Introduction to Statistical Limit Theory

CHAPMAN & HALL/CRC
Texts in Statistical Science Series

Series Editors

Bradley P. Carlin, *University of Minnesota, USA*
Julian J. Faraway, *University of Bath, UK*
Martin Tanner, *Northwestern University, USA*
Jim Zidek, *University of British Columbia, Canada*

Texts in Statistical Science

Introduction to Statistical Limit Theory

Alan M. Polansky

Northern Illinois University
Dekalb, Illinois, USA

CRC Press
Taylor & Francis Group
Boca Raton London New York

CRC Press is an imprint of the
Taylor & Francis Group, an **informa** business

A CHAPMAN & HALL BOOK

CRC Press
Taylor & Francis Group
6000 Broken Sound Parkway NW, Suite 300
Boca Raton, FL 33487-2742

First issued in paperback 2019

© 2011 by Taylor and Francis Group, LLC
CRC Press is an imprint of Taylor & Francis Group, an Informa business

No claim to original U.S. Government works

ISBN-13: 978-1-4200-7660-8 (hbk)
ISBN-13: 978-0-367-38313-8 (pbk)

Library of Congress Cataloging-in-Publication Data

Polansky, Alan M.
    Introduction to statistical limit theory / Alan M. Polansky.
        p. cm. -- (Chapman & Hall/CRC texts in statistical science)
    "A Chapman & Hall book."
    Includes bibliographical references and index.
    ISBN 978-1-4200-7660-8 (hardback)
    1. Limit theorems (Probability theory) I. Title.

QA273.67.P65 2011
519.2--dc22
                                                                            2010045942

Visit the Taylor & Francis Web site at
http://www.taylorandfrancis.com

and the CRC Press Web site at
http://www.crcpress.com

TO

CATHERINE AND ALTON

AND

ADLER, AGATHA, FLUFFY, AND KINSEY

AND THE MEMORY OF

FLATNOSE, HOLMES, LISA, MILHOUSE, MIRROR,
RALPH, AND SEYMOUR

Contents

Preface

The motivation for writing a new book is heavily dependent on the books that are currently available in the area you wish to present. In many cases one can make the case that a new book is needed when there are no books available in that area. In other cases one can make the argument that the current books are out of date or are of poor quality. I wish to make it clear from the onset that I do not have this opinion of any of the books on statistical asymptotic theory that are currently available, or have been available in the past. In fact, I find myself humbled as I search my book shelves and find the many well written books on this subject. Indeed, I feel strongly that some of the best minds in statistics have written books in this area. This leaves me to the task of explaining why I found it necessary to write this book on asymptotic statistical theory.

Many students of statistics are finding themselves more specialized away from mathematical theory in favor of newer and more complex statistical methods that necessarily require specialized study. Inevitably, this has led to a diminished focus of pure mathematics. However, this does not diminish the need for a good understanding of asymptotic theory. Good students of statistics, for example, should know exactly what the central limit theorem says and what exactly it means. They should be able to understand the assumptions required for the theory to work. There are many modern methods that are complex enough so that they cannot be justified directly, as therefore must be justified from an asymptotic viewpoint. Students should have a good understanding as to what such a justification guarantees and what it does not. Students should also have a good understanding of what can go wrong.

Asymptotic theory is mathematical in nature. Over the years I have helped many students understand results from asymptotic theory, often as part of their research or classwork. In this time I began to realize that the decreased exposure to mathematical theory, particularly in real analysis, was making it much more difficult for students to understand this theory.

I wrote this book with the goal of explaining as much of the background material as I could, while still keeping a reasonable presentation and length. The reader will not find a detailed review of the whole of real analysis, and other important subjects from mathematics. Instead the reader will find sufficient background of those subjects that are important for the topic at hand, along with references which the reader may explore for a more detailed understanding. I have also attempted to present a much more detailed account of the modes of convergence of random variables, distributions, and moments than can be found in many other texts. This creates a firm foundation for the applications that appear in the book, along with further study that students may do on their own.

As I began the job of writing this book, I recalled a quote from one of my favorite authors, Sir Arthur Conan Doyle in his Sherlock Holmes adventure "The Dancing Men." Mr. Holmes is conversing with Dr. Watson and has just presented him with one of his famous deductions that leaves Dr. Watson startled by Holmes' deductive abilities. Mr. Holmes replies as follows:

> "You see, my dear Watson,"—he propped his test-tube in the rack, and began to lecture with the air of a professor addressing his class—"it is not really difficult to construct a series of inferences, each dependent upon its predecessor and each simple in itself. If, after doing so, one simply knocks out all the central inferences and presents one's audience with the starting-point and the conclusion, one may produce a startling, though possibly a meretricious, effect."

A mathematical theorem essentially falls into this category in the sense that it is a series of assumptions followed by a logical result. The central inferences have been removed, sometimes producing a somewhat startling result. When one uses the theorem as a tool to obtain further results, the central inferences are usually unimportant. If one wishes to truly *understand* the result, one must understand the central inferences, which is usually called the "proof" of the theorem. As this book is meant to be either a textbook for a course, or a reference book for someone unfamiliar with many of the results within, I feel duty bound to not produce startling results in this book, but to include all of the central inferences of each result in such a way that the simple logical progression is clear to the reader. As a result, most of the proofs presented in this book are very detailed.

I have included proofs to many of the results in the book. Nearly all of these arguments come from those before me, but I have attempted to include detailed explanations for the results while pointing out important tools and techniques along the way. I have attempted to give due credit to the sources I used for the proofs. If I have left anyone out, please accept my apologies and be assured that it was not on purpose, but out of my ignorance. I have also not steered away from the more complicated proofs in this field. These proofs often contain valuable techniques that a user of asymptotic theory should be familiar with. For many of the standard results there are several proofs that one can choose from. I have not always chosen the shortest or the "slickest" method of proof. In my choice of proofs I weighed the length of the argument along with its pedagogical value and how well the proof provides insight into the final result. This is particularly true when complicated or strange looking assumptions are part of the result. I have attempted to point out in the proofs where these conditions come from.

This book is mainly concerned with providing a detailed introduction into the common modes of convergence and their related tools used in statistics. However, it is also useful to consider how these results can be applied to several common areas of statistics. Therefore, I have included several chapters that deal more with the application of the theory developed in the first part of the book. These applications are not an exhaustive offering by any stretch of the imagination, and to be sure an exhaustive offering would have enlarged the book to the size of a phone book from a large metropolitan area. Therefore, I have attempted to include a few topics whose deeper understanding benefits greatly from asymptotic theory and whose applications provide illustrative examples of the theory developed earlier.

Many people have helped me along the way in developing the ideas behind this book and implementing them on paper. Bob Stern and David Grubbs at CRC/Chapman & Hall have been very supportive of this project and have put up with my countless delays. There were several students who took a course from me based on an early draft of this book: Devrim Bilgili, Arpita Chatterjee, Ujjwal Das, Santu Ghosh, Priya Kohli, and Suchitrita Sarkar. They all provided me with useful comments, found numerous typographical errors, and asked intelligent questions, which helped develop the book from its early phase to a more coherent and complete document. I would also like to thank Qian Dong, Lingping Liu, and Kristin McCullough who studied from a later version of the book and were able to point out several typographical errors and places where there could be some improvement in the presentation. I also want thank Sanjib Basu who helped me by answering numerous questions about Bayesian theory.

My friends and colleagues who have supported me through the years also deserve a special note of thanks: My advisor, Bill Schucany, has also been a constant supporter of my activities. He recently announced his retirement and I wish him all the best. I also want to thank Bob Mason, Youn-Min Chou, Dick Gunst, Wayne Woodward, Bennie Pierce, Pat Gerard, Michael Ernst, Carrie Helmig, Donna Lynn, and Jane Eesley for their support over the years.

Jeff Reynolds, my bandmate, my colleague, my student, and my friend, has always shown me incredible support. He has been a constant source for reality checks, help, and motivation. We have faced many battles together, many of them quite enjoyable: "My brother in arms."

There is always family, and mine is the best. My wife Catherine and my son Alton have been there for me all throughout this process. I love and cherish both of them; I could not have completed this project without their support, understanding, sacrifices, and patience. I also wish to thank my extended family; My Mom and Dad, Kay and Al Polansky who celebrated their 60th wedding anniversary in 2010; my brother Gary and his famiy: Marcia, Kara, Krista, Mandy, Nellie and Jack; my brother Dale and his new family, which we welcomed in the summer of 2009: Jennifer and Sydney; and my wife's family: Karen, Mike, Ginnie, Christopher, Jonathan, Jackie, Frank, and Mila.

Finally, I would like to thank those who always provide me with the diversions I need during such a project as this. Matt Groening, who started a little show called *The Simpsons* the year before I started graduate school, will never know how much joy his creation has brought to my life. I also wish to thank David Silverman who visited Southern Methodist University while I was there working on my Ph.D.; he drew me a wonderful Homer Simpson, which hangs on my wall to this day. Jebediah was right: "A noble spirit embiggens the smallest man." There was also those times when what I needed was a good polka. For this I usually turned to the music of Carl Finch and Jeffrey Barnes of Brave Combo. See you at Westfest!

Much has happened during the time while I was writing this book. I has very saddened to hear of the passing Professor Erich Lehmann on September 12, 2009, at the age of 91. One cannot understate the contributions of Professor Lehmann on the field of statistics, and particularly on much of the material presented in this book. I did not know Professor Lehmann personally, but did meet him once at the Joint Statistical Meetings where he was kind enough to sign a copy of the new edition of his book *Theory of Point Estimation*. His classic books, which have always had a special place on my bookshelf since beginning my career as a graduate student, have been old and reliable friends for many years. On July 8, 2010, David Blackwell, the statistician and mathematician who wrote many groundbreaking papers on probability and game theory passed away as well. Besides his many contributions to statistics, Professor Blackwell also held the distinction of being the first African American scholar to be admitted to the National Academy of Sciences and was the first African American tenured professor at Berkeley. He too will be missed.

During the writing of this book I also passed my 40th birthday and began to think about the turmoil and fear that pervaded my birth year of 1968. What a time it must have been to bring a child into the world. There must have been few hopes and many fears. I find myself looking at my own child Alton and wondering what the world will have in store for him. It has again been a time of turmoil and fear, and I can only hope that humanity begins to heed the words of Dr. Martin Luther King:

"We must learn to live together as brothers or perish together as fools."

Peace to All,

Alan M. Polansky
Creston, Illinois, USA

CHAPTER 1

Sequences of Real Numbers and Functions

K. felt slightly abandoned as, probably observed by the priest, he walked by himself between the empty pews, and the size of the cathedral seemed to be just at the limit of what a man could bear.

The Trial by Franz Kafka

1.1 Introduction

The purpose of this chapter is to introduce much of the mathematical limit theory used throughout the reminder of the book. For many readers, many of the topics will consist of review material, while other topics may be quite new. The study of asymptotic properties in statistics relies heavily on results and concepts from real analysis and calculus. As such, we begin with a review of limits of sequences of real numbers and sequences of real functions. This is followed by the development of what will be a most useful tool, Taylor's Theorem. We then introduce the concept of asymptotic expansions, a topic that may be new to some readers and is vitally important to many of the results treated later in the book. Of particular importance is the introduction of the asymptotic order notation, which is popular in modern research in probability and statistics. The related topic of inversion of asymptotic expansions is then briefly introduced.

1.2 Sequences of Real Numbers

An infinite sequence of real numbers given by $\{x_n\}_{n=1}^{\infty}$ is specified by the function $x_n : \mathbb{N} \to \mathbb{R}$. That is, for each $n \in \mathbb{N}$, the sequence has a real value x_n. The set $\mathbb{N}$ is usually called the *index set*. In this case the domain of the function is countable, and the sequence is usually thought to evolve sequentially through the increasing values in $\mathbb{N}$. For example, the simple harmonic sequence specified by $x_n = n^{-1}$ has the values

$$x_1 = 1, x_2 = \tfrac{1}{2}, x_3 = \tfrac{1}{3}, x_4 = \tfrac{1}{4}, \ldots,$$

while the simple alternating sequence $x_n = (-1)^n$ has values

$$x_1 = -1, x_2 = 1, x_3 = -1, x_4 = 1, \ldots.$$

One can also consider real sequences of the form x_t with an uncountable domain such as the real line which is specified by a function $x_t : \mathbb{R} \to \mathbb{R}$. Such sequences are essentially just real functions. This section will consider only real sequences whose index set is $\mathbb{N}$.

The asymptotic behavior of such sequences is often of interest. That is, what general conclusions can be made about the behavior of the sequence as n becomes very large? In particular, do the values in the sequence appear to "settle down" and become arbitrarily close to a single number $x \in \mathbb{R}$ as $n \to \infty$? For example, the sequence specified by $x_n = n^{-1}$ appears to become closer and closer to 0 as n becomes very large. If a sequence has this type of property, then the sequence is said to *converge* to x as $n \to \infty$, or that the *limit* of x_n as $n \to \infty$ is x, usually written as

$$\lim_{n \to \infty} x_n = x,$$

or as $x_n \to x$ as $n \to \infty$. To decide whether a given sequence has this type of behavior, a mathematical definition of the convergence or limiting concept is required. The most common definition is given below.

Definition 1.1. *Let $\{x_n\}_{n=1}^{\infty}$ be a sequence of real numbers and let $x \in \mathbb{R}$ be a real number. Then x_n converges to x as $n \to \infty$, written as $x_n \to x$ as $n \to \infty$, or as*

$$\lim_{n \to \infty} x_n = x,$$

if and only if for every $\varepsilon > 0$ there exists $n_\varepsilon \in \mathbb{N}$ such that $|x_n - x| < \varepsilon$ for every $n \geq n_\varepsilon$.

This definition ensures the behavior described above. Specify any distance $\varepsilon > 0$ to x, and all of the terms in a convergent sequence will eventually be closer than that distance to x.

Example 1.1. Consider the harmonic sequence defined by $x_n = n^{-1}$ for all $n \in \mathbb{N}$. This sequence appears to monotonically become closer to zero as n increases. In fact, it can be proven that the limit of this sequence is zero. Let $\varepsilon > 0$. Then there exists $n_\varepsilon \in \mathbb{N}$ such that $n_\varepsilon^{-1} < \varepsilon$, which can be seen by taking n_ε to be any integer greater than ε^{-1}. It follows that any $n \geq n_\varepsilon$ will also have the property that $n^{-1} < \varepsilon$. Therefore, according to Definition 1.1, the sequence $x_n = n^{-1}$ converges to zero, or

$$\lim_{n \to \infty} n^{-1} = 0.$$

∎

For the real number system there is an equivalent development of the concept of a limit that is dependent on the concept of a *Cauchy* sequence.

Definition 1.2. *Let $\{x_n\}_{n=1}^{\infty}$ be a sequence of real numbers. The sequence is a Cauchy sequence if for every $\varepsilon > 0$ there is an integer n_ε such that $|x_n - x_m| < \varepsilon$ for every $n > n_\varepsilon$ and $m > n_\varepsilon$.*

In general, not every Cauchy sequence converges to a limit. For example, not every Cauchy sequence of rational numbers has a rational limit. See Example 6.9 of Sprecher (1970). There are, however, some spaces where Cauchy sequence has a unique limit. Such spaces are said to be *complete*. The real number system is an example of a complete space.

Theorem 1.1. *Every Cauchy sequence of real numbers has a unique limit.*

The advantage of using Cauchy sequences is that we sometimes only need to show the existence of a limit and Theorem 1.1 can be used to show that a real sequence has a limit even if we do not know what the limit may be.

Simple algebraic transformations can be applied to convergent sequences with the resulting limit being subject to the same transformation. For example, adding a constant to each term of a convergent sequence results in a convergent sequence whose limit equals the limit of the original sequence plus the constant. A similar results applies to sequences that have been multiplied by a constant.

Theorem 1.2. *Let $\{x_n\}_{n=1}^{\infty}$ be a sequence of real numbers such that*

$$\lim_{n\to\infty} x_n = x,$$

and let $c \in \mathbb{R}$ be a constant. Then

$$\lim_{n\to\infty} (x_n + c) = x + c,$$

and

$$\lim_{n\to\infty} cx_n = cx.$$

Proof. We will prove the first result. The second result is proven in Exercise 6. Let $\{x_n\}_{n=1}^{\infty}$ be a sequence of real numbers that converges to x, let c be a real constant, and let $\varepsilon > 0$. Definition 1.1 and the convergence of the sequence $\{x_n\}_{n=1}^{\infty}$ implies that there exists a positive integer n_ε such that $|x_n - x| < \varepsilon$ for all $n \geq n_\varepsilon$. Now consider the sequence $\{x_n + c\}_{n=1}^{\infty}$. Note that for $\varepsilon > 0$ we have that $|(x_n + c) - (x + c)| = |x_n - x|$. Therefore, $|(x_n + c) - (x + c)|$ is also less than ε for all $n \geq n_\varepsilon$ and the result follows from Definition 1.1. $\quad\square$

Theorem 1.2 can be generalized to continuous transformations of convergent sequences.

Theorem 1.3. *Let $\{X_n\}_{n=1}^{\infty}$ be a sequence of real numbers such that*

$$\lim_{n\to\infty} x_n = x,$$

and let f be a function that is continuous at x. Then

$$\lim_{n\to\infty} f(x_n).$$

A proof of Theorem 1.3 can be found in Section 2.2 of Buck (1965).

In many cases we may consider combining two or more convergent sequences through a simple algebraic transformation. For example, we might consider a sequence whose value is equal to the sum of the corresponding terms of two other convergent sequences. This new sequence is also convergent, and has a limit equal to the sum of the limits of the two sequences. Similar results also apply to the product and ratio of two convergent sequences.

Theorem 1.4. *Let $\{x_n\}_{n=1}^{\infty}$ and $\{y_n\}_{n=1}^{\infty}$ be sequences of real numbers such that*

$$\lim_{n \to \infty} x_n = x,$$

and

$$\lim_{n \to \infty} y_n = y.$$

Then

$$\lim_{n \to \infty} (x_n + y_n) = x + y,$$

and

$$\lim_{n \to \infty} x_n y_n = xy.$$

If, in addition to the assumptions above, $y_n \neq 0$ for all $n \in \mathbb{N}$ and $y \neq 0$, then

$$\lim_{n \to \infty} \frac{x_n}{y_n} = \frac{x}{y}.$$

Proof. Only the first result will be proven here. The remaining results are proven as part of Exercise 6. Let $\{x_n\}_{n=1}^{\infty}$ and $\{y_n\}_{n=1}^{\infty}$ be convergent sequences with limits x and y respectively. Let $\varepsilon > 0$, then Definition 1.1 implies that there exists integers $n_{\varepsilon,x}$ and $n_{\varepsilon,y}$ such that $|x_n - x| < \varepsilon/2$ for all $n \geq n_{\varepsilon,x}$ and $|y_n - y| < \varepsilon/2$ for all $n \geq n_{\varepsilon,y}$. Now, note that Theorem A.18 implies that

$$|(x_n + y_n) - (x + y)| = |(x_n - x) + (y_n - y)| \leq |x_n - x| + |y_n - y|.$$

Let $n_{\varepsilon} = \max\{n_{\varepsilon,x}, n_{\varepsilon,y}\}$ so that $|x_n - x| < \varepsilon/2$ and $|y_n - y| < \varepsilon/2$ for all $n \geq n_{\varepsilon}$. Therefore $|(x_n + y_n) - (x + y)| < \varepsilon/2 + \varepsilon/2 = \varepsilon$ for all $n \geq n_{\varepsilon}$ and the result follows from Definition 1.1. ☐

The focus of our study of limits so far has been for convergent sequences. Not all sequences of real numbers are convergent, and the limits of non-convergent real sequences do not exist.

Example 1.2. Consider the alternating sequence defined by $x_n = (-1)^n$ for all $n \in \mathbb{N}$. It is intuitively clear that the alternating sequence does not "settle down" at all, or that it does not converge to any real number. To prove this, let $l \in \mathbb{R}$ be any real number. Take $0 < \varepsilon < \max\{|l - 1|, |l + 1|\}$, then for any $n' \in \mathbb{N}$ there will exist at least one $n'' > n'$ such that $|x_n - l| > \varepsilon$. Hence, this sequence does not have a limit. ∎

While a non-convergent sequence does not have a limit, the asymptotic behavior of non-convergent sequences can be described to a certain extent by considering the asymptotic behavior of the upper and lower bounds of the sequence. Let $\{x_n\}_{n=1}^{\infty}$ be a sequence of real numbers, then $u \in \mathbb{R}$ is an *upper bound* for $\{x_n\}_{n=1}^{\infty}$ if $x_n \leq u$ for all $n \in \mathbb{N}$. Similarly, $l \in \mathbb{R}$ is a *lower bound* for $\{x_n\}_{n=1}^{\infty}$ if $x_n \geq l$ for all $n \in \mathbb{N}$. The *least upper bound* of $\{x_n\}_{n=1}^{\infty}$ is $u_l \in \mathbb{R}$ if u_l is an upper bound for $\{x_n\}_{n=1}^{\infty}$ and $u_l \leq u$ for any upper bound u of $\{x_n\}_{n=1}^{\infty}$. The least upper bound will be denoted by

$$u_l = \sup_{n \in \mathbb{N}} x_n,$$

and is often called the *supremum* of the sequence $\{x_n\}_{n=1}^{\infty}$. Similarly, the *greatest lower bound* of $\{x_n\}_{n=1}^{\infty}$ is $l_u \in \mathbb{R}$ if l_u is a lower bound for $\{x_n\}_{n=1}^{\infty}$ and $l_u \geq l$ for any lower bound l of $\{x_n\}_{n=1}^{\infty}$. The greatest lower bound of $\{x_n\}_{n=1}^{\infty}$ will be denoted by

$$l_u = \inf_{n \in \mathbb{N}} x_n,$$

and is often called the *infimum* of $\{x_n\}_{n=1}^{\infty}$. It is a property of the real numbers that any sequence that has a lower bound also has a greatest lower bound and that any sequence that has an upper bound also has a least upper bound. See Page 33 of Royden (1988) for further details. The asymptotic behavior of non-convergent sequences can be studied in terms of how the supremum and infimum of a sequence behaves as $n \to \infty$. That is, we can consider for example the asymptotic behavior of the upper limit of a sequence $\{x_n\}_{n=1}^{\infty}$ by calculating

$$\lim_{n \to \infty} \sup_{k \geq n} x_k.$$

Note that the sequence

$$\left\{ \sup_{k \geq n} x_k \right\}_{n=1}^{\infty}, \tag{1.1}$$

is a monotonically decreasing sequence of real numbers. The limiting value of this sequence should occur at its smallest value, or in mathematical terms

$$\lim_{n \to \infty} \sup_{k \geq n} x_k = \inf_{k \in \mathbb{N}} \sup_{k \geq n} x_k. \tag{1.2}$$

From the discussion above it is clear that if the sequence $\{x_n\}_{n=1}^{\infty}$ is bounded, then the sequence given in Equation (1.1) is also bounded and the limit in Equation (1.2) will always exist. A similar concept can be developed to study the asymptotic behavior of the infimum of a sequence as well.

Definition 1.3. *Let $\{x_n\}_{n=1}^{\infty}$ be a sequence of real numbers. The limit supremum of $\{x_n\}_{n=1}^{\infty}$ is*

$$\limsup_{n \to \infty} x_n = \inf_{n \in \mathbb{N}} \sup_{k \geq n} x_k,$$

and the limit infimum of $\{x_n\}_{n=1}^{\infty}$ is

$$\liminf_{n \to \infty} x_n = \sup_{n \in \mathbb{N}} \inf_{k \geq n} x_k.$$

If

$$\liminf_{n \to \infty} x_n = \limsup_{n \to \infty} x_n = c \in \mathbb{R},$$

then the limit of $\{x_n\}_{n=1}^{\infty}$ exists and is equal to c.

The usefulness of the limit supremum and the limit infimum can be demonstrated through some examples.

Example 1.3. Consider the sequence $\{x_n\}_{n=1}^{\infty}$ where $x_n = n^{-1}$ for all $n \in \mathbb{N}$. Note that

$$\sup_{k \geq n} x_n = \sup_{k \geq n} k^{-1} = n^{-1},$$

since n^{-1} is the largest element in the sequence. Therefore,

$$\limsup_{n \to \infty} x_n = \inf_{k \in \mathbb{N}} \sup_{k \geq n} x_k = \inf_{n \in \mathbb{N}} n^{-1} = 0.$$

Note that zero is a lower bound of n^{-1} since $n^{-1} > 0$ for all $n \in \mathbb{N}$. Further, zero is the greatest lower bound since there exists an $n_\varepsilon \in \mathbb{N}$ such that $n_\varepsilon^{-1} < \varepsilon$ for any $\varepsilon > 0$. Similar arguments can be used to show that

$$\liminf_{n \to \infty} x_n = \sup_{n \in \mathbb{N}} \inf_{k \geq n} x_k = \sup_{n \in \mathbb{N}} 0 = 0.$$

Therefore Definition 1.3 implies that

$$\lim_{n \to \infty} n^{-1} = \liminf_{n \to \infty} n^{-1} = \limsup_{n \to \infty} n^{-1} = 0,$$

and the sequence is convergent. ∎

Example 1.4. Consider the sequence $\{x_n\}_{n=1}^{\infty}$ where $x_n = (-1)^n$ for all $n \in \mathbb{N}$. Note that

$$\sup_{k \geq n} x_k = \sup_{k \geq n}(-1)^k = 1,$$

so that,

$$\limsup_{n \to \infty} x_n = \inf_{k \in \mathbb{N}} \sup_{k \geq n} x_k = \inf_{n \in \mathbb{N}} 1 = 1.$$

Similarly,

$$\inf_{k \geq n} x_k = \inf_{k \geq n}(-1)^k = -1,$$

so that

$$\liminf_{n \to \infty} x_n = \sup_{n \in \mathbb{N}} \inf_{k \geq n} x_k = \sup_{n \in \mathbb{N}} -1 = -1.$$

In this case it is clear that

$$\liminf_{n \to \infty} x_n \neq \limsup_{n \to \infty} x_n,$$

so that, as shown in Example 1.2, the limit of the sequence does not exist. The limit infimum and limit supremum indicate the extent of the limit of the variation of $\{x_n\}_{n=1}^{\infty}$ as $n \to \infty$. ∎

Example 1.5. Consider the sequence $\{x_n\}_{n=1}^{\infty}$ where $x_n = (-1)^n(1 + n^{-1})$ for all $n \in \mathbb{N}$. Note that

$$\sup_{k \geq n} x_k = \sup_{k \geq n}(-1)^k(1 + k^{-1}) = \begin{cases} 1 + n^{-1} & \text{if } n \text{ is even,} \\ 1 + (n+1)^{-1} & \text{if } n \text{ is odd.} \end{cases}$$

In either case,

$$\inf_{n\in\mathbb{N}}(1+n^{-1}) = \inf_{n\in\mathbb{N}}[1+(n+1)^{-1}] = 1,$$

so that

$$\limsup_{n\to\infty} x_n = 1.$$

Similarly,

$$\inf_{k\geq n} x_k = \inf_{k\geq n}(-1)^k(1+k^{-1}) = \begin{cases} -(1+n^{-1}) & \text{if } n \text{ is odd,} \\ -[1+(n+1)^{-1}] & \text{if } n \text{ is even.} \end{cases}$$

and

$$\sup_{n\in\mathbb{N}} -(1+n^{-1}) = \sup_{n\in\mathbb{N}} -[1+(n+1)^{-1}] = -1,$$

so that

$$\liminf_{n\to\infty} x_n = -1.$$

As in Example 1.4

$$\liminf_{n\to\infty} x_n \neq \limsup_{n\to\infty} x_n,$$

so that the limit does not exist. Note that this sequence has the same asymptotic behavior on its upper and lower bounds as the much simpler sequence in Example 1.4. ∎

The properties of the limit supremum and limit infimum are similar to those of the limit with some notable exceptions.

Theorem 1.5. *Let $\{x_n\}_{n=1}^{\infty}$ be a sequence of real numbers. Then*

$$\inf_{n\in\mathbb{N}} x_n \leq \liminf_{n\to\infty} x_n \leq \limsup_{n\to\infty} x_n \leq \sup_{n\in\mathbb{N}} x_n,$$

and

$$\limsup_{n\to\infty}(-x_n) = -\liminf_{n\to\infty} x_n.$$

Proof. The second property will be proven here. The first property is proven in Exercise 10. Let $\{x_n\}_{n=1}^{\infty}$ be a sequence of real numbers. Note that the negative of any lower bound of $\{x_n\}_{n=1}^{\infty}$ is an upper bound of the sequence $\{-x_n\}_{n=1}^{\infty}$. To see why, let l be a lower bound of $\{x_n\}_{n=1}^{\infty}$. Then $l \leq x_n$ for all $n \in \mathbb{N}$. Multiplying each side of the inequality by -1 yields $-l \geq -x_n$ for all $n \in \mathbb{N}$. Therefore $-l$ is an upper bound of $\{-x_n\}_{n=1}^{\infty}$, and it follows that the negative of the greatest lower bound of $\{x_n\}_{n=1}^{\infty}$ is the least upper bound of $\{-x_n\}_{n=1}^{\infty}$. That is

$$-\inf_{k\geq n} x_k = \sup_{k\geq n} -x_k, \tag{1.3}$$

and

$$-\sup_{k\geq n} x_k = \inf_{k\geq n} -x_k. \tag{1.4}$$

Therefore, an application of Equations (1.3) and (1.4) implies that

$$\limsup_{n\to\infty}(-x_n) = \inf_{n\in\mathbb{N}}\sup_{k\geq n}(-x_k) =$$

$$\inf_{n\in\mathbb{N}}\left(-\inf_{k\geq n}x_k\right) = -\sup_{n\in\mathbb{N}}\left(\inf_{k\geq n}x_k\right) = -\liminf_{n\to\infty}x_n.$$

□

Combining the limit supremum and infimum of two or more sequences is also possible, though the results are not as simple as in the case of limits of convergent sequences.

Theorem 1.6. *Let $\{x_n\}_{n=1}^{\infty}$ and $\{y_n\}_{n=1}^{\infty}$ be sequences of real numbers.*

1. *If $x_n \leq y_n$ for all $n \in \mathbb{N}$, then*

$$\limsup_{n\to\infty} x_n \leq \limsup_{n\to\infty} y_n,$$

and

$$\liminf_{n\to\infty} x_n \leq \liminf_{n\to\infty} y_n.$$

2. *If*

$$\left|\limsup_{n\to\infty} x_n\right| < \infty,$$

and

$$\left|\liminf_{n\to\infty} x_n\right| < \infty,$$

then

$$\liminf_{n\to\infty} x_n + \liminf_{n\to\infty} y_n \leq \liminf_{n\to\infty}(x_n + y_n),$$

and

$$\limsup_{n\to\infty}(x_n + y_n) \leq \limsup_{n\to\infty} x_n + \limsup_{n\to\infty} y_n.$$

3. *If $x_n > 0$ and $y_n > 0$ for all $n \in \mathbb{N}$,*

$$0 < \limsup_{n\to\infty} x_n < \infty,$$

and

$$0 < \limsup_{n\to\infty} y_n < \infty,$$

then

$$\limsup_{n\to\infty} x_n y_n \leq \left(\limsup_{n\to\infty} x_n\right)\left(\limsup_{n\to\infty} y_n\right).$$

Proof. Part of the first property is proven here. The remaining properties are proven in Exercises 12 and 13. Suppose that $\{x_n\}_{n=1}^{\infty}$ and $\{y_n\}_{n=1}^{\infty}$ are real sequences such that $x_n \leq y_n$ for all $n \in \mathbb{N}$. Then $x_k \leq y_k$ for all $k \in \{n, n+1, \ldots\}$. This implies that any upper bound for $\{y_n, y_{n+1}, \ldots, \}$ is also an upper bound for $\{x_n, x_{n+1}, \ldots\}$. It follows that the least upper bound for $\{x_n, x_{n+1}, \ldots\}$ is less than or equal to the least upper bound for $\{y_n, y_{n+1}, \ldots, \}$. That is

$$\sup_{k \geq n} x_k \leq \sup_{k \geq n} y_k. \tag{1.5}$$

Because k is arbitrary in the above argument, the property in Equation (1.5) holds for all $k \in \mathbb{N}$. Now, note that a lower bound for the sequence

$$\left\{ \sup_{k \geq 1} x_k \right\}_{k=1}^{n}, \tag{1.6}$$

is also a lower bound for

$$\left\{ \sup_{k \geq 1} y_k \right\}_{k=1}^{n}, \tag{1.7}$$

due to the property in Equation (1.5). Therefore, the greatest lower bound for the sequence in Equation (1.6) is also a lower bound for the sequence in Equation (1.7). Hence, it follows that the greatest lower bound for the sequence in Equation (1.6) is bounded above by the lower bound for the sequence in Equation (1.7). That is

$$\limsup_{n \to \infty} x_n = \inf_{n \in \mathbb{N}} \sup_{k \geq n} x_k \leq \inf_{n \in \mathbb{N}} \sup_{k \geq n} y_k = \limsup_{n \to \infty} y_n.$$

The property for the limit infimum is proven in Exercise 11. □

Example 1.6. Consider two sequences of real numbers $\{x_n\}_{n=1}^{\infty}$ and $\{y_n\}_{n=1}^{\infty}$ given by $x_n = (-1)^n$ and $y_n = (1 + n^{-1})$ for all $n \in \mathbb{N}$. Using Definition 1.3 it can be shown that

$$\limsup_{n \to \infty} (-1)^n = 1$$

and

$$\lim_{n \to \infty} y_n = \limsup_{n \to \infty} y_n = 1.$$

Therefore,

$$\left(\limsup_{n \to \infty} x_n \right) \left(\limsup_{n \to \infty} y_n \right) = 1.$$

It was shown in Example 1.5 that

$$\limsup_{n \to \infty} x_n y_n = 1.$$

Therefore, in this case,

$$\limsup_{n \to \infty} x_n y_n = \left(\limsup_{n \to \infty} x_n \right) \left(\limsup_{n \to \infty} y_n \right).$$

∎

Example 1.7. Consider two sequences of real numbers $\{x_n\}_{n=1}^{\infty}$ and $\{y_n\}_{n=1}^{\infty}$ where $x_n = \frac{1}{2}[(-1)^n - 1]$ and $y_n = (-1)^n$ for all $n \in \mathbb{N}$. Using Definition 1.3 it can be shown that

$$\limsup_{n \to \infty} x_n = 0,$$

and

$$\limsup_{n \to \infty} y_n = 1,$$

so that

$$\left(\limsup_{n \to \infty} x_n \right) \left(\limsup_{n \to \infty} y_n \right) = 0.$$

Now the sequence $x_n y_n$ can be defined as $x_n y_n = \frac{1}{2}[1 - (-1)^n]$, for all $n \in \mathbb{N}$, so that

$$\limsup_{n \to \infty} x_n y_n = 1,$$

which would apparently contradict Theorem 1.6, except for the fact that the assumption

$$\limsup_{n \to \infty} x_n > 0,$$

is violated.   ∎

Example 1.8. Consider two sequences of real numbers $\{x_n\}_{n=1}^{\infty}$ and $\{y_n\}_{n=1}^{\infty}$ where $x_n = (-1)^n$ and $y_n = (-1)^{n+1}$ for all $n \in \mathbb{N}$. From Definition 1.3 it follows that

$$\limsup_{n \to \infty} x_n = \limsup_{n \to \infty} y_n = 1.$$

The sequence $x_n y_n$ can be defined by $x_n y_n = -1$ for all $n \in \mathbb{N}$, so that

$$\limsup_{n \to \infty} x_n y_n = -1.$$

Therefore, in this case

$$\limsup_{n \to \infty} x_n y_n < \left(\limsup_{n \to \infty} x_n \right) \left(\limsup_{n \to \infty} y_n \right).$$

∎

One sequence of interest is specified by $x_n = (1 + n^{-1})^n$, which will be used in later chapters. The computation of the limit of this sequence is slightly more involved and specifically depends on the definition of limit given in Definition 1.3, and as such is an excellent example of using the limit infimum and supremum to compute a limit.

Theorem 1.7. *Define the sequence $\{x_n\}_{n=1}^{\infty}$ as $x_n = (1 + n^{-1})^n$. Then*

$$\lim_{n \to \infty} x_n = e.$$

Proof. The standard proof of this result, such as this one which is adapted from Sprecher (1970), is based on the Theorem A.22. Using Theorem A.22, note that when $n \in \mathbb{N}$ is fixed

$$(1 + n^{-1})^n = \sum_{i=0}^{n} \binom{n}{i} n^{-i} (1)^{n-i} = \sum_{i=0}^{n} \binom{n}{i} n^{-i}. \tag{1.8}$$

Consider the ith term in the series on the right hand side of Equation (1.8), and note that

$$\binom{n}{i} n^{-i} = \frac{(n-i)!}{i!(n-i)!} \prod_{j=0}^{i-1} \frac{n-j}{n} \le \frac{1}{i!}.$$

Therefore it follows that

$$(1 + n^{-1})^n \le \sum_{i=0}^{n} \frac{1}{i!} \le e,$$

for every $n \in \mathbb{N}$. The second inequality comes from the fact that

$$e = \sum_{i=0}^{\infty} \frac{1}{i!},$$

where all of the terms in the sequence are positive. It then follows that

$$\limsup_{n \to \infty} (1 + n^{-1})^n \le e.$$

See Theorem 1.6. Now suppose that $m \in \mathbb{N}$ such that $m \le n$. Then

$$(1 + n^{-1})^n = \sum_{i=0}^{n} \binom{n}{i} n^{-i} \ge \sum_{i=0}^{m} \binom{n}{i} n^{-i},$$

since each term in the sum is positive. Note that for fixed i,

$$\lim_{n \to \infty} \binom{n}{i} n^{-i} = \lim_{n \to \infty} \frac{1}{i!} \prod_{j=0}^{i-1} \frac{n-j}{n} = \frac{1}{i!}.$$

Therefore, using the same argument as above, it follows that

$$\liminf_{n \to \infty} (1 + n^{-1})^n \ge \sum_{i=0}^{m} \frac{1}{i!}.$$

Letting $m \to \infty$ establishes the inequality

$$\liminf_{n \to \infty} (1 + n^{-1})^n \ge e.$$

Therefore it follows that

$$\limsup_{n \to \infty} (1 + n^{-1})^n \le e \le \liminf_{n \to \infty} (1 + n^{-1})^n.$$

The result of Theorem 1.5 then implies that

$$\limsup_{n \to \infty} (1 + n^{-1})^n = e = \liminf_{n \to \infty} (1 + n^{-1})^n,$$

so that Definition 1.3 implies that

$$\lim_{n \to \infty} (1 + n^{-1})^n = e.$$

□

Another limiting result that is used in this book is *Stirling's Approximation*, which is used to approximate the factorial operation.

Theorem 1.8 (Stirling).

$$\lim_{n\to\infty}\frac{n^n(2n\pi)^{1/2}}{\exp(n)n!}=1.$$

A proof of Theorem 1.8 can be found in Slomson (1991). Theorem 1.8 implies that when n is large,

$$\frac{n^n(2n\pi)^{1/2}}{\exp(n)n!}\simeq 1,$$

and hence one can approximate $n!$ with $n^n\exp(-n)(2n\pi)^{1/2}$.

Example 1.9. Let n and k be positive integers such that $k\leq n$. The number of combinations of k items selected from a set of n items is

$$\binom{n}{k}=\frac{n!}{k!(n-k)!}.$$

Theorem 1.8 implies that when both n and k are large, we can approximate $\binom{n}{k}$ as

$$\begin{aligned}\binom{n}{k}&\simeq\frac{n^n\exp(-n)(2n\pi)^{1/2}}{k^k\exp(-k)(2k\pi)^{1/2}(n-k)^{n-k}\exp(k-n)[2(n-k)\pi]^{1/2}}\\&=(2\pi)^{-1/2}n^{n+1/2}k^{-k-1/2}(n-k)^{k-n-1/2}.\end{aligned}$$

For example, $\binom{10}{5}=252$ exactly, while the approximation yields $\binom{10}{5}\simeq258.4$, giving a relative error of 2.54%. The approximation improves as both n and k increase. For example, $\binom{20}{10}=184756$ exactly, while the approximation yields $\binom{20}{10}\simeq187078.973$, giving a relative error of 1.26%. ∎

1.3 Sequences of Real Functions

The convergence of sequences of functions will play an important role throughout this book due to the fact that sequences of random variables are really just sequences of functions. This section will review the basic definitions and results for convergent sequences of functions from a calculus based viewpoint. More sophisticated results, such as those based on measure theory, will be introduced in Chapter 3.

The main difference between studying the convergence properties of a sequence of real numbers and a sequence of real valued functions is that there is not a single definition that characterizes a convergent sequence of functions. This section will study two basic types of convergence: pointwise convergence and uniform convergence.

Definition 1.4. *Let $\{f_n(x)\}_{n=1}^{\infty}$ be a sequence of real valued functions. The sequence converges pointwise to a real function f if*

$$\lim_{n\to\infty}f_n(x)=f(x)\qquad(1.9)$$

for all $x \in \mathbb{R}$. We will represent this property as $f_n \xrightarrow{pw} f$ as $n \to \infty$.

In Definition 1.4 the problem of convergence is reduced to looking at the convergence properties of the real sequence of numbers given by $\{f_n(x)\}_{n=1}^{\infty}$ when x is a fixed real number. Therefore, the limit used in Equation (1.9) is the same limit that is defined in Definition 1.1.

Example 1.10. Consider a sequence of functions $\{f_n(x)\}_{n=1}^{\infty}$ defined on the unit interval by $f_n(x) = 1 + n^{-1}x^2$ for all $n \in \mathbb{N}$. Suppose that $x \in [0,1]$ is a fixed real number. Then

$$\lim_{n \to \infty} f_n(x) = \lim_{n \to \infty} (1 + n^{-1}x^2) = 1,$$

regardless of the value of $x \in [0,1]$. Therefore $f_n \xrightarrow{pw} f$ as $n \to \infty$.   ∎

Example 1.11. Let $\{f_n(x)\}_{n=1}^{\infty}$ be a sequence of real valued functions defined on the unit interval $[0,1]$ as $f_n(x) = x^n$ for all $n \in \mathbb{N}$. First suppose that $x \in [0,1)$ and note that

$$\lim_{n \to \infty} f_n(x) = \lim_{n \to \infty} x^n = 0.$$

However, if $x = 1$ then

$$\lim_{n \to \infty} f_n(x) = \lim_{n \to \infty} 1 = 1.$$

Therefore $f_n \xrightarrow{pw} f$ as $n \to \infty$ where $f_n(x) = \delta\{x; \{1\}\}$, and δ is the indicator function defined by

$$\delta\{x; A\} = \begin{cases} 1 & if x \in A, \\ 0 & if x \notin A. \end{cases}$$

Note that this example also demonstrates that the limit of a sequence of continuous functions need not also be continuous.   ∎

Because the definition of pointwise convergence for sequences of functions is closely related to Definition 1.1, the definition for limit for real sequences, many of the properties of limits also hold for sequences of functions that converge pointwise. For example, if $\{f_n(x)\}_{n=1}^{\infty}$ and $\{g_n(x)\}_{n=1}^{\infty}$ are sequences of functions that converge to the functions f and g, respectively, then the sequence $\{f_n(x) + g_n(x)\}_{n=1}^{\infty}$ converges pointwise to $f(x) + g(x)$. See Exercise 14.

A different approach to defining convergence for sequences of real functions requires not only that the sequence of functions converge pointwise to a limiting function, but that the convergence must be uniform in $x \in \mathbb{R}$. That is, if $\{f_n\}_{n=1}^{\infty}$ is a sequence of functions that convergence pointwise to a function f, we further require that the rate of convergence of $f_n(x)$ to $f(x)$ as $n \to \infty$ does not depend on $x \in \mathbb{R}$.

Definition 1.5. *A sequence of functions $\{f_n(x)\}_{n=1}^{\infty}$ converges uniformly to a function $f(x)$ as $n \to \infty$ if for every $\varepsilon > 0$ there exists an integer n_ε such that $|f_n(x) - f(x)| < \varepsilon$ for all $n \geq n_\varepsilon$ and $x \in \mathbb{R}$. This type of convergence will be represented as $f_n \xrightarrow{u} f$ as $n \to \infty$.*

Example 1.12. Consider once again the sequence of functions $\{f_n(x)\}_{n=1}^\infty$ given by $f_n(x) = 1 + n^{-1}x^2$ for all $n \in \mathbb{N}$ and $x \in [0,1]$. It was shown in Example 1.10 that $f_n \xrightarrow{pw} f = 1$ as $n \to \infty$ on $[0,1]$. Now we investigate whether this convergence is uniform or not. Let $\varepsilon > 0$, and note that for $x \in [0,1]$,

$$|f_n(x) - f(x)| = |1 + n^{-1}x^2 - 1| = |n^{-1}x^2| \leq n^{-1},$$

for all $x \in [0,1]$. Take $n_\varepsilon = \varepsilon^{-1} + 1$ and we have that $|f_n(x) - f(x)| < \varepsilon$ for all $n > n_\varepsilon$. Because n_ε does not depend on x, we have that $f_n \xrightarrow{u} f$, as $n \to \infty$ on $[0,1]$. ∎

Example 1.13. Consider the sequence of functions $\{f_n(x)\}_{n=1}^\infty$ given by $f_n(x) = x^n$ for all $x \in [0,1]$ and $n \in \mathbb{N}$. It was shown in Example 1.11 that $f_n \xrightarrow{pw} f$ as $n \to \infty$ where $f_n(x) = \delta\{x; \{1\}\}$ on $[0,1]$. Consider $\varepsilon \in (0,1)$. For any value of $n \in \mathbb{N}$, $|f_n(x) - 0| < \varepsilon$ on $(0,1)$ when $x^n < \varepsilon$ which implies that $n > \log(\varepsilon)/\log(x)$, where we note that $\log(x) < 0$ when $x \in (0,1)$. Hence, such a bound on n will always depend on x since $\log(x)$ is unbounded in the interval $(0,1)$, and therefore the sequence of functions $\{f_n\}_{n=1}^\infty$ does not converge uniformly to f as $n \to \infty$. ∎

Example 1.12 demonstrates one characteristic of sequences of functions that are uniformly convergent, in that the limit of a sequence of uniformly convergent continuous functions must also be continuous.

Theorem 1.9. *Suppose that $\{f_n(x)\}_{n=1}^\infty$ is a sequence of functions on a subset R of $\mathbb{R}$ that converge uniformly to a function f. If each f_n is continuous at a point $x \in R$, then f is also continuous at x.*

A proof of Theorem 1.9 can be found in Section 9.4 of Apostol (1974). Note that uniform convergence is a sufficient, but not a necessary condition, for the limit function to be continuous, as is demonstrated by Example 1.13. An alternate view of uniformly convergent sequences of functions can be defined in terms of Cauchy sequences, as shown below.

Theorem 1.10. *Suppose that $\{f_n(x)\}_{n=1}^\infty$ is a sequence of functions on a subset R of $\mathbb{R}$. There exists a function f such that $f_n \xrightarrow{u} f$ as $n \to \infty$ if and only if for every $\varepsilon > 0$ there exists $n_\varepsilon \in \mathbb{N}$ such that $|f_n(x) - f_m(x)| < \varepsilon$ for all $n > n_\varepsilon$ and $m > n_\varepsilon$, for every $x \in R$.*

A proof of Theorem 1.10 can be found in Section 9.5 of Apostol (1974).

Another important property of sequences of functions $\{f_n(x)\}_{n=1}^\infty$ is whether a limit and an integral can be exhanged. That is, let a and b be real constants that do not depend on n such that $a < b$. Does it necessarily follow that

$$\lim_{n\to\infty} \int_a^b f_n(x)dx = \int_a^b \lim_{n\to\infty} f_n(x)dx = \int_a^b f(x)dx?$$

An example can be used to demonstrate that such an exchange is not always justified.

Example 1.14. Consider a sequence of real functions defined by $f_n(x) = 2^n \delta\{x; (2^{-n}, 2^{-(n-1)})\}$ for all $n \in \mathbb{N}$. The integral of f_n is given by

$$\int_{-\infty}^{\infty} f_n(x)dx = \int_{2^{-n}}^{2^{-(n-1)}} 2^n dx = 1,$$

for all $n \in \mathbb{N}$. Therefore

$$\lim_{n \to \infty} \int_{-\infty}^{\infty} f_n(x)dx = 1.$$

However, for each $x \in \mathbb{R}$,

$$\lim_{n \to \infty} f_n(x) = 0,$$

so that

$$\int_{-\infty}^{\infty} \lim_{n \to \infty} f_n(x) = \int_{-\infty}^{\infty} 0 dx = 0.$$

Therefore, in this case

$$\lim_{n \to \infty} \int_a^b f_n(x)dx \neq \int_a^b \lim_{n \to \infty} f_n(x)dx.$$

■

While Example 1.14 shows it is not always possible to interchange a limit and an integral, there are some instances where the change is allowed. One of the most useful of these cases occurs when the sequence of functions $\{f_n(x)\}_{n=1}^{\infty}$ is dominated by an integrable function, that is, a function whose integral exists and is finite. This result is usually called the *Dominated Convergence Theorem*.

Theorem 1.11 (Lebesgue). *Let $\{f_n(x)\}_{n=1}^{\infty}$ be a sequence of real functions. Suppose that $f_n \xrightarrow{pw} f$ as $n \to \infty$ for some real valued function f, and that there exists a real function g such that*

$$\int_{-\infty}^{\infty} |g(x)|dx < \infty,$$

and $|f_n(x)| \leq g(x)$ for all $x \in \mathbb{R}$ and $n \in \mathbb{N}$. Then

$$\int_{-\infty}^{\infty} |f(x)|dx < \infty,$$

and

$$\lim_{n \to \infty} \int_{-\infty}^{\infty} f_n(x)dx = \int_{-\infty}^{\infty} \lim_{n \to \infty} f_n(x)dx = \int_{-\infty}^{\infty} f(x)dx.$$

A proof of this result can be found in Chapter 9 of Sprecher (1970). Example 1.14 demonstrated a situation where the interchange between an integral and a limit is not justified. As such, the sequence of functions in Example 1.14 must violate the assumptions of Theorem 1.11 in some way. The main assumption in Theorem 1.11 is that the sequence of functions $\{f_n(x)\}_{n=1}^{\infty}$ is dominated by an integrable function g. This is the assumption that is violated in Example 1.14. Consider the function defined by

$$g(x) = \sum_{n=1}^{\infty} 2^n \delta\{x; (2^{-n}, 2^{-(n-1)})\}.$$

This function dominates $f_n(x)$ for all $n \in \mathbb{N}$ in that $|f_n(x)| \leq g(x)$ for all $x \in \mathbb{R}$ and $n \in \mathbb{N}$. But note that

$$\int_{-\infty}^{\infty} g(x)dx = \sum_{n=1}^{\infty} \int_{-\infty}^{\infty} 2^n \delta\{x; (2^{-n}, 2^{-(n-1)})\} = \sum_{n=1}^{\infty} 1 = \infty. \qquad (1.10)$$

Therefore, g is not an integrable function. Could there be another function that dominates the sequence $\{f_n(x)\}_{n=1}^{\infty}$ that is integrable? Such a function would have to be less than g for at least some values of x in order to make the integral in Equation (1.10) finite. This is not possible because $g(x) = f_n(x)$ when $x \in (2^{-n}, 2^{-(n-1)})$, for all $n \in \mathbb{N}$. Therefore, there is not an integrable function g that dominates the sequence $\{f_n(x)\}_{n=1}^{\infty}$ for all $n \in \mathbb{N}$.

A common application of Theorem 1.11 in statistical limit theory is to show that the convergence of the density of a sequence of random variables also implies that the distribution function of the sequence converges.

Example 1.15. Consider a sequence of real functions $\{f_n(x)\}_{n=1}^{\infty}$ defined by

$$f_n(x) = \begin{cases} (1 + n^{-1})\exp[-x(1 + n^{-1})] & \text{for } x > 0, \\ 0 & \text{for } x \leq 0, \end{cases}$$

for all $n \in \mathbb{N}$. The pointwise limit of the sequence of functions is given by

$$\lim_{n \to \infty} f_n(x) = \lim_{n \to \infty} (1 + n^{-1})\exp[-x(1 + n^{-1})] = \exp(-x),$$

for all $x > 0$. Now, note that

$$|f_n(x)| = (1 + n^{-1})\exp[-x(1 + n^{-1})] \leq 2\exp(-x),$$

for all $n \in \mathbb{N}$. Therefore, if we define

$$g(x) = \begin{cases} 2\exp(-x) & \text{for } x > 0, \\ 0, & \text{for } x \leq 0 \end{cases}$$

we have that $|f_n(x)| \leq g(x)$ for all $x \in \mathbb{R}$ and $n \in \mathbb{N}$. Now

$$\int_{-\infty}^{\infty} g(x)dx = \int_0^{\infty} 2\exp(-x)dx = 2 < \infty,$$

so that g is an integrable function. Theorem 1.11 then implies that

$$\lim_{n\to\infty} \int_0^x f_n(t)dt = \lim_{n\to\infty} \int_0^x (1+n^{-1})\exp[-t(1+n^{-1})]dt$$

$$= \int_0^x \lim_{n\to\infty} (1+n^{-1})\exp[-t(1+n^{-1})]dt$$

$$= \int_0^x \exp(-t)dt$$

$$= 1 - \exp(-x).$$

∎

A related result that allows for the interchange of a limit and an integral is based on the assumption that the sequence of functions is monotonically increasing or decreasing to a limiting function.

Theorem 1.12 (Lebesgue's Monotone Convergence Theorem). *Let $\{f_n(x)\}_{n=1}^{\infty}$ be a sequence of real functions that are monotonically increasing to f on $\mathbb{R}$. That is $f_i(x) \leq f_j(x)$ for all $x \in \mathbb{R}$ when $i < j$, for positive integers i and j and $f_n \xrightarrow{pw} f$ as $n \to \infty$ on $\mathbb{R}$. Then*

$$\lim_{n\to\infty} \int_{-\infty}^{\infty} f_n(x)dx = \int_{-\infty}^{\infty} \lim_{n\to\infty} f_n(x)dx = \int_{-\infty}^{\infty} f(x)dx. \qquad (1.11)$$

It is important to note that the integrals in Equation (1.11) may be infinite, unless the additional assumption that f is integrable is added. The result is not just limited to monotonically increasing sequences of functions. The corollary below provides a similar result for monotonically decreasing functions.

Corollary 1.1. *Let $\{f_n(x)\}_{n=1}^{\infty}$ be a sequence of non-negative real functions that are monotonically decreasing to f on $\mathbb{R}$. That is $f_i(x) \geq f_j(x)$ for all $x \in \mathbb{R}$ when $i < j$, for positive integers i and j and $f_n \xrightarrow{pw} f$ as $n \to \infty$ on $\mathbb{R}$. If f_1 is integrable then*

$$\lim_{n\to\infty} \int_{-\infty}^{\infty} f_n(x)dx = \int_{-\infty}^{\infty} \lim_{n\to\infty} f_n(x)dx = \int_{-\infty}^{\infty} f(x)dx.$$

Proofs of Theorem 1.12 and Corollary 1.1 can be found in Gut (2005) or Sprecher (1970).

Example 1.16. The following setup is often used when using arguments that rely on the truncation of random variables. Let g be an integrable function on $\mathbb{R}$ and define a sequence of functions $\{f_n(x)\}_{n=1}^{\infty}$ as $f_n(x) = g(x)\delta\{|x|; (0, n)\}$. It follows that for all $x \in \mathbb{R}$ that

$$\lim_{n\to\infty} f_n(x) = \lim_{n\to\infty} g(x)\delta\{|x|; (0, n)\} = g(x),$$

since for each $x \in \mathbb{R}$ there exists an integer n_x such that $f_{n_x}(x) = g(x)$ for all $n \geq n_x$. Noting that $\{f_n(x)\}_{n=1}^{\infty}$ is a monotonically increasing sequence of functions allows us to use Theorem 1.12 to conclude that

$$\lim_{n\to\infty} \int_{-\infty}^{\infty} f_n(x)dx = \int_{-\infty}^{\infty} \lim_{n\to\infty} f_n(x)dx = \int_{-\infty}^{\infty} g(x)dx.$$

∎

1.4 The Taylor Expansion

Taylor's Theorem, and the associated approximation of smooth functions, are crucial elements of asymptotic theory in both mathematics and statistics. In particular, the Central Limit Theorem is based on this result. Adding to the usefulness of this idea is the fact that the general method can be extended to vector functions and spaces of functions. Throughout this book various forms of Taylor's Theorem will serve as important tools in many diverse situations.

Before proceeding to the main result, it is useful to consider some motivation of the form of the most common use of Taylor's Theorem, which is the approximation of smooth functions near a known value of the function. Consider a generic function f that is reasonably smooth. That is, f has no jumps or kinks. This can be equivalently stated in terms of the derivatives of f. In this example it will be assumed that f has at least three continuous derivatives in a sufficiently large neighborhood of x.

Suppose that the value of the function f is known at a point x, but the value at $x+\delta$ is desired, where δ is some positive small real number. The sign of δ is not crucial to this discussion, but restricting δ to be positive will simplify the use of the figures that illustrate this example. If f is linear, that is $f(x) = a + bx$ for some real constants a and b, then it follows that

$$f(x + \delta) = f(x) + \delta f'(x). \tag{1.12}$$

Now suppose that f is not linear, but is approximately linear in a small neighborhood of x. This will be true of all reasonably smooth functions as long as the neighborhood is small enough. In this case, the right hand side of Equation (1.12) can be used to approximate $f(x + \delta)$. That is

$$f(x + \delta) \simeq f(x) + \delta f'(x). \tag{1.13}$$

See Figure 1.1. One can observe from Figure 1.1 that the quality of the approximation will depend on many factors. It is clear, at least visually, that the approximation becomes more accurate as $\delta \to 0$. This is generally true and can be proven using the continuity of f. It is also clear from Figure 1.1 that the accuracy of the approximation will also depend on the curvature of f. The less curvature f has, the more accurate the approximation will be. Hence, the accuracy of the approximation depends on $|f''|$ in a neighborhood of x as well. The accuracy will also depend on higher derivatives when they exist and are non-zero. To consider the accuracy of this approximation more closely, define

$$E_1(x, \delta) = f(x + \delta) - f(x) - \delta f'(x),$$

Figure 1.1 *The linear approximation of a curved function. The solid line indicates the form of the function f and the dotted line shows the linear approximation of $f(z + \delta)$ given by $f(z + \delta) \simeq f(z) + \delta f'(z)$ for the point z indicated on the graph.*

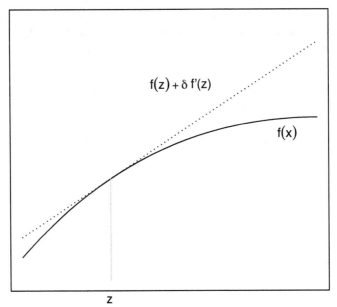

$$f(z) + \delta\, f'(z)$$

$$f(x)$$

$$z$$

the error of the approximation as a function of both x and δ. Note that using Theorem A.3 yields

$$f(x + \delta) - f(x) = \int_x^{x+\delta} f'(t)dt,$$

and

$$\delta f'(x) = f'(x) \int_x^{x+\delta} dt,$$

so that

$$E_1(x, \delta) = \int_x^{x+\delta} [f'(t) - f'(x)]dt.$$

An application of Theorem A.4 yields

$$
\begin{aligned}
E_1(x, \delta) &= -(x + \delta - t)[f'(t) - f'(x)]\big|_x^{x+\delta} + \int_x^{x+\delta} (x + \delta - t)f''(t)dt \\
&= \int_x^{x+\delta} (x + \delta - t)f''(t)dt,
\end{aligned}
$$

which establishes the role of the second derivative in the error of the approximation.

Note that if f has a high degree of curvature, with no change in the direction of the concavity of f, the absolute value of the integral in $E_1(x, \delta)$ will be large. This indicates that the function continues turning away from the linear approximation as shown in Figure 1.1. If f has an inflection point in the interval $(x, x + \delta)$, the direction on the concavity will change and the integral will become smaller. The change in the sign of f'' indicates that the function is turning back toward the linear approximation and the error will decrease. See Figure 1.2.

A somewhat simpler form of the error term $E_1(x, \delta)$ can be obtained through an application of the Theorem A.5, as

$$
\begin{aligned}
E_1(x, \delta) &= \int_x^{x+\delta} (x + \delta - t) f''(t) dt \\
&= f''(\xi) \int_x^{x+\delta} (x + \delta - t) dt \\
&= \tfrac{1}{2} f''(\xi) \delta^2
\end{aligned}
$$

for some $\xi \in [x, x + \delta]$. The exact value of ξ will depend on x, δ, and f.

If f is a quadratic polynomial it follows that $f(x + \delta)$ can be written as

$$
f(x + \delta) = f(x) + \delta f'(x) + \tfrac{1}{2} \delta^2 f''(x). \tag{1.14}
$$

See Exercise 21. As in the case of the linear approximation, if f is not a quadratic polynomial then the right hand side of Equation (1.14) can be used to approximate $f(x + \delta)$. Note that in the case where f is linear, the approximation would revert back to the linear approximation given in Equation (1.13) and the result would be exact. In the case where the right hand side of Equation (1.14) is an approximation, it is logical to consider whether the quadratic approximation is better than the linear approximation given in Equation (1.13). Obviously the linear approximation could be better in some specialized cases for specific values of δ even when f is not linear. See Figure 1.2.

Assume that $f'''(x)$ exists, is finite, and continuous in the interval $(x, x + \delta)$. Define

$$
E_2(x, \delta) = f(x + \delta) - f(x) - \delta f'(x) - \tfrac{1}{2} \delta^2 f''(x). \tag{1.15}
$$

Note that the first three terms on the right hand side of Equation (1.15) equals $E_1(x, \delta)$ so that

$$
E_2(x, \delta) = E_1(x, \delta) - \tfrac{1}{2} \delta^2 f''(x) = \int_x^{x+\delta} (x + \delta - t) f''(t) dt - \tfrac{1}{2} \delta^2 f''(x).
$$

Now note that

$$
\tfrac{1}{2} \delta^2 f''(x) = f''(x) \int_x^{x+\delta} (x + \delta - t) dt,
$$

Figure 1.2 *The linear approximation of a curved function. The solid line indicates the form of the function f and the dotted line shows the linear approximation of $f(z + \delta)$ given by $f(z + \delta) \simeq f(z) + \delta f'(z)$ for the point z indicated on the graph. In this case, the concavity of f changes and the linear approximation becomes more accurate again for larger values of δ, for the range of the values plotted in the figure.*

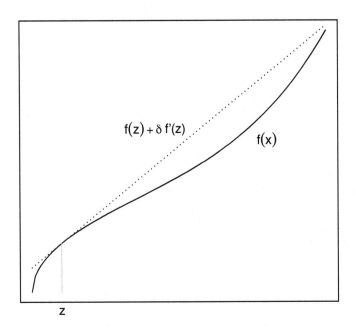

so that

$$E_2(x, \delta) = \int_x^{x+\delta} (x + \delta - t)[f''(t) - f''(x)]dt.$$

An application of Theorem A.4 yields

$$E_2(x, \delta) = \tfrac{1}{2} \int_x^{x+\delta} (x + \delta - t)^2 f'''(t)dt,$$

which indicates that the third derivative is the essential determining factor in the accuracy of this approximation. As with $E_1(x, \delta)$, the error term can be restated in a somewhat simpler form by using an application of Theorem A.5. That is,

$$
\begin{aligned}
E_2(x, \delta) &= \int_x^{x+\delta} (x + \delta - t)f'''(t)dt \\
&= f'''(\xi) \int_x^{x+\delta} (x + \delta - t)^2 dt \\
&= \tfrac{1}{6}\delta^3 f'''(\xi),
\end{aligned}
$$

for some $\xi \in [x, x + \delta]$, where the value of ξ will depend on x, δ and f. Note that δ is cubed in $E_2(x, \delta)$ as opposed to being squared in $E_1(x, \delta)$. This would imply, depending on the relative sizes of f'' and f''' in the interval $[x, x + \delta]$, that $E_2(x, \delta)$ will generally be smaller than $E_1(x, \delta)$ for small values of δ. In fact, one can note that

$$\frac{E_2(x, \delta)}{E_1(x, \delta)} = \frac{\frac{1}{6}\delta^3 f'''[\xi_2(\delta)]}{\frac{1}{2}\delta^2 f''[\xi_1(\delta)]} = \frac{\delta f'''[\xi_2(\delta)]}{3 f''[\xi_1(\delta)]}.$$

The values $\xi_1(\delta)$ and $\xi_2(\delta)$ are the values of ξ for $E_1(x, \delta)$ and $E_2(x, \delta)$, respectively, written here as a function of δ to emphasize the fact that these values change as $\delta \to 0$. In fact, as $\delta \to 0$, $\xi_1(\delta) \to x$ and $\xi_2(\delta) \to x$. Now assume that the derivatives are continuous so that $f''[\xi_1(\delta)] \to f''(x)$ and $f'''[\xi_1(\delta)] \to f'''(x)$ as $\delta \to 0$ and that $|f'''(x)/f''(x)| < \infty$. Then

$$\lim_{\delta \to 0} \frac{E_2(x, \delta)}{E_1(x, \delta)} = \lim_{\delta \to 0} \frac{\delta f'''[\xi_2(\delta)]}{f''[\xi_1(\delta)]} = 0. \tag{1.16}$$

Hence, under these conditions, the error from the quadratic approximation will be dominated by the error from the linear approximation.

If f is a sufficiently smooth function, ensured by the existence of a required number of derivatives, then the process described above can be iterated further to obtain potentially smaller error terms when δ is small. This results in what is usually known as Taylor's Theorem.

Theorem 1.13 (Taylor). *Let f be a function that has $p + 1$ bounded and continuous derivatives in the interval $(x, x + \delta)$. Then*

$$f(x + \delta) = \sum_{k=0}^{p} \frac{\delta^k f^{(k)}(x)}{k!} + E_p(x, \delta),$$

where

$$E_p(x, \delta) = \frac{1}{p!} \int_x^{x+\delta} (x + \delta - t)^p f^{(p+1)}(t)dt = \frac{\delta^{p+1} f^{(p+1)}(\xi)}{(p + 1)!},$$

for some $\xi \in (x, x + \delta)$.

For a proof of Theorem 1.13 see Exercises 24 and 25. What is so special about the approximation that is obtained using Theorem 1.13? Aside from the motivation given earlier in this section, consider taking the first derivative of the pth order approximation at $\delta = 0$, which is

$$\frac{d}{d\delta} \sum_{k=0}^{p} \frac{\delta^k f^{(k)}(x)}{k!} \bigg|_{\delta=0} = \frac{d}{d\delta} \left[f(x) + \delta f'(x) + \sum_{k=2}^{p} \frac{\delta^k f^{(k)}(x)}{k!} \right] \bigg|_{\delta=0} = f'(x).$$

Hence, the approximating function has the same derivative at x as the actual function $f(x)$. In general it can be shown that

$$\frac{d^j}{d\delta^j} \sum_{k=0}^{p} \frac{\delta^k f^{(k)}(x)}{k!} \bigg|_{\delta=0} = f^{(j)}(x), \tag{1.17}$$

for $j = 1, \ldots, p$. Therefore, the pth-order approximating function has the same derivatives of order $1, \ldots, p$ as the actual function $f(x)$ at the point x. A proof of Equation (1.17) is given in Exercise 27.

Note that an alternative form of the expansion given in Theorem 1.13 can be obtained by setting $y = x + \delta$ and $x = y_0$ so that $\delta = y - x = y - y_0$ and the expansion has the form

$$f(y) = \sum_{k=0}^{p} \frac{(y - y_0)^k f^{(k)}(y_0)}{k!} + E_p(y, y_0), \qquad (1.18)$$

where

$$E_p(y, y_0) = \frac{(y - y_0)^{p+1} f^{(p+1)}(\xi)}{(p+1)!},$$

and ξ is between y and y_0. The expansion given in Equation (1.18) is usually called the expansion of f around the point y_0.

Example 1.17. Suppose we wish to approximate the exponential function for positive arguments near zero. That is, we wish to approximate $\exp(\delta)$ for small values of δ. A simple approximation for these values may be useful since the exponential function does not have a simple closed form from which it can be evaluated. Several approximations based on Theorem 1.13 will be considered in detail. For the first approximation, take $p = 1$ in Theorem 1.13 to obtain $\exp(x+\delta) = \exp(x)+\delta\exp(x)+E_1(x,\delta)$ so that when $x = 0$ the approximation simplifies to $\exp(\delta) = 1 + \delta + E_1(\delta)$. The error term is now written only as a function of δ since the value of x is now fixed. Similarly, when $p = 2$ and $p = 3$, Theorem 1.13 yields the approximations $\exp(\delta) = 1 + \delta + \frac{1}{2}\delta^2 + E_2(\delta)$ and $\exp(\delta) = 1 + \delta + \frac{1}{2}\delta^2 + \frac{1}{6}\delta^3 + E_3(\delta)$, respectively. The simple polynomial approximations given by Theorem 1.13 in this case are due to the simple form of the derivative of the exponential function. The exponential function, along with these three approximations are plotted in Figure 1.3. One can observe from the figure that all of the approximations, even the linear one, do well for very small values of δ. This is due to the fact that all of the error terms converge to zero as $\delta \to 0$. As more terms are added to the Taylor expansion, the approximation gets better, and has relatively smaller error even for larger values of δ. For example, the absolute error for the cubic approximation for $\delta = 1$ is smaller than that of the linear approximation at $\delta = 0.3$.

To emphasize the difference in the behavior of the error terms for each of the approximations, the size of the absolute error for each approximation has been plotted in Figure 1.4. The large differences in the absolute error for each of the three approximations is quite apparent from Figure 1.4, as well as the fact that all three error terms converge to zero as $\delta \to 0$.

The relative absolute errors of each approximation are plotted in Figure 1.5. One can visually observe the effect that is derived in Equation (1.16). All three of the relative errors converge to zero as $\delta \to 0$, demonstrating that the

Figure 1.3 *The exponential function (solid line) and three approximations based on Theorem 1.13 using p = 1 (dashed line), p = 2 (dotted line) and p = 3 (dash-dot line).*

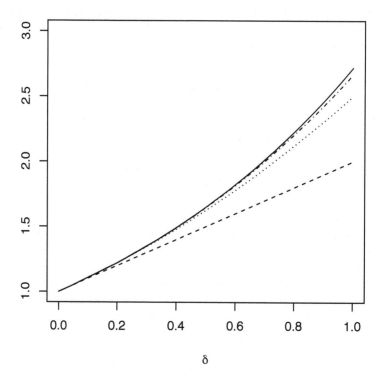

absolute errors from the higher order polynomial approximations are dominated by the absolute errors from the lower order polynomial approximations as $\delta \to 0$. One can also observe from Figure 1.5 that error of the quadratic approximation relative to that of the linear approximation is much larger than that of the absolute error of the cubic approximation relative to that of the linear approximation. ∎

Example 1.18. Consider the distribution function of a N(0, 1) distribution given by

$$\Phi(x) = \int_{-\infty}^{x} (2\pi)^{-1/2} \exp(-t^2/2)dt.$$

We would like to approximate $\Phi(x)$, an integral that has no simple closed form, with a simple function for values of x near 0. As in the previous example, three approximations based on Theorem 1.13 will be considered. Applying Theorem 1.13 to $\Phi(x + \delta)$ with $p = 1$ yields the approximation

$$\Phi(x + \delta) = \Phi(x) + \delta\Phi'(x) + E_1(x, \delta) = \Phi(x) + \delta\phi(x) + E_1(x, \delta),$$

Figure 1.4 *The absolute error for approximating the exponential function using the three approximations based on Theorem 1.13 with $p = 1$ (dashed line), $p = 2$ (dotted line) and $p = 3$ (dash-dot line).*

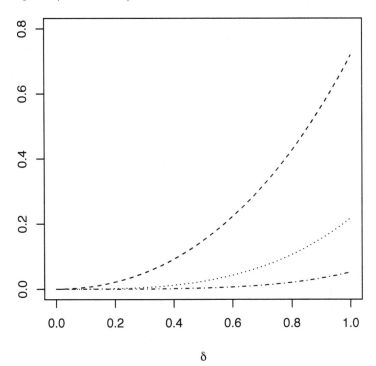

where $\phi(x) = \Phi'(x)$ is the density function of a $N(0,1)$ distribution. Setting $x = 0$ yields

$$\Phi(\delta) = \tfrac{1}{2} + \delta\phi(0) + E_1(\delta) = \tfrac{1}{2} + \delta(2\pi)^{-1/2} + E_1(\delta).$$

With $p = 2$ and $x = 0$, Theorem 1.13 yields

$$\Phi(\delta) = \tfrac{1}{2} + (2\pi)^{-1/2}\delta + \tfrac{1}{2}\phi'(0)\delta^2 + E_2(\delta),$$

where

$$\phi'(0) = \left.\frac{d}{dx}\phi(x)\right|_{x=0} = \left.-x(2\pi)^{-1/2}\exp(-x^2/2)\right|_{x=0} = \left.-x\phi(x)\right|_{x=0} = 0.$$

Hence, the quadratic approximation is the same as the linear one. This indicates that the linear approximation is more accurate in this case than what would usually be expected. The cubic approximation has the form

$$\Phi(\delta) = \tfrac{1}{2} + (2\pi)^{-1/2}\delta + \tfrac{1}{2}\phi'(0)\delta^2 + \tfrac{1}{6}\phi''(0)\delta^3 + E_3(\delta), \tag{1.19}$$

Figure 1.5 *The absolute relative errors for approximating the exponential function using the three approximations based on Theorem 1.13 with $p = 1$, $p = 2$ and $p = 3$. The relative errors are $|E_2(\delta)/E_1(\delta)|$ (solid line), $|E_3(\delta)/E_1(\delta)|$ (dashed line) and $|E_3(\delta)/E_2(\delta)|$ (dotted line).*

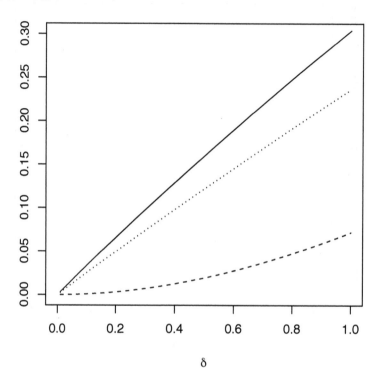

where

$$\phi''(0) = \frac{d}{dx} - x\phi(x)\Big|_{x=0} = (x^2 - 1)\phi(x)\Big|_{x=0} = -(2\pi)^{-1/2}.$$

Hence, the cubic approximation has the form

$$\Phi(\delta) = \tfrac{1}{2} + \delta\phi(0) + E_1(\delta) = \tfrac{1}{2} + \delta(2\pi)^{-1/2} - \tfrac{1}{6}(2\pi)^{-1/2}\delta^3 + E_3(\delta).$$

The linear and quadratic approximations are plotted in Figure 1.6. It is again clear that both approximations are accurate for very small values of δ. The cubic approximation does relatively well for $\delta \in [0, 1]$, but quickly becomes worse for larger values of δ. Note further that the approximations do not provide a valid distribution function. The linear approximation quickly exceeds one and the cubic approximation is not a non-decreasing function. Most approximations for distribution functions have a limited range where the approximation is both accurate, and provides a valid distribution function. The decreased error of the linear approximation when $x = 0$ is due to the fact that the stan-

Figure 1.6 *The standard normal distribution function (solid line) and two approxi-mations based on Theorem 1.13 using p = 1 (dashed line) and p = 3 (dotted line).*

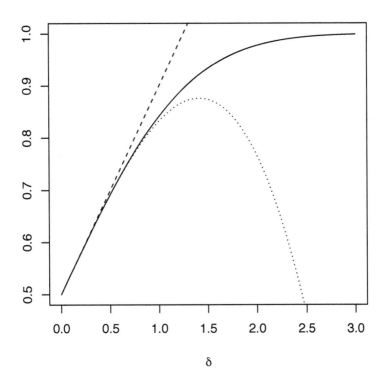

dard normal distribution function is nearly linear at $x = 0$. This increased accuracy does not hold for other values of x, as the term $\frac{1}{2}\phi'(x)\delta^2$ is only zero at $x = 0$. What is the purpose of the cubic term in this approximation? While the normal distribution function is nearly linear at around 0, there is some curvature present. In fact, the concavity of $\Phi(x)$ changes at $x = 0$. The linear term, whose concavity does not change, is unable to capture this behavior. However, a cubic term can. Note that the cubic term is negative when $\delta > 0$ and is positive when $\delta < 0$. This allows the cubic approximation to capture the slight curvature around $x = 0$ that the linear approximation cannot. ■

From Example 1.18 it is clear that the derivatives of the standard normal density have a specific form in that they are all a polynomial multiplied by the standard normal density. The polynomial multipliers, called *Hermite poly-nomials* are quite useful and will be used later in the book.

Definition 1.6. *Let $\phi(x)$ be the density of a standard normal random variable.*

The kth derivative of $\phi(x)$ can be expressed as

$$\frac{d^k}{dx^k}\phi(x) = (-1)^k H_k(x)\phi(x),$$

where $H_k(x)$ is a kth-order polynomial in x called the kth Hermite polynomial. That is,

$$H_k(x) = \frac{(-1)^k \phi^{(k)}(x)}{\phi(x)}.$$

Hermite polynomials are an example of a set of orthogonal polynomials. See Exercise 33. Hermite polynomials also have many interesting properties including a simple recurrence relation between successive polynomials.

Theorem 1.14. *Let $H_k(x)$ be the kth Hermite polynomial, then*

1. For any positive integer k,

$$H_k(x) = \sum_{i=0}^{\lfloor k/2 \rfloor} (-1)^i \frac{(2i)!}{2^i i!} \binom{k}{2i} x^{k-2i},$$

where $\lfloor k/2 \rfloor$ is the greatest integer less than or equal to $k/2$.

2. For any integer $k \geq 2$, $H_k(x) = xH_{k-1}(x) - (k-1)H_{k-2}(x)$.

3. For any positive integer k,

$$\frac{d}{dx}H_k(x) = kH_{k-1}(x).$$

For a proof of Theorem 1.14 see Exercises 30, 31, and 32.

1.5 Asymptotic Expansions

An asymptotic expansion is an approximation of a function usually written as a partial series that gains accuracy usually as a certain parameter approaches a specified value, usually zero or infinity. Asymptotic expansions have already been encountered in Section 1.4 where the Taylor expansion was used. The general form of an asymptotic expansion is

$$f(x,\delta) = d_0(x) + \delta d_1(x) + \delta^2 d_2(x) + \cdots + \delta^p d_p(x) + E_p(x,\delta). \qquad (1.20)$$

This approximation is asymptotic in the sense that the accuracy of the approximation gets better as $\delta \to 0$ for fixed values of x. This type of expansion is usually called a pth-order expansion since the highest power of δ in the approximation is p. The functions $d_1(x), \ldots, d_p(x)$ are functions of x only, and do not depend on δ. The power of δ need not be integers, but the power of δ should go up by a set amount for each additional term. For example, many asymptotic expansions may have the form

$$f(x,\delta) = d_0(x) + \delta^{1/2}d_1(x) + \delta d_2(x) + \cdots + \delta^{p/2}d_p(x) + E_p(x,\delta),$$

which is written in terms of powers of the square root of δ. Asymptotic expansions are also often defined in terms of the general form

$$f(x,\delta) = f_0(x)[1 + \delta d_1(x) + \cdots + \delta^p d_p(x) + E_p(x,\delta)], \qquad (1.21)$$

for approximating $f(x,\delta)$ around a leading term $f_0(x)$. See Chapter 3 of Barndorff-Nielsen and Cox (1989). An essential issue when dealing with asymptotic expansion is the fact that adding terms to an expansion, for example going from a pth-order expansion to a $(p+1)^{\text{st}}$-order expansion, does not necessarily increase the accuracy of the corresponding approximation for fixed δ. In fact, the expansion may not even have a finite sum as $p \to \infty$. The accuracy of the expansion is an asymptotic property. That is, the error term $E_{p+1}(x,\delta)$ will usually be dominated by $E_p(x,\delta)$ only in the limit, or as $\delta \to 0$.

Example 1.19. Consider the function $f(x) = x^{-1}$. Suppose we wish to approximate $f(x+\delta)$ when $x = 1$ and δ is small. The first few derivatives of $f(x)$ are $f'(x) = -x^{-2}$, $f''(x) = 2x^{-3}$, and $f'''(x) = -6x^{-4}$. An analysis of these derivatives suggests the general form $f^{(k)}(x) = (-1)^k (k!) x^{-(k+1)}$. Therefore, the kth term of the Taylor expansion of $f(1+\delta)$ is

$$\frac{\delta^k f^{(k)}(1)}{k!} = (-1)^k \delta^k.$$

Hence, Theorem 1.13 implies

$$(1+\delta)^{-1} = \sum_{k=0}^{p}(-1)^k \delta^k + E_p(\delta). \qquad (1.22)$$

When δ is small, specifically when δ is fixed so that $0 \le \delta < 1$, it follows that

$$(1+\delta)^{-1} = \lim_{p\to\infty} \sum_{k=0}^{p}(-1)^k \delta^k = \sum_{k=0}^{\infty}(-1)^k \delta^k,$$

where the properties of the geometric series have been used. See Theorem A.23. However, if δ is fixed so that $\delta > 1$ then

$$\sum_{k=0}^{p}(-1)^{k+1}\delta^k \to \infty,$$

as $p \to \infty$. Therefore, for fixed values of p, the asymptotic expansion given in Equation 1.22 becomes more accurate as $\delta \to 0$, but for fixed δ the asymptotic expansion may become less accurate as $p \to \infty$. ∎

The asymptotic expansions typically encountered in statistical applications are based on an increasing sample size. These types of expansions are usually of the form

$$f(x,n) = d_0(x) + n^{-1}d_1(x) + n^{-2}d_2(x) + \cdots + n^{-p}d_p(x) + E_p(x,n),$$

or

$$f(x,n) = d_0(x) + n^{-1/2}d_1(x) + n^{-1}d_2(x) + \cdots + n^{-p/2}d_p(x) + E_p(x,n),$$

as $n \to \infty$. This form of the expansions is obtained by setting $\delta = n^{-1}$ in the previous expansions, where $\delta \to 0$ as $n \to \infty$. In many applications the function f is a density or distribution function that is being approximated for large sample sizes, or as $n \to \infty$. The function $d_0(x)$ is usually a known density or distribution function that the function f converges to as $n \to \infty$. In many cases $d_0(x)$ is the standard normal density or distribution function.

Example 1.20. Let $X_1, \ldots, X_n$ be a set of independent and identically distributed random variables following a BETA(α, β) distribution, and let $\bar{X}_n$ denote the sample mean. The distribution of $Z_n = n^{1/2}[\bar{X} - \mu(\alpha, \beta)]/\sigma(\alpha, \beta)$ is approximately N$(0, 1)$ when n is large, where $\mu(\alpha, \beta) = \alpha(\alpha + \beta)^{-1}$ and

$$\sigma(\alpha, \beta) = \left[\frac{\alpha\beta}{(\alpha + \beta)^2(\alpha + \beta + 1)} \right]^{1/2}.$$

In fact, more precise information about the asymptotic distribution of Z can be obtained using an *Edgeworth expansion*. These expansions will be covered in great detail later, but for the current presentation assume that the density of Z has the asymptotic expansion

$$f(z) = \phi(z) - \tfrac{1}{6}n^{-1/2}\phi(z)H_3(z)\kappa_3(\alpha, \beta) + R_2(z, n),$$

where

$$\kappa_3(\alpha, \beta) = \frac{2(\beta - \alpha)(\alpha + \beta + 1)^{1/2}}{(\alpha + \beta + 2)(\alpha\beta)^{1/2}},$$

which is the coefficient of skewness of the BETA(α, β) distribution, and $H_3(z)$ is the third-order Hermite polynomial from Definition 1.6 . The error term for this expansion is not the same as given by Theorem 1.13, and its form will not be considered at this time, other than the fact that

$$\lim_{n \to \infty} n^{1/2} R_2(z, n) = 0,$$

as $n \to \infty$. Therefore, the density converges pointwise to the standard normal density $\phi(z)$ as $n \to \infty$. In fact, one can use the first-order term

$$\tfrac{1}{6}n^{-1/2}\phi(z)H_3(z)\kappa_3(\alpha, \beta),$$

to more precisely describe the shape of this density when n is large. ∎

The exact form of the error terms of asymptotic expansions are typically not of interest. However, the asymptotic behavior of these errors as $\delta \to 0$ or as $n \to \infty$ terms is important. For example, in Section 1.4 it was argued that when certain conditions are met, the error term $E_2(x, \delta)$ from Theorem 1.13 is dominated as $\delta \to 0$ by $E_1(x, \delta)$, when x is fixed. Because the asymptotic behavior of the error term, and not its exact form, is important, a specific type of notation has been developed to symbolize the limit behavior of these sequences. *Asymptotic order notation* is a simple type of shorthand that indicates the asymptotic behavior of a sequence with respect to another sequence.

Definition 1.7. Let $\{x_n\}_{n=1}^{\infty}$ and $\{y_n\}_{n=1}^{\infty}$ be real sequences.

1. *The notation $x_n = o(y_n)$ as $n \to \infty$ means that*

$$\lim_{n \to \infty} \frac{x_n}{y_n} = 0.$$

That is, the sequence $\{x_n\}_{n=1}^{\infty}$ is dominated by $\{y_n\}_{n=1}^{\infty}$ as $n \to \infty$.

2. *The notation $x_n = O(y_n)$ as $n \to \infty$ means that the sequence $\{|x_n/y_n|\}_{n=1}^{\infty}$ remains bounded as $n \to \infty$.*

3. *The notation $x_n \asymp y_n$ means that*

$$\lim_{n \to \infty} \frac{x_n}{y_n} = 1,$$

or that the two sequence are asymptotically equivalent as $n \to \infty$.

The O-notation is often called *Backmann-Landau* notation. See Landau (1974). The asymptotic order notation is often used to replace the error term in an asymptotic expansion. For example, the error term of the expansion $f(x) = d_0(x) + n^{-1/2}d_1(x) + E_1(x, n)$ may have the property that $E_1(x, n) = O(n^{-1})$ as $n \to \infty$. In this case the expansion is often rewritten as $f(x) = d_0(x) + n^{-1/2}d_1(x) + O(n^{-1})$ as $n \to \infty$. The term $O(n^{-1})$ in the expansion represents an error sequence that has the property that when it is multiplied by n, the ratio remains bounded as $n \to \infty$. Convention usually stipulates that the sign of the error term is always positive when using asymptotic order notation, regardless of the sign of the actual sequence. It is always important to remember that the actual error sequence usually does not have the *exact* form specified by the asymptotic order notation. Rather, the notation provides a simple form for the *asymptotic* behavior of the sequence. Because the order notation is asymptotic in nature, it must be qualified by the limit that is being taken. In this book any limiting notation in the sample size n will be understood to be qualified by the statement $n \to \infty$, unless specifically noted otherwise. Similarly, any limiting notation in a real value δ will be understood to be qualified by the statement $\delta \to 0$, unless specifically noted otherwise. For example, $E(\delta) = o(\delta)$ will automatically be taken to mean

$$\lim_{\delta \to 0} \frac{E(\delta)}{\delta} = 0,$$

or that $E(\delta)$ is dominated by δ as $\delta \to 0$. We will usually continue to qualify these types of statements to emphasize their importance, but it is common in statistical literature for this notation to be understood.

Another important note on the order notation is that the asymptotic behavior of sequences is not unique. That is, if we suppose that the sequence a_n is $O(n^{-1})$ as $n \to \infty$, it then follows that the sequence can also be characterized as $O(n^{-1/2})$ and $O[(n + 1)^{-1}]$ as $n \to \infty$. To see why this is true, we refer to Definition 1.7 which tells us that $a_n = O(n^{-1})$ as $n \to \infty$ means than the sequence $|na_n|$ remains bounded for all $n \in \mathbb{N}$. Because $|n^{1/2}a_n| \leq |na_n|$ for all $n \in \mathbb{N}$, it follows that the sequence $|n^{1/2}a_n|$ remains bounded for all $n \in \mathbb{N}$

and therefore $a_n = O(n^{-1/2})$ as $n \to \infty$. A similar argument can be used to establish the fact that $a_n = O[(n+1)^{-1}]$ as $n \to \infty$.

Example 1.21. Consider two real sequences $\{a_n\}_{n=1}^{\infty}$ and $\{b_n\}_{n=1}^{\infty}$ defined by $a_n = (n-1)^{-1}$ and $b_n = (n+1)^{-1}$ for all $n \in \mathbb{N}$. Note that

$$\lim_{n\to\infty} \frac{a_n}{b_n} = 1,$$

so that $a_n \asymp b_n$ as $n \to \infty$. The fact that the sequence a_n/b_n remains bounded as $n \to \infty$ also implies that $a_n = O(b_n)$ and $b_n = O(a_n)$ as $n \to \infty$. In practice we would usually pick the simpler form of the sequence and conclude that $a_n = O(n^{-1})$ and $b_n = O(n^{-1})$ as $n \to \infty$. The key idea for many error sequences is that they converge to zero as $n \to \infty$. In these cases the asymptotic order notation allows us to study how quickly such sequences converge to zero. Hence, the results that $a_n = O(n^{-1})$ and $b_n = O(n^{-1})$ are usually interpreted by concluding that the sequences $\{a_n\}_{n=1}^{\infty}$ and $\{b_n\}_{n=1}^{\infty}$ converge to zero at the same rate as each other, and at the same rate as n^{-1}. Now consider the limits

$$\lim_{n\to\infty} n^{1/2} a_n = \lim_{n\to\infty} n^{1/2} b_n = 0.$$

This indicates that $a_n = o(n^{-1/2})$ and $b_n = o(n^{-1/2})$ as $n \to \infty$, and the conclusion is then that the sequences $\{a_n\}_{n=1}^{\infty}$ and $\{b_n\}_{n=1}^{\infty}$ converge to zero at a faster rate than $n^{-1/2}$. To emphasize the fact that these representations are not unique, we can also conclude that $a_n = o(n^{-1/4})$ and $a_n = o(n^{-1/256})$ as $n \to \infty$ as well, with similar conclusions for the sequence $\{b_n\}_{n=1}^{\infty}$. Note finally that any sequence that converges to zero, including the sequences $\{a_n\}_{n=1}^{\infty}$ and $\{b_n\}_{n=1}^{\infty}$, are also $o(1)$ as $n \to \infty$. ∎

The main tool that we have encountered for deriving asymptotic expansions is given by Theorem 1.13 (Taylor), which provided fairly specific forms of the error terms in the expansion. These error terms can also be written in terms of the asymptotic order notation to provide a simple asymptotic form of the errors.

Theorem 1.15. *Let f be a function that has $p+1$ bounded and continuous derivatives in the interval $(x, x + \delta)$. Then*

$$f(x+\delta) = \sum_{k=0}^{p} \frac{\delta^k f^{(k)}(x)}{k!} + E_p(x, \delta),$$

where $E_p(x, \delta) = O(\delta^{p+1})$ and $E_p(x, \delta) = o(\delta^p)$ as $\delta \to 0$.

Proof. We will prove that $E_p(x, \delta) = O(\delta^{p+1})$ as $\delta \to 0$. The fact that $E_p(x, \delta) = o(\delta^p)$ is proven in Exercise 34. From Theorem 1.13 we have that

$$E_p(x, \delta) = \frac{\delta^{p+1} f^{(p+1)}(\xi)}{(p+1)!},$$

for some $\xi \in (x, x + \delta)$. Hence, the sequence $E_p(x, \delta)/\delta^{p+1}$ has the form

$$\frac{f^{(p+1)}(\xi)}{(p+1)!},$$

which depends on δ through the value of ξ. The assumption that f has $p+1$ bounded and continuous derivatives in the interval $(x, x + \delta)$ ensures that this sequence remains bounded for all $\xi \in (x, x + \delta)$. Hence it follows from Definition 1.7 that $E_p(x, \delta) = O(\delta^{p+1})$ as $\delta \to 0$. $\qquad\square$

Example 1.22. Consider the asymptotic expansions developed in Example 1.17, that considered approximating the function $\exp(\delta)$ as $\delta \to 0$. Theorem 1.15 can be applied to these results to conclude that $\exp(\delta) = 1 + \delta + O(\delta^2)$, $\exp(\delta) = 1 + \delta + \frac{1}{2}\delta^2 + O(\delta^3)$, and $\exp(\delta) = 1 + \delta + \frac{1}{2}\delta^2 + \frac{1}{6}\delta^3 + O(\delta^4)$, as $\delta \to 0$. This allows us to easily evaluate the asymptotic properties of the error sequences. In particular, the asymptotically most accurate approximation has an error term that converges to zero at the same rate as δ^4. Alternatively, we could also apply Theorem 1.15 to these approximations to conclude that $\exp(\delta) = 1 + \delta + o(\delta)$, $\exp(\delta) = 1 + \delta + \frac{1}{2}\delta^2 + o(\delta^2)$, and

$$\exp(\delta) = 1 + \delta + \frac{1}{2}\delta^2 + \frac{1}{6}\delta^3 + o(\delta^3),$$

as $\delta \to 0$. Hence, the asymptotically most accurate approximation considered here has an error term that converges to 0 at a rate faster than δ^3. $\qquad\blacksquare$

Example 1.23. Consider the asymptotic expansions developed in Example 1.18 that approximated the standard normal distribution function near zero. The first and second-order approximations coincide in this case so that we can conclude using Theorem 1.15 that $\Phi(\delta) = \frac{1}{2} + \delta(2\pi)^{-1/2} + O(\delta^3)$ and $\Phi(\delta) = \frac{1}{2} + \delta(2\pi)^{-1/2} + o(\delta^2)$ as $\delta \to 0$. The third order approximation has the forms $\Phi(\delta) = \frac{1}{2} + \delta(2\pi)^{-1/2} + \frac{1}{6}\delta^3(2\pi)^{-1/2} + O(\delta^4)$ and $\Phi(\delta) = \frac{1}{2} + \delta(2\pi)^{-1/2} + \frac{1}{6}\delta^3(2\pi)^{-1/2} + o(\delta^3)$ as $\delta \to 0$. $\qquad\blacksquare$

There are other methods besides the Taylor expansion which can also be used to generate asymptotic expansions. A particular method that is useful for approximating integral functions of a certain form is based on Theorem A.4. Integral functions with an exponential type form often fall into this category, and the normal integral is a particularly interesting example.

Example 1.24. Consider the problem of approximating the tail probability function of a standard normal distribution given by

$$\bar{\Phi}(z) = \int_z^\infty \phi(t)dt,$$

for large values of z, or as $z \to \infty$. To apply integration by parts, first note that from Definition 1.6 it follows that $\phi'(t) = -H_1(t)\phi(t) = -t\phi(t)$ or equivalently $\phi(t) = -\phi'(t)/t$. Therefore

$$\bar{\Phi}(z) = -\int_z^\infty t^{-1}\phi'(t)dt.$$

A single application of Theorem A.4 yields

$$-\int_z^\infty t^{-1}\phi'(t)dt = \lim_{t\to\infty}[-t^{-1}\phi(t)] + z^{-1}\phi(z) - \int_z^\infty t^{-2}\phi(t)dt$$

$$= z^{-1}\phi(z) - \int_z^\infty t^{-2}\phi(t)dt. \tag{1.23}$$

The integral of the right hand side of Equation (1.23) can be rewritten as

$$-\int_z^\infty t^{-2}\phi(t)dt = \int_z^\infty t^{-3}\phi'(t)dt,$$

which suggests an additional application of integration by parts to yield

$$\int_z^\infty t^{-3}\phi'(t)dt = -z^{-3}\phi(z) + \int_z^\infty 3t^{-4}\phi(t)dt.$$

Therefore, an approximation for $\bar{\Phi}(z)$ can be written as

$$\bar{\Phi}(z) = z^{-1}\phi(z) - z^{-3}\phi(z) + \tilde{E}_2(z),$$

where

$$\tilde{E}_2(z) = \int_z^\infty 3t^{-4}\phi(t)dt.$$

Note that the terms of this approximation are not in the form of the asymptotic expansion of Equation (1.20), but are in the form of Equation (1.21), written as

$$\bar{\Phi}(z) = \phi(z)[z^{-1} - z^{-3} + E_2(z)],$$

where

$$E_2(z) = \frac{1}{\phi(z)}\int_z^\infty 3t^{-4}\phi(t)dt.$$

To evaluate the asymptotic behavior of the error term $E_2(z)$, note that iterating the process described above one more time yields

$$\int_z^\infty 3t^{-4}\phi(t)dt = 3z^{-5}\phi(z) - \int_z^\infty 15t^{-6}\phi(t)dt,$$

so that

$$E_2(z) = 3z^{-5} - \frac{1}{\phi(z)}\int_z^\infty 15t^{-6}\phi(t)dt.$$

Now

$$z^5 E_2(z) = 3 - \frac{z^5}{\phi(z)}\int_z^\infty 15t^{-6}\phi(t)dt.$$

The first term is bounded, and noting that $\phi(t)$ is a decreasing function for $t > 0$ it follows that

$$\frac{z^5}{\phi(z)}\int_z^\infty 15t^{-6}\phi(t)dt \le \frac{z^5}{\phi(z)}\int_z^\infty 15t^{-6}\phi(z)dt = z^5\int_z^\infty 15t^{-6}dt = 3.$$

Therefore $z^5 E_2(z)$ remains bounded and positive for all $z > 0$, and it follows that $E_2(z) = O(z^{-5})$ as $z \to \infty$. This process can be iterated further by applying integration by parts to the error term $E_2(z)$ which will result in an error term that is $O(z^{-7})$ as $z \to \infty$. Barndorff-Nielsen and Cox (1989) point out several interesting properties of the resulting asymptotic expansion, including the fact that if the process is continued the resulting sequence has alternating signs, and that each successive approximation provides a lower or upper bound for the true value $\bar{\Phi}(z)$. Moreover, the infinite sequence that results from continuing the expansion indefinitely is divergent when z is fixed. See Example 3.1 of Barndorff-Nielsen and Cox (1989) for further details. ∎

The approximation of exponential type integrals can be generalized using the LaPlace expansion. See Section 3.3 of Barndorff-Nielsen and Cox (1989), Chapter 5 of Copson (1965), Chapter 4 of De Bruijn (1958), and Erdélyi (1956) for further details on this and other techniques for deriving asymptotic expansions. Example 1.24 also suggests that it might be of interest to study how asymptotic relations relate to integration and differentiation. General results for integration can be established as shown in the theorem below.

Theorem 1.16. *Suppose $f(z) = O[g(z)]$ as $z \to \infty$ and that*

$$\int_z^\infty |g(t)|dt < \infty,$$

for all $z \in \mathbb{R}$. Then

$$\int_z^\infty f(t)dt = O\left(\int_z^\infty |g(t)|dt\right),$$

as $z \to \infty$.

Proof. The assumption that $f(z) = O[g(z)]$ as $z \to \infty$ implies that $|f(z)/g(z)|$ remains bounded as $z \to \infty$, or that there exists a positive real number $b < \infty$ such that $|f(z)| \leq b|g(z)|$ as $z \to \infty$. Therefore,

$$\left|\int_z^\infty f(t)dt\right| \leq \int_z^\infty |f(t)|dt \leq b\int_z^\infty |g(t)|dt,$$

as $z \to \infty$, which implies that

$$\frac{\left|\int_z^\infty f(t)dt\right|}{\int_z^\infty |g(t)|dt} \leq b < \infty,$$

as $z \to \infty$, which yields the result.   □

More general theorems on the relationship between integration and the asymptotic order notation can be developed as well. See, for example, Section 1.1 of Erdélyi (1956). It is important to note that it is generally not permissible to exchange a derivative and an asymptotic order relation, though some results are possible if additional assumptions can be made. For example, the following result is based on the development of Section 7.3 of De Bruijn (1958), which contains a proof of the result.

Theorem 1.17. *Let g be a real function that is integrable over a finite interval and define*

$$G(t) = \int_0^t g(x)dx.$$

If g is non-decreasing and $G(t) \asymp (\alpha+1)^{-1}t^{\alpha+1}$ as $t \to \infty$, then $g(t) \asymp t^{\alpha}$ as $t \to \infty$.

Example 1.25. Consider the function $G(t) = t^3 + t^2$ and note that $G(t) \asymp t^3$ as $t \to \infty$. Knowing the exact form of $G(t)$ in this case allows us to compute the derivative using direct calculations to be $g(t) = 3t^2 + 2t$ so that $g(t) \asymp 3t^2$ as $t \to \infty$. Therefore, differentiating the asymptotic rate is permissible here. In fact, if we did not know the exact form of $g(t)$, but knew that g is a non-decreasing function, the same rate could be obtained from Theorem 1.17. Note also that $G(t) = O(t^3)$ and $g(t) = O(t^2)$ here as $t \to \infty$. ∎

Example 1.26. Consider the function $G(t) = t^{1/2} + \sin(t)$ and note that $G(t) = O(t^{1/2})$ as $t \to \infty$. Direct differentiation yields $g(t) = \frac{1}{2}t^{-1/2} + \cos(t)$, but in this case it is not true that $g(t) = O(\frac{1}{2}t^{-1/2})$ as $t \to \infty$. Indeed, note that

$$\frac{\frac{1}{2}t^{-1/2} + \cos(t)}{t^{-1/2}} = 1 + t^{1/2}\cos(t),$$

does not remain bounded as $t \to \infty$. The reason that differentiation is not applicable here is due to the cyclic nature of $G(t)$. As $t \to \infty$ the $t^{1/2}$ term dominates the $\sin(t)$ term in $G(t)$, so this periodic pattern is damped out in the limit. However, the $t^{-1/2}$ term, which converges to 0, in $g(t)$ is dominated by the $\cos(t)$ term as $t \to \infty$ so the periodic nature of the function results. Note that Theorem 1.17 is not applicable in this case as $g(t)$ is not strictly increasing. ∎

Other common operations, such as multiplication and addition, are permissible with order relations subject to a set of simple rules. As a motivational example, consider asymptotic expansions in n for two functions given by $f(x) = d_0(x) + O(n^{-1/2})$ and $g(y) = h_0(y) + O(n^{-1/2})$. Suppose that we are interested in approximating the product $f(x)g(y)$. Multiplying the two asymptotic expansions yields an expansion of the form

$$f(x)g(y) = d_0(x)h_0(y) + O(n^{-1/2})h_0(y) + d_0(x)O(n^{-1/2}) + O(n^{-1/2})O(n^{-1/2}).$$

Here, the notation $O(n^{-1/2})h_0(y)$ indicates that the error sequence from approximating f with $d_0(x)$ that has the asymptotic property that the sequence is $O(n^{-1/2})$ as $n \to \infty$, is multiplied by $h_0(y)$. What is the asymptotic behavior of the resulting sequence? This question can be answered by appealing to Definition 1.7. Assume that $h_0(y)$ is finite for the value of y we are interested in and is constant with respect to n. Let $E_0(x, n)$ be the error term from this expansion that has the property that $E_0(x, n) = O(n^{-1/2})$ as $n \to \infty$. Then it follows that the sequence $n^{1/2}E_0(x, n)h_0(y)$ remains bounded for all $n \in \mathbb{N}$ because the sequence $n^{1/2}E_0(x, n)$ is guaranteed to remain bounded by Definition 1.7 and the fact that $h_0(y)$ is finite and does not depend on n. Hence it follows that $O(n^{-1/2})h_0(y) = O(n^{-1/2})$ as $n \to \infty$. Since $h_0(y)$ is finite for all $n \in \mathbb{N}$ it follows that $h_0(y) = O(1)$ and we have proved that if we multiply a sequence that is $O(n^{-1/2})$ by a sequence that is $O(1)$ we obtain a sequence that is $O(n^{-1/2})$ as $n \to \infty$. This type of behavior is generalized in Theorem 1.18.

Theorem 1.18. *Let* $\{a_n\}_{n=1}^{\infty}$, $\{b_n\}_{n=1}^{\infty}$, $\{c_n\}_{n=1}^{\infty}$, *and* $\{d_n\}_{n=1}^{\infty}$ *be real sequences.*

1. *If* $a_n = o(b_n)$ *and* $c_n = o(d_n)$ *as* $n \to \infty$ *then* $a_n c_n = o(b_n d_n)$ *as* $n \to \infty$.
2. *If* $a_n = o(b_n)$ *and* $c_n = O(d_n)$ *as* $n \to \infty$ *then* $a_n c_n = o(b_n d_n)$ *as* $n \to \infty$.
3. *If* $a_n = O(b_n)$ *and* $c_n = O(d_n)$ *as* $n \to \infty$ *then* $a_n c_n = O(b_n d_n)$ *as* $n \to \infty$.

Proof. We will prove the first result, leaving the proofs of the remaining results as Exercise 37. Definition 1.7 implies that

$$\lim_{n \to \infty} \left| \frac{a_n}{b_n} \right| = 0,$$

and

$$\lim_{n \to \infty} \left| \frac{c_n}{d_n} \right| = 0.$$

Therefore, from Theorem 1.4, it follows that

$$\lim_{n \to \infty} \left| \frac{a_n c_n}{b_n d_n} \right| = \lim_{n \to \infty} \left| \frac{a_n}{b_n} \right| \lim_{n \to \infty} \left| \frac{a_n}{b_n} \right| = 0,$$

and hence $a_n b_n = o(c_n d_n)$ as $n \to \infty$. $\qquad \square$

Theorem 1.18 essentially yields two types of results. First, one can observe the multiplicative effect of the asymptotic behavior of the sequences. Second, one can also observe the dominating effect of sequences that have o-type behavior over those with O-type behavior, in that the product of a sequence with o-type behavior with a sequence that has O-type behavior yields a sequence with o-type behavior. The reason for this dominance comes from the fact that the product of a bounded sequence with a sequence that converges to zero, also converges to zero.

Returning to the discussion on the asymptotic expansion for the product of the functions $f(x)g(y)$, it is now clear from Theorem 1.18 that the form of the asymptotic expansion for $f(x)g(y)$ can be written as

$$f(x)g(y) = d_0(x)d_0'(y) + O(n^{-1/2}) + O(n^{-1/2}) + O(n^{-1}).$$

The next step in simplifying this asymptotic expansion is to consider the behavior of the sum of the sequences that are $O(n^{-1/2})$ and $O(n^{-1})$ as $n \to \infty$. Define error terms $E_1(n)$, $E_1'(n)$ and $E_2(n)$ such that $E_1(n) = O(n^{-1/2})$, $E_1'(n) = O(n^{-1/2})$ and $E_2(n) = O(n^{-1})$ as $n \to \infty$. Then

$$n^{1/2}[E_1(n) + E_1'(n) + E_2(n)] = n^{1/2}E_1(n) + n^{1/2}E_1(n)' + n^{1/2}E_2(n).$$

The fact that the first two sequences in the sum remain bounded for all $n \in \mathbb{N}$ follows directly from Definition 1.7. Because the third sum in the sequence is $O(n^{-1})$ as $n \to \infty$ it follows that $nE_2(n)$ remains bounded for all $n \in \mathbb{N}$. Because $n^{1/2}E_2(n) \le nE_2(n)$ for all $n \in \mathbb{N}$, it follows that $n^{1/2}E_2(n)$ remains bounded for all $n \in \mathbb{N}$. Therefore $E_1(n) + E_1'(n) + E_2(n) = O(n^{-1/2})$ as $n \to \infty$ and the asymptotic expansion for $f(x)g(y)$ can be written as $f(x)g(y) = d_0(x)d_0'(y) + O(n^{-1/2})$ as $n \to \infty$. Is it possible that the error sequence converges to zero at a faster rate than $n^{-1/2}$? Such a result cannot be found using the assumptions that are given because the error sequences $E_1(n)$ and $E_1'(n)$ are only guaranteed to remain bounded when multiplied by $n^{1/2}$, and not any larger sequence in n. This type of result is generalized in Theorem 1.19.

Theorem 1.19. *Consider two real sequences $\{a_n\}_{n=1}^{\infty}$ and $\{b_n\}_{n=1}^{\infty}$ and positive real numbers k and m where $k \le m$. Then*

1. *If $a_n = o(n^{-k})$ and $b_n = o(n^{-m})$ as $n \to \infty$, then $a_n + b_n = o(n^{-k})$ as $n \to \infty$.*

2. *If $a_n = O(n^{-k})$ and $b_n = O(n^{-m})$ as $n \to \infty$, then $a_n + b_n = O(n^{-k})$ as $n \to \infty$.*

3. *If $a_n = O(n^{-k})$ and $b_n = o(n^{-m})$ as $n \to \infty$, then $a_n + b_n = O(n^{-k})$ as $n \to \infty$.*

4. *If $a_n = o(n^{-k})$ and $b_n = O(n^{-m})$ as $n \to \infty$, then $a_n + b_n = O(n^{-k})$ as $n \to \infty$.*

Proof. Only the first result will be proven, leaving the proofs of the remaining results the subject of Exercise 38. Suppose $a_n = o(n^{-k})$ and $b_n = o(n^{-m})$ as $n \to \infty$ and consider the sequence $n^k(a_n + b_n)$. Because $a_n = o(n^{-k})$ it follows that $n^k a_n \to 0$ as $n \to \infty$. Similarly, $n^k b_n \to 0$ as $n \to \infty$ due to the fact that $|n^k b_n| \le |n^m b_n| \to 0$ as $n \to \infty$. It follows that $n^k(a_n + b_n) \to 0$ as $n \to \infty$ which yields the result. $\square$

Example 1.27. In Example 1.20 it was established that the density of $Z = n^{1/2}[\bar{X} - \mu(\alpha, \beta)]/\sigma(\alpha, \beta)$, where $\bar{X}_n$ is the sample mean from a sample of size n from a BETA(α, β) distribution, has asymptotic expansion

$$f(x) = \phi(x) - \tfrac{1}{6}n^{-1/2}\phi(x)H_3(x)\kappa_3(\alpha, \beta) + R_2(x, n).$$

It will be shown later that $R_2(x, n) = O(n^{-1})$ as $n \to \infty$. In some applications $\kappa_3(\alpha, \beta)$ is not known exactly and is replaced by a sequence $\hat{\kappa}_3(\alpha, \beta)$ where $\hat{\kappa}_3(\alpha, \beta) = \kappa_3(\alpha, \beta) + O(n^{-1/2})$ as $n \to \infty$. Theorems 1.18 and 1.19 can be employed to yield

$$
\begin{aligned}
\hat{f}(x) &= \phi(x) - \tfrac{1}{6}n^{-1/2}\phi(x)H_3(x)\hat{\kappa}_3(\alpha, \beta) + O(n^{-1}) \\
&= \phi(x) - \tfrac{1}{6}n^{-1/2}\phi(x)H_3(x)[\kappa_3(\alpha, \beta) + O(n^{-1/2})] + O(n^{-1}) \\
&= \phi(x) - \tfrac{1}{6}n^{-1/2}\phi(x)H_3(x)\kappa_3(\alpha, \beta) + O(n^{-1}), \quad\quad (1.24)
\end{aligned}
$$

as $n \to \infty$. Therefore, it is clear that replacing $\kappa_3(\alpha, \beta)$ with $\hat{\kappa}_3(\alpha, \beta)$ does not change the asymptotic order of the error in the asymptotic expansion. That is $|\hat{f}(x) - f(x)| = O(n^{-1})$ as $n \to \infty$, for a fixed value of $x \in \mathbb{R}$. Note that it is not proper to conclude that $\hat{f}(x) = f(x)$, even though both functions have asymptotic expansions of the form $\phi(x) - \frac{1}{6}n^{-1/2}\phi(x)H_3(x)\kappa_3(\alpha, \beta) + O(n^{-1})$. It is clear from the development given in Equation (1.24) that the error terms of the two expansions differ, even though they are both $O(n^{-1})$ as $n \to \infty$. ∎

The final result of this section will provide a more accurate representation of the approximation given by Stirling's approximation to factorials given in Theorem 1.8 by specifying the asymptotic behavior of the error of this approximation.

Theorem 1.20. $n! = n^n \exp(-n)(2n\pi)^{1/2}[1 + O(n^{-1})]$ *as $n \to \infty$.*

For a proof of Theorem 1.20 see Example 3.5 of Barndorff-Nielsen and Cox (1989). The theory of asymptotic expansions and divergent series is far more expansive than has been presented in this brief overview. The material presented in this section is sufficient for understanding the expansion theory used in the rest of this book. Several book length treatments of this topic can be consulted for further information. These include Barndorff-Nielsen and Cox (1989), Copson (1965), De Bruijn (1958), Erdélyi (1956), and Hardy (1949). Some care must be taken when consulting some references on asymptotic expansions as many presentations are for analytic functions in the complex domain. The theoretical properties of asymptotic expansions for these functions can differ greatly in some cases than for real functions.

1.6 Inversion of Asymptotic Expansions

Suppose that f is a function with asymptotic expansion given by

$$f(x, n) = d_0(x) + n^{-1/2}d_1(x) + n^{-1}d_2(x) + \cdots$$
$$+ n^{-p/2}d_p(x) + O(n^{-(p+1)/2}), \quad (1.25)$$

as $n \to \infty$, and we wish to obtain an asymptotic expansion for the inverse of the function with respect to x in terms of powers of $n^{-1/2}$. That is, we wish to obtain an asymptotic expansion for a point $x_{a,n}$ such that $f(x_{a,n}) = a + O(n^{-(p+1)/2})$, as $n \to \infty$. Note that from the onset we work under the assumption that the inverse will not be exact, or that we would be able to find $x_{a,n}$ such that $f(x_{a,n}) = a$ exactly. This is due to the error term whose behavior is only known asymptotically. This section will begin by demonstrating a method that is often useful for finding such an inverse for an asymptotic expansion of the type given in Equation (1.25). The method can easily be adapted to asymptotic expansions that are represented in other forms, such as powers of n^{-1}, $\delta^{1/2}$ and δ.

To begin the process, assume that $x_{a,n}$ has an asymptotic expansion of the form that matches the asymptotic expansion of $f(x, n)$. That is, assume that

$$x_{a,n} = v_0(a) + n^{-1/2}v_1(a) + n^{-1}v_2(a) + \cdots + n^{-p/2}v_p(a) + O(n^{-(p+1)/2}),$$

as $n \to \infty$. Now, substitute the asymptotic expansion for $x_{a,n}$ for x in the asymptotic expansion for $f(x, n)$ given in Equation (1.25), and use Theorem 1.13, and other methods, to obtain an expansion for the expression in terms of the powers of $n^{-1/2}$. That is, find functions $r_0, \ldots, r_p$, such that

$$f(x_{a,n}, n) = r_0 + n^{-1/2}r_1 + \cdots + n^{-p/2}r_p + O(n^{-(p+1)/2}),$$

as $n \to \infty$, where r_k is a function of a, and $v_0, \ldots, v_k$, for $k = 1, \ldots, p$. That is $r_k = r_k(a; v_0, \ldots, v_k)$, for $k = 1, \ldots, p$. Now note that $f(x_a, n)$ is also equal to $a + O(n^{-(p+1)/2})$ as $n \to \infty$. Equating the coefficients of the powers of n implies that $v_0, \ldots, v_p$ should satisfy $r_0(a; v_0) = a$ and $r_k(a; v_0, \ldots, v_k) = 0$ for $k = 1, \ldots, p$. Therefore, solving this system of equations will result in expressions for $v_0, \ldots, v_p$, and an asymptotic expansion for $x_{a,n}$ is obtained.

As presented, this method is somewhat ad hoc, and we have not provided any general guidelines as to when the method is applicable and what happens in cases such as when the inverse is not unique. For the problems encountered in this book the method is generally reliable. For a rigorous justification of the method see De Bruijn (1958).

Example 1.28. Example 1.20 showed that the density of $Z_n = n^{1/2}(\bar{X}_n - \mu(\alpha, \beta))/\sigma(\alpha, \beta)$, where $\bar{X}_n$ is computed from a sample of size n from a BETA(α, β) distribution, has asymptotic expansion

$$f_n(z) = \phi(z) + \tfrac{1}{6}n^{-1/2}H_3(z)\phi(z)\kappa_3(\alpha, \beta) + O(n^{-1}), \qquad (1.26)$$

as $n \to \infty$. In this example we will consider finding an asymptotic expansion for the quantile of this distribution. Term by term integration of the expansion in Equation (1.26) with respect to z yields an expansion for the distribution function of Z_n given by

$$F_n(z) = \Phi(z) - \tfrac{1}{6}n^{-1/2}H_2(z)\phi(z)\kappa_3(\alpha, \beta) + O(n^{-1}),$$

as $n \to \infty$, where it is noted that from Definition 1.6 it follows that the integral of $-H_3(z)\phi(z)$, which is the third derivative of the standard normal density, is given by $H_2(z)\phi(z)$, which is the second derivative of the standard normal density. We assume that the integration of the error term with respect to z does not change the order of the error term. This actually follows from the fact that the error term can be shown to be uniform in z. Denote the αth quantile of F_n as $f_{\alpha,n}$ and assume that $f_{\alpha,n}$ has an asymptotic expansion of the form $f_{\alpha,n} = v_0(\alpha) + n^{-1/2}v_1(\alpha) + O(n^{-1})$, as $n \to \infty$. To obtain $v_0(\alpha)$ and $v_1(\alpha)$ set $F_n(f_{\alpha,n}) = \alpha + O(n^{-1})$, which is the property that $f_{\alpha,n}$ should have to be the αth quantile of F_n up to order $O(n^{-1})$, as $n \to \infty$. Therefore, it follows that $F_n[v_0(\alpha) + n^{-1/2}v_1(\alpha) + O(n^{-1})] = \alpha + O(n^{-1})$, or equivalently

$$\Phi[v_0(\alpha) + n^{-1/2}v_1(\alpha) + O(n^{-1})] - \tfrac{1}{6}n^{-1/2}\phi[v_0(\alpha) + n^{-1/2}v_1(\alpha) + O(n^{-1})] \times$$
$$H_2[v_0(\alpha) + n^{-1/2}v_1(\alpha) + O(n^{-1})]\kappa_3(\alpha, \beta) = \alpha + O(n^{-1}), \quad (1.27)$$

as $n \to \infty$. Now expand each term in Equation (1.27) using Theorem 1.13 and the related theorems in Section 1.5. Applying Theorem 1.13 the standard normal distribution function yields

$$\Phi[v_0(\alpha) + n^{-1/2}v_1(\alpha) + O(n^{-1})] =$$
$$\Phi[v_0(\alpha)] + [n^{-1/2}v_1(\alpha) + O(n^{-1})]\phi[v_0(\alpha)] + O(n^{-1}) =$$
$$\Phi[v_0(\alpha)] + n^{-1/2}v_1(\alpha)\phi[v_0(\alpha)] + O(n^{-1}),$$

as $n \to \infty$. To expand the second term in Equation (1.27), first apply Theorem 1.13 to the standard normal density. This yields

$$\phi[v_0(\alpha) + n^{-1/2}v_1(\alpha) + O(n^{-1})] = \phi[v_0(\alpha)] + O(n^{-1/2}),$$

as $n \to \infty$, where all terms of order $n^{-1/2}$ are absorbed into the error term. Direct evaluation of $H_2[z_\alpha + n^{-1/2}v_1(\alpha) + O(n^{-1})]$ using Definition 1.6 can be used to show that

$$H_2[v_0(\alpha) + n^{-1/2}v_1(\alpha) + O(n^{-1})] = H_2[v_0(\alpha)] + O(n^{-1/2}),$$

as $n \to \infty$. Therefore, the second term in Equation (1.27) has the form $-\tfrac{1}{6}n^{-1/2}\kappa_3(\alpha, \beta)H_2[v_0(\alpha)]\phi[v_0(\alpha)] + O(n^{-1})$. Combining these results yields

$$F(f_{\alpha,n}) = \Phi[v_0(\alpha)] + n^{-1/2}\phi[v_0(\alpha)]\{v_1(\alpha) - \tfrac{1}{6}\kappa_3(\alpha, \beta)H_2[v_0(\alpha)]\} + O(n^{-1}),$$

as $n \to \infty$. Using the notation of this section we have that $r_0(\alpha; v_0) = \Phi[v_0(\alpha)]$ and $r_1(\alpha; v_0, v_1) = \phi[v_0(\alpha)]\{v_1(\alpha) - \tfrac{1}{6}\kappa_3(\alpha, \beta)H_2[v_0(\alpha)]\}$. Setting $r_0(\alpha, v_0) = \Phi[v_0(\alpha)] = \alpha$ implies that $v_0(\alpha) = z_\alpha$, the αth quantile of a $N(0,1)$ distribution. Similarly, setting $r_1(\alpha; v_0, v_1) = \phi[v_0(\alpha)]\{v_1(\alpha) - \tfrac{1}{6}\kappa_3(\alpha, \beta)H_2[v_0(\alpha)]\} = 0$ implies that $v_1(\alpha) = \tfrac{1}{6}\kappa_3(\alpha, \beta)H_2[v_0(\alpha)] = \tfrac{1}{6}\kappa_3(\alpha, \beta)H_2(z_\alpha)$. Therefore, an asymptotic expansion for the αth quantile of F is given by

$$f_{\alpha,n} = v_0(\alpha) + n^{-1/2}v_1(\alpha) + O(n^{-1}) = z_\alpha + n^{-1/2}\tfrac{1}{6}\kappa_3(\alpha, \beta)H_2(z_\alpha) + O(n^{-1}),$$

as $n \to \infty$. ∎

It should be noted that if closed forms for the derivatives of f are known, then it can be easier to derive an asymptotic expansion for the inverse of a function $f(x)$ using Theorem 1.13 or Theorem 1.15 directly. The derivatives of the inverse of $f(x)$ are required for this approach. The following result from calculus can be helpful with this calculation.

Theorem 1.21. *Assume that g is a strictly increasing and continuous real function on an interval $[a, b]$ and let h be the inverse of g. If the derivative of g exists and is non-zero at a point $x \in (a, b)$ then the derivative of h also* *exists and is non-zero at the corresponding point $y = g(x)$ and*

$$\frac{d}{dy}h(y) = \left[\frac{d}{dx}g(x)\Big|_{x=h(y)} \right]^{-1}.$$

A proof of Theorem 1.21 can be found in Section 6.20 of Apostol (1967). Note the importance of the monotonicity condition in Theorem 1.21, which ensures that the function g has a unique inverse. Further, the restriction that g is strictly increasing implies that the derivative of the inverse will be positive.

Example 1.29. Consider the standard normal distribution function $\Phi(z)$, and suppose we wish to obtain an asymptotic expansion for the standard normal quantile function $\Phi^{-1}(\alpha)$ for values of α near $\frac{1}{2}$. Theorem 1.15 implies that

$$\Phi^{-1}(\alpha + \delta) = \Phi^{-1}(\alpha) + \delta\frac{d}{d\alpha}\Phi^{-1}(\alpha) + O(\delta^2),$$

as $\delta \to 0$. Noting that $\Phi(t)$ is monotonically increasing and that $\Phi'(t) \neq 0$ for all $t \in \mathbb{R}$, we can apply Theorem 1.21 to $\Phi^{-1}(\alpha)$ to find that

$$\frac{d}{d\alpha}\Phi^{-1}(\alpha) = \left[\frac{d}{dz}\Phi(z)\Big|_{z=\Phi^{-1}(\alpha)} \right]^{-1} = \frac{1}{\phi(z_\alpha)}.$$

Therefore a one-term asymptotic expansion for the standard normal quantile function is given by

$$\Phi^{-1}(\alpha + \delta) = z_\alpha + \frac{\delta}{\phi(z_\alpha)} + O(\delta^2),$$

as $\delta \to 0$. Now take $\alpha = \frac{1}{2}$ as in Example 1.18 to yield $z_{\frac{1}{2}+\delta} = \delta(2\pi)^{1/2} + O(\delta^2)$, as $\delta \to 0$. ∎

1.7 Exercises and Experiments

1.7.1 Exercises

1. Let $\{x_n\}_{n=1}^\infty$ be a sequence of real numbers defined by

$$x_n = \begin{cases} -1 & n = 1 + 3(k-1), k \in \mathbb{N}, \\ 0 & n = 2 + 3(k-1), k \in \mathbb{N}, \\ 1 & n = 3 + 3(k-1), k \in \mathbb{N}. \end{cases}$$

Compute

$$\liminf_{n\to\infty} x_n,$$

and

$$\limsup_{n\to\infty} x_n.$$

Determine if the limit of x_n as $n \to \infty$ exists.

2. Let $\{x_n\}_{n=1}^{\infty}$ be a sequence of real numbers defined by

$$x_n = \frac{n}{n+1} - \frac{n+1}{n},$$

for all $n \in \mathbb{N}$. Compute

$$\liminf_{n\to\infty} x_n,$$

and

$$\limsup_{n\to\infty} x_n.$$

Determine if the limit of x_n as $n \to \infty$ exists.

3. Let $\{x_n\}_{n=1}^{\infty}$ be a sequence of real numbers defined by $x_n = n^{(-1)^n - n}$ for all $n \in \mathbb{N}$. Compute

$$\liminf_{n\to\infty} x_n,$$

and

$$\limsup_{n\to\infty} x_n.$$

Determine if the limit of x_n as $n \to \infty$ exists.

4. Let $\{x_n\}_{n=1}^{\infty}$ be a sequence of real numbers defined by $x_n = n2^{-n}$, for all $n \in \mathbb{N}$. Compute

$$\liminf_{n\to\infty} x_n,$$

and

$$\limsup_{n\to\infty} x_n.$$

Determine if the limit of x_n as $n \to \infty$ exists.

5. Each of the sequences given below converges to zero. Specify the smallest value of n_ε so that $|x_n| < \varepsilon$ for every $n > n_\varepsilon$ as a function of ε.

a. $x_n = n^{-2}$
b. $x_n = n(n+1)^{-1} - 1$
c. $x_n = [\log(n+1)]^{-1}$
d. $x_n = 2(n^2 + 1)^{-1}$

6. Let $\{x_n\}_{n=1}^{\infty}$ and $\{y_n\}_{n=1}^{\infty}$ be sequences of real numbers such that

$$\lim_{n\to\infty} x_n = x,$$

and

$$\lim_{n\to\infty} y_n = y.$$

a. Prove that if $c \in \mathbb{R}$ is a constant, then

$$\lim_{n\to\infty} cx_n = cx.$$

b. Prove that

$$\lim_{n\to\infty} (x_n + y_n) = x + y.$$

c. Prove that

$$\lim_{n \to \infty} x_n y_n = xy.$$

d. Prove that

$$\lim_{n \to \infty} \frac{x_n}{y_n} = \frac{x}{y},$$

where $y_n \neq 0$ for all $n \in \mathbb{N}$ and $y \neq 0$.

7. Let $\{x_n\}_{n=1}^{\infty}$ and $\{y_n\}_{n=1}^{\infty}$ be sequences of real numbers such that $x_n \leq y_n$ for all $n \in \mathbb{N}$. Prove that if the limit of the two sequences exist, then

$$\lim_{n \to \infty} x_n \leq \lim_{n \to \infty} y_n.$$

8. Let $\{x_n\}_{n=1}^{\infty}$ and $\{y_n\}_{n=1}^{\infty}$ be sequences of real numbers such that

$$\lim_{n \to \infty} (x_n + y_n) = s,$$

and

$$\lim_{n \to \infty} (x_n - y_n) = d.$$

Prove that

$$\lim_{n \to \infty} x_n y_n = \tfrac{1}{4}(s^2 - d^2).$$

9. Find the supremum and infimum limits for each sequence given below.

a. $x_n = (-1)^n (1 + n^{-1})$

b. $x_n = (-1)^n$

c. $x_n = (-1)^n n$

d. $x_n = n^2 \sin^2(\tfrac{1}{2} n \pi)$

e. $x_n = \sin(n)$

f. $x_n = (1 + n^{-1}) \cos(n\pi)$

g. $x_n = \sin(\tfrac{1}{2} n \pi) \cos(\tfrac{1}{2} n \pi)$

h. $x_n = (-1)^n n (1 + n)^{-n}$

10. Let $\{x_n\}_{n=1}^{\infty}$ be a sequence of real numbers.

a. Prove that

$$\inf_{n \in \mathbb{N}} x_n \leq \liminf_{n \to \infty} x_n \leq \limsup_{n \to \infty} x_n \leq \sup_{n \in \mathbb{N}} x_n.$$

b. Prove that

$$\liminf_{n \to \infty} x_n = \limsup_{n \to \infty} x_n = l,$$

if and only if

$$\lim_{n \to \infty} x_n = l.$$

11. Let $\{x_n\}_{n=1}^{\infty}$ and $\{y_n\}_{n=1}^{\infty}$ be a sequences of real numbers such that $x_n \leq y_n$ for all $n \in \mathbb{N}$. Prove that

$$\liminf_{n \to \infty} x_n \leq \liminf_{n \to \infty} y_n.$$

12. Let $\{x_n\}_{n=1}^{\infty}$ and $\{y_n\}_{n=1}^{\infty}$ be a sequences of real numbers such that

$$\left| \limsup_{n \to \infty} x_n \right| < \infty,$$

and

$$\left| \limsup_{n \to \infty} y_n \right| < \infty,$$

Then prove that

$$\liminf_{n \to \infty} x_n + \liminf_{n \to \infty} y_n \leq \liminf_{n \to \infty}(x_n + y_n),$$

and

$$\limsup_{n \to \infty}(x_n + y_n) \leq \limsup_{n \to \infty} x_n + \limsup_{n \to \infty} y_n.$$

13. Let $\{x_n\}_{n=1}^{\infty}$ and $\{y_n\}_{n=1}^{\infty}$ be a sequences of real numbers such that $x_n > 0$ and $y_n > 0$ for all $n \in \mathbb{N}$,

$$0 < \limsup_{n \to \infty} x_n < \infty,$$

and

$$0 < \limsup_{n \to \infty} y_n < \infty.$$

Prove that

$$\limsup_{n \to \infty} x_n y_n \leq \left(\limsup_{n \to \infty} x_n \right) \left(\limsup_{n \to \infty} y_n \right).$$

14. Let $\{f_n(x)\}_{n=1}^{\infty}$ and $\{g_n(x)\}_{n=1}^{\infty}$ be sequences of real valued functions that converge pointwise to the real functions f and g, respectively.

a. Prove that $cf_n \xrightarrow{pw} cf$ as $n \to \infty$ where c is any real constant.

b. Prove that $f_n + c \xrightarrow{pw} f + c$ as $n \to \infty$ where c is any real constant.

c. Prove that $f_n + g_n \xrightarrow{pw} f + g$ as $n \to \infty$.

d. Prove that $f_n g_n \xrightarrow{pw} fg$ as $n \to \infty$.

e. Suppose that $g_n(x) > 0$ and $g(x) > 0$ for all $x \in \mathbb{R}$. Prove that $f_n/g_n \xrightarrow{pw} f/g$ as $n \to \infty$.

15. Let $\{f_n(x)\}_{n=1}^{\infty}$ and $\{g_n(x)\}_{n=1}^{\infty}$ be sequences of real valued functions that converge uniformly on $\mathbb{R}$ to the real functions f and g as $n \to \infty$, respectively. Prove that $f_n + g_n \xrightarrow{u} f + g$ on $\mathbb{R}$ as $n \to \infty$.

16. Let $\{f_n(x)\}_{n=1}^{\infty}$ be a sequence of real functions defined by $f_n(x) = \frac{1}{2}n\delta\{x; (n - n^{-1}, n + n^{-1})\}$ for all $n \in \mathbb{N}$.

a. Prove that

$$\lim_{n\to\infty} f_n(x) = 0$$

for all $x \in \mathbb{R}$, and hence conclude that

$$\int_{-\infty}^{\infty} \lim_{n\to\infty} f_n(x)dx = 0.$$

b. Compute

$$\lim_{n\to\infty} \int_{-\infty}^{\infty} f_n(x)dx.$$

Does this match the result derived above?

c. State whether Theorem 1.11 applies to this case, and use it to explain the results you found.

17. Let $\{f_n(x)\}_{n=1}^{\infty}$ be a sequence of real functions defined by $f_n(x) = (1 + n^{-1})\delta\{x; (0, 1)\}$ for all $n \in \mathbb{N}$.

a. Prove that

$$\lim_{n\to\infty} f_n(x) = \delta\{x; (0, 1)\}$$

for all $x \in \mathbb{R}$, and hence conclude that

$$\int_{-\infty}^{\infty} \lim_{n\to\infty} f_n(x)dx = 1.$$

b. Compute

$$\lim_{n\to\infty} \int_{-\infty}^{\infty} f_n(x)dx.$$

Does this match the result you found above?

c. State whether Theorem 1.11 applies to this case, and use it to explain the results you found above.

18. Let $g(x) = \exp(-|x|)$ and define a sequence of functions $\{f_n(x)\}_{n=1}^{\infty}$ as $f_n(x) = g(x)\delta\{|x|; (n, \infty)\}$, for all $n \in \mathbb{N}$.

a. Calculate

$$f(x) = \lim_{n\to\infty} f_n(x),$$

for each fixed $x \in \mathbb{R}$.

b. Calculate

$$\lim_{n\to\infty} \int_{-\infty}^{\infty} f_n(x)dx,$$

and

$$\int_{-\infty}^{\infty} f(x)dx.$$

Is the exchange of the limit and the integral justified in this case? Why or why not?

19. Define a sequence of functions $\{f_n(x)\}_{n=1}^{\infty}$ as $f_n(x) = n^2 x(1-x)^n$ for $x \in \mathbb{R}$ and for all $n \in \mathbb{N}$.

    a. Calculate

$$f(x) = \lim_{n \to \infty} f_n(x),$$

    for each fixed $x \in \mathbb{R}$.

    b. Calculate

$$\lim_{n \to \infty} \int_{-\infty}^{\infty} f_n(x)dx,$$

    and

$$\int_{-\infty}^{\infty} f(x)dx.$$

    Is the exchange of the limit and the integral justified in this case? Why or why not?

20. Define a sequence of functions $\{f_n(x)\}_{n=1}^{\infty}$ as $f_n(x) = n^2 x(1-x)^n$ for $x \in [0,1]$. Determine whether

$$\lim_{n \to \infty} \int_0^1 f_n(x)dx = \int_0^1 \lim_{n \to \infty} f_n(x)dx.$$

21. Suppose that f is a quadratic polynomial. Prove that for $\delta \in \mathbb{R}$,

$$f(x + \delta) = f(x) + \delta f'(x) + \tfrac{1}{2}\delta^2 f''(x).$$

22. Suppose that f is a cubic polynomial. Prove that for $\delta \in \mathbb{R}$,

$$f(x + \delta) = f(x) + \delta f'(x) + \tfrac{1}{2}\delta^2 f''(x) + \tfrac{1}{6}\delta^3 f'''(x).$$

23. Prove that if f is a polynomial of degree p then

$$f(x + \delta) = \sum_{i=1}^{p} \frac{\delta^i f^{(i)}}{i!}.$$

24. Prove Theorem 1.13 using induction. That is, assume that

$$E_1(x, \delta) = \int_x^{x+\delta} (x + \delta - t)f''(t)dt,$$

which has been shown to be true, and that

$$E_p(x, \delta) = \frac{1}{p!} \int_x^{x+\delta} (x + \delta - t)^p f^{(p+1)}(t)dt,$$

and show that these imply

$$E_{p+1}(x, \delta) = \frac{1}{(p+1)!} \int_x^{x+\delta} (x + \delta - t)^p f^{(p+2)}(t)dt.$$

25. Given that $E_p(x, \delta)$ from Theorem 1.13 can be written as

$$E_p(x, \delta) = \frac{1}{p!} \int_x^{x+\delta} (x + \delta - t)^p f^{(p+1)}(t) dt,$$

show that $E_p(x, \delta) = \delta^{p+1} f^{(p+1)}(\xi)/(p+1)!$ for some $\xi \in [x, x + \delta]$.

26. Use Theorem 1.13 with $p = 1, 2$ and 3 to find approximations for each of the functions listed below for small values of δ.

a. $f(\delta) = 1/(1 + \delta)$
b. $f(\delta) = \sin^2(\pi/4 + \delta)$
c. $f(\delta) = \log(1 + \delta)$
d. $f(\delta) = (1 + \delta)/(1 - \delta)$

27. Prove that the pth-order Taylor expansion of a function $f(x)$ has the same derivatives of order $1, \ldots, p$ as $f(x)$. That is, show that

$$\frac{d^j}{d\delta^j} \sum_{k=0}^p \frac{\delta^k f^{(k)}(x)}{k!} \bigg|_{\delta=0} = f^{(j)}(x),$$

for $j = 1, \ldots, p$. What assumptions are required for this result to be true?

28. Show that by taking successive derivatives of the standard normal density that $H_3(x) = x^3 - 3x$, $H_4(x) = x^4 - 6x^2 + 3$ and $H_5(x) = x^5 - 10x^3 + 15x$.

29. Use Theorem 1.13 (Taylor) to find fourth and fifth order polynomials that are approximations to the standard normal distribution function $\Phi(x)$. Is there a difference between the approximations? What can be said in general about two consecutive even and odd order polynomial approximations of $\Phi(x)$? Prove your conjecture using the results of Theorem 1.14.

30. Prove Part 1 of Theorem 1.14 using induction. That is, prove that for any non-negative integer k,

$$H_k(x) = \sum_{i=0}^{\lfloor k/2 \rfloor} (-1)^i \frac{(2i)!}{2^i i!} \binom{k}{2i} x^{k-2i},$$

where $\lfloor k/2 \rfloor$ is the greatest integer less than or equal to $k/2$. It may prove useful to use the result of Exercise 32.

31. Prove Part 2 of Theorem 1.14. That is, prove that for any non-negative integer $k \geq 2$,

$$H_k(x) = x H_{k-1}(x) - (k - 1) H_{k-2}(x).$$

The simplest approach is to use Definition 1.6.

32. Prove Part 3 of Theorem 1.14 using only Definition 1.6. That is, prove that for any non-negative integer k,

$$\frac{d}{dx} H_k(x) = k H_{k-1}(x).$$

Do not use the result of Part 1 of Theorem 1.14.

33. The Hermite polynomials are often called a set of *orthogonal polynomials*. Consider the Hermite polynomials up to a specified order d. Let $\mathbf{h}_k$ be a vector in $\mathbb{R}^d$ whose elements correspond to the coefficients of the Hemite polynomial $H_k(x)$. That is, for example, $\mathbf{h}'_1 = (1, 0, 0, 0 \cdots 0)$, $\mathbf{h}'_2 = (0, 1, 0, 0 \cdots 0)$, and $\mathbf{h}'_3 = (-1, 0, 1, 0 \cdots 0)$. Then the polynomials $H_i(x)$ and $H_j(x)$ are said to be orthogonal if $\mathbf{h}'_i \mathbf{h}_j = 0$. Show that the first six Hermite polynomials are all orthogonal to one another.

34. In Theorem 1.15 prove that $E_p(x, \delta) = o(\delta^p)$, as $\delta \to 0$.

35. Consider approximating the normal tail integral

$$\bar{\Phi}(z) = \int_z^\infty \phi(t)dt,$$

for large values of z using integration by parts as discussed in Example 1.24. Use repeated integration by parts to show that

$$\bar{\Phi}(z) = z^{-1}\phi(z) - z^{-3}\phi(z) + 3z^{-5}\phi(z) - 15z^{-7}\phi(z) + O(z^{-9}),$$

as $z \to \infty$.

36. Using integration by parts, show that the exponential integral

$$\int_z^\infty t^{-1}e^{-t}dt,$$

has asymptotic expansion

$$z^{-1}e^{-z} - z^{-2}e^{-z} + 2z^{-3}e^{-z} - 6z^{-4}e^{-z} + O(z^{-5}),$$

as $z \to \infty$.

37. Prove the second and third results of Theorem 1.18. That is, let $\{a_n\}_{n=1}^\infty$, $\{b_n\}_{n=1}^\infty$, $\{c_n\}_{n=1}^\infty$, and $\{d_n\}_{n=1}^\infty$ be real sequences.

  a. Prove that if $a_n = o(b_n)$ and $c_n = O(d_n)$ as $n \to \infty$ then $a_n b_n = o(c_n d_n)$ as $n \to \infty$.

  b. Prove that if $a_n = O(b_n)$ and $c_n = O(d_n)$ as $n \to \infty$ then $a_n b_n = O(c_n d_n)$ as $n \to \infty$.

38. Prove the remaining three results of Theorem 1.19. That is, consider two real sequences $\{a_n\}_{n=1}^\infty$ and $\{b_n\}_{n=1}^\infty$ and positive integers k and m where $k \leq m$. Then

  a. Suppose $a_n = O(n^{-k})$ and $b_n = O(n^{-m})$ as $n \to \infty$. Then prove that $a_n + b_n = O(n^{-k})$ as $n \to \infty$.

  b. Suppose $a_n = O(n^{-k})$ and $b_n = o(n^{-m})$ as $n \to \infty$. Then prove that $a_n + b_n = O(n^{-k})$ as $n \to \infty$.

  c. Suppose $a_n = o(n^{-k})$ and $b_n = O(n^{-m})$ as $n \to \infty$. Then prove that $a_n + b_n = O(n^{-k})$ as $n \to \infty$.

39. For each specified pair of functions $G(t)$ and $g(t)$, determine the value of α and c so that $G(t) \asymp ct^{\alpha-1}$ as $t \to \infty$ and determine if there is a function $g(t) \asymp dt^{\alpha}$ for some d as $t \to \infty$ where c and d are real constants. State whether Theorem 1.17 is applicable in each case.

   a. $G(t) = 2t^4 + t$

   b. $G(t) = t + t^{-1}$

   c. $G(t) = t^2 + \cos(t)$

   d. $G(t) = t^{1/2} + \cos(t)$

40. Consider a real function f that can be approximated with the asymptotic expansion

$$f_n(x) = \pi x + \tfrac{1}{2}n^{-1/2}\pi^2 x^{1/2} - \tfrac{1}{3}n^{-1}\pi^3 x^{1/4} + O(n^{-3/2}),$$

as $n \to \infty$, uniformly in x, where x is assumed to be positive. Use the first method demonstrated in Section 1.6 to find an asymptotic expansion with error $O(n^{-3/2})$ as $n \to \infty$ for x_a where $f(x_a) = a + O(n^{-3/2})$ as $n \to \infty$.

41. Consider the problem of approximating the function $\sin(x)$ and its inverse for values of x near 0.

   a. Using Theorem 1.15 show that $\sin(\delta) = \delta - \tfrac{1}{6}\delta^3 + O(\delta^4)$ as $\delta \to 0$.

   b. Using Theorems 1.15 and the known derivatives of the inverse sine function show that $\sin^{-1}(\delta) = \delta + \tfrac{1}{6}\delta^3 + O(\delta^4)$ as $\delta \to 0$.

   c. Recompute the first term of the expansion found in Part (b) using Theorem 1.21. Do they match? What restrictions on x are required to apply Theorem 1.21?

1.7.2 Experiments

1. Refer to the three approximations derived for each of the four functions in Exercise 26. For each function use R to construct a line plot of the function, along with the three approximations versus δ on a single plot. The lines corresponding to each approximation and the original function should be different, that is, the plots should look like the one given in Figure 1.3. You may need to try several ranges of δ to find one that provides a good indication of the behavior of each approximation. What do these plots suggest about the errors of the three approximations?

2. Refer to the three approximations derived for each of the four functions in Exercise 26. For each function use R to construct a line plot of the error terms $E_1(x, \delta)$, $E_2(x, \delta)$ and $E_3(x, \delta)$ versus δ on a single plot. The lines corresponding to each error function should be different so that the plots should look like the one given in Figure 1.4. What do these plots suggest about the errors of the three approximations?

3. Refer to the three approximations derived for each of the four functions in Exercise 26. For each function use R to construct a line plot of the error terms $E_2(x, \delta)$ and $E_3(x, \delta)$ relative to the error term $E_1(x, \delta)$. That is, for each function, plot $E_2(x, \delta)/E_1(x, \delta)$ and $E_3(x, \delta)/E_1(x, \delta)$ versus δ. The lines corresponding to each relative error function should be different. What do these plots suggest about the relative error rates?

4. Consider the approximation for the normal tail integral $\bar{\Phi}(z)$ studied in Example 1.24 given by

$$\bar{\Phi}(z) \simeq z^{-1}\phi(z)(1 - z^{-2} + 3z^{-4} - 15z^{-6} + 105z^{-8}).$$

A slight rearrangement of the approximation implies that

$$\frac{z\bar{\Phi}(z)}{\phi(z)} \simeq 1 - z^{-2} + 3z^{-4} - 15z^{-6} + 105z^{-8}.$$

Define $S_1(z) = 1 - z^{-2}$, $S_2(z) = 1 - z^{-2} + 3z^{-4}$, $S_3(z) = 1 - z^{-2} + 3z^{-4} - 15z^{-6}$ and $S_4(z) = 1 - z^{-2} + 3z^{-4} - 15z^{-6} + 105z^{-8}$, which are the successive approximations of $z\bar{\Phi}(z)/\phi(z)$. Using R, compute $z\bar{\Phi}(z)/\phi(z)$, $S_1(z)$, $S_2(z)$, $S_3(z)$, and $S_4(z)$ for $z = 1, \ldots, 10$. Comment on which approximation performs best for each value of z and whether the approximations become better as z becomes larger.

CHAPTER 2

Random Variables and Characteristic Functions

Self-control means wanting to be effective at some random point in the infinite radiations of my spiritual existence.

Franz Kafka

2.1 Introduction

This chapter begins with a short review of probability measures and random variables. A sound formal understanding of random variables is crucial to have a complete understanding of much of the asymptotic theory that follows. Inequalities are also very useful in asymptotic theory, and the second section of this chapter reviews several basic inequalities for both probabilities and expectations, as well as some more advanced results that will have specific applications later in the book. The next section develops some limit theory that is useful for working with probabilities of sequences of events, including the Borel-Cantelli lemmas. We conclude the chapter with a review of moment generating functions, characteristic functions, and cumulant generating functions. Moment generating functions and characteristic functions are often a useful surrogate for distributions themselves. While the moment generating function may be familiar to many readers, the characteristic function may not, due to the need for some complex analysis. However, the extra effort required to use the characteristic function is worthwhile as many of the results presented later in the book are more useful when derived using characteristic functions.

2.2 Probability Measures and Random Variables

Consider an *experiment*, an action that selects a point from a set Ω called a sample space. The point that is selected is called the *outcome* of the experiment. In probability and statistics the focus is usually on *random* or *stochastic* experiments where the point that is selected cannot be predicted with absolute certainty, except in very specialized cases. Subsets of the sample space are called *events*, and are taken to be members of a collection of subsets of Ω called a σ-field.

Definition 2.1. *Let Ω be a set, then $\mathcal{F}$ is a σ-field of subsets of Ω if $\mathcal{F}$ has the following properties:*

1. $\emptyset \in \mathcal{F}$ and $\Omega \in \mathcal{F}$.
2. $A \in \mathcal{F}$ implies that $A^c \in \mathcal{F}$.
3. If $A_i \in \mathcal{F}$ for $i \in \mathbb{N}$ then

$$\bigcup_{i=1}^{\infty} A_i \in \mathcal{F} \quad and \quad \bigcap_{i=1}^{\infty} A_i \in \mathcal{F}.$$

In some cases a σ-field will be *generated* from a sample space Ω. This σ-field is the smallest σ-field that contains the events in Ω. The term smallest in this case means that this σ-field is a subset of any other σ-field that contains the events in Ω. For further information about σ-fields, their generators, and their use in probability theory see Section 1.2 of Gut (2005) or Section 2.1 of Pollard (2002).

The experiment selects an outcome in Ω according to a *probability measure* P, which is a set function that maps $\mathcal{F}$ to $\mathbb{R}$.

Definition 2.2. *Let Ω be a sample space and $\mathcal{F}$ be a σ-algebra of subsets of Ω. A set function $P : \mathcal{F} \to \mathbb{R}$ is a probability measure if P satisfies the axioms of probability set forth by Kolmogorov (1933). The axioms are:*

Axiom 1: $P(A) \geq 0$ for every $A \in \mathcal{F}$.

Axiom 2: $P(\Omega) = 1$.

Axiom 3: If $\{A_i\}_{i=1}^{\infty}$ is a sequence of mutually exclusive events in $\mathcal{F}$, then

$$P\left(\bigcup_{i=1}^{\infty} A_i\right) = \sum_{i=1}^{\infty} P(A_i).$$

The term mutually exclusive refers to the property that A_i and A_j are disjoint, or that $A_i \cap A_j = \emptyset$ for all $i \neq j$.

From Definition 2.2 it is clear that there are three elements that are required to assign a set of probabilities to outcomes from an experiment: the sample space Ω, which identifies the possible outcomes of the experiment; the σ-field $\mathcal{F}$, which identifies which events in Ω that the probability measure is able to compute probabilities for; and the probability measure P, which assigns probabilities to the events in $\mathcal{F}$. These elements are often collected together in a triple $(\Omega, \mathcal{F}, P)$, called a *probability space*.

When the sample space of an experiment is $\mathbb{R}$, or an interval subset of $\mathbb{R}$, the σ-field used to define the probability space is usually generated from the open subsets of the sample space.

Definition 2.3. *Let Ω be a sample space. The Borel σ-field corresponding to Ω is the σ-field generated by the collection of open subsets of Ω. The Borel σ-field generated by Ω is denoted by $\mathcal{B}\{\Omega\}$.*

In the case where $\Omega = \mathbb{R}$, it can be shown that $\mathcal{B}\{\mathbb{R}\}$ can be generated from simpler collections of events. In particular, $\mathcal{B}\{\mathbb{R}\}$ can be generated from the collection of intervals $\{(-\infty, b] : b \in \mathbb{R}\}$, with a similar result when Ω is a subset of $\mathbb{R}$ such as $\Omega = [0, 1]$. Other simple collections of intervals can also be used to generate $\mathcal{B}\{\mathbb{R}\}$. See Section 3.3 of Gut (2005) for further details.

The main purpose of this section is to introduce the concept of a random variable. Random variables provide a convenient way of referring to events within a sample space that often have simple interpretations with regard to the underlying experiment. Intuitively, random variables are often thought of as mathematical variables that are subject to random behavior. This informal way of thinking about random variables may be helpful to understand certain concepts in probability theory, but a true understanding, especially with regard to statistical limit theorems, comes from the formal mathematical definition below.

Definition 2.4. *Let $(\Omega, \mathcal{F}, P)$ be a probability space, X be a function that maps Ω to $\mathbb{R}$, and $\mathcal{B}$ be a σ-algebra of subsets of $\mathbb{R}$. The function X is a random variable if $X^{-1}(B) = \{\omega \in \Omega : X(\omega) \in B\} \in \mathcal{F}$, for all $B \in \mathcal{B}$.*

Note that according to Definition 2.4, there is actually nothing random about a random variable. When the experiment is performed an element of Ω is chosen at random according to the probability measure P. The role of the random variable is to map this outcome to the real line. Therefore, the output of the random variable is random, but the mapping itself is not. The restriction that $X^{-1}(B) = \{\omega \in \Omega : X(\omega) \in B\}$, for all $B \in \mathcal{B}$ assures that probability of the inverse mapping can be calculated.

Events written in terms of random variables are interpreted by selecting outcomes from the sample space Ω that satisfy the event. That is, if $A \in \mathcal{B}$ then the event $\{X \in A\}$ is equivalent to the event that consists of all outcomes $\omega \in \Omega$ such that $X(\omega) \in A$. This allows for the computation of probabilities of events written in terms of random variables. That is, $P(X \in A) = P(\omega : X(\omega) \in A)$, where it is assumed that the event will be empty when A is not a subset of the range of the function X. Random variables need not be one-to-one functions, but in the case where X is a one-to-one function and $a \in \mathbb{R}$ the computation simplifies to $P(X = a) = P[X^{-1}(a)]$.

Example 2.1. Consider the simple experiment where a fair coin is flipped three times, and the sequence of flips is observed. The elements of the sample space will be represented by triplets containing the symbols H_i, signifying that the ith flip is heads, and T_i, signifying that the ith flip is tails. The order of the symbols in the triplet signify the order in which the outcomes are observed. For example, the event $H_1 T_2 H_3$ corresponds to the event that heads was observed first, then tails, then heads again. The sample space for this experiment is given by

$$\Omega = \{T_1 T_2 T_3, T_1 T_2 H_3, T_1 H_2 T_3, H_1 T_2 T_3,$$
$$H_1 H_2 T_3, H_1 T_2 H_3, T_1 H_2 H_3, H_1 H_2 H_3\}.$$

Because the coin is fair, the probability measure P on this sample space is uniform so that each outcome has a probability of $\frac{1}{8}$. A suitable σ-field for Ω is given by the power set of Ω. Now consider a random variable X defined as

$$X(\omega) = \begin{cases} 0 & \text{if } \omega = \{T_1 T_2 T_3\}, \\ 1 & \text{if } \omega \in \{T_1 T_2 H_3, T_1 H_2 T_3, H_1 T_2 T_3\}, \\ 2 & \text{if } \omega \in \{H_1 H_2 T_3, H_1 T_2 H_3, T_1 H_2 H_3\}, \\ 3 & \text{if } \omega = \{H_1 H_2 H_3\}. \end{cases}$$

Hence, the random variable X counts the number of heads in the three flips of the coin. For example, the event that two heads are observed in the three flips of the coin can be represented by the event $\{X = 2\}$. The probability of this event is computed by considering all of the outcomes in the original sample space that satisfy this event. That is

$$P(X = 2) = P[\omega \in \Omega : X(\omega) = 2] = P(H_1 H_2 T_3, H_1 T_2 H_3, T_1 H_2 H_3) = \frac{3}{8}.$$

Because the inverse image of each possible value of X is in $\mathcal{F}$, we are always able to compute the probability of the corresponding inverse image.   ∎

Example 2.2. Consider a countable sample space of the form

$$\Omega = \{H_1, T_1 H_2, T_1 T_2 H_3, T_1 T_2 T_3 H_4, \ldots, \},$$

that corresponds to the experiment of flipping a coin repeatedly until the first heads is observed where the notation of Example 2.1 has been used to represent the possible outcomes of the experiment. If the coin is fair then the probability measure P defined as

$$P(\omega) = \begin{cases} \frac{1}{2}, & \omega = \{H_1\}, \\ 2^{-k}, & \omega = \{T_1 T_2 \cdots T_{k-1} H_k\}, \\ 0 & \text{otherwise}, \end{cases}$$

is appropriate for the described experiment with the σ-algebra $\mathcal{F}$ defined to be the power set of Ω. A useful random variable to define for this experiment is to set $X(H_1) = 1$ and $X(\omega) = k$ when $\omega = \{T_1 T_2 \cdots T_{k-1} H_k\}$, which counts the number of flips until the first head is observed.   ∎

Example 2.3. Consider the probability space $(\Omega, \mathcal{F}, P)$ where $\Omega = (0, 1]$, $\mathcal{F}$ is the Borel sets on the interval $(0, 1]$, and P is Lebesgue measure on $(0, 1]$. The measure P assigns an interval that is contained in $(0, 1]$ of length l the probability l. Probabilities of other events are obtained using the three axioms of Definition 2.2. Define a random variable $X(\omega) = -\log(\omega)$ for all $\omega \in (0, 1]$ so that the range of X is the positive real line. Let $0 < a < b < \infty$ and consider computing the probability that $X \in (a, b)$. Working with the inverse image, we find that

$$
\begin{aligned}
P[X \in (a, b)] &= P[\omega : X(\omega) \in (a, b)] \\
&= P[\omega : a < -\log(\omega) < b] \\
&= P[\omega : \exp(-b) < \omega < \exp(-a)] \\
&= \exp(-b) - \exp(-a),
\end{aligned}
$$

where we have used the monotonicity of the log function. This results in an EXPONENTIAL(1) random variable. ∎

Note that the inverse image of an event written in terms of a random variable may be empty if there is not a non-empty event in $\mathcal{F}$ that is mapped to the event in question. This does not present a problem as $\emptyset$ is guaranteed to be a member of $\mathcal{F}$ by Definition 2.1. Definition 2.2 then implies that the probability of the corresponding event is zero.

Functions of random variables can also be shown to be random variables themselves as long as the function has certain properties. According to Definition 2.4, a random variable is a function that maps the sample space ω to the real line in such a way that the inverse image of a Borel set is in $\mathcal{F}$. Suppose that X is a random variable and g is a real function. It is clear that the composition $g[X(\omega)]$ maps ω to $\mathbb{R}$. But for $g(X)$ to be a random variable we also require that

$$
[g(X)]^{-1}(B) = \{\omega \in \Omega : g[X(\omega)] \in B\} \in \mathcal{F},
$$

for all $B \in \mathcal{B}\{\mathbb{R}\}$. This can be guaranteed by requiring that the inverse image from g of a Borel set will always be a Borel set. That is, we require that

$$
g^{-1}(B) = \{b \in \mathbb{R} : g(b) \in B\} \in \mathcal{B}\{\mathbb{R}\},
$$

for all $B \in \mathcal{B}\{\mathbb{R}\}$. We will call such functions *Borel functions*. This development yields the following result.

Theorem 2.1. *Let X be a random variable and g be a Borel function. Then $g(X)$ is a random variable.*

The development of random vectors follows closely the development of random variables. Let $(\Omega, \mathcal{F}, P)$ be a probability space and let $X_1, \ldots, X_d$ be a set of random variables defined on $(\Omega, \mathcal{F}, P)$. We can construct a d-dimensional vector $\mathbf{X}$ by putting $X_1, \ldots, X_d$ into a $d \times 1$ array as $\mathbf{X} = (X_1, \ldots, X_d)'$. What does this construction represent? According to Definition 2.4, each X_i is a function that maps Ω to $\mathbb{R}$ in such a way that $X^{-1}(B) = \{\omega \in \Omega : X(\omega) \in B\} \in \mathcal{F}$ for all $B \in \mathcal{B}\{\mathbf{R}\}$. In the case of $\mathbf{X}$, a point $\omega \in \Omega$ will be mapped to $\mathbf{X}(\omega) = [X_1(\omega), \ldots, X_d(\omega)]' \in \mathbb{R}^d$. Therefore, $\mathbf{X}$ is a function that maps Ω to $\mathbb{R}^d$. However, in order for $\mathbf{X}$ to be a random vector we must be assured that we can compute the probability of events that are written in terms of the random vector. That is, we need to be able to compute the probability of the inverse mapping of any reasonable subset of $\mathbb{R}^d$. Reasonable subsets will be those that come from a σ-field on $\mathbb{R}^d$.

Definition 2.5. *Let $(\Omega, \mathcal{F}, P)$ be a probability space, $\mathbf{X}$ be a function that maps Ω to $\mathbb{R}^d$, and $\mathcal{B}^d$ be a σ-algebra of subsets of $\mathbb{R}^d$. The function $\mathbf{X}$ is a d-dimensional random vector if $\mathbf{X}^{-1}(B) = \{\omega \in \Omega : \mathbf{X}(\omega) \in B\} \in \mathcal{F}$, for all $B \in \mathcal{B}^d$.*

The usual σ-field used for random vectors is the Borel sets on $\mathbb{R}^d$, denoted by $\mathcal{B}\{\mathbb{R}^d\}$.

Example 2.4. Consider a probability space $(\Omega, \mathcal{F}, P)$ where $\Omega = (0, 1) \times (0, 1)$ is the unit square and P is a bivariate extension of Lebesgue measure. That is, if R is a rectangle of the form

$$R = \{(\omega_1, \omega_2) : \underline{\omega}_1 \leq \omega_1 \leq \overline{\omega}_1, \underline{\omega}_2 \leq \omega_1 \leq \overline{\omega}_2\},$$

where $\underline{\omega}_1 \leq \overline{\omega}_1$ and $\underline{\omega}_2 \leq \overline{\omega}_2$ then $P(R) = (\overline{\omega}_1 - \underline{\omega}_1)(\overline{\omega}_2 - \underline{\omega}_1)$, which corresponds to the area of the rectangle. Let $\mathcal{F} = \mathcal{B}\{(0, 1) \times (0, 1)\}$ and in general define $P(B)$ to be the area of B for any $B \in \mathcal{B}\{(0, 1) \times (0, 1)\}$. It is shown in Exercise 3 that this is a probability measure. A simple bivariate random variable defined on $(\Omega, \mathcal{F}, P)$ is $\mathbf{X}(\omega) = (\omega_1, \omega_2)'$ which would return the ordered pair from the sample space. Another choice, $\mathbf{X}(\omega) = [-\log(\omega_1), -\log(\omega_2)]'$ yields a bivariate EXPONENTIAL random vector with independent EXPONENTIAL(1) marginal distributions. ∎

As with the case of random variables there is the technical question as to whether a function of a random vector is also a random vector. The development leading to Theorem 2.1 can be extended to random vectors by considering a function $g : \mathbb{R}^d \to \mathbb{R}^q$ that has the property that

$$g^{-1}(B) = \{\mathbf{b} \in \mathbb{R}^d : g(\mathbf{b}) \in B\} \in \mathcal{B}\{\mathbb{R}^d\},$$

for all $B \in \mathcal{B}\{\mathbb{R}^q\}$. As with the univariate case, we shall call such a function as simply a Borel function, and we get a parallel result to Theorem 2.1.

Theorem 2.2. *Let $\mathbf{X}$ be a d-dimensional random vector and $g : \mathbb{R}^d \to \mathbb{R}^q$ be a Borel function. Then $g(\mathbf{X})$ is a q-dimensional random vector.*

If $\mathbf{X}' = (X_1, \ldots, X_d)$ is a random vector in $\mathbb{R}^b$ then some examples of functions of $\mathbf{X}$ that are random variables include $\min\{X_1, \ldots, X_d\}$ and $\max\{X_1, \ldots, X_d\}$, as well as each of the components of $\mathbf{X}$. Further, it follows that if $\{X_n\}_{n=1}^\infty$ is a sequence of random variables the

$$\inf_{n \in \mathbb{N}} X_n,$$

and

$$\sup_{n \in \mathbb{N}} X_n,$$

are also random variables. See Section 2.1.1 of Gut (2005) for further details.

2.3 Some Important Inequalities

In many applications in asymptotic theory, inequalities provide useful upper or lower bounds on probabilities and expectations. These bounds are often helpful in proving important properties. This section briefly reviews many of the inequalities that are used later in the book. We begin with some basic inequalities that are based on probability measures.

Theorem 2.3. *Let A and B be events in a probability space $(\Omega, \mathcal{F}, P)$. Then*

1. *If $A \subset B$ then $P(A) \leq P(B)$.*
2. *$P(A \cap B) \leq P(A)$*

Proof. Note that if $A \subset B$ then B can be partitioned into two mutually exclusive events as $B = A \cup (A^c \cap B)$. Axiom 3 of Definition 2.2 then implies that $P(B) = P(A) + P(A^c \cap B)$. Axiom 1 of Definition 2.2 then yields the result by noting that $P(A^c \cap AB) \geq 0$. The second property follows from the first by noting that $A \cap B \subset A$. $\qquad\qquad\square$

When events are not necessarily mutually exclusive, the *Bonferroni Inequality* is useful for obtaining an upper bound on the probability of the union of the events. In the special case were the events are mutually exclusive, Axiom 3 of Definition 2.2 applies and an equality results.

Theorem 2.4 (Bonferroni). *Let $\{A_i\}_{i=1}^n$ be a sequence of events from a probability space $(\Omega, \mathcal{F}, P)$. Then*

$$P\left(\bigcup_{i=1}^n A_i\right) \leq \sum_{i=1}^n P(A_i).$$

Theorem 2.4 is proven in Exercise 4.

The moments of random variables play an important role in obtaining bounds for probabilities. The first result given below is rather simple, but useful, and can be proven by noting that the variance of a random variable is always non-negative.

Theorem 2.5. *Consider a random variable X where $V(X) < \infty$. Then $V(X) \leq E(X^2)$.*

Markov's Theorem is a general result that places a bound on the tail probabilities of a random variable using the fact that a certain set of moments of the random variable are finite. Essentially the result states that only so much probability can be in the tails of the distribution of a random variable X when $E(|X|^r) < \infty$ for some $r > 0$.

Theorem 2.6 (Markov). *Consider a random variable X where $E(|X|^r) < \infty$ for some $r > 0$ and let $\delta > 0$. Then $P(|X| > \delta) \leq \delta^{-r} E(|X|^r)$.*

Proof. Assume for simplicity that X is a continuous random variable with distribution function F. If $\delta > 0$ then

$$
\begin{aligned}
E(|X|^r) &= \int_{-\infty}^{\infty} |x|^r dF(x) \\
&= \int_{\{x:|x|\leq\delta\}} |x|^r dF(x) + \int_{\{x:|x|>\delta\}} |x|^r dF(x) \\
&\geq \int_{\{x:|x|>\delta\}} |x|^r dF(x),
\end{aligned}
$$

since

$$
\int_{\{x:|x|\leq\delta\}} |x|^r dF(x) \geq 0.
$$

Note that $|x|^r \geq \delta^r$ within the set $\{x : |x| > \delta\}$, so that

$$
\int_{\{x:|x|>\delta\}} |x|^r dF(x) \geq \int_{\{x:|x|>\delta\}} \delta^r dF(x) = \delta^r P(|X| > \delta).
$$

Therefore $E(|X|^r) \geq \delta^r P(|X| > \delta)$ and the result follows. □

Tchebysheff's Theorem is an important special case of Theorem 2.6 that restates the tail probability in terms of distance from the mean when the variance of the random variable is finite.

Theorem 2.7 (Tchebysheff). *Consider a random variable X with $E(X) = \mu$ and $V(X) = \sigma^2 < \infty$. Then $P(|X - \mu| > \delta) \leq \delta^{-2}\sigma^2$.*

Theorem 2.7 is proven in Exercise 6. Several inequalities for expectations are also used in this book. This first result is a direct consequence of Theorems A.16 and A.18.

Theorem 2.8. *Let $0 < r \leq 1$ and suppose that X and Y are random variables such that $E(|X|^r) < \infty$ and $E(|Y|^r) < \infty$. Then $E(|X + Y|^r) \leq E(|X|^r) + E(|Y|^r)$.*

Minkowski's Inequality provides a similar result to Theorem 2.8, expect that the power r now exceeds one.

Theorem 2.9 (Minkowski). *Let $X_1, \ldots, X_n$ be random variables such that $E(|X_i|^r) < \infty$ for $i = 1, \ldots, n$. Then for $r \geq 1$,*

$$
\left[E\left(\left| \sum_{i=1}^{n} X_i \right|^r \right) \right]^{1/r} \leq \sum_{i=1}^{n} [E(|X_i|^r)]^{1/r}.
$$

Note that Theorem A.18 is a special case of Theorems 2.8 and 2.9 when r is take to be one. Theorem 2.9 can be proven using Theorem A.18 and *Hölder's Inequality*, given below.

Theorem 2.10 (Hölder). *Let X and Y be random variables such that $E|X|^p < \infty$ and $E|Y|^q < \infty$ where p and q are real numbers such that $p^{-1} + q^{-1} = 1$. Then*

$$
E(|XY|) \leq [E(|X|^p)]^{1/p} [E(|Y|^q)]^{1/q}.
$$

Figure 2.1 *An example of a convex function.*

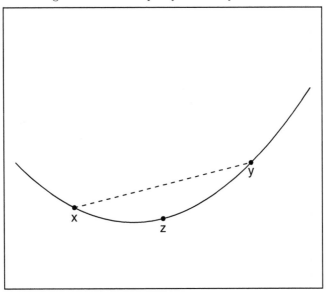

For proofs of Theorems 2.9 and 2.10, see Section 3.2 of Gut (2005).

A more general result is *Jensen's Inequality*, which is based on the properties of *convex functions*.

Definition 2.6. *Let f be a real function such that $f[\lambda x + (1 - \lambda)y] \leq \lambda f(x) + (1 - \lambda)f(y)$ for all $x \in \mathbb{R}$, $y \in \mathbb{R}$ and $\lambda \in (0, 1)$, then f is a convex function. If the inequality is strict then the function is strictly convex. If the function $-f(x)$ is convex, then the function $f(x)$ is concave.*

Note that if $\lambda = 0$ then $\lambda f(x) + (1 - \lambda)f(y) = f(y)$, and if $\lambda = 1$ it follows that $\lambda f(x) + (1 - \lambda)f(y) = f(x)$. Noting that $\lambda f(x) + (1 - \lambda)f(y) = \lambda[f(x) - f(y)] + f(y)$ we see that when x and y are fixed, $\lambda f(x) + (1 - \lambda)f(y) = f(y)$ is a linear function of λ. Therefore, $\lambda f(x) + (1 - \lambda)f(y) = f(y)$ represents the line that connects $f(x)$ to $f(y)$. Hence, convex functions are ones such that $f(z)$ is always below the line connecting $f(x)$ and $f(y)$ for all real numbers x, y, and z such that $x < z < y$. Figure 2.1 gives an example of a convex function while Figure 2.2 gives an example of a function that is not convex. This property of convex functions allows us to develop the following inequality for expectations of convex functions of a random variable.

Theorem 2.11 (Jensen). *Let X be a random variable and let g be a convex function, then $E[g(X)] \geq g[E(X)]$. If g is strictly convex then $E[g(X)] > g[E(X)]$.*

For a proof of Theorem 2.11, see Section 5.3 of Fristedt and Gray (1997).

Figure 2.2 *An example of function that is not convex.*

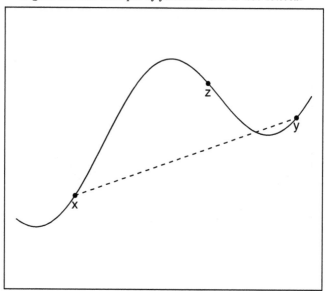

Example 2.5. Let X be a random variable such that $E(|X|^r) < \infty$, for some $r > 0$. Let s be a real number such that $0 < s < r$. Since $r/s > 1$ it follows that $f(x) = x^{r/s}$ is a convex function and therefore Theorem 2.11 implies that $[E(|X|^s)]^{r/s} \leq E[(|X|^s)^{r/s}] = E(|X|^r) < \infty$, so that it follows that $E(|X|^s) < \infty$. This establishes the fact that the existence of higher order moments implies the existence of lower order moments. ∎

The following result establishes a bound for the absolute expectation of the sum of truncated random variables. Truncation is often a useful tool that can be applied when some of the moments of a random variable do not exist. If the random variables are truncated at some finite value, then all of the moments of the truncated random variables must exist, and tools such as Theorem 2.7 can be used.

Theorem 2.12. Let $X_1, \dots, X_n$ be a set of independent and identically distributed random variables such that $E(X_1) = 0$. Then, for any $\varepsilon > 0$,

$$\left| E\left(\sum_{i=1}^{n} X_i \delta\{|X_i|; [0, \varepsilon]\} \right) \right| \leq n E(|X_1| \delta\{|X_1|; (\varepsilon, \infty)\}).$$

Proof. Note that for all $\omega \in \Omega$ and $i \in \{1, \dots, n\}$ we have that

$$X_i(\omega) = X_i(\omega)\delta\{|X_i(\omega)|; [0, \varepsilon]\} + X_i(\omega)\delta\{|X_i(\omega)|; (\varepsilon, \infty)\},$$

so that

$$E(X) = E(X_i\delta\{|X_i|; [0, \varepsilon]\}) + E(X_i\delta\{|X_i|; (\varepsilon, \infty)\}).$$

Because $E(X_i) = 0$ for all $i \in \{1, \ldots, n\}$ we have that

$$E(X_i \delta\{|X_i|; [0, \varepsilon]\}) = -E(X_i \delta\{|X_i|; (\varepsilon, \infty)\}). \tag{2.1}$$

Now

$$
\begin{aligned}
\left| E\left(\sum_{i=1}^{n} X_i \delta\{|X_i|; [0, \varepsilon]\} \right) \right| &= \left| \sum_{i=1}^{n} E(X_i \delta\{|X_i|; [0, \varepsilon]\}) \right| \\
&\leq \sum_{i=1}^{n} |E(X_i \delta\{|X_i|; [0, \varepsilon]\})| \\
&\leq \sum_{i=1}^{n} |E(X_i \delta\{|X_i|; (\varepsilon, \infty)\})| \\
&\leq \sum_{i=1}^{n} E(|X_i| \delta\{|X_i|; (\varepsilon, \infty)\}) \\
&= nE(|X_1| \delta\{|X_1|; (\varepsilon, \infty)\}),
\end{aligned}
$$

where the first inequality follows from Theorem A.18, the second inequality follows from Equation (2.1), and the third inequality follows from Theorem A.6. $\qquad \square$

The following inequality is an analog of Theorem 2.7 for the case when the random variable of interest is a sum or an average. While the bound has the same form as Theorem 2.7, the event in the probability concerns a much stronger event in terms of the maximal value of the random variable. The result is usually referred to as *Kolmogorov's Maximal Inequality*.

Theorem 2.13 (Kolmogorov). *Let $\{X_n\}_{n=1}^{\infty}$ be a sequence of independent random variables where $E(X_n) = 0$ and $V(X_n) < \infty$ for all $n \in \mathbb{N}$. Let*

$$S_n = \sum_{i=1}^{n} X_i.$$

Then, for any $\varepsilon > 0$,

$$P\left(\max_{i \in \{1, \ldots, n\}} |S_i| > \varepsilon \right) \leq \varepsilon^{-2} V(S_n).$$

Proof. The proof provided here is fairly standard for this result, though this particular version runs most closely along what is shown in Gut (2005) and Sen and Singer (1993). Let $\varepsilon > 0$ and define $S_0 \equiv 0$ and define a sequence of events $\{A_i\}_{i=0}^{n}$ as

$$A_0 = \{|S_i| \leq \varepsilon; i \in \{0, 1, \ldots, n\}\},$$

and

$$A_i = \{|S_k| \leq \varepsilon; k \in \{0, 1, \ldots, i-1\}\} \cap \{|S_i| > \varepsilon\},$$

for $i \in \{1, \ldots, n\}$. The essential idea to deriving the bound is based on the fact that

$$\left\{ \max_{i \in \{1,\ldots,n\}} |S_i| > \varepsilon \right\} = \bigcup_{i=1}^{n} A_i, \tag{2.2}$$

where the events in the sequence $\{A_i\}_{i=0}^{n}$ are mutually exclusive. We now consider S_n to be a random variable that maps some sample space Ω to $\mathbb{R}$. Equation (2.2) implies that

$$\bigcup_{i=1}^{n} A_i \subset \Omega,$$

so that

$$S_n^2(\omega) \geq S_n^2(\omega)\delta \left\{ \omega; \bigcup_{i=1}^{n} A_i \right\} = \sum_{i=1}^{n} S_n^2(\omega)\delta\{\omega; A_i\},$$

for all $\omega \in \Omega$, where we have used the fact that the events in the sequence $\{A_i\}_{i=0}^{n}$ are mutually exclusive. Therefore, Theorem A.16 implies that

$$E(S_n^2) \geq \sum_{i=1}^{n} E(S_n^2 \delta\{A_i\}),$$

where we have suppressed the ω argument in the indicator function. Note that we can write S_n as $(S_n - S_i) + S_i$ and therefore $S_n^2 = S_i^2 + 2S_i(S_n - S_i) + (S_n - S_i)^2$. Hence

$$
\begin{aligned}
E(S_n^2) &\geq \sum_{i=1}^{n} E\{[S_i^2 + 2S_i(S_n - S_i) + (S_n - S_i)^2]\delta\{A_i\}\} \\
&\geq \sum_{i=1}^{n} E\{[S_i^2 + 2S_i(S_n - S_i)]\delta\{A_i\}\} \\
&= \sum_{i=1}^{n} E(S_i^2 \delta\{A_i\}) + 2\sum_{i=1}^{n} E[S_i(S_n - S_i)\delta\{A_i\}],
\end{aligned}
$$

where the second inequality is due to the fact that $[S_n(\omega) - S_i(\omega)]^2 \geq 0$ for all $\omega \in \Omega$. Now note that

$$S_n - S_i = \sum_{k=i+1}^{n} X_k,$$

is independent of $S_i \delta\{A_i\}$ because the event A_i and the sum S_i depend only on $X_1, \ldots, X_i$. Therefore

$$E[S_i(S_n - S_i)\delta\{A_i\}] = E[(S_n - S_i)]E(S_i\delta\{A_i\}) =$$

$$E\left(\sum_{k=i+1}^{n} X_k\right) E(S_i\delta\{A_i\}) = 0,$$

since $E(X_i) = 0$ for all $i \in \mathbb{N}$. Therefore,

$$E(S_n^2) \geq \sum_{i=1}^{n} E(S_i^2 \delta\{A_i\}) \geq \varepsilon^2 \sum_{i=1}^{n} E(\delta\{A_i\}). \qquad (2.3)$$

To obtain the second inequality in Equation (2.3), note that when $\omega \in A_i$ then $|S_i| > \varepsilon$, and therefore $S_i^2(\omega)\delta\{\omega; A_i\} \geq \varepsilon^2$. When $\omega \notin A_i$ we have that $S_i^2(\omega)\delta\{\omega; A_i\} = 0$. It follows that $S_i^2(\omega)\delta\{\omega; A_i\} \geq \varepsilon^2\delta\{\omega; A_i\}$, for all $\omega \in \Omega$ and hence Theorem A.16 implies that $E(S_i^2\delta\{A_i\}) \geq \varepsilon^2 E(\delta\{A_i\})$. Now note that

$$\varepsilon^2 \sum_{i=1}^{n} E(\delta\{A_i\}) = \varepsilon^2 \sum_{i=1}^{n} P(A_i) = \varepsilon^2 P\left(\bigcup_{i=1}^{n} A_i\right) = \varepsilon^2 P\left(\max_{i\in\{1,\dots,n\}} |S_i| > \varepsilon\right).$$

Therefore, we have established that

$$E(S_n^2) \geq \varepsilon^2 P\left(\max_{i\in\{1,\dots,n\}} |S_i| > \varepsilon\right),$$

which yields the final result. $\square$

The exchange of a limit and an expectation is equivalent to the problem of exchanging a limit and an integral or sum. Therefore, concepts such as monotone or dominated convergence play a role in determining when such an exchange can take place. A related result is *Fatou's Lemma*, which provides inequalities concerning such exchanges for sequences of random variables. The result is also related to the problem of whether the convergence of a sequence of random variables implies the convergence of the corresponding moments, which we shall consider in detail in Chapter 5.

Theorem 2.14 (Fatou). *Let $\{X_n\}_{n=1}^{\infty}$ be a sequence of non-negative random variables and suppose there exist random variables L and U such that $P(L \leq X_n \leq U) = 1$ for all $n \in \mathbb{N}$, $E(|L|) < \infty$, and $E(|U|) < \infty$. Then*

$$E\left(\liminf_{n\to\infty} X_n\right) \leq \liminf_{n\to\infty} E(X_n) \leq \limsup_{n\to\infty} E(X_n) \leq E\left(\limsup_{n\to\infty} X_n\right).$$

A proof of Theorem 2.14 can be found in Section 2.5 of Gut (2005).

2.4 Some Limit Theory for Events

The concept of a limit can also be applied to sequences of events. Conceptually, the limiting behavior of sequences of events is more abstract than it is with sequences of real numbers. We will begin by defining the limit supremum and infimum for sequences of events. Recall from Definition 1.3 that the limit supremum of a sequence of real numbers $\{x_n\}_{n=1}^{\infty}$ is given by

$$\limsup_{n\to\infty} x_n = \inf_{n\in\mathbb{N}} \sup_{k\geq n} x_k.$$

To convert this concept to sequences of events note that the supremum of a sequence of real numbers is the least upper bound of the sequence. In terms of events we will say that an event A is an upper bound of an event B if $B \subset A$. For a sequence of events $\{A_n\}_{n=1}^{\infty}$ in $\mathcal{F}$, the supremum will be defined to be the smallest event that contains the sequence, which is the least upper bound of the sequence. That is, we will define

$$\sup_{n \in \mathbb{N}} A_k = \bigcup_{n=1}^{\infty} A_n.$$

Note that any event smaller than this union will not be an upper bound of the sequence. Following similar arguments, the infimum of a sequence of real numbers is defined to be the greatest lower bound. Therefore, the infimum of a sequence of events is the largest event that is contained by all of the events in the sequence. Hence, the infimum of a sequence of events $\{A_n\}_{n=1}^{\infty}$ is defined as

$$\inf_{n \in \mathbb{N}} A_n = \bigcap_{n=1}^{\infty} A_n.$$

These concepts can be combined to define the limit supremum and the limit infimum of a sequence of events.

Definition 2.7. *Let $\{A_n\}_{n=1}^{\infty}$ be a sequence of events in a σ-field $\mathcal{F}$ generated from a sample space Ω. Then*

$$\liminf_{n \to \infty} A_n = \sup_{n \in \mathbb{N}} \inf_{k \geq n} A_n = \bigcup_{n=1}^{\infty} \bigcap_{k=n}^{\infty} A_k,$$

and

$$\limsup_{n \to \infty} A_n = \inf_{n \in \mathbb{N}} \sup_{k \geq n} A_n = \bigcap_{n=1}^{\infty} \bigcup_{k=n}^{\infty} A_k.$$

As with sequences of real numbers, the limit of a sequence of events exists when the limit infimum and limit supremum of the sequence agree.

Definition 2.8. *Let $\{A_n\}_{n=1}^{\infty}$ be a sequence of events in a σ-field $\mathcal{F}$ generated from a sample space Ω. If*

$$A = \liminf_{n \to \infty} A_n = \limsup_{n \to \infty} A_n,$$

then the limit of the sequence $\{A_n\}_{n=1}^{\infty}$ exists and

$$\lim_{n \to \infty} A_n = A.$$

If

$$\liminf_{n \to \infty} A_n \neq \limsup_{n \to \infty} A_n,$$

then the limit of the sequence $\{A_n\}_{n=1}^{\infty}$ does not exist.

Example 2.6. Consider the probability space $(\Omega, \mathcal{F}, P)$ where $\Omega = (0, 1)$,

$\mathcal{F} = \mathcal{B}\{(0,1)\}$, and the sequence of events $\{A_n\}_{n=1}^{\infty}$ is defined by $A_n = (\frac{1}{3} - (3n)^{-1}, \frac{2}{3} + (3n)^{-1})$ for all $n \in \mathbb{N}$. Now

$$\inf_{k \geq n} A_n = \bigcap_{k=n}^{\infty} (\tfrac{1}{3} - (3k)^{-1}, \tfrac{2}{3} + (3k)^{-1}) = [\tfrac{1}{3}, \tfrac{2}{3}],$$

which does not depend on n so that by Definition 2.7

$$\liminf_{n \to \infty} A_n = \bigcup_{n=1}^{\infty} \bigcap_{k=n}^{\infty} (\tfrac{1}{3} - (3k)^{-1}, \tfrac{2}{3} + (3k)^{-1}) = \bigcup_{n=1}^{\infty} [\tfrac{1}{3}, \tfrac{2}{3}] = [\tfrac{1}{3}, \tfrac{2}{3}].$$

Similarly, by Definition 2.7

$$
\begin{aligned}
\limsup_{n \to \infty} A_n &= \bigcap_{n=1}^{\infty} \bigcup_{k=n}^{\infty} (\tfrac{1}{3} - (3k)^{-1}, \tfrac{2}{3} + (3k)^{-1}) \\
&= \bigcap_{n=1}^{\infty} (\tfrac{1}{3} - (3n)^{-1}, \tfrac{2}{3} + (3n)^{-1}) \\
&= [\tfrac{1}{3}, \tfrac{2}{3}],
\end{aligned}
$$

which matches the limit infimum. Therefore, by Definition 2.8,

$$\lim_{n \to \infty} A_n = [\tfrac{1}{3}, \tfrac{2}{3}].$$

∎

Example 2.7. Consider the probability space $(\Omega, \mathcal{F}, P)$ where $\Omega = (0,1)$, $\mathcal{F} = \mathcal{B}\{(0,1)\}$, and the sequence of events $\{A_n\}_{n=1}^{\infty}$ is defined by $A_n = (\frac{1}{2} + (-1)^n \frac{1}{4}, 1)$ for all $n \in \mathbb{N}$. Definition 2.7 implies that

$$\liminf_{n \to \infty} A_n = \bigcup_{n=1}^{\infty} \bigcap_{k=n}^{\infty} (\tfrac{1}{2} + (-1)^k \tfrac{1}{4}, 1) = \bigcup_{n=1}^{\infty} (\tfrac{3}{4}, 1) = (\tfrac{3}{4}, 1).$$

Similarly, Definition 2.7 implies that

$$\limsup_{n \to \infty} A_n = \bigcap_{n=1}^{\infty} \bigcup_{k=n}^{\infty} (\tfrac{1}{2} + (-1)^k \tfrac{1}{4}, 1) = \bigcap_{n=1}^{\infty} (\tfrac{1}{4}, 1) = (\tfrac{1}{4}, 1).$$

In this case the limit of the sequence of events $\{A_n\}_{n=1}^{\infty}$ does not exist.   ∎

The sequence of events studied in Example 2.6 has a property that is very important in the theory of limits of sequences of events. In that example, $A_{n+1} \subset A_n$ for all $n \in \mathbb{N}$, which corresponds to a monotonically increasing sequence of events. The computation of the limits of such sequences is simplified by this structure.

Theorem 2.15. *Let $\{A_n\}_{n=1}^{\infty}$ be a sequence of events from a σ-field $\mathcal{F}$ of subsets of a sample space Ω.*

1. If $A_n \subset A_{n+1}$ for all $n \in \mathbb{N}$ then the sequence is monotonically increasing *and has limit*

$$\lim_{n\to\infty} A_n = \bigcup_{n=1}^{\infty} A_n.$$

2. *If $A_{n+1} \subset A_n$ for all $n \in \mathbb{N}$ then the sequence is monotonically decreasing and has limit*

$$\lim_{n\to\infty} A_n = \bigcap_{n=1}^{\infty} A_n.$$

Proof. We will prove the first result. The second result is proven in Exercise 7. Let $\{A_n\}_{n=1}^{\infty}$ be a sequence of monotonically increasing events from $\mathcal{F}$. That is $A_n \subset A_{n+1}$ for all $n \in \mathbb{N}$. Then Definition 2.7 implies

$$\liminf_{n\to\infty} A_n = \bigcup_{n=1}^{\infty} \bigcap_{k=n}^{\infty} A_k.$$

Because $A_n \subset A_{n+1}$ it follows that

$$\bigcap_{k=n}^{\infty} A_k = A_n,$$

and therefore

$$\liminf_{n\to\infty} A_n = \bigcup_{n=1}^{\infty} A_n.$$

Similarly, Definition 2.7 implies that

$$\limsup_{n\to\infty} A_n = \bigcap_{n=1}^{\infty} \bigcup_{k=n}^{\infty} A_k.$$

The monotonicity of the sequence implies that

$$\bigcup_{k=1}^{\infty} A_k = \bigcup_{k=n}^{\infty} A_k,$$

for all $n \in \mathbb{N}$. Therefore,

$$\limsup_{n\to\infty} A_n = \bigcap_{n=1}^{\infty} \bigcup_{k=1}^{\infty} A_k = \bigcup_{k=1}^{\infty} A_k.$$

Therefore, Definition 2.8 implies

$$\lim_{n\to\infty} A_n = \bigcup_{n=1}^{\infty} A_n.$$

$\square$

It is important to note that the limits of the sequences are members of $\mathcal{F}$.

For monotonically increasing sequences this follows from the fact that count-able unions of events in $\mathcal{F}$ are also in $\mathcal{F}$ by Definition 2.1. For monotonically decreasing sequences note that by Theorem A.2,

$$\left(\bigcap_{n=1}^{\infty} A_n\right)^c = \bigcup_{n=1}^{\infty} A_n^c. \tag{2.4}$$

The complement of each event is a member of $\mathcal{F}$ by Defintion 2.1 and there-fore the countable union is also a member of $\mathcal{F}$. Because the limits of these sequences of events are members of $\mathcal{F}$, the probabilities of the limits can also be computed. In particular, Theorem 2.16 shows that the limit of the probabil-ities of the sequence and the probability of the limit of the sequence coincide.

Theorem 2.16. *Let $\{A_n\}_{n=1}^{\infty}$ be a sequence of events from a σ-field $\mathcal{F}$ of subsets of a sample space Ω. If the sequence $\{A_n\}_{n=1}^{\infty}$ is either monotonically increasing or decreasing then*

$$\lim_{n\to\infty} P(A_n) = P\left(\lim_{n\to\infty} A_n\right).$$

Proof. To prove this result break the sequence up into mutually exclusive events and use Definition 2.2. If $\{A_n\}_{n=1}^{\infty}$ is a sequence of monotonically in-creasing events then define $B_{n+1} = A_{n+1} \cap A_n^c$ for $n \in \mathbb{N}$ where B_1 is defined to be A_1. Note that the sequence $\{B_n\}_{n=1}^{\infty}$ is defined so that

$$\bigcup_{i=1}^{n} A_i = \bigcup_{i=1}^{n} B_i,$$

for all $n \in \mathbb{N}$ and hence

$$\bigcup_{n=1}^{\infty} A_n = \bigcup_{n=1}^{\infty} B_n. \tag{2.5}$$

Definition 2.2 implies that

$$P(A_n) = \sum_{i=1}^{n} P(B_i),$$

for all $n \in \mathbb{N}$. Therefore, taking the limit of each side of the equation as $n \to \infty$ yields

$$\lim_{n\to\infty} P(A_n) = \sum_{n=1}^{\infty} P(B_n).$$

Definition 2.2, Equation (2.5), and Theorem 2.15 then imply

$$\sum_{n=1}^{\infty} P(B_n) = P\left(\bigcup_{n=1}^{\infty} B_n\right) = P\left(\bigcup_{n=1}^{\infty} A_n\right) = P\left(\lim_{n\to\infty} A_n\right),$$

which proves the result for monotonically increasing events. For monotonically decreasing events take the complement as shown in Equation (2.4) and note that the resulting sequence is monotonically increasing. The above result is then applied to this resulting sequence of monotonically increasing events. $\quad\Box$

The *Borel-Cantelli Lemmas* relate the probability of the limit supremum of a sequence of events to the convergence of the sum of the probabilities of the events in the sequence.

Theorem 2.17 (Borel and Cantelli). *Let $\{A_n\}_{n=1}^{\infty}$ be a sequence of events. If*

$$\sum_{n=1}^{\infty} P(A_n) < \infty,$$

then

$$P\left(\limsup_{n \to \infty} A_n\right) = 0.$$

Proof. From Definition 2.7

$$P\left(\limsup_{n \to \infty} A_n\right) = P\left(\bigcap_{n=1}^{\infty} \bigcup_{m=n}^{\infty} A_m\right) \leq P\left(\bigcup_{m=n}^{\infty} A_m\right),$$

for each $n \in \mathbb{N}$, where the inequality follows from Theorem 2.3 and the fact that

$$\bigcap_{n=1}^{\infty} \bigcup_{m=n}^{\infty} A_m \subset \bigcup_{m=n}^{\infty} A_m.$$

Now, Theorem 2.4 (Bonferroni) implies that

$$P\left(\limsup_{n \to \infty} A_n\right) \leq \sum_{m=n}^{\infty} P(A_m),$$

for each $n \in \mathbb{N}$. The assumed condition

$$\sum_{n=1}^{\infty} P(A_n) < \infty,$$

implies that

$$\lim_{n \to \infty} \sum_{m=n}^{\infty} P(A_m) = 0,$$

so that Theorem 2.16 implies that

$$P\left(\limsup_{n \to \infty} A_n\right) = 0.$$

$\square$

The usual interpretation of Theorem 2.17 relies on considering how often, in an asymptotic sense, the events in the sequence occur. Recall that the event

$$\bigcup_{m=n}^{\infty} A_m,$$

occurs when at least one event in the sequence $A_n, A_{n+1}, \ldots$ occurs. Now, if the event

$$\bigcap_{n=1}^{\infty} \bigcup_{m=n}^{\infty} A_m,$$

occurs, then at least one event in the sequence $A_n, A_{n+1}, \ldots$ occurs *for every* $n \in \mathbb{N}$. This means that whenever one of the events A_n occurs in the sequence, we are guaranteed with probability one that $A_{n'}$ will occur for some $n' > n$. That is, if the event

$$\limsup_{n \to \infty} A_n = \bigcap_{n=1}^{\infty} \bigcup_{m=n}^{\infty} A_m,$$

occurs, then events in the sequence occur infinitely often. This event is usually represented as $\{A_n \text{ i.o.}\}$. In this context, Theorem 2.17 implies that if

$$\sum_{n=1}^{\infty} P(A_n) < \infty,$$

then the probability that A_n occurs infinitely often is zero. That is, there will exist an $n' \in \mathbb{N}$ such that none of the events in the sequence $\{A_{n'+1}, A_{n'+2}, \ldots\}$ will occur, with probability one. The second Borel and Cantelli Lemma relates the divergence of the sum of the probabilities of the events in the sequence to the case where A_n occurs infinitely often with probability one. This result only applies to the case where the events in the sequence are independent.

Theorem 2.18 (Borel and Cantelli). *Let $\{A_n\}_{n=1}^{\infty}$ be a sequence of independent events. If*

$$\sum_{n=1}^{\infty} P(A_n) = \infty,$$

then

$$P\left(\limsup_{n \to \infty} A_n\right) = P(A_n \text{ i.o.}) = 1.$$

Proof. We use the method of Billingsley (1986) to prove this result. Note that by Theorem A.2

$$\left(\limsup_{n \to \infty} A_n\right)^c = \left(\bigcap_{n=1}^{\infty} \bigcup_{m=n}^{\infty} A_m\right)^c = \bigcup_{n=1}^{\infty} \bigcap_{m=n}^{\infty} A_m^c.$$

Therefore, if we can prove that

$$P\left(\bigcup_{n=1}^{\infty} \bigcap_{m=n}^{\infty} A_m^c\right) = 0,$$

then the result will follow. In fact, note that if we are able to show that

$$P\left(\bigcap_{m=n}^{\infty} A_m^c\right) = 0, \tag{2.6}$$

for each $n \in \mathbb{N}$, then Theorem 2.4 implies that

$$P\left(\bigcup_{n=1}^{\infty} \bigcap_{m=n}^{\infty} A_m^c\right) \leq \sum_{n=1}^{\infty} P\left(\bigcap_{m=n}^{\infty} A_m^c\right) = 0.$$

Therefore, it suffices to show that the property in Equation (2.6) is true. We will first consider a finite part of the intersection in Equation (2.6). The independence of the $\{A_n\}_{n=1}^{\infty}$ sequence implies that

$$P\left(\bigcap_{m=n}^{n+k} A_m^c\right) = \prod_{m=n}^{n+k} P(A_m^c).$$

Now, note that $1 - x \leq \exp(-x)$ for all positive and real x, so that

$$P(A_m^c) = 1 - P(A_m) \leq \exp[-P(A_m)],$$

for all $m \in \{n, n+1, \ldots\}$. Therefore,

$$P\left(\bigcap_{m=n}^{n+k} A_m^c\right) \leq \prod_{m=n}^{n+k} \exp[-P(A_m)] = \exp\left[-\sum_{m=n}^{n+k} P(A_m)\right]. \qquad (2.7)$$

We wish to take the limit of both sides of Equation (2.7), which means that we need to evaluate the limit of the sum on the right hand side. By supposition we have that

$$\sum_{m=1}^{\infty} P(A_m) = \sum_{m=1}^{n-1} P(A_m) + \sum_{m=n}^{\infty} P(A_m) = \infty,$$

and

$$\sum_{m=1}^{n-1} P(A_m) \leq n - 1 < \infty,$$

since $P(A_m) \leq 1$ for each $m \in \mathbb{N}$. Therefore, it follows that

$$\sum_{m=n}^{\infty} P(A_m) = \infty, \qquad (2.8)$$

for each $n \in \mathbb{N}$. Finally, to take the limit, note that

$$\left\{\bigcap_{m=n}^{n+k} A_m^c\right\}_{k=1}^{\infty},$$

is a monotonically decreasing sequence of events so that Theorem 2.16 implies that

$$\lim_{k \to \infty} P\left(\bigcap_{m=n}^{n+k} A_m^c\right) = P\left(\lim_{k \to \infty} \bigcap_{m=n}^{n+k} A_m^c\right) = P\left(\bigcap_{m=n}^{\infty} A_m^c\right).$$

But Equation (2.8) implies that

$$\lim_{k \to \infty} P\left(\bigcap_{m=n}^{n+k} A_m^c\right) \leq \exp\left[-\lim_{k \to \infty} \sum_{m=n}^{n+k} P(A_m)\right] = 0,$$

and the result follows. □

Note that Theorem 2.18 requires the sequence of events to be independent, whereas Theorem 2.17 does not. However, Theorem 2.17 is still true under the assumption of independence, and combining the results of Theorems 2.17 and 2.18 under this assumption yields the following *zero-one law*.

Corollary 2.1. *Let $\{A_n\}_{n=1}^{\infty}$ be a sequence of independent events. Then*

$$P\left(\limsup_{n\to\infty} A_n\right) = \begin{cases} 0 & \text{if } \sum_{n=1}^{\infty} P(A_n) < \infty, \\ 1 & \text{if } \sum_{n=1}^{\infty} P(A_n) = \infty. \end{cases}$$

Results such as the one given in Corollary 2.1 are called zero-one laws because the probability of the event of interest can only take on the values zero and one.

Example 2.8. Let $\{U_n\}_{n=1}^{\infty}$ be a sequence of independent random variables where U_n has a UNIFORM$\{1, 2, \ldots, n\}$ distribution for all $n \in \mathbb{N}$. Define a sequence of events $\{A_n\}_{n=1}^{\infty}$ as $A_n = \{U_n = 1\}$ for all $n \in \mathbb{N}$. Note that

$$\sum_{n=1}^{\infty} P(A_n) = \sum_{n=1}^{\infty} n^{-1} = \infty,$$

so that Corollary 2.1 implies that

$$P\left(\limsup_{n\to\infty} A_n\right) = P(A_n \text{ i.o.}) = 1.$$

Therefore, each time we observe $\{U_n = 1\}$, we are guaranteed to observe $\{U_{n'} = 1\}$ for some $n' \geq n$, with probability one. Hence, the event A_n will occur an infinite number of times in the sequence, with probability one. Now suppose that U_n has a UNIFORM$\{1, 2, \ldots, n^2\}$ distribution for all $n \in \mathbb{N}$, and assume that A_n has the same definition as above. In this case

$$\sum_{n=1}^{\infty} P(A_n) = \sum_{n=1}^{\infty} n^{-2} = \frac{\pi^2}{6},$$

so that Corollary 2.1 implies that

$$P\left(\limsup_{n\to\infty} A_n\right) = P(A_n \text{ i.o.}) = 0.$$

In this case there will be a last occurrence of an event in the sequence A_n with probability one. That is, there will exist an integer n' such that $\{U_n = 1\}$ will not be observed for all $n > n'$, with probability one. That is, the event $\{U_n = 1\}$ will occur in this sequence only a finite number of times, with probability one. Squaring the size of the sample space in this case creates too many opportunities for events other than those events in the sequence A_n to occur as $n \to \infty$ for an event A_n to ever occur again after a certain point. ■

2.5 Generating and Characteristic Functions

Let X be a random variable with distribution function F. Many of the characteristics of the behavior of a random variable can be studied by considering the moments of X.

Definition 2.9. *The kth moment of a random variable X with distribution function F is*

$$\mu'_k = E(X^k) = \int_{-\infty}^{\infty} x^k dF(x), \tag{2.9}$$

provided

$$\int_{-\infty}^{\infty} |x|^k dF(x) < \infty.$$

The kth central moment of a random variable X with distribution function F is

$$\mu_k = E[(X - \mu'_1)^k] = \int_{-\infty}^{\infty} (x - \mu'_1)^k dF(x), \tag{2.10}$$

provided

$$\int_{-\infty}^{\infty} |x - \mu'_1|^k dF(x) < \infty.$$

The integrals used in Definition 2.9 are Lebesgue-Stieltjes integrals, which can be applied to any random variable, discrete or continuous.

Definition 2.10. *Let X be a random variable with distribution function F. Let g be any real function. Then the integral*

$$\int_{-\infty}^{\infty} g(x) dF(x),$$

will be interpreted as a Lebesgue-Stieltjes integral. That is, if X is continuous with density f then

$$\int_{-\infty}^{\infty} g(x) dF(x) = \int_{-\infty}^{\infty} g(x) f(x) dx,$$

provided

$$\int_{-\infty}^{\infty} |g(x)| f(x) dx < \infty.$$

If X is a discrete random variable that takes on values in the set $\{x_1, x_2, \ldots\}$ with probability distribution function f then

$$\int_{-\infty}^{\infty} g(x) dF(x) = \sum_{i=1}^{n} g(x_i) f(x_i),$$

provided

$$\sum_{i=1}^{n} |g(x_i)| f(x_i) < \infty.$$

by

$$
\begin{aligned}
\mu'_k(\theta) &= \int_0^\infty x^k f(x)\{1 + \theta \sin[2\pi \log(x)]\}dx \\
&= \int_0^\infty x^k f(x)dx + \theta \int_0^\infty x^k f(x) \sin[2\pi \log(x)]dx \\
&= \mu'_k + \theta \int_0^\infty x^k f(x) \sin[2\pi \log(x)]dx,
\end{aligned}
$$

where μ'_k is the kth moment of $f(x)$. Using the same change of variable as above, it follows that

$$
\int_0^\infty x^k f(x) \sin[2\pi \log(x)]dx =
$$
$$
(2\pi)^{-1/2} \exp(\tfrac{1}{2}k^2) \int_{-\infty}^\infty \exp(-\tfrac{1}{2}u^2) \sin[2\pi(u + k)]du =
$$
$$
(2\pi)^{-1/2} \exp(\tfrac{1}{2}k^2) \int_{-\infty}^\infty \exp(-\tfrac{1}{2}u^2) \sin(2\pi u)du = 0,
$$

for each $k \in \mathbb{N}$. Hence, $\mu'_k(\theta) = \mu'_k$ for all $\theta \in [-1, 1]$, and we have demonstrated that this family of distributions all have the same sequence of moments. Therefore, the moment sequence of this distribution does not uniquely identify this distribution. A plot of the lognormal density along with the density given in Equation (2.13) when $\theta = 1$ is given in Figure 2.3. This example is based on one from Heyde (1963). ∎

The *moment generating function* is a function that contains the information about all of the moments of a distribution. As suggested above, under certain conditions the moment generating function can serve as a surrogate for a distribution. This is useful in some cases where the distribution function itself may not be convenient to work with.

Definition 2.11. *Let X be a random variable with distribution function F. The moment generating function of X, or equivalently of F, is*

$$
m(t) = E[\exp\{tX\}] = \int_{-\infty}^\infty \exp(tx)dF(x),
$$

provided

$$
\int_{-\infty}^\infty exp(tx)dF(x) < \infty
$$

for $|t| < b$ for some $b > 0$.

Example 2.11. Let X be a BINOMIAL(n, p) random variable. The moment

Figure 2.3 *The lognormal density (solid line) and the density given in Equation (2.13) with $\theta = 1$ (dashed line). Both densities have the same moment sequence.*

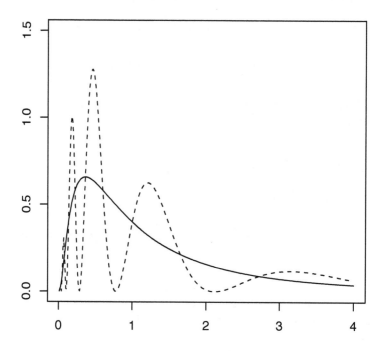

generating function of X is given by

$$
\begin{aligned}
m(t) &= \int_{-\infty}^{\infty} \exp(tx) dF(x) \\
&= \sum_{x=0}^{n} \exp(tx) \binom{n}{x} p^x (1-p)^{n-x} \\
&= \sum_{x=0}^{n} \binom{n}{x} [p \exp(t)]^x (1-p)^{n-x} \\
&= [1 - p + p \exp(t)]^n,
\end{aligned}
$$

where the final inequality results from Theorem A.22.   ■

Example 2.12. Let X be an EXPONENTIAL(β) distribution. Suppose that

$|t| < \beta^{-1}$, then the moment generating function of X is given by

$$
\begin{aligned}
m(t) &= \int_{-\infty}^{\infty} \exp(tx)dF(x) \\
&= \int_{0}^{\infty} \exp(tx)\beta^{-1}\exp(-\beta^{-1}x)dx \\
&= \beta^{-1}\int_{0}^{\infty} \exp[-x(\beta^{-1}-t)]dx. \quad (2.14)
\end{aligned}
$$

Note that the integral in Equation (2.14) diverges unless the restriction $|t| < \beta^{-1}$ is employed. Now consider the change of variable $v = x(\beta^{-1} - t)$ so that $dx = (\beta^{-1} - t)^{-1}dv$. The moment generating function is then given by

$$
m(t) = \beta^{-1}(\beta^{-1} - t)^{-1}\int_{0}^{\infty} \exp(-v)dv = (1 - t\beta)^{-1}.
$$

∎

When the convergence condition given in Definition 2.11 holds then all of the moments of X are finite, and the moment generating function contains information about all of the moments of F. To observe why this is true, consider the Taylor series for $\exp(tX)$ given by

$$
\exp(tX) = \sum_{i=0}^{\infty} \frac{(tX)^i}{i!}.
$$

Taking the expectation of both sides yields

$$
m(t) = E[\exp(tX)] = \sum_{i=0}^{\infty} \frac{t^i \mu_i'}{i!}, \quad (2.15)
$$

where we have assumed that the exchange between the infinite sum and the expectation is permissible. Note that the coefficient of t^i, divided by $i!$, is equal to the ith moment of F. As indicated by its name, the moment generating function can be used to generate the moments of the corresponding distribution. The operation is outlined in Theorem 2.21, which can be proven using a standard induction argument. See Problem 16.

Theorem 2.21. *Let X be a random variable that has moment generating function $m(t)$ that converges on some radius $|t| \leq b$ for some $b > 0$. Then*

$$
\mu_k' = \frac{d^k m(t)}{dt^k}\bigg|_{t=0}.
$$

Example 2.13. Suppose that X is a BINOMIAL(n,p) random variable. It was shown in Example 2.11 that the moment generating function of X is

$m(t) = [1 - p + p\exp(t)]^n$. To find the mean of X use Theorem 2.21 to obtain

$$
\begin{aligned}
E(X) &= \left.\frac{dm(t)}{dt}\right|_{t=0} \\
&= \left.\frac{d}{dt}[1 - p + p\exp(t)]^n\right|_{t=0} \\
&= \left. np\exp(t)[1 - p + p\exp(t)]^{n-1}\right|_{t=0} \\
&= np.
\end{aligned}
$$

∎

Of equal importance is the ability of the moment generating function to uniquely identify the distribution of the random variable. This is what allows one to use the moment generating function of a random variable as a surrogate for the distribution of the random variable. In particular, the Central Limit Theorem can be proven by showing that the limiting moment generating function of a particular sum of random variables matches the moment generating function of the normal distribution.

Theorem 2.22. *Let F and G be two distribution functions all of whose moments exist, and whose moment generating functions are $m_F(t)$ and $m_G(t)$, respectively. If $m_F(t)$ and $m_G(t)$ exist and $m_F(t) = m_G(t)$ for all $|t| \leq b$, for some $b > 0$ then $F(x) = G(x)$ for all $x \in \mathbb{R}$.*

We do not formally prove Theorem 2.22, but to informally see why the result would be true one can compare Equation (2.15) to Equation (2.12). When the moment generating function exists then expansion given in Theorem 2.20 must converge for some radius of convergence. Therefore, Theorems 2.19 and 2.20 imply that the distribution is uniquely characterized by its moment sequence. Note that since Equation (2.15) is a polynomial in t whose coefficients are functions of the moments, the two moment generating functions will be equal only when all of the moments are equal, which will mean that the two distributions are equal.

Another useful result relates the distribution of a function of a random variable to the moment generating function of the random variable.

Theorem 2.23. *Let X be a random variable with distribution function F. If g is a real function then the moment generating function of $g(X)$ is*

$$
m(t) = E\{\exp[tg(X)]\} = \int_{-\infty}^{\infty} \exp[tg(x)] dF(x),
$$

provided $m(t) < \infty$ when $|t| < b$ for some $b > 0$.

The results of Theorems 2.22 and 2.23 can be combined to identify the distributions of transformed random variables.

Example 2.14. Suppose X is an EXPONENTIAL(1) random variable. Example 2.12 showed that the moment generating function of X is $m_X(t) =$

$(1-t)^{-1}$ when $|t| < 1$. Now consider a new random variable $Y = \beta X$ where $\beta \in \mathbb{R}$. Theorem 2.23 implies that the moment generating function of Y is $m_Y(t) = E\{\exp[t(\beta X)]\} = E\{\exp[(t\beta)X]\} = m_X(t\beta)$ as long as $t\beta < 1$ or equivalently $t < \beta^{-1}$. Evaluating the moment generating function of X at $t\beta$ yields $m_Y(t) = (1-t\beta)^{-1}$, which is the moment generating function of an Ex-PONENTIAL(β) random variable. Theorem 2.22 can be used to conclude that if X is an EXPONENTIAL(1) random variable then βX is an EXPONENTIAL(β) random variable. ∎

Example 2.15. Let Z be a $N(0,1)$ random variable and define a new random variable $Y = Z^2$. Theorem 2.23 implies that the moment generating function of Y is given by

$$
\begin{aligned}
m_Y(t) &= E[\exp(tZ^2)] \\
&= \int_{-\infty}^{\infty} (2\pi)^{-1/2} \exp(-\tfrac{1}{2}z^2 + tz^2) dz \\
&= \int_{-\infty}^{\infty} (2\pi)^{-1/2} \exp[-\tfrac{1}{2}z^2(1-2t)] dz.
\end{aligned}
$$

Assuming that $1 - 2t > 0$, or that $|t| < \tfrac{1}{2}$ so that integral converges, use the change of variable $v = z(1-2t)^{1/2}$ to obtain

$$
m_Y(t) = (1-2t)^{-1/2} \int_{-\infty}^{\infty} (2\pi)^{-1/2} \exp(-\frac{1}{2}v^2) dv = (1-2t)^{-1/2}.
$$

Note that $m_Y(t)$ is the moment generating function of a CHI-SQUARED(1) random variable. Therefore, Theorem 2.22 implies that if Z is a $N(0,1)$ random variable then Z^2 is a CHI-SQUARED(1) random variable. ∎

In Example 2.14 it is clear that the moment generating functions of linear transformations of random variables can be evaluated directly without having to re-evaluate the integral in Definition 2.11. This result can be generalized.

Theorem 2.24. *Let X be a random variable with moment generating function $m_X(t)$ that exists and is finite for $|t| < b$ for some $b > 0$. Suppose that Y is a new random variable defined by $Y = \alpha X + \beta$ where α and β are real constants. Then the moment generating function of Y is $m_Y(t) = \exp(t\beta)m_X(\alpha t)$ provided $|\alpha t| < b$.*

Theorem 2.24 is proven in Exercise 23. Another important property relates the moment generating function of the sum of independent random variables to the moment generating functions of the individual random variables.

Theorem 2.25. *Let $X_1, \ldots, X_n$ be a sequence of independent random variables where X_i has moment generating function $m_i(t)$ for $i = 1, \ldots, n$. Suppose that $m_1(t), \ldots, m_n(t)$ all exist and are finite when $|t| < b$ for some $b > 0$. If*

$$
S_n = \sum_{i=1}^{n} X_i,
$$

then the moment generating function of S_n is

$$m_{S_n}(t) = \prod_{k=1}^{n} m_k(t),$$

when $|t| < b$. If $X_1, \ldots, X_n$ are identically distributed with moment generating function $m(t)$ then $m_{S_n}(t) = m^n(t)$ provided $|t| < b$.

Proof. The moment generating function of S_n is given by

$$
\begin{aligned}
m_{S_n}(t) &= E[\exp(tS_n)] = E\left[\exp\left(t\sum_{i=1}^{n} X_i\right)\right] = E\left[\exp\left(\sum_{i=1}^{n} tX_i\right)\right] \\
&= E\left[\prod_{i=1}^{n} \exp(tX_i)\right] = \prod_{i=1}^{n} E[\exp(tX_i)] = \prod_{i=1}^{n} m_i(t),
\end{aligned}
$$

where the interchange of the product and the expectation follows from the independence of the random variables. The second result follows by setting $m_1(t) = \cdots = m_n(t) = m(t)$. $\qquad\square$

Example 2.16. Suppose that $X_1, \ldots, X_n$ are a set of independent $N(0, 1)$ random variables. The moment generating function of X_k is $m(t) = \exp(\frac{1}{2}t^2)$ for all $t \in \mathbb{R}$. Theorem 2.25 implies that the moment generating function of

$$S_n = \sum_{i=1}^{n} X_i,$$

is $m_{S_n}(t) = m^n(t) = \exp(\frac{1}{2}nt^2)$, which is the moment generating function of a $N(0, n)$ random variable. Therefore, Theorem 2.22 implies that the sum of n independent $N(0, 1)$ random variables is a $N(0, n)$ random variable. $\qquad\blacksquare$

The moment generating function can be a useful tool in asymptotic theory, but has the disadvantage that it does not exist for many distributions of interest. A function that has many similar properties to the moment generating function is the *characteristic function*. Using the characteristic function requires a little more work as it is based on some ideas from complex analysis. However, the benefits far outweigh this inconvenience as the characteristic function always exists, can be used to generate moments when they exist, and also uniquely identifies a distribution.

Definition 2.12. *Let X be a random variable with distribution function F. The characteristic function of X, or equivalently of F, is*

$$\psi(t) = E[\exp(itX)] = \int_{-\infty}^{\infty} \exp(itx)dF(x). \tag{2.16}$$

In the case where F has density f, the characteristic function is equivalent to

$$\psi(t) = \int_{-\infty}^{\infty} \exp(itx)f(x)dx. \tag{2.17}$$

One may recognize the integral in Equation (2.17) as the *Fourier transformation* of the function f. Such a transformation exists for other functions beside densities, but for the special case of density functions the Fourier transformation is called the characteristic function. Similarly, the integral in Equation (2.16) is called the *Fourier–Stieltjes* transformation of F. These transformations are defined for all bounded measures, and in the special case of a measure that is normed to one, we again call the transformation a characteristic function.

There is no provision for a radius of convergence in Definition 2.12 because the integral always exists. To see this note that

$$\left| \int_{-\infty}^{\infty} \exp(itx) dF(x) \right| \le \int_{-\infty}^{\infty} |\exp(itx)| dF(x),$$

by Theorem A.6. Now Definitions A.5 and A.6 (Euler) imply that

$$|\exp(itx)| = |\cos(tx) + i\sin(tx)| = [\cos^2(tx) + \sin^2(tx)]^{1/2} = 1.$$

Therefore, it follows that

$$\left| \int_{-\infty}^{\infty} \exp(itx) dF(x) \right| \le \int_{-\infty}^{\infty} dF(x) = 1.$$

Before giving some examples on deriving characteristic functions for specific distributions it is worth pointing out that complex integration is not required to find characteristic functions, since we are not integrating over the complex plane. In fact, Definition A.6 implies that the characteristic function of a random variable that has distribution function F is given by

$$\psi(t) = \int_{-\infty}^{\infty} \exp(itx) dF(x) =$$

$$\int_{-\infty}^{\infty} \cos(tx) dF(x) + i \int_{-\infty}^{\infty} \sin(tx) dF(x), \quad (2.18)$$

where both of the integrals in Equation (2.18) are of real functions integrated over the real line.

Example 2.17. Let X be a BINOMIAL(n, p) random variable. The characteristic function of X is given by

$$
\begin{aligned}
\psi(t) &= \int_{-\infty}^{\infty} \exp(itx) dF(x) \\
&= \sum_{x=0}^{n} \exp(itx) \binom{n}{x} p^x (1-p)^{n-x} \\
&= \sum_{x=0}^{n} \binom{n}{x} [p\exp(it)]^x (1-p)^{n-x} \\
&= [1 - p + p\exp(it)]^n,
\end{aligned}
$$

where the final equality results from Theorem A.22. ∎

Example 2.18. Let X be an EXPONENTIAL(β) distribution. The characteristic function of X is given by

$$
\begin{aligned}
\psi(t) &= \int_{-\infty}^{\infty} \exp(itx) dF(x) \\
&= \int_{0}^{\infty} \exp(itx) \beta^{-1} \exp(-\beta^{-1}x) dx \\
&= \beta^{-1} \int_{0}^{\infty} \cos(tx) \exp(-\beta^{-1}x) + i\beta^{-1} \int_{0}^{\infty} \sin(tx) \exp(-\beta^{-1}x) dx.
\end{aligned}
$$

Now, using standard results from calculus, it follows that

$$
\int_{0}^{\infty} \cos(tx) \exp(-\beta^{-1}x) dx = \frac{\beta^{-1}}{\beta^{-2} + t^2},
$$

and

$$
\int_{0}^{\infty} \sin(tx) \exp(-\beta^{-1}x) dx = \frac{t}{\beta^{-2} + t^2}.
$$

Therefore

$$
\begin{aligned}
\psi(t) &= \frac{1}{\beta} \left(\frac{\beta^{-1}}{\beta^{-2} + t^2} + \frac{it}{\beta^{-2} + t^2} \right) = \frac{1 + it\beta}{1 + t^2\beta^2} \\
&= \frac{1 + it\beta}{(1 + it\beta)(1 - it\beta)} = \frac{1}{1 - it\beta}.
\end{aligned}
$$

In both of the previous two examples the characteristic function turns out to be a function whose range is in the complex field. This is not always the case, as there are some circumstances under which the characteristic function is a real valued function.

Theorem 2.26. *Let X be a random variable. The characteristic function of X is real valued if and only if X has the same distribution as $-X$.*

Theorem 2.26 is proven in Exercise 22. Note that the condition that X has the same distribution as $-X$ implies that $P(X \geq x) = P(X \leq -x)$ for all $x > 0$, which is equivalent to the case where X has a symmetric distribution about the origin. For example, a random variable with a N(0, 1) distribution has a real valued characteristic function, whereas a random variable with a non-symmetric distribution like a GAMMA distribution has a complex valued characteristic function. Note that Theorem 2.26 requires that X have a distribution that is symmetric about the origin. That is, if X has a N(μ, σ) distribution where $\mu \neq 0$, then the characteristic function of X is complex valued.

As with the moment generating function, the characteristic function uniquely characterizes the distribution of a random variable, though the characteristic function can be used in more cases as there is no need to consider potential convergence issues with the characteristic function.

Theorem 2.27. *Let F and G be two distribution functions whose characteristic functions are $\psi_F(t)$ and $\psi_G(t)$ respectively. If $\psi_F(t) = \psi_G(t)$ for all $t \in \mathbb{R}$ then $F(x) = G(x)$ for all $x \in \mathbb{R}$.*

Inversion provides a method for deriving the distribution function of a random variable directly from its characteristic function. The two main results used in this book focus on inversion of characteristic functions for certain continuous and discrete distributions.

Theorem 2.28. *Consider a random variable X with characteristic function $\psi(t)$. If*

$$\int_{-\infty}^{\infty} |\psi(t)| dt < \infty,$$

then X has an absolutely continuous distribution F with a bounded and continuous density f given by

$$f(x) = \frac{1}{2\pi} \int_{-\infty}^{\infty} \exp(-itx)\psi(t) dt.$$

A proof of Theorem 2.28 can be found in Chapter 4 of Gut (2005).

Example 2.19. Consider a random variable X that has characteristic function $\psi(t) = \exp(-\frac{1}{2}t^2)$, which follows the condition given in Theorem 2.28. Therefore, X has an absolutely continuous distribution function F that has a continuous and bounded density given by

$$
\begin{aligned}
f(x) &= \frac{1}{2\pi} \int_{-\infty}^{\infty} \exp(-itx)\exp(-\tfrac{1}{2}t^2) dt \\
&= \frac{1}{2\pi} \int_{-\infty}^{\infty} \exp[-\tfrac{1}{2}(t^2 + 2itx)] dt \\
&= \frac{1}{2\pi} \int_{-\infty}^{\infty} \exp[-\tfrac{1}{2}(t + ix)^2] \exp(-\tfrac{1}{2}x^2) dt \\
&= \frac{1}{2\pi} \exp(-\tfrac{1}{2}x^2) \int_{-\infty}^{\infty} \exp[-\tfrac{1}{2}(t + ix)^2] dt.
\end{aligned}
$$

Note that $(2\pi)^{-1/2}$ multiplied by the integrand corresponds to a $\mathrm{N}(-ix, 1)$ distribution, and the corresponding integral is one. Therefore

$$f(x) = (2\pi)^{-1/2} \exp(-\tfrac{1}{2}x^2),$$

which is the density of a $\mathrm{N}(0,1)$ random variable. ∎

For discrete distributions we will focus on random variables that take on values on a regular lattice. That is, for random variables X that take on values in the set $\{kd + l : k \in \mathbb{Z}\}$ for some $d > 0$ and $l \in \mathbb{R}$. Many of the common discrete distributions, such as the BINOMIAL, GEOMETRIC, and POISSON, have supports on a regular lattice with $d = 1$ and $l = 0$.

Theorem 2.29. *Consider a random variable X that takes on values in the set*

$\{kd + l : k \in \mathbb{Z}\}$ for some $d > 0$ and $l \in \mathbb{R}$. Suppose that X has characteristic function $\psi(t)$, then for any $x \in \{kd + l : k \in \mathbb{Z}\}$,

$$P(X = x) = \frac{d}{2\pi} \int_{-\pi/d}^{\pi/d} \exp(-itx)\psi(t)dt.$$

Example 2.20. Let X be a discrete random variable that takes on values in the set $\{0, 1, \dots, n\}$ where $n \in \mathbb{N}$ is fixed so that $d = 1$ and $l = 0$. If X has characteristic function $\psi(t) = [1 - p + p\exp(it)]^n$ where $p \in (0, 1)$ and $x \in \{1, \dots, n\}$, then Theorem 2.29 implies that

$$
\begin{aligned}
P(X = x) &= \frac{1}{2\pi} \int_{-\pi}^{\pi} \exp(-itx)[1 - p + p\exp(it)]^n dt \\
&= \frac{1}{2\pi} \int_{-\pi}^{\pi} \exp(-itx) \sum_{k=0}^{n} \binom{n}{k} p^k \exp(itk)(1 - p)^{n-k} dt \\
&= \frac{1}{2\pi} \sum_{k=0}^{n} \binom{n}{k} p^k (1 - p)^{n-k} \int_{-\pi}^{\pi} \exp[it(k - x)]dt,
\end{aligned}
$$

where Theorem A.22 has been used to expand the polynomial. There are two distinct cases to consider for the value of the index k. When $k = x$ the exponential function becomes one and the expression simplifies to

$$\frac{1}{2\pi} \binom{n}{x} p^x (1 - p)^{n-x} \int_{-\pi}^{\pi} dt = \binom{n}{x} p^x (1 - p)^{n-x}.$$

When $k \neq x$ the integral expression in the sum can be calculated as

$$\int_{-\pi}^{\pi} \exp[it(k - x)]dt = \int_{-\pi}^{\pi} \cos[t(k - x)]dt + i \int_{-\pi}^{\pi} \sin[t(k - x)]dt.$$

The second integral is zero because $\sin[-t(k - x)] = -\sin[(t(k - x)]$. For the first integral we note that $(k - x) \in \mathbb{N}$ and therefore

$$\int_{-\pi}^{\pi} \cos[t(k - x)]dt = 2 \int_{0}^{\pi} \cos[t(k - x)]dt = 0,$$

since the range of the integral over the cosine function is an integer multiple of π. Therefore, the integral expression is zero when $k \neq x$ and we have that

$$P(X = x) = \binom{n}{x} p^x (1 - p)^{n-x},$$

which corresponds to the probability of a BINOMIAL(n, p) random variable. ∎

Due to the similar form of the definition of the characteristic function to the moment generating function, the two functions have similar properties. In particular, the characteristic function of a random variable can also be used to obtain moments of the random variable if they exist. We first establish that when the associated moments exist, the characteristic function can be approximated by partial expansions whose terms correspond to the Taylor series for the exponential function.

Theorem 2.30. *Let X be a random variable with characteristic function ψ. Then if $E(|X|^n) < \infty$ for some $n \in \mathbb{N}$, then*

$$\left| \psi(t) - \sum_{k=0}^{n} \frac{(it)^k E(X^k)}{k!} \right| \leq E\left(\frac{2|t|^n |X|^n}{n!} \right),$$

and

$$\left| \psi(t) - \sum_{k=0}^{n} \frac{(it)^k E(X^k)}{k!} \right| \leq E\left[\frac{|t|^{n+1} |X|^{n+1}}{(n+1)!} \right].$$

Proof. We will prove the first statement. The second statement follows using similar arguments. Theorem A.11 implies that for $y \in \mathbb{R}$,

$$\left| \exp(iy) - \sum_{k=0}^{n} \frac{(iy)^k}{k!} \right| \leq \frac{2|y|^n}{n!}. \tag{2.19}$$

Let $y = tX$ where $t \in \mathbb{R}$ and note then that

$$\begin{aligned}
\left| \psi(t) - \sum_{k=0}^{n} \frac{(it)^k E(X^k)}{k!} \right| &= \left| E[\exp(itX)] - \sum_{k=0}^{n} \frac{E[(itX)^k]}{k!} \right| \\
&\leq E\left[\left| \exp(itX) - \sum_{k=0}^{n} \frac{(itX)^k}{k!} \right| \right] \\
&\leq E\left(\frac{2|t|^n |X|^n}{n!} \right),
\end{aligned}$$

where the final inequality follows from the bound in Equation (2.19). $\qquad\square$

The results given in Theorem 2.30 allow us to determine the asymptotic behavior of the error that is incurred by approximating a characteristic function with the corresponding partial series. This, in turn, allows us to obtain a method by which the moments of a distribution can be obtained from a characteristic function.

Theorem 2.31. *Let X be a random variable with characteristic function ψ. If $E(|X|^n) < \infty$ for some $n \in \{1, 2, \ldots\}$ then $\psi^{(k)}$ exists, is uniformly continuous for $k \in \{1, 2, \ldots, n\}$,*

$$\psi(t) = 1 + \sum_{k=1}^{n} \frac{\mu'_k (it)^k}{k!} + o(|t|^n),$$

as $t \to 0$ and

$$\left. \frac{d^k \psi(t)}{dt^k} \right|_{t=0} = i^k \mu'_k.$$

A proof of Theorem 2.31 is the subject of Exercise 33. Note that the characteristic function allows one to find the finite moments of distributions that may not have all finite moments, as opposed to the moment generating function which does not exist when *any* of the moments of a random variable are infinite.

Example 2.21. Let X be a $N(0,1)$ random variable, which has characteristic function $\psi(t) = \exp(-\frac{1}{2}t^2)$. Taking a first derivative yields

$$\frac{d}{dt}\psi(t)\bigg|_{t=0} = \frac{d}{dt}\exp(-\frac{1}{2}t^2)\bigg|_{t=0} = -t\exp(-\frac{1}{2}t^2)\bigg|_{t=0} = 0,$$

which is the mean of the $N(0,1)$ distribution. $\blacksquare$

The characteristic function also has a simple relationship with linear transformations of a random variable.

Theorem 2.32. *Suppose that X is a random variable with characteristic function $\psi(t)$. Let $Y = \alpha X + \beta$ where α and β are real constants. Then the characteristic function of Y is $\psi_Y(t) = \exp(it\beta)\psi(t\alpha)$.*

Theorem 2.32 is proven in Exercise 24. As with moment generating functions, the characteristic function of the sum of independent random variables is the product of the characteristic functions.

Theorem 2.33. *Let $X_1, \ldots, X_n$ be a sequence of independent random variables where X_i has characteristic function $\psi_i(t)$, for $i = 1, \ldots, n$, then the characteristic function of*

$$S_n = \sum_{i=1}^{n} X_i,$$

is

$$\psi_{S_n}(t) = \prod_{i=1}^{n} \psi_i(t).$$

If $X_1, \ldots, X_n$ are identically distributed with characteristic function $\psi(t)$ then the characteristic function of S_n is $\psi_{S_n}(t) = \psi^n(t)$.

Theorem 2.33 is proven in Exercise 25.

A function that is related to the moment generating function is the *cumulant generating function*, defined below.

Definition 2.13. *Let X be a random variable with moment generating function $m(t)$ defined on a radius of convergence $|t| < b$ for some $b > 0$. The cumulant generating function of X is given by $c(t) = \log[m(t)]$.*

The usefulness of the cumulant generating function may not be apparent from the definition given above, though one can immediately note that when the moment generating function of a random variable exists, the cumulant generating function will also exist and will uniquely characterize the distribution of the random variable as the moment generating function does. This follows from the fact that the cumulant generating function is a one-to-one function of the moment generating function. As indicated by its name, $c(t)$ can be used to generate the *cumulants* of a random variable, which are related to the moments of a random variable. Before defining the cumulants of a random variable, some expansion theory is required to investigate the structure of the cumulant generating function more closely.

We begin by assuming that the moment generating function $m(t)$ is defined on a radius of convergence $|t| < b$ for some $b > 0$. As shown in Equation (2.15), the moment generating function can be written as

$$m(t) = 1 + \mu_1' t + \tfrac{1}{2}\mu_2' t^2 + \tfrac{1}{6}\mu_3' t^3 + \cdots + \frac{\mu_n' t^n}{k!} + O(t^n), \qquad (2.20)$$

as $t \to 0$. An application of Theorem 1.15 to the logarithmic function can be used to show that

$$\log(1 + \delta) = \delta - \tfrac{1}{2}\delta^2 + \tfrac{1}{3}\delta^3 - \tfrac{1}{4}\delta^4 + \cdots + \tfrac{1}{n}(-1)^{n+1}\delta^n + O(\delta^n), \qquad (2.21)$$

as $\delta \to 0$. Substituting

$$\delta = \sum_{i=1}^{n} \frac{\mu_i' t^i}{i!} + O(t^n), \qquad (2.22)$$

into Equation (2.21) yields a polynomial in t of the same form as given in Equation (2.20), but with different coefficients. That is, the cumulant generating function has the form

$$c(t) = \kappa_1 t + \tfrac{1}{2}\kappa_2 t^2 + \tfrac{1}{6}\kappa_3 t^3 + \cdots + \frac{\kappa_n t^n}{n!} + R_c(t), \qquad (2.23)$$

where the coefficients κ_i are called the *cumulants* of X, and $R_c(t)$ is an error term that depends on t whose order is determined below. Note that since $c(t)$ and $m(t)$ have the same form, the cumulants can be generated from the cumulant generating function in the same way that moments can be generated from the moment generating function through Theorem 2.21. That is,

$$\kappa_i = \left. \frac{d^i c(t)}{dt^i} \right|_{t=0}.$$

Matching the coefficients of t^i in Equations (2.21) and (2.23) yields expressions for the cumulants of X in terms of the moments of X. For example, matching the coefficients of t yields the relation $\mu_1' t = \kappa_1 t$ so that the first cumulant is equal to μ_1', the mean. The remaining cumulants are not equal to the corresponding moments. Matching the coefficients of t^2 yields

$$\tfrac{1}{2}\kappa_2 t^2 = \tfrac{1}{2}\mu_2' t^2 - \tfrac{1}{2}(\mu_1')^2 t^2,$$

so that $\kappa_2 = \mu_2' - (\mu_1')^2$, the variance of X. Similarly, matching the coefficients of t^3 yields

$$\tfrac{1}{6}\kappa_3 t^3 = \tfrac{1}{6}\mu_3' t^3 - \tfrac{1}{2}\mu_1'\mu_2' t^3 + \tfrac{1}{3}(\mu_1')^3 t^3,$$

so that $\kappa_3 = \mu_3' - 3\mu_1'\mu_2' + 2(\mu_1')^3$. In this form this expression may not seem familiar, but note that

$$E[(X - \mu_1')^3] = \mu_3' - 3\mu_1'\mu_2' + 2(\mu_1')^3,$$

so that κ_3 is the skewness of X. It can be similarly shown that κ_4 is the kurtosis of X. For further cumulants, see Exercises 34 and 35, and Chapter 3 of Kendall and Stuart (1977).

The reminder term $R_c(t)$ from the expansion given in Equation (2.23) will now be quantified. Consider the powers of t given in the expansion for δ in Equation (2.22) when δ is substituted into Equation (2.23). The reminder term from the linear term is $O(t^n)$ as $t \to 0$. When the expansion given for δ is squared, there will be terms that range in powers of t from t^2 to t^{2n}. All of the powers of t less than $n+1$ have coefficients that are used to identify the first k cumulants in terms of the moments of X. All the remaining terms are $O(t^n)$ as $t \to 0$, assuming that all of the moments are finite, which must follow if the moment generating function of X converges on a radius of convergence. This argument can be applied to the remaining terms to obtain the result

$$c(t) = \sum_{i=1}^{n} \frac{\kappa_i t^i}{i!} + O(t^n), \tag{2.24}$$

as $t \to \infty$.

Example 2.22. Suppose X has a $N(\mu, \sigma^2)$ distribution. The moment generating function of X is given by $m(t) = \exp(\mu t + \frac{1}{2}\sigma^2 t^2)$. From Definition 2.13, the cumulant generating function of X is

$$c(t) = \log[m(t)] = \mu t + \frac{1}{2}\sigma^2 t^2.$$

Matching this cumulant generating function to the general form given in Equation (2.24) implies the $N(\mu, \sigma^2)$ distribution has cumulants

$$\kappa_i = \begin{cases} \mu & i = 1, \\ \sigma^2 & i = 2 \\ 0 & i = 3, 4, \ldots \end{cases}$$

∎

In some cases additional calculations are required to obtain the necessary form of the cumulant generating function.

Example 2.23. Suppose that X has an EXPONENTIAL(β) distribution. The moment generating function of X is $m(t) = (1 - \beta t)^{-1}$ when $|t| < \beta^{-1}$ so that the cumulant generating function of X is $c(t) = -\log(1 - \beta t)$. This cumulant generating function is not in the form given in Equation (2.24) so the cumulants cannot be obtained by directly observing $c(t)$. However, when $|\beta t| < 1$ it follows from the Taylor expansion given in Equation (2.21) that

$$\begin{aligned} \log(1 - \beta t) &= (-\beta t) - \tfrac{1}{2}(-\beta t)^2 + \tfrac{1}{3}(-\beta t)^3 + \cdots \\ &= -\beta t - \tfrac{1}{2}\beta^2 t^2 - \tfrac{1}{3}\beta^3 t^3 + \cdots . \end{aligned}$$

Therefore $c(t) = \beta t + \frac{1}{2}\beta^2 t^2 + \frac{1}{3}\beta^3 t^3 + \cdots$, which is now in the form of Equation (2.24). It follows that the ith cumulant of X can be found by solving

$$\frac{\kappa_i t^i}{i!} = \frac{\beta^i t^i}{i},$$

for $t \neq 0$. This implies that the ith cumulant of X is $\kappa_i = (i-1)!\beta^i$ for $i \in \mathbb{N}$. Alternatively, one could also find the cumulants by differentiating the cumulant generating function. For example, the first cumulant is given by

$$
\begin{aligned}
\left.\frac{d}{dt}c(t)\right|_{t=0} &= \left.\frac{d}{dt}[-\log(1-\beta t)]\right|_{t=0} \\
&= \beta(1-\beta t)^{-1}\big|_{t=0} \\
&= \beta.
\end{aligned}
$$

The remaining cumulants can be found by taking additional derivatives. ∎

Cumulant generating functions are particularly easy to work with for sums of independent random variables.

Theorem 2.34. *Let $X_1, \ldots, X_n$ be a sequence of independent random variables where X_i has cumulant generating function $c_i(t)$ for $i = 1, \ldots, n$. Then the cumulant generating function of*

$$
S_n = \sum_{i=1}^{n} X_i,
$$

is

$$
c_{S_n}(t) = \sum_{i=1}^{n} c_i(t).
$$

If $X_1, \ldots, X_n$ are also identically distributed with cumulant generating function $c(t)$ then the cumulant generating function of S_n is $nc(t)$.

Theorem 2.34 is proven in Exercise 36. The fact that the cumulant generating functions add for sums of independent random variables implies that the coefficients of $t^i/i!$ add as well. Therefore the ith cumulant of the sum of independent random variables is equal to the sum of the corresponding cumulants of the individual random variables. This result gives some indication as to why cumulants are often preferable to work with, as the moments or central moments of a sum of independent random variables can be a complex function of the individual moments. For further information about cumulants see Barndorff-Nielsen and Cox (1989), Gut (2005), Kendall and Stuart (1977), and Severini (2005).

2.6 Exercises and Experiments

2.6.1 Exercises

1. Verify that $\mathcal{F} = \{\emptyset, \omega_1, \omega_2, \omega_3, \omega_1 \cup \omega_2, \omega_1 \cup \omega_3, \omega_2 \cup \omega_3, \omega_1 \cup \omega_2 \cup \omega_3\}$ is a σ-field containing the sets ω_1, ω_2, and ω_3. Prove that this is the smallest possible σ-field containing these sets by showing that $\mathcal{F}$ is no longer a σ-field if any of the sets are eliminated from $\mathcal{F}$.

2. Let $\{A_n\}_{n=1}^{\infty}$ be a sequence of monotonically increasing events from a σ-field $\mathcal{F}$ of subsets of a sample space Ω. Prove that the sequence $\{A_n^c\}_{n=1}^{\infty}$ is a monotonically decreasing sequence of events from $\mathcal{F}$.

3. Consider a probability space $(\Omega, \mathcal{F}, P)$ where $\Omega = (0,1) \times (0,1)$ is the unit square and P is a bivariate extension of Lebesgue measure. That is, if R is a rectangle of the form

$$R = \{(\omega_1, \omega_2) : \underline{\omega}_1 \le \omega_1 \le \overline{\omega}_1, \underline{\omega}_2 \le \omega_1 \le \overline{\omega}_2\},$$

where $\underline{\omega}_1 \le \overline{\omega}_1$ and $\underline{\omega}_2 \le \overline{\omega}_2$ then $P(R) = (\overline{\omega}_1 - \underline{\omega}_1)(\overline{\omega}_2 - \underline{\omega}_1)$, which corresponds to the area of the rectangle. Let $\mathcal{F} = \mathcal{B}\{(0,1) \times (0,1)\}$ and in general define $P(B)$ to be the area of B for any $B \in \mathcal{B}\{(0,1) \times (0,1)\}$. Prove that P is a probability measure.

4. Prove Theorem 2.4. That is, prove that

$$P\left(\bigcup_{i=1}^{n} A_i\right) \le \sum_{i=1}^{n} P(A_i).$$

The most direct approach is based on mathematical induction using the general addition rule to prove the basis and the induction step. The general addition rule states that for any two events A_1 and A_2, $P(A_1 \cup A_2) = P(A_1) + P(A_2) - P(A_1 \cap A_2)$.

5. Prove Theorem 2.6 (Markov) for the case when X is a discrete random variable on $\mathbb{N}$ with probability distribution function $p(x)$.

6. Prove Theorem 2.7 (Tchebysheff). That is, prove that if X is a random variable such that $E(X) = \mu$ and $V(X) = \sigma^2 < \infty$, then $P(|X - \mu| > \delta) \le \delta^{-2}\sigma^2$.

7. Let $\{A_n\}_{n=1}^{\infty}$ be a sequence of events from a σ-field $\mathcal{F}$ of subsets of a sample space Ω. Prove that if $A_{n+1} \subset A_n$ for all $n \in \mathbb{N}$ then the sequence has limit

$$\lim_{n \to \infty} A_n = \bigcap_{n=1}^{\infty} A_n.$$

8. Let $\{A_n\}_{n=1}^{\infty}$ be a sequence of events from $\mathcal{F}$, a σ-field on the sample space $\Omega = (0,1)$, defined by

$$A_n = \begin{cases} (\frac{1}{3}, \frac{2}{3}) & \text{if } n \text{ is even,} \\ (\frac{1}{4}, \frac{3}{4}) & \text{if } n \text{ is odd,} \end{cases}$$

for all $n \in \mathbb{N}$. Compute

$$\liminf_{n \to \infty} A_n,$$

$$\limsup_{n \to \infty} A_n,$$

and determine if the limit of the sequence $\{A_n\}_{n=1}^{\infty}$ exists.

9. Let $\{A_n\}_{n=1}^{\infty}$ be a sequence of events from $\mathcal{F}$, a σ-field on the sample space $\Omega = \mathbb{R}$, defined by $A_n = (-1 - n^{-1}, 1 + n^{-1})$ for all $n \in \mathbb{N}$. Compute
$$\liminf_{n \to \infty} A_n,$$
$$\limsup_{n \to \infty} A_n,$$
and determine if the limit of the sequence $\{A_n\}_{n=1}^{\infty}$ exists.

10. Let $\{A_n\}_{n=1}^{\infty}$ be a sequence of events from $\mathcal{F}$, a σ-field on the sample space $\Omega = (0,1)$, defined by
$$A_n = \begin{cases} B & \text{if } n \text{ is even,} \\ B^c & \text{if } n \text{ is odd,} \end{cases}$$
for all $n \in \mathbb{N}$ where B is a fixed member of $\mathcal{F}$. Compute
$$\liminf_{n \to \infty} A_n,$$
$$\limsup_{n \to \infty} A_n,$$
and determine if the limit of the sequence $\{A_n\}_{n=1}^{\infty}$ exists.

11. Let $\{A_n\}_{n=1}^{\infty}$ be a sequence of events from $\mathcal{F}$, a σ-field on the sample space $\Omega = \mathbb{R}$, defined by
$$A_n = \begin{cases} [\frac{1}{2}, \frac{1}{2} + n^{-1}) & \text{if } n \text{ is even,} \\ (\frac{1}{2} - n^{-1}, \frac{1}{2}] & \text{if } n \text{ is odd,} \end{cases}$$
for all $n \in \mathbb{N}$. Compute
$$\liminf_{n \to \infty} A_n,$$
$$\limsup_{n \to \infty} A_n,$$
and determine if the limit of the sequence $\{A_n\}_{n=1}^{\infty}$ exists.

12. Consider a probability space $(\Omega, \mathcal{F}, P)$ where $\Omega = (0,1)$, $\mathcal{F} = \mathcal{B}\{(0,1)\}$ and P is Lebesgue measure on $(0,1)$. Let $\{A_n\}_{n=1}^{\infty}$ be a sequence of events in $\mathcal{F}$ defined by $A_n = (0, \frac{1}{2}(1 + n^{-1}))$ for all $n \in \mathbb{N}$. Show that
$$\lim_{n \to \infty} P(A_n) = P\left(\lim_{n \to \infty} A_n\right).$$

13. Consider tossing a fair coin repeatedly and define H_n to be the event that the nth toss of the coin yields a head. Prove that
$$P(\limsup_{n \to \infty} H_n) = 1,$$
and interpret this result in terms of how often the event occurs.

14. Consider the case where $\{A_n\}_{n=1}^{\infty}$ is a sequence of independent events that all have the same probability $p \in (0,1)$. Prove that
$$P(\limsup_{n \to \infty} A_n) = 1,$$
and interpret this result in terms of how often the event occurs.

15. Let $\{U_n\}_{n=1}^{\infty}$ be a sequence of independent UNIFORM$(0,1)$ random variables. For each definition of A_n given below, calculate

$$P\left(\limsup_{n\to\infty} A_n\right).$$

a. $A_n = \{U_n < n^{-1}\}$ for all $n \in \mathbb{N}$.

b. $A_n = \{U_n < n^{-3}\}$ for all $n \in \mathbb{N}$.

c. $A_n = \{U_n < \exp(-n)\}$ for all $n \in \mathbb{N}$.

d. $A_n = \{U_n < 2^{-n}\}$ for all $n \in \mathbb{N}$.

16. Let X be a random variable that has moment generating function $m(t)$ that converges on some radius $|t| \leq b$ for some $b > 0$. Using induction, prove that

$$\mu_k' = \left.\frac{d^k m(t)}{dt^k}\right|_{t=0}.$$

17. Let X be a POISSON(λ) random variable.

a. Prove that the moment generating function of X is $\exp[\lambda \exp(t) - 1]$.

b. Prove that the characteristic function of X is $\exp[\lambda \exp(it) - 1]$.

c. Using the moment generating function, derive the first three moments of X. Repeat the process using the characteristic function.

18. Let Z be a N$(0,1)$ random variable.

a. Prove that the moment generating function of X is $\exp[-\frac{1}{2}t^2]$.

b. Prove that the characteristic function of X is also $\exp[-\frac{1}{2}t^2]$.

c. Using the moment generating function, derive the first three moments of X. Repeat the process using the characteristic function.

19. Let Z be a N$(0,1)$ random variable and define $X = \mu + \sigma Z$ for some $\mu \in \mathbb{R}$ and $0 < \sigma < \infty$. Using the fact that X is a N(μ, σ^2) random variable, derive the moment generating function and the characteristic function of a N(μ, σ^2) random variable.

20. Let X be a N(μ, σ^2) random variable. Using the moment generating function, derive the first three moments of X. Repeat the process using the characteristic function.

21. Let X be a UNIFORM(α, β) random variable.

a. Prove that the moment generating function of X is $[t(\beta - \alpha)]^{-1}[\exp(t\beta) - \exp(t\alpha)]$.

b. Prove that the characteristic function of X is $[it(\beta - \alpha)]^{-1}[\exp(it\beta) - \exp(it\alpha)]$.

c. Using the moment generating function, derive the first three moments of X. Repeat the process using the characteristic function.

22. Let X be a random variable. Prove that the characteristic function of X is real valued if and only if X has the same distribution as $-X$.

23. Prove Theorem 2.24. That is, suppose that X is a random variable with moment generating function $m_X(t)$ that exists and is finite for $|t| < b$ for some $b > 0$. Suppose that Y is a new random variable defined by $Y = \alpha X + \beta$ where α and β are real constants. Prove that the moment generating function of Y is $m_Y(t) = \exp(t\beta) m_X(\alpha t)$ provided $|\alpha t| < b$.

24. Prove Theorem 2.32. That is, suppose that X is a random variable with characteristic function $\psi(t)$. Let $Y = \alpha X + \beta$ where α and β are real constants. Prove that the characteristic function of Y is $\psi_Y(t) = \exp(it\beta)\psi(\alpha t)$.

25. Prove Theorem 2.33. That is, suppose that $X_1, \ldots, X_n$ be a sequence of independent random variables where X_i has characteristic function $\psi_i(t)$, for $i = 1, \ldots, n$. Prove that the characteristic function of

$$ S_n = \sum_{i=1}^{n} X_i, $$

is

$$ \psi_{S_n}(t) = \prod_{i=1}^{n} \psi_i(t). $$

Further, prove that if $X_1, \ldots, X_n$ are identically distributed with characteristic function $\psi(t)$ then the characteristic function of S_n is $\psi_{S_n}(t) = \psi^n(t)$.

26. Let $X_1, \ldots, X_n$ be a sequence of independent random variables where X_i has a GAMMA(α_i, β) distribution for $i = 1, \ldots, n$. Let

$$ S_n = \sum_{i=1}^{n} X_i. $$

Find the moment generating function of S_n, and identify the corresponding distribution of the random variable.

27. Suppose that X is a discrete random variable that takes on non-negative integer values and has characteristic function $\psi(t) = \exp\{\theta[\exp(it) - 1]\}$. Use Theorem 2.29 to find the probability that X equals k where $k \in \{0, 1, \ldots\}$.

28. Suppose that X is a discrete random variable that takes on the values $\{0, 1\}$ and has characteristic function $\psi(t) = \cos(t)$. Use Theorem 2.29 to find the probability that X equals k where $k \in \{0, 1\}$.

29. Suppose that X is a discrete random variable that takes on positive integer values and has characteristic function

$$ \psi(t) = \frac{p \exp(it)}{1 - (1-p)\exp(it)}. $$

Use Theorem 2.29 to find the probability that X equals k where $k \in \{1, 2, \ldots\}$.

30. Suppose that X is a continuous random variable that takes on real values and has characteristic function $\psi(t) = \exp(-|t|)$. Use Theorem 2.28 to find the density of X.

31. Suppose that X is a continuous random variable that takes on values in $(0,1)$ and has characteristic function $\psi(t) = [\exp(it) - 1]/it$. Use Theorem 2.28 to find the density of X.

32. Suppose that X is a continuous random variable that takes on positive real values and has characteristic function $\psi(t) = (1 - \theta it)^{-\alpha}$. Use Theorem 2.28 to find the density of X.

33. Let X be a random variable with characteristic function ψ. Suppose that $E(|X|^n) < \infty$ for some $n \in \{1, 2, \ldots\}$ and that $\psi^{(k)}$ exists and is uniformly continuous for $k \in \{1, 2, \ldots, n\}$.

    a. Prove that
    $$\psi(t) = 1 + \sum_{k=1}^{n} \frac{\mu_k'(it)^k}{k!} + o(|t|^n),$$
    as $t \to 0$.

    b. Prove that
    $$\left. \frac{d^k \psi(t)}{dt^k} \right|_{t=0} = i^k \mu_k'.$$

34. a. Prove that $\kappa_4 = \mu_4' - 4\mu_3'\mu_1' - 3(\mu_2')^2 + 12\mu_2'(\mu_1')^2 - 6(\mu_1')^4$.

    b. Prove that $\kappa_4 = \mu_4 - 3\mu_2^2$, which is often called the *kurtosis* of a random variable.

    c. Suppose that X is an EXPONENTIAL(θ) random variable. Compute the fourth cumulant of X.

35. a. Prove that
    $$\kappa_5 = \mu_5' - 5\mu_4'\mu_1' - 10\mu_3'\mu_2' + 20\mu_3'(\mu_1')^2 + 30(\mu_2')^2\mu_1' - 60\mu_2'(\mu_1')^3 + 24(\mu_1')^5.$$

    b. Prove that $\kappa_5 = \mu_5 - 10\mu_2\mu_3$.

    c. Suppose that X is an EXPONENTIAL(θ) random variable. Compute the fifth cumulant of X.

36. Prove Theorem 2.34. That is, suppose that $X_1, \ldots, X_n$ be a sequence of independent random variables where X_i has cumulant generating function $c_i(t)$ for $i = 1, \ldots, n$. Then prove that the cumulant generating function of
    $$S_n = \sum_{i=1}^{n} X_i,$$
    is
    $$c_{S_n}(t) = \sum_{i=1}^{n} c_i(t).$$

37. Suppose that X is a POISSON(λ) random variable, so that the moment generating function of X is $m(t) = \exp\{\lambda[\exp(t) - 1]\}$. Find the cumulant generating function of X, and put it into the form given in Equation (2.24). Using the form of the cumulant generating function, find a general form for the cumulants of X.

38. Suppose that X is a GAMMA(λ) random variable, so that the moment generating function of X is $m(t) = (1-t\beta)^{-\alpha}$. Find the cumulant generating function of X, and put it into the form given in Equation (2.24). Using the form of the cumulant generating function, find a general form for the cumulants of X.

39. Suppose that X is a LAPLACE(α, β) random variable, so that the moment generating function of X is $m(t) = (1 - t^2\beta^2)^{-1}\exp(t\alpha)$ when $|t| < \beta^{-1}$. Find the cumulant generating function of X, and put it into the form given in Equation (2.24). Using the form of the cumulant generating function, find a general form for the cumulants of X.

40. One consequence of defining the cumulant generating function in terms of the moment generating function is that the cumulant generating function will not exist any time the moment generating function does not. An alternate definition of the cumulant generating function is defined in terms of the characteristic function. That is, if X has characteristic function $\psi(t)$, then the cumulant generating function can be defined as $c(t) = \log[\psi(t)]$.

   a. Assume all of the cumulants (and moments) of X exist. Prove that the coefficient of $(it)^k/k!$ for the cumulant generating function defined using the characteristic function is the kth cumulant of X. You may want to use Theorem 2.31.

   b. Find the cumulant generating function of a random variable X that has a CAUCHY($0, 1$) distribution based on the fact that the characteristic function of X is $\psi(t) = \exp(|t|)$. Use the form of this cumulant generating function to argue that the cumulants of X do not exist.

2.6.2 Experiments

1. For each of the distributions listed below, use R to compute $P(|X - \mu| > \delta)$ and compare the result to the bound given by Theorem 2.7 as $\delta^{-2}\sigma^2$ for $\delta = \frac{1}{2}, 1, \frac{3}{2}, 2$. Which distributions become closest to achieving the bound? What are the properties of these distributions?

   a. N($0, 1$)

   b. T(3)

   c. GAMMA($1, 1$)

   d. UNIFORM($0, 1$)

2. For each distribution listed below, plot the corresponding characteristic function of the density as a function of t if the characteristic function is real-valued, or as a function of t on the complex plane if the function is complex-valued. Describe each characteristic function. Are there any properties of the associated random variables that have an apparent effect on the properties of the characteristic function? See Section B.3 for details on plotting complex functions in the complex plane.

   a. BERNOULLI$(\frac{1}{2})$

   b. BINOMIAL$(5, \frac{1}{2})$

   c. GEOMETRIC$(\frac{1}{2})$

   d. POISSON(2)

   e. UNIFORM$(0, 1)$

   f. EXPONENTIAL(2)

   g. CAUCHY$(0, 1)$

3. For each value of μ and σ listed below, plot the characteristic function of the corresponding $N(\mu, \sigma^2)$ distribution as a function of t in the complex plane. Describe how the changes in the parameter values affect the properties of the corresponding characteristic function. This will require a three-dimensional plot. See Section B.3 for further details.

   a. $\mu = 0$, $\sigma = 1$

   b. $\mu = 1$, $\sigma = 1$

   c. $\mu = 0$, $\sigma = 2$

   d. $\mu = 1$, $\sigma = 2$

4. *Random walks* are a special type of discrete stochastic process that are able to change from one state to any adjacent state according to a conditional probability distribution. This experiment will investigate the properties of random walks in one, two, and three dimensions.

   a. Consider a sequence of random variables $\{X_n\}_{n=1}^{\infty}$ where $X_1 = 0$ with probability one and $P(X_n = k | X_{n-1} = x) = \frac{1}{2}\delta\{k; \{x - 1, x + 1\}\}$, for $n = 2, 3, \ldots$. Such a sequence is known as a symmetric one-dimensional random walk. Write a program in R that simulates such a sequence of length 1000 and keeps track of the number of times the origin is visited after the initial start of the sequence. Use this program to repeat the experiment 100 times to estimate the average number of visits to the origin for a sequence of length 1000.

   b. Consider a sequence of two-dimensional random vectors $\{\mathbf{Y}_n\}_{n=1}^{\infty}$ where $\mathbf{Y}_1' = (0, 0)$ with probability one and $P(\mathbf{Y}_n = \mathbf{y}_n | \mathbf{Y}_{n-1} = \mathbf{y}) = \frac{1}{4}\delta\{\mathbf{y}_n - \mathbf{y}; D\}$, for $n = 2, 3, \ldots$ where $D = \{(1, 0), (0, 1), (-1, 0), (0, -1)\}$. Such a sequence is known as a symmetric two-dimensional random walk. Write a program in R that simulates such a sequence of length 1000 and keeps track of the number of times the origin is visited after the initial start of the sequence. Use this program to repeat the experiment 100 times to estimate the average number of visits to the origin for a sequence of length 1000.

c. Consider a sequence of three-dimensional random vectors $\{\mathbf{Z}_n\}_{n=1}^{\infty}$ where $\mathbf{Z}_1' = (0,0,0)$ with probability one and $P(\mathbf{Z}_n = \mathbf{z}_n | \mathbf{Z}_{n-1} = \mathbf{z}) = \frac{1}{6}\delta\{\mathbf{z}_n - \mathbf{z}; D\}$, for $n = 2, 3, \ldots$ where

$$D = \{(1,0,0), (0,1,0), (0,0,1), (-1,0,0), (0,-1,0), (0,0,-1)\}.$$

Such a sequence is known as a symmetric three-dimensional random walk. Write a program in R that simulates such a sequence of length 1000 and keeps track of the number of times the origin is visited after the initial start of the sequence. Use this program to repeat the experiment 100 times to estimate the average number of visits to the origin for a sequence of length 1000.

d. The theory of Markov chains can be used to show that with probability one, the one and two-dimensional symmetric random walks will return to the origin infinitely often, whereas the three-dimensional random walk will not. Discuss, as much as is possible, the properties of the probability of the process visiting the origin at step n for each case in terms of Theorem 2.17. Note that Theorem 2.18 cannot be applied to this case because the events in the sequence are dependent. Discuss whether your simulation results provide evidence about this behavior. Do you think that a property such as this could ever be verified empirically?

CHAPTER 3

Convergence of Random Variables

The man from the country has not expected such difficulties: the law should always be accessible for everyone, he thinks, but as he now looks more closely at the gatekeeper in his fur coat, at his large pointed nose and his long, thin, black Tartars beard, he decides that it would be better to wait until he gets permission to go inside.

Before the Law by Franz Kafka

3.1 Introduction

Let $\{X_n\}_{n=1}^{\infty}$ be a sequence of random variables and let X be some other random variable. Under what conditions is it possible to say that X_n converges to X as $n \to \infty$? That is, is it possible to define a limit for a sequence of random variables so that the statement

$$\lim_{n \to \infty} X_n = X,$$

has a well defined mathematical meaning? The answer, or answers, to this question arise from a detailed consideration of the mathematical structure of random variables. There are two ways to conceptualize random variables. From an informal viewpoint one can view the sequence $\{X_n\}_{n=1}^{\infty}$ as a sequence of quantities that are random, and this random behavior somehow depends on the index n. The quantity X is also random. It is clear in this context that the mathematical definition of the limit of a sequence of real numbers could only be applied to an observed sequence of these random variables. That is, if we observed $X_n = x_n$ for all $n \in \mathbb{N}$ and $X = x$, then it is easy to ascertain whether

$$\lim_{n \to \infty} x_n = x,$$

using Definition 1.1. But what can be concluded about the sequence before the random variables are observed? It is clear that the usual mathematical definition of convergence and limit will not suffice, and that a new view of convergence will need to be established for random variables. In some sense we wish X_n and X to act the same when n is large, but how can this be guaranteed? It is clear that probability should play a role, but what exactly should the role be? For example, we could insist that the limit of the sequence

101

$\{X_n\}_{n=1}^{\infty}$ match X with probability one, which would provide the definition that X_n converges to X as $n \to \infty$ if

$$P\left(\lim_{n\to\infty} X_n = X\right) = 1.$$

Another viewpoint might insist that X_n and X be arbitrarily close with a probability converging to one. That is, one could say that X_n converges to X as $n \to \infty$ if for every $\varepsilon > 0$,

$$\lim_{n\to\infty} P(|X_n - X| < \varepsilon) = 1.$$

Now there are two definitions of convergence of random variables to contend with, which yields several more questions. Are the two notions of convergence equivalent? Are there other competing definitions which should be considered?

Fortunately, a more formal analysis of this problem reveals a fully established mathematical framework for this problem. Recall from Section 2.2 that random variables are not themselves random quantities, but are functions that map points in the sample space to the real line. This indicates that $\{X_n\}_{n=1}^{\infty}$ is really a sequence of functions and that X is a possible limiting function. Well established methods that originate in real analysis and measure theory yield definitions for the convergence of functions as well as many other relevant properties that will be used throughout this book. From the mathematician's point of view, statistical limit theory may seem like a specialized application of real analysis and measure theory to problems that specifically involve a probability (normed) measure. However, the application is not trivial, and the viewpoint of problems considered by statisticians may considerably differ from those that interest pure mathematicians.

The purpose of this chapter is to introduce the concept of convergence of random variables within a statistical framework. The presentation in this chapter will stress the mathematical viewpoint of these developments, as this viewpoint provides the best understanding of the true nature of the modes of convergence, their properties, and how they relate to one another. The chapter begins by considering convergence in probability, which corresponds to the second of the definitions considered above.

3.2 Convergence in Probability

We now formally develop the second definition of convergence of random variables that was proposed informally in Section 3.1.

Definition 3.1. Let $\{X_n\}_{n=1}^{\infty}$ be a sequence of random variables. The sequence converges in probability to a random variable X if for every $\varepsilon > 0$,

$$\lim_{n\to\infty} P(|X_n - X| \geq \varepsilon) = 0.$$

This relationship is represented by $X_n \xrightarrow{p} X$ as $n \to \infty$.

A simple application of the compliment rule provides an equivalent condition for convergence in probability. In particular, $X_n \xrightarrow{p} X$ as $n \to \infty$ if for every $\varepsilon > 0$,

$$\lim_{n \to \infty} P(|X_n - X| < \varepsilon) = 1.$$

Either representation of the condition implies the same idea behind this mode of convergence. That is, X_n should be close to X with a high probability for large values of n. While the sequence $\{|X_n - X|\}_{n=1}^{\infty}$ is a sequence of random variables indexed by n, the sequence of probabilities $\{P(|X_n - X| > \varepsilon)\}_{n=1}^{\infty}$ is a sequence of real constants indexed by $n \in \mathbb{N}$. Hence, Definition 1.1 can be applied to the latter sequence. Therefore an equivalent definition of convergence in probability is that $X_n \xrightarrow{p} X$ as $n \to \infty$ if for every $\varepsilon > 0$ and $\delta > 0$ there exists an integer $n_\delta \in \mathbb{N}$ such that $P(|X_n - X| > \varepsilon) < \delta$ for all $n \geq n_\delta$.

Another viewpoint of convergence in probability can be motivated by Definition 2.4. A random variable is a measureable mapping from a sample space Ω to $\mathbb{R}$. Hence the sequence random variables $\{X_n\}_{n=1}^{\infty}$ is really a sequence of functions and the random variable X is a limiting function. Convergence in probability requires that for any $\varepsilon > 0$ the sequence of functions must be within an ε-band of the function X over a set of points from the sample space whose probability increases to one as $n \to \infty$. Equivalently, the set of points for which the sequence of functions is not within the ε-band must decrease to zero as $n \to \infty$. See Figure 3.1.

Example 3.1. Let Z be a $N(0,1)$ random variable and let $\{X_n\}_{n=1}^{\infty}$ be a sequence of random variables such that $X_n = Z + n^{-1}$ for all $n \in \mathbb{N}$. Let $\varepsilon > 0$, then

$$P(|X_n - Z| \geq \varepsilon) = P(|Z + n^{-1} - Z| \geq \varepsilon) = P(n^{-1} \geq \varepsilon),$$

for all $n \in \mathbb{N}$. There exists an $n_\varepsilon \in \mathbb{N}$ such that $n^{-1} < \varepsilon$ for all $n \geq n_\varepsilon$ so that $P(n^{-1} > \varepsilon) = 0$ for all $n \geq n_\varepsilon$. Therefore,

$$\lim_{n \to \infty} P(|X_n - Z| \geq \varepsilon) = 0,$$

and it follows from Definition 3.1 that $X_n \xrightarrow{p} Z$ as $n \to \infty$. ∎

Example 3.2. Suppose that $\hat{\theta}_n$ is an unbiased estimator of θ, that is $E(\hat{\theta}_n) = \theta$ for all values of θ within the parameter space. In many cases the standard error, or equivalently, the variance of $\hat{\theta}_n$ converges to zero as $n \to \infty$. That is,

$$\lim_{n \to \infty} V(\hat{\theta}_n) = 0.$$

Under these conditions note that Theorem 2.7 (Tchebysheff) implies that for any $\varepsilon > 0$ $P(|\hat{\theta}_n - \theta| > \varepsilon) \leq \varepsilon^{-2} V(\hat{\theta}_n)$. The limiting condition on the variance of $\hat{\theta}_n$ and Definition 2.2 imply that

$$0 \leq \lim_{n \to \infty} P(|\hat{\theta}_n - \theta| > \varepsilon) \leq \lim_{n \to \infty} \varepsilon^{-2} V(\hat{\theta}_n) = 0,$$

Figure 3.1 *Convergence in probability from the viewpoint of convergence of functions.*
The solid line represents the random variable X and the dashed lines represent an ε-
band around the function where the horizontal axis is the sample space Ω. The dotted
line represents a random variable X_n. Convergence in probability requires that more
of the function X_n, with respect to the probability measure on Ω, be within the ε band
as n becomes larger.

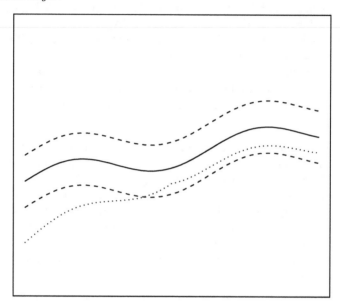

since ε is constant with respect to n. Therefore,

$$\lim_{n\to\infty} P(|\hat{\theta}_n - \theta| \geq \varepsilon) = 0,$$

and Definition 3.1 implies that $\hat{\theta}_n \xrightarrow{p} \theta$ as $n \to \infty$. In estimation theory this
property is called *consistency*. That is, $\hat{\theta}_n$ is a *consistent* estimator of θ. A
special case of this result applies to the sample mean. Suppose that $X_1, \ldots, X_n$
are a set of independent and identically distributed random variables from a
distribution with mean θ and finite variance σ^2. The sample mean $\bar{X}_n$ is
an unbiased estimator of θ with variance $n^{-1}\sigma^2$ which converges to zero as
$n \to \infty$ as long as $\sigma^2 < \infty$. Therefore it follows that $\bar{X}_n \xrightarrow{p} \theta$ as $n \to \infty$ and
the sample mean is a consistent estimator of θ. This result is a version of what
are known as *Laws of Large Numbers*. In particular, this result is known as
the *Weak Law of Large Numbers*. Various results of this type can be proven
under many different conditions. In particular, it will be shown in Section 3.6
that the condition that the variance is finite can be relaxed. This result can be
visualized with the aid of simulated data. Consider simulating samples from a
$N(0, 1)$ distribution of size $n = 5, 10, 15, \ldots, 250$, where the sample mean $\bar{X}_n$

Figure 3.2 *The results of a small simulation demonstrating convergence in probability due to the weak law of large numbers. Each line represents a sequence of sample means computed on a sequence of independent* $N(0, 1)$ *random variables. The means were computed when* $n = 5, 10, \ldots, 250$. *An* ε*-band of size* $\varepsilon = 0.10$ *has been placed around the limiting value.*

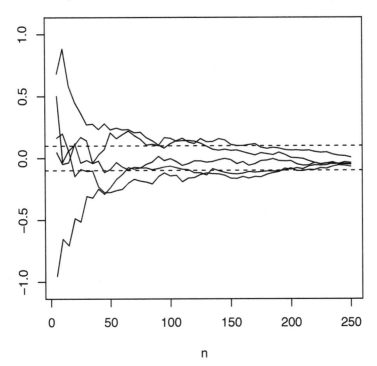

is computed on each sample. The Weak Law of Large Numbers states that these sample means should converge in probability to 0 as $n \to \infty$. Figure 3.2 shows the results of five such simulated sequences. An ε-band has been plotted around 0. Note that all of the sequences generally become closer to 0 as n becomes larger, and that there is a point where all of the sequences are within the ε-band. Remember that the definition of convergence in probability is a result for random sequences. This does not mean that all such sequences will be within the ε-band for a given sample size, only that the probability that the sequences are within the ε-band converges to one as $n \to \infty$. This can also be observed from the fact that the individual sequences do not monotonically converge to 0 as n becomes large. There are random fluctuations in all of the sequences, but the overall behavior of the sequence does become closer to 0 as n becomes large. ∎

Example 3.3. Let $\{c_n\}_{n=1}^{\infty}$ be a sequence of real constants where

$$\lim_{n \to \infty} c_n = c,$$

for some constant $c \in \mathbb{R}$. Let $\{X_n\}_{n=1}^{\infty}$ be a sequence of random variables with a degenerate distribution at c_n for all $n \in \mathbb{N}$. That is $P(X_n = c_n) = 1$ for all $n \in \mathbb{N}$. Let $\varepsilon > 0$, then

$$P(|X_n - c| \geq \varepsilon) = P(|X_n - c| \geq \varepsilon | X_n = c_n)P(X_n = c_n) = P(|c_n - c| \geq \varepsilon).$$

Definition 1.1 implies that for any $\varepsilon > 0$ there exists an $n_\varepsilon \in \mathbb{N}$ such that $|c_n - c| < \varepsilon$ for all $n > n_\varepsilon$. Therefore $P(|c_n - c| > \varepsilon) = 0$ for all $n > n_\varepsilon$, and it follows that

$$\lim_{n \to \infty} P(|X_n - c| \geq \varepsilon) = \lim_{n \to \infty} P(|c_n - c| \geq \varepsilon) = 0.$$

Therefore, by Definition 3.1 it follows that $X_n \xrightarrow{P} c$ as $n \to \infty$. ∎

Example 3.4. Let $X_1, \ldots, X_n$ be a set of independent and identically distributed random variables from a UNIFORM$(\theta, \theta + 1)$ distribution for some $\theta \geq 0$. Let $X_{(1)}$ be the first sample order statistic. That is

$$X_{(1)} = \min\{X_1, \ldots, X_n\}.$$

The distribution function of $X_{(1)}$ can be found by using the fact that if $X_{(1)} \geq t$ for some $t \in \mathbb{R}$, then $X_i \geq t$ for all $i = 1, \ldots, n$. Therefore, the distribution function of $X_{(1)}$ is given by

$$F(t) = P(X_{(1)} \leq t) = 1 - P(X_{(1)} > t)$$
$$= 1 - P\left(\bigcap_{i=1}^{n} \{X_i > t\}\right) = 1 - \prod_{i=1}^{n} P(X_i > t).$$

If $t \in (\theta, \theta + 1)$ then $P(X_i > t) = 1 + \theta - t$ so that the distribution function of $X_{(1)}$ is

$$F(t) = \begin{cases} 0 & \text{for } t < \theta, \\ 1 - (1 + \theta - t)^n & \text{for } t \in (\theta, \theta + 1), \\ 1 & \text{for } t > \theta + 1. \end{cases}$$

Let $\varepsilon > 0$ and consider the inequality $|X_{(1)} - \theta| \leq \varepsilon$. If $\varepsilon \geq 1$ then $|X_{(1)} - \theta| \leq \varepsilon$ with probability one because $X_{(1)} \in (\theta, \theta + 1)$ with probability one. If $\varepsilon \in (0, 1)$ then

$$
\begin{aligned}
P(|X_{(1)} - \theta| < \varepsilon) &= P(-\varepsilon < X_{(1)} - \theta < \varepsilon) \\
&= P(\theta - \varepsilon < X_{(1)} < \theta + \varepsilon) \\
&= P(\theta < X_{(1)} < \theta + \varepsilon) \\
&= F(\theta + \varepsilon) \\
&= 1 - (1 - \varepsilon)^n,
\end{aligned}
$$

where the fact that $X_{(1)}$ must be greater than θ has been used. Therefore

$$\lim_{n \to \infty} P(|X_{(1)} - \theta| < \varepsilon) = 1,$$

since $0 < 1 - \varepsilon < 1$. Definition 3.1 then implies that $X_{(1)} \xrightarrow{p} \theta$ as $n \to \infty$, or that $X_{(1)}$ is a consistent estimator of θ.   ■

3.3 Stronger Modes of Convergence

In Section 3.1 we considered another reasonable interpretation of the limit of a sequence of random variables, which essentially requires that the sequence converge with probability one. This section formalizes this definition and studies how this type of convergence is related to convergence in probability. We also introduce another concept of convergence of random variables that is based on modifying the definition of convergence in probability.

Definition 3.2. *Let $\{X_n\}_{n=1}^{\infty}$ be a sequence of random variables. The sequence converges almost certainly to a random variable X if*

$$P\left(\lim_{n \to \infty} X_n = X\right) = 1. \tag{3.1}$$

This relationship is represented by $X_n \xrightarrow{a.c.} X$ as $n \to \infty$.

To better understand this type of convergence it is sometimes helpful to rewrite Equation (3.1) as

$$P\left[\omega : \lim_{n \to \infty} X_n(\omega) = X(\omega)\right] = 1.$$

Hence, almost certain convergence requires that the set of all ω for which $X_n(\omega)$ converges to $X(\omega)$, have probability one. Note that the limit used in Equation (3.1) is the usual limit for a sequence of constants given in Definition 1.1, as when ω is fixed, $X_n(\omega)$ is a sequence of constants. See Figure 3.3.

Example 3.5. Consider the sample space $\Omega = [0, 1]$ with probability measure P such that ω is chosen according to a UNIFORM$[0, 1]$ distribution on the Borel σ-field $\mathcal{B}[0, 1]$. Define a sequence $\{X_n\}_{n=1}^{\infty}$ of random variables as $X_n(\omega) = \delta\{\omega; [0, n^{-1})\}$. Let $\omega \in [0, 1]$ be fixed and note that there exists an $n_\omega \in \mathbb{N}$ such that $n^{-1} < \omega$ for all $n \geq n_\omega$. Therefore $X_n(\omega) = 0$ for all $n \geq n_\omega$, and it follows that for this value of ω

$$\lim_{n \to \infty} X_n(\omega) = 0.$$

The exception is when $\omega = 0$ for which $X_n(\omega) = 1$ for all $n \in \mathbb{N}$. Therefore

$$P\left(\lim_{n \to \infty} X_n = 0\right) = P[\omega \in (0, 1)] = 1,$$

and it follows from Definition 3.2 that $X_n \xrightarrow{a.c.} 0$ as $n \to \infty$.   ■

Figure 3.3 *Almost certain convergence is characterized by the point-wise convergence of the random variable sequence to the limiting random variable. In this figure, the limiting random variable is represented by the function given by the solid black line, and random variables in the sequence are represented by the functions plotted with the dotted line. The horizontal axis represents the sample space. For a fixed value of $\omega \in \Omega$, the corresponding sequence is simply a sequence of real constants as represented by the black points.*

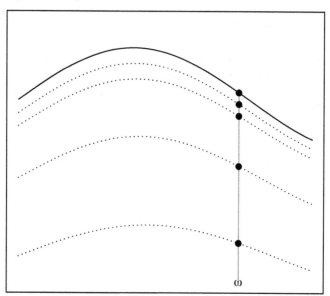

Definition 3.2 can be difficult to apply in practice, and is not always useful when studying the properties of almost certain convergence. By applying Definition 1.1 to the limit inside the probability in Equation (3.1), an equivalent definition that relates almost certain convergence to a statement that is similar to the one used in Definition 3.1 can be obtained.

Theorem 3.1. *Let $\{X_n\}_{n=1}^{\infty}$ be a sequence of random variables. Then X_n converges almost certainly to a random variable X as $n \to \infty$ if and only if for every $\varepsilon > 0$,*

$$\lim_{n\to\infty} P(|X_m - X| < \varepsilon \text{ for all } m \geq n) = 1. \tag{3.2}$$

Proof. This result is most easily proven by rewriting the definition of a limit using set operations. This is the method used by Halmos (1950), Serfling (1980), and many others. To prove the equivalence, consider the set

$$A = \left\{ \omega : \lim_{n\to\infty} X_n(\omega) = X(\omega) \right\}.$$

Definition 1.1 implies that

$$A = \{\omega : \text{for every } \varepsilon \text{ there exists } n \in \mathbb{N} \text{ such that}$$
$$|X_m(\omega) - X(\omega)| < \varepsilon \text{ for all } m \geq n\} \quad (3.3)$$

Since the condition in the event on the right hand side of Equation (3.3) must be true for every $\varepsilon > 0$, we have that

$$A = \bigcap_{\varepsilon > 0} \{\omega : \text{ there exists } n \in \mathbb{N} \text{ such that}$$
$$|X_m(\omega) - X(\omega)| < \varepsilon \text{ for all } m \geq n\} \quad (3.4)$$

The condition within each event of the intersection on the right hand side of Equation (3.4) needs only to be true for at least one $n \in \mathbb{N}$. Therefore

$$A = \bigcap_{\varepsilon > 0} \bigcup_{n=1}^{\infty} \{\omega : |X_m(\omega) - X(\omega)| < \varepsilon \text{ for all } m \geq n\}.$$

Now consider $0 < \varepsilon < \delta$, then we have that

$$\bigcup_{n=1}^{\infty} \{\omega : |X_m(\omega) - X(\omega)| < \varepsilon \text{ for all } m \geq n\} \subset$$
$$\bigcup_{n=1}^{\infty} \{\omega : |X_m(\omega) - X(\omega)| < \delta \text{ for all } m \geq n\}. \quad (3.5)$$

This implies that the sequence of events within the intersection on the right hand side of Equation (3.4) is monotonically decreasing as $\varepsilon \to 0$. Therefore, Theorem 2.15 implies that

$$A = \lim_{\varepsilon \to 0} \bigcup_{n=1}^{\infty} \{\omega : |X_m(\omega) - X(\omega)| < \varepsilon \text{ for all } m \geq n\}.$$

Similarly, note that

$$\{\omega : |X_m(\omega) - X(\omega)| < \varepsilon \text{ for all } m \geq n+1\} \subset$$
$$\{\omega : |X_m(\omega) - X(\omega)| < \varepsilon \text{ for all } m \geq n\}$$

so that the sequence of events within the union on the right hand side of Equation (3.4) is monotonically increasing as $n \to \infty$. Therefore, Theorem 2.15 implies that

$$A = \lim_{\varepsilon \to 0} \lim_{n \to \infty} \{\omega : |X_m(\omega) - X(\omega)| < \varepsilon \text{ for all } m \geq n\}.$$

Therefore, Theorem 2.16 implies that

$$P\left[\omega : \lim_{n \to \infty} X_n(\omega) = X(\omega)\right] =$$
$$\lim_{\varepsilon \to 0} \lim_{n \to \infty} P[\omega : |X_m(\omega) - X(\omega)| < \varepsilon \text{ for all } m \geq n].$$

Now, suppose that for every $\varepsilon > 0$,

$$\lim_{n \to \infty} P[\omega : |X_m(\omega) - X(\omega)| < \varepsilon \text{ for all } m \geq n] = 1.$$

It then follows that

$$\lim_{\varepsilon \to 0} \lim_{n \to \infty} P[\omega : |X_m(\omega) - X(\omega)| < \varepsilon \text{ for all } m \geq n] = 1$$

and hence

$$P\left(\lim_{n \to \infty} X_n = X\right) = 1,$$

so that $X_n \xrightarrow{a.c.} X$ as $n \to \infty$. Now suppose that $X_n \xrightarrow{a.c.} X$ as $n \to \infty$ and let $\varepsilon > 0$ and note that Equation (3.5) implies that

$$1 = P\left(\lim_{n \to \infty} X_n = X\right) =$$

$$\lim_{\varepsilon \to 0} \lim_{n \to \infty} P[\omega : |X_m(\omega) - X(\omega)| < \varepsilon \text{ for all } m \geq n] \leq$$

$$\lim_{n \to \infty} P[\omega : |X_m(\omega) - X(\omega)| < \varepsilon \text{ for all } m \geq n],$$

so that

$$\lim_{n \to \infty} P[\omega : |X_m(\omega) - X(\omega)| < \varepsilon \text{ for all } m \geq n] = 1,$$

and the result is proven. □

Example 3.6. Suppose that $\{U_n\}_{n=1}^{\infty}$ is a sequence of independent UNI-FORM$(0,1)$ random variables and let $U_{(1,n)}$ be the smallest order statistic of $U_1, \ldots, U_n$ defined by $U_{(1,n)} = \min\{U_1, \ldots, U_n\}$. Let $\varepsilon > 0$, then because $U_{(1,n)} \geq 0$ with probability one, it follows that

$$P(|U_{(1,m)} - 0| < \varepsilon \text{ for all } m \geq n) = P(U_{(1,m)} < \varepsilon \text{ for all } m \geq n)$$
$$= P(U_{(1,n)} < \varepsilon),$$

where the second equality follows from the fact that if $U_{(1,n)} < \varepsilon$ then $U_{(1,m)} < \varepsilon$ for all $m \geq n$. Similarly, if $U_{(1,m)} < \varepsilon$ for all $m \geq n$ then $U_{(1,n)} < \varepsilon$ so that it follows that the two events are equivalent. Now note that the independence of the random variables in the sequence implies that when $\varepsilon < 1$,

$$P(U_{(1,n)} < \varepsilon) = 1 - P(U_{(1,n)} \geq \varepsilon) = 1 - \prod_{k=1}^{n} P(U_k \geq \varepsilon) = 1 - (1 - \varepsilon)^n.$$

Therefore, it follows that

$$\lim_{n \to \infty} P(U_{(1,m)} < \varepsilon \text{ for all } m \geq n) = \lim_{n \to \infty} 1 - (1 - \varepsilon)^n = 1.$$

When $\varepsilon \geq 1$ the probability equals one for all $n \in \mathbb{N}$. Hence Theorem 3.1 implies that $U_{(1,n)} \xrightarrow{a.c.} 0$ as $n \to \infty$. ∎

Theorem 3.1 is also useful in beginning to understand the relationship between convergence in probability and almost certain convergence.

Theorem 3.2. *Let $\{X_n\}_{n=1}^{\infty}$ be a sequence of random variables that converge almost certainly to a random variable X as $n \to \infty$. Then $X_n \xrightarrow{p} X$ as $n \to \infty$.*

Proof. Suppose that $\{X_n\}_{n=1}^{\infty}$ is a sequence of random variables that converge almost certainly to a random variable X as $n \to \infty$. Then Theorem 3.1 implies that for every $\varepsilon > 0$,

$$\lim_{n\to\infty} P(|X_m - X| < \varepsilon \text{ for all } m \geq n) = 1.$$

Now note that

$$\{\omega : |X_m(\omega) - X(\omega)| < \varepsilon \text{ for all } m \geq n\} \subset \{\omega : |X_n(\omega) - X(\omega)| < \varepsilon\},$$

so that is follows that

$$P(|X_m(\omega) - X(\omega)| < \varepsilon \text{ for all } m \geq n) \leq P(|X_n(\omega) - X(\omega)| < \varepsilon),$$

for each $n \in \mathbb{N}$. Therefore

$$\lim_{n\to\infty} P(|X_m(\omega) - X(\omega)| < \varepsilon \text{ for all } m \geq n) \leq \lim_{n\to\infty} P(|X_n(\omega) - X(\omega)| < \varepsilon),$$

which implies

$$\lim_{n\to\infty} P(|X_n(\omega) - X(\omega)| < \varepsilon) = 1.$$

Therefore Definition 3.1 implies that $X_n \xrightarrow{p} X$ as $n \to \infty$. □

Theorem 3.2 makes it clear that almost certain convergence is potentially a stronger concept of convergence of random variables than convergence in probability. However, the question remains as to whether they may be equivalent. The following example demonstrates that convergence in probability is actually a weaker concept of convergence than almost certain convergence by identifying a sequence of random variables that converges in probability, but does not converge almost certainly.

Example 3.7. This example is a version of a popular example used before by Serfling (1980), Royden (1988), and Halmos (1974). Consider a sample space $\Omega = [0, 1]$ with a uniform probability measure P. That is, the probability associated with the interval $[a, b] \subset \Omega$ is $b - a$. Let $m(n)$ be a sequence of intervals of the form $[i/k, (i+1)/k]$ for $i = 0, \ldots, k-1$ and $k \in \mathbb{N}$, where $m(1) = [0, 1]$, $m(2) = [0, 1/2]$, $m(3) = [1/2, 1]$, $m(4) = [0, 1/3]$, $m(5) = [1/3, 2/3]$, $m(6) = [2/3, 1]$, ... See Figure 3.4. Define a sequence of random variables on this sample space as $X_n(\omega) = \delta\{\omega; m(n)\}$ for $\omega \in [0, 1]$. Let $\varepsilon > 0$, and note that since $X_n(\omega)$ is 1 only on the interval $m(n)$ that $P(|X_n| < \varepsilon) \leq 1 - k(n)$, where $k(n)$ is a sequence that converges to 0, as $n \to \infty$. It is then clear that for any $\varepsilon > 0$,

$$\lim_{n\to\infty} P(|X_n| < \varepsilon) = 1,$$

so that it follows that $X_n \xrightarrow{p} 0$ as $n \to \infty$. Now consider any fixed $\omega \in [0, 1]$. Note that for any $n \in \mathbb{N}$ there exists $n' \in \mathbb{N}$ and $n'' \in \mathbb{N}$ such that $n' > n$, $n'' > n$, $X_{n'}(\omega) = 1$, and $X_{n''}(\omega) = 0$. Hence the limit of the sequence $X_n(\omega)$ does not exist for any fixed $\omega \in [0, 1]$. Since ω is arbitrary it follows that

$$\{\omega : \lim_{n\to\infty} X_n(\omega) = 0\} = \emptyset,$$

Figure 3.4 *The first ten subsets of the unit interval used in Example 3.7.*

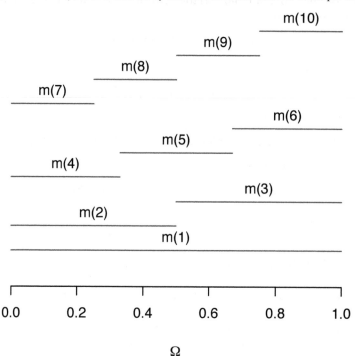

$$P\left[\lim_{n\to\infty} X_n(\omega) = 0\right] = 0.$$

and, therefore,

Therefore, this sequence does not converge almost certainly to 0. In fact, the probability that X_n converges at all is zero. Note the fundamental difference between the two modes of convergence demonstrated by this example. Convergence in probability requires that X_n be arbitrarily close to X for a set of ω whose probability limits to one, while almost certain convergence requires that the set of ω for which $X_n(\omega)$ converges to X have probability one. This latter set does not depend on n. ■

While almost certain convergence requires a stronger condition on the sequence of random variables than convergence in probability does, Theorem 3.2 indicates that there are sequences that converge in probability that also converge almost certainly. Therefore, some sequences that converge in probability have additional properties that allow these sequences to meet the stronger requirements of a sequence that converges almost certainly. The quantification of the properties of such sequences motivates an additional concept of convergence defined by Hsu and Robbins (1947).

Definition 3.3. *Let $\{X_n\}_{n=1}^{\infty}$ be a sequence of random variables. The sequence converges completely to a random variable X if for every $\varepsilon > 0$,*

$$\sum_{n=1}^{\infty} P(|X_n - X| > \varepsilon) < \infty. \tag{3.6}$$

This relationship is represented by $X_n \xrightarrow{c} X$ as $n \to \infty$.

Complete convergence is stronger than convergence in probability since convergence in probability only requires that for every $\varepsilon > 0$ that $P(|X_n - X| > \varepsilon)$ converge to zero as $n \to \infty$. Complete convergence not only requires that $P(|X_n - X| > \varepsilon)$ converge to zero, it must do it at a rate fast enough to ensure that the infinite sum

$$\sum_{n=1}^{\infty} P(|X_n - X| > \varepsilon),$$

converges. For example, if $P(|X_n - X| > \varepsilon) = n^{-1}$ then the sequence of random variables $\{X_n\}_{n=1}^{\infty}$ would converge in probability to X, but not completely, as $n \to \infty$. If $P(|X_n - X| > \varepsilon) = n^{-2}$ then the sequence of random variables $\{X_n\}_{n=1}^{\infty}$ would converge in probability and completely to X, as $n \to \infty$.

Example 3.8. Let U be a UNIFORM$(0,1)$ random variable and define a sequence of random variables $\{X_n\}_{n=1}^{\infty}$ such that $X_n = \delta\{U; (0, n^{-2})\}$. Let $\varepsilon > 0$, then

$$P(X_n > \varepsilon) = \begin{cases} 0 & \text{if } \varepsilon \geq 1 \\ n^{-2} & \text{if } \varepsilon < 1. \end{cases}$$

Therefore, for every $\varepsilon > 0$,

$$\sum_{n=1}^{\infty} P(X_n > \varepsilon) \leq \sum_{n=1}^{\infty} n^{-2} < \infty,$$

so that it follows from Defintion 3.3 that $X_n \xrightarrow{c} 0$ as $n \to \infty$. ∎

Example 3.9. Let $\{\hat{\theta}_n\}_{n=1}^{\infty}$ be a sequence of random variables such that $E(\hat{\theta}_n) = c$ for all $n \in \mathbb{N}$ where $c \in \mathbb{R}$ is a constant that does not depend on n. Suppose further that $V(\hat{\theta}_n) = n^{-2}\tau$ where τ is a positive and finite constant that does not depend on n. Under these conditions, note that for any $\varepsilon > 0$, Theorem 2.7 implies that

$$P(|\hat{\theta}_n - c| > \varepsilon) \leq \varepsilon^{-2} V(X_n) = n^{-2} \varepsilon^{-2} \tau.$$

Therefore, for every $\varepsilon > 0$,

$$\sum_{n=1}^{\infty} P(|\hat{\theta}_n - c| > \varepsilon) \leq \sum_{n=1}^{\infty} n^{-2} \varepsilon^{-2} \tau \leq \varepsilon^{-2} \tau \sum_{n=1}^{\infty} n^{-2} < \infty,$$

where we have used the fact that ε and τ do not depend on n. Therefore, $\hat{\theta}_n \xrightarrow{c} c$ as $n \to \infty$. ∎

While complete convergence is sufficient to ensure convergence in probability, we must still investigate the relationship between complete convergence and almost certain convergence. As it turns out, complete convergence also implies almost certain convergence, and is one of the strongest concepts of convergence of random variables that we will study in this book.

Theorem 3.3. *Let* $\{X_n\}_{n=1}^{\infty}$ *be a sequence of random variables that converges completely to a random variable* X *as* $n \to \infty$. *Then* $X_n \xrightarrow{a.c.} X$ *as* $n \to \infty$.

Proof. There are several approaches to proving this result, including one based on Theorems 2.17 and 2.18. See Exercise 13. The approach we use here is used by Serfling (1980). Suppose that $X_n \xrightarrow{c} X$ as $n \to \infty$. This method of proof shows that the complement of the event in Equation (3.2) has limiting probability zero, which in turn proves that the event has limiting probability one, and the almost certain convergence of the sequence the results from Theorem 3.1. Note that the complement of $\{|X_m(\omega) - X(\omega)| < \varepsilon$ for all $m \geq n\}$ contains all $\omega \in \Omega$ where $|X_m(\omega) - X(\omega)| \geq \varepsilon$ for at least one $m \geq n$. That is,

$$\{|X_m(\omega) - X(\omega)| < \varepsilon \text{ for all } m \geq n\}^c =$$
$$\{|X_m(\omega) - X(\omega)| \geq \varepsilon \text{ for some } m \geq n\}.$$

Now let $\varepsilon > 0$, and note that

$$P(\omega : |X_m(\omega) - X(\omega)| > \varepsilon \text{ for some } m \geq n) =$$
$$P\left(\bigcup_{m=n}^{\infty} \{|X_m(\omega) - X(\omega)| > \varepsilon\} \right),$$

because $|X_m(\omega) - X(\omega)| > \varepsilon$ must be true for at least one $m \geq n$. Theorem 2.4 then implies that

$$P\left(\bigcup_{m=n}^{\infty} \{|X_m(\omega) - X(\omega)| > \varepsilon\} \right) \leq \sum_{m=n}^{\infty} P(\omega : |X_m(\omega) - X(\omega)| > \varepsilon).$$

Definition 3.3 implies that for every $\varepsilon > 0$

$$\sum_{n=1}^{\infty} P(\omega : |X_n(\omega) - X(\omega)| > \varepsilon) < \infty.$$

In order for this sum to converge, the limit

$$\lim_{n\to\infty} \sum_{m=n}^{\infty} P(\omega : |X_m(\omega) - X(\omega)| > \varepsilon) = 0,$$

must hold. Therefore, Definition 2.2 implies that for every $\varepsilon > 0$

$$\lim_{n\to\infty} P(\omega : |X_m(\omega) - X(\omega)| > \varepsilon \text{ for some } m \geq n) = 0.$$

Because the probabilities of an event and its complement always add to one,

the probability of the complement of the event given earlier must converge to
one. That is,

$$\lim_{n \to \infty} P(\omega : |X_m(\omega) - X(\omega)| \le \varepsilon \text{ for all } m \ge n) = 1.$$

Theorem 3.1 then implies that $X_n \xrightarrow{a.c.} X$ as $n \to \infty$. $\square$

The result in Theorem 3.3, coupled with Definition 3.3, essentially implies that
if a sequence $\{X_n\}_{n=1}^{\infty}$ converges in probability to X as $n \to \infty$ at a sufficiently
fast rate, then the sequence will also converge almost certainly to X. The fact
that complete convergence is not equivalent to almost certain convergence is
established by Example 3.5. The sequence of random variables is shown in that
example to converge almost certainly, but because $P(\omega : |X_n(\omega) - X(\omega)| > \varepsilon) = n^{-1}$ if $\varepsilon < 1$, the sequence does not converge completely.

There are some conditions under which almost certain convergence and complete convergence are equivalent.

Theorem 3.4. *Let $\{X_n\}_{n=1}^{\infty}$ be a sequence of independent random variables,
and let c be a real constant. If $X_n \xrightarrow{a.c.} c$ as $n \to \infty$ then $X_n \xrightarrow{c} c$ as $n \to \infty$.*

Proof. Suppose that $X_n \xrightarrow{a.c.} c$ as $n \to \infty$ where c is a real constant. Then,
for every $\varepsilon > 0$

$$\lim_{n \to \infty} P(|X_m - c| \le \varepsilon \text{ for all } m \ge n) = 1,$$

or equivalently

$$\lim_{n \to \infty} P(|X_m - c| > \varepsilon \text{ for at least one } m \ge n) = 0.$$

Note that

$$\lim_{n \to \infty} P(|X_m - c| > \varepsilon \text{ for at least one } m \ge n) =$$

$$\lim_{n \to \infty} P\left(\bigcup_{m=n}^{\infty} \{|X_m - c| > \varepsilon\} \right) = P\left(\bigcap_{n=1}^{\infty} \bigcup_{m=n}^{\infty} \{|X_m - c| > \varepsilon\} \right) =$$

$$P\left(\limsup_{n \to \infty} \{|X_n - c| > \varepsilon\} \right), \quad (3.7)$$

where the second equality follows from Theorem 2.16 and the fact that

$$\left\{ \bigcup_{m=n}^{\infty} \{|X_m - c| > \varepsilon\} \right\}_{n=1}^{\infty},$$

is a monotonically decreasing sequence of events. Now note that since $\{X_n\}_{n=1}^{\infty}$
is a sequence of independent random variables it follows that $\{|X_n - c| > \varepsilon\}_{n=1}^{\infty}$
is a sequence of independent random variables for every $\varepsilon > 0$. Therefore,
Corollary 2.1 implies that since

$$P\left(\limsup_{n \to \infty} \{|X_n - c| > \varepsilon\} \right) = 0,$$

for every $\varepsilon > 0$, then

$$\sum_{n=1}^{\infty} P(|X_n - c| > \varepsilon) < \infty,$$

for every $\varepsilon > 0$. Definition 3.3 then implies that $X_n \xrightarrow{c} c$ as $n \to \infty$. □

Note that Theorem 3.4 obtains an equivalence between almost certain convergence and convergence in probability. When a sequence converges in probability at a fast enough rate to a constant, convergence in probability and almost certain convergence are equivalent. Such a result was the main motivation of Hsu and Robbins (1947). Note further that convergence to a constant plays an important role in the proof and the application of Corollary 2.1. If $X_n \xrightarrow{c} X$ as $n \to \infty$, then the sequence is not independent and Corollary 2.1 cannot be applied to the sequence of events $\{|X_n - X| > \varepsilon\}_{n=1}^{\infty}$. But when $X_n \xrightarrow{c} c$ as $n \to \infty$ the sequence is independent and Corollary 2.1 can be applied.

As with convergent sequences of real numbers, subsequences of convergent sequences of random variables can play an important role in the development of asymptotic theory.

Theorem 3.5. *Let $\{X_n\}_{n=1}^{\infty}$ be a sequence of random variables that converge in probability to a random variable X. Then there exists a non-decreasing sequence of positive integers $\{n_k\}_{k=1}^{\infty}$ such that $X_{n_k} \xrightarrow{c} X$ and $X_{n_k} \xrightarrow{a.c.} X$ as $k \to \infty$.*

For a proof of Theorem 3.5, see Chapter 5 of Gut (2005).

Example 3.10. Let $\{U_n\}_{n=1}^{\infty}$ be a sequence of independent UNIFORM$(0, 1)$ random variables and define a sequence of random variables $\{X_n\}_{n=1}^{\infty}$ as $X_n = \delta\{U_n; (0, n^{-1})\}$ so that $X_n \xrightarrow{P} 0$ as $n \to \infty$. Let $\varepsilon > 0$ and note that when $\varepsilon < 1$ it follows that $P(|X_n - 0| > \varepsilon) = n^{-1}$ and therefore

$$\sum_{n=1}^{\infty} P(|X_n - 0| > \varepsilon) = \sum_{n=1}^{\infty} n^{-1} = \infty.$$

Hence, X_n does not converge completely to 0 as $n \to \infty$. Now define a non-decreasing sequence of positive integers $n_k = k^2$. In this case $P(|X_{n_k} - 0| < \varepsilon) = k^{-2}$ and therefore

$$\sum_{k=1}^{\infty} P(X_{n_k} - 0| > \varepsilon) = \sum_{k=1}^{\infty} k^{-2} < \infty.$$

Therefore $X_{n_k} \xrightarrow{c} 0$ as $k \to \infty$. Hence, we have found a subsequence that converges completely to zero. ■

It is important to note that Theorem 3.5 is not constructive in that it does not identify a particular subsequence that will converge completely to the random variable of interest. Theorem 3.5 is an existence result in that it merely guarantees the existence of at least one subsequence that converges completely. Such a result may not be seen as being useful at first, but there are several important applications of results such as these. It is sometimes possible to prove properties for the entire sequence by working with properties of the subsequence.

Example 3.11. This example, which is based on Theorem 3.5 of Gut (2005), highlights how convergent subsequences can be used to prove properties of the entire sequence. Let $\{X_n\}_{n=1}^{\infty}$ be a sequence of monotonically increasing random variables that converge in probability to a random variable X as $n \to \infty$. That is, $P(X_n < X_{n+1}) = 1$ for all $n \in \mathbb{N}$ and $X_n \overset{p}{\to} X$ as $n \to \infty$. Note that this also implies that $P(X_n < X) = 1$ for all $n \in \mathbb{N}$. See Exercise 14. This example demonstrates that the monotonicity of the sequence, along with the convergence in probability, is enough to conclude that X_n converges almost certainly to X as $n \to \infty$. Theorem 3.5 implies that there exists a sequence of monotonically increasing integers $\{n_k\}_{k=1}^{\infty}$ such that $X_{n_k} \overset{a.c.}{\to} X$ as $k \to \infty$. Theorem 3.1 then implies that for every $\varepsilon > 0$

$$\lim_{k \to \infty} P(|X_{n_{k'}} - X| \leq \varepsilon \text{ for all } k' \geq k) = 1.$$

We now appeal to the monotonicity of the sequence. Let $M_1 = \{\omega : X_n(\omega) \leq X_{n+1}(\omega) \text{ for all } n \in \mathbb{N}\}$, $M_2 = \{\omega : X_n(\omega) \leq X(\omega) \text{ for all } n \in \mathbb{N}\}$, and $M = M_1 \cup M_2$. By assumption $P(M_1) = P(M_2) = 1$. Theorems 2.4 and A.2 imply that $P(M^c) = P(M_1^c \cap M_2^c) \leq P(M_1^c) + P(M_2^c) = 0$ so that $P(M) = 1$. Suppose that $\omega \in M$ and

$$\omega \in \{|X_{n_{k'}}(\omega) - X(\omega)| \leq \varepsilon \text{ for all } k' \geq k\}.$$

The monotonicity of the sequence implies that $|X_{n_{k'}}(\omega) - X(\omega)| = X(\omega) - X_{n_{k'}}(\omega)$ so that $|X_{n_{k'}}(\omega) - X(\omega)| \leq \varepsilon$ implies that $X(\omega) - X_{n_{k'}}(\omega) < \varepsilon$. Note further that because the sequence is monotonically increasing then if $X(\omega) - X_{n_{k'}}(\omega) < \varepsilon$ it follows that $X(\omega) - X_n(\omega) < \varepsilon$ for all $n \geq n_k'$ and $\omega \in M$. In fact, the two events are equivalent so that it follows that

$$\lim_{n \to \infty} P(|X_n - X| \geq \varepsilon \text{ for all } m \geq n) = 1,$$

and therefore $X_n \overset{a.c.}{\to} X$ as $n \to \infty$. ∎

Further applications of Theorem 3.5 can be found in Simmons (1971). Another application is given in Exercise 19.

3.4 Convergence of Random Vectors

This section will investigate how the three modes of convergence studied in Sections 3.2 and 3.3 can be applied to random vectors. Let $\{\mathbf{X}_n\}_{n=1}^{\infty}$ be a sequence of d-dimensional random vectors and let $\mathbf{X}$ be another d-dimensional random vector. For an arbitrary d-dimensional vector $\mathbf{x}' = (x_1, \ldots, x_d) \in \mathbb{R}^d$

let $\|\mathbf{x}\|$ be the usual vector norm in d-dimensional Euclidean space defined by

$$\|\mathbf{x}\| = \left(\sum_{i=1}^{d} x_i^2 \right)^{1/2}.$$

When $d = 1$ the norm reduces to the absolute value of x, that is $\|x\| = |x|$. Therefore, we can generalize the one-dimensional requirement that $|X_n(\omega) - X(\omega)| > \varepsilon$ to $\|\mathbf{X}_n(\omega) - \mathbf{X}(\omega)\| > \varepsilon$ in the d-dimensional case.

Definition 3.4. *Let $\{\mathbf{X}_n\}_{n=1}^{\infty}$ be a sequence of d-dimensional random vectors and let $\mathbf{X}$ be another d-dimensional random vector.*

1. *The sequence $\{\mathbf{X}_n\}_{n=1}^{\infty}$ convergences in probability to $\mathbf{X}$ as $n \to \infty$ if for every $\varepsilon > 0$*

$$\lim_{n \to \infty} P(\|\mathbf{X}_n - \mathbf{X}\| \geq \varepsilon) = 0.$$

2. *The sequence $\{\mathbf{X}_n\}_{n=1}^{\infty}$ convergences almost certainly to $\mathbf{X}$ as $n \to \infty$ if*

$$P\left(\lim_{n \to \infty} \mathbf{X}_n = \mathbf{X} \right) = 1.$$

3. *The sequence $\{\mathbf{X}_n\}_{n=1}^{\infty}$ convergences completely to $\mathbf{X}$ as $n \to \infty$ if for every $\varepsilon > 0$*

$$\sum_{n=1}^{\infty} P(\|\mathbf{X}_n - \mathbf{X}\| \geq \varepsilon) < \infty.$$

While Definition 3.4 is intuitively appealing, it is not easy to apply in practice. The definition can be simplified by finding an equivalent condition on the convergence of the individual elements of the random vector. Recall from Defintion 2.5 that $\mathbf{X}$ is a random vector if it is a measureable function that maps the sample space Ω to $\mathcal{B}\{\mathbb{R}^d\}$. That is $\mathbf{X}^{-1}(B) = \{\omega \in \Omega : \mathbf{X}(\omega) \in B\} \in \mathcal{F}$, for all $B \in \mathcal{B}\{\mathbb{R}^d\}$. Note that $\mathbf{X}' = (X_1, \ldots, X_d)$ where X_i is a measureable function that maps Ω to $\mathbb{R}$ for $i = 1, \ldots, d$. That is $X_i^{-1}(B) = \{\omega \in \Omega : X(\omega) \in B\} \in \mathcal{F}$ for all $B \in \mathcal{B}\{\mathbb{R}\}$. Therefore a d-dimensional random vector is made up of d one-dimensional random variables. Similarly, let $\mathbf{X}_n' = (X_{1n}, \ldots, X_{dn})$ have the same measureability properties for each of the components for all $n \in \mathbb{N}$.

Suppose $\varepsilon > 0$, and consider the inequality $\|\mathbf{X}_n(\omega) - \mathbf{X}(\omega)\| < \varepsilon$ for a specified $\omega \in \Omega$. The inequality implies that $|X_{in}(\omega) - X_i(\omega)| < \varepsilon$ for $i = 1, \ldots, d$. Therefore,

$$\{\omega : \|\mathbf{X}_n(\omega) - \mathbf{X}(\omega)\| < \varepsilon\} \subset \bigcap_{i=1}^{d} \{\omega : |X_{in}(\omega) - X_i(\omega)| < \varepsilon\}. \qquad (3.8)$$

This turns out to be the essential relationship required to establish that the convergence of a random vector is equivalent to the convergence of the individual elements of the random vector.

Theorem 3.6. *Let $\{\mathbf{X}_n\}_{n=1}^{\infty}$ be a sequence of d-dimensional random vectors and let $\mathbf{X}$ be another d-dimensional random vector where $\mathbf{X}' = (X_1, \ldots, X_d)$ and $\mathbf{X}'_n = (X_{1,n} \ldots, X_{d,n})$ for all $n \in \mathbb{N}$.*

1. $\mathbf{X}_n \xrightarrow{P} \mathbf{X}$ *as $n \to \infty$ if and only if $X_{k,n} \xrightarrow{P} X_k$ as $n \to \infty$ for all $k \in \{1, \ldots, d\}$.*

2. $\mathbf{X}_n \xrightarrow{a.c.} \mathbf{X}$ *as $n \to \infty$ if and only if $X_{k,n} \xrightarrow{a.c.} X_k$ as $n \to \infty$ for all $k \in \{1, \ldots, d\}$.*

Proof. We will prove this result for convergence in probability. The remaining result is proven in Exercise 15. Suppose that $\mathbf{X}_n \xrightarrow{P} \mathbf{X}$ as $n \to \infty$. Then, from Definition 3.4 is follows that for every $\varepsilon > 0$

$$\lim_{n \to \infty} P(\|\mathbf{X}_n - \mathbf{X}\| \leq \varepsilon) = 1,$$

which in turn implies that

$$\lim_{n \to \infty} P\left(\bigcap_{i=1}^{d} \{\omega : |X_{i,n}(\omega) - X_i(\omega)| \leq \varepsilon\} \right) = 1,$$

where we have used the relationship in Equation (3.8). Now let $k \in \{1, \ldots, d\}$. Theorem 2.3 implies that since

$$\bigcap_{i=1}^{d} \{\omega : |X_{i,n}(\omega) - X_i(\omega)| \leq \varepsilon\} \subset \{\omega : |X_{k,n}(\omega) - X_k(\omega)| \leq \varepsilon\},$$

for all $n \in \mathbb{N}$ it follows that,

$$P\left(\bigcap_{i=1}^{d} \{\omega : |X_{i,n}(\omega) - X_i(\omega)| \leq \varepsilon\} \right) \leq P(\omega : |X_{k,n}(\omega) - X_k(\omega)| \leq \varepsilon),$$

for all $n \in \mathbb{N}$. Therefore,

$$\lim_{n \to \infty} P\left(\bigcap_{i=1}^{d} \{\omega : |X_{i,n}(\omega) - X_i(\omega)| \leq \varepsilon\} \right)$$
$$\leq \lim_{n \to \infty} P(\omega : |X_{k,n}(\omega) - X_k(\omega)| \leq \varepsilon).$$

Definition 2.2 then implies that

$$\lim_{n \to \infty} P(\omega : |X_{k,n}(\omega) - X_k(\omega)| \leq \varepsilon) = 1,$$

and Defintion 3.1 implies that $X_{k,n} \xrightarrow{P} X_k$ as $n \to \infty$. Because k is arbitrary in the above arguments, we have proven that $\mathbf{X}_n \xrightarrow{P} \mathbf{X}$ as $n \to \infty$ implies that $X_{k,n} \xrightarrow{P} X_k$ as $n \to \infty$ for all $k \in \{1, \ldots, d\}$, or that the convergence of the random vector in probability implies the convergence in probability of the elements of the random vector.

Now suppose that $X_{k,n} \xrightarrow{p} X_k$ as $n \to \infty$ for all $k \in \{1, \ldots, d\}$, and let $\varepsilon > 0$ be given. Then Definition 3.1 implies that

$$\lim_{n \to \infty} P(\omega : |X_{k,n}(\omega) - X_k(\omega)| > d^{-1}\varepsilon) = 0,$$

for $k \in \{1, \ldots, d\}$. Theorem 2.4 implies that

$$P\left(\bigcup_{i=1}^{d} \{\omega : |X_{i,n}(\omega) - X_i(\omega)| > d^{-1}\varepsilon\}\right) \leq$$

$$\sum_{i=1}^{d} P(\omega : |X_{i,n}(\omega) - X_i(\omega)| > d^{-1}\varepsilon),$$

for each $n \in \mathbb{N}$. Therefore,

$$\lim_{n \to \infty} P\left(\bigcup_{i=1}^{d} \{\omega : |X_{i,n}(\omega) - X_i(\omega)| > d^{-1}\varepsilon\}\right) \leq$$

$$\lim_{n \to \infty} \sum_{i=1}^{d} P(\omega : |X_{i,n}(\omega) - X_i(\omega)| > d^{-1}\varepsilon) = 0.$$

Now,

$$\bigcap_{i=1}^{d} \{\omega : |X_{i,n}(\omega) - X_i(\omega)| \leq d^{-1}\varepsilon\} \subset \{\omega : |X_{i,n}(\omega) - X_i(\omega)| \leq \varepsilon\},$$

so that

$$P\left(\bigcap_{i=1}^{d} \{\omega : |X_{i,n}(\omega) - X_i(\omega)| \leq d^{-1}\varepsilon\}\right) \leq P(\omega : |X_{i,n}(\omega) - X_i(\omega)| \leq \varepsilon).$$

Hence, Theorem A.2 then implies that

$$\lim_{n \to \infty} P(\omega : \|\mathbf{X}_n(\omega) - \mathbf{X}(\omega)\| \leq \varepsilon) \geq$$

$$\lim_{n \to \infty} P\left(\bigcap_{i=1}^{d} \{\omega : |X_{i,n}(\omega) - X_i(\omega)| \leq \varepsilon\}\right) = 1.$$

Therefore Definition 3.4 implies that $\mathbf{X}_n \xrightarrow{p} \mathbf{X}$ as $n \to \infty$, and the result is proven. $\qquad\square$

Example 3.12. As a simple example of these results consider a set of independent and identically distributed random variables $X_1, \ldots, X_n$ having distribution F with mean μ and another set of independent and identically distributed random variables $Y_1, \ldots, Y_n$ having distribution G with mean ν. Let $\bar{X}_n$ and $\bar{Y}_n$ be the sample means of the two samples, respectively. The results from Example 3.2 can be used to show that $\bar{X}_n \xrightarrow{p} \mu$ and $\bar{Y}_n \xrightarrow{p} \nu$ as $n \to \infty$ as long as F and G both have finite variances. Now define $\mathbf{Z}_i' = (X_i, Y_i)$

for $i = 1, \ldots, n$ and

$$\bar{\mathbf{Z}}_n = n^{-1} \sum_{i=1}^{n} \mathbf{Z}_i = \begin{bmatrix} \bar{X}_n \\ \bar{Y}_n \end{bmatrix}.$$

Theorem 3.6 implies that $\bar{\mathbf{Z}}_n \xrightarrow{p} \boldsymbol{\theta}$ as $n \to \infty$ where $\boldsymbol{\theta}' = (\mu, \nu)$. Therefore, the result of Example 3.2 has been extended to the bivariate case. Note that we have not assumed anything specific about the joint distribution of X_i and Y_i, other than that the covariance must be finite which follows from the assumption that the variances of the marginal distributions are finite. ∎

Example 3.13. Let $\{U_n\}_{n=1}^{\infty}$ be a sequence of independent UNIFORM$[0,1]$ random variables and define a sequence of random vectors $\{\mathbf{X}_n\}_{n=1}^{\infty}$ such that $\mathbf{X}_n' = (X_{1,n}, X_{2,n}, X_{3,n})$ for all $n \in \mathbb{N}$, where $X_{1,n} = \delta\{U_n; (0, n^{-1})\}$, $X_{2,n} = \frac{1}{2}\delta\{U_n; (0, 1 - n^{-1})\}$, and $X_{3,n} = \delta\{U_n; (0, 1 - n^{-2})\}$. From Definition 3.2 it follows that $X_{1,n} \xrightarrow{a.c.} 0$, $X_{2,n} \xrightarrow{a.c.} \frac{1}{2}$, and $X_{3,n} \xrightarrow{a.c.} 1$ as $n \to \infty$. Therefore, Theorem 3.6 implies that $\mathbf{X}_n \xrightarrow{a.c.} (0, \frac{1}{2}, 1)'$ as $n \to \infty$. ∎

Example 3.14. Let $\{X_n\}_{n=1}^{\infty}$, $\{Y_n\}_{n=1}^{\infty}$, and $\{Z_n\}_{n=1}^{\infty}$ be sequences of random variables that converge in probability to the random variables X, Y, and Z, respectively as $n \to \infty$. Suppose that X has a $\mathrm{N}(\theta_x, \sigma_x^2)$ distribution, Y has a $\mathrm{N}(\theta_y, \sigma_y^2)$ distribution, and Z has a $\mathrm{N}(\theta_z, \sigma_z^2)$ distribution, where X, Y and Z are independent. Define a sequence of random vectors $\{\mathbf{W}\}_{n=1}^{\infty}$ as $\mathbf{W}_n = (X_n, Y_n, Z_n)'$ for all $n \in \mathbb{N}$ and let $\mathbf{W} = (X, Y, Z)$. Then Theorem 3.6 implies that $\mathbf{W}_n \xrightarrow{p} \mathbf{W}$ as $n \to \infty$ where the independence of the components of $\mathbf{W}$ imply that $\mathbf{W}$ has a $\mathrm{N}(\boldsymbol{\mu}, \boldsymbol{\Sigma})$ distribution where $\boldsymbol{\mu} = (\theta_x, \theta_y, \theta_z)'$ and

$$\boldsymbol{\Sigma} = \begin{bmatrix} \sigma_x^2 & 0 & 0 \\ 0 & \sigma_y^2 & 0 \\ 0 & 0 & \sigma_z^2 \end{bmatrix}.$$

∎

3.5 Continuous Mapping Theorems

When considering a sequence of constants, a natural result that arises is that if a sequence converges to a point $c \in \mathbb{R}$, and we apply a continuous real function g to the sequence, then it follows that the transformed sequence converges to $g(c)$. This is the result given in Theorem 1.3. In some sense, this is the essential property of continuous functions. Extension of this result to sequences of random variables can prove to be very useful. For example, suppose that we are able to prove that under certain conditions the sample variance is a consistent estimator of the population variance as the sample size increases to ∞. That is, the sample variance converges in probability to the population variance. Can we automatically conclude that the square root of the sequence converges in probability to the square root of the limit? That is, does it follows that the sample standard deviation converges in probability to the population standard deviation, or that, the sample standard deviation

is a consistent estimator of the population standard deviation? As we show
in this section, such operations are usually permissible under the additional
assumption that the transforming function g is continuous with probability
one with respect to the distribution of the limiting random variable. We begin
by considering the simple case where a sequence of random variables converges
to a real constant. In this case the transformation need only be continuous at
the constant.

Theorem 3.7. *Let $\{X_n\}_{n=1}^{\infty}$ be a sequence of random variables, c be a real
constant, and g be a Borel function that is continuous as c.*

1. *If $X_n \xrightarrow{a.c.} c$ as $n \to \infty$, then $g(X_n) \xrightarrow{a.c.} g(c)$ as $n \to \infty$.*

2. *If $X_n \xrightarrow{p} c$ as $n \to \infty$, then $g(X_n) \xrightarrow{p} g(c)$ as $n \to \infty$.*

Proof. We will prove the second result of the theorem, leaving the proof of the
first part as Exercise 18. Suppose that $X_n \xrightarrow{p} c$ as $n \to \infty$, so that Definition
3.1 implies that for every $\delta > 0$

$$\lim_{n \to \infty} P(|X_n - c| < \delta) = 1.$$

Because g is a continuous function at the point c, it follows that for every
$\varepsilon > 0$ there exists a real number δ_ε such that $|x - c| < \delta_\varepsilon$ implies that
$|g(x) - g(c)| < \varepsilon$, where x is a real number. This in turn implies that for each
$n \in \mathbb{N}$ that

$$\{\omega \in \Omega : |X_n(\omega) - c| < \delta_\varepsilon\} \subset \{\omega \in \Omega : |g[X_n(\omega)] - g(c)| < \delta_\varepsilon\},$$

and therefore Theorem 2.3 implies that for each $n \in \mathbb{N}$

$$P(|X_n - c| < \delta_\varepsilon) \leq P(|g(X_n) - g(c)| < \varepsilon).$$

Therefore, for every $\varepsilon > 0$,

$$\lim_{n \to \infty} P(|g(X_n) - g(c)| < \varepsilon) \geq \lim_{n \to \infty} P(|X_n - c| < \delta_\varepsilon) = 1,$$

and therefore $g(X_n) \xrightarrow{p} g(c)$ as $n \to \infty$. $\square$

Example 3.15. Let $\{X_n\}_{n=1}^{\infty}$ be a sequence of independent random variables
where X_n has a POISSON(θ) distribution. Let $\bar{X}_n$ be the sample mean computed on $X_1, \ldots, X_n$. Example 3.2 implies that $\bar{X}_n \xrightarrow{p} \theta$ as $n \to \infty$. If we
wish to find a consistent estimator of the standard deviation of X_n which is
$\theta^{1/2}$ we can consider $\bar{X}_n^{1/2}$. Theorem 3.7 implies that since the square root
transformation is continuous at θ if $\theta > 0$ that $\bar{X}_n^{1/2} \xrightarrow{p} \theta^{1/2}$ as $n \to \infty$.   ∎

Example 3.16. Let $\{X_n\}_{n=1}^{\infty}$ be a sequence of independent random variables
where X_n has a N$(0, 1)$ distribution. Let $\bar{X}_n$ be the sample mean computed
on $X_1, \ldots, X_n$. Example 3.2 implies that $\bar{X}_n \xrightarrow{p} 0$ as $n \to \infty$. Consider the
function $g(x) = \delta\{x; \{0\}\}$ and note that $P[g(\bar{X}_n) = 0] = P(\bar{X}_n \neq 0) = 1$ for
all $n \in \mathbb{N}$. Therefore, it is clear that $g(\bar{X}_n) \xrightarrow{p} 0$ as $n \to \infty$. However, note
that $g(0) = 1$ so that in this case $g(\bar{X}_n)$ does not converge in probability to
$g(0)$. This, of course, is due to the discontinuity at 0 in the function g.   ∎

Theorem 3.7 can be extended to the case where $\bar{X}_n$ converges to a random variable X, instead of a real constant using essentially the same argument of proof, if we are willing to assume that g is uniformly continuous. This is due to the fact that when g is uniformly continuous we have that for every $\varepsilon > 0$ that there exists a real number $\delta_\varepsilon > 0$ such that $|x - y| < \delta_\varepsilon$ implies $|g(x) - g(y)| < \varepsilon$ no matter what value of x and y are considered. This implies that

$$P(|X_n - X| \le \delta_\varepsilon) \le P(|g(X_n) - g(X)| \le \varepsilon),$$

for all $n \in \mathbb{N}$, and the corresponding results follow. However, the assumption of uniform continuity turns out to be needlessly strong, and in fact can be weakened to the assumption that the transformation is continuous with probability one with respect to the distribution of the limiting random variable.

Theorem 3.8. *Let $\{X_n\}_{n=1}^\infty$ be a sequence of random variables, X be a random variable, and g be a Borel function on $\mathbb{R}$. Let $C(g)$ be the set of continuity points of g and suppose that $P[X \in C(g)] = 1$.*

1. *If $X_n \xrightarrow{a.c.} X$ as $n \to \infty$, then $g(X_n) \xrightarrow{a.c.} g(X)$ as $n \to \infty$.*

2. *If $X_n \xrightarrow{p} X$ as $n \to \infty$, then $g(X_n) \xrightarrow{p} g(X)$ as $n \to \infty$.*

Proof. We will prove the first result in this case. See Exercise 19 for proof of the second result. Definition 3.2 implies that if $X_n \xrightarrow{a.c.} X$ as $n \to \infty$ then

$$P\left[\omega : \lim_{n \to \infty} X_n(\omega) = X(\omega) \right] = 1.$$

Let

$$N = \left\{ \omega \in \Omega : \lim_{n \to \infty} X_n(\omega) = X(\omega) \right\},$$

and note that by assumption $P(N) = P[C(g)] = 1$. Consider $\omega \in N \cap C(g)$. For such ω is follows from Theorem 1.3 that

$$\lim_{n \to \infty} g[X_n(\omega)] = g[X(\omega)].$$

Noting that $P[N \cap C(g)] = 1$ yields the result.   □

Example 3.17. Let $\{X_n\}_{n=1}^\infty$ be a sequence of random variables where $X_n = Z + n^{-1}$ for all $n \in \mathbb{N}$, and Z is a $N(0, 1)$ random variable. As was shown in Example 3.1, we have that $X_n \xrightarrow{p} Z$ as $n \to \infty$. The transformation $g(x) = x^2$ is continuous with probability one with respect to the normal distribution and hence it follows from Theorem 3.8 that $X_n^2 \xrightarrow{p} Y = Z^2$ where Y is a random variable with a CHISQUARED(1) distribution.   ∎

The extension of Theorem 3.8 to the case of multivariate transformations of random vectors is almost transparent. The arguments simply consist of applying Theorem 3.8 element-wise to the random vectors and appealing to Theorem 3.6 to obtain the convergence of the corresponding random vectors. Note that we also are using the fact that a Borel function of a random vector is also a random vector from Theorem 2.2.

Theorem 3.9. *Let $\{\mathbf{X}_n\}$ be a sequence of d-dimensional random vectors, $\mathbf{X}$ be a d-dimensional random vector, and $g : \mathbb{R}^d \to \mathbb{R}^q$ be a Borel function. Let $C(g)$ be the set of continuity points of g and suppose that $P[\mathbf{X} \in C(g)] = 1$.*

1. *If $\mathbf{X}_n \xrightarrow{a.c.} \mathbf{X}$ as $n \to \infty$, then $g(\mathbf{X}_n) \xrightarrow{a.c.} g(\mathbf{X})$ as $n \to \infty$.*

2. *If $\mathbf{X}_n \xrightarrow{p} \mathbf{X}$ as $n \to \infty$, then $g(\mathbf{X}_n) \xrightarrow{p} g(\mathbf{X})$ as $n \to \infty$.*

Theorem 3.9 is proven in Exercise 20.

Example 3.18. Suppose that $X_1, \ldots, X_n$ be a sample from $N(\mu, \sigma)$ distribution. Example 3.2 shows that $\bar{X}_n$ is a consistent estimator of μ and Exercise 7 shows that the sample variance S_n^2 is a consistent estimator of σ^2. Define a parameter vector $\boldsymbol{\theta} = (\mu, \sigma^2)'$ along with a sequence of random vectors given by $\hat{\boldsymbol{\theta}}_n = (\bar{X}_n, S_n^2)'$ for all $n \in \mathbb{N}$, as a sequence of estimators of $\boldsymbol{\theta}$. Theorem 3.6 implies that $\hat{\boldsymbol{\theta}}_n \xrightarrow{p} \boldsymbol{\theta}$ as $n \to \infty$. The αth quantile of a $N(\mu, \sigma)$ distribution is given by $\mu + \sigma z_\alpha$, where z_α is the αth quantile of a $N(0, 1)$ distribution. This quantile can be estimated from the sample with $\bar{X}_n + S_n z_\alpha$, which is a continuous transformation of the sequence of random vectors $\{\hat{\boldsymbol{\theta}}_n\}_{n=1}^\infty$. Therefore, Theorem 3.9 implies that $\bar{X}_n + S_n z_\alpha \xrightarrow{p} \mu + \sigma z_\alpha$ as $n \to \infty$, or that the estimator is consistent. ∎

Example 3.19. Consider the result of Example 3.14 where a sequence of random vectors $\{\mathbf{W}_n\}_{n=1}^\infty$ was created that converged in probability to a random vector $\mathbf{W}$ that has a $N(\boldsymbol{\mu}, \boldsymbol{\Sigma})$ distribution where $\boldsymbol{\mu} = (\theta_x, \theta_y, \theta_z)'$ and

$$\boldsymbol{\Sigma} = \begin{bmatrix} \sigma_x^2 & 0 & 0 \\ 0 & \sigma_y^2 & 0 \\ 0 & 0 & \sigma_z^2 \end{bmatrix}.$$

Let $\mathbf{1}$ be a 3×1 vector where each element is equal to 1. Then Theorem 3.9 implies that $\mathbf{1}'\mathbf{W}_n \xrightarrow{p} \mathbf{1}'\mathbf{W}$ where $\mathbf{1}'\mathbf{W}$ has a $N(\theta_x + \theta_y + \theta_z, \sigma_x^2 + \sigma_y^2 + \sigma_z^2)$ distribution. In the case where $\boldsymbol{\mu} = \mathbf{0}$ and $\boldsymbol{\Sigma} = \mathbf{I}$, then Theorem 3.9 also implies that $\mathbf{W}_n'\mathbf{W}_n$ converges to the random variable $\mathbf{W}'\mathbf{W}$ which has a CHISQUARED(3) distribution.

3.6 Laws of Large Numbers

Example 3.2 discussed some general conditions under which an estimator $\hat{\theta}_n$ of a parameter θ converges in probability to θ as $n \to \infty$. In the special case where the estimator is the sample mean calculated from a sequence of independent and identically distributed random variables, Example 3.2 states that the sample mean will converge in probability to the population mean as long as the variance of the population is finite. This result is often called the *Weak Law of Large Numbers*. The purpose of this section is to explore other versions of this result. In particular we will consider alternate sets of conditions under which the result remains the same. We will also consider under what conditions the result can be strengthened to the *Strong Law of*

Large Numbers, for which the sample mean converges almost certainly to the population mean. The first result given in Theorem 3.10 below shows that the assumption that the variance of the population is finite can be removed as long as the mean of the population exists and is finite.

Theorem 3.10 (Weak Law of Large Numbers). *Let $X_1, \ldots, X_n$ be a set of independent and identically distributed random variables from a distribution F with finite mean θ and let $\bar{X}_n$ be the sample mean computed on the random variables. Then $\bar{X}_n \xrightarrow{p} \theta$ as $n \to \infty$.*

Proof. Because there is no assumption that the variance of X_1 is finite, Theorem 2.7 cannot be used directly to establish the result. Instead, we will apply Theorem 2.7 to a version of X_1 whose values are truncated at a finite value. This will insure that the variance of the truncated random variables is finite. The remainder of the proof is then concerned with showing that the results obtained for the truncated random variables can be translated into equivalent results for the original random variables.

This proof is based on one used in Section 6.3 of Gut (2005) with some modifications. A simpler proof, based on characteristic functions, will be given in Chapter 3, where some additional results will provide a simpler argument. However, the proof given here is worth considering as several important concepts will be used that will also prove useful later.

For simplicity we begin by assuming that $\theta = 0$. Define a truncated version of X_k as $Y_k = X_k \delta\{|X_k|; [0, n\varepsilon^3]\}$ for all $k \in \mathbb{N}$ where $\varepsilon > 0$ is arbitrary. That is, Y_k will be equal to X_k when $|X_k| \leq n\varepsilon^3$, but will be zero otherwise. Define the partial sums

$$S_n = n^{-1} \sum_{k=1}^{n} X_k,$$

and

$$T_n = n^{-1} \sum_{k=1}^{n} Y_k.$$

We will now consider the asymptotic relationship between S_n and $E(T_n)$. Definition 2.2 implies that

$$P(|S_n - E(T_n)| > n\varepsilon) =$$
$$P(\{|S_n - E(T_n)| > n\varepsilon\} \cap A) + P(\{|S_n - E(T_n)| > n\varepsilon\} \cap A^c), \quad (3.9)$$

where

$$A = \bigcap_{k=1}^{n} \{|X_k| \leq n\varepsilon^3\}.$$

Considering the first term in Equation (3.9) we note that if the event A is true, then $S_n = T_n$ because none of the random variables are truncated. That is, the events $\{|S_n - E(T_n)| > n\varepsilon^3\} \cap A$ and $\{|T_n - E(T_n)| > n\varepsilon^3\} \cap A$ are the same. Therefore

$$P(\{|S_n - E(T_n)| > n\varepsilon\} \cap A) = P(\{|T_n - E(T_n)| > n\varepsilon\} \cap A)$$
$$\leq P(|T_n - E(T_n)| > n\varepsilon),$$

where the inequality follows from Theorem 2.3. Regardless of the distribution of S_n, the variance of T_n must be finite due to that fact that the support of Y_k is finite for all $k \in \{1, \ldots, n\}$. Therefore, Theorem 2.7 can be applied to T_n to obtain

$$P(|T_n - E(T_n)| > n\varepsilon) \leq \frac{V(T_n)}{n^2\varepsilon^2}.$$

Now

$$V(T_n) = V\left(\sum_{k=1}^{n} Y_k\right) = nV(Y_1).$$

Therefore, Theorem 2.5 implies

$$P(|T_n - E(T_n)| > n\varepsilon) \leq (n\varepsilon^2)^{-1}V(Y_1) \leq n^{-1}\varepsilon^{-2}E(Y_1^2).$$

Now, recall that $Y_1 = X_1\delta\{|X_1|; [0, n\varepsilon^3]\}$ so that

$$E(Y_1^2) = E(X_1^2\delta\{|X_1|; [0, n\varepsilon^3]\})$$
$$\leq E[(n\varepsilon^3)|X_1|\delta\{|X_1|; [0, n\varepsilon^3]\}]$$
$$= n\varepsilon^3 E(|X_1|\delta\{|X_1|; [0, n\varepsilon^3]\})$$
$$\leq n\varepsilon^3 E(|X_1|),$$

where we have used Theorem A.7 and the fact that

$$X_1^2(\omega)\delta\{|X_1|; [0, n\varepsilon^3]\} \leq n\varepsilon^3|X_1(\omega)|\delta\{|X_1|; [0, n\varepsilon^3]\},$$

and

$$|X_1(\omega)|\delta\{|X_1|; [0, n\varepsilon^3]\} \leq |X_1(\omega)|,$$

for all $\omega \in \Omega$. Therefore

$$P(|T_n - E(T_n)| > n\varepsilon) \leq n^{-1}\varepsilon^{-2}(n\varepsilon^3)E(|X_1|) = \varepsilon E(|X_1|). \qquad (3.10)$$

To find a bound on the second term in Equation (3.9) first note that

$$A^c = \left(\bigcap_{k=1}^{n}\{|X_k| \leq n\varepsilon^3\}\right)^c = \bigcup_{k=1}^{n}\{|X_k| \leq n\varepsilon^3\}^c = \bigcup_{k=1}^{n}\{|X_k| > n\varepsilon^3\},$$

which follows from Theorem A.2. Therefore, Theorems 2.3 and 2.4 and the fact that the random variables are identically distributed imply

$$P(\{|S_n - E(T_n)| > n\varepsilon^3\} \cap A^c) \leq P\left(\bigcup_{k=1}^{n}\{|X_k| > n\varepsilon^3\}\right)$$
$$\leq \sum_{k=1}^{n} P(|X_k| > n\varepsilon^3)$$
$$= nP(|X_1| > n\varepsilon^3). \qquad (3.11)$$

Combining the results of Equations (3.9)–(3.11) implies that

$$P(|S_n - E(T_n)| > n\varepsilon) \leq \varepsilon E(|X_1|) + nP(|X_1| > n\varepsilon^3).$$

Let G be the distribution of $|X_1|$, then note that Theorem A.7 implies that

$$nP(|X_1| > n\varepsilon^3) = n \int_{n\varepsilon^3}^{\infty} dG(t) = \varepsilon^{-3} \int_{n\varepsilon^3}^{\infty} n\varepsilon^3 dG(t) \leq \varepsilon^{-3} \int_{n\varepsilon^3}^{\infty} t \, dG(t).$$

Since we have assumed that $E(|X_1|) < \infty$, it follows that

$$\lim_{n \to \infty} \int_{n\varepsilon^3}^{\infty} t \, dG(t) = 0.$$

Therefore, Theorem 1.6 implies that

$$\limsup_{n \to \infty} P(|S_n - E(T_n)| > n\varepsilon) \leq \varepsilon E(|X_1|).$$

We use the limit supremum in the limit instead of the usual limit since we do not yet know whether the sequence converges or not. Equivalently, we have shown that

$$\limsup_{n \to \infty} P(|n^{-1}S_n - n^{-1}E(T_n)| > \varepsilon) \leq \varepsilon E(|X_1|).$$

Because $\varepsilon > 0$ is arbitrary it follows that we have shown that $n^{-1}S_n - n^{-1}E(T_n) \xrightarrow{P} 0$ as $n \to \infty$. See Exercise 22. It is tempting to want to conclude that $n^{-1}S_n \xrightarrow{P} n^{-1}E(T_n)$ as $n \to \infty$, but this is not permissible since the limit value in such a statement depends on n. To finish the proof we note that Theorem 2.12 implies that

$$
\begin{aligned}
|E(T_n)| &= \left| E\left(\sum_{k=1}^{n} Y_k \right) \right| \\
&= \left| E\left(\sum_{k=1}^{n} X_k \delta\{|X_k|; [0, n\varepsilon^3]\} \right) \right| \\
&\leq nE(|X_1|\delta\{|X_1|; (n\varepsilon^3, \infty)\}).
\end{aligned}
$$

Therefore

$$\lim_{n \to \infty} n^{-1}|E(T_n)| \leq \lim_{n \to \infty} E(|X_1|\delta\{|X_1|; (n\varepsilon^3, \infty)\}) = 0,$$

since for every $\omega \in \Omega$

$$\lim_{n \to \infty} |X_1(\omega)|\delta\{|X_1(\omega)|; (n\varepsilon^3, \infty)\} = 0.$$

Theorem 3.8 then implies that $n^{-1}S_n \xrightarrow{P} 0$ as $n \to \infty$, and the result is proven for $\theta = 0$. If $\theta \neq 0$ one can simply use this same proof for the transformed random variables $X_k - \theta$ to make the conclusion that $n^{-1}S_n - \theta \xrightarrow{P} 0$ as $n \to \infty$, or equivalently that $n^{-1}S_n \xrightarrow{P} \theta$ as $n \to \infty$. ☐

Example 3.20. Suppose that $\{X_n\}_{n=1}^{\infty}$ is a sequence of independent random

variables where X_n has a T(2) distribution. The variance of X_n does not exist, but Theorem 3.10 still applies to this case and we can still therefore conclude that $\bar{X}_n \xrightarrow{p} 0$ as $n \to \infty$.   ∎

The *Strong Law of Large Numbers* keeps the same essential result as Theorem 3.10 except that the mode of convergence is strengthened from convergence in probability to almost certain convergence. The path to this stronger result requires slightly more complicated mathematics, and we will therefore develop some intermediate results before presenting the final result and its proof. The general approach used here is the development used by Feller (1971). Somewhat different approaches to this result can be found in Gut (2005), Gnedenko (1962), and Sen and Singer (1993), though the basic ideas are essentially the same.

Theorem 3.11. *Let* $\{X_n\}_{n=1}^{\infty}$ *be a sequence of independent random variables where* $E(X_n) = 0$ *for all* $n \in \mathbb{N}$ *and*

$$\sum_{n=1}^{\infty} E(X_n^2) < \infty.$$

Then S_n *converges almost certainly to a limit* S.

Proof. Let $\varepsilon > 0$, and consider the probability

$$P\left(\max_{n \leq k \leq m} |S_k - S_n| > \varepsilon\right) = P\left(\max_{n < k \leq m} \left|\sum_{j=n+1}^{k} X_j\right| > \varepsilon\right).$$

Theorem 2.13 implies that

$$P\left(\max_{n < k \leq m} \left|\sum_{j=n+1}^{k} X_j\right| > \varepsilon\right) \leq \varepsilon^{-2} \sum_{k=n+1}^{m} V(X_k)$$

$$\leq \varepsilon^{-2} \sum_{k=n+1}^{\infty} V(X_k), \quad (3.12)$$

where the second inequality is due to the fact that the terms of the sequence are non-negative. Now, the right hand side of Equation (3.12) does not depend on m, so that we can take the limit of the left hand side as $m \to \infty$. Theorem 2.16 the implies that

$$P\left(\sup_{k \geq n} |S_k - S_n| > \varepsilon\right) = P\left(\bigcup_{k=n}^{\infty} \{|S_k - S_n| > \varepsilon\}\right) \leq \varepsilon^{-2} \sum_{k=n+1}^{\infty} V(X_k).$$

Now take the limit as $n \to \infty$ to obtain

$$\lim_{n \to \infty} P\left(\sup_{k \geq n} |S_k - S_n| > \varepsilon\right) \leq \lim_{n \to \infty} \varepsilon^{-2} \sum_{k=n+1}^{\infty} V(X_k) = 0, \quad (3.13)$$

where the limit on the right hand side of Equation (3.13) follows from the assumption that

$$\sum_{n=1}^{\infty} E(X_n^2) < \infty.$$

It follows from Equation (3.12) that the sequence $\{S_n\}_{n=1}^{\infty}$ is a Cauchy sequence with probability one, or by Theorem 1.1, $\{S_n\}_{n=1}^{\infty}$ has a limit with probability one. Therefore $S_n \xrightarrow{a.c.} S$ as $n \to \infty$ for some S. $\qquad\square$

Theorem 3.11 actually completes much of the work we need to prove the Strong Law of Large Numbers in that we now know that the sum converges almost certainly to a limit. However, the assumption on the variance of X_n is quite strong and we will need to find a way to weaken this assumption. The method will be the same as used in the proof of Theorem 3.10 in that we will use truncated random variables. In order to apply the result in Theorem 3.11 to these truncated random variables a slight generalization of the result is required.

Corollary 3.1. *Let $\{X_n\}_{n=1}^{\infty}$ be a sequence of independent random variables where $E(X_n) = 0$ for all $n \in \mathbb{N}$. Let $\{b_n\}_{n=1}^{\infty}$ be a monotonically increasing sequence of real numbers such that $b_n \to \infty$ as $n \to \infty$. If*

$$\sum_{n=1}^{\infty} b_n^{-2} E(X_n^2) < \infty,$$

then

$$\sum_{n=1}^{\infty} b_n^{-1} X_n,$$

converges almost certainly to some limit and $b_n^{-1} S_n \xrightarrow{a.c.} 0$ as $n \to \infty$.

Proof. The first result is obtained directly from Theorem 3.11 using $\{b_n X_n\}_{n=1}^{\infty}$ as the sequence of random variables of interest. In that case we require

$$\sum_{n=1}^{\infty} E[(b_n^{-1} X_n)^2] = \sum_{n=1}^{\infty} b_n^{-2} E(X_n^2) < \infty,$$

which is the condition that is assumed. To prove the second result see Exercise 23. $\qquad\square$

The final result required to prove the Strong Law of Large Numbers is a condition on the existence of the mean of a random variable.

Theorem 3.12. *Let $\{X_n\}_{n=1}^{\infty}$ be a sequence of random variables where X_n has distribution function F for all $n \in \mathbb{N}$. Then for any $\varepsilon > 0$*

$$\sum_{n=1}^{\infty} P(|X_n| > n\varepsilon) < \infty,$$

if and only if $E(|X_n|) < \infty$.

A proof of Theorem 3.12 can be found in Section VII.8 of Feller (1971). We now have enough tools to consider the result of interest.

Theorem 3.13. *Let* $\{X_n\}_{n=1}^{\infty}$ *be a sequence of independent and identically distributed random variables. If* $E(X_n)$ *exists and is equal to* θ *then* $\bar{X}_n \xrightarrow{a.c.} \theta$.

Proof. Suppose that $E(X_n)$ exists and equals θ. We will consider the case when $\theta = 0$. The case when $\theta \neq 0$ can be proven using the same methodology used at the end of the proof of Theorem 3.10. Define two new sequences of random variables $\{Y_n\}_{n=1}^{\infty}$ and $\{Z_n\}_{n=1}^{\infty}$ as $Y_n = X_n \delta\{|X_n|; [0,n]\}$ and $Z_n = X_n \delta\{|X_n|; (n,\infty)\}$. Hence, it follows that

$$\bar{X}_n = n^{-1} \sum_{k=1}^{n} X_k = n^{-1} \sum_{k=1}^{n} Y_k + n^{-1} \sum_{k=1}^{n} Z_k.$$

Because $E(X_n)$ exists, Theorem 3.12 implies that for every $\varepsilon > 0$,

$$\sum_{n=1}^{\infty} P(|X_n| > n\varepsilon) < \infty.$$

Therefore, for $\varepsilon = 1$, $P(|X_n| > n) = P(X_n = Z_n) = P(Z_n \neq 0)$. Hence

$$\sum_{n=1}^{\infty} P(Z_n \neq 0) < \infty,$$

and Theorem 2.17 then implies that $P(\{Z_n \neq 0\} \text{ i.o.}) = 0$. This means that with probability one there will only be a finite number of times that $Z_n \neq 0$ over all the values of $n \in \mathbb{N}$, which implies that the sum

$$\sum_{n=1}^{\infty} Z_n,$$

will be finite with probability one. Hence

$$P\left(\lim_{n\to\infty} n^{-1} \sum_{k=1}^{n} Z_k = 0\right) = 1,$$

and therefore

$$n^{-1} \sum_{k=1}^{n} Z_k \xrightarrow{a.c.} 0,$$

as $n \to \infty$. The convergence behavior of the sum

$$\sum_{k=1}^{n} Y_k,$$

will be studied with the aid of Corollary 3.1. To apply this result we must show that

$$\sum_{n=1}^{\infty} n^{-1} E(Y_n^2) < \infty.$$

First note that since Y_n is the version of X_n truncated at n and $-n$ we have that

$$E(Y_n^2) = \int_{-\infty}^{\infty} x^2 \delta\{|x|; [0, n]\} dF(x) = \int_{-n}^{n} x^2 dF(x) = \sum_{k=1}^{n} \int_{R_k} x^2 dF(x),$$

where $R_k = \{x \in \mathbb{R} : k - 1 \le |x| < k\}$ and F is the distribution function of X_n. Hence

$$\sum_{n=1}^{\infty} n^{-2} E(Y_n^2) = \sum_{n=1}^{\infty} \sum_{k=1}^{n} n^{-2} \int_{R_k} x^2 dF(x).$$

Note that the set of pairs (n, k) for which $n \in \{1, 2, \ldots\}$ and $k \in \{1, 2, \ldots, n\}$ are the same set of pairs for which $k \in \{1, 2, \ldots\}$ and $n \in \{k, k+1, \ldots\}$. This allows us to change the order in the double sum as

$$\sum_{n=1}^{\infty} n^{-2} E(Y_n^2) = \sum_{k=1}^{\infty} \sum_{n=k}^{\infty} n^{-2} \int_{R_k} x^2 dF(x) = \sum_{k=1}^{\infty} \left[\int_{R_k} x^2 dF(x) \left(\sum_{n=k}^{\infty} n^{-2} \right) \right],$$

Now

$$\sum_{n=k}^{\infty} n^{-2} \le 2k^{-1},$$

so that

$$\sum_{n=1}^{\infty} n^{-2} E(Y_n^2) \le \sum_{k=1}^{\infty} \int_{R_k} 2x^2 k^{-1} dF(x).$$

But when $x \in R_k$ we have that $|x| < k$ so that

$$\sum_{k=1}^{\infty} \int_{R_k} 2x^2 k^{-1} dF(x) \le \sum_{k=1}^{\infty} \int_{R_k} 2k|x|k^{-1} dF(x) \le 2 \sum_{k=1}^{\infty} \int_{R_k} |x| dF(x) < \infty,$$

where the last inequality follows from our assumptions. Therefore we have shown that

$$\sum_{n=1}^{\infty} n^{-2} E(Y_n^2) < \infty.$$

Now, consider the centered sequence of random variables $Y_n - E(Y_n)$, which have mean zero. The fact that $E\{[Y_n - E(Y_n)]^2\} \le E(Y_n^2)$ implies that

$$\sum_{n=1}^{\infty} n^{-2} E\{[Y_n - E(Y_n)]^2\} < \infty.$$

We are now in the position where Corollary 3.1 can be applied to the centered sequence, which allows us to conclude that

$$n^{-1} \sum_{k=1}^{n} [Y_k - E(Y_k)] = \left[n^{-1} \sum_{k=1}^{n} Y_k - E \left(n^{-1} \sum_{k=1}^{n} Y_k \right) \right] \xrightarrow{a.c.} 0,$$

as $n \to \infty$, leaving us to evaluate the asymptotic behavior of

$$E\left(n^{-1}\sum_{k=1}^{n} Y_k\right) = n^{-1}\sum_{k=1}^{n} E(Y_k).$$

This is the same problem encountered in the proof of Theorem 3.10, and the same solution based on Theorem 2.12 implies that

$$\lim_{n\to\infty} n^{-1}\sum_{k=1}^{n} E(Y_k) = 0.$$

Therefore, Theorem 3.9 implies that

$$n^{-1}\sum_{k=1}^{n} Y_k \xrightarrow{a.c.} 0,$$

as $n \to \infty$.   □

Aside from the stronger conclusion about the mode of convergence about the sample mean, there is another major difference between the Weak Law of Large Numbers and the Strong Law of Large Numbers. Section VII.8 of Feller (1971) actually shows that the existence of the mean of X_n is both a necessary and sufficient condition to assure the almost certain convergence of the sample mean. As it turns out, the existence of the mean is not a necessary condition for a properly centered sample mean to converge in probability to a limit.

Theorem 3.14. *Let $\{X_n\}_{n=1}^{\infty}$ be a sequence of independent random variables each having a common distribution F. Then $\bar{X}_n - E(X_1\delta\{|X_1|; [0,n]\}) \xrightarrow{p} 0$ as $n \to \infty$ if and only if*

$$\lim_{n\to\infty} nP(|X_1| > n) = 0.$$

A proof of Theorem 3.14 can be found in Section 6.4 of Gut (2005). It is important to note that in Theorem 3.14 that the result does not imply that the sample mean converges to any value in this case. Rather, the conclusion is that the difference between the normal mean and the truncated mean converges to zero as $n \to \infty$. In the special case where the distribution of X_1 is symmetric about zero, $\bar{X}_n \xrightarrow{p} 0$ as $n \to \infty$, but if the mean of X_1 does not exist then $\bar{X}_n$ is not converging to the population mean. We could, however, conclude that $\bar{X}_n$ converges in probability to the population median as $n \to \infty$ in this special case. Note further that the condition that $nP(|X_1| > n) \to 0$ as $n \to \infty$ is both necessary and sufficient to ensure this convergence. This implies that there are cases where the convergence does not take place.

Example 3.21. Let $\{X_n\}_{n=1}^{\infty}$ be a sequence of independent and identically distributed random variables from a CAUCHY$(0,1)$ distribution. The mean of the distribution does not exist, and further it can be shown that $nP(|X_1| > n) \to 2\pi^{-1}$ as $n \to \infty$, so that the condition of Theorem 3.14 does not hold. Therefore, even though the distribution of X_1 is symmetric about zero, the

Figure 3.5 *The results of a small simulation demonstrating the behavior of sample means computed from a* CAUCHY$(0,1)$ *distribution. Each line represents a sequence of sample means computed on a sequence of independent* CAUCHY$(0,1)$ *random variables. The means were computed when* $n = 5, 10, \ldots, 250$.

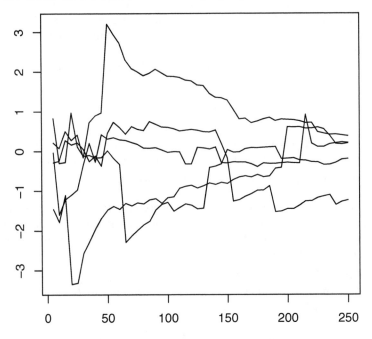

sample mean will not converge to zero. To observe the behavior of the sample mean in this case see Figure 3.5, where five realizations of the sample mean have been plotted for $n = 5, 10, \ldots, 250$. Note that the values of the mean do not appear to be settling down as we observed in Figure 3.2. ∎

Example 3.22. Let $\{X_n\}_{n=1}^{\infty}$ be a sequence of independent and identically distributed random variables from a continuous distribution with density

$$f(x) = \begin{cases} \frac{1}{2}x^{-2} & |x| > 1 \\ 0 & |x| \leq 1. \end{cases}$$

We first note that

$$\tfrac{1}{2}\int_1^n xf(x)dx = \tfrac{1}{2}\int_1^n x^{-1}dx = \tfrac{1}{2}\log(n),$$

so that

$$\int_1^n xf(x)dx \to \infty,$$

as $n \to \infty$, and therefore the mean of X_1 does not exist. Checking the condition in Theorem 3.14 we have that, due to the symmetry of the density,

$$nP(|X_1| > n) = 2n \int_n^\infty dF = \int_n^\infty x^{-2}dx = 1.$$

Therefore, Theorem 3.14 implies that $\bar{X}_n - E(X_1\delta\{|X_1|; [0,n]\})$ does not converge in probability to the truncated mean, which in this case is zero due to symmetry. One the other hand, if we modify the tails of the density so that they drop off at a slightly faster rate, then we can achieve convergence. For example, consider the density suggested in Section 6.4 of Gut (2005), given by

$$f(x) = \begin{cases} cx^2 \log(|x|) & |x| > 2 \\ 0 & |x| \le 2, \end{cases}$$

where c is a normalizing constant. In this case it can be shown that $nP(|X_1| > n) \to 0$ as $n \to \infty$, but that the mean does not exist. However, we can still conclude that $\bar{X}_n \xrightarrow{P} 0$ as $n \to \infty$ due to Theorem 3.14. ∎

The Laws of Large Numbers given by Theorems 3.10 and 3.13 provide a characterization of the limiting behavior of the sample mean as the sample size $n \to \infty$. The *Law of the Iterated Logarithm* provides information about the extreme fluctuations of the sample mean as $n \to \infty$.

Theorem 3.15 (Hartman and Wintner). *Let $\{X_n\}_{n=1}^\infty$ be a sequence of independent random variables each having a common distribution F such that $E(X_n) = \mu$ and $V(X_n) = \sigma^2 < \infty$. Then*

$$P\left(\limsup_{n\to\infty} \frac{n^{1/2}(\bar{X}_n - \mu)}{\{2\sigma^2 \log[\log(n)]\}^{1/2}} = 1\right) = 1,$$

and

$$P\left(\liminf_{n\to\infty} \frac{n^{1/2}(\bar{X}_n - \mu)}{\{2\sigma^2 \log[\log(n)]\}^{1/2}} = -1\right) = 1.$$

A proof of Theorem 3.15 can be found in Section 8.3 of Gut (2005). Theorem 3.15 shows that the extreme fluctuations of the sequence of random variables given by $\{Z_n\}_{n=1}^\infty = \{n^{1/2}\sigma^{-1}(\bar{X}_n - \mu)\}_{n=1}^\infty$ are about the same size as $\{2\log[\log(n)]\}^{1/2}$. More precisely, let $\varepsilon > 0$ and consider the interval

$$I_{\varepsilon,n} = [-(1+\varepsilon)\{2\log[\log(n)]\}^{1/2}, (1+\varepsilon)\{2\log[\log(n)]\}^{1/2}],$$

then all but a finite number of values in the sequence $\{Z_n\}_{n=1}^\infty$ will be contained in $I_{\varepsilon,n}$ with probability one. On the other hand if we define the interval

$$J_{\varepsilon,n} = [-(1-\varepsilon)\{2\log[\log(n)]\}^{1/2}, (1-\varepsilon)\{2\log[\log(n)]\}^{1/2}],$$

then $Z_n \notin J_{n,\varepsilon}$ an infinite number of times with probability one.

Example 3.23. Theorem 3.15 is somewhat difficult to visualize, but some simulated results can help. In Figure 3.6 we have plotted a realization of $n^{1/2}\bar{X}_n$ for $n = 1,\ldots,500$, where the population is $N(0,1)$, along with its

Figure 3.6 *A simulated example of the behavior indicated by Theorem 3.15. The solid line is a realization of $n^{1/2}\bar{X}_n$ for a sample of size n from a $N(0,1)$ distribution with $n = 1, \ldots, 500$. The dotted line indicates the extent of the envelope $\pm\{2\log[\log(n)]\}^{1/2}$ and the dashed line indicates the extreme fluctuations of $n^{1/2}\bar{X}_n$.*

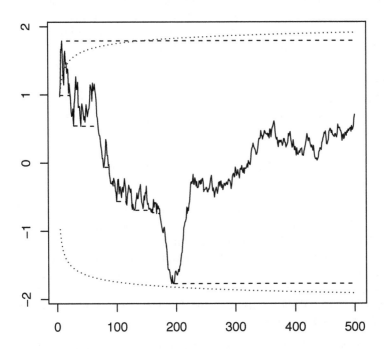

extreme fluctuations and the limits $\pm\{2\log[\log(n)]\}^{1/2}$. One would not expect the extreme fluctuations of $n^{1/2}\bar{X}_n$ to exactly follow the limits given by $\pm\{2\log[\log(n)]\}^{1/2}$, but note in our realization that the general shape of the fluctuations does follow the envelope fairly well as n becomes larger. ■

3.7 The Glivenko–Cantelli Theorem

In many settings statistical inference is based on the *nonparametric* framework, which attempts to make as few assumptions as possible about the underlying process that produces a set of observations. A major assumption in many methods of statistical inference is that the population has a certain distribution. For example, statistical inference on population means and variances often assumes that the underlying population is at least approximately normal. If the assumption about the normality of the population is not true, then there can be an effect on the reliability of the associated methods of statistical inference.

The *empirical distribution function* provides a method for estimating a distribution function F based only the assumptions that the observed data consist of independent and identically distributed random variables following the distribution F. This estimator is useful in two ways. First, the reasonableness of assumptions about the distribution of the sample can be studied using this estimator. Second, if no assumption about the distribution is to be made, the empirical distribution can often be used in place of the unknown distribution to obtain approximate methods for statistical inference. This section introduces the empirical distribution function and studies many of the asymptotic properties of the estimate.

Definition 3.5. *Let $X_1, \ldots, X_n$ be a set of independent and identically distributed random variables from a distribution F. The empirical distribution function of $X_1, \ldots, X_n$ is*

$$\hat{F}_n(t) = n^{-1} \sum_{i=1}^{n} \delta\{X_i; (-\infty, t]\}.$$

It is useful to consider the structure of the estimate of F proposed in Definition 3.5 in some detail. For an observed sample $X_1, \ldots, X_n$, the empirical distribution function is a step function that has steps of size n^{-1} at each of the observed sample values, under the assumption that each of the sample values is unique. Therefore, if we consider the empirical distribution function conditional on the observed sample $X_1, \ldots, X_n$, the empirical distribution function corresponds to a discrete distribution that has a probability of n^{-1} at each of the observations in the sample.

The properties of the empirical distribution function can be studied in two ways. For a fixed point $t \in \mathbb{R}$ we can consider $\hat{F}_n(t)$ as an estimator of $F(t)$, at that point. Alternatively we can consider the entire function $\hat{F}_n$ over the real line simultaneously as an estimate of the function F. We begin by considering the point-wise viewpoint first. Finite sample properties follow from the fact that when $t \in \mathbb{R}$ is fixed,

$$P(\delta\{X_i; (-\infty, t]\} = 1) = P(X_i \in (-\infty, t]) = P(X_i \leq t) = F(t).$$

Similarly, $P(\delta\{X_i; (-\infty, t]\} = 0) = 1 - F(t)$. Therefore $\delta\{X_i; (-\infty, t]\}$ is a BERNOULLI$[F(t)]$ random variable for each $i = 1, \ldots, n$, and hence

$$n\hat{F}(t) = \sum_{i=1}^{n} \delta(X_i; (-\infty, t]),$$

is a BINOMIAL$[n, F(t)]$ random variable. Using this fact, it can be proven that for a fixed value of $t \in \mathbb{R}$, $\hat{F}_n(t)$ is an unbiased estimator of $F(t)$ with standard error $n^{-1/2}\{F(t)[1 - F(t)]\}^{1/2}$. The empirical distribution function is also point-wise consistent.

Theorem 3.16. *Let $X_1, \ldots, X_n$ be a set of independent and identically distributed random variables from a distribution F, and let $\hat{F}_n$ be the empir-*

ical distribution function computed on $X_1, \ldots, X_n$. *Then for each* $t \in \mathbb{R}$, $\hat{F}_n(t) \xrightarrow{a.c.} F(t)$ *as* $n \to \infty$.

The next step in our development is to extend the consistency result of Theorem 3.16 to the entire empirical distribution function. That is, we wish to conclude that $\hat{F}_n$ is a consistent estimator of F, or that $\hat{F}_n$ convergences almost certainly to F as $n \to \infty$. This differs from the previous result in that we wish to show that the random function $\hat{F}_n$ becomes arbitrarily close to F with probability one as $n \to \infty$. Therefore, we require a measure of distance between two distribution functions. Many distance functions, or *metrics*, can be defined on the space of distribution functions. For examples, see Young (1988). A common metric for comparing two distribution functions in statistical inference is based on the *supremum metric*.

Theorem 3.17. *Let* F *and* G *be two distribution functions. Then,*

$$d_\infty(F, G) = \|F - G\|_\infty = \sup_{t \in \mathbb{R}} |F(t) - G(t)|$$

is a metric in the space of distribution functions called the supremum metric.

A stronger result can actually be proven in that the metric defined in Theorem 3.17 is actually a metric over the space of all functions. Now that a metric in the space of distribution functions has been defined, it is relevant to ascertain whether the empirical distribution function is a consistent estimator of F with respect to this metric. That is, we would conclude that $\hat{F}_n$ converges almost certainly to F as $n \to \infty$ if,

$$P\left(\lim_{n \to \infty} \|\hat{F}_n - F\| = 0\right) = 1,$$

or equivalently that $\|\hat{F}_n - F\|_\infty \xrightarrow{a.c.} 0$ as $n \to \infty$.

Theorem 3.18 (Glivenko and Cantelli). *Let* $X_1, \ldots, X_n$ *be a set of independent and identically distributed random variables from a distribution* F, *and let* $\hat{F}_n$ *be the empirical distribution function computed on* $X_1, \ldots, X_n$. *Then* $\|\hat{F}_n - F\|_\infty \xrightarrow{a.c.} 0$.

Proof. For a fixed value of $t \in \mathbb{R}$, Theorem 3.16 implies that $\hat{F}_n(t) \xrightarrow{a.c.} F(t)$ as $n \to \infty$. The result we wish to prove states that the maximum difference between $\hat{F}_n$ and F also converges to 0 as $n \to \infty$, a stronger result. Rather than attempt to quantify the behavior of the maximum difference directly, we will instead prove that $\hat{F}_n(t) \xrightarrow{a.c.} F(t)$ uniformly in t as $n \to \infty$. We will follow the method of proof used by van der Vaart (1998). Alternate approaches can be found in Sen and Singer (1993) and Serfling (1980). The result was first proven under various conditions by Glivenko (1933) and Cantelli (1933).

Let $\varepsilon > 0$ be given. Then, there exists a partition of $\mathbb{R}$ given by $-\infty = t_0 < t_1 < \ldots < t_k = \infty$ such that

$$\lim_{t \uparrow t_i} F(t) - F(t_{i-1}) < \varepsilon,$$

for some $k \in \mathbb{N}$. We will begin by arguing that such a partition exists. First consider the endpoints of the partition t_1 and t_{k-1}. Because F is a distribution function we know that

$$\lim_{t \to t_0} F(t) = \lim_{t \to -\infty} F(t) = 0,$$

and hence there must be a point t_1 such that $F(t_1) < \varepsilon$. Similarly, since

$$\lim_{t \to t_k} F(t) = \lim_{t \to \infty} F(t) = 1,$$

it follows that there must be a point t_{k-1} such that $F(t_{k-1})$ is within ε of $F(t_k) = 1$. If the distribution function is continuous with an interval (a, b), then the definition of continuity implies that there must exist two points t_i and t_{i-1} such that $F(t_i) - F(t_{i-1}) < \varepsilon$. Noting that for points where F is continuous we have that

$$\lim_{t \uparrow t_i} F(t) = F(t_i),$$

which shows that the partition exists on any interval (a, b) where the distribution function is continuous.

Now consider an interval (a, b) where there exists a point $t' \in (a, b)$ such that F has a jump of size δ' at t'. That is,

$$\delta' = F(t') - \lim_{t \uparrow t'} F(t).$$

There is no problem if $\delta' < \varepsilon$, for a specific value of ε, but this cannot be guaranteed for every $\varepsilon > 0$. However, the partition can still be created by setting one of the points of the partition exactly at t'. First, consider the case where $t_i = t'$. Considering the limit of $F(t)$ as $t \uparrow t'$ results in the value of F at t' if F was *left continuous* at t'. It then follows that there must exist a point t_{i-1} such that

$$\lim_{t \uparrow t'} F(t) - F(t_{i-1}) < \varepsilon.$$

See Figure 3.7. In the case where $t_{i+1} = t'$, the property follows from the fact that F is always right continuous, and is therefore continuous on the interval (t_{i+1}, b) for some b. Therefore, there does exist a partition with the indicated property. The partition is finite because the range of F, which is $[0, 1]$, is bounded.

Now consider $t \in (t_{i-1}, t_i)$ for some $i \in \{0, \ldots, k\}$ and note that because $\hat{F}_n$ and F are non-decreasing it follows that

$$\hat{F}_n(t) \leq \lim_{t \uparrow t_i} \hat{F}_n(t),$$

and

$$F(t) \geq F(t_{i-1}) > \lim_{t \uparrow t_i} F(t) - \varepsilon,$$

so that is follows that

$$\hat{F}_n(t) - F(t) \leq \lim_{t \uparrow t_i} \hat{F}_n(t) - \lim_{t \uparrow t_i} F(t) + \varepsilon.$$

Similar computations can be used to show that $\hat{F}_n(t) - F(t) \geq \hat{F}_n(t_{i-1}) - F(t_{i-1}) - \varepsilon$.

We already know that $\hat{F}_n(t) \xrightarrow{a.c.} F(t)$ for every $t \in \mathbb{R}$. However, because the partition $t_0 < t_1 < \ldots < t_k$ is finite, it follows that $\hat{F}_n(t) \xrightarrow{a.c.} F(t)$ uniformly on the partition $t_0 < t_1 < \ldots < t_k$. That is, for every $\varepsilon > 0$, there exists a positive integer n_ε such that $|\hat{F}_n(t) - F(t)| < \varepsilon$ for all $n \geq n_\varepsilon$ and $t \in \{t_0, t_1, \ldots, t_k\}$, with probability one. To prove this we need only find $n_{\varepsilon,t}$ such that $|\hat{F}_n(t) - F(t)| < \varepsilon$ for all $n \geq n_{\varepsilon,t}$ with probability one and assign $n_\varepsilon = \max\{n_{\varepsilon,t_0}, \ldots, n_{\varepsilon,t_k}\}$.

This implies that for every $\varepsilon > 0$ and $t \in (t_{i-1}, t_i)$ that there is a positive integer n_ε such that

$$\hat{F}_n(t) - F(t) \leq \lim_{t \uparrow t_i} \hat{F}_n(t) - \lim_{t \uparrow t_i} F(t) + \varepsilon \leq 2\varepsilon,$$

and

$$\hat{F}_n(t) - F(t) \geq \hat{F}_n(t_{i-1}) - F(t_{i-1}) - \varepsilon \geq -2\varepsilon.$$

Hence, for every $\varepsilon > 0$ and $t \in (t_{i-1}, t_i)$ that there is a positive integer n_ε such that $|\hat{F}_n(t) - F(t)| < 2\varepsilon$ for all $n \geq n_\varepsilon$, with probability one. Noting that the value of n_ε does not depend on i, we have proven that for every $\varepsilon > 0$ there is a positive integer n_ε such that $|\hat{F}_n(t) - F(t)| < 2\varepsilon$ for every $n \geq n_\varepsilon$ and $t \in \mathbb{R}$, with probability one. Therefore, $\hat{F}_n(t)$ converges uniformly to $F(t)$ with probability one. This uniform convergence implies that

$$P\left(\lim_{n \to \infty} \sup_{t \in \mathbb{R}} |\hat{F}_n(t) - F(t)| = 0\right) = 1,$$

or that

$$\sup_{t \in \mathbb{R}} |\hat{F}_n(t) - F(t)| \xrightarrow{a.c.} 0,$$

as $n \to \infty$.   □

Example 3.24. Let $\{X_n\}_{n=1}^\infty$ be a sequence of independent and identically distributed random variables each having an EXPONENTIAL(θ) distribution. Let $\hat{F}_n$ be the empirical distribution function computed on $X_1, \ldots, X_n$. We can visualize the convergence of the $\hat{F}_n$ to F as described in Theorem 3.18 by looking at the results of a small simulation. In Figures 3.8–3.10 we compare the empirical distribution function computed on simulated samples of size $n = 5$, 25 and 50 from an EXPONENTIAL(θ) distribution to the true distribution function, which in this case is given by $F(t) = [1 - \exp(-t)]\delta\{t; (0, \infty)\}$. One can observe in Figures 3.8–3.10 that the empirical distribution function becomes uniformly closer to the true distribution function as the sample size becomes larger. Theorem 3.18 guarantees that this type of behavior will occur with probability one.   ■

Figure 3.7 *Constructing the partition used in the proof of Theorem 3.18 when there is a discontinuity in the distribution function. By locating a partition point at the jump point (grey line), the continuity of the distribution function to the left of this point can be used to find a point (dotted line) such that the difference between the distribution function at these two points does not exceed ε for any specified ε > 0.*

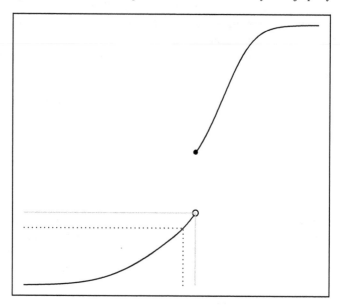

3.8 Sample Moments

Let $X_1, \ldots, X_n$ be a sequence of independent and identically distributed random variables from a distribution F and consider the problem of estimating the kth moment of F defined in Definition 2.9 as

$$\mu_k' = \int_{-\infty}^{\infty} x^k dF(x). \tag{3.14}$$

We will assume for the moment that

$$\int_{-\infty}^{\infty} |x|^m dF(x) < \infty, \tag{3.15}$$

for some $m \geq k$, with further emphasis on this assumption to be considered later. The empirical distribution function introduced in Section 3.7 provides a nonparametric estimate of F based on a sample $X_1, \ldots, X_n$. Substituting the empirical distribution function into Equation (3.14) we obtain a nonparametric estimate of the kth moment of F given by

$$\hat{\mu}_k' = \int_{-\infty}^{\infty} x^k d\hat{F}_n(x) = n^{-1} \sum_{n=1}^{\infty} X_i^k,$$

Figure 3.8 *The empirical distribution function computed on a simulated sample of size $n = 5$ from an* EXPONENTIAL(θ) *distribution (solid line) compared to the actual distribution function (dashed line).*

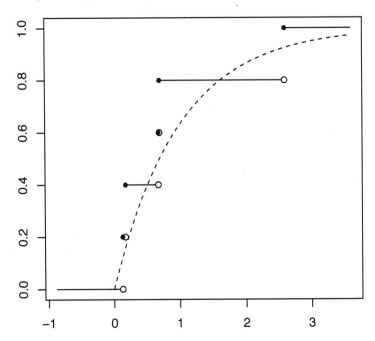

where the integral is evaluated using Definition 2.10. This estimate is known as the kth sample moment. The properties of this estimate are detailed below.

Theorem 3.19. *Let $X_1, \ldots, X_n$ be a set of independent and identically distributed random variables from a distribution F.*

1. *If $E(|X_1|^k) < \infty$ then $\hat{\mu}'_k$ is an unbiased estimator of μ'_k.*

2. *If $E(|X_1|^{2k}) < \infty$ then the standard error of $\hat{\mu}'_k$ is $n^{-1/2}(\mu'_{2k} - \mu'^2_k)^{1/2}$.*

3. *If $E(|X_1|^k) < \infty$ then $\hat{\mu}'_k \xrightarrow{a.c.} \mu'_k$ as $n \to \infty$.*

For a proof of Theorem 3.19 see Exercise 28.

The kth central moment can be handled in a similar way. From Definition 2.9 the kth central moment of F is

$$\mu_k = \int_{-\infty}^{\infty} (x - \mu'_1)^k dF(x), \tag{3.16}$$

where we will again use the assumption in Equation (3.15) that $E(|X|^m) < \infty$ for a value of m to be determined later. Substituting the empirical distribution

Figure 3.9 *The empirical distribution function computed on a simulated sample of size $n = 25$ from an* EXPONENTIAL(θ) *distribution (solid line) compared to the actual distribution function (dashed line).*

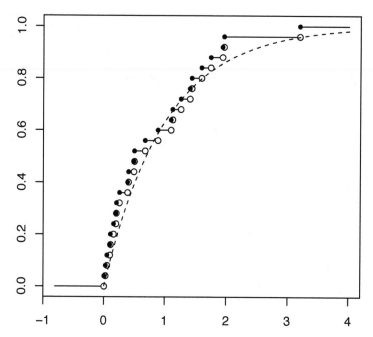

function for F in Equation (3.16) provides an estimate of μ_k given by

$$\hat{\mu}_k = \int_{-\infty}^{\infty} \left[x - \int_{-\infty}^{\infty} t d\hat{F}_n(t) \right]^k d\hat{F}_n(x) = n^{-1} \sum_{i=1}^{n} (X_i - \hat{\mu}_1')^k.$$

This estimate has a more complex structure than that of the kth sample moment which makes the bias and standard error more difficult to obtain. One result which makes this job slightly easier is given below.

Theorem 3.20. *Let $\{X_n\}_{n=1}^{\infty}$ be a sequence of independent and identically distributed random variables such that $E(X_1) = 0$ and $E(|X_n|^k) < \infty$ for some $k \geq 2$. Then*

$$E\left(\left| \sum_{i=1}^{n} X_i \right|^k \right) = O(n^{-k/2}),$$

as $n \to \infty$.

A proof of Theorem 3.20 can be found in Chapter 19 of Loéve (1977). A similar result is given in Lemma 9.2.6.A of Serfling (1980).

Theorem 3.21. *Let $X_1, \ldots, X_n$ be a set of independent and identically distributed random variables from a distribution F.*

Figure 3.10 *The empirical distribution function computed on a simulated sample of size $n = 50$ from an* EXPONENTIAL(θ) *distribution (solid line) compared to the actual distribution function (dashed line).*

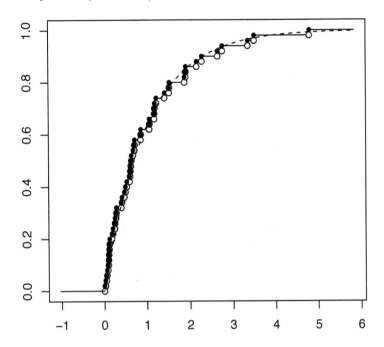

1. If $E(|X_1|^k) < \infty$ then the bias of $\hat{\mu}_k$ as an estimator of μ_k is

$$\tfrac{1}{2}n^{-1}k(k-1)\mu_{k-2}\mu_2 - n^{-1}k\mu_k + O(n^{-2}),$$

as $n \to \infty$.

2. If $E(|X_1|^{2k}) < \infty$ then the variance of $\hat{\mu}_k$ is

$$n^{-1}(\mu_{2k} - \mu_k^2 - 2k\mu_{k-1}\mu_{k+1} + k^2\mu_2\mu_{k-1}^2) + O(n^{-2}),$$

as $n \to \infty$.

3. If $E(|X_1|^k) < \infty$ then $\hat{\mu}_k \xrightarrow{a.c.} \mu_k$ as $n \to \infty$.

Proof. We follow the method of Serfling (1980) to prove the first result. We begin by noting that

$$
\begin{aligned}
\hat{\mu}_k &= n^{-1} \sum_{i=1}^{n} (X_i - \hat{\mu}_1')^k \\
&= n^{-1} \sum_{i=1}^{n} [(X_i - \mu_1') + (\mu_1' - \hat{\mu}_1')]^k \\
&= n^{-1} \sum_{i=1}^{n} \sum_{j=0}^{k} \binom{k}{j} (\mu_1' - \hat{\mu}_1')^j (X_i - \mu_1')^{k-j} \\
&= \sum_{j=0}^{k} \binom{k}{j} (\mu_1' - \hat{\mu}_1')^j \left[n^{-1} \sum_{i=1}^{n} (X_i - \mu_1')^{k-j} \right].
\end{aligned}
$$

Therefore,

$$
E(\hat{\mu}_k) = \sum_{j=0}^{k} \binom{k}{j} E \left\{ (\mu_1' - \hat{\mu}_1')^j \left[n^{-1} \sum_{i=1}^{n} (X_i - \mu_1')^{k-j} \right] \right\}. \tag{3.17}
$$

Now, the $j = 0$ term in the sum in Equation (3.17) equals

$$
E \left[n^{-1} \sum_{i=1}^{n} (X_i - \mu_1')^k \right] = \mu_k.
$$

It follows that the bias of $\hat{\mu}_k$ is given by

$$
\sum_{j=1}^{k} \binom{k}{j} E \left\{ (\mu_1' - \hat{\mu}_1')^j \left[n^{-1} \sum_{i=1}^{n} (X_i - \mu_1')^{k-j} \right] \right\}. \tag{3.18}
$$

The first term of this sum is given by

$$
\binom{k}{1} E \left\{ (\mu_1' - \hat{\mu}_1') \left[n^{-1} \sum_{i=1}^{n} (X_i - \mu_1')^{k-1} \right] \right\} =
$$

$$
kn^{-2} \sum_{i=1}^{n} \sum_{j=1}^{n} E[(\mu_1' - X_i)(X_j - \mu_1')^{k-1}].
$$

Note that when $i \neq j$ that the two terms in the sum are independent, and therefore the expectation of the product is the product of the expectations. In all of these cases $E(X_i - \mu_1') = 0$ and the term vanishes. When $i = j$ the term in the sum equals $-E[(X_i - \mu_1')^k] = -\mu_k$. Combining these results implies that the first term of Equation (3.18) is $-n^{-1} k \mu_k$. The second term ($j = 2$)

of Equation (3.18) equals

$$\binom{k}{2} E\left\{ (\mu_1' - \hat{\mu}_1')^2 \left[n^{-1} \sum_{i=1}^{n} (X_i - \mu_1')^{k-2} \right] \right\} =$$

$$\tfrac{1}{2} k(k-1) n^{-3} E\left\{ \left[\sum_{i=1}^{n} (\mu_1' - X_i) \right]^2 \sum_{j=1}^{n} (X_j - \mu_1')^{k-2} \right\} =$$

$$\tfrac{1}{2} k(k-1) n^{-3} \sum_{i=1}^{n} \sum_{j=1}^{n} \sum_{l=1}^{n} E[(\mu_1' - X_i)(\mu_1' - X_j)(X_l - \mu_1')^{k-2}]. \quad (3.19)$$

When i differs from j and l the expectation is zero due to independence. When $i = j$, the sum in Equation (3.19) becomes

$$\tfrac{1}{2} k(k-1) n^{-3} \sum_{i=1}^{n} \sum_{j=1}^{n} E[(\mu_1' - X_i)^2 (X_j - \mu_1')^{k-2}]. \quad (3.20)$$

When $i \neq j$ the two terms in the sum in Equation (3.20) are independent therefore

$$E[(\mu_1' - X_i)^2 (X_j - \mu_1')^{k-2}] = E[(\mu_1' - X_i)^2] E[(X_j - \mu_1')^{k-2}] = \mu_2 \mu_{k-2}.$$

When $i = j$ the term in the sum in Equation (3.20) is given by

$$E[(\mu_1' - X_i)^2 (X_i - \mu_1')^{k-2}] = E[(X_i - \mu_1')^{k}] = \mu_k.$$

Therefore, the sum in Equation (3.20) equals

$$\tfrac{1}{2} k(k-1) n^{-3} [n(n-1)\mu_2 \mu_{k-2} + n\mu_k] = \tfrac{1}{2} k(k-1) n^{-1} \mu_2 \mu_{k-2} + O(n^{-2}),$$

as $n \to \infty$. When $j = 3$, the term in the sum in Equation (3.18) can be shown to be $O(n^{-2})$ as $n \to \infty$. See Theoretical Exercise 29. To obtain the behavior of the remaining terms we note that the jth term in the sum in Equation (3.18) has the form

$$\binom{k}{j} \left[n^{-1} \sum_{i=1}^{n} (\mu_1' - X_i) \right]^j \left[n^{-1} \sum_{i=1}^{n} (X_i - \mu_1')^{k-j} \right].$$

Now apply Theorem 2.10 (Hölder) to the two sums to yield

$$\left| E\left\{ \left[n^{-1}\sum_{i=1}^{n}(\mu_1' - X_i) \right]^{j} \left[n^{-1}\sum_{i=1}^{n}(X_i - \mu_1')^{k-j} \right] \right\} \right| \leq$$

$$\left\{ E\left[\left| \left(n^{-1}\sum_{i=1}^{n}(\mu_1' - X_i) \right)^{j} \right|^{k/j} \right] \right\}^{j/k}$$

$$\times \left\{ E\left[\left| n^{-1}\sum_{i=1}^{n}(X_i - \mu_1')^{k-j} \right|^{k/(k-j)} \right] \right\}^{(k-j)/k} . \quad (3.21)$$

Applying Theorem 3.20 to the first term of Equation (3.21) implies

$$\left\{ E\left[\left| \left(n^{-1}\sum_{i=1}^{n}(\mu_1' - X_i) \right)^{j} \right|^{k/j} \right] \right\}^{j/k} =$$

$$\left\{ E\left[\left| n^{-1}\sum_{i=1}^{n}(\mu_1' - X_i) \right|^{k} \right] \right\}^{j/k} = [O(n^{-k/2})]^{j/k} = O(n^{-j/2}),$$

as $n \to \infty$. To simplify the second term in Equation (3.21), apply Theorem 2.9 (Minkowski's Inequality) to find that

$$\left\{ E\left[\left| n^{-1}\sum_{i=1}^{n}(X_i - \mu_1')^{k-j} \right|^{k/(k-j)} \right] \right\}^{(k-j)/k} \leq$$

$$n^{-1}\sum_{i=1}^{n}\{E[|(X_i - \mu_1')^{k-j}|^{k/(k-j)}]\}^{(k-j)/k} =$$

$$n^{-1}\sum_{i=1}^{n}\{E[|X_i - \mu_1'|^{k}]\}^{(k-j)/k} = E(|X_1 - \mu_1'|^{k})^{(k-j)/k} = O(1), \quad (3.22)$$

as $n \to \infty$. Therefore Theorem 1.18 implies that the expression in Equation (3.21) is of order $O(1)O(n^{-j/2}) = O(n^{-2})$ as $n \to \infty$ when $j \geq 4$. Combining these results implies that

$$E(\hat{\mu}_k) = -n^{-1}k\mu_k + \tfrac{1}{2}n^{-1}k(k-1)\mu_2\mu_{k-2} + O(n^{-2}),$$

as $n \to \infty$, which yields the result. The second result is proven in Exercise 30 and the third result is proven in Exercise 31. □

Example 3.25. Let $X_1, \ldots, X_n$ be a sequence of independent and identically distributed random variables from a distribution F with finite fourth moment.

Then Theorem 3.21 implies that the sample variance given by

$$\hat{\mu}_n = n^{-1} \sum_{i=1}^{n} (X_i - \bar{X}_n)^2,$$

is consistent with bias

$$n^{-1}\mu_0\mu_2 - n^{-1}2\mu_2 + O(n^{-2}) = -n^{-1}\mu_2 + O(n^{-2}),$$

as $n \to \infty$. Similarly, the variance is given by

$$n^{-1}(\mu_4 - \mu_2^2 - 4\mu_1\mu_3 + 4\mu_2\mu_1^2) + O(n^{-2}) = n^{-1}(\mu_4 - \mu_2^2) + O(n^{-2}),$$

as $n \to \infty$, where we note that $\mu_1 = 0$ by definition. In fact, a closer analysis of these results indicate that the error terms are identically zero in this case. See Exercise 7 for an alternate approach to determining the corresponding results for the unbiased version of the sample variance. ∎

3.9 Sample Quantiles

This section investigates under what conditions a sample quantile provides a consistent estimate of the corresponding population quantile. Let X be a random variable with distribution function

$$F(x) = P(X \le x) = \int_{-\infty}^{x} dF(t).$$

The pth quantile of X is defined to be $\xi_p = F^{-1}(p) = \inf\{x : F(x) \ge p\}$. The population quantile as defined above is always unique, even though the inverse of the distribution function may not be unique in every case. There are three essential examples to consider. In the case where $F(x)$ is continuous and strictly increasing in the neighborhood of the quantile, then the distribution function has a unique inverse in that neighborhood and it follows that $F(\xi_p) = F[F^{-1}(p)] = p$. The continuity of the function in this case also guarantees that $F(\xi_p-) = F(\xi_p) = p$. That is, the quantile ξ_p can be seen as the unique solution to the equation $F(\xi_p-) = F(\xi_p) = p$ with respect to ξ_p. See Figure 3.11. In the case where a discontinuous jump occurs at the quantile, the distribution function does not have an inverse in the sense that $F[F^{-1}(p)] = p$. The quantile ξ_p as defined above is located at the jump point. In this case $F(\xi_p-) < p < F(\xi_p)$, and once again the quantile can be defined to be the unique solution to the equation $F(\xi_p-) < p < F(\xi_p)$ with respect to ξ_p. See Figure 3.12. In the last case F is continuous in a neighborhood of the quantile, but is not increasing. In this case, due to the continuity of F in the neighborhood of the quantile, $F(\xi_p-) = p = F(\xi_p)$. However, the difference in this case is that the quantile is not the unique solution to the equation $F(\xi_p-) = p = F(\xi_p)$ in that any point in the non-increasing neighborhood of the quantile will also be a solution to this equation. See Figure 3.13. Therefore, for the first two situations (Figures 3.11 and 3.12) there is

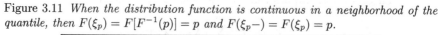

Figure 3.11 *When the distribution function is continuous in a neighborhood of the quantile, then $F(\xi_p) = F[F^{-1}(p)] = p$ and $F(\xi_p-) = F(\xi_p) = p$.*

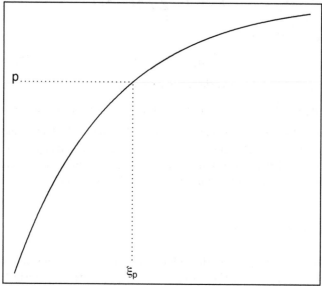

Figure 3.12 *When the distribution function has a discontinuous jump in at the quantile, then $F(\xi_p-) < p < F(\xi_p)$.*

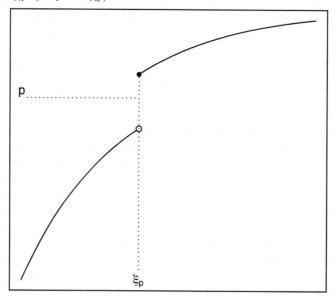

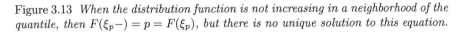

Figure 3.13 *When the distribution function is not increasing in a neighborhood of the quantile, then $F(\xi_p-) = p = F(\xi_p)$, but there is no unique solution to this equation.*

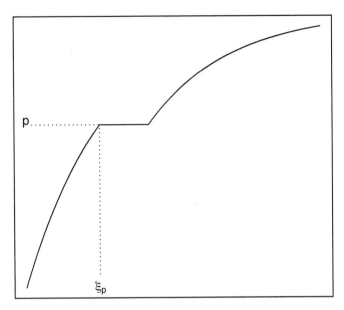

a unique solution to the equation $F(\xi_p-) \leq p \leq F(\xi_p)$ that corresponds to the quantile ξ_p as defined earlier. The third situation, where there is not a unique solution, is problematical as we shall see in the development below. A few other facts about distribution functions and quantiles will be also helpful in establishing the results of this section.

Theorem 3.22. *Let F be a distribution function and let $\xi_p = F^{-1}(p) = \inf\{x : F(x) \geq p\}$. Then*

1. $\xi_p = F^{-1}(p)$ *is a non-decreasing function of $p \in (0,1)$.*
2. $\xi_p = F^{-1}(p)$ *is left-continuous function of $p \in (0,1)$.*
3. $F^{-1}[F(x)] \leq x$ *for all $x \in \mathbb{R}$.*
4. $F[F^{-1}(p)] \geq p$ *for all $p \in (0,1)$.*
5. $F(x) \geq p$ *if and only if $x \geq F^{-1}(p)$.*

Proof. To prove Part 1 we consider $p_1 \in (0,1)$ and $p_2 \in (0,1)$ such that $p_1 < p_2$. Then we have that $\xi_{p_1} = F^{-1}(p_1) = \inf\{x : F(x) > p_1\}$. The key idea to this proof is establishing that $\xi_{p_1} = F^{-1}(p_1) = \inf\{x : F(x) > p_1\} \leq \inf\{x : F(x) > p_2\} = \xi_{p_2}$. The reason for this follows from the fact that F is a non-decreasing function. Therefore, the smallest value of x such that $F(x) > p_2$ must be at least as large as the smallest value of x such that $F(x) > p_1$. $\square$

Figure 3.14 *Example of computing a sample quantile when* $p = kn^{-1}$. *In this example, the empirical distribution function for a* UNIFORM$(0,1)$ *sample of size* $n = 5$ *is plotted and we wish to estimate* $\xi_{0.6}$. *In this case* $p = 0.6 = kn^{-1}$ *where* $k = 3$ *so that the estimate corresponds to the third order statistic.*

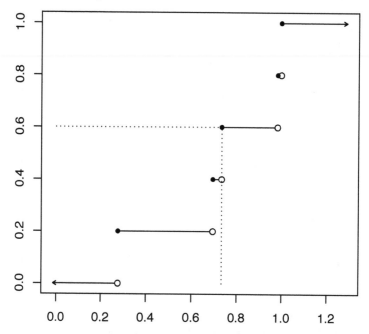

Now suppose that $X_1, \ldots, X_n$ is a set of independent and identically distributed random variables following the distribution F and we wish to estimate ξ_p, the pth quantile of F. If no assumptions can be made about F, a logical estimator would be the corresponding quantile of the empirical distribution function from Definition 3.5. That is, let $\hat{\xi}_p = \inf\{x : \hat{F}_n(x) \geq p\}$, where $\hat{F}_n(x)$ is the empirical distribution function computed on the sample $X_1, \ldots, X_n$. There are two specific cases that can occur when using this estimate of a quantile. The empirical distribution function is a step function with steps of size n^{-1} at each of the observed values in the sample. Therefore, if $p = kn^{-1}$ for some $k \in \{1, \ldots, n\}$ then $\hat{\xi}_p = \inf\{x : \hat{F}_n(x) \geq kn^{-1}\} = X_{(k)}$, the kth order statistic of the sample. For an example of this case, see Figure 3.14. If there does not exist a $k \in \{1, \ldots, n\}$ such that $p = kn^{-1}$ then there will be a value of k such that $(k-1)n^{-1} < p < kn^{-1}$ and therefore p will be between steps of the empirical distribution function. In this case $\hat{\xi}_p = \inf\{x : \hat{F}_n(x) \geq (k-1)n^{-1}\} = X_{(k)}$ as well. For an example of this case, see Figure 3.15.

Theorem 3.23. *Let* $\{X_n\}_{n=1}^{\infty}$ *be a sequence of independent random variables*

Figure 3.15 *Example of computing a sample quantile when $p \neq kn^{-1}$. In this example, the empirical distribution function for a* UNIFORM$(0, 1)$ *sample of size $n = 5$ is plotted and we wish to estimate the median $\xi_{0.5}$. In this case there is not a value of k such that $p = 0.5 = kn^{-1}$, but when $k = 3$ we have that $(k-1)n^{-1} < p < kn^{-1}$ and therefore the estimate of the median corresponds to the third order statistic.*

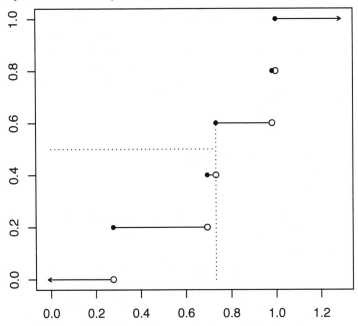

with common distribution F. Suppose that $p \in (0, 1)$ and that ξ_p is the unique solution to $F(\xi_p-) \leq p \leq F(\xi_p)$. Then $\hat{\xi}_p \xrightarrow{a.c.} \xi_p$ as $n \to \infty$.

Proof. Let $\varepsilon > 0$ and note that the assumption that $F(\xi_p-) \leq p \leq F(\xi_p)$ implies that $F(\xi_p - \varepsilon) < p < F(\xi_p + \varepsilon)$. Now, Theorem 3.16 implies that $\hat{F}_n(x) \xrightarrow{a.c.} F(x)$ for every $x \in \mathbb{R}$ so that $\hat{F}_n(\xi_p - \varepsilon) \xrightarrow{a.c.} F(\xi_p - \varepsilon)$ and $\hat{F}_n(\xi_p + \varepsilon) \xrightarrow{a.c.} F(\xi_p + \varepsilon)$ as $n \to \infty$. Theorem 3.1 implies that for every $\delta > 0$

$$\lim_{n \to \infty} P[|\hat{F}_m(\xi_p - \varepsilon) - F(\xi_p - \varepsilon)| < \delta \text{ for all } m \geq n] = 1,$$

and

$$\lim_{n \to \infty} P[|\hat{F}_m(\xi_p + \varepsilon) - F(\xi_p + \varepsilon)| < \delta \text{ for all } m \geq n] = 1.$$

Now, take δ small enough so that

$$\lim_{n \to \infty} P[\hat{F}_m(\xi_p - \varepsilon) < p < \hat{F}_m(\xi_p + \varepsilon) \text{ for all } m \geq n] = 1.$$

Theorem 3.22 then implies that

$$\lim_{n \to \infty} P(\xi_p - \varepsilon < \hat{F}_m^{-1}(p) < \xi_p + \varepsilon \text{ for all } m \geq n) = 1,$$

but note that $\hat{F}_m^{-1}(p) = \hat{\xi}_{p,n}$ so that

$$\lim_{n\to\infty} P(\xi_p - \varepsilon < \hat{\xi}_{p,n} < \xi_p + \varepsilon \text{ for all } m \geq n) = 1,$$

which in turn implies that $\hat{\xi}_{p,n} \xrightarrow{a.c.} \xi_p$ as $n \to \infty$.   □

Example 3.26. Let $X_1, \ldots, X_n$ be a set of independent and identically distributed random variables, each having a $N(\mu, \sigma^2)$ distribution. It was shown in Example 3.18 that a consistent estimator of the αth sample quantile is given by $\bar{X}_n + \hat{\sigma}_n z_\alpha$ where z_α is the αth quantile of a $N(0,1)$ distribution. In the special case where we are interested in estimating the third quartile, which corresponds to the 0.75 quantile, this estimate approximately equals $\bar{X}_n + \frac{1349}{2000}\hat{\sigma}_n$. An alternative consistent estimate is given by Theorem 3.23 as the third sample quartile.   ■

3.10 Exercises and Experiments

3.10.1 Exercises

1. Let $\{X_n\}_{n=1}^{\infty}$ be a sequence of independent random variables where X_n is a GAMMA(α, β) random variable with $\alpha = n$ and $\beta = n^{-1}$ for $n \in \mathbb{N}$. Prove that $X_n \xrightarrow{p} 1$ as $n \to \infty$.

2. Let Z be a $N(0,1)$ random variable and let $\{X_n\}_{n=1}^{\infty}$ be a sequence of random variables such that $X_n = Y_n + Z$ where Y_n is a $N(n^{-1}, n^{-1})$ random variable for all $n \in \mathbb{N}$. Prove that $X_n \xrightarrow{p} Z$ and $n \to \infty$.

3. Consider a sequence of independent random variables $\{X_n\}_{n=1}^{\infty}$ where X_n has a BINOMIAL$(1, \theta)$ distribution. Prove that the estimator

$$\hat{\theta}_n = n^{-1} \sum_{k=1}^{n} X_k$$

   is a consistent estimator of θ.

4. Let U be a UNIFORM$(0,1)$ random variable and define a sequence of random variables $\{X_n\}_{n=1}^{\infty}$ as $X_n = \delta\{U; (0, n^{-1})\}$. Prove that $X_n \xrightarrow{p} 0$ as $n \to \infty$.

5. Let $\{c_n\}_{n=1}^{\infty}$ be a sequence of real constants such that

$$\lim_{n\to\infty} c_n = c,$$

   for some constant $c \in \mathbb{R}$. Let $\{X_n\}_{n=1}^{\infty}$ be a sequence of random variables such that $P(X_n = c_n) = 1 - n^{-1}$ and $P(X_n = 0) = n^{-1}$. Prove that $X_n \xrightarrow{p} c$ as $n \to \infty$.

6. Let $X_1, \ldots, X_n$ be a set of independent and identically distributed random variables from a shifted exponential density of the form

$$f(x) = \begin{cases} \exp[-(x - \theta)] & \text{for } x \geq \theta \\ 0 & \text{for } x < \theta. \end{cases}$$

Let $X_{(1)} = \min\{X_1, \ldots, X_n\}$. Prove that $X_{(1)} \xrightarrow{p} \theta$ as $n \to \infty$.

7. Let $X_1, \ldots, X_n$ be a set of independent and identically distributed random variables from a distribution F with variance μ_2 where $E(|X_1|^4) < \infty$. Let $\bar{X}_n$ be the sample mean and S_n^2 be the unbiased version of the sample variance.

a. Prove that the sample variance can be rewritten as

$$S_n^2 = \frac{1}{2n(n-1)} \sum_{i=1}^{n} \sum_{j=1}^{n} (X_i - X_j)^2.$$

b. Prove that S_n^2 is an unbiased estimator of μ_2, that is, prove that $E(S_n^2) = \mu_2$ for all $\mu_2 > 0$.

c. Prove that the variance of $\hat{\mu}_2$ is

$$V(\hat{\mu}_2) = n^{-1}(\mu_4 - \tfrac{n-3}{n-1}\mu_2^2).$$

d. Use the results derived above to prove that $\hat{\mu}_2$ is a consistent estimator of μ_2. That is, prove that $\hat{\mu}_2 \xrightarrow{p} \mu_2$ as $n \to \infty$.

e. Relate the results observed here with the results given in Theorem 3.21.

8. Consider a sequence of independent random variables $\{X_n\}_{n=1}^{\infty}$ where X_n has probability distribution function

$$f_n(x) = \begin{cases} 2^{-(n+1)} & x = -2^{n(1-\varepsilon)}, 2^{n(1-\varepsilon)} \\ 1 - 2^{-n} & x = 0 \\ 0 & \text{elsewhere,} \end{cases}$$

where $\varepsilon > \frac{1}{2}$ (Sen and Singer, 1993).

a. Compute the mean and variance of X_n.

b. Let

$$\bar{X}_n = n^{-1} \sum_{k=1}^{n} X_k,$$

for all $n \in \mathbb{N}$. Compute the mean and variance of $\bar{X}_n$.

c. Prove that $\bar{X}_n \xrightarrow{p} 0$ as $n \to \infty$.

9. Let $\{X_n\}_{n=1}^{\infty}$ be a sequence of random variables such that

$$\lim_{n \to \infty} E(|X_n - c|) = 0,$$

for some $c \in \mathbb{R}$. Prove that $X_n \xrightarrow{p} c$ as $n \to \infty$. *Hint: Use Theorem 2.6.*

10. Let U be a UNIFORM$[0, 1]$ random variable and let $\{X_n\}_{n=1}^{\infty}$ be a sequence of random variables such that

$$X_n = \delta\{U; (0, \tfrac{1}{2} - [2(n+1)]^{-1})\} + \delta\{U; (\tfrac{1}{2} + [2(n+1)]^{-1}, 1)\},$$

for all $n \in \mathbb{N}$. Prove that $X_n \xrightarrow{a.c.} 1$ as $n \to \infty$.

11. Let $\{X_n\}_{n=1}^{\infty}$ be a sequence of independent random variables where X_n has probability distribution function

$$f(x) = \begin{cases} 1 - n^{-1} & x = 0 \\ n^{-1} & x = n^{\alpha} \\ 0 & \text{elsewhere,} \end{cases}$$

where $\alpha \in \mathbb{R}$.

 a. For what values of α does $X_n \xrightarrow{P} 0$ as $n \to \infty$?
 b. For what values of α does $X_n \xrightarrow{a.c.} 0$ as $n \to \infty$?
 c. For what values of α does $X_n \xrightarrow{c} 0$ as $n \to \infty$?

12. Let $X_1, \ldots, X_n$ be a set of independent and identically distributed random variables following the distribution F. Prove that for a fixed value of $t \in \mathbb{R}$, the empirical distribution function $\hat{F}_n(t)$ is an unbiased estimator of $F(t)$ with standard error $n^{-1/2}\{F(t)[1 - F(t)]\}^{1/2}$.

13. Prove Theorem 3.3 using the theorems of Borel and Cantelli. That is, let $\{X_n\}_{n=1}^{\infty}$ be a sequence of random variables that converges completely to a random variable X as $n \to \infty$. Then prove that $X_n \xrightarrow{a.c.} X$ as $n \to \infty$ using Theorems 2.17 and 2.18.

14. Let $\{X_n\}_{n=1}^{\infty}$ be a sequence of monotonically increasing random variables that converge in probability to a random variable X. That is, $P(X_n < X_{n+1}) = 1$ for all $n \in \mathbb{N}$ and $X_n \xrightarrow{P} X$ as $n \to \infty$. Prove that $P(X_n < X) = 1$ for all $n \in \mathbb{N}$.

15. Prove Part 2 of Theorem 3.6. That is, let $\{\mathbf{X}_n\}_{n=1}^{\infty}$ be a sequence of d-dimensional random vectors and let $\mathbf{X}$ be another d-dimensional random vector. and prove that $\mathbf{X}_n \xrightarrow{a.c.} \mathbf{X}$ as $n \to \infty$ if and only if $X_{kn} \xrightarrow{a.c.} X_k$ as $n \to \infty$ for all $k \in \{1, \ldots, d\}$.

16. Let $\{\mathbf{X}_n\}_{n=1}^{\infty}$ be a sequence of random vectors that converge almost certainly to a random vector $\mathbf{X}$ as $n \to \infty$. Prove that for every $\varepsilon > 0$ it follows that

$$\lim_{n \to \infty} P(\|\mathbf{X}_m - \mathbf{X}\| < \varepsilon \text{ for all } m \geq n) = 1.$$

Prove the converse as well.

17. Let $\{U_n\}_{n=1}^{\infty}$ be a sequence of independent and identically distributed UNIFORM$(0, 1)$ random variables and let $U_{(n)}$ be the largest order statistic of $U_1, \ldots, U_n$. That is, $U_{(n)} = \max\{U_1, \ldots, U_n\}$. Prove that $U_{(n)} \xrightarrow{a.c.} 1$ as $n \to \infty$.

18. Prove the first part of Theorem 3.7. That is let $\{X_n\}_{n=1}^{\infty}$ be a sequence of random variables, c be a real constant, and g be a Borel function on $\mathbb{R}$ that is continuous at c. Prove that if $X_n \xrightarrow{a.c.} c$ as $n \to \infty$, then $g(X_n) \xrightarrow{a.c.} g(c)$ as $n \to \infty$.

19. Prove the second part of Theorem 3.8. That is let $\{X_n\}_{n=1}^{\infty}$ be a sequence of random variables, X be a random variable, and g be a Borel function on $\mathbb{R}$. Let $C(g)$ be the set of continuity points of g and suppose that $P[X \in C(g)] = 1$. Prove that if $X_n \xrightarrow{p} X$ as $n \to \infty$, then $g(X_n) \xrightarrow{p} g(X)$ as $n \to \infty$. *Hint: Prove by contradiction using Theorem 3.5.*

20. Let $\{\mathbf{X}_n\}$ be a sequence of d-dimensional random vectors, $\mathbf{X}$ be a d-dimensional random vector, and $g : \mathbb{R}^d \to \mathbb{R}^q$ be a Borel function. Let $C(g)$ be the set of continuity points of g and suppose that $P[\mathbf{X} \in C(g)] = 1$.

   a. Prove that if $\mathbf{X}_n \xrightarrow{a.c.} \mathbf{X}$ as $n \to \infty$, then $g(\mathbf{X}_n) \xrightarrow{a.c.} g(\mathbf{X})$ as $n \to \infty$.

   b. Prove that if $\mathbf{X}_n \xrightarrow{p} \mathbf{X}$ as $n \to \infty$, then $g(\mathbf{X}_n) \xrightarrow{p} g(\mathbf{X})$ as $n \to \infty$.

21. Let $\{X_n\}_{n=1}^{\infty}$, $\{Y_n\}_{n=1}^{\infty}$, and $\{Z_n\}_{n=1}^{\infty}$ be independent sequences of random variables that converge in probability to the random variables X, Y, and Z, respectively. Suppose that X, Y, and Z are independent and that each of these random variables has a $N(0,1)$ distribution.

   a. Let $\{\mathbf{W}_n\}_{n=1}^{\infty}$ be a sequence of three-dimensional random vectors defined by $\mathbf{W}_n = (X_n, Y_n, Z_n)'$ and let $\mathbf{W} = (X, Y, Z)'$. Prove that $\mathbf{W}_n \xrightarrow{p} \mathbf{W}$ as $n \to \infty$. Identify the distribution of $\mathbf{W}$. Would you be able to completely identify this distribution if the random variables X, Y, and Z are not independent? What additional information would be required?

   b. Prove that $\frac{1}{3}\mathbf{1}'\mathbf{W}_n \xrightarrow{p} \frac{1}{3}\mathbf{1}'\mathbf{W}$ and identify the distribution of $\frac{1}{3}\mathbf{1}'\mathbf{W}$.

22. Let $\{X_n\}_{n=1}^{\infty}$ be a sequence of random variables. Suppose that for every $\varepsilon > 0$ we have that
$$\limsup_{n \to \infty} P(|X_n| > \varepsilon) \le c\varepsilon,$$
where c is a finite real constant. Prove that $X_n \xrightarrow{p} 0$ as $n \to \infty$.

23. A result from calculus is *Kronecker's Lemma*, which states that if $\{b_n\}_{n=1}^{\infty}$ is a monotonically increasing sequence of real numbers such that $b_n \to \infty$ as $n \to \infty$, then the convergence of the series
$$\sum_{n=1}^{\infty} b_n x_n$$
implies that
$$\lim_{n \to \infty} b_n^{-1} \sum_{k=1}^{n} x_k = 0.$$
Use Kronecker's Lemma to prove the second result in Corollary 3.1. That is, let $\{X_n\}_{n=1}^{\infty}$ be a sequence of independent random variables where $E(X_n) = 0$ for all $n \in \mathbb{N}$. Prove that if
$$\sum_{n=1}^{\infty} b_n^{-2} E(X_n^2) < \infty,$$
then $b_n^{-1} S_n \xrightarrow{a.c.} 0$ as $n \to \infty$.

24. Let $\{X_n\}_{n=1}^{\infty}$ be a sequence of independent and identically distributed random variables from a CAUCHY$(0,1)$ distribution. Prove that the mean of the distribution does not exist, and further prove that it can be shown that $nP(|X_1| > n) \to 2\pi^{-1}$ as $n \to \infty$, so that the condition of Theorem 3.14 does not hold.

25. Let $\{X_n\}_{n=1}^{\infty}$ be a sequence of independent and identically distributed random variables from a density of the form

$$f(x) = \begin{cases} cx^2 \log(|x|) & |x| > 2 \\ 0 & |x| \le 2, \end{cases}$$

where c is a normalizing constant. Prove that $nP(|X_1| > n) \to 0$ as $n \to \infty$, but that the mean does not exist. Hence, we can still conclude that $\bar{X}_n \overset{p}{\to} 0$ as $n \to \infty$ due to Theorem 3.14.

26. Prove Theorem 3.17. That is, let F and G be two distribution functions. Show that

$$\|F - G\|_{\infty} = \sup_{t \in \mathbb{R}} |F(t) - G(t)|$$

is a metric in the space of distribution functions.

27. In the proof of Theorem 3.18, verify that $\hat{F}_n(t) - F(t) \ge \hat{F}_n(t) - F(t_{i-1}) - \varepsilon$.

28. Let $X_1, \ldots, X_n$ be a set of independent and identically distributed random variables from a distribution F.

   a. Prove that if $E(|X_1|^k) < \infty$ then $\hat{\mu}'_k$ is an unbiased estimator of μ'_k.

   b. Prove that if $E(|X_1|^{2k}) < \infty$ then the standard error of $\hat{\mu}'_k$ is $n^{-1/2}(\mu'_{2k} - \mu'^2_k)^{1/2}$.

   c. Prove that if $E(|X_1|^k) < \infty$ then $\hat{\mu}'_k \overset{a.c.}{\longrightarrow} \mu'_k$ as $n \to \infty$.

29. Let $X_1, \ldots, X_n$ be a set of independent and identically distributed random variables from a distribution F where $E(|X_1|^k) < \infty$. Consider the sum

$$\sum_{j=1}^{k} \binom{k}{j} E\left\{ (\mu'_1 - \hat{\mu}'_1)^j \left[n^{-1} \sum_{i=1}^{n} (X_i - \mu'_1)^{k-j} \right] \right\}.$$

Prove that when $j = 3$ the term in the sum is $O(n^{-2})$ as $n \to \infty$.

30. Let $X_1, \ldots, X_n$ be a set of independent and identically distributed random variables from a distribution F with $E(|X_1|^{2k}) < \infty$.

   a. Prove that

$$E(\hat{\mu}_k^2) = \mu_k^2 + n^{-1}[\mu_{2k} - \mu_k^2 - 2k(\mu_k^2 + \mu_{k-1}\mu_{k+1}) + k^2\mu_{k-1}^2\mu_2 +$$
$$k(k-1)\mu_k\mu_{k-2}\mu_2] + O(n^{-2}),$$

as $n \to \infty$.

b. Use this result to prove that

$$V(\hat{\mu}_k) = n^{-1}(\mu_{2k} - \mu_k^2 - 2k\mu_{k-1}\mu_{k+1} + k^2\mu_2\mu_{k-1}^2) + O(n^{-2}),$$

as $n \to \infty$.

31. Let $X_1, \ldots, X_n$ be a set of independent and identically distributed random variables from a distribution F with $E(|X_1|^k) < \infty$. Prove that $\hat{\mu}_k \xrightarrow{a.c.} \mu_k$ as $n \to \infty$.

3.10.2 Experiments

1. Consider an experiment that flips a fair coin 100 times. Define an indicator random variable B_n so that

$$B_k = \begin{cases} 1 & \text{if the } k^{\text{th}} \text{ flip is heads} \\ 0 & \text{if the } k^{\text{th}} \text{ flip is tails.} \end{cases}$$

The proportion of heads up to the nth flip is then

$$\hat{p}_n = n^{-1} \sum_{k=1}^{n} B_k.$$

Run a simulation that will repeat this experiment 25 times. On the same set of axes, plot $\hat{p}_n$ versus n for $n = 1, \ldots, 100$ with the points connected by lines, for each replication of the experiment. For comparison, plot a horizontal line at $\frac{1}{2}$, the true probability of flipping heads. Comment on the outcome of the experiments. What results are being demonstrated by this set of experiments? How do these results related to what is usually called the frequency method for computing probabilities?

2. Write a program in R that generates a sample $X_1, \ldots, X_n$ from a specified distribution F, computes the empirical distribution function of $X_1, \ldots, X_n$, and plots both the empirical distribution function and the specified distribution function F on the same set of axes. Use this program with $n = 5, 10, 25, 50$, and 100 to demonstrate the consistency of the empirical distribution function given by Theorem 3.16. Repeat this experiment for each of the following distributions: $N(0, 1)$, BINOMIAL$(10, 0.25)$, CAUCHY$(0, 1)$, and GAMMA$(2, 4)$.

3. Write a program in R that generates a sample $X_1, \ldots, X_n$ from a specified distribution F and computes the sample mean $\bar{X}_n$. Use this program with $n = 5, 10, 25, 50, 100$, and 1000 and plot the sample size against $\bar{X}_n$. Repeat the experiment five times, and plot all the results on a single set of axes. Produce the plot described above for each of the following distributions $N(0, 1)$, T(1), and T(2). For each distribution state whether the Strong Law of Large Numbers or the Weak Law of Large Numbers regulates the behavior of $\bar{X}_n$. What differences in behavior are observed on the plots?

4. Write a program in R that generates independent UNIFORM$(0, 1)$ random variables $U_1, \ldots, U_n$. Define two sequences of random variables $X_1, \ldots, X_n$ and $Y_1, \ldots, Y_n$ as $X_k = \delta\{U_k; (0, k^{-1})\}$ and $Y_k = \delta\{U_k; (0, k^{-2})\}$. Plot $X_1, \ldots, X_n$ and $Y_1, \ldots, Y_n$ against $k = 1, \ldots, n$ on the same set of axes for $n = 25$. Is it apparent from the plot that $X_n \overset{p}{\to} 0$ as $n \to \infty$ but that $Y_n \overset{c}{\nrightarrow} 0$ as $n \to \infty$? Repeat this process five times to get an idea of the average behavior in each plot.

5. Write a program in R that generates a sample $X_1, \ldots, X_n$ from a specified distribution F, computes the empirical distribution function of $X_1, \ldots, X_n$, computes the maximum distance between $\hat{F}_n$ and F, and computes the location of the maximum distance between $\hat{F}_n$ and F. Use this program with $n = 5, 10, 25, 50$, and 100 and plot the sample size versus the maximum distance to demonstrate Theorem 3.18. Separately, plot the location of the maximum distance between $\hat{F}_n$ and F against the sample size. Is there an area where the maximum tends to stay, or does it tend to occur where F has certain properties? Repeat this experiment for each of the following distributions: N$(0, 1)$, BINOMIAL$(10, 0.25)$, CAUCHY$(0, 1)$, and GAMMA$(2, 4)$.

6. Write a program in R that generates a sample from a population with distribution function

$$F(x) = \begin{cases} 0 & x < -1 \\ 1 + x & -1 \leq x < -\frac{1}{2} \\ \frac{1}{2} & -\frac{1}{2} \leq x < \frac{1}{2} \\ 1 - x & \frac{1}{2} \leq x < 1 \\ 1 & x \geq 1. \end{cases}$$

This distribution is UNIFORM on the set $[-1, -\frac{1}{2}] \cup [\frac{1}{2}, 1]$. Use this program to generate samples of size $n = 5, 10, 25, 50, 100, 500$, and 1000. For each sample compute the sample median $\hat{\xi}_{0.5}$. Repeat this process five times and plot the results on a single set of axes. What effect does the flat area of the distribution have on the convergence of the sample median? For comparison, repeat the entire experiment but compute $\hat{\xi}_{0.75}$ instead.

CHAPTER 4

Convergence of Distributions

"Ask them then," said the deputy director. "It's not that important," said K., although in that way his earlier excuse, already weak enough, was made even weaker. As he went, the deputy director continued to speak about other things.

The Trial by Franz Kafka

4.1 Introduction

In statistical inference it is often the case that we are not interested in whether a random variable converges to another specific random variable, rather we are just interested in the distribution of the limiting random variable. Statistical hypothesis testing provides a good example of this situation. Suppose that we have a random sample from a distribution F with mean μ, and we wish to test some hypothesis about μ. The most common test statistic to use in this situation is $Z_n = n^{1/2}\hat{\sigma}_n^{-1}(\hat{\mu}_n - \mu_0)$ where $\hat{\mu}_n$ and $\hat{\sigma}_n$ are the sample mean and standard deviation, respectively. The value μ_0 is a constant that is specified by the null hypothesis. In order to derive a statistical hypothesis test for the null hypothesis based on this test statistic we need to know the distribution of Z_n when the null hypothesis is true, which in this case we will take to be the condition that $\mu = \mu_0$. If the parametric form of F is not known explicitly, then this distribution can be approximated using the Central Limit Theorem, which states that Z_n approximately has a $N(0,1)$ distribution when n is large and $\mu = \mu_0$. See Section 4.4. This asymptotic result does not identify a specific random variable Z that Z_n converges to as $n \to \infty$. There is no need because all we are interested in is the distribution of the limiting random variable. This chapter will introduce and study a type of convergence that only specifies the distribution of the random variable of interest as $n \to \infty$.

4.2 Weak Convergence of Random Variables

We wish to define a type of convergence for a sequence of random variables $\{X_n\}_{n=1}^\infty$ to a random variable X that focuses on how the distribution of the random variables in the sequence converge to the distribution of X as $n \to \infty$. To define this type of convergence we will consider how the distribution functions of the random variables in the sequence $\{X_n\}_{n=1}^{\infty}$ converge to the distribution function of X.

If all of the random variables in the sequence $\{X_n\}_{n=1}^{\infty}$ and the random variable X were all continuous then it might make sense to consider the densities of the random variables. Or, if all of the random variables were discrete we could consider the convergence of the probability distribution functions of the random variables. Such approaches are unnecessarily restrictive. In fact, some of the more interesting examples of convergence of distributions of sequences of random variables are for sequences of discrete random variables that have distributions that converge to a continuous distribution as $n \to \infty$. By defining the mode of convergence in terms of distribution functions, which are defined for all types of random variables, our definition will allow for such results.

Definition 4.1. *Let $\{X_n\}_{n=1}^{\infty}$ be a sequence of random variables where X_n has distribution function F_n for all $n \in \mathbb{N}$. Then X_n converges in distribution to a random variable X with distribution function F as $n \to \infty$ if*

$$\lim_{n \to \infty} F_n(x) = F(x),$$

for all $x \in C(F)$, the set of points where F is continuous. This relationship will be represented by $X_n \xrightarrow{d} X$ as $n \to \infty$.

It is clear from Definition 4.1 that the concept of convergence in distribution literally requires that the distribution of the random variables in the sequence converge to a distribution that matches the distribution of the limiting random variable. Convergence in distribution is also often called *weak convergence* since the random variables play a secondary role in Definition 4.1. In fact, the concept of convergence in distribution can be defined without them.

Definition 4.2. *Let $\{F_n\}_{n=1}^{\infty}$ be a sequence of distribution functions. Then F_n converges weakly to F as $n \to \infty$ if*

$$\lim_{n \to \infty} F_n(x) = F(x),$$

for all $x \in C(F)$. This relationship will be represented by $F_n \rightsquigarrow F$ as $n \to \infty$.

Definitions 4.1 and 4.2 do not require that the point-wise convergence of the distribution functions take place where there are discontinuities in the limiting distribution function. It turns out that requiring point-wise convergence for all real values is too strict a requirement in that there are situations where the distribution obviously converges to a limit, but the stricter requirement is not met.

Example 4.1. Consider a sequence of random variables $\{X_n\}_{n=1}^{\infty}$ whose distribution function has the form $F_n(x) = \delta\{x; [n^{-1}, \infty)\}$. That is, $P(X_n = n^{-1}) = 1$ for all $n \in \mathbb{N}$. In the limit we would expect $X_n \xrightarrow{d} X$ as $n \to \infty$ where X has distribution function $F(x) = \delta\{x; [0, \infty)\}$. That is, $P(X = 0) = 1$. If we consider any $x \in (-\infty, 0)$ it is clear that

$$\lim_{n \to \infty} F_n(x) = \lim_{n \to \infty} \delta\{x; [n^{-1}, \infty)\} = 0 = F(x).$$

Similarly, if $x \in (0, \infty)$ we have that there exists a value $n_x \in \mathbb{N}$ such that $n^{-1} < x$ for all $n \geq n_x$. Therefore, for any $x \in (0, \infty)$ we have that

$$\lim_{n \to \infty} F_n(x) = \lim_{n \to \infty} \delta\{x; [n^{-1}, \infty)\} = 1 = F(x).$$

Now consider what occurs at $x = 0$. For every $n \in \mathbb{N}$, $F_n(x) = 0$ so that

$$\lim_{n \to \infty} F_n(x) = \delta\{x; [n^{-1}, \infty)\} = 0 \neq F(x).$$

Therefore, if we insist that $F_n \rightsquigarrow F$ as $n \to \infty$ only if

$$\lim_{n \to \infty} F_n(x) = F(x),$$

for *all* $x \in \mathbb{R}$, we would not be able to conclude that the convergence takes place in this instance. However, Definitions 4.1 and 4.2 do not have this strict requirement, and noting that $0 \notin C(F)$ in this case allows us to conclude that $F_n \rightsquigarrow F$, or $X_n \xrightarrow{d} X$, as $n \to \infty$. ∎

The two definitions of convergence in distribution arise in different applications. In the first application we have a sequence of random variables and we wish to determine the distribution of the limiting random variable. The Central Limit Theorem is an example of this application. In the second application we are only interested in the limit of a sequence of distribution functions. Approximation results, such as the normal approximation to the binomial distribution, are examples of this application. Some examples of both applications are given below.

Example 4.2. Let $\{X_n\}_{n=1}^{\infty}$ be a sequence of independent NORMAL$(0, 1 + n^{-1})$ random variables, and let X be a NORMAL$(0, 1)$ random variable. Taking the limit of the distribution function of X_n as $n \to \infty$ yields

$$\lim_{n \to \infty} \int_{-\infty}^{x} [2\pi(1 + n^{-1})]^{-1/2} \exp\{-\tfrac{1}{2}(1 + n^{-1})^{-1}t^2\} dt =$$

$$\lim_{n \to \infty} \int_{-\infty}^{x(1+n^{-1})^{-1/2}} (2\pi)^{-1/2} \exp\{-\tfrac{1}{2}v^2\} dv =$$

$$\int_{-\infty}^{x} (2\pi)^{-1/2} \exp\{-\tfrac{1}{2}v^2\} dv = \Phi(x),$$

where the change of variable $v = t(1 + n^{-1})^{-1/2}$ has been used. This limit is valid for all $x \in \mathbb{R}$, and since the distribution function of X is given by $\Phi(x)$, it follows from Definition 4.1 that $X_n \xrightarrow{d} X$ as $n \to \infty$. Note that we are required to verify that the limit is valid for all $x \in \mathbb{R}$ in this case because $C(\Phi) = \mathbb{R}$. ∎

Example 4.3. Let $\{X_n\}_{n=1}^{\infty}$ be a sequence of independent random variables where the distribution function of X_n is

$$F_n(x) = \begin{cases} 1 - \exp(\theta - x) & \text{for } x \in [\theta, \infty), \\ 0 & \text{for } x \in (-\infty, \theta). \end{cases}$$

Define a new sequence of random variables given by $Y_n = \min\{X_1, \ldots, X_n\}$ for all $n \in \mathbb{N}$. The distribution function of Y_n can be found by noting that $Y_n > y$ if and only if $X_k > y$ for all $k \in \{1, \ldots, n\}$. Therefore, for $y \in [\theta, \infty)$,

$$
\begin{aligned}
P(Y_n \leq y) &= 1 - P(Y_n > y) \\
&= 1 - P\left(\bigcap_{i=1}^{n} \{X_i > y\}\right) \\
&= 1 - \prod_{i=1}^{n} P(X_i > y) \\
&= 1 - \prod_{i=1}^{n} \exp(\theta - y) \\
&= 1 - \exp[n(\theta - y)].
\end{aligned}
$$

Further, note that if $y \in (-\infty, \theta)$, then $P(Y_n \leq y)$ is necessarily zero due to the fact that $P(X_n \leq \theta) = 0$ for all $n \in \mathbb{N}$. Therefore, the distribution function of Y_n is

$$
G_n(y) = \begin{cases} 1 - \exp[n(\theta - y)] & \text{for } y \in [\theta, \infty), \\ 0 & \text{for } y \in (-\infty, \theta). \end{cases}
$$

For $y \in [\theta, \infty)$ we have that

$$
\lim_{n \to \infty} G_n(y) = \lim_{n \to \infty} \{1 - \exp[n(\theta - y)]\} = 1.
$$

For $y \in (-\infty, \theta)$ the distribution function $G_n(y)$ is zero for all $n \in \mathbb{N}$. Therefore, it follows from Definition 4.2 that $G_n \rightsquigarrow G$ as $n \to \infty$ where $G(y) = \delta(y; [\theta, \infty))$. In terms of random variables it follows that

$$
Y_n = \min\{X_1, \ldots, X_n\} \xrightarrow{d} Y,
$$

as $n \to \infty$ where $P(Y = \theta) = 1$, a degenerate distribution at θ. In subsequent discussions will use the simpler notation $Y_n \xrightarrow{d} \theta$ as $n \to \infty$ where it is understood that convergence in distribution to a constant indicates convergence to a degenerate random variable at that constant. ∎

Example 4.4. This example motivates the POISSON approximation to the BINOMIAL distribution. Consider a sequence of random variables $\{X_n\}_{n=1}^{\infty}$ where X_n has a BINOMIAL$(n, n^{-1}\lambda)$ distribution where λ is a positive constant. Let x be a positive real value. The distribution function of X_n is given by

$$
F_n(x) = P(X_n \leq x) = \begin{cases} 0 & \text{for } x < 0, \\ \sum_{i=0}^{\lfloor x \rfloor} \binom{n}{i} \left(\frac{\lambda}{n}\right)^i \left(1 - \frac{\lambda}{n}\right)^{n-i} & \text{for } 0 \leq x \leq n \\ 1 & \text{for } x > n. \end{cases}
$$

Consider the limiting behavior of the term

$$\binom{n}{i}\left(\frac{\lambda}{n}\right)^i\left(1-\frac{\lambda}{n}\right)^{n-i},$$

as $n \to \infty$ when i is a fixed integer between 0 and n. Some care must be taken with this limit to keep a single term from diverging while the remaining terms converge to zero, resulting in an indeterminate form. Therefore, note that

$$\binom{n}{i}\left(\frac{\lambda}{n}\right)^i\left(1-\frac{\lambda}{n}\right)^{n-i} = \frac{n!}{i!(n-i)!}\left(\frac{\lambda}{n}\right)^i\left(1-\frac{\lambda}{n}\right)^{-i}\left(1-\frac{\lambda}{n}\right)^n$$

$$= \frac{n(n-1)\cdots(n-i+1)}{i!}\lambda^i(n-\lambda)^{-i}\left(1-\frac{\lambda}{n}\right)^n$$

$$= \frac{\lambda^i}{i!}\left(1-\frac{\lambda}{n}\right)^n\prod_{k=1}^{i}\frac{n-k+1}{n-\lambda}.$$

Now

$$\lim_{n\to\infty}\prod_{k=1}^{i}\frac{n-k+1}{n-\lambda} = 1,$$

and Theorem 1.7 implies that

$$\lim_{n\to\infty}\left(1-\frac{\lambda}{n}\right)^n = \exp(-\lambda).$$

Therefore, it follows that

$$\lim_{n\to\infty}\binom{n}{i}\left(\frac{\lambda}{n}\right)^i\left(1-\frac{\lambda}{n}\right)^{n-i} = \frac{\lambda^i\exp(-\lambda)}{i!},$$

for each $i \in \{0,1,\ldots,n\}$, and hence for $0 \le x \le n$ we have that

$$\lim_{n\to\infty}F_n(x) = \lim_{n\to\infty}\sum_{i=0}^{\lfloor x\rfloor}\binom{n}{i}\left(\frac{\lambda}{n}\right)^i\left(1-\frac{\lambda}{n}\right)^{n-i} = \sum_{i=0}^{\lfloor x\rfloor}\frac{\lambda^i\exp(-\lambda)}{i!}, \qquad (4.1)$$

which is the distribution function of a POISSON(λ) random variable. Therefore, Definition 4.1 implies that we have proven that $X_n \overset{d}{\to} X$ as $n \to \infty$ where X is a POISSON(λ) random variable. It is also clear that from the limit exhibited in Equation (4.1), that when n is large we may make use of the approximation $P(X_n \le x) \simeq P(X \le x)$. Hence, we have proven that under certain conditions, probabilities of a BINOMIAL($n, n^{-1}\lambda$) random variable can be approximated with POISSON(λ) probabilities. ∎

An important question related to the convergence of distributions is under what conditions that limiting distribution is a valid distribution function. That is, suppose that $\{F_n\}_{n=1}^{\infty}$ is a sequence of distribution functions such that

$$\lim_{n\to\infty}F_n(x) = F(x),$$

for all $x \in C(F)$ for some function $F(x)$. Must $F(x)$ necessarily be a distribution function? Standard mathematical arguments from calculus can be used to show that $F(x) \in [0,1]$ for all $x \in \mathbb{R}$ and that $F(x)$ is non-decreasing. See Exercise 6. We do not have to worry about the right-continuity of F as we do not necessarily require convergence at the points of discontinuity of F, so that we can make these points right-continuous if we wish. The final property that is required for F to be a valid distribution function is that

$$\lim_{x \to \infty} F(x) = 1,$$

and

$$\lim_{x \to -\infty} F(x) = 0.$$

The example given below shows that these properties do not always follow for the limiting distribution.

Example 4.5. Lehmann (1999) considers a sequence of random variables $\{X_n\}_{n=1}^{\infty}$ such that the distribution function of X_n is given by

$$F_n(x) = \begin{cases} 0 & x < 0 \\ 1 - p_n, & 0 \le x < n, \\ 1 & x \ge n, \end{cases}$$

where $\{p_n\}_{n=1}^{\infty}$ is a sequence of real numbers such that

$$\lim_{n \to \infty} p_n = p.$$

If $p = 0$ then

$$\lim_{n \to \infty} F_n(x) = F(x) = \begin{cases} 0 & x < 0 \\ 1 & x \ge 0, \end{cases}$$

which is a degenerate distribution at zero. However, if $0 < p < 1$ then the limiting function is given by

$$F(x) = \begin{cases} 0 & x < 0, \\ 1 - p & x \ge 0. \end{cases}$$

It is clear in this case that

$$\lim_{x \to \infty} F(x) = 1 - p < 1.$$

Therefore, even though $F_n \rightsquigarrow F$ as $n \to \infty$, it does not follow that F is a valid distribution function. ∎

In Example 4.5 we observed a situation where a sequence of distribution functions converged to a function that was not a valid distribution function. The problem in this case was that the sequence of distribution functions has a non-zero mass located at a point that diverged to ∞ as $n \to \infty$, leaving the upper tail of the distribution function less than one. It turns out that this is the essential problem which keeps us from concluding that a limiting function is a valid distribution function. Sequences that do not have this problem are said to be *bounded in probability*.

Definition 4.3. *Let $\{X_n\}_{n=1}^{\infty}$ be a sequence of random variables. The sequence is bounded in probability if for every $\varepsilon > 0$ there exists an $x_\varepsilon \in \mathbb{R}$ and $n_\varepsilon \in \mathbb{N}$ such that $P(|X_n| \leq x_\varepsilon) > 1 - \varepsilon$ for all $n > n_\varepsilon$.*

Example 4.6. Reconsider the situation in Example 4.5 where $\{X_n\}_{n=1}^{\infty}$ is a sequence of random variables such that the distribution function of X_n is given by

$$F_n(x) = \begin{cases} 0 & x < 0 \\ 1 - p_n, & 0 \leq x < n, \\ 1 & x \geq n, \end{cases}$$

where $\{p_n\}_{n=1}^{\infty}$ is a sequence of real numbers such that

$$\lim_{n \to \infty} p_n = p.$$

If we first consider the case where $p = 0$, then for every $\varepsilon > 0$, we need only use Definition 1.1 and find a value n_ε such that $p_n < \varepsilon$ for all $n \geq n_\varepsilon$. For this value of n, it will follow that $P(|X_n| \leq 0) \geq 1 - \varepsilon$ for all $n > n_\varepsilon$, and by Definition 4.3, the sequence is bounded in probability. On the other hand, consider the case where $p > 0$ and we set a value of ε such that $0 < \varepsilon < p$. Let x be a positive real value. For any $n > x$ we have the property that $P(|X_m| \leq x) = 1 - p \leq 1 - \varepsilon$ for all $m > n$. Therefore, it is not possible to find the value of x required in Definition 4.3, and the sequence is not bounded in probability. ∎

In Examples 4.5 and 4.6 we found that when the sequence in question was bounded in probability, the corresponding limiting distribution function was a valid distribution function. When the sequence was not bounded in probability, the limiting distribution function was not a valid distribution function. Hence, for that example, the property that the sequence is bounded in probability is equivalent to the condition that the limiting distribution function is valid. This property is true in general.

Theorem 4.1. *Let $\{X_n\}_{n=1}^{\infty}$ be a sequence of random variables where X_n has distribution function F_n for all $n \in \mathbb{N}$. Suppose that $F_n \rightsquigarrow F$ as $n \to \infty$ where F may or may not be a valid distribution function. Then,*

$$\lim_{x \to -\infty} F(x) = 0,$$

and

$$\lim_{x \to \infty} F(x) = 1,$$

if and only if the sequence $\{X_n\}_{n=1}^{\infty}$ is bounded in probability.

Proof. We will first assume that the sequence $\{X_n\}_{n=1}^{\infty}$ is bounded in probability and that $X_n \xrightarrow{d} X$ as $n \to \infty$. Let $\varepsilon > 0$. Because $\{X_n\}_{n=1}^{\infty}$ is bounded in probability there exist $x_{\varepsilon/2} \in \mathbb{R}$ and $n_{\varepsilon/2} \in \mathbb{N}$ such that $P(|X_n| \leq x_{\varepsilon/2}) \geq$

$1 - \varepsilon/2$ for all $n > n_{\varepsilon/2}$. This implies that $F_n(x) \geq 1 - \varepsilon/2$ and because $F_n(x) \leq 1$ for all $x \in \mathbb{R}$ and $n \in \mathbb{N}$, it follows that $|F_n(x) - 1| < \varepsilon/2$ for all $x > x_{\varepsilon/2}$ and $n > n_{\varepsilon/2}$. Because $X_n \overset{d}{\to} X$ as $n \to \infty$ it follows that there exists $n'_{\varepsilon/2} \in \mathbb{N}$ such that $|F_n(x) - F(x)| < \varepsilon/2$ for all $n > n'_{\varepsilon/2}$. Let $n_\varepsilon = \max\{n_{\varepsilon/2}, n'_{\varepsilon/2}\}$. Theorem A.18 implies that

$$|F(x) - 1| \leq |F(x) - F_n(x)| + |F_n(x) - 1| \leq \varepsilon,$$

for all $x \in C(F)$ such that $x > x_{\varepsilon/2}$ and $n > n_{\varepsilon/2}$. Hence, it follows that $|F(x) - 1| \leq \varepsilon$ for all $x \in C(F)$ such that $x > x_{\varepsilon/2}$. It follows then that

$$\lim_{x \to \infty} F(x) = 1.$$

Similar arguments can be used to calculate the limit of F as $x \to -\infty$. See Exercise 7. To prove the converse, see Exercise 8. □

The convergence of sequences that are not bounded in probability can still be studied in a more general framework called *vague convergence*. See Section 5.8.3 of Gut (2005) for further information on this type of convergence.

Recall that a quantile function is defined as the inverse of the distribution function, where we have used $\xi_p = F^{-1}(p) = \inf\{x : F(x) \geq p\}$ as the inverse in cases where there is not a unique inverse. Therefore, associated with each sequence of distribution functions $\{F_n\}_{n=1}^\infty$, is a sequence of quantile functions $\{F_n^{-1}(t)\}_{n=1}^\infty$. The weak convergence of the distribution functions to a distribution function F as $n \to \infty$ implies the convergence of the quantile functions as well. However, as distribution functions are not required to converge at every point, a similar result holds for the convergence of the quantile functions.

Theorem 4.2. *Let $\{F_n\}_{n=1}^\infty$ be a sequence of distribution functions that converge weakly to a distribution function F as $n \to \infty$. Let $\{F_n^{-1}\}_{n=1}^\infty$ be the corresponding sequence of quantile functions and let F^{-1} be the quantile function corresponding to F. Define N to be the set of points where F_n^{-1} does not converge pointwise to F^{-1}. That is*

$$N = (0, 1) \setminus \left\{ t \in (0, 1) : \lim_{n \to \infty} F_n^{-1}(t) = F^{-1}(t) \right\}.$$

Then N is countable.

A proof of Theorem 4.2 can be found in Section 1.5.6 of Serfling (1980). As Theorem 4.2 implies, there may be as many as a countable number of points where the convergence does not take place. Certainly for the points where the distribution function does not converge, we cannot expect the quantile function to necessarily converge at the inverse of those points. However, the result is not specific about at which points the convergence may not take place, and other points may be included in the set N as well, such as the inverse of points that occur where the distribution functions are not increasing. On the other hand, there may be cases where the convergence of the quantile functions may occur at all points in $(0, 1)$, as the next example demonstrates.

Example 4.7. Let $\{X_n\}_{n=1}^{\infty}$ be a sequence of random variables where X_n is an EXPONENTIAL$(\theta + n^{-1})$ random variable for all $n \in \mathbb{N}$ where θ is a positive real constant. Let X be an EXPONENTIAL(θ) random variable. It can be shown that $X_n \xrightarrow{d} X$ as $n \to \infty$. See Exercise 2. The quantile function associated with X_n is given by $F_n^{-1}(t) = -(\theta + n^{-1})^{-1} \log(1 - t)$ for all $n \in \mathbb{N}$ and $t \in (0, 1)$. Similarly, the quantile function associated with X is given by $F^{-1}(t) = -\theta^{-1} \log(1 - t)$. Let $t \in (0, 1)$ and note that

$$\lim_{n \to \infty} F_n^{-1}(t) = \lim_{n \to \infty} -(\theta + n^{-1})^{-1} \log(1 - t) = -\theta^{-1} \log(1 - t) = F^{-1}(t).$$

Therefore, the convergence of the quantile functions holds for all $t \in (0, 1)$, and therefore $N = \emptyset$. ∎

We also present an example where the convergence of the quantile function does not hold at some points $t \in (0, 1)$.

Example 4.8. Let $\{X_n\}_{n=1}^{\infty}$ be a sequence of random variables where X_n has a BERNOULLI$[\frac{1}{2} + (n + 2)^{-1}]$ distribution for all $n \in \mathbb{N}$ and let X be a BERNOULLI$(\frac{1}{2})$ random variable. It can be shown that $X_n \xrightarrow{d} X$ as $n \to \infty$. See Exercise 4. The distribution function of X_n is given by

$$F_n(x) = \begin{cases} 0 & x < 0 \\ \frac{1}{2} - (n + 2)^{-1} & 0 \le x < 1 \\ 1 & x \ge 1, \end{cases}$$

and the distribution function of X is given by

$$F(x) = \begin{cases} 0 & x < 0 \\ \frac{1}{2} & 0 \le x < 1 \\ 1 & x \ge 1. \end{cases}$$

Therefore, $F_n^{-1}(\frac{1}{2}) = 1$ for all $n \in \mathbb{N}$ but $F^{-1}(\frac{1}{2}) = 0$. Therefore,

$$\lim_{n \to \infty} F_n^{-1}(\tfrac{1}{2}) \ne F(\tfrac{1}{2}),$$

and hence $\frac{1}{2} \in N$. ∎

Another important question is whether the expectations of random variables that convergence in distribution to a random variable also converge. The answer to this question is not straightforward, and as we will show in Chapter 5, weak convergence is not enough to ensure that the corresponding moments will also converge. In this section we will develop some limited results along this line of inquiry that will be specifically developed to result in an equivalence between the convergence of certain expectations and weak convergence. We begin by with the result of Helly and Bray, which concerns expectations truncated to a finite real interval.

Theorem 4.3 (Helly and Bray). *Let $\{F_n\}_{n=1}^{\infty}$ be a sequence of distribution functions and let g be a function that is continuous on $[a, b]$ where $-\infty <$*

$a < b < \infty$ and a and b are continuity points of a distribution function F. If $F_n \rightsquigarrow F$ as $n \to \infty$, then

$$\lim_{n\to\infty} \int_a^b g(x)dF_n(x) = \int_a^b g(x)dF(x).$$

Proof. We follow the development of this result given in Sen and Singer (1993). Let $\varepsilon > 0$ and consider a partition of the interval $[a, b]$ given by $a = x_0 < x_1 < \cdots < x_m < x_{m+1} = b$. We will assume that x_k is a continuity point of F for all $k \in \{0, 1, \ldots, m+1\}$ and that $x_{k+1} - x_k < \varepsilon$ for $k \in \{0, 1, \ldots, m\}$. We will form a step function to approximate g on the interval $[a, b]$. Define $g_m(x) = g[\frac{1}{2}(x_k + x_{k+1})]$ whenever $x \in (x_k, x_{k+1})$ and note that because $a = x_0 < \cdots < x_{m+1} = b$ it follows that for every $m \in \mathbb{N}$ and $x \in [a, b]$ we can write $g_m(x)$ as

$$g_m(x) = \sum_{k=0}^{m} g[\tfrac{1}{2}(x_k + x_{k+1})]\delta\{x; (x_k, x_{k+1})\}.$$

Now, repeated application of Theorem A.18 implies that

$$\left| \int_a^b g(x)dF_n(x) - \int_a^b g(x)dF(x) \right| \leq \left| \int_a^b g(x)dF_n(x) - \int_a^b g_m(x)dF_n(x) \right|$$

$$+ \left| \int_a^b g_m(x)dF_n(x) - \int_a^b g_m(x)dF(x) \right|$$

$$+ \left| \int_a^b g_m(x)dF(x) - \int_a^b g(x)dF(x) \right|.$$

To bound the first term we note that since $x_{k+1} - x_k < \varepsilon$ for all $k \in \{0, \ldots, m\}$ it then follows that if $x \in (x_k, x_{k+1})$ then $|x - \frac{1}{2}(x_k + x_{k+1})| < \varepsilon$. Because g is a continuous function it follows that there exists $\eta_\varepsilon > 0$ such that $|g_m(x) - g(x)| = |g[\frac{1}{2}(x_k + x_{k+1})] - g(x)| < \eta_\varepsilon$, for all $x \in (x_k, x_{k+1})$. Therefore, there exists $\delta_\varepsilon > 0$ such that

$$\sup_{x\in[a,b]} |g_m(x) - g(x)| < \tfrac{1}{3}\delta_\varepsilon.$$

Now, Theorem A.6 implies that

$$\left| \int_a^b g(x)dF_n(x) - \int_a^b g_m(x)dF_n(x) \right| = \left| \int_a^b [g(x) - g_m(x)]dF_n(x) \right|$$

$$\leq \int_a^b |g(x) - g_m(x)|dF_n(x)$$

$$\leq \sup_{x\in[a,b]} |g(x) - g_m(x)| \int_a^b dF_n(x)$$

$$\leq \tfrac{1}{3}\delta_\varepsilon.$$

Hence, this term can be made arbitrarily small by choosing ε to be small, or equivalently by choosing m to be large. For the second term we note that since g_m is a step function, we have that

$$\int_a^b g_m(x)dF_n(x) = \sum_{k=0}^{m} g_m(x)[F_n(x_{k+1}) - F_n(x_k)]$$

$$= \sum_{k=0}^{m} g[\tfrac{1}{2}(x_{k+1} - x_k)][F_n(x_{k+1}) - F_n(x_k)],$$

where the second equality follows from the definition of g_m. Similarly

$$\int_a^b g_m(x)dF(x) = \sum_{k=0}^{m} g[\tfrac{1}{2}(x_{k+1} - x_k)][F(x_{k+1}) - F(x_k)].$$

Therefore,

$$\int_a^b g_m(x)dF_n(x) - \int_a^b g_m(x)dF(x) =$$

$$\sum_{k=0}^{m} g[\tfrac{1}{2}(x_{k+1} - x_k)][F_n(x_{k+1}) - F(x_{k+1}) - F_n(x_k) + F(x_k)].$$

Now, since $F_n \rightsquigarrow F$ as $n \to \infty$ and x_k and x_{k+1} are continuity points of F, it follows from Defintion 4.1 that

$$\lim_{n \to \infty} F_n(x_k) = F(x_k),$$

for all $k \in \{0, 1, \ldots, m+1\}$. Therefore, it follows that

$$\lim_{n \to \infty} \left| \int_a^b g_m(x)dF_n(x) - \int_a^b g_m(x)dF(x) \right| = 0.$$

The third term can also be bounded by $\frac{1}{3}\delta_\varepsilon$ using similar arguments to those above. See Exercise 19. Since all three terms can be made smaller than $\frac{1}{3}\delta_\varepsilon$ for any $\delta_\varepsilon > 0$, it follows that

$$\lim_{n \to \infty} \left| \int_a^b g(x)dF_n(x) - \int_a^b g(x)dF(x) \right| < \delta_\varepsilon,$$

for any $\delta_\varepsilon > 0$ and therefore it follows that

$$\lim_{n \to \infty} \left| \int_a^b g(x)dF_n(x) - \int_a^b g(x)dF(x) \right| = 0,$$

which completes the proof. $\qquad\square$

The restriction that the range of the integral is bounded can be weakened if we are willing to assume that the function of interest is instead bounded. This result is usually called the extended or generalized theorem of Helly and Bray.

Theorem 4.4 (Helly and Bray). *Let g be a continuous and bounded function and let $\{F_n\}_{n=1}^{\infty}$ be a sequence of distribution functions such that $F_n \rightsquigarrow F$ as $n \to \infty$, where F is a distribution function. Then,*

$$\lim_{n \to \infty} \int_{-\infty}^{\infty} g(x)dF_n(x) = \int_{-\infty}^{\infty} g(x)dF(x).$$

Proof. Once again, we will use the method of proof from Sen and Singer (1993). This method of proof breaks up the integrals in the difference

$$\int_{-\infty}^{\infty} g(x)dF_n(x) - \int_{-\infty}^{\infty} g(x)dF(x),$$

into two basic parts. In the first part, the integrals are integrated over a finite range, and hence Theorem 4.3 (Helly and Bray) can be used to show that the difference converges to zero. The second part corresponds to the integrals of the leftover tails of the range. These differences will be made arbitrarily small by appealing to both the assumed boundedness of the function g and the fact that F is a distribution function. To begin, let $\varepsilon > 0$ and let

$$\tilde{g} = \sup_{x \in \mathbb{R}} |g(x)|,$$

where $\tilde{g} < \infty$ by assumption. Let a and b be continuity points of F. Repeated use of Theorem A.18 implies that

$$\left| \int_{-\infty}^{\infty} g(x)dF_n(x) - \int_{-\infty}^{\infty} g(x)dF(x) \right| \leq \left| \int_{-\infty}^{a} g(x)dF_n(x) - \int_{-\infty}^{a} g(x)dF(x) \right|$$

$$+ \left| \int_{a}^{b} g(x)dF_n(x) - \int_{a}^{b} g(x)dF(x) \right|$$

$$+ \left| \int_{b}^{\infty} g(x)dF_n(x) - \int_{b}^{\infty} g(x)dF(x) \right|.$$

Theorem 4.3 (Helly and Bray) can be applied to the second term as long as a and b are finite constants that do not depend on n, to obtain

$$\lim_{n \to \infty} \left| \int_{a}^{b} g(x)dF_n(x) - \int_{a}^{b} g(x)dF(x) \right| = 0.$$

The find bounds for the remaining two terms, we note that since F is a distribution function it follows that

$$\lim_{x \to \infty} F(x) = 1,$$

and

$$\lim_{x \to -\infty} F(x) = 0.$$

Therefore, Definition 1.1 implies that for every $\delta > 0$ there exist finite continuity points $a < b$ such that

$$\int_{-\infty}^{a} dF(x) = F(a) < \delta,$$

and

$$\int_b^\infty dF(x) = 1 - F(b) < \delta.$$

Therefore, for these values of a and b it follows that

$$\left| \int_{-\infty}^a g(x)dF_n(x) \right| \le \left| \int_{-\infty}^a \tilde{g} dF_n(x) \right| = \tilde{g} \int_{-\infty}^a dF_n(x) = \tilde{g} F_n(a).$$

For the choice of a described above we have that

$$\lim_{n \to \infty} \tilde{g} F_n(a) = \tilde{g} F(a) < \tilde{g}\delta,$$

since a is a continuity point of F, where δ can be chosen so that $\tilde{g}\delta < \varepsilon$. Therefore,

$$\lim_{n \to \infty} \tilde{g} F_n(a) < \varepsilon.$$

Similarly, for this same choice of a we have that

$$\int_{-\infty}^a g(x)dF(x) < \tilde{g}\delta,$$

and therefore Theorem A.18 implies that

$$\lim_{n \to \infty} \left| \int_{-\infty}^a g(x)dF_n(x) - \int_{-\infty}^a g(x)dF(x) \right| \le$$
$$\lim_{n \to \infty} \left| \int_{-\infty}^a g(x)dF_n(x) \right| + \left| \int_{-\infty}^a g(x)dF(x) \right| < 2\varepsilon,$$

for every $\varepsilon > 0$. Therefore, it follows that

$$\lim_{n \to \infty} \left| \int_{-\infty}^a g(x)dF_n(x) - \int_{-\infty}^a g(x)dF(x) \right| = 0.$$

Similarly, it can be shown that

$$\lim_{n \to \infty} \left| \int_b^\infty g(x)dF_n(x) - \int_b^\infty g(x)dF(x) \right| = 0,$$

and the result follows. See Exercise 9. □

The limit property in Theorem 4.4 (Helly and Bray) can be further shown to be equivalent to the weak convergence of the sequence of distribution functions, thus characterizing weak convergence. Another characterization of weak convergence is based on the convergence of the characteristic functions corresponding to the sequence of distribution functions. However, in order to prove this equivalence, we require another of Helly's Theorems.

Theorem 4.5 (Helly). *Let $\{F_n\}_{n=1}^\infty$ be a sequence of non-decreasing functions that are uniformly bounded. Then the sequence $\{F_n\}_{n=1}^\infty$ contains at least one subsequence $\{F_{n_m}\}_{n=1}^\infty$ where $\{n_m\}_{m=1}^\infty$ is an increasing sequence in $\mathbb{N}$ such that $F_{n_m} \rightsquigarrow F$ as $m \to \infty$ where F is a non-decreasing function.*

A proof of Theorem 4.5 can be found in Section 37 of Gnedenko (1962). Theorem 4.5 is somewhat general in that it deals with sequences that may or may not be distribution functions. The result does apply to sequences of distribution functions since they are uniformly bounded between zero and one. However, as our discussion earlier in this chapter suggests, the limiting function F may not be a valid distribution function, even when the sequence is comprised of distribution functions. There are two main potential problems in this case. The first is that F need not be right continuous. But, since weak convergence is defined on the continuity points of F, it follows that we can always define F in such a way that F is right continuous at the points of discontinuity of F, without changing the weak convergence properties of the sequence. See Exercise 10. The second potential problem is that F may not have the proper limits as $x \to \infty$ and $x \to -\infty$. This problem cannot be addressed without further assumptions. See Theorem 4.1. It turns out that the convergence of the corresponding distribution functions provides the additional assumptions that are required. With this assumption, the result given below provides other cromulent methods for assessing weak convergence.

Theorem 4.6. *Let $\{F_n\}_{n=1}^{\infty}$ be a sequence of distribution functions and let $\{\psi_n\}_{n=1}^{\infty}$ be a sequence of characteristic functions such that ψ_n is the characteristic function of F_n for all $n \in \mathbb{N}$. Let F be a distribution function with characteristic function ψ. The following three statements are equivalent:*

1. *$F_n \rightsquigarrow F$ as $n \to \infty$.*

2. *For each $t \in \mathbb{R}$,*
$$\lim_{n \to \infty} \psi_n(t) = \psi(t).$$

3. *For each bounded and continuous function g,*
$$\lim_{n \to \infty} \int g(x) dF_n(x) = \int g(x) dF(x).$$

Proof. We begin by showing that Conditions 1 and 3 are equivalent. The fact that Condition 1 implies Condition 3 follows directly from Theorem 4.4 (Helly and Bray). To prove that Condition 3 implies Condition 1 we follow the method of proof given by Serfling (1980). Let $\varepsilon > 0$ and let $t \in \mathbb{R}$ be a continuity point of F. Consider a function g given by

$$g(x) = \begin{cases} 1 & x \leq t, \\ 1 - \varepsilon^{-1}(x - t) & t < x < t + \varepsilon \\ 0 & x \geq t + \varepsilon. \end{cases}$$

Note that g is continuous and bounded. See Figure 4.1. Therefore, Condition 3 implies that

$$\lim_{n \to \infty} \int_{-\infty}^{\infty} g(x) dF_n(x) = \int_{-\infty}^{\infty} g(x) dF(x).$$

Now, because $g(x) = 0$ when $x \geq t + \varepsilon$, it follows that

$$\int_{-\infty}^{\infty} g(x)dF_n(x) = \int_{-\infty}^{t} dF_n(x) + \int_{t}^{t+\varepsilon} g(x)dF_n(x).$$

Since $g(x) \geq 0$ for all $t \in [t, t+\varepsilon]$ it follows that

$$\int_{-\infty}^{\infty} g(x)dF_n(x) \geq \int_{-\infty}^{t} dF_n(x) = F_n(t),$$

for all $n \in \mathbb{N}$. Therefore, Theorem 1.6 implies that

$$\limsup_{n\to\infty} F_n(t) \leq \lim_{n\to\infty} \int_{-\infty}^{\infty} g(x)dF_n(x) = \int_{-\infty}^{\infty} g(x)dF(x),$$

where we have used the limit superior because we do not know if the limit of $F_n(t)$ converges or not. Now

$$
\begin{aligned}
\int_{-\infty}^{\infty} g(x)dF(x) &= \int_{-\infty}^{t} dF(x) + \int_{t}^{t+\varepsilon} g(x)dF(x) \\
&\leq \int_{-\infty}^{t} dF(x) + \int_{t}^{t+\varepsilon} dF(x) \\
&= F(t+\varepsilon),
\end{aligned}
$$

since $g(x) \leq 1$ for all $x \in [t, t+\varepsilon]$. Therefore, we have shown that

$$\limsup_{n\to\infty} F_n(t) \leq F(t+\varepsilon).$$

Similar arguments can be used to show that

$$\liminf_{n\to\infty} F_n(t) \geq F(t-\varepsilon).$$

See Exercise 11. Hence, we have shown that

$$F(t-\varepsilon) \leq \liminf_{n\to\infty} F_n(t) \leq \limsup_{n\to\infty} F_n(t) \leq F(t+\varepsilon),$$

for every $\varepsilon > 0$. Since t is a continuity point of F, this implies that

$$\liminf_{n\to\infty} F_n(t) = \limsup_{n\to\infty} F_n(t)$$

or equivalently that

$$\lim_{n\to\infty} F_n(t) = F(t),$$

by Definition 1.3, when t is a continuity point of F. Therefore, Definition 4.2 implies that $F_n \rightsquigarrow F$ as $n \to \infty$. Hence, we have shown that Conditions 1 and 3 are equivalent.

To prove that Conditions 1 and 2 are equivalent we use the method of proof given by Gnendenko (1962). We will first show that Condition 1 implies Condition 2. Therefore, let us assume that $\{F_n\}_{n=1}^{\infty}$ is a sequence of distribution functions that converge weakly to a distribution function F as $n \to \infty$. Define $g(x) = \exp(itx)$ where t is a constant and note that $g(x)$ is continuous and bounded. Therefore, we use the equivalence between Conditions 1 and 3 to conclude that

$$\lim_{n\to\infty} \psi_n(t) = \int_{-\infty}^{\infty} \exp(itx)dF_n(x) = \int_{-\infty}^{\infty} \exp(itx)dF(x) = \psi(t),$$

for all $t \in \mathbb{R}$. Therefore, Condition 2 follows. The converse is somewhat more difficult to prove because we need to account for the fact that F may not be a distribution function. To proceed, we assume that

$$\lim_{n\to\infty} \psi_n(t) = \psi(t),$$

for all $t \in \mathbb{R}$. We first prove that $\{F_n\}_{n=1}^{\infty}$ converges to a distribution function. We then finish the proof by showing that the characteristic function of F must be $\psi(t)$. To obtain the weak convergence we begin by concluding from Theorem 4.5 (Helly) that there is a subsequence $\{F_{n_m}\}_{m=1}^{\infty}$, where $\{n_m\}_{m=1}^{\infty}$ is an increasing sequence in $\mathbb{N}$, that converges weakly to some non-decreasing function F. From the discussion following Theorem 4.5, we know that we can assume that F is a right continuous function. The proof that F has the correct limit properties is rather technical and is somewhat beyond the scope of this book. For a complete argument see Gnendenko (1962). In turn, we shall for the rest of this argument, assume that F has the necessary properties to be a distribution function. It follows from the fact that Condition 1 implies Condition 2 that the characteristic function ψ must correspond to the distribution function F. To complete the proof, we must now show that the sequence $\{F_n\}_{n=1}^{\infty}$ converges weakly to F as $n \to \infty$. We will use a proof by contradiction. That is, let us suppose that the sequence $\{F_n\}_{n=1}^{\infty}$ does not converge weakly to F as $n \to \infty$. In this case we would be able to find a sequence of integers $\{c_n\}_{n=1}^{\infty}$ such that F_{c_n} converges weakly to some distribution function G that differs from F at at least one point of continuity. However, as we have stated above, G must have a characteristic function equal to $\psi(t)$ for all $t \in \mathbb{R}$. But Theorem 2.27 implies that G must be the same distribution function as F, thereby contradicting our assumption, and hence it follows that $\{F_n\}_{n=1}^{\infty}$ converges weakly to F as $n \to \infty$. $\qquad\square$

Example 4.9. Let $\{X_n\}_{n=1}^{\infty}$ be a sequence of random variables where X_n has a $N(0, 1 + n^{-1})$ distribution for all $n \in \mathbb{N}$ and let X be a $N(0,1)$ random variable. Hence X_n has characteristic function $\psi_n(t) = \exp[-\frac{1}{2}t^2(1 + n^{-1})]$ and X has characteristic function $\psi(t) = \exp(-\frac{1}{2}t^2)$. Let $t \in \mathbb{R}$ and note that

$$\lim_{n\to\infty} \psi_n(t) = \lim_{n\to\infty} \exp[-\tfrac{1}{2}t^2(1 + n^{-1})] = \exp(-\tfrac{1}{2}t^2) = \psi(t).$$

Therefore, Theorem 4.6 implies that $X_n \overset{d}{\to} X$ as $n \to \infty$. $\qquad\blacksquare$

Example 4.10. Let $\{X_n\}_{n=1}^{\infty}$ be a sequence of random variables where X_n has distribution F_n for all $n \in \mathbb{N}$. Suppose the X_n converges in distribution to a random variable X with distribution F as $n \to \infty$. Now let $g(x) = x\delta\{x; (-\delta, \delta)\}$, for a specified $0 < \delta < \infty$, which is a bounded and continuous

Figure 4.1 *The bounded and continuous function used in the proof of Theorem 4.6.*

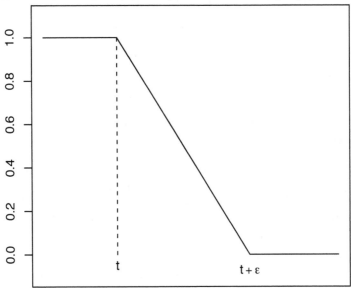

function. Then, it follows from Theorem 4.6 that

$$\lim_{n\to\infty} E(X_n\delta\{X_n;(-\delta,\delta)\}) = \lim_{n\to\infty} \int g(x)dF_n(x)$$

$$= \int g(x)dF(x)$$

$$= E(X\delta\{X;(-\delta,\delta)\}).$$

Thus we have shown that trimmed expectations of sequences of random variables that converge in distribution also converge.   ∎

It is important to note that we cannot take $g(x) = x$ in Theorem 4.6 because this function is not bounded. Therefore, Theorem 4.6 cannot be used to conclude that

$$\lim_{n\to\infty} E(X_n) = E(X),$$

whenever $X_n \overset{d}{\to} X$ as $n \to \infty$. In fact, such a result is not true in general, though Theorem 4.6 does provide some important clues as to what conditions may be required to make such a conclusion.

Example 4.11. Let $\{X_n\}_{n=1}^{\infty}$ be a sequence of random variables where X_n has distribution F_n for all $n \in \mathbb{N}$. Suppose the X_n converges in distribution to a random variable X with distribution F as $n \to \infty$. Assume in this case that there exists an interval $(-a, a)$ where $0 < a < \infty$ such that $P[X_n \in (-a, a)] = P[X \in (-a, a)] = 1$ for all $n \in \mathbb{N}$. Define $g(x) = x\delta\{x;(-\delta,\delta)\}$ as before where we specify δ such that $a < \delta < \infty$. Once again, this is a bounded and continuous function. It follows from Theorem 4.6 that

$$\lim_{n\to\infty} E(X_n \delta\{X_n; (-\delta, \delta)\}) = E(X\delta\{X; (-\delta, \delta)\}),$$

but in this case we note that $E(X_n) = E(X_n\delta\{X_n; (-\delta, \delta)\})$ and $E(X) = E(X\delta\{X; (-\delta, \delta)\})$. Therefore, in this case we have proven that

$$\lim_{n\to\infty} E(X_n) = E(X),$$

under the assumption that the random variables stay within a bounded subset of the real line.   ∎

Note that the boundedness of the sequence of random variables is not a necessary condition for the expectations of sequences of random variables that converge in distribution to also converge, though it is sufficient. We will return to this topic in Chapter 5 where we will develop equivalent conditions to the convergence of the expectations.

Noting that the convergence detailed in Definitions 4.1 and 4.2 is pointwise in terms of the sequence of distribution functions, it may be somewhat surprising that the convergence can be shown to be uniform as well. This is due to the special properties associated with distribution functions in that they are bounded and non-decreasing.

Theorem 4.7 (Pólya). *Suppose that $\{F_n\}_{n=1}^{\infty}$ is a sequence of distribution functions such that $F_n \rightsquigarrow F$ as $n \to \infty$ where F is a continuous distribution function. Then*

$$\lim_{n\to\infty} \sup_{t\in\mathbb{R}} |F_n(t) - F(t)| = 0.$$

The proof of Theorem 4.7 essentially follows along the same lines as that of Theorem 3.18 where we considered the uniform convergence of the empirical distribution function. See Exercise 12.

When considering the modes of convergence studied thus far, it would appear conceptually that convergence in distribution is a rather weak concept in that it does not require that the random variables X_n and X should be close to one another when n is large. Only the corresponding distributions of the sequence need coincide with the distribution of X as $n \to \infty$. Therefore, it would seem that any of the modes of convergence studied in Chapter 3, which all require that the random variables X_n and X coincide in some sense in the limit, would also require that distributions of the random variables would also coincide. This should essentially guarantee convergence in distribution for these sequences. It suffices to prove this property for convergence in probability, the weakest of the modes of convergence studied in Chapter 3.

Theorem 4.8. *Let $\{X_n\}_{n=1}^{\infty}$ be a sequence of random variables that converge in probability to a random variable X as $n \to \infty$. Then $X_n \xrightarrow{d} X$ as $n \to \infty$.*

Proof. Let $\varepsilon > 0$, the distribution function of X_n be $F_n(x) = P(X_n \leq x)$ for all $n \in \mathbb{N}$, and F denote the distribution function of X. Then

$$
\begin{aligned}
F_n(x) &= P(X_n \leq x) \\
&= P(\{X_n \leq x\} \cap \{|X_n - X| < \varepsilon\}) + P(\{X_n \leq x\} \cap \{|X_n - X| \geq \varepsilon\}) \\
&\leq P(\{X_n \leq x\} \cap \{|X_n - X| < \varepsilon\}) + P(\{|X_n - X| \geq \varepsilon\}),
\end{aligned}
$$

by Theorem 2.3. Now note that

$$
\{X_n \leq x\} \cap \{|X_n - X| < \varepsilon\} \subset \{X \leq x + \varepsilon\},
$$

so that Theorem 2.3 implies

$$
P(\{X_n \leq x\} \cap \{|X_n - X| < \varepsilon\}) \leq P(X \leq x + \varepsilon),
$$

and therefore,

$$
F_n(x) \leq P(X \leq x + \varepsilon) + P(|X_n - X| \geq \varepsilon).
$$

Hence, without assuming that the limit exists, we can use Theorem 1.6 to show that

$$
\begin{aligned}
\limsup_{n \to \infty} F_n(x) &\leq \limsup_{n \to \infty} P(X \leq x + \varepsilon) + \limsup_{n \to \infty} P(|X_n - X| \geq \varepsilon) \\
&= P(X \leq x + \varepsilon) = F(x + \varepsilon),
\end{aligned}
$$

where the second term in the sum converges to zero because $X_n \overset{P}{\to} X$ as $n \to \infty$. It can similarly be shown that

$$
\liminf_{n \to \infty} F_n(x) \geq F(x - \varepsilon).
$$

See Exercise 13. Suppose that $x \in C(F)$. Since $\varepsilon > 0$ is arbitrary, we have shown that

$$
F(x) = F(x-) \leq \liminf_{n \to \infty} F_n(x) \leq \limsup_{n \to \infty} F_n(x) \leq F(x+) = F(x).
$$

Therefore, it follows that

$$
\liminf_{n \to \infty} F_n(x) = \limsup_{n \to \infty} F_n(x) = F(x),
$$

and Definition 1.3 implies that

$$
\lim_{n \to \infty} F_n(x) = F(x),
$$

for all $x \in C(F)$. Therefore, it follows that $X_n \overset{d}{\to} X$ as $n \to \infty$. $\square$

Given the previous discussion it would be quite reasonable to conclude that the converse of Theorem 4.8 is not true. This can be proven using the example below.

Example 4.12. Let $\{X_n\}_{n=1}^\infty$ be a sequence of random variables where $X_n = (-1)^n Z$ where Z has a $\mathrm{N}(0,1)$ distribution for all $n \in \mathbb{N}$. Because Z and $-Z$ both have a $\mathrm{N}(0,1)$ distribution, it follows that $X_n \overset{d}{\to} Z$ as $n \to \infty$. But note that for $\varepsilon > 0$ it follows that $P(|X_n - Z| \leq \varepsilon) = P(|2Z| \leq \varepsilon)$, which is a non-zero constant for all odd valued $n \in \mathbb{N}$. Therefore, it does not follow that

$$\lim_{n \to \infty} P(|X_n - Z| \leq \varepsilon) = 0,$$

and therefore the sequence $\{X_n\}_{n=1}^{\infty}$ does not converge in probability to Z as $n \to \infty$. ∎

Additional assumptions may be added to a sequence that converges in distribution that imply that the sequence converges in probability as well. One such condition occurs when the sequence converges in distribution to a degenerate distribution.

Theorem 4.9. *Let $\{X_n\}_{n=1}^{\infty}$ be a sequence of random variables that converge in distribution to a real constant c as $n \to \infty$. Then $X_n \xrightarrow{p} c$ as $n \to \infty$.*

Proof. Let $\varepsilon > 0$ and suppose that $X_n \xrightarrow{d} c$ as $n \to \infty$. Denote the distribution function of X_n by F_n for all $n \in \mathbb{N}$. The limiting distribution function F is given by $F(x) = \delta\{x; [c, \infty)\}$ so that $C(F) = \mathbb{R} \setminus \{c\}$. Then

$$P(|X_n - c| \leq \varepsilon) = P(c - \varepsilon \leq X_n \leq c + \varepsilon) = F_n(c + \varepsilon) - F_n[(c - \varepsilon)-].$$

Noting that $c + \varepsilon$ and $(c - \varepsilon)-$ are both elements of $C(F)$ for all $\varepsilon > 0$ implies that

$$
\begin{aligned}
\lim_{n \to \infty} P(|X_n - c| \leq \varepsilon) &= \lim_{n \to \infty} F_n(c + \varepsilon) - F_n[(c - \varepsilon)-] \\
&= F(c + \varepsilon) - F[(c - \varepsilon)-] \\
&= 1 - 0 = 1.
\end{aligned}
$$

where we have used the fact that $F[(c - \varepsilon)-] = F(c - \varepsilon)$ since $(c - \varepsilon) \in C(F)$. Therefore, Definition 3.1 implies that $X_n \xrightarrow{p} X$ as $n \to \infty$. □

While a sequence that converges in distribution may not always converge in probability, it turns out that there does exist a sequence of random variables with the same convergence in distribution properties, that also converge almost certainly. This result is known as the *Skorokhod Representation Theorem.*

Theorem 4.10 (Skorokhod). *Let $\{X_n\}_{n=1}^{\infty}$ be a sequence of random variables that converge in distribution to a random variable X. Then there exists a sequence of random variables $\{Y_n\}_{n=1}^{\infty}$ and a random variable Y defined on a probability space $(\Omega, \mathcal{F}, P)$, where $\Omega = [0, 1]$, $\mathcal{F} = \mathcal{B}\{[0, 1]\}$, and P is a continuous uniform probability measure on Ω, such that*

1. *X and Y have the same distribution.*
2. *X_n and Y_n have the same distribution for all $n \in \mathbb{N}$.*
3. *$Y_n \xrightarrow{a.c.} Y$ as $n \to \infty$.*

Proof. This proof is based on the development of Serfling (1980). Like many existence proofs, this one is constructive in nature. Let F be the distribution function of X and F_n be the distribution function of X_n for all $n \in \mathbb{R}$. Define the random variable $Y : [0,1] \to \mathbb{R}$ as $Y(\omega) = F^{-1}(\omega)$. Similarly define $Y_n(\omega) = F_n^{-1}(\omega)$ for all $n \in \mathbb{N}$. We now prove that these random variables have the properties listed earlier. We begin by proving that X and Y have the same distribution. The distribution function of Y is given by $G(y) = P[\omega : Y(\omega) \leq y] = P[\omega : F^{-1}(\omega) \leq y]$. Now Theorem 3.22 implies that $P[\omega : F^{-1}(\omega) \leq y] = P[\omega : \omega \leq F(y)]$. Because P is a uniform probability measure, it follows that $G(y) = P[\omega : \omega \leq F(y)] = F(y)$, and therefore we have proven that X and Y have the same distribution. Using similar arguments it can be shown that X_n and Y_n have the same distribution for all $n \in \mathbb{N}$. Hence, it remains for us to show that Y_n converges almost certainly to Y as $n \to \infty$. Note that the sequence of random variables $\{Y_n\}_{n=1}^{\infty}$ is actually the sequence $\{F_n^{-1}\}_{n=1}^{\infty}$, which is a sequence of quantile functions corresponding to a sequence of distribution functions that converge weakly. Theorem 4.2 implies that if we collect together all $\omega \in [0,1]$ such that the sequence $\{Y_n(\omega)\}_{n=1}^{\infty}$ does not converge pointwise to $Y(\omega) = F^{-1}(\omega)$ we get a countable set, which has probability zero with respect to the continuous probability measure P on Ω. Therefore, Definition 3.2 implies that $Y_n \xrightarrow{a.c.} Y$ as $n \to \infty$. $\qquad\square$

The most common example of using Theorem 4.10 arises when proving that continuous functions of a sequence of random variables that converge weakly also converge. See Theorem 4.12 and the corresponding proof.

We often encounter problems that concern a sequence of random variables that converge in distribution and we perturb the sequence with another sequence that also has some convergence properties. In these cases it is important to determine how such perturbations affect the convergence of the sequence. For example, we might know that a sequence of random variables $\{X_n\}_{n=1}^{\infty}$ converges in distribution to a random variable Z that has a $\mathrm{N}(\mu, 1)$ distribution. We may wish to standardize this sequence, but may not know μ. Suppose we have a consistent estimator of μ. That is, we can compute $\hat{\mu}_n$ such that $\hat{\mu}_n \xrightarrow{p} \mu$ as $n \to \infty$. Given this information, is it possible to conclude that the sequence $\{X_n - \hat{\mu}_n\}_{n=1}^{\infty}$ converges in distribution to a standard normal distribution? *Slutsky's Theorem* is a result that considers both additive and multiplicative perturbations of sequences that converge in distribution.

Theorem 4.11 (Slutsky). *Let $\{X_n\}_{n=1}^{\infty}$ be a sequence of random variables that converge weakly to a random variable X. Let $\{Y_n\}_{n=1}^{\infty}$ be a sequence of random variables that converge in probability to a real constant c. Then,*

1. $X_n + Y_n \xrightarrow{d} X + c$ *as* $n \to \infty$.

2. $X_n Y_n \xrightarrow{d} cX$ *as* $n \to \infty$.

3. $X_n / Y_n \xrightarrow{d} X/c$ *as* $n \to \infty$ *as long as* $c \neq 0$.

Proof. The first result will be proven here. The remaining results are proven in Exercises 14 and 15. Denote the distribution function of $X_n + Y_n$ to be G_n

and let F be the distribution function of X. Let $\varepsilon > 0$ and set x such that $x - c \in C(F)$ and $x + \varepsilon - c \in C(F)$. Then

$$
\begin{aligned}
G_n(x) &= P(X_n + Y_n \leq x) \\
&= P(\{X_n + Y_n \leq x\} \cap \{|Y_n - c| \leq \varepsilon\}) + \\
&\quad P(\{X_n + Y_n \leq x\} \cap \{|Y_n - c| > \varepsilon\}).
\end{aligned}
$$

Now, Theorem 2.3 implies that

$$
P(\{X_n + Y_n \leq x\} \cap \{|Y_n - c| \leq \varepsilon\}) \leq P(X_n + c \leq x + \varepsilon),
$$

and

$$
P(\{X_n + Y_n \leq x\} \cap \{|Y_n - c| > \varepsilon\}) \leq P(|Y_n - c| > \varepsilon).
$$

Therefore,

$$
G_n(x) \leq P(X_n + c \leq x + \varepsilon) + P(|Y_n - c| > \varepsilon),
$$

and Theorem 1.6 implies that

$$
\begin{aligned}
\limsup_{n \to \infty} G_n(x) &\leq \limsup_{n \to \infty} P(X_n + c \leq x + \varepsilon) + \limsup_{n \to \infty} P(|Y_n - c| > \varepsilon) \\
&= P(X \leq x + \varepsilon - c),
\end{aligned}
$$

where we have used the fact that $X_n \xrightarrow{d} X$ and $Y_n \xrightarrow{p} c$ as $n \to \infty$. Similarly, it can be shown that

$$
P(X_n \leq x - \varepsilon - c) \leq G_n(x) + P(|Y_n - c| > \varepsilon).
$$

See Exercise 16. Therefore, Theorem 1.6 implies that

$$
\liminf_{n \to \infty} P(X_n \leq x - \varepsilon - c) \leq \liminf_{n \to \infty} G_n(x) + \liminf_{n \to \infty} P(|Y_n - c| > \varepsilon)
$$

so that

$$
\liminf_{n \to \infty} G_n(x) \geq P(X \leq x - \varepsilon - c).
$$

Because $\varepsilon > 0$ is arbitrary, we have proven that

$$
\limsup_{n \to \infty} G_n(x) \leq P(X \leq x - c) \leq \liminf_{n \to \infty} G_n(x),
$$

which can only occur when

$$
\liminf_{n \to \infty} G_n(x) = \limsup_{n \to \infty} G_n(x) = \lim_{n \to \infty} G_n(x) = P(X \leq x - c).
$$

The result then follows by noting that $P(X \leq x - c) = F(x - c)$ is the distribution function of the random variable $X + c$. $\square$

Example 4.13. Suppose that $\{Z_n\}_{n=1}^{\infty}$ is a sequence of random variables such that $Z_n/\sigma \xrightarrow{d} Z$ as $n \to \infty$ where Z is a $N(0,1)$ random variable. Suppose that σ is unknown, but can be estimated by a consistent estimator $\hat{\sigma}_n$. That is $\hat{\sigma}_n \xrightarrow{p} \sigma$ as $n \to \infty$. Noting that $\sigma \xrightarrow{d} \sigma$ as $n \to \infty$, it follows from Part 3 of Theorem 4.11 that $\sigma/\hat{\sigma}_n \xrightarrow{d} 1$ as $n \to \infty$. Theorem 4.9 then implies that $\sigma/\hat{\sigma}_n \xrightarrow{p} 1$ as $n \to \infty$, so that Part 2 of Theorem 4.11 can be used to show that $Z_n/\hat{\sigma}_n \xrightarrow{d} Z$ as $n \to \infty$. This type of argument is used in later sections to justify estimating a variance on a set of standardized random variables that are asymptotically NORMAL. ∎

Example 4.14. Let $\{X_n\}_{n=1}^{\infty}$ be a sequence of random variables such that X_n has a GAMMA(α_n, β_n) distribution where $\{\alpha_n\}_{n=1}^{\infty}$ and $\{\beta_n\}_{n=1}^{\infty}$ are sequences of positive real numbers such that $\alpha_n \to \alpha$ and $\beta_n \to \beta$ as $n \to \infty$, some some positive real numbers α and β. Let $\hat{\alpha}_n$ and $\hat{\beta}_n$ be consistent estimators of α and β in the sense that $\hat{\alpha}_n \xrightarrow{p} \alpha$ and $\hat{\beta}_n \xrightarrow{p} \beta$ as $n \to \infty$. It can be shown that $X_n \xrightarrow{d} X$ as $n \to \infty$ where X has a GAMMA(α, β) distribution. See Exercise 3. Part 3 of Theorem 4.11 implies that $X_n/\hat{\beta}_n \xrightarrow{d} X/\beta$ as $n \to \infty$ where X/β has a GAMMA$(\alpha, 1)$ distribution. Similarly, an additional application of Part 1 of Theorem 4.11 implies that if we define a new random variable $Y_n = X_n/\hat{\beta}_n - \alpha_n$ for all $n \in \mathbb{N}$, then $Y_n \xrightarrow{d} Y$ as $n \to \infty$ where Y has a shifted GAMMA$(\alpha, 1)$ distribution with density

$$f(y) = \tfrac{1}{\Gamma(\alpha)}(y+\alpha)^{\alpha-1}\exp[-(y+\alpha)]\delta\{y;(\alpha,\infty)\}.$$

∎

More general transformations of sequences of weakly convergent transformations also converge as well, under the condition that the transformation is continuous with respect to the limiting distribution.

Theorem 4.12. *Let $\{X_n\}_{n=1}^{\infty}$ be a sequence of random variables that converge in distribution to a random variable X as $n \to \infty$. Let g be a Borel function on $\mathbb{R}$ and suppose that $P[X \in C(g)] = 1$. Then $g(X_n) \xrightarrow{d} g(X)$ as $n \to \infty$.*

Proof. In this proof we follow the method of Serfling (1980) which translates the problem into one concerning almost certain convergence using Theorem 4.10 (Skorokhod), where the convergence of the transformation is known to hold. The result is then translated back to weak convergence using the fact that almost certain convergence implies convergence in distribution. To proceed, let us suppose that $\{X_n\}_{n=1}^{\infty}$ is a sequence of random variables that converge in distribution to a random variable X as $n \to \infty$. Theorem 4.10 then implies that there exists a sequence of random variables $\{Y_n\}_{n=1}^{\infty}$ such that X_n and Y_n have the same distribution for all $n \in \mathbb{N}$ where $Y_n \xrightarrow{a.c.} Y$ as $n \to \infty$, and Y has the same distribution as X. It then follows from Theorem 3.8 that $g(Y_n) \xrightarrow{a.c.} g(Y)$ as $n \to \infty$ as long as g is continuous with probability one with respect to the distribution of Y. To show this, let $D(g)$ denote the set of discontinuities of the function g. Let P be the probability measure from the measure space used to define X and P^* be the probability measure from the measure space used to define Y. Then it follows that since X and Y have the same distribution, that $P^*[Y \in D(g)] = P[X \in D(g)] = 0$. Therefore, it follows that $g(Y_n) \xrightarrow{a.c.} g(Y)$ as $n \to \infty$. Theorems 3.2 and 4.8 then imply that $g(Y_n) \xrightarrow{d} g(Y)$ as $n \to \infty$. But $g(Y_n)$ has the same distribution as $g(X_n)$

for all $n \in \mathbb{R}$ and $g(Y)$ has the same distribution as $g(X)$. Therefore, it follows that $g(X_n) \xrightarrow{d} g(X)$ as $n \to \infty$ as well, and the result is proven. $\qquad\square$

Proofs based on Theorem 4.10 (Skorokhod) can sometimes be confusing and indeed one must be careful with their application. It may seem that Theorem 4.10 could be used to prove that any result that holds for almost certain convergence also holds for convergence in distribution. This of course is not true. The key step in the proof of Theorem 4.12 is that we can translate the desired result on the random variables that converge almost certainly back to a parallel result for weak convergence. In the proof of Theorem 4.12 this is possible because the two sets of random variables have the same distributions.

Example 4.15. Let $\{X_n\}_{n=1}^{\infty}$ be a sequence of independent $N(0, 1+n^{-1})$ random variables and let X be a $N(0,1)$ random variable. We have already shown in Example 4.2 that $X_n \xrightarrow{d} X$ as $n \to \infty$. Consider the function $g(x) = x^2$, which is continuous with probability one with respect to a $N(0,1)$ distribution. It follows from Theorem 4.12 that the sequence $\{Y_n\}_{n=1}^{\infty}$ converges in distribution to a random variable Y, where $Y_n = g(X_n) = X_n^2$ for all $n \in \mathbb{R}$ and $Y = g(X) = X^2$. Using the fact that Y has a CHISQUARED(1) distribution, it follows that we have proven that $\{X_n^2\}_{n=1}^{\infty}$ converges in distribution to a CHISQUARED(1) distribution as $n \to \infty$. $\qquad\blacksquare$

Example 4.16. Let $\{X_n\}_{n=1}^{\infty}$ be a sequence of random variables where X_n has an EXPONENTIAL(θ_n) distribution for all $n \in \mathbb{N}$ where $\{\theta_n\}_{n=1}^{\infty}$ is a sequence of positive real numbers such that $\theta_n \to \theta$ as $n \to \infty$. Note that $X_n \xrightarrow{d} X$ as $n \to \infty$ where X has an EXPONENTIAL(1) distribution. Consider the transformation $g(x) = \log(x)$ and note that g is continuous with probability one with respect to the random variable X. Therefore, Theorem 4.12 implies that $\log(X_n) \xrightarrow{d} \log(X)$ where $\log(X)$ has a UNIFORM$(0,1)$ distribution. $\qquad\blacksquare$

4.3 Weak Convergence of Random Vectors

The extension of the concept of convergence in distribution to the case of random vectors is relatively straightforward in that the univariate definition is directly generalized to the multivariate definition. While the generalization follows directly, there are some important consequences of moving to the multivariate setting which must be studied carefully.

Definition 4.4. *Let $\{\mathbf{X}_n\}_{n=1}^{\infty}$ be a sequence of d-dimensional random vectors where $\mathbf{X}_n$ has distribution function F_n for all $n \in \mathbb{N}$. Then $\mathbf{X}_n$ converges in distribution to a d-dimensional random vector $\mathbf{X}$ with distribution function F if*

$$\lim_{n \to \infty} F_n(\mathbf{x}) = F(\mathbf{x}),$$

for all $\mathbf{x} \in C(F)$.

While the definition of convergence in distribution is essentially the same as the univariate case, there are some hidden differences that arise due to the fact that the distribution functions in this case are functions of d-dimensional vectors. This is important because in the univariate case the set $C(F)$ corresponds to all real x such that $P(X = x) = 0$. This is not necessarily true in the multivariate case as demonstrated in the example below.

Example 4.17. This example is based on a discussion in Lehmann (1999). Consider a discrete bivariate random vector $\mathbf{X}$ with probability distribution function

$$f(\mathbf{x}) = \begin{cases} \frac{1}{2} & \mathbf{x} \in \{(0, -1)', (-1, 0)'\} \\ 0 & \text{elsewhere.} \end{cases} \tag{4.2}$$

The distribution function of $\mathbf{X}$ is $F(\mathbf{x}) = P(\mathbf{X} \leq \mathbf{x})$, where the inequality between the vectors is interpreted element-wise. Therefore, the distribution function is given by

$$F(\mathbf{x}) = \begin{cases} 0 & x_1 < 0, x_2 < 0; x_1 < -1, x_2 > 0 \text{ or } x_1 > 0, x_2 < -1 \\ \frac{1}{2} & -1 \leq x_1 < 0, x_2 \geq 0 \text{ or } x_1 \geq 0, -1 \leq x_2 < 0 \\ 1 & x_1 \geq 0, x_2 \geq 0 \end{cases}$$

where $\mathbf{x} = (x_1, x_2)'$. The probability contours of this distribution function are given in Figure 4.2. It is clear from the figure that the point $\mathbf{x} = (0, 0)'$ is a point of discontinuity of the distribution function, yet $P[\mathbf{x} = (0, 0)'] = 0$. ■

Continuity points can be found by examining the probability of the boundary of the set $\{\mathbf{t} : \mathbf{t} \leq \mathbf{x}\}$ where we continue to use the convention in this book that inequalities between vectors are interpreted pointwise.

Theorem 4.13. *Let $\mathbf{X}$ be a d-dimensional random vector with distribution function F and let $B(\mathbf{x}) = \{\mathbf{t} \in \mathbb{R}^d : \mathbf{t} \leq \mathbf{x}\}$. Then a point $\mathbf{x} \in \mathbb{R}^d$ is a continuity point of F if and only if $P[\mathbf{X} \in \partial B(\mathbf{x})] = 0$.*

A proof of Theorem 4.13 can be found in Section 5.1 of Lehmann (1999).

Example 4.18. Consider the random vector with distribution function introduced in Example 4.17. Consider once again the point $\mathbf{x}' = (0, 0)$ which has boundary set

$$\begin{aligned} \partial B(\mathbf{x}) &= \partial \{\mathbf{t} \in \mathbb{R}^2 : \mathbf{t} \leq \mathbf{0}\} \\ &= \{(x_1, x_2) \in \mathbb{R}^2 : x_1 = 0, x_2 \leq 0\} \cup \\ &\quad \{(x_1, x_2) \in \mathbb{R}^2 : x_1 \leq 0, x_2 = 0\}. \end{aligned}$$

Now, from Equation 4.2 it follows that $P[\partial B(\mathbf{x})] = P(\{(0, -1), (-1, 0)\}) = 1$. Therefore, Theorem 4.13 implies that $\mathbf{x}' = (0, 0)$ is not a continuity point. ■

When the limiting distribution is continuous, the problem of proving weak convergence for random vectors simplifies greatly.

Example 4.19. Let $\{\mathbf{X}_n\}_{n=1}^{\infty}$ be a sequence of d-dimensional random vectors

Figure 4.2 *Probability contours of the discrete bivariate distribution function from Example 4.17. The dotted lines indicate the location of discrete steps in the distribution function, with the height of the steps being indicated on the plot. It is clear from this plot that the point $(0,0)$ is a point of discontinuity of the distribution function.*

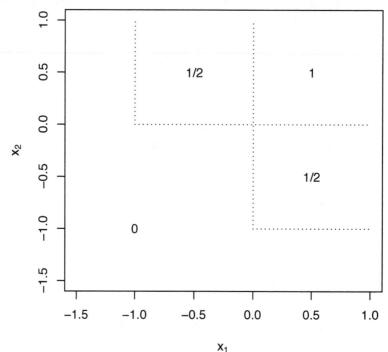

such that $\mathbf{X}_n$ has a $\mathbf{N}(\boldsymbol{\mu}_n, \boldsymbol{\Sigma}_n)$ distribution where $\{\boldsymbol{\mu}_n\}_{n=1}^{\infty}$ is a sequence of d-dimensional means such that

$$\lim_{n \to \infty} \boldsymbol{\mu}_n = \boldsymbol{\mu},$$

for some $\boldsymbol{\mu} \in \mathbb{R}^d$ and $\{\boldsymbol{\Sigma}_n\}_{n=1}^{\infty}$ is a sequence of $d \times d$ positive definite covariance matrices such that

$$\lim_{n \to \infty} \boldsymbol{\Sigma}_n = \boldsymbol{\Sigma},$$

for some $d \times d$ positive definite covariance matrix $\boldsymbol{\Sigma}$. It follows that $\mathbf{X}_n \xrightarrow{d} \mathbf{X}$ as $n \to \infty$ where $\mathbf{X}$ has a $\mathbf{N}(\boldsymbol{\mu}, \boldsymbol{\Sigma})$ distribution. To show this we note that

$$F_n(\mathbf{x}) = \int_{B(\mathbf{x})} (2\pi)^{-d/2} |\boldsymbol{\Sigma}_n|^{-1/2} \exp[-\tfrac{1}{2}(\mathbf{t} - \boldsymbol{\mu}_n)'\boldsymbol{\Sigma}_n^{-1}(\mathbf{t} - \boldsymbol{\mu}_n)]d\mathbf{t},$$

and that

$$F(\mathbf{x}) = \int_{B(\mathbf{x})} (2\pi)^{-d/2} |\boldsymbol{\Sigma}|^{-1/2} \exp[-\tfrac{1}{2}(\mathbf{t} - \boldsymbol{\mu})'\boldsymbol{\Sigma}^{-1}(\mathbf{t} - \boldsymbol{\mu})]d\mathbf{t},$$

where $B(\mathbf{x})$ is defined in Theorem 4.13. Because the limiting distribution is continuous, $P[\mathbf{X} \in \partial B(\mathbf{x})] = 0$ for all $\mathbf{x} \in \mathbb{R}^d$. To show that weak convergence follows, we note that $F_n(\mathbf{x})$ can be written as

$$F_n(\mathbf{x}) = \int_{\boldsymbol{\Sigma}_n^{-1/2}[B(\mathbf{x})-\boldsymbol{\mu}_n]} (2\pi)^{-d/2} \exp(-\tfrac{1}{2}\mathbf{t}'\mathbf{t})dt,$$

and similarly

$$F(\mathbf{x}) = \int_{\boldsymbol{\Sigma}^{-1/2}[B(\mathbf{x})-\boldsymbol{\mu}]} (2\pi)^{-d/2} \exp(-\tfrac{1}{2}\mathbf{t}'\mathbf{t})d\mathbf{t}.$$

In both cases we have used the shorthand $\mathbf{A}B(\mathbf{x}) + \mathbf{c}$ to represent the linear transformation $\{\mathbf{A}\mathbf{t}+\mathbf{c} : \mathbf{t} \in B(\mathbf{x})\}$. Now it follows that $\boldsymbol{\Sigma}_n^{-1/2}[B(\mathbf{x}-\boldsymbol{\mu}_n)] \to \boldsymbol{\Sigma}^{-1/2}[B(\mathbf{x}-\boldsymbol{\mu})]$ for all $\mathbf{x} \in \mathbb{R}^d$. Therefore, it follows that

$$\lim_{n\to\infty} F_n(\mathbf{x}) = F(\mathbf{x}),$$

for all $\mathbf{x} \in \mathbb{R}^d$ and, hence, it follows from Definition 4.4 that $\mathbf{X}_n \xrightarrow{d} \mathbf{X}$ as $n \to \infty$. ∎

If a sequence of d-dimensional distribution functions $\{F_n\}_{n=1}^{\infty}$ converges weakly to a distribution function F as $n \to \infty$, then Defintion 4.4 implies that

$$\lim_{n\to\infty} F_n(\mathbf{x}) = F(\mathbf{x}),$$

as long as $\mathbf{x}$ is a continuity point of F. Let $\mathbf{X}_n$ be a d-dimensional random vector with distribution function F_n for all $n \in \mathbb{N}$ and $\mathbf{X}$ have distribution function F. Then the fact that $F_n \leadsto F$ as $n \to \infty$ implies that

$$\lim_{n\to\infty} P(\mathbf{X}_n \leq \mathbf{x}) = P(\mathbf{X} \leq \mathbf{x}),$$

as long as $\mathbf{x}$ is a continuity point of F. Using Theorem 4.13, this property can also be written equivalently as

$$\lim_{n\to\infty} P[\mathbf{X}_n \in B(\mathbf{x})] = P[\mathbf{X} \in B(\mathbf{x})],$$

as long as $P[\mathbf{X} \in \partial B(\mathbf{x})] = 0$. This type of result can be extended to any subset of $\mathbb{R}^d$.

Theorem 4.14. *Let $\{\mathbf{X}_n\}_{n=1}^{\infty}$ be a sequence of d-dimensional random vectors where $\mathbf{X}_n$ has distribution function F_n for all $n \in \mathbb{N}$ and let $\mathbf{X}$ be a d-dimensional random vector with distribution function F. Suppose that $F_n \leadsto F$ as $n \to \infty$ and let B be any subset of $\mathbb{R}^d$. If $P(\mathbf{X} \in \partial B) = 0$, then,*

$$\lim_{n\to\infty} P(\mathbf{X}_n \in B) = P(\mathbf{X} \in B).$$

Example 4.20. Consider the setup of Example 4.19 where $\{\mathbf{X}_n\}_{n=1}^{\infty}$ is a sequence of d-dimensional random vectors such that $\mathbf{X}_n$ has a $\mathbf{N}(\boldsymbol{\mu}_n, \boldsymbol{\Sigma}_n)$ distribution and $\mathbf{X}_n \xrightarrow{d} \mathbf{X}$ as $n \to \infty$ where $\mathbf{X}$ has a $\mathbf{N}(\boldsymbol{\mu}, \boldsymbol{\Sigma})$ distribution. Define a region

$$E(\alpha) = \{\mathbf{x} \in \mathbb{R}^d : (\mathbf{x} - \boldsymbol{\mu})'\boldsymbol{\Sigma}^{-1}(\mathbf{x} - \boldsymbol{\mu}) \leq \chi_{d;\alpha}^2\}$$

where $\chi^2_{d;\alpha}$ is the α quantile of a CHISQUARED(d) distribution. The boundary region of $E(\alpha)$ is given by

$$\partial E(\alpha) = \{\mathbf{x} \in \mathbb{R}^d : (\mathbf{x} - \boldsymbol{\mu})'\boldsymbol{\Sigma}^{-1}(\mathbf{x} - \boldsymbol{\mu}) = \chi^2_{d;\alpha}\},$$

which is an ellipsoid in $\mathbb{R}^d$. Now

$$P[\mathbf{X} \in \partial E(\alpha)] = P[(\mathbf{X} - \boldsymbol{\mu})'\boldsymbol{\Sigma}^{-1}(\mathbf{X} - \boldsymbol{\mu}) = \chi^2_{d;\alpha}] = 0,$$

since $\mathbf{X}$ is a continuous random vector. It then follows from Theorem 4.14 that

$$\lim_{n\to\infty} P[\mathbf{X}_n \in E(\alpha)] = P[\mathbf{X} \in E(\alpha)] = \alpha.$$

∎

The result of Theorem 4.14 is actually part of a larger result that generalizes Theorem 4.6 to the multivariate case.

Theorem 4.15. *Let $\{\mathbf{X}_n\}_{n=1}^{\infty}$ be a sequence of d-dimensional random vectors where $\mathbf{X}_n$ has distribution function F_n for all $n \in \mathbb{N}$ and let $\mathbf{X}$ be a d-dimensional random vector with distribution function F. Then the following statements are equivalent.*

1. *$F_n \rightsquigarrow F$ as $n \to \infty$.*

2. *For any bounded and continuous function g,*

$$\lim_{n\to\infty} \int_{\mathbb{R}^d} g(\mathbf{x}) dF_n(\mathbf{x}) = \int_{\mathbb{R}^d} g(\mathbf{x}) dF(\mathbf{x}).$$

3. *For any closed set of $C \subset \mathbb{R}^d$,*

$$\limsup_{n\to\infty} P(\mathbf{X}_n \in C) = P(\mathbf{X} \in C).$$

4. *For any open set of $G \subset \mathbb{R}^d$,*

$$\liminf_{n\to\infty} P(\mathbf{X}_n \in G) = P(\mathbf{X} \in G).$$

5. *For any set B where $P(\mathbf{X} \in \partial B) = 0$,*

$$\lim_{n\to\infty} P(\mathbf{X}_n \in B) = P(\mathbf{X} \in B).$$

Proof. We shall follow the method of proof given by Billingsley (1986), which first shows the equivalence on Conditions 2–5, and then proves that Condition 5 is equivalent to Condition 1. We begin by proving that Condition 2 implies Condition 3. Let C be a closed subset of $\mathbb{R}^d$. Define a metric $\Delta(\mathbf{x}, C)$ that measures the distance between a point $\mathbf{x} \in \mathbb{R}^d$ and the set C as the smallest distance between $\mathbf{x}$ and any point in C. That is,

$$\Delta(\mathbf{x}, C) = \inf_{\mathbf{c} \in C}\{||\mathbf{x} - \mathbf{c}||\}.$$

In order to effectively use Condition 2 we need to define a bounded and continuous function. Therefore, define

$$h_k(t) = \begin{cases} 1 & \text{if } t < 0, \\ 1 - tk & \text{if } 0 \leq t \leq k^{-1}, \\ 0 & \text{if } k^{-1} \leq t. \end{cases}$$

Let $g_k(\mathbf{x}) = h_k[\Delta(\mathbf{x}, C)]$. It follows that the function is continuous and bounded between zero and one for all $k \in \mathbb{N}$. Now, suppose that $\mathbf{x} \in C$ so that $\Delta(\mathbf{x}, C) = 0$ and hence,

$$\lim_{k \to \infty} h_k[\Delta(\mathbf{x}, C)] = \lim_{k \to \infty} 1 = 1,$$

since $h_k(0) = 1$ for all $k \in \mathbb{N}$. On the other hand, if $\mathbf{x} \notin C$ so that $\Delta(\mathbf{x}, C) > 0$, then there exists a positive integer $k_{\mathbf{x}}$ such that $k^{-1} < \Delta(\mathbf{x}, C)$ for all $k > k_{\mathbf{x}}$. Therefore, it follows that

$$\lim_{k \to \infty} h_k[\Delta(\mathbf{x}, C)] = 0.$$

Hence, Definition 1.4 implies that $g_k \xrightarrow{pw} \delta\{\mathbf{x}; C\}$ as $k \to \infty$. It further follows that $g_k(\mathbf{x}) \geq \delta\{\mathbf{x}; C\}$ for all $\mathbf{x} \in \mathbb{R}^d$. Therefore, Theorem 1.6 implies

$$\begin{aligned} \limsup_{n \to \infty} P(\mathbf{X}_n \in C) &= \limsup_{n \to \infty} \int_C dF_n(\mathbf{x}) \\ &= \limsup_{n \to \infty} \int_{\mathbb{R}^d} \delta\{\mathbf{x}; c\} dF_n(\mathbf{x}) \\ &\leq \limsup_{n \to \infty} \int_{\mathbb{R}^d} g_k(\mathbf{x}) dF_n(\mathbf{x}) \\ &= \int_{\mathbb{R}^d} g_k(\mathbf{x}) dF(\mathbf{x}), \end{aligned}$$

where the final equality follows from Condition 2. It can further be proven that g_k converges monotonically (decreasing) to $\delta\{\mathbf{x}; C\}$ as $n \to \infty$ so that Theorem 1.12 (Lebesgue) implies that

$$\lim_{k \to \infty} \int_{\mathbb{R}^d} g_k(\mathbf{x}) dF(\mathbf{x}) = \int_{\mathbb{R}^d} g(\mathbf{x}) dF(\mathbf{x}) = \int_{\mathbb{R}^d} \delta\{\mathbf{x}; c\} dF(\mathbf{x}) = P(\mathbf{X} \in C).$$

Therefore, we have proven that

$$\limsup_{n \to \infty} P(\mathbf{X}_n \in C) = P(\mathbf{X} \in C),$$

which corresponds to Condition 3. For a proof that Condition 3 implies Condition 4, see Exercise 20. We will now prove that Conditions 3 and 4 imply Condition 5. Let A be a subset of $\mathbb{R}^d$ such that $P(\mathbf{X} \in \partial A) = 0$. Let $A^\circ = A \setminus \partial A$ be the open interior of A and let $A^- = A \cup \partial A$ be the closure of A. From Condition 3 and the fact that $A^\circ \subset A$, we have

$$P(\mathbf{X} \in A^\circ) \leq \liminf_{n \to \infty} P(\mathbf{X}_n \in A^\circ) \leq \liminf_{n \to \infty} P(\mathbf{X}_n \in A).$$

Similary, Condition 2 and the fact that $A \subset A^-$ implies

$$\limsup_{n\to\infty} P(\mathbf{X}_n \in A) \leq \limsup_{n\to\infty} P(\mathbf{X}_n \in A^-) \leq P(\mathbf{X} \in A^-).$$

Theorem 1.5 implies then that

$$P(\mathbf{X} \in A^\circ) \leq \liminf_{n\to\infty} P(\mathbf{X}_n \in A) \leq \limsup_{n\to\infty} P(\mathbf{X}_n \in A) \leq P(\mathbf{X} \in A^-).$$

But Condition 5 assumes that $P(\mathbf{X} \in \partial A) = 0$ so that $P(\mathbf{X} \in A^\circ) = P(\mathbf{X} \in A^-)$. Therefore, Definition 1.3 implies that

$$\liminf_{n\to\infty} P(\mathbf{X}_n \in A) = \limsup_{n\to\infty} P(\mathbf{X}_n \in A) = \lim_{n\to\infty} P(\mathbf{X}_n \in A) = P(\mathbf{X} \in A),$$

and Condition 5 is proven. We now prove that Condition 5 implies Condition 2, from which we can conclude that Conditions 2–5 are equivalent. Suppose that g is continuous and bounded so that $|g(\mathbf{x})| < b$ for all $\mathbf{x} \in \mathbb{R}^d$ from some $b > 0$. Let $\varepsilon > 0$ and define a partition of $[-b, b]$ given by $a_0 < a_1 < \cdots < a_m$ where $a_k - a_{k-1} < \varepsilon$ for all $k = 1, \ldots, m$. We will assume that $P[g(\mathbf{X}) = a_k] = 0$ for all $k = 0, \ldots, m$. Let $A_k = \{\mathbf{x} \in \mathbb{R}^d : a_{k-1} < g(\mathbf{x}) \leq a_k\}$ for $k = 0, \ldots, m$. From the way the sets $A_0, \ldots, A_m$ are constructed it follows that

$$
\begin{aligned}
\sum_{k=1}^{m} \left[\int_{A_k} g(\mathbf{x}) dF_n(\mathbf{x}) - a_k P(\mathbf{X}_n \in A_k) \right] &= \sum_{k=1}^{m} \int_{A_k} [g(\mathbf{x}) - a_k] dF_n(\mathbf{x}) \\
&\leq \sum_{k=1}^{m} \int_{A_k} \varepsilon \, dF_n(\mathbf{x}) \\
&= \varepsilon \sum_{k=1}^{m} \int_{A_k} dF_n(\mathbf{x}) \\
&= \varepsilon,
\end{aligned}
$$

since

$$\bigcup_{k=0}^{m} A_k = \{\mathbf{x} \in \mathbb{R}^d : |g(\mathbf{x})| \leq b\},$$

and $P[|g(\mathbf{X}_n)| \leq b] = 1$. This same property then implies that

$$\int_{\mathbb{R}^d} g(\mathbf{x}) dF_n(\mathbf{x}) - \sum_{k=0}^{m} a_k P(\mathbf{X}_n \in A_k) \leq \varepsilon.$$

It can similarly be shown that

$$\int_{\mathbb{R}^d} g(\mathbf{x}) dF_n(\mathbf{x}) - \sum_{k=0}^{m} a_k P(\mathbf{X}_n \in A_k) \geq -\varepsilon,$$

so that it follows that

$$\left| \int_{\mathbb{R}^d} g(\mathbf{x}) dF_n(\mathbf{x}) - \sum_{k=0}^{m} a_k P(\mathbf{X}_n \in A_k) \right| \leq \varepsilon.$$

See Exercise 21. The same arguments can be used with F_n replaced by F and $\mathbf{X}_n$ replaced by $\mathbf{X}$ to obtain

$$\left| \int_{\mathbb{R}^d} g(\mathbf{x}) dF(\mathbf{x}) - \sum_{k=0}^m a_k P(\mathbf{X} \in A_k) \right| \leq \varepsilon.$$

Now, it can be shown that $P(\mathbf{X} \in \partial A_k) = 0$ for $k = 0, \ldots, m$. See Section 29 of Billingsley (1986). Therefore, Condition 5 implies that when m is fixed,

$$\lim_{n \to \infty} \sum_{k=0}^m a_k P(\mathbf{X}_n \in A_k) = \sum_{k=0}^m a_k P(\mathbf{X} \in A_k).$$

Therefore, it follows that

$$\left| \lim_{n \to \infty} \int_{\mathbb{R}^d} g(\mathbf{x}) dF_n(\mathbf{x}) - \int_{\mathbb{R}^d} g(\mathbf{x}) dF(\mathbf{x}) \right| \leq 2\varepsilon.$$

Because $\varepsilon > 0$ is arbitrary, it follows that

$$\lim_{n \to \infty} \int_{\mathbb{R}^d} g(\mathbf{x}) dF_n(\mathbf{x}) = \int_{\mathbb{R}^d} g(\mathbf{x}) dF(\mathbf{x}),$$

and Condition 2 follows. We will finally show that Condition 5 implies Condition 1. To show this define sets $B(\mathbf{x}) = \{\mathbf{t} \in \mathbb{R}^d : \mathbf{t} \leq \mathbf{x}\}$. Theorem 4.13 implies that $\mathbf{x}$ is a continuity point of F if and only if $P[\mathbf{X} \in \partial B(\mathbf{x})] = 0$. Therefore, if $\mathbf{x}$ is a continuity point of F, we have from Condition 5 that

$$\lim_{n \to \infty} F_n(\mathbf{x}) = \lim_{n \to \infty} P[\mathbf{X} \in B(\mathbf{x})] = P[\mathbf{X} \in B(\mathbf{x})] = F(\mathbf{x}).$$

Therefore, it follows that $F_n \rightsquigarrow F$ as $n \to \infty$ and Condition 1 follows. It remains to show that Condition 1 implies one of Conditions 2–5. This proof is beyond the scope of this book, and can be found in Section 29 of Billingsley (1986). □

In the case of convergence of random variables it followed that the convergence of the individual elements of a random vector was equivalent to the convergence of the random vector itself. The same equivalence does not hold for convergence in distribution of random vectors. This is due to the fact that a set of marginal distributions do not uniquely determine the joint distribution of a random vector.

Example 4.21. Let $\{X_n\}_{n=1}^\infty$ be a sequence of random variables such that X_n has a $N(0, 1 + n^{-1})$ distribution for all $n \in \mathbb{N}$. Similarly, let $\{Y_n\}_{n=1}^\infty$ be a sequence of random variables such that Y_n has a $N(0, 1 + n^{-1})$ distribution for all $n \in \mathbb{N}$. Now consider the random vector $\mathbf{Z}_n = (X_n, Y_n)'$ for all $n \in \mathbb{N}$. Of interest is the limiting behavior of the sequence $\{\mathbf{Z}_n\}_{n=1}^\infty$. That is, is there a unique random vector $\mathbf{Z}$ such that $\mathbf{Z}_n \xrightarrow{d} \mathbf{Z}$ as $n \to \infty$? The answer is no, unless further information is known about the joint behavior of X_n and Y_n. For example, if X_n and Y_n are independent for all $n \in \mathbb{N}$, then $\mathbf{Z}$ is a $N(\mathbf{0}, \mathbf{I})$ distribution, where $\mathbf{I}$ is the identity matrix. On the other hand, suppose that

$\mathbf{Z}_n = \mathbf{C}_n \mathbf{W}$ where $\mathbf{W}$ is a $\mathbf{N}(\mathbf{0}, \boldsymbol{\Sigma})$ random vector where

$$\boldsymbol{\Sigma} = \begin{bmatrix} 1 & \tau \\ \tau & 1 \end{bmatrix},$$

where τ is a constant such that $\tau \in (0, 1)$, and $\{\mathbf{C}_n\}_{n=1}^{\infty}$ is a sequence of 2×2 matrices defined by

$$\mathbf{C}_n = \begin{bmatrix} (1 + n^{-1})^{1/2} & 0 \\ 0 & (1 + n^{-1})^{1/2} \end{bmatrix}.$$

In this case $\mathbf{Z}_n \xrightarrow{d} \mathbf{Z}$ as $n \to \infty$ where $\mathbf{Z}$ has a $\mathbf{N}(\mathbf{0}, \boldsymbol{\Sigma})$ distribution. Hence, there are an infinite number of choices of $\mathbf{Z}$ depending on the covariance between X_n and Y_n. There are even more choices for the limiting distribution because the limiting joint distribution need not even be multivariate NORMAL, due to the fact that the joint distribution of two normal random variables need not be multivariate NORMAL.   ∎

The converse of this property is true. That is, if a sequence of random vectors converge in distribution to another random vector, then all of the elements in the sequence of random vectors must also converge to the elements of the limiting random vector. This result follows from Theorem 4.14.

Corollary 4.1. *Let* $\{\mathbf{X}_n\}_{n=1}^{\infty}$ *be a sequence of d-dimensional random vectors that converge in distribution to a random vector* $\mathbf{X}$ *as* $n \to \infty$. *Let* $\mathbf{X}'_n = (X_{n1}, \ldots, X_{nd})$ *and* $\mathbf{X}' = (X_1, \ldots, X_d)$, *then* $X_{nk} \xrightarrow{d} X_k$ *as* $n \to \infty$ *for all* $k \in \{1, \ldots, d\}$.

For a proof of Corollary 4.1, see Exercise 22.

The convergence of random vectors was simplified to the univariate case using Theorem 3.6. As Example 4.21 demonstrates, the same simplification is not applicable to the convergence of distributions. The Cramér-Wold Theorem does provide a method for reducing the convergence in distribution of random vectors to the univariate case. Before presenting this result some preliminary setup is required. The result depends on multivariate characteristic functions, which are defined below.

Definition 4.5. *Let* $\mathbf{X}$ *be a d-dimensional random vector. The characteristic function of* $\mathbf{X}$ *is given by* $\psi(\mathbf{t}) = E[\exp(i\mathbf{t}'\mathbf{X})]$ *for all* $\mathbf{t} \in \mathbb{R}^d$.

Example 4.22. Suppose that $X_1, \ldots, X_d$ are independent standard normal random variables and let $\mathbf{X}' = (X_1, \ldots, X_d)$. Suppose that $\mathbf{t}' = (t_1, \ldots, t_d)$, then Definition 4.5 implies that the characteristic function of $\mathbf{X}$ is given by

$$
\begin{aligned}
\psi(\mathbf{t}) &= E[\exp(it'\mathbf{X})] \\
&= E\left[\exp\left(\sum_{k=1}^{n} it_k X_k\right)\right] \\
&= E\left[\prod_{k=1}^{n} \exp(it_k X_k)\right] \\
&= \prod_{k=1}^{n} E[\exp(it_k X_k)] \\
&= \prod_{k=1}^{n} \exp(-\tfrac{1}{2}t_k^2),
\end{aligned}
$$

where we have used the independence assumption and the fact that the characteristic function of a standard normal random variable is $\exp(-\tfrac{1}{2}t^2)$. Therefore, it follows that

$$
\psi(\mathbf{t}) = \exp\left(-\tfrac{1}{2}\sum_{k=1}^{n} t_k^2\right) = \exp(-\tfrac{1}{2}\mathbf{t}'\mathbf{t}).
$$

∎

As with the univariate case, characteristic functions uniquely identify a distribution, and in particular, a convergent sequence of characteristic functions is equivalent to the weak convergence of the corresponding random vectors or distribution functions. This result is the multivariate generalization of the equivalence of Conditions 1 and 2 of Theorem 4.6.

Theorem 4.16. *Let $\{\mathbf{X}_n\}_{n=1}^{\infty}$ be a sequence of d-dimensional random vectors where $\mathbf{X}_n$ has characteristic function $\psi(\mathbf{t})$ for all $n \in \mathbb{N}$. Let $\mathbf{X}$ be a d-dimensional random vector with characteristic function $\psi(\mathbf{t})$. Then $\mathbf{X}_n \xrightarrow{d} \mathbf{X}$ as $n \to \infty$ if and only if*

$$
\lim_{n\to\infty} \psi_n(\mathbf{t}) = \psi(\mathbf{t}),
$$

for every $\mathbf{t} \in \mathbb{R}^d$.

A proof of Theorem 4.16 can be found in Section 29 of Billingsley (1986).

Example 4.23. Let $\{\mathbf{X}_n\}_{n=1}^{\infty}$ be a sequence of d-dimensional random vectors where $\mathbf{X}_n$ has a $\mathbf{N}_d(\boldsymbol{\mu}_n, \boldsymbol{\Sigma}_n)$ distribution for all $n \in \mathbb{N}$. Suppose that $\boldsymbol{\mu}_n$ is a sequence of d-dimensional vectors defined as $\boldsymbol{\mu}_n = n^{-1}\mathbf{1}$, where $\mathbf{1}$ is a d-dimensional vector of the form $\mathbf{1}' = (1,1,\ldots,1)$. Further suppose that $\boldsymbol{\Sigma}_n$ is a sequence of $d \times d$ covariance matrices of the form $\boldsymbol{\Sigma}_n = \mathbf{I}+n^{-1}(\mathbf{J}-\mathbf{I})$, where $\mathbf{I}$ is the $d \times d$ identity matrix and $\mathbf{J}$ is a $d \times d$ matrix of ones. The characteristic function of $\mathbf{X}_n$ is given by $\psi(\mathbf{t}) = \exp(it'\boldsymbol{\mu}_n - \tfrac{1}{2}\mathbf{t}'\boldsymbol{\Sigma}_n\mathbf{t})$. Note that

$$
\lim_{n\to\infty} \boldsymbol{\mu}_n = \mathbf{0},
$$

and

$$\lim_{n \to \infty} \Sigma_n = I,$$

where 0 is a $d \times 1$ vector of the form $0' = (0, 0, \ldots, 0)$. Therefore, it follows that for every $t \in \mathbb{R}^d$ that

$$\lim_{n \to \infty} \psi(t) = \lim_{n \to \infty} \exp(it'\mu_n - \tfrac{1}{2}t'\Sigma_n t) = \exp(-\tfrac{1}{2}t't),$$

which is the characteristic function of a $N_d(0, I)$ distribution. Therefore, it follows from Theorem 4.16 that $X_n \overset{d}{\to} X$ as $n \to \infty$ where X is a $N_d(0, I)$ random vector. ∎

We are now in a position to present the theorem of Cramér and Wold, which reduces the task of proving that a sequence of random vectors converge weakly to another random vector to the univariate case by considering the convergence all possible linear combinations of the components of the random vectors.

Theorem 4.17 (Cramér and Wold). *Let $\{X_n\}_{n=1}^{\infty}$ be a sequence of random vectors in $\mathbb{R}^d$ and let X be a d-dimensional random vector. Then $X_n \overset{d}{\to} X$ as $n \to \infty$ if and only if $v'X_n \overset{d}{\to} v'X$ as $n \to \infty$ for all $v \in \mathbb{R}^d$.*

Proof. We will follow the proof of Serfling (1980). Let us first suppose that for any $v \in \mathbb{R}^d$ that $v'X_n \overset{d}{\to} vX$ as $n \to \infty$. Theorem 4.6 then implies that the characteristic function of $v'X_n$ converges to the characteristic function of $v'X$. Let $\psi_n(t)$ be the characteristic function of X_n for all $n \in \mathbb{N}$. The characteristic function of $v'X_n$ is then given by $E[\exp(itv'X_n)] = E\{\exp[i(tv')X_n]\} = \psi_n(tv)$ by Definition 4.5. Similarly, if $\psi(t)$ is the characteristic function of X, then the characteristic function of $v'X$ is given by $\psi(tv)$. Theorem 4.6 then implies that if $v'X_n \overset{d}{\to} v'X$ as $n \to \infty$ for all $v \in \mathbb{R}^d$, then

$$\lim_{n \to \infty} \psi_n(tv) = \psi(tv),$$

for all $v \in \mathbb{R}^d$ and $t \in \mathbb{R}$. This is equivalent to concluding that

$$\lim_{n \to \infty} \psi_n(u) = \psi(u),$$

for all $u \in \mathbb{R}^d$. Therefore, Theorem 4.16 implies that $X_n \overset{d}{\to} X$ as $n \to \infty$. For a proof of the converse, see Exercise 23. □

Example 4.24. Let $\{X_n\}_{n=1}^{\infty}$ and $\{Y_n\}_{n=1}^{\infty}$ be sequences of random variables where X_n has a $N(\mu_n, \sigma_n^2)$ distribution, Y_n has a $N(\nu_n, \tau_n^2)$ distribution, and X_n is independent of Y_n for all $n \in \mathbb{N}$. We assume that $\{\mu_n\}_{n=1}^{\infty}$ and $\{\nu_n\}_{n=1}^{\infty}$ are sequences of real numbers such that $\mu_n \to \mu$ and $\nu_n \to \nu$ as $n \to \infty$ for some real numbers μ and ν. Similarly, assume that $\{\sigma_n\}_{n=1}^{\infty}$ and $\{\tau_n\}_{n=1}^{\infty}$ are sequences of positive real numbers such that $\sigma_n \to \sigma$ and $\tau_n \to \tau$ as $n \to \infty$ for some positive real numbers σ and τ. Let v_1 and v_2 be arbitrary real numbers, then $E(v_1 X_n + v_2 Y_n) = v_1 \mu_n + v_2 \nu_n$ and $V(v_1 X_n + v_2 Y_n) = v_1^2 \sigma_n^2 + v_2^2 \tau_n^2$, for all $n \in \mathbb{N}$. It follows that $v_1 X_n + v_2 Y_n$ has a $N(v_1 \mu_n + v_2 \nu_n, v_1^2 \sigma_n^2 + v_2^2 \tau_n^2)$

distribution for all $n \in \mathbb{N}$, due to the assumed independence between X_n and Y_n. Similarly $v_1 X + v_2 Y$ has a $N(v_1 \mu + v_2 \nu, v_1^2 \sigma^2 + v_2^2 \tau^2 + 2v_1 v_2 \gamma)$ distribution. Example 4.2 implies that $v_1 X_n + v_2 Y_n \xrightarrow{d} v_1 X + v_2 Y$ as $n \to \infty$ for all $v_1 \in \mathbb{R}$ and $v_2 \in \mathbb{R}$. See Exercise 24 for further details on this conclusion. Now let $\mathbf{Z}'_n = (X_n, Y_n)$ for all $n \in \mathbb{N}$ and let $\mathbf{Z}' = (X, Y)$. Because v_1 and v_2 are arbitrary, Theorem 4.17 implies that $\mathbf{Z}_n \xrightarrow{d} \mathbf{Z}$ as $n \to \infty$. ∎

Example 4.25. Let $\{\mathbf{X}_n\}_{n=1}^{\infty}$ be a sequence of d_x-dimensional random vectors that converge in distribution to a random vector $\mathbf{X}$ as $n \to \infty$, and let $\{\mathbf{Y}_n\}_{n=1}^{\infty}$ be a sequence of d_y-dimensional random vectors that converge in probability to a constant vector $\mathbf{y}$ as $n \to \infty$. Let $d = d_x + d_y$ and consider the random vector defined by $\mathbf{Z}'_n = (\mathbf{X}'_n, \mathbf{Y}'_n)$ for all $n \in \mathbb{N}$ and similarly define $\mathbf{Z}' = (\mathbf{X}', \mathbf{y}')$. Let $\mathbf{v} \in \mathbb{R}^d$ be an arbitrary vector that can be partitioned as $\mathbf{v}' = (\mathbf{v}'_1, \mathbf{v}'_2)$ where $\mathbf{v}_1 \in \mathbb{R}^{d_x}$ and $\mathbf{v}_2 \in \mathbb{R}^{d_y}$. Now $\mathbf{v}'\mathbf{Z}_n = \mathbf{v}'_1 \mathbf{X}_n + \mathbf{v}'_2 \mathbf{Y}_n$ and $\mathbf{v}'\mathbf{Z} = \mathbf{v}'_1 \mathbf{X} + \mathbf{v}'_2 \mathbf{y}$. Theorem 4.17 implies that $\mathbf{v}'_1 \mathbf{X}_n \xrightarrow{d} \mathbf{v}'_1 \mathbf{X}$ as $n \to \infty$ and Theorem 3.9 implies that $\mathbf{v}'_2 \mathbf{Y}_n \xrightarrow{p} \mathbf{v}'_2 \mathbf{y}$ as $n \to \infty$. Theorem 4.11 (Slutsky) then implies that $\mathbf{v}'_1 \mathbf{X}_n + \mathbf{v}'_2 \mathbf{Y}_n \xrightarrow{d} \mathbf{v}'_1 \mathbf{X} + \mathbf{v}'_2 \mathbf{y}$ as $n \to \infty$. Because $\mathbf{v}$ is arbitrary, Theorem 4.17 implies that $\mathbf{Z}_n \xrightarrow{d} \mathbf{Z}$ as $n \to \infty$. ∎

Linear functions are not the only function of convergent sequences of random vectors that converge in distribution. The result of Theorem 4.12 can be generalized to Borel functions of sequences of random vectors that converge weakly as long as the function is continuous with respect to the distribution of the limiting random vector.

Theorem 4.18. *Let $\{\mathbf{X}_n\}_{n=1}^{\infty}$ be a sequence of d-dimensional random vectors that converge in distribution to a random vector $\mathbf{X}$ as $n \to \infty$. Let g be a Borel function that maps $\mathbb{R}^d$ to $\mathbb{R}^m$ and suppose that $P[\mathbf{X} \in C(g)] = 1$. Then $g(\mathbf{X}_n) \xrightarrow{d} g(\mathbf{X})$ as $n \to \infty$.*

Theorem 4.18 can be proven using an argument that parallels the proof of Theorem 4.12 (Continuous Mapping Theorem) using a multivariate version of Theorem 4.10 (Skorokhod).

Example 4.26. Let $\{\mathbf{Z}_n\}_{n=1}^{\infty}$ be a sequence of bivariate random variables that converge in distribution to a bivariate random variable $\mathbf{Z}$ as $n \to \infty$ where $\mathbf{Z}$ as a $\mathbf{N}(\mathbf{0}, \mathbf{I})$ distribution. Let $g(\mathbf{z}) = g(z_1, z_2) = z_1/z_2$ which is continuous except on the line $z_2 = 0$. Therefore, $P[\mathbf{Z} \in C(g)] = 1$ and Theorem 4.12 implies that $g(\mathbf{Z}_n) \xrightarrow{d} g(\mathbf{Z})$ as $n \to \infty$ where $g(\mathbf{Z})$ has a CAUCHY$(0,1)$ distribution. ∎

Example 4.27. Let $\{R_n\}_{n=1}^{\infty}$ be a sequence of random variables where $R_n \xrightarrow{d} R$ as $n \to \infty$ where R^2 has a CHISQUARED(2) distribution. Let $\{T_n\}_{n=1}^{\infty}$ be a sequence of random variables where $T_n \xrightarrow{d} T$ as $n \to \infty$ where T has a UNIFORM$(0, 2\pi)$ distribution. Assume that the bivariate sequence of random vectors $\{(R_n, T_n)\}_{n=1}^{\infty}$ converge in distribution to the random vector (R, T) as $n \to \infty$, where R and T are independent of one another. Consider the transformation $g : \mathbb{R}^2 \to \mathbb{R}^2$ defined by $g(R, T) = (R\cos(T), R\sin(T))$, which is continuous with probability one with respect to the joint distribution of R and T. Then it follows from Theorem 4.12 that the sequence $\{g(R_n, T_n)\}_{n=1}^{\infty}$ converges in distribution to a bivariate normal distribution with mean vector $\mathbf{0}$ and covariance matrix $\mathbf{I}$. ∎

Example 4.28. Let $\{X_n\}_{n=1}^{\infty}$ and $\{Y_n\}_{n=1}^{\infty}$ be sequences of random variables that converge in distribution to random variables X and Y as $n \to \infty$, respectively. Can we conclude necessarily that $X_n + Y_n \xrightarrow{d} X + Y$ as $n \to \infty$ based on Theorem 4.12? Without further information, we cannot make such a conclusion. The reason is that the weak convergence of the sequences $\{X_n\}_{n=1}^{\infty}$ and $\{Y_n\}_{n=1}^{\infty}$ do not imply the weak convergence of the associated sequence of random vectors $\{(X_n, Y_n)\}_{n=1}^{\infty}$ and Theorem 4.12 requires that the random vectors converge in distribution. Therefore, without further information about the two sequences, and the convergence behavior, no conclusion can be made. For example, if X_n and Y_n converge to independent normal random variables X and Y, then the conclusion does follow. But for example, take X_n to have a standard normal distribution and $Y_n = -X_n$ so that Y_n also has a standard normal distribution. In this case X and Y are standard normal distributions, and if we assume that X and Y where independent then $X + Y$ would have a $N(0, 2)$ distribution whereas $X_n + Y_n$ is a degenerate distribution at 0 for all $n \in \mathbb{N}$, that converges to a degenerate distribution at 0 as $n \to \infty$. Therefore, we would have to know about the relationship between X_n and Y_n in order to draw the correct conclusion in this case. ∎

As in the univariate case, we are often interested in perturbed sequences of random vectors that have some weak convergence property. The results of Theorem 4.11 generalize to the multivariate case.

Theorem 4.19. *Let $\{\mathbf{X}_n\}_{n=1}^{\infty}$ be a sequence of d-dimensional random vectors that converge in distribution to a random vector $\mathbf{X}$ as $n \to \infty$. Let $\{\mathbf{Y}_n\}_{n=1}^{\infty}$ be a sequence of d-dimensional random vectors that converge in probability to a constant vector $\mathbf{y}$ and let $\{\mathbf{Z}_n\}_{n=1}^{\infty}$ be a sequence of d-dimensional random vectors that converge in probability to a constant vector $\mathbf{z}$ as $n \to \infty$. Then $diag(\mathbf{Y}_n)\mathbf{X}_n + \mathbf{Z}_n \xrightarrow{d} diag(\mathbf{y})\mathbf{X} + \mathbf{z}$ as $n \to \infty$, where $diag(\mathbf{y})$ is a $d \times d$ diagonal matrix whose diagonal values equal the elements of $\mathbf{y}$.*

Proof. We will use the method of proof suggested by Lehmann (1999). Define a sequence of $3d$-dimensional random variables $\{\mathbf{W}_n\}_{n=1}^{\infty}$ where $\mathbf{W}_n' = (\mathbf{X}_n', \mathbf{Y}_n', \mathbf{Z}_n')$ for all $n \in \mathbb{N}$ and similarly define $\mathbf{W}' = (\mathbf{X}', \mathbf{y}', \mathbf{z}')$. Theorem 3.6 and Example 4.25 imply that $\mathbf{W}_n \xrightarrow{d} \mathbf{W}$ as $n \to \infty$. Now define $g(\mathbf{w}) = g(\mathbf{x}, \mathbf{y}, \mathbf{z}) = diag(\mathbf{y})\mathbf{x} + \mathbf{z}$ which is an everywhere continuous function so that $P[\mathbf{W} \in C(g)] = 1$. Theorem 4.18 implies that $g(\mathbf{W}_n) = diag(\mathbf{Y}_n)\mathbf{X}_n + \mathbf{Z}_n \xrightarrow{d} g(\mathbf{W}) = diag(\mathbf{y})\mathbf{X} + \mathbf{z}$ as $n \to \infty$. □

Example 4.29. Let $\{\mathbf{X}_n\}_{n=1}^{\infty}$ be a sequence of d-dimensional random vectors such that $n^{1/2}\mathbf{X}_n \xrightarrow{d} \mathbf{Z}$ as $n \to \infty$ where $\mathbf{Z}$ is a $\mathbf{N}_p(\mathbf{0}, \mathbf{I})$ random vector.

Suppose that $\{\mathbf{Y}_n\}_{n=1}^{\infty}$ is any sequence of random vectors that converges in probability to a vector $\boldsymbol{\theta}$. Then Theorem 4.19 implies that $n^{1/2}\mathbf{X}_n + \mathbf{Y}_n \xrightarrow{d} \mathbf{W}$ as $n \to \infty$ where $\mathbf{W}$ has a $\mathbf{N}_p(\boldsymbol{\theta}, \mathbf{I})$ distribution. ∎

4.4 The Central Limit Theorem

The central limit theorem, as the name given to it by G. Pólya in 1920 implies, is the key asymptotic result in statistics. The result in some form has existed since 1733 when De Moivre proved the result for a sequence of independent and identically distributed BERNOULLI(θ) random variables. In some sense, one can question how far the field of statistics could have progressed without this essential result. It is the Central Limit Theorem, in its various forms, that allow us to construct approximate normal tests and confidence intervals for unknown means when the sample size is large. Without such a result we would be required to develop tests for each possible population. The result allows us to approximate BINOMIAL probabilities under certain circumstances when the number of Bernoulli experiments is large. These probabilities, with such an approximation, would have been very difficult to compute, especially before the advent of the digital computer. Another key attribute of the Central Limit Theorem is its widespread applicability. The Central Limit Theorem, with appropriate modifications, applies not only to the case of independent and identically distributed random variables, but can also be applied to dependent sequences of variables, sequences that have varying distributions, and other cases as well. Finally, the power of the normal approximation is additionally quite important. When the parent population is not too far from normality, then the Central Limit Theorem provides quite accurate approximations to the distribution of the sample mean, even when the sample size is quite small.

In this section we will introduce the simplest form of the central limit theorem which applies to sequences of independent and identically distributed random variables with finite mean and variance and present the usual proof which is based on limits of characteristic functions. We will also present the simple form of the multivariate version of the central limit theorem. We will revisit this topic with much more detail in Chapter 6 where we consider several generalizations of this result.

Theorem 4.20 (Lindeberg and Lévy). *Let $\{X_n\}_{n=1}^{\infty}$ be a sequence of independent and identically distributed random variables such that $E(X_n) = \mu$ and $V(X_n) = \sigma^2 < \infty$ for all $n \in \mathbb{N}$, then $Z_n = n^{1/2}\sigma^{-1}(\bar{X}_n - \mu) \xrightarrow{d} Z$ as $n \to \infty$ where Z has a $N(0,1)$ distribution.*

Proof. We will begin by assuming, without loss of generality, that $\mu = 0$ and $\sigma = 1$. We can do this since if μ and σ are known, we can always transform the sequence $\{X_n\}_{n=1}^{\infty}$ as $\sigma^{-1}(X_n - \mu)$ to get a sequence of random variables with zero mean and variance equal to one. When $\mu = 0$ and $\sigma = 1$ the random variable of interest is

$$Z_n = n^{1/2}\bar{X}_n = n^{-1/2}\sum_{k=1}^{n} X_k.$$

The general method of proof involves computing the characteristic function of Z_n and then showing that this characteristic function converges to that of a standard normal random variable as $n \to \infty$. An application of Theorem 4.6 then completes the proof. Suppose that X_n has characteristic function $\psi(t)$ for all $n \in \mathbb{N}$. Because the random variables in the sequence $\{X_n\}_{n=1}^{\infty}$ are independent and identically distributed, Theorem 2.33 implies that the characteristic function of the sum of $X_1, \ldots, X_n$ is $\psi^n(t)$ for all $n \in \mathbb{N}$. An application of Theorem 2.32 then implies that the characteristic function of Z_n equals $\psi^n(n^{-1/2}t)$ for all $n \in \mathbb{N}$. By assumption the variance σ^2 is finite, and hence the second moment is also finite and Theorem 2.31 implies that $\psi(t) = 1 - \frac{1}{2}t^2 + o(t^2)$ as $t \to 0$. Therefore, the characteristic function of Z_n equals

$$\psi^n(n^{-1/2}t) = [1 - \tfrac{1}{2}n^{-1}t^2 + o(n^{-1}t^2)]^n = [1 - \tfrac{1}{2}n^{-1}t^2]^n + o(n^{-1}t^2).$$

The second inequality can be justified using Theorem A.22. See Exercise 26. Therefore, for fixed $t \in \mathbb{R}$, Theorem 1.7 implies that

$$\lim_{n\to\infty} \psi^n(n^{-1/2}t) = \lim_{n\to\infty}[1 - \tfrac{1}{2}n^{-1}t^2]^n + o(n^{-1}t^2) = \exp(-\tfrac{1}{2}t^2),$$

which is the characteristic function of a standard normal random variable. Therefore, Theorem 4.6 implies that $n^{1/2}\bar{X}_n \xrightarrow{d} Z$ as $n \to \infty$ where Z is a standard normal random variable. $\qquad\square$

Example 4.30. Suppose that $\{B_n\}_{n=1}^{\infty}$ be a sequence of independent and identically distributed BERNOULLI(θ) random variables. Let $\bar{B}_n$ be the sample mean computed on $B_1, \ldots, B_n$, which in this case will correspond to the sample proportion. Theorem 4.20 then implies that $n^{1/2}\sigma^{-1}(\bar{B}_n - \mu) \xrightarrow{d} Z$ as $n \to \infty$ where Z is a standard normal random variable. In this case $\mu = E(B_1) = \theta$ and $\sigma^2 = V(B_1) = \theta(1 - \theta)$ so that the result above is equivalent to $n^{1/2}(\bar{B}_n - \theta)[\theta(1 - \theta)]^{-1/2} \xrightarrow{d} Z$ as $n \to \infty$. This implies that when n is large that

$$P\{n^{1/2}(\bar{B}_n - \theta)[\theta(1 - \theta)]^{-1/2} \le t\} \simeq \Phi(t),$$

or equivalently that

$$P\left\{\sum_{k=1}^{n} B_k \le \theta + n^{1/2}t[\theta(1 - \theta)]^{1/2}\right\} \simeq \Phi(t),$$

which is also equivalent to

$$P\left(\sum_{k=1}^{n} B_k \le t\right) \simeq \Phi\{(t - \theta)[n\theta(1 - \theta)]^{-1/2}\}.$$

Figure 4.3 *The distribution function of a* BINOMIAL(n, θ) *distribution and a* N$[n\theta, n\theta(1 - \theta)]$ *distribution when* $n = 5$ *and* $\theta = \frac{1}{4}$.

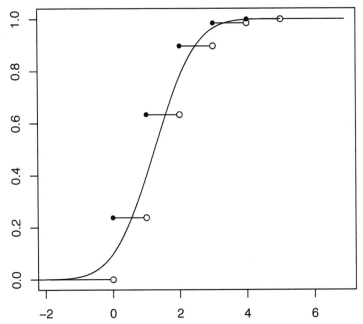

The left hand side of this last equation is a BINOMIAL probability and the right hand side is a NORMAL probability. Therefore, this last equation provides a method for approximating BINOMIAL probabilities with NORMAL probabilities under the condition that n is large. Figures 4.3 and 4.4 compare the BINOMIAL(n, p) and N$[n\theta, n\theta(1 - \theta)]$ distribution functions for $n = 5$ and $n = 10$ when $\theta = \frac{1}{4}$. One can observe that the NORMAL distribution function, though continuous, does capture the general shape of the BINOMIAL distribution function, with the approximation improving as n becomes larger.   ∎

Example 4.31. Let $\{X_n\}_{n=1}^{\infty}$ be a sequence of independent and identically distributed random variables where X_n has a UNIFORM$(0, 1)$ distribution, so that it follows that $E(X_n) = \frac{1}{2}$ and $V(X_n) = \frac{1}{12}$ for all $n \in \mathbb{N}$. Therefore, Theorem 4.20 implies that $Z_n = (12n)^{1/2}(\bar{X}_n - \frac{1}{2}) \xrightarrow{d} Z$ as $n \to \infty$ where Z is a N$(0, 1)$ random variable. The power of Theorem 4.20 is particularly impressive in this case. When $n = 2$ the true distribution of Z_n is a TRIANGULAR distribution, which already begins to show a somewhat more normal shape when compared to the population. Figures 4.5 to 4.6 show histograms based on 100,000 simulated samples of Z_n when $n = 5$ and 10. Again, it is apparent how quickly the normal approximation becomes accurate, particularly near the center of the distribution, even for smaller values of n. In fact, a simple

Figure 4.4 *The distribution function of a* BINOMIAL(n, θ) *distribution and a* N$[n\theta, n\theta(1 - \theta)]$ *distribution when $n = 10$ and $\theta = \frac{1}{4}$.*

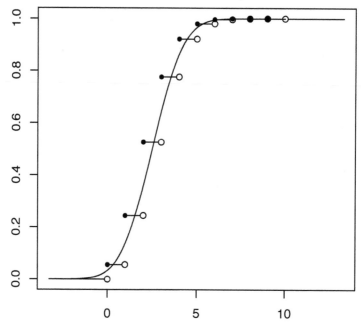

algorithm for generating approximate normal random variables in computer simulations is to generate independent random variables $U_1, \ldots, U_{12}$ where U_k has a UNIFORM$(0, 1)$ distribution and then compute

$$Z = \left(\sum_{k=1}^{12} U_k\right) - 6,$$

which has an approximate N$(0, 1)$ distribution by Theorem 4.20. This algorithm, however, is not highly recommended as there are more efficient and accurate methods to generate standard normal random variables. See Section 5.3 of Bratley, Fox and Schrage (1987). ∎

In the usual setting in Theorem 4.20, the random variable Z_n can be written as $n^{1/2}\sigma^{-1}(\bar{X}_n - \mu)$. From this formulation it is clear that if $\sigma < \infty$, as required by the assumptions of Theorem 4.20, then it follows that

$$\lim_{n\to\infty} n^{-1/2}\sigma = 0.$$

Note that if Z_n is to have a normal distribution in the limit then it must follow that $\bar{X}_n$ must approach μ at the same rate. That is, the result of the Theorem 4.20 implies that $\bar{X}_n$ is a consistent estimator of μ.

Theorem 4.21. *Suppose that $\{X_n\}_{n=1}^{\infty}$ is a sequence of random variables*

Figure 4.5 *Density histogram of* 100,000 *simulated values of* $Z_n = (12n)^{1/2}(\bar{X}_n - \frac{1}{2})$ *for samples of size* $n = 5$ *from a* UNIFORM$(0,1)$ *distribution compared to a* N$(0,1)$ *density.*

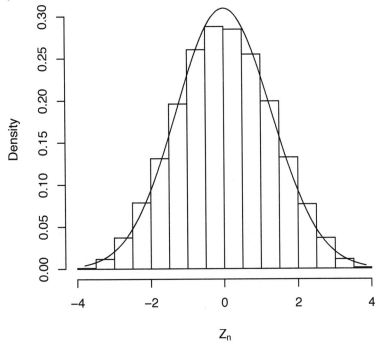

such that $\sigma_n^{-1}(X_n - \mu) \xrightarrow{d} Z$ *as* $n \to \infty$ *where* Z *is a* N$(0,1)$ *random variable and* $\{\sigma_n\}_{n=1}^{\infty}$ *is a sequence of real numbers such that*

$$\lim_{n \to \infty} \sigma_n = \sigma \in \mathbb{R},$$

and $\mu \in \mathbb{R}$. *Then* $X_n \xrightarrow{p} \mu$ *if and only if* $\sigma = 0$.

Proof. Let us first assume that $\sigma = 0$, which implies that $\sigma_n \xrightarrow{p} 0$ as $n \to \infty$. Part 2 of Theorem 4.11 (Slutsky) implies that $\sigma_n \sigma_n^{-1}(X_n - \mu) = X_n - \mu \xrightarrow{d} 0$ as $n \to \infty$. Theorem 4.9 then implies that $X_n - \mu \xrightarrow{p} 0$ as $n \to \infty$, which is equivalent to the result that $X_n \xrightarrow{p} \mu$ as $n \to \infty$. On the other hand if we assume that $X_n \xrightarrow{p} \mu$ as $n \to \infty$, then again we automatically conclude that $X_n - \mu \xrightarrow{p} 0$. Suppose that $\sigma \neq 0$ and let us find a contradiction. Since $\sigma \neq 0$ it follows that $\sigma_n / \sigma \xrightarrow{p} 1$ as $n \to \infty$ and therefore Theorem 4.11 implies that $\sigma_n \sigma^{-1} \sigma_n^{-1}(X_n - \mu) = X_n - \mu \xrightarrow{d} Z$ as $n \to \infty$ where Z is a N$(0,1)$ random variable. This is a contradiction since we know that $X_n - \mu \xrightarrow{p} 0$ as $n \to \infty$. Therefore, σ cannot be non-zero. $\square$

Figure 4.6 *Density histogram of* 100,000 *simulated values of* $Z_n = (12n)^{1/2}(\bar{X}_n - \frac{1}{2})$ *for samples of size* $n = 10$ *from a* UNIFORM$(0,1)$ *distribution compared to a* N$(0,1)$ *density.*

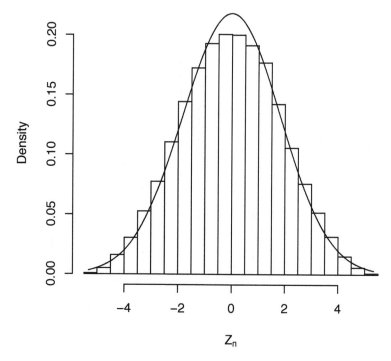

Example 4.32. Suppose that $\{B_n\}_{n=1}^{\infty}$ be a sequence of independent and identically distributed BERNOULLI(θ) random variables. In Example 4.30 it was shown that $Z_n = \sigma_n^{-1}(\bar{B}_n - \theta) \xrightarrow{d} Z$ as $n \to \infty$ where Z is a standard normal random variable and $\sigma_n = n^{-1/2}[\theta(1-\theta)]^{1/2}$. Since

$$\lim_{n\to\infty} \sigma_n = \lim_{n\to\infty} n^{-1/2}[\theta(1-\theta)]^{1/2} = 0,$$

if follows from Theorem 4.21 that $\bar{B}_n \xrightarrow{p} \theta$ as $n \to \infty$. ∎

Example 4.33. Let $\{X_n\}_{n=1}^{\infty}$ be a sequence of independent and identically distributed random variables where X_n has a UNIFORM$(0,1)$ distribution. In Example 4.31 it was shown that $Z_n = \sigma_n^{-1}(\bar{X}_n - \frac{1}{2}) \xrightarrow{d} Z$ as $n \to \infty$ where Z is a standard normal random variable and where $\sigma_n = (12n)^{-1/2}$. Since

$$\lim_{n\to\infty} \sigma_n = \lim_{n\to\infty} (12n)^{-1/2} = 0,$$

if follows from Theorem 4.21 that $\bar{X}_n \xrightarrow{p} \frac{1}{2}$ as $n \to \infty$. ∎

The multivariate version of Theorem 4.20 is a direct generalization to the case of a sequence of independent and identically distributed random vectors from a distribution that has a covariance matrix with finite elements.

Theorem 4.22. *Let $\{\mathbf{X}_n\}_{n=1}^{\infty}$ be a sequence of independent and identically distributed d-dimensional random vectors such that $E(\mathbf{X}_n) = \boldsymbol{\mu}$ and $V(\mathbf{X}_n) = \boldsymbol{\Sigma}$, where the covariance matrix $\boldsymbol{\Sigma}$ has elements that are all finite. Then $n^{1/2}\boldsymbol{\Sigma}^{-1/2}(\bar{\mathbf{X}}_n - \boldsymbol{\mu}) \overset{d}{\to} \mathbf{Z}$ as $n \to \infty$ where $\mathbf{Z}$ is a d-dimensional $\mathbf{N}(\mathbf{0}, \mathbf{I})$ random variable.*

The proof of Theorem 4.22 is similar to that of the univariate case given by Theorem 4.20, the only real difference being that the multivariate characteristic function is used.

Example 4.34. Consider a sequence of discrete bivariate random variables $\{\mathbf{X}_n\}_{n=1}^{\infty}$ where $\mathbf{X}_n$ has probability distribution

$$
f(\mathbf{x}) = \begin{cases} \theta & \mathbf{x}' = (1, 0, 0) \\ \eta & \mathbf{x}' = (0, 1, 0) \\ 1 - \theta - \eta & \mathbf{x}' = (0, 0, 1) \\ 0 & \text{otherwise,} \end{cases}
$$

for all $n \in \mathbb{N}$ where θ and η are parameters such that $0 < \theta < 1$, $0 < \eta < 1$ and $0 < \theta + \eta < 1$. The mean vector of $\mathbf{X}_n$ is given by $E(\mathbf{X}_n) = \boldsymbol{\mu} = (\theta, \eta, 1 - \theta - \eta)'$. The covariance matrix of $\mathbf{X}_n$ is given by

$$
V(\mathbf{X}_n) = \boldsymbol{\Sigma} = \begin{bmatrix} \theta(1-\theta) & -\theta\eta & -\theta(1-\theta-\eta) \\ -\theta\eta & \eta(1-\eta) & -\eta(1-\theta-\eta) \\ -\theta(1-\theta-\eta) & -\eta(1-\theta-\eta) & (\theta+\eta)(1-\theta-\eta) \end{bmatrix}.
$$

Theorem 4.22 implies that $n^{1/2}\boldsymbol{\Sigma}^{-1/2}(\bar{\mathbf{X}}_n - \boldsymbol{\mu}) \overset{d}{\to} \mathbf{Z}$ as $n \to \infty$ where $\mathbf{Z}$ is a three dimensional $\mathbf{N}(\mathbf{0}, \mathbf{I})$ random vector. That is, when n is large, it follows that

$$
P[n^{1/2}\boldsymbol{\Sigma}^{-1/2}(\bar{\mathbf{X}}_n - \boldsymbol{\mu}) \leq \mathbf{t}] \simeq \boldsymbol{\Phi}(\mathbf{t}),
$$

where $\boldsymbol{\Phi}$ is the multivariate distribution function of a $\mathbf{N}(\mathbf{0}, \mathbf{I})$ random vector. Equivalently, we have that

$$
P\left(\sum_{k=1}^{n} \mathbf{X}_k \leq \mathbf{t}\right) \simeq \boldsymbol{\Phi}[n^{-1/2}\boldsymbol{\Sigma}^{-1/2}(\mathbf{t} - \boldsymbol{\mu})],
$$

which results in a NORMAL approximation to a three dimensional MULTINOMIAL distribution. ∎

4.5 The Accuracy of the Normal Approximation

Let $\{X_n\}_{n=1}^{\infty}$ be a sequence of independent and identically distributed random variables from a distribution F such that $E(X_n) = \mu$ and $V(X_n) = \sigma^2 < \infty$ for all $n \in \mathbb{N}$. Theorem 4.20 implies that the standardized sample mean given by $Z_n = \sigma^{-1}n^{1/2}(\bar{X}_n - \mu)$ converges in distribution to a random variable Z that has a N(0,1) distribution as $n \to \infty$. That is,

$$\lim_{n\to\infty} P(Z_n \le z) = \Phi(z),$$

for all $z \in \mathbb{R}$. Definition 1.1 then implies that for large values of n, we can use the approximation $P(Z_n \le z) \simeq \Phi(z)$. One of the first concerns when one uses any approximation should be about the accuracy of the approximation. In the case of the normal approximation given by Theorem 4.20, we are interested in how well the normal distribution approximates probabilities of the standardized sample mean and how the quality of the approximation depends on n and on the parameters of the distribution of the random variables in the sequence. In some cases it is possible to study these effects using direct calculation.

Example 4.35. Let $\{X_n\}_{n=1}^{\infty}$ be a sequence of independent and identically distributed random variables where X_n has a BERNOULLI(θ), distribution. In this case it is well known that the sum S_n has a BINOMIAL(n,θ) distribution. The distribution of the standardized sample mean, $Z_n = n^{1/2}\theta^{-1/2}(1 - \theta)^{-1/2}(\bar{X}_n-\theta)$ then has a BINOMIAL(n,θ) distribution that has been scaled so that Z_n has support $\{n^{1/2}\theta^{-1/2}(1-\theta)^{-1/2}(0-\theta), n^{1/2}\theta^{-1/2}(1-\theta)^{-1/2}(n^{-1} - \theta), \ldots, n^{1/2}\theta^{-1/2}(1-\theta)^{-1/2}(1-\theta)\}$. That is,

$$P[Z_n = n^{1/2}\theta^{-1/2}(1-\theta)^{-1/2}(kn^{-1} - \theta)] = \binom{n}{k}\theta^k(1-\theta)^{n-k},$$

for $k \in \{1,\ldots,n\}$. As shown in Example 4.30, Theorem 4.20 implies that $Z_n \xrightarrow{d} Z$ as $n \to \infty$ where Z is a N(0,1) random variable. This means that for large n, we have the approximation $P(Z_n \le z) \simeq \Phi(z)$. Because the distribution of the standardized mean is known in this case we can assess the accuracy of the normal approximation directly. For example, when $\theta = \frac{1}{4}$ and $n = 5$ we have that the Kolmogorov distance between $P(Z_n \le t)$ and $\Phi(t)$ is 0.2346. See Figure 4.7. Similarly, when $\theta = \frac{1}{2}$ and $n = 10$ we have that the Kolmogorov distance between $P(Z_n \le t)$ and $\Phi(t)$ is 0.1230. See Figure 4.8. A more complete table of comparisons is given in Table 4.1. It is clear from the table that both n and θ affect the accuracy of the normal approximation. As n increases, the Kolmogorov distance becomes smaller. This is guaranteed by Theorem 4.7. However, another effect can be observed from Table 4.1. The approximation becomes progressively worse as θ approaches zero. This is due to the fact that the binomial distribution becomes more skewed as θ approaches zero. The normal approximation requires larger sample sizes to overcome this skewness. ∎

Example 4.36. Let $\{X_n\}_{n=1}^{\infty}$ be a sequence of independent and identically distributed random variables where X_n has a EXPONENTIAL(θ) distribution. In this case it is well known that the sum S_n has a GAMMA(n,θ) distribution, and therefore the standardized mean $Z_n = n^{1/2}\theta^{-1}(\bar{X}_n - \theta)$ has a translated

Figure 4.7 *The normal approximation to the* BINOMIAL$(5, \frac{1}{4})$ *distribution. The largest difference is indicated by the dotted line.*

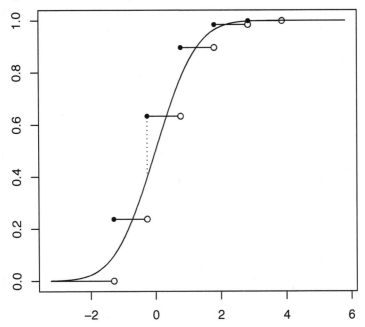

Table 4.1 *The normal approximation of the* BINOMIAL(n, θ) *distribution for* $n = 5, 10, 25, 50, 100$ *and* $\theta = 0.01, 0.05, 0.10, 0.25, 0.50$. *The value reported is the Kolmogorov distance between the scaled binomial distribution function and the normal distribution function.*

			θ		
n	0.01	0.05	0.10	0.25	0.50
5	0.5398	0.4698	0.3625	0.2347	0.1726
10	0.5291	0.3646	0.2361	0.1681	0.1230
25	0.4702	0.2331	0.1677	0.1071	0.0793
50	0.3664	0.1677	0.1161	0.0758	0.0561
100	0.2358	0.1160	0.0832	0.0535	0.0398

Figure 4.8 *The normal approximation to the* BINOMIAL$(10, \frac{1}{2})$ *distribution. The largest difference is indicated by the dotted line.*

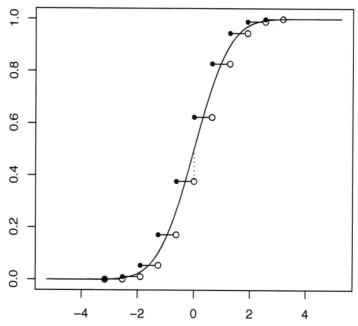

GAMMA$(n, n^{-1/2})$ distribution given by

$$P(Z_n \le z) = P[n^{1/2}\theta^{-1}(\bar{X}_n - \theta) \le z] \tag{4.3}$$

$$= \int_{-n^{1/2}}^{z} \frac{n^{1/2}}{\Gamma(n)}(t + n^{1/2})^{n-1} \exp[-n^{1/2}(t + n^{1/2})]dt. \tag{4.4}$$

See Exercise 27. Theorem 4.20 implies that $Z_n \xrightarrow{d} Z$ as $n \to \infty$ where Z is a $N(0,1)$ distribution, and therefore for large n, we have the approximation $P(Z_n \le z) \simeq \Phi(z)$. Because the exact density of Z_n is known in this case, we can compute the Kolmogorov distance between $P(Z_n \le z)$ and $\Phi(z)$ to get an overall view of the accuracy of this approximation. Note that in this case the distribution of Z_n does not depend on θ, so that we need only consider the sample size n for our calculations. For example, when $n = 2$, the Kolmogorov distance between $P(Z_n \le z)$ and $\Phi(z)$ is approximately 0.0945, and when $n = 5$ the Kolmogorov distance between $P(Z_n \le z)$ and $\Phi(z)$ is approximately 0.0596. See Figures 4.9 and 4.10. A plot of the Kolmogorov distance against n is given in Figure 4.11. We observe once again that the distance decreases with n, as required by Theorem 4.7. ∎

It is not always possible to perform this type of calculation directly. In many cases the exact distribution of the standardized sample mean is not known,

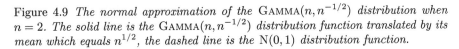

Figure 4.9 *The normal approximation of the* GAMMA$(n, n^{-1/2})$ *distribution when* $n = 2$. *The solid line is the* GAMMA$(n, n^{-1/2})$ *distribution function translated by its mean which equals* $n^{1/2}$, *the dashed line is the* N$(0, 1)$ *distribution function.*

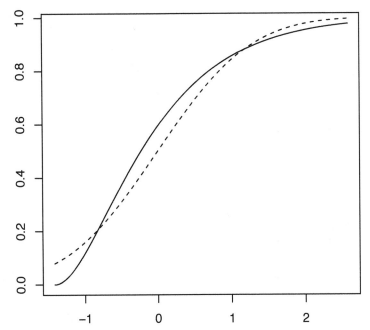

or may not have a simple form. In these cases one can often use simulations to approximate the behavior of the normal approximation.

Example 4.37. Let $\{\{X_{n,m}\}_{m=1}^{n}\}_{n=1}^{\infty}$ be a triangular array of N$(0, 1)$ random variables that are mutually independent both within and between rows. Define a new sequence of random variables $\{Y_n\}_{n=1}^{\infty}$ where

$$Y_n = \max\{X_{n,1}, \ldots, X_{n,n}\} - \min\{X_{n,1}, \ldots, X_{n,n}\}$$

for all $n \in \mathbb{N}$. The distribution function of Y_n has the form

$$P(Y_n \leq y) = n \int_{-\infty}^{\infty} [\Phi(t + y) - \Phi(t)]^{n-1} \phi(t) dt, \qquad (4.5)$$

for $y > 0$. See Arnold, Balakrishnan and Nagaraja (1993). Now let $\bar{Y}_n$ be the sample mean of $Y_1, \ldots, Y_n$, and let $Z_n = n^{1/2}\sigma(\bar{Y}_n - \mu)$ where $\mu = E(Y_n)$ and $\sigma^2 = V(Y_n)$. While the distribution function in Equation (4.5) can be computed numerically with some ease, the distributions of $\bar{Y}_n$ and Z_n are not so simple to compute and approximations based on simulations are usually easier. ∎

When applying the normal approximation in practice, the distribution F may

Figure 4.10 *The normal approximation of the* GAMMA$(n, n^{-1/2})$ *distribution when* $n = 5$*. The solid line is the* GAMMA$(n, n^{-1/2})$ *distribution function translated by its mean which equals* $n^{1/2}$*, the dashed line is the* N$(0, 1)$ *distribution function.*

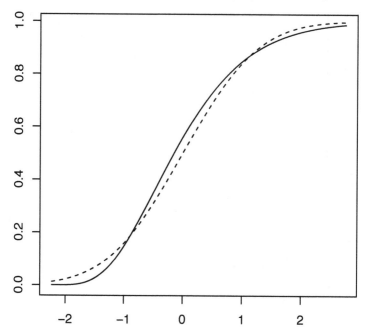

not be known. This problem prevents one from studying the accuracy of the normal approximation using either direct calculation or simulations. In this case one needs to be able to study the accuracy of the normal approximation in such a way that it does not depend of the distribution of the random variables in the sequence. One may be surprised to learn that there are theoretical results which can be used in this case. Specifically, one can find universal bounds on the accuracy of the normal approximation that do not depend on the distribution of the random sequence. For the case we study in this section we will have to make some further assumptions about the distribution F. Specifically, we will need to know something about the third moment of F.

Developing these bounds depends heavily on the theory of characteristic functions. For a given distribution F, it is not always an easy matter to obtain analytic results about the sum, or average, of n random variables having that distribution since convolutions are based on integrals or sums. However, the characteristic function of a sum or average is much easier to compute because of the results in Theorems 2.32 and 2.33. Because of the uniqueness of characteristic functions guaranteed by Theorem 2.27 it may not be surprising that

Figure 4.11 *The Kolmorogov distance between $P(Z_n \leq z)$ and $\Phi(z)$ for different values of n where Z_n is the standardized sample mean computed on n EXPONENTIAL(θ) random variables.*

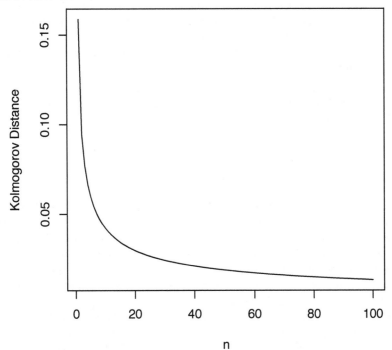

we are able to study differences between distributions by studying differences between their corresponding distribution functions. The main tool for studying these differences is based on what is commonly known as the *smoothing theorem*, a version of which is given below.

Theorem 4.23 (Smoothing Theorem). *Suppose that F is a distribution function with characteristic function ψ such that*

$$\int_{-\infty}^{\infty} t\, dF(t) = 0,$$

and

$$\int_{-\infty}^{\infty} t^2\, dF(t) = 1.$$

Let G be a differentiable distribution function with characteristic function ζ such that $\zeta'(0) = 0$. Then for $T > 0$,

$$\pi|F(x) - G(x)| \leq \int_{-T}^{T} \frac{|\psi(t) - \zeta(t)|}{|t|}\, dt + 24T^{-1} \sup_{t \in \mathbb{R}} |G'(t)|.$$

A proof of Theorem 4.23 can be found in Section 2.5 of Kolassa (2006). The form of the Smoothing Theorem given in Theorem 4.23 allows us to bound the difference between distribution functions based on an integral of the corresponding characteristic functions. A more general version of this result will be considered later in the book. Theorem 4.23 allows us to find a universal bound for the accuracy of the normal approximation that depends only on the absolute third moment of F. This famous result is known as the theorem of Berry and Esseen.

Theorem 4.24 (Berry and Esseen). *Let $\{X_n\}_{n=1}^{\infty}$ be a sequence of independent and identically distributed random variables from a distribution such that $E(X_n) = 0$ and $V(X_n) = 1$. If $\rho = E(|X_n|^3) < \infty$ then*

$$\sup_{t \in \mathbb{R}} \left| P(n^{1/2} \bar{X}_n \leq t) - \Phi(t) \right| \leq n^{-1/2} B\rho, \tag{4.6}$$

for all $n \in \mathbb{N}$ where B is a constant that does not depend on n.

Proof. The proof of this result is based on Theorem 4.23, and follows the method of proof given by Feller (1971) and Kolassa (2006). To avoid certain overly technical arguments, we will make the assumption in this proof that the distribution of the random variables is symmetric, which results in a real valued characteristic function. The general proof follows the same overall path. See Section 6.2 of Gut (2005) for further details. We will assume that $B = 3$ and first dispense with the case where $n < 10$. To simplify notation, let X be a random variable following the distribution F. Note that Theorem 2.11 (Jensen) implies that $[E(|X|^2)]^{3/2} \leq E(|X|^3) = \rho$, and by assumption we have that $E(|X|^2) = 1$. Therefore, it follows that $\rho \geq 1$. Now, when $B = 3$, the bound given in Equation (4.6) has the property $n^{-1/2} B\rho = 2n^{-1/2}\rho$ where $\rho \geq 1$. Hence, if $n^{1/2} \leq 3$, or equivalently if $n < 10$, it follows that $nB\rho \geq 1$. Since the difference between two distribution functions is always bounded above by one, it follows that the result is true without any further calculation when $B = 3$ and $n < 10$. For the remainder of the proof we will consider only the case where $n \geq 10$.

Let $\psi(t)$ be the characteristic function of X_n, and focusing our attention on the characteristic function of the standardized mean, let

$$Z_n = n^{1/2} \bar{X}_n = n^{-1/2} \sum_{k=1}^{n} X_k.$$

Theorems 2.32 and 2.33 imply that the characteristic function of Z_n is given by $\psi^n(n^{-1/2}t)$. Letting F_n denote the distribution function of Z_n, Theorem 4.23 implies that

$$|F_n(t) - \Phi(x)| \leq \pi^{-1} \int_{-T}^{T} |t|^{-1} |\psi^n(n^{-1/2}t) - \exp(-\tfrac{1}{2}t^2)| dt$$

$$+ 24T^{-1}\pi^{-1} \sup_{t \in \mathbb{R}} |\phi(t)|. \tag{4.7}$$

The bound of integration used in Equation (4.7) is chosen to be $T = \frac{4}{3}\rho^{-1}n^{1/2}$, where it follows that since $\rho \geq 1$, we have that $T \leq \frac{4}{3}n^{1/2}$. Now,

$$\sup_{t\in\mathbb{R}} |\phi(t)| = (2\pi)^{-1/2} < \tfrac{2}{5},$$

so that the second term can be bounded as

$$24T^{-1}\pi^{-1}\sup_{t\in\mathbb{R}} |\phi(t)| < \tfrac{48}{5}T^{-1}\pi^{-1}.$$

Note that $\frac{48}{5} = 9.6$, matching the bound given in Feller (1971). In order to find a bound on the integral in Equation (4.7) we will use Theorem A.10, which states that for any two real numbers ξ and γ we have that

$$|\xi^n - \gamma^n| \leq n|\xi - \gamma|\zeta^{n-1}, \tag{4.8}$$

if $|\xi| \leq \zeta$ and $|\gamma| \leq \zeta$. We will use this result with $\xi = \psi(n^{-1/2}t)$ and $\gamma = \exp(-\frac{1}{2}n^{-1}t^2)$, and we therefore need to find a value for ζ. To begin finding such a bound, we first note that

$$|\psi(t) - 1 + \tfrac{1}{2}t^2| = \left| \int_{-\infty}^{\infty} \exp(itx)dF(x) - \int_{-\infty}^{\infty} dF(x) - \int_{-\infty}^{\infty} itxdF(x) + \tfrac{1}{2}t^2\int_{-\infty}^{\infty} x^2dF(x) \right|,$$

where we have used the fact that our assumptions imply that

$$\int_{-\infty}^{\infty} \exp(itx)dF(x) = \psi(t),$$

$$\int_{-\infty}^{\infty} dF(x) = 1,$$

$$\int_{-\infty}^{\infty} xdF(x) = 0,$$

and

$$\int_{-\infty}^{\infty} x^2dF(x) = 1.$$

Theorem A.12 can be applied to real integrals of complex functions so that we have that

$$|\psi(t) - 1 + \tfrac{1}{2}t^2| \leq \int_{-\infty}^{\infty} \left|\exp(itx) - 1 - itx + \tfrac{1}{2}t^2x^2\right| dF(x). \tag{4.9}$$

Now, Theorem A.11 implies that

$$\left|\exp(itx) - 1 - itx + \tfrac{1}{2}t^2x^2\right| \leq \tfrac{1}{6}|tx|^3, \tag{4.10}$$

so that it follows that

$$\int_{-\infty}^{\infty} \left| \exp(itx) - 1 - itx + \tfrac{1}{2}t^2 x^2 \right| dF(x) \le$$

$$\int_{-\infty}^{\infty} \tfrac{1}{6}|tx|^3 dF(x) = \tfrac{1}{6}|t|^3 \int_{-\infty}^{\infty} x^3 dF(x) = \tfrac{1}{6}|t|^3 \rho.$$

Now we use the assumption that F has a symmetric distribution about the origin. Note that $\rho = E(|X|^3)$, and is not the skewness of F, so that $\rho > 0$ as long as F is not a degenerate distribution at the origin. In this case, Theorem 2.26 implies that $\psi(t)$ is real valued and that $|\psi(t) - 1 + \tfrac{1}{2}t^2| \le \tfrac{1}{6}\rho|t|^3$ implies that $\psi(t) - 1 + \tfrac{1}{2}t^2 \le \tfrac{1}{6}\rho|t|^3$, or that

$$\psi(t) \le 1 - \tfrac{1}{2}t^2 + \tfrac{1}{6}\rho|t|^3. \qquad (4.11)$$

Note that such a comparison would not make sense if $\psi(t)$ were complex valued. When $\psi(t)$ is real-valued we have that $\psi(t) \ge 0$ for all $t \in \mathbb{R}$. Also, if $1 - \tfrac{1}{2}t^2 > 0$, or equivalently if $t^2 < 2$, it follows that the right hand side of Equation (4.11) is positive. Hence, if $(n^{-1/2}t)^2 < 2$ we have that

$$\begin{aligned} |\psi(n^{-1/2}t)| &\le 1 - \tfrac{1}{2}(n^{-1/2}t)^2 + \tfrac{1}{6}\rho|n^{-1/2}t|^3 \\ &= 1 - \tfrac{1}{2}n^{-1}t^2 + \tfrac{1}{6}n^{-3/2}\rho|t|^3. \end{aligned}$$

Now, if we assume that t is also within our proposed region of integration, that is that $|t| \le T = \tfrac{4}{3}\rho^{-1}n^{1/2}$ it follows that

$$\tfrac{1}{6}\rho n^{-3/2}|t|^3 = \tfrac{1}{6}\rho n^{-3/2}|t|^2|t| \le \tfrac{4}{18}n^{-1}t^2,$$

and therefore $|\psi(n^{-1/2}t)| \le 1 - \tfrac{5}{18}n^{-1}t^2$. From Theorem A.21 we have that $\exp(x) \ge 1+x$ for all $x \in \mathbb{R}$, so that it follows that $1 - \tfrac{5}{18}n^{-1}t^2 \le \exp(-n^{-1}\tfrac{5}{18}t^2)$ for all $t \in \mathbb{R}$. Therefore, we have established that $|\psi(n^{-1/2}t)| \le \exp(-n^{-1}\tfrac{5}{18}t^2)$, for all $|t| \le T$, the bound of integration established earlier. It then follows that $|\psi(n^{-1/2}t)|^{n-1} \le \exp[-\tfrac{5}{18}n^{-1}(n-1)t^2]$. But note that $\tfrac{5}{18}n^{-1}(n-1) \ge \tfrac{1}{4}$ for $n \ge 10$ so that $|\psi(n^{-1/2}t)|^{n-1} \le \exp(-\tfrac{1}{4}t^2)$ when $n \ge 10$. We will use this bound for ζ in Equation (4.8) to find that

$$|\psi^n(n^{-1/2}t) - \exp(-\tfrac{1}{2}t^2)| \le n|\psi(n^{-1/2}t) - \exp(-\tfrac{1}{2}t^2)||\psi(n^{-1/2}t)|^{n-1} \le$$

$$n|\psi(n^{-1/2}t) - \exp(-\tfrac{1}{2}t^2)|\exp(-\tfrac{1}{4}t^2), \quad (4.12)$$

when $n \ge 10$. Now add and subtract $1 - \tfrac{1}{2}n^{-1}t^2$ inside the absolute value on the right hand side of Equation 4.12 and apply Theorem A.18 to yield

$$n|\psi(n^{-1/2}t) - \exp(-\tfrac{1}{2}t^2)| \le n|\psi(n^{-1/2}t) - 1 + \tfrac{1}{2}n^{-1}t^2| +$$

$$n|1 - \tfrac{1}{2}n^{-1}t^2 - \exp(-\tfrac{1}{2}n^{-1}t^2)|. \quad (4.13)$$

The first term on the right hand side of Equation (4.13) can be bounded using Equations (4.9) and (4.10) to find that

$$|\psi(n^{-1/2}t) - 1 + \tfrac{1}{2}n^{-1}t^2| \le \tfrac{1}{6}n^{-3/2}\rho|t|^3.$$

For the second term on the right hand side of Equation (4.13) we use Theorem A.21 to find that

$$\left|\exp(-\tfrac{1}{2}n^{-1}t^2) - 1 + \tfrac{1}{2}n^{-1}t^2\right| \le \tfrac{1}{8}n^{-2}t^4.$$

Therefore,

$$n|\psi(n^{-1/2}t) - \exp(-\tfrac{1}{2}t^2)| \le \tfrac{1}{6}n^{-1/2}\rho|t|^3 + \tfrac{1}{8}n^{-1}t^4. \qquad (4.14)$$

Now, using Equation (4.8), the integrand in Equation (4.7) can be bounded as

$$|t|^{-1}|\psi^n(n^{-1/2}t) - \exp(-\tfrac{1}{2}t^2)| \le$$
$$n|t|^{-1}|\psi(n^{-1/2}t) - \exp(-\tfrac{1}{2}n^{-1}t^2)|\exp(-\tfrac{1}{4}t^2) \le$$
$$(\tfrac{1}{6}n^{-1/2}\rho|t|^2 + \tfrac{1}{8}n^{-1}|t|^3)\exp(-\tfrac{1}{4}t^2), \qquad (4.15)$$

where the second inequality follows from Equation (4.14). If we multiply and divide the bound in Equation (4.15) by $T = \tfrac{4}{3}\rho^{-1}n^{1/2}$, we get that the bound equals

$$T^{-1}(\tfrac{2}{9}t^2 + \tfrac{1}{6}\rho^{-1}n^{-1/2}|t|^3)\exp(-\tfrac{1}{4}t^2).$$

Recalling that $\rho \ge 1$ and that $n^{1/2} > 3$ when $n \ge 10$, we have that $\rho^{-1}n^{-1/2} \le \tfrac{1}{3}$, so that

$$T^{-1}(\tfrac{2}{9}t^2 + \tfrac{1}{6}\rho^{-1}n^{-1/2}|t|^3)\exp(-\tfrac{1}{4}t^2) \le T^{-1}(\tfrac{2}{9}t^2 + \tfrac{1}{18}|t|^3)\exp(-\tfrac{1}{4}t^2).$$

This function is non-negative and integrable over $\mathbb{R}$ and, therefore, we can bound the integral in Equation (4.7) by

$$\pi^{-1}\int_{-\infty}^{\infty}|t|^{-1}|\psi^n(n^{-1/2}t) - \exp(-\tfrac{1}{2}t^2)|dt + 24T^{-1}\pi^{-1}\sup_{t\in\mathbb{R}}|\phi(t)| \le$$
$$(\pi T)^{-1}\int_{-\infty}^{\infty}(\tfrac{2}{9}t^2 + \tfrac{1}{18}|t|^3)\exp(-\tfrac{1}{4}t^2)dt + \tfrac{48}{5}T^{-1}\pi^{-1}.$$

Now

$$T^{-1}\tfrac{2}{9}\int_{-\infty}^{\infty}t^2\exp(-\tfrac{1}{4}t^2)dt = \tfrac{2}{3}\pi^{1/2}\rho n^{-1/2},$$

and

$$T^{-1}\tfrac{1}{18}\int_{-\infty}^{\infty}|t|^3\exp(-\tfrac{1}{4}t^2)dt = \tfrac{2}{3}\rho n^{-1/2}.$$

See Exercise 28. Similarly, $\tfrac{48}{5}T^{-1} = \tfrac{36}{5}\rho n^{-1/2}$. Therefore, it follows that

$$\pi^{-1}\int_{-\infty}^{\infty}|t|^{-1}|\psi^n(n^{-1/2}t) - \exp(-\tfrac{1}{2}t^2)|dt + 24T^{-1}\pi^{-1}\sup_{t\in\mathbb{R}}|\phi(t)| \le$$
$$\pi^{-1}\rho n^{-1/2}(\tfrac{2}{3}\pi^{1/2} + \tfrac{2}{3} + \tfrac{36}{5}) \le \tfrac{136}{15}\pi^{-1}\rho n^{-1/2},$$

where we have used the fact that $\pi^{1/2} \le \tfrac{9}{5}$ for the second inequality. Now, to finish up, note that $\pi^{-1} \le \tfrac{135}{408}$ so that we finally have the conclusion that from Equation (4.7) that

$$|F_n(t) - \Phi(x)| \leq \tfrac{135}{408}\tfrac{136}{15}\rho n^{-1/2} = \tfrac{18360}{6120}\rho n^{-1/2} = 3\rho n^{-1/2},$$

which completes the proof.   □

Note that the more general case where $E(X_n) = \mu$ and $V(X_n) = \sigma^2$ can be addressed by applying Theorem 4.24 to the standardized sequence $Z_n = \sigma^{-1}(X_n - \mu)$ for all $n \in \mathbb{N}$, yielding the result below.

Corollary 4.2. *Let* $\{X_n\}_{n=1}^{\infty}$ *be a sequence of independent and identically distributed random variables from a distribution such that* $E(X_n) = \mu$ *and* $V(X_n) = \sigma^2$. *If* $E(|X_n - \mu|^3) < \infty$ *then*

$$\sup_{t \in \mathbb{R}} \left| P[n^{1/2}\sigma^{-1}(\bar{X}_n - \mu) \leq t] - \Phi(t) \right| \leq n^{-1/2} B\sigma^{-3} E(|X_n - \mu|^3)$$

for all $n \in \mathbb{N}$ *where* B *is a constant that does not depend on* n.

Corollary 4.2 is proven in Exercise 29. The bound given in Theorem 4.24 and Corollary 4.2 were first derived by Berry (1941) and Esseen (1942). Extension to the case of non-identical distributions has been studied by Esseen (1945). The exact value of the constant B specified in Theorem 4.24 and Corollary 4.2 is not known, though there has been a considerable amount of research devoted to the topic of finding upper and lower bounds for B. The proof of Theorem 4.24 uses the constant $B = 3$. Esseen's original value for the constant is 7.59. Esseen and Wallace also showed that $B = 2.9$ and $B = 2.05$ works as well in unpublished works. See page 26 of Kolassa (2006). The recent best upper bounds for B have been shown to be 0.7975 by van Beek (1972) and 0.7655 by Shiganov (1986). Chen (2002) provides further refinements of B. Further information about this constant can be found in Petrov (1995) and Zolotarev (1986). A lower bound of 0.4097 is given by Esseen (1956), using a type of BERNOULLI population for the sequence of random variables. A lower bound based on similar arguments is derived in Example 4.38 below.

Example 4.38. Lower bounds for the constant B used in Theorem 4.24 and Corollary 4.2 can be found by computing the observed distance between the standardized distribution of the mean and the standard normal distribution for specific examples. Any such distance must provide a lower bound for the maximum distance, and therefore can be used to provide a lower bound for B. The most useful examples provide a distance that is a multiple of $n^{-1/2}$ so that a lower bound for B can be derived that does not depend on n. For example, Petrov (2000) considers the case where $\{X_n\}_{n=1}^{\infty}$ is a sequence of independent and identically distributed random variables where X_n has probability distribution function

$$f(x) = \begin{cases} \tfrac{1}{2} & x \in \{-1, 1\}, \\ 0 & \text{otherwise.} \end{cases}$$

In this case we have that $E(X_n) = 0$, $V(X_n) = 1$, and $E(|X_n|^3) = 1$ so that

Theorem 4.24 implies that

$$\sup_{t \in \mathbb{R}} \left| P(n^{-1/2}\bar{X}_n \le t) - \Phi(t) \right| \le n^{-1/2}B. \tag{4.16}$$

In this case the $\frac{1}{2}(X_n+1)$ has a BERNOULLI($\frac{1}{2}$) distribution so that $\frac{1}{2}(S_n+n)$ has a BINOMIAL($n, \frac{1}{2}$) distribution. Therefore, it follows that

$$P[\tfrac{1}{2}(S_n + n) = k] = P(S_n = 2k - n) = \binom{n}{k}2^{-n},$$

for $k \in \{0, 1, \dots, n\}$. Suppose n is an even integer, then

$$P(n^{-1/2}\bar{X}_n = 0) = P(S_n = 0) = P[\tfrac{1}{2}(S_n + n) = \tfrac{n}{2}] = \binom{n}{\frac{n}{2}}2^{-n}.$$

Now apply Theorem 1.20 (Stirling) to the factorial operators in the combination to find

$$
\begin{aligned}
\binom{n}{\frac{n}{2}} &= \frac{n!}{(\frac{n}{2})!(\frac{n}{2})!} \\
&= \frac{n^n (2n\pi)^{1/2} \exp(\frac{n}{2}) \exp(\frac{n}{2})[1 + o(1)]}{n\pi(\frac{n}{2})^{n/2}(\frac{n}{2})^{n/2} \exp(n)[1 + o(1)][1 + o(1)]} \\
&= 2^n (\tfrac{2}{n\pi})^{1/2}[1 + o(1)],
\end{aligned}
$$

as $n \to \infty$. Therefore it follows that $P(n^{-1/2}\bar{X}_n = 0) \simeq (\frac{2}{n\pi})^{1/2}$, or more accurately $P(n^{-1/2}\bar{X}_n = 0) = (\frac{2}{n\pi})^{1/2}[1+o(1)]$ as $n \to \infty$. It follows that the distribution function $P(n^{-1/2}\bar{X}_n \le x)$ has a jump of size $(\frac{2}{n\pi})^{-1/2}[1 + o(1)]$ at $x = 0$. Therefore, in a neighborhood of $x = 0$, $P(n^{-1/2}\bar{X}_n \le x)$ cannot be approximated by a continuous with error less than $\frac{1}{2} \times (\frac{2}{n\pi})^{1/2}[1 + o(1)] = (2n\pi)^{-1/2}[1 + o(1)]$ as $n \to \infty$. From Equation 4.16, this suggests that $B \ge (2\pi)^{-1/2} \simeq 0.3989$. ∎

Example 4.39. Let $\{X_n\}_{n=1}^{\infty}$ be a sequence of independent and identically distributed random variables where X_n has a BERNOULLI(θ) distribution as discussed in Example 4.35. Therefore, in this case, $\mu = E(X_n) = \theta$, $\sigma^2 = V(X_n) = \theta(1 - \theta)$, and $E(|X_n - \theta|^3) = |0 - \theta|^3(1 - \theta) + |1 - \theta|^3\theta = \theta(1 - \theta)[\theta^2 + (1 - \theta)^2]$. Therefore, Corollary 4.2 implies that

$$\sup_{t \in \mathbb{R}} \left| P\{n^{1/2}[\theta(1 - \theta)]^{-1/2}(\bar{X}_n - \theta) \le t\} - \Phi(t) \right| \le$$

$$n^{-1/2}B[\theta(1 - \theta)]^{-1/2}[\theta^2 + (1 - \theta)^2],$$

where $B = 0.7655$ can be used as the constant. This bound is given for the cases studied in Table 4.1 in Table 4.2. Note that in each of the cases studied, the actual error given in Table 4.1 is lower than the error given by Corollary 4.2. ∎

Table 4.2 *Upper bounds on the error of the normal approximation of the* BINOMIAL(n, θ) *distribution for* n = 5, 10, 25, 50, 100 *and* θ = 0.01, 0.05, 0.10, 0.25, 0.50 *provided by the Corollary 4.2 with* $B = 0.7655$.

			θ		
n	0.01	0.05	0.10	0.25	0.50
5	3.3725	1.4215	0.9357	0.4941	0.3423
10	2.3847	1.0052	0.6617	0.3494	0.2421
25	1.5082	0.6357	0.4185	0.221	0.1531
50	1.0665	0.4495	0.2959	0.1563	0.1083
100	0.7541	0.3179	0.2092	0.1105	0.0765

4.6 The Sample Moments

Let $\{X_n\}_{n=1}^{\infty}$ be a sequence of independent and identically distributed random variables and let μ_k' be the kth moment defined in Definition 2.9 and let $\hat{\mu}_k'$ be the kth sample moment defined in Section 3.8 as

$$\hat{\mu}_k' = n^{-1} \sum_{i=1}^{n} X_i^k.$$

Note that for a fixed value of k, $\{X_n^k\}_{n=1}^{\infty}$ is a sequence of independent and identically distributed random variables with $E(X_n^k) = \mu_k'$ and $V(X_n^k) = \mu_{2k}' - (\mu_k')^2$, so that Theorem 4.20 (Lindeberg and Lévy) implies that if $\mu_{2k}' < \infty$ then $n^{1/2}(\hat{\mu}_k' - \mu_k')[\mu_{2k}' - (\mu_k')^2]^{-1/2} \xrightarrow{d} Z$ as $n \to \infty$ where Z is a $N(0,1)$ random variable.

This general argument can be extended to the joint distribution of any set of sample moments, as long as the required moments conditions are met.

Theorem 4.25. *Let* $\{X_n\}_{n=1}^{\infty}$ *be a sequence of independent and identically distributed random variables. Let* $\boldsymbol{\mu}_d = (\mu_1', \ldots, \mu_d')'$, $\hat{\boldsymbol{\mu}}_d = (\hat{\mu}_1', \ldots, \hat{\mu}_d')'$ *and assume that* $\mu_{2d}' < \infty$. *Then* $n^{1/2} \boldsymbol{\Sigma}^{-1/2}(\hat{\boldsymbol{\mu}}_n - \boldsymbol{\mu}) \xrightarrow{d} \mathbf{Z}$ *as* $n \to \infty$ *where* $\mathbf{Z}$ *is a d-dimensional* $\mathbf{N}(\mathbf{0}, \mathbf{I})$ *random variable where the* $(i,j)^{\text{th}}$ *element of* $\boldsymbol{\Sigma}$ *is* $\mu_{i+j}' - \mu_i' \mu_j'$, *for* $i = 1, \ldots, d$ *and* $j = 1, \ldots, d$.

Proof. Define a sequence of d-dimensional random vectors $\{\mathbf{Y}_n\}_{n=1}^{\infty}$ as $\mathbf{Y}_n' = (X_n, X_n^2, \ldots, X_n^d)$ for all $n \in \mathbb{N}$ and note that

$$\bar{\mathbf{Y}}_n = n^{-1} \sum_{k=1}^{n} Y_k = \begin{bmatrix} n^{-1} \sum_{k=1}^{n} X_k \\ n^{-1} \sum_{k=1}^{n} X_k^2 \\ \vdots \\ n^{-1} \sum_{k=1}^{n} X_k^d \end{bmatrix} = \begin{bmatrix} \hat{\mu}_1' \\ \hat{\mu}_2' \\ \vdots \\ \hat{\mu}_d' \end{bmatrix} = \hat{\boldsymbol{\mu}}_d.$$

Taking the expectation of $\bar{\mathbf{Y}}_n$ elementwise implies that $E(\mathbf{Y}_n) = \boldsymbol{\mu}_d$. Let $\boldsymbol{\Sigma}$

be the covariance matrix of $\mathbf{Y}_n$ so that the $(i,j)^{\text{th}}$ element of Σ is given by $C(X_n^i, X_n^j) = E(X_n^i X_n^j) - E(X_n^i)E(X_n^j) = \mu'_{i+j} - \mu'_i \mu'_j$ for $i = 1, \ldots, d$ and $j = 1, \ldots, d$. The result now follows by applying Theorem 4.22 to the sequence $\{\mathbf{Y}_n\}_{n=1}^{\infty}$. $\qquad\qquad\qquad\qquad\qquad\qquad\qquad\qquad\qquad\qquad\qquad\qquad\quad$ $\square$

Example 4.40. Let $\{X_n\}_{n=1}^{\infty}$ be a sequence of independent and identically distributed random variables from a distribution where $\mu'_4 < \infty$. Theorem 4.25 implies that $n^{1/2}\Sigma^{-1/2}(\hat{\mu}_2 - \mu_2) \xrightarrow{d} \mathbf{Z}$ as $n \to \infty$ where $\mathbf{Z}$ is a two-dimensional $N(\mathbf{0}, \mathbf{I})$ random vector, where $\mu'_2 = (\mu'_1, \mu'_2)$ and

$$\Sigma = \begin{bmatrix} \mu'_2 - (\mu'_1)^2 & \mu'_3 - \mu'_1 \mu'_2 \\ \mu'_3 - \mu'_1 \mu'_2 & \mu'_4 - (\mu'_2)^2 \end{bmatrix}.$$

Suppose that the distribution of $\{X_n\}_{n=1}^{\infty}$ is $N(\mu, \sigma^2)$. Then $\mu'_1 = \mu$, $\mu'_2 = \sigma^2 + \mu^2$, $\mu'_3 = \mu^3 + 3\mu\sigma^2$, $\mu'_4 = \mu^4 + 6\mu^2\sigma^2 + 3\sigma^4$, and

$$\Sigma = \begin{bmatrix} \sigma^2 & 2\mu\sigma^2 \\ 2\mu\sigma^2 & 4\mu^2\sigma^2 + 2\sigma^4 \end{bmatrix}.$$

In the special case of a standard normal distribution where $\mu = 0$ and $\sigma^2 = 1$ the covariance matrix simplifies to

$$\Sigma = \begin{bmatrix} 1 & 0 \\ 0 & 2 \end{bmatrix}.$$

$\blacksquare$

4.7 The Sample Quantiles

In Section 3.9 we proved that the sample quantiles converge almost certainly to the population quantiles under some assumptions on the local behavior of the distribution function in a neighborhood of the quantile. In this section we establish that sample quantiles also have an asymptotic NORMAL distribution. Of interest in this case is the fact that the results again depend on local properties of the distribution function, in particular, the derivative of the distribution function at the point of the quantile. This result differs greatly from the case of the sample moments whose asymptotic NORMALITY depends on global properties of the distribution, which in that case was dependent on the moments of the distribution. The main result given below establishes NORMAL limits for some specific forms of probabilities involving sample quantiles. These will then be used to establish the asymptotic NORMALITY of the sample quantiles under certain additional assumptions.

Theorem 4.26. Let $\{X_n\}_{n=1}^{\infty}$ be a sequence of independent random variables that have a common distribution F. Let $p \in (0,1)$ and suppose that F is continuous at ξ_p. Then,

1. If $F'(\xi_p-)$ exists and is positive then for $x < 0$,

$$\lim_{n \to \infty} P\{n^{1/2} F'(\xi_p-)(\hat{\xi}_{p,n} - \xi_p)[p(1-p)]^{-1/2} \le x\} = \Phi(x).$$

2. *If $F'(\xi_p+)$ exists and is positive then for $x > 0$,*

$$\lim_{n\to\infty} P\{n^{1/2}F'(\xi_p+)(\hat{\xi}_{p,n} - \xi_p)[p(1-p)]^{-1/2} \le x\} = \Phi(x).$$

3. *In any case*

$$\lim_{n\to\infty} P[n^{1/2}(\hat{\xi}_{p,n} - \xi_p) \le 0] = \Phi(0) = \tfrac{1}{2}.$$

Proof. Fix $t \in \mathbb{R}$ and let v be a normalizing constant whose specific value will be specified later in the proof. Define $G_n(t) = P[n^{1/2}v^{-1}(\hat{\xi}_{pn} - \xi_p) \le t]$, which is the standardized distribution of the pth sample quantile. Now,

$$\begin{aligned}
G_n(t) &= P[n^{1/2}v^{-1}(\hat{\xi}_{pn} - \xi_p) \le t] \\
&= P(\hat{\xi}_{pn} \le \xi_p + tvn^{-1/2}) \\
&= P[\hat{F}_n(\hat{\xi}_{pn}) \le \hat{F}_n(\xi_p + tvn^{-1/2})]
\end{aligned}$$

where $\hat{F}_n$ is the empirical distribution function computed on $X_1, \ldots, X_n$ and the last equality follows from the fact that the $\hat{F}_n$ is a non-decreasing function. Theorem 3.22 implies that

$$\begin{aligned}
P[\hat{F}_n(\hat{\xi}_{pn}) \le \hat{F}_n(\xi_p + tvn^{-1/2})] &= P[p \le \hat{F}_n(\xi_p + tvn^{-1/2})] \\
&= P[np \le n\hat{F}_n(\xi_p + tvn^{-1/2})].
\end{aligned}$$

The development in Section 3.7 implies that $n\hat{F}_n(\xi_p + tvn^{-1/2})$ has a BI-NOMIAL$[n, F(\xi_p + tvn^{-1/2})]$ distribution. Let $\theta = F(\xi_p + tvn^{-1/2})$ and note that

$$\begin{aligned}
G_n(t) &= P[n\hat{F}_n(\xi_p + tvn^{-1/2}) \ge np] \\
&= P[n\hat{F}_n(\xi_p + tvn^{-1/2}) - n\theta \ge np - n\theta] \\
&= P\left[\frac{n^{1/2}[\hat{F}_n(\xi_p + tvn^{-1/2}) - n\theta]}{[\theta(1-\theta)]^{1/2}} \ge \frac{n^{1/2}(p - \theta)}{[\theta(1-\theta)]^{1/2}} \right], \quad (4.17)
\end{aligned}$$

where the random variable $n^{1/2}[\hat{F}_n(\xi_p + tvn^{-1/2}) - n\theta][\theta(1-\theta)]^{-1/2}$ is a standardized BINOMIAL(n, θ) random variable. Now let us consider the case when $t = 0$. When $t = 0$ is follows that $\theta = F(\xi_p + tvn^{-1/2}) = F(\xi_p) = p$ and $n^{1/2}(p - \theta)[\theta(1-\theta)]^{-1/2} = 0$ as long as $p \in (0, 1)$. Theorem 4.20 (Lindeberg and Lévy) then implies that

$$\lim_{n\to\infty} P\left[\frac{n^{1/2}[\hat{F}_n(\xi_p + tvn^{-1/2}) - n\theta]}{[\theta(1-\theta)]^{1/2}} \ge 0 \right] = 1 - \Phi(0) = \tfrac{1}{2},$$

which proves Statement 3. Note that the normalizing constant v does not enter into this result as it is cancelled out when $t = 0$. To prove the remaining statements, note that

$$\Phi(t) - G_n(t) = \Phi(t) - P\left\{\frac{n^{1/2}[n\hat{F}_n(\xi_p + tvn^{-1/2}) - n\theta]}{[\theta(1-\theta)]^{1/2}} \geq \frac{n^{1/2}(p-\theta)}{[\theta(1-\theta)]^{1/2}}\right\} =$$

$$P\left\{\frac{n^{1/2}[n\hat{F}_n(\xi_p + tvn^{-1/2}) - n\theta]}{[\theta(1-\theta)]^{1/2}} < \frac{n^{1/2}(p-\theta)}{[\theta(1-\theta)]^{1/2}}\right\} - 1 + \Phi(t) =$$

$$P\left\{\frac{n^{1/2}[n\hat{F}_n(\xi_p + tvn^{-1/2}) - n\theta]}{[\theta(1-\theta)]^{1/2}} < \frac{n^{1/2}(p-\theta)}{[\theta(1-\theta)]^{1/2}}\right\} - 1 + \Phi(t)$$

$$- \Phi\left\{\frac{n^{1/2}(p-\theta)}{[\theta(1-\theta)]^{1/2}}\right\} + \Phi\left\{\frac{n^{1/2}(p-\theta)}{[\theta(1-\theta)]^{1/2}}\right\}$$

$$= P\left\{\frac{n^{1/2}[n\hat{F}_n(\xi_p + tvn^{-1/2}) - n\theta]}{[\theta(1-\theta)]^{1/2}} < \frac{n^{1/2}(p-\theta)}{[\theta(1-\theta)]^{1/2}}\right\} - \Phi\left\{\frac{n^{1/2}(p-\theta)}{[\theta(1-\theta)]^{1/2}}\right\}$$

$$+ \Phi(t) - \Phi\left\{\frac{n^{1/2}(\theta-p)}{[\theta(1-\theta)]^{1/2}}\right\}.$$

To obtain a bound on the first difference we use Theorem 4.24 (Berry and Esseen) which implies that

$$\sup_{t\in\mathbb{R}}\left|P\left\{\frac{n^{1/2}[n\hat{F}_n(\xi_p + tvn^{-1/2}) - n\theta]}{[\theta(1-\theta)]^{1/2}} < t\right\} - \Phi(t)\right| \leq n^{-1/2}B\gamma\tau^{-3},$$

where B is a constant that does not depend on n, and

$$\gamma = \theta|1-\theta|^3 + (1-\theta)|-\theta^3| = \theta(1-\theta)[(1-\theta)^2 + \theta^2],$$

and $\tau^2 = \theta(1-\theta)$. Therefore,

$$n^{-1/2}B\gamma\tau^{-3} = \frac{B\theta(1-\theta)[(1-\theta)^2 + \theta^2]}{n^{1/2}\theta^{3/2}(1-\theta)^{3/2}} = Bn^{-1/2}\frac{(1-\theta)^2 + \theta^2}{\theta^{1/2}(1-\theta)^{1/2}}.$$

Therefore, it follows that

$$\left|P\left\{\frac{n^{1/2}[n\hat{F}_n(\xi_p + tvn^{-1/2}) - n\theta]}{[\theta(1-\theta)]^{1/2}} < \frac{n^{1/2}(p-\theta)}{[\theta(1-\theta)]^{1/2}}\right\} - \Phi\left\{\frac{n^{1/2}(p-\theta)}{[\theta(1-\theta)]^{1/2}}\right\}\right|$$

$$\leq Bn^{-1/2}\frac{(1-\theta)^2 + \theta^2}{\theta^{1/2}(1-\theta)^{1/2}}$$

and, hence,

$$|\Phi(t) - G_n(t)| \leq Bn^{-1/2}\frac{(1-\theta)^2 + \theta^2}{\theta^{1/2}(1-\theta)^{1/2}} + \left|\Phi(t) - \Phi\left\{\frac{n^{1/2}(\theta-p)}{[\theta(1-\theta)]^{1/2}}\right\}\right|.$$

To complete the arguments we must investigate the limiting behavior of θ. Note that $\theta = F(\xi_p + tvn^{-1/2})$, which is a function of n. Because we have assumed that F is continuous at ξ_p it follows that for a fixed value of $t \in \mathbb{R}$,

$$\lim_{n\to\infty} \theta = \lim_{n\to\infty} F(\xi_p + n^{-1/2}vt) = F(\xi_p) = p,$$

and therefore it follows that

$$\lim_{n \to \infty} \theta(1 - \theta) = p(1 - p).$$

Now note that

$$\frac{n^{1/2}(\theta - p)}{[\theta(1 - \theta)]^{1/2}} = \frac{n^{1/2}[F(\xi_p + n^{-1/2}vt) - p]}{[\theta(1 - \theta)]^{1/2}} =$$

$$\frac{tv}{[\theta(1 - \theta)]^{1/2}} \frac{F(\xi_p + n^{-1/2}vt) - p}{tvn^{-1/2}},$$

where we have used the fact that $F(\xi_p) = p$. Noting that the second term of the last expression is in the form of a derivative, we have that if $t > 0$

$$\lim_{n \to \infty} \frac{n^{1/2}(\theta - p)}{[\theta(1 - \theta)]^{1/2}} = \lim_{n \to \infty} \frac{tv}{[\theta(1 - \theta)]^{1/2}} F'(\xi_p+).$$

Therefore, if we choose the normalizing constant as $v = [p(1 - p)]^{1/2}/F'(\xi_p+)$ we have that

$$\lim_{n \to \infty} \frac{n^{1/2}(\theta - p)}{[\theta(1 - \theta)]^{1/2}} = t.$$

Therefore, when $t > 0$ we have that

$$\lim_{n \to \infty} |G_n(t) - \Phi(t)| \le \lim_{n \to \infty} Bn^{-1/2} \frac{(1 - \theta)^2 + \theta^2}{\theta^{1/2}(1 - \theta)^{1/2}} +$$

$$\lim_{n \to \infty} \left| \Phi(t) - \Phi \left\{ \frac{n^{1/2}(\theta - p)}{[\theta(1 - \theta)]^{1/2}} \right\} \right| = 0.$$

Hence, we have shown that

$$\lim_{n \to \infty} P[n^{1/2}v^{-1}(\hat{\xi}_{pn} - \xi_p) \le t] = \Phi(t),$$

for all $t > 0$, which proves Statement 2. Similar arguments are used to prove Statement 3. See Exercise 30. □

The result of Theorem 4.26 simplifies when we are able to make additional assumptions about the structure of F. In the first case we assume that F is differentiable at the point ξ_p and that $F(\xi_p) > 0$.

Corollary 4.3. *Let $\{X_n\}_{n=1}^{\infty}$ be a sequence of independent random variables that have a common distribution F. Let $p \in (0, 1)$ and suppose that F is differentiable at ξ_p and that $F(\xi_p) > 0$. Then,*

$$\frac{n^{1/2}F'(\xi_p)(\hat{\xi}_{pn} - \xi_p)}{[p(1 - p)]^{1/2}} \xrightarrow{d} Z$$

as $n \to \infty$ where Z is a $N(0, 1)$ random variable.

Corollary 4.3 is proven in Exercise 31. An additional simplification of the results occurs when F has a density f in a neighborhood of ξ_p.

Corollary 4.4. *Let $\{X_n\}_{n=1}^{\infty}$ be a sequence of independent random variables that have a common distribution F. Let $p \in (0,1)$ and suppose that F has density f in a neighborhood of ξ_p and that f is positive and continuous at ξ_p. Then,*

$$\frac{n^{1/2}f(\xi_p)(\hat{\xi}_{pn} - \xi_p)}{[p(1-p)]^{1/2}} \xrightarrow{d} Z,$$

as $n \to \infty$ where Z is a $N(0,1)$ random variable.

Corollary 4.4 is proven in Exercise 32. Corollary 4.4 highlights the difference referred to earlier between the asymptotic NORMALITY of a sample moment and a sample quantile. Let $X_1, \ldots, X_n$ be a set of independent and identically distributed random variables from a distribution F that has a density f. The sample mean has a standard error equal to

$$n^{-1/2}\left\{\int_{-\infty}^{\infty}\left[u - \int_{-\infty}^{\infty} t\,dF(t)\right]^2 dF(u)\right\}^{1/2}.$$

This is a global property of F in that many distributions with vastly different local properties may have the same variance. On the other hand, the asymptotic standard error of the sample median is given by $\frac{1}{2}[f(\xi_{1/2})]^{-1}$. This standard error depends only on the behavior of the density f near the population median, and is therefore a local property. Note in fact that the standard error is inversely related to the density at the median. This is due to the fact that the sample median is determined by the values from the sample that are closest to the middle of the data, and these values will tend to be centered around the sample median on average. If $f(\xi_{1/2})$ is large, then there will tend to be a large amount of data concentrated around the population median, which will provide a less variable estimate. On the other hand, if $f(\xi_{1/2})$ is small, then there will tend to be little data concentrated around the population median, and hence the sample median will be more variable.

Example 4.41. Suppose that $\{X_n\}_{n=1}^{\infty}$ is a sequence of independent random variables that have a common distribution F with positive density f at $\xi_{1/2}$. Then Corollary 4.4 implies that $2f(\xi_{1/2})(\hat{\xi}_{1/2,n} - \xi_{1/2}) \xrightarrow{d} Z$ as $n \to \infty$ where Z is a $N(0,1)$ random variable. In the special case where f is a $N(\mu, \sigma^2)$ density we have that $f(\xi_{1/2}) = f(\mu) = (2\pi\sigma^2)^{1/2}$ and hence $2^{1/2}(\pi\sigma^2)^{1/2}(\hat{\xi}_{1/2,n} - \xi_{1/2}) \xrightarrow{d} Z$ as $n \to \infty$ where Z is a $N(0,1)$ random variable. Note that even when the population is normal, the finite sample distribution of the sample median is not normal, but is asymptotically NORMAL. Now, consider a bimodal NORMAL mixture of the form

$$f(x) = 2^{-3/2}\pi^{-1/2}\left\{\exp[-2(x + \tfrac{3}{2})^2] + \exp[-2(x - \tfrac{3}{2})^2]\right\}..$$

The mean and median of this distribution is zero, and the variance is $\frac{5}{2}$. The asymptotic standard error of the median is

$$[2f(\xi_{1/2})]^{-1} = [2^{1/2}\pi^{-1/2}\exp(-\tfrac{9}{2})]^{-1} = 2^{-1/2}\pi^{1/2}\exp(\tfrac{9}{2}).$$

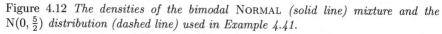

Figure 4.12 *The densities of the bimodal* NORMAL *(solid line) mixture and the* $N(0, \frac{5}{2})$ *distribution (dashed line) used in Example 4.41.*

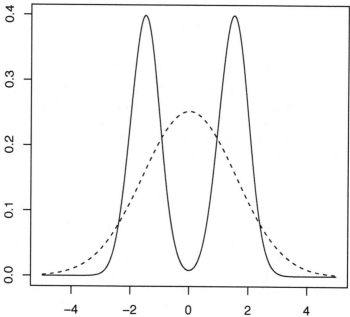

Let us compare this standard error with that of the sample median based on a sample of size n from a $N(0, \frac{5}{2})$ distribution, which has the same median, mean and variance as the bimodal NORMAL mixture considered earlier. In this case the asymptotic standard error is given by

$$[2f(\xi_{1/2})]^{-1} = [2(5\pi)^{-1/2}]^{-1} = \tfrac{1}{2}(5\pi)^{1/2},$$

so that the ratio of the asymptotic standard error of the sample median for the bimodal NORMAL mixture relative to that of the $N(0, \frac{5}{2})$ distribution is

$$\frac{2^{-1/2}\pi^{1/2}\exp(\frac{9}{2})}{\tfrac{1}{2}(5\pi)^{1/2}} = (\tfrac{2}{5})^{1/2}\exp(\tfrac{9}{2}) \simeq 56.9.$$

Therefore, the sample median has a much larger asymptotic standard error for estimating the median of the bimodal NORMAL mixture. This is because the density of the mixture distribution is much lower near the location of the median. See Figure 4.12. ∎

Example 4.42. Suppose that $\{X_n\}_{n=1}^{\infty}$ is a sequence of independent random

variables that have a common distribution F given by

$$F(x) = \begin{cases} 0 & x < 0, \\ \frac{1}{2}x & 0 \leq x < \frac{1}{2}, \\ \frac{1}{2}(3x - 1) & \frac{1}{2} \leq x < 1, \\ 1 & x \geq 1. \end{cases}$$

This distribution function is plotted in Figure 4.13. It is clear that for this distribution $\xi_{1/4} = \frac{1}{2}$, but note that $F'(\xi_{1/4}-) = \frac{1}{2}$ and $F'(\xi_{1/4}+) = \frac{3}{2}$ so that the derivative of F does not exist at $\xi_{1/4}$. This means that $\hat{\xi}_{1/4,n}$ does not have an asymptotically normal distribution, but according to Theorem 4.26, probabilities of appropriately standardized functions of $\hat{\xi}_{1/4,n}$ can be approximated by normal probabilities. In particular, Theorem 4.26 implies that

$$\lim_{n \to \infty} P[3^{-1/2} n^{1/2} (\hat{\xi}_{1/4,n} - \xi_{1/4}) \leq t] = \Phi(t),$$

for $t < 0$,

$$\lim_{n \to \infty} P[3^{1/2} n^{1/2} (\hat{\xi}_{1/4,n} - \xi_{1/4}) \leq t] = \Phi(t),$$

for $t > 0$, and

$$\lim_{n \to \infty} P[n^{1/2} (\hat{\xi}_{1/4,n} - \xi_{1/4}) \leq 0] = \frac{1}{2}.$$

It is clear from this result that the distribution of $\hat{\xi}_{1/4}$ has a longer tail below $\xi_{1/4}$ and a shorter tail above $\xi_{1/4}$. This is due to the fact that there is less density, and hence less data, that will typically be observed on average below $\xi_{1/4}$. See Experiment 6. ∎

It is important to note that the regularity conditions are crucially important to these results. For an example of what can occur when the regularity conditions are not met, see Koenker and Bassett (1984).

4.8 Exercises and Experiments

4.8.1 Exercises

1. Let $\{X_n\}_{n=1}^{\infty}$ be a sequence of random variables such that X_n has a UNI-FORM$\{0, n^{-1}, 2n^{-2}, \ldots, 1\}$ distribution for all $n \in \mathbb{N}$. Prove that $X_n \xrightarrow{d} X$ as $n \to \infty$ where X has a UNIFORM$[0, 1]$ distribution.

2. Let $\{X_n\}_{n=1}^{\infty}$ be a sequence of random variables where X_n is an EXPO-NENTIAL$(\theta + n^{-1})$ random variable for all $n \in \mathbb{N}$ where θ is a positive real constant. Let X be an EXPONENTIAL(θ) random variable. Prove that $X_n \xrightarrow{d} X$ as $n \to \infty$.

3. Let $\{X_n\}_{n=1}^{\infty}$ be a sequence of random variables such that for each $n \in \mathbb{N}$, X_n has a GAMMA(α_n, β_n) distribution where $\{\alpha_n\}_{n=1}^{\infty}$ and $\{\beta_n\}_{n=1}^{\infty}$ are sequences of positive real numbers such that $\alpha_n \to \alpha$ and $\beta_n \to \beta$ as

Figure 4.13 *The distribution function considered in Example 4.42. Note that the derivative of the distribution function does not exist at the point $\frac{1}{2}$, which equals $\xi_{1/4}$ for this population. According to Theorem 4.26, this is enough to ensure that the asymptotic distribution of the sample quantile is not normal.*

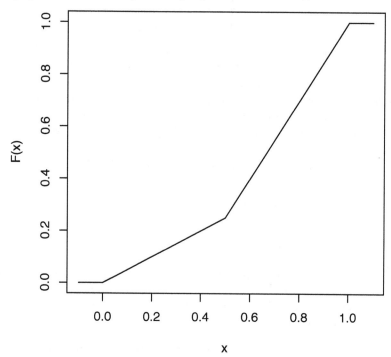

$n \to \infty$, some some positive real numbers α and β. Prove that $X_n \xrightarrow{d} X$ as $n \to \infty$ where X has a GAMMA(α, β) distribution.

4. Let $\{X_n\}_{n=1}^{\infty}$ be a sequence of random variables where for each $n \in \mathbb{N}$, X_n has an BERNOULLI$[\frac{1}{2}+(n+2)^{-1}]$ distribution, and let X be a BERNOULLI$(\frac{1}{2})$ random variable. Prove that $X_n \xrightarrow{d} X$ as $n \to \infty$.

5. Let $\{X_n\}$ be a sequence of independent and identically distributed random variables where the distribution function of X_n is

$$F_n(x) = \begin{cases} 1 - x^{-\theta} & \text{for } x \in (1, \infty), \\ 0 & \text{for } x \in (-\infty, 1]. \end{cases}$$

Define a new sequence of random variables given by

$$Y_n = n^{-1/\theta} \max\{X_1, \ldots, X_n\},$$

for all $n \in \mathbb{N}$.

a. Prove that the distribution function of Y_n is

$$G_n(y) = \begin{cases} [1 - (nx^\theta)^{-1}]^n & x > 1 \\ 0 & \text{for } x \le 1. \end{cases}$$

b. Consider the distribution function of a random variable Y given by

$$G(y) = \begin{cases} \exp(-x^{-\theta}) & x > 1 \\ 0 & \text{for } x \le 1. \end{cases}$$

Prove that $G_n \rightsquigarrow G$ as $n \to \infty$.

6. Suppose that $\{F_n\}_{n=1}^\infty$ is a sequence of distribution functions such that

$$\lim_{n \to \infty} F_n(x) = F(x),$$

for all $x \in \mathbb{R}$ for some function $F(x)$. Prove the following properties of $F(x)$.

a. $F(x) \in [0, 1]$ for all $x \in \mathbb{R}$.

b. $F(x)$ is a non-decreasing function.

7. Let $\{X_n\}_{n=1}^\infty$ be a sequence of random variables that converge in distribution to a random variable X where X_n has distribution function F_n for all $n \in \mathbb{N}$ and X has distribution function F, which may or may not be a valid distribution function. Prove that if the sequence $\{X_n\}_{n=1}^\infty$ is bounded in probability then

$$\lim_{x \to -\infty} F(x) = 0.$$

8. Let $\{X_n\}_{n=1}^\infty$ be a sequence of random variables that converge in distribution to a random variable X where X_n has distribution function F_n for all $n \in \mathbb{N}$ and X has a valid distribution function F. Prove that the sequence $\{X_n\}_{n=1}^\infty$ is bounded in probability.

9. Let g be a continuous and bounded function and let $\{F_n\}_{n=1}^\infty$ be a sequence of distribution functions such that $F_n \rightsquigarrow F$ as $n \to \infty$ where F is a distribution function. Prove that when b is a finite continuity point of F, that

$$\lim_{n \to \infty} \left| \int_b^\infty g(x) dF_n(x) - \int_b^\infty g(x) dF(x) \right| = 0.$$

10. Consider the sequence of distribution functions $\{F_n\}_{n=1}^\infty$ where

$$F_n(x) = \begin{cases} 0 & x < 0 \\ \frac{1}{2} + (n+2)^{-1} & 0 \le x < 1 \\ 1 - (n+2)^{-1} & x \ge 1. \end{cases}$$

a. Specify a function G such that $F_n \rightsquigarrow G$ as $n \to \infty$ and G is a right continuous function.

b. Specify a function G such that $F_n \leadsto G$ as $n \to \infty$ and G is a left continuous function.

c. Specify a function G such that $F_n \leadsto G$ as $n \to \infty$ and G is neither right continuous nor left continuous.

11. Let $\{F_n\}_{n=1}^{\infty}$ be a sequence of distribution functions and let F be a distribution function such that for each bounded and continuous function g,

$$\lim_{n \to \infty} \int_{-\infty}^{\infty} g(x) dF_n(x) = \int_{-\infty}^{\infty} g(x) dF(x).$$

Prove that if $\varepsilon > 0$ and t is a continuity point of F, then

$$\liminf_{n \to \infty} F_n(t) \geq F(t - \varepsilon).$$

12. Let $\{F_n\}_{n=1}^{\infty}$ be a sequence of distribution functions that converge in distribution to a distribution function F as $n \to \infty$. Prove that

$$\lim_{n \to \infty} \sup_{x \in \mathbb{R}} |F_n(x) - F(x)| = 0.$$

Hint: Adapt the proof of Theorem 3.18 to the current case.

13. In the context of the proof of Theorem 4.8 prove that

$$\liminf_{n \to \infty} F_n(x) \geq F(x - \varepsilon).$$

14. Prove the second result of Theorem 4.11. That is, let $\{X_n\}_{n=1}^{\infty}$ be a sequence of random variables that converge weakly to a random variable X. Let $\{Y_n\}_{n=1}^{\infty}$ be a sequence of random variables that converge in probability to a real constant c. Prove that $X_n Y_n \xrightarrow{d} cX$ as $n \to \infty$.

15. Prove the third result of Theorem 4.11. That is, let $\{X_n\}_{n=1}^{\infty}$ be a sequence of random variables that converge weakly to a random variable X. Let $\{Y_n\}_{n=1}^{\infty}$ be a sequence of random variables that converge in probability to a real constant $c \neq 0$. Prove that $X_n/Y_n \xrightarrow{d} X/c$ as $n \to \infty$.

16. In the context of the proof of the first result of Theorem 4.11, prove that

$$P(X_n \leq x - \varepsilon - c) \leq G_n(x) + P(|Y_n - c| > \varepsilon).$$

17. Use Theorem 4.11 to prove that if $\{X_n\}_{n=1}^{\infty}$ is a sequence of random variables that converge in probability to a random variable X as $n \to \infty$, then $X_n \xrightarrow{d} X$ as $n \to \infty$.

18. Use Theorem 4.11 to prove that if $\{X_n\}_{n=1}^{\infty}$ is a sequence of random variables that converge in distribution to a real constant c as $n \to \infty$, then $X_n \xrightarrow{P} c$ as $n \to \infty$.

19. In the context of the proof of Theorem 4.3, prove that

$$\lim_{n \to \infty} \left| \int_a^b g_m(x) dF(x) - \int_a^b g(x) dF(x) \right| < \tfrac{1}{3}\delta_\varepsilon,$$

for any $\delta_\varepsilon > 0$.

20. Let $\{\mathbf{X}_n\}_{n=1}^{\infty}$ be a sequence of d-dimensional random vectors where $\mathbf{X}_n$ has distribution function F_n for all $n \in \mathbb{N}$ and let $\mathbf{X}$ be a d-dimensional random vector with distribution function F. Prove that if for any closed set of $C \subset \mathbb{R}^d$,

$$\limsup_{n\to\infty} P(\mathbf{X}_n \in C) = P(\mathbf{X} \in C),$$

then for any open set of $G \subset \mathbb{R}^d$,

$$\liminf_{n\to\infty} P(\mathbf{X}_n \in G) = P(\mathbf{X} \in G).$$

Hint: Let $C = \mathbb{R}^d \setminus G$ and show that C is closed.

21. Let $\mathbf{X}$ be a d-dimensional random vector with distribution function F. Let $g : \mathbb{R}^d \to \mathbb{R}$ be a continuous function such that $|g(\mathbf{x})| \leq b$ for a finite real value b for all $\mathbf{x} \in \mathbb{R}$. Let $\varepsilon > 0$ and define a partition of $[-b, b]$ given by $a_0 < a_1 < \cdots < a_m$ where $a_k - a_{k-1} < \varepsilon$ for all $k = 1, \ldots, m$ and let $A_k = \{\mathbf{x} \in \mathbb{R}^d : a_{k-1} < f(\mathbf{x}) \leq a_k\}$ for $k = 0, \ldots, m$. Prove that

$$\sum_{k=1}^{m} \left[\int_{A_k} g(\mathbf{x}) dF(\mathbf{x}) - a_k P(\mathbf{X} \in A_k) \right] \geq -\varepsilon.$$

22. Let $\{\mathbf{X}_n\}_{n=1}^{\infty}$ be a sequence of d-dimensional random vectors that converge in distribution to a random vector $\mathbf{X}$ as $n \to \infty$. Let $\mathbf{X}_n' = (X_{n1}, \ldots, X_{nd})$ and $\mathbf{X}' = (X_1, \ldots, X_d)$. Prove that if $\mathbf{X}_n \xrightarrow{d} \mathbf{X}$ as $n \to \infty$ then $X_{nk} \xrightarrow{d} X_k$ as $n \to \infty$ for all $k \in \{1, \ldots, d\}$.

23. Prove the converse part of the proof of Theorem 4.17. That is, let $\{\mathbf{X}_n\}_{n=1}^{\infty}$ be a sequence of d-dimensional random vectors and let $\mathbf{X}$ be a d-dimensional random vector. Prove that if $\mathbf{X}_n \xrightarrow{d} \mathbf{X}$ as $n \to \infty$ then $\mathbf{v}'\mathbf{X}_n \xrightarrow{d} \mathbf{v}'\mathbf{X}$ as $n \to \infty$ for all $\mathbf{v} \in \mathbb{R}^d$.

24. Let $\{X_n\}_{n=1}^{\infty}$ and $\{Y_n\}_{n=1}^{\infty}$ be sequences of random variables where X_n has a $N(\mu_n, \sigma_n^2)$ distribution and Y_n has a $N(\nu_n, \tau_n^2)$ and $\nu_n \to \nu$ as $n \to \infty$ for some real numbers μ and ν. Assume that $\sigma_n \to \sigma$ and $\tau_n \to \tau$ as $n \to \infty$ for some positive real numbers σ and τ, and that X_n and Y_n are independent for all $n \in \mathbb{N}$. Let v_1 and v_2 be arbitrary real numbers. Then it follows that $v_1 X_n + v_2 Y_n$ has a $N(v_1\mu_n + v_2\nu_n, v_1^2\sigma_n^2 + v_2^2\tau_n^2)$ distribution for all $n \in \mathbb{N}$. Similarly $v_1 X + v_2 Y$ has a $N(v_1\mu + v_2\nu, v_1^2\sigma^2 + v_2^2\tau^2)$ distribution. Provide the details for the argument that $v_1 X_n + v_2 Y_n \xrightarrow{d} v_1 X + v_2 Y$ as $n \to \infty$ for all $v_1 \in \mathbb{R}$ and $v_2 \in \mathbb{R}$. In particular, consider the cases where $v_1 = 0$ and $v_2 \neq 0$, $v_1 \neq 0$ and $v_2 = 0$, and finally $v_1 = v_2 = 0$.

25. Let $\{X_n\}_{n=1}^{\infty}$ and $\{Y_n\}_{n=1}^{\infty}$ be sequences of random variables that converge in distribution as $n \to \infty$ to the random variables X and Y, respectively. Suppose that X_n has a $N(0, 1 + n^{-1})$ distribution for all $n \in \mathbb{N}$ and that Y_n has a $N(0, 1 + n^{-1})$ distribution for all $n \in \mathbb{N}$.

a. Identify the distributions of the random variables X and Y.

b. Find some conditions under which we can conclude that $X_n Y_n^{-1} \xrightarrow{d} Z$ as $n \to \infty$ where Z has a CAUCHY$(0, 1)$ distribution.

c. Find some conditions under which we can conclude that $X_n Y_n^{-1} \xrightarrow{d} W$ as $n \to \infty$ where W is a degenerate distribution at one.

26. In the context of the proof of Theorem 4.20, use Theorem A.22 to prove that $[1 - \frac{1}{2}n^{-1}t^2 + o(n^{-1})]^n = [1 - \frac{1}{2}n^{-1}t^2] + o(n^{-1})$ as $n \to \infty$ for fixed t.

27. Let $\{X_n\}_{n=1}^{\infty}$ be a sequence of independent and identically distributed random variables where X_n has a EXPONENTIAL(θ) distribution. Prove that the standardized sample mean $Z_n = n^{1/2}\theta^{-1}(\bar{X}_n - \theta)$ has a translated GAMMA$(n, n^{-1/2})$ distribution given by

$$
\begin{aligned}
P(Z_n \le z) &= P[n^{1/2}\theta^{-1}(\bar{X}_n - \theta) \le z] \\
&= \int_{-n^{1/2}}^{z} \frac{n^{1/2}}{\Gamma(n)}(t + n^{1/2})^{n-1}\exp[-n^{1/2}(t + n^{1/2})]dt.
\end{aligned}
$$

28. Prove that

$$
T^{-1}\frac{2}{9}\int_{-\infty}^{\infty} t^2 \exp(-\tfrac{1}{4}t^2)dt = \tfrac{2}{3}\pi^{1/2}\rho n^{-1/2},
$$

and

$$
T^{-1}\frac{1}{18}\int_{-\infty}^{\infty} |t|^3 \exp(-\tfrac{1}{4}t^2)dt = \tfrac{2}{3}\rho n^{-1/2},
$$

where $T = \frac{4}{3}\rho^{-1}n^{1/2}$.

29. Use Theorem 4.24 to prove Corollary 4.2.

30. Prove Statement 3 of Theorem 4.26.

31. Prove Corollary 4.3. That is, let $\{X_n\}_{n=1}^{\infty}$ be a sequence of independent random variables that have a common distribution F. Let $p \in (0, 1)$ and suppose that F is differentiable at ξ_p and that $F(\xi_p) > 0$. Then, prove that

$$
\frac{n^{1/2}F'(\xi_p)(\hat{\xi}_{pn} - \xi_p)}{[p(1-p)]^{1/2}} \xrightarrow{d} Z,
$$

as $n \to \infty$ where Z is a N$(0, 1)$ random variable.

32. Prove Corollary 4.4. That is, let $\{X_n\}_{n=1}^{\infty}$ be a sequence of independent random variables that have a common distribution F. Let $p \in (0, 1)$ and suppose that F has density f in a neighborhood of ξ_p and that f is positive and continuous at ξ_p. Then, prove that

$$
\frac{n^{1/2}f(\xi_p)(\hat{\xi}_{pn} - \xi_p)}{[p(1-p)]^{1/2}} \xrightarrow{d} Z,
$$

as $n \to \infty$ where Z is a N$(0, 1)$ random variable.

4.8.2 Experiments

1. Write a program in R that simulates b samples of size n from an EXPONEN-
   TIAL(1) distribution. For each of the b samples compute the minimum value
   of the sample. When the b samples have been simulated, a histogram of the
   b sample minimums should be produced. Run this simulation for $n = 5$,
   10, 25, 50, and 100 with $b = 10{,}000$ and discuss the resulting histograms in
   terms of the theoretical result of Example 4.3.

2. Write a program in R that simulates a sample of size b from a BINO-
   MIAL(n, n^{-1}) distribution and a sample of size b from a POISSON(1) distribution. For each sample, compute the relative frequencies associated with
   each of the values $\{1, \ldots, n\}$. Notice that it is possible that not all of the
   sample from the POISSON distribution will be used. Run this simulation for
   $n = 10$, 25, 50, and 100 with $b = 10{,}000$ and discuss how well the relative
   frequencies from each sample compare in terms of the theoretical result
   given in Example 4.4.

3. Write a program in R that generates a sample of size b from a specified
   distribution F_n (specified below) that weakly converges to a distribution F
   as $n \to \infty$. Compute the Kolmogorov distance between F and the empirical
   distribution from the sample. Repeat this for $n = 5$, 10, 25, 50, 100, 500,
   and 1000 with $b = 10{,}000$ for each of the cases given below. Discuss the
   behavior of the Kolmogorov distance and n becomes large in terms of the
   result of Theorem 4.7 (Pólya). Be sure to notice the specific assumptions
   that are a part of Theorem 4.7.

   a. F_n corresponds to a NORMAL$(n^{-1}, 1 + n^{-1})$ distribution and F corre-
      sponds to a NORMAL$(0, 1)$ distribution.

   b. F_n corresponds to a GAMMA$(1 + n^{-1}, 1 + n^{-1})$ distribution and F cor-
      responds to a GAMMA$(1, 1)$ distribution.

   c. F_n corresponds to a BINOMIAL(n, n^{-1}) distribution and F corresponds
      to a POISSON(1) distribution.

4. Write a program in R that generates b samples of size n from a $\mathbf{N}(\mathbf{0}, \boldsymbol{\Sigma}_n)$
   distribution where
   $$\boldsymbol{\Sigma}_n = \begin{bmatrix} 1 + n^{-1} & n^{-1} \\ n^{-1} & 1 + n^{-1} \end{bmatrix}.$$
   Transform each of the b samples using the bivariate transformation
   $$g(x_1, x_2) = \tfrac{1}{2}x_1 + \tfrac{1}{4}x_2,$$
   and produce a histogram of the resulting transformed values. Run this
   simulation for $n = 10$, 25, 50, and 100 with $b = 10{,}000$ and discuss how this
   histogram compares to what would be expected for large n as regulated by
   the underlying theory.

5. Write a program in R that generates b samples of size n from a specified
   distribution F. For each sample compute the statistic $Z_n = n^{1/2}\sigma^{-1}(\bar{X}_n -$

μ) where μ and σ correspond to the mean and standard deviation of the specified distribution F. Produce of histogram of the b observed values of Z_n. Run this simulation for $n = 10, 25, 50$, and 100 with $b = 10{,}000$ for each of the distributions listed below and discuss how these histograms compare to what would be expected for large n as regulated by the underlying theory given by Theorem 4.20.

a. F corresponds to a $\mathrm{N}(0,1)$ distribution.

b. F corresponds to an EXPONENTIAL(1) distribution.

c. F corresponds to a GAMMA$(2,2)$ distribution.

d. F corresponds to a UNIFORM$(0,1)$ distribution.

e. F corresponds to a BINOMIAL$(10, \frac{1}{2})$ distribution.

f. F corresponds to a BINOMIAL$(10, \frac{1}{10})$ distribution.

6. Write a program in R that simulates b samples of size n from a distribution that has distribution function

$$
F(x) = \begin{cases}
0 & x < 0 \\
\frac{1}{2}x & 0 \le x < \frac{1}{2} \\
\frac{1}{2}(3x - 1) & \frac{1}{2} \le x < 1 \\
1 & x \ge 1.
\end{cases}
$$

For each sample, compute the sample quantile $\hat{\xi}_{1/4}$. When the b samples have been simulated, a histogram of the b sample values of $\hat{\xi}_{1/4}$ should be produced. Run this simulation for $n = 5, 10, 25, 50$, and 100 with $b = 10{,}000$ and discuss the resulting histograms in terms of the theoretical result of Example 4.42.

CHAPTER 5

Convergence of Moments

And at such a moment, his solitary supporter, Karl comes along wanting to give him a piece of advice, but instead only shows that all is lost.

Amerika by Franz Kafka

5.1 Convergence in r^{th} Mean

Consider an arbitrary probability measure space $(\Omega, \mathcal{F}, P)$ and let $\mathcal{X}_r$ be the collection of all possible random variables X that map Ω to $\mathbb{R}$ subject to the restriction that $E(|X|^r) < \infty$. One can note that $\mathcal{X}_r$ is a *vector space* if we introduce the operators $\oplus$ and $\otimes$ such that $X \oplus Y$ is the function $X(\omega) + Y(\omega)$ for all $\omega \in \Omega$ when $X \in \mathcal{X}_r$ and $Y \in \mathcal{X}_r$, and for a scalar $a \in \mathbb{R}$, $a \otimes X$ is the function $aX(\omega)$ for all $\omega \in \Omega$. See Chapter 1 of Halmos (1958) and Exercise 1. By defining a function $||X||_r$ for $X \in \mathcal{X}_r$ as

$$||X||_r = \left[\int_\Omega |X(\omega)|^r dP(\omega) \right]^{1/r} = [E(|X|^r)]^{1/r},$$

it can be shown that $\mathcal{X}_r$ is a *normed vector space*. See Chapter 3 of Halmos (1958) and Exercise 2. In such a normed vector space we are able to define the distance between two elements as $d_r(X, Y) = ||X - Y||_r = [E(|X - Y|^r)]^{1/r}$. This development suggests defining a mode of convergence for random variables by requiring that the distance between a sequence of random variables and the limiting random variable converge to zero.

Definition 5.1. *Let $\{X_n\}_{n=1}^\infty$ be a sequence of random variables. Then X_n converges to a random variable X in r^{th} mean for a specified $r > 0$ if*

$$\lim_{n \to \infty} E(|X_n - X|^r) = 0.$$

This type of convergence will be represented as $X_n \xrightarrow{r} X$ as $n \to \infty$.

Therefore, convergence in r^{th} mean is equivalent to requiring that the distance $d_r(X_n, X)$ converge to zero as $n \to \infty$. A different interpretation of this mode of convergence is based on interpreting the expectation from a statistical viewpoint. That is, X_n converges in r^{th} mean to X if the expected absolute difference between the two random variables converges to zero as $n \to \infty$.

Figure 5.1 *Convergence in r^{th} mean when $r = 1$ and P is a uniform probability measure. The criteria that $E(|X_n - X|)$ converges to zero as $n \to \infty$ implies that the area between X and X_n, as indicated by the shaded area between the two random variables on the plot, converges to zero as $n \to \infty$.*

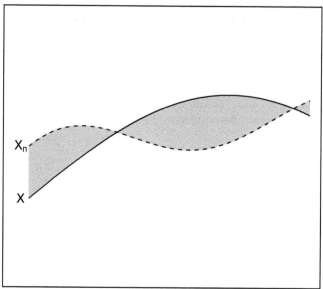

When $r = 1$ the absolute difference between X_n and X corresponds to the area between X_n and X, weighted with respect to the probability measure P. Hence $X_n \xrightarrow{r=1} X$ as $n \to \infty$ when the area between X_n and X, weighted by the probability measure P, converges to zero as $n \to \infty$. When P is a uniform probability measure, this area is the usual geometric area between X_n and X. See Figure 5.1. When $r = 2$ then convergence in r^{th} mean is usually called *convergence in quadratic mean*. We will represent this case as $X_n \xrightarrow{qm} X$ as $n \to \infty$. When $r = 1$ then convergence in r^{th} mean is usually called *convergence in absolute mean*. We will represent this case as $X_n \xrightarrow{am} X$ as $n \to \infty$.

Example 5.1. Consider the probability space $(\Omega, \mathcal{F}, P)$ where $\Omega = [0, 1]$, $\mathcal{F} = \mathcal{B}\{[0, 1]\}$ and P is a uniform probability measure on $[0, 1]$. Define a random variable X as $X(\omega) = \delta\{\omega; [0, 1]\}$ and a sequence of random variables $\{X_n\}_{n=1}^{\infty}$ as $X_n(\omega) = n^{-1}(n - 1 + \omega)$ for all $\omega \in [0, 1]$ and $n \in \mathbb{N}$. Then $E(|X_n - X|) = (2n)^{-1}$ and therefore

$$\lim_{n \to \infty} E(|X_n - X|) = \lim_{n \to \infty} (2n)^{-1} = 0.$$

See Figure 5.2. Hence, Definition 5.1 implies that $X_n \xrightarrow{am} X$ as $n \to \infty$. ∎

Example 5.2. Suppose that $\{X_n\}_{n=1}^{\infty}$ is a sequence of independent random

Figure 5.2 *Convergence in r^{th} mean when $r = 1$ in Example 5.1 with $X(\omega) = \delta\{\omega; [0,1]\}$ and $X_n(\omega) = n^{-1}(n-1+\omega)$ for all $\omega \in [0,1]$ and $n \in \mathbb{N}$. In this case $E(|X_n - X|) = (2n)^{-1}$ corresponds to the triangular area between X_n and X. The area for $n = 3$ is shaded on the plot.*

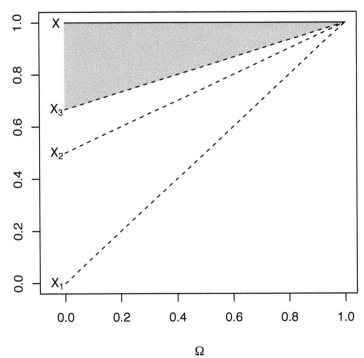

$$\Omega$$

variables from a common distribution that has mean μ and variance $\sigma^2 < \infty$. Let $\bar{X}_n$ be the sample mean. Then,

$$\lim_{n\to\infty} E(|\bar{X}_n - \mu|^2) = \lim_{n\to\infty} V(\bar{X}_n) = \lim_{n\to\infty} n^{-1}\sigma^2 = 0.$$

Therefore, Definition 5.1 implies that $\bar{X}_n \xrightarrow{qm} \mu$ as $n \to \infty$. Are we able to also conclude that $\bar{X}_n \xrightarrow{r=4} \mu$ as $n \to \infty$? This would require that

$$\lim_{n\to\infty} E(|\bar{X}_n - \mu|^4) = 0.$$

Note that

$$E(|\bar{X}_n - \mu|^4) = E(\bar{X}_n^4) - 4\mu E(\bar{X}_n^3) + 6\mu^2 E(\bar{X}_n^2) - 4\mu^3 E(\bar{X}_n) + \mu^4.$$

This expectation can be computed explicitly, though it is not necessary to do so here to see that $E(|\bar{X}_n - \mu|^4)$ depends on $E(X_n^4)$, which is not guaranteed to be finite by the assumption $\sigma^2 < \infty$. Therefore, we cannot conclude that

$\bar{X}_n \xrightarrow{r=4} \mu$ as $n \to \infty$ without further assumptions on the moments of the distribution of X_n. ∎

Example 5.3. Let $\{B_n\}_{n=1}^{\infty}$ be a sequence of independent random variables where B_n has a BERNOULLI(θ) distribution for all $n \in \mathbb{N}$. Let $\bar{B}_n$ be the sample mean computed on $B_1, \ldots, B_n$, which in this case corresponds to the sample proportion, which is an unbiased estimator of θ. Are we able to conclude that $n^{1/2}\{\log[\log(n)]\}^{-1/2}(\bar{B}_n - \theta)\} \xrightarrow{qm} 0$ as $n \to \infty$? To evaluate this possibility we compute

$$
\begin{aligned}
E\left(\left|n^{1/2}\{\log[\log(n)]\}^{-1/2}(\bar{B}_n - \theta) - 0\right|^2\right) &= \frac{n}{\log[\log(n)]} E[(\bar{B}_n - \theta)^2] \\
&= \frac{n}{\log[\log(n)]} V(\bar{B}_n) \\
&= \frac{n\theta(1 - \theta)}{n \log[\log(n)]} \\
&= \frac{\theta(1 - \theta)}{\log[\log(n)]}.
\end{aligned}
$$

Now, note that

$$
\lim_{n \to \infty} \frac{\theta(1 - \theta)}{\log[\log(n)]} = 0,
$$

so that Definition 5.1 implies that $n^{1/2}\{\log[\log(n)]\}^{-1/2}(\bar{B}_n - \theta)\} \xrightarrow{qm} 0$ as $n \to \infty$. ∎

The following result establishes a hierarchy in the convergence of means in that convergence of higher order means implies the convergence of lower order means.

Theorem 5.1. *Let $\{X_n\}_{n=1}^{\infty}$ be a sequence of random variables that converge in r^{th} mean to a random variable X as $n \to \infty$. Suppose that s is a real number such that $0 < s < r$. Then $X_n \xrightarrow{s} X$ as $n \to \infty$.*

Proof. Suppose that $X_n \xrightarrow{r} X$ as $n \to \infty$. Definition 5.1 implies that

$$
\lim_{n \to \infty} E(|X_n - X|^r) = 0.
$$

We note that, as in Example 2.5, that $x^{r/s}$ is a convex function so that Theorem 2.11 (Jensen) implies that

$$
\lim_{n \to \infty} [E(|X_n - X|^s)]^{r/s} \leq \lim_{n \to \infty} E[(|X_n - X|^s)^{r/s}] = \lim_{n \to \infty} E(|X_n - X|^r) = 0,
$$

which implies that

$$
\lim_{n \to \infty} E(|X_n - X|^s) = 0.
$$

Definition 5.1 then implies that $X_n \xrightarrow{s} X$ as $n \to \infty$. ☐

Example 5.4. Suppose that $\{X_n\}_{n=1}^{\infty}$ is a sequence of independent random

variables from a common distribution that has mean μ and variance σ^2, such that $E(|X_n|^4) < \infty$. Recall from Example 5.2 that

$$E(|\bar{X}_n - \mu|^4) = E(\bar{X}_n^4) - 4\mu E(\bar{X}_n^3) + 6\mu^2 E(\bar{X}_n^2) - 3\mu^4. \qquad (5.1)$$

Calculations in Example 5.2 also show that $E(\bar{X}_n^2) = n^{-1}\sigma^2 + \mu^2$. To obtain $E(\bar{X}_n^3)$ we first note that

$$
\begin{aligned}
E(\bar{X}_n^3) &= E\left[\left(n^{-1}\sum_{k=1}^{n} X_k\right)^3\right] \\
&= E\left(n^{-3}\sum_{i=1}^{n}\sum_{j=1}^{n}\sum_{k=1}^{n} X_i X_j X_k\right) \\
&= n^{-3}\sum_{i=1}^{n}\sum_{j=1}^{n}\sum_{k=1}^{n} E(X_i X_j X_k). \qquad (5.2)
\end{aligned}
$$

The sum in Equation 5.2 has n^3 terms in all, which can be partitioned as follows. There are n terms where $i = j = k$ for which the expectation in the sum has the form $E(X_i^3) = \gamma$, where γ will be used to denote the third moment. There are $\binom{3}{2}n(n-1)$ terms where two of the indices are the same, while the other is different. For these terms the expectation has the form $E(X_i^2)E(X_j) = \mu^3 + \mu\sigma^2$. Finally, there are $n(n-1)(n-2)$ terms where i, j, and k are all unequal. In these cases the expectation has the form $E(X_i)E(X_j)E(X_k) = \mu^3$. Therefore, it follows that

$$
\begin{aligned}
E(\bar{X}_n^3) &= n^{-3}[n\gamma + 3n(n-1)(\mu^3 + \mu\sigma^2) + n(n-1)(n-2)\mu^3] \\
&= A(n) + n^{-2}(n-1)(n-2)\mu^3,
\end{aligned}
$$

where $A(n) = O(n^{-1})$ as $n \to \infty$. One can find $E(\bar{X}_n^4)$ using the same basic approach. That is

$$E(\bar{X}_n^4) = n^{-4}\sum_{i=1}^{n}\sum_{j=1}^{n}\sum_{k=1}^{n}\sum_{m=1}^{n} E(X_i X_j X_k X_m). \qquad (5.3)$$

An analysis of the number of terms of each type in Equation (5.3) yields

$$
\begin{aligned}
E(\bar{X}_n^4) &= n^{-4}[n\lambda + 4n(n-1)\mu\gamma + 6n(n-1)(n-2)\mu^2(\mu^2 + \sigma^2) \\
&\quad + n(n-1)(n-2)(n-3)\mu^4] \\
&= B(n) + n^{-3}(n-1)(n-2)(n-3)\mu^4,
\end{aligned}
$$

where $\lambda = E(X_i^4)$ and $B(n) = O(n^{-1})$ as $n \to \infty$. See Exercise 5. Substituting the results of Equation (5.2) and Equation (5.3) into Equation (5.1) implies

that

$$
\begin{aligned}
E(|\bar{X}_n - \mu|^4) &= B(n) + n^{-3}(n-1)(n-2)(n-3)\mu^4 - \\
&\quad 4\mu[A(n) + n^{-2}(n-1)(n-2)\mu^3] + 6\mu^2[n^{-1}\sigma^2 + \mu^2] - 3\mu^4 \\
&= C(n) + \mu^4[n^{-3}(n-1)(n-2)(n-3) - \\
&\quad 4n^{-2}(n-1)(n-2) + 3],
\end{aligned}
$$

where $C(n) = O(n^{-1})$ as $n \to \infty$. Note that

$$
\lim_{n\to\infty} n^{-3}(n-1)(n-2)(n-3) - 4n^{-2}(n-1)(n-2) = -3,
$$

so that it follows that

$$
\lim_{n\to\infty} E(|\bar{X}_n - \mu|^4) = 0,
$$

and therefore Definition 5.1 implies that $\bar{X}_n \xrightarrow{r=4} \mu$ as $n \to \infty$. Theorem 5.1 also allows us to also conclude that $\bar{X}_n \xrightarrow{r=3} \mu$, $\bar{X}_n \xrightarrow{qm} \mu$, and $\bar{X}_n \xrightarrow{am} \mu$, as $n \to \infty$. Note of course that the converse is not true. If $E(|X_n|^4) = \infty$ but $E(|X_n|^2) < \infty$ it still follows that $\bar{X}_n \xrightarrow{qm} \mu$ and $\bar{X}_n \xrightarrow{am} \mu$, as $n \to \infty$, but the conclusion that $\bar{X}_n \xrightarrow{r=4} \mu$ is no longer valid. ∎

The dependence of the definition of rth mean convergence on the concept of expectation makes it difficult to easily determine how this mode of convergence fits in with the other modes of convergence studied up to this point. We begin to determine these relationships by considering convergence in probability.

Theorem 5.2. Let $\{X_n\}_{n=1}^\infty$ be a sequence of random variables that converge in r^{th} mean to a random variable X as $n \to \infty$ for some $r > 0$. Then $X_n \xrightarrow{p} X$ as $n \to \infty$.

Proof. Let $\varepsilon > 0$ and suppose that $X_n \xrightarrow{r} X$ as $n \to \infty$ for some $r > 0$. Theorem 2.6 (Markov) implies that

$$
\lim_{n\to\infty} P(|X_n - X| > \varepsilon) \leq \lim_{n\to\infty} \varepsilon^r E(|X_n - X|^r) = 0.
$$

Definition 3.1 then implies that $X_n \xrightarrow{p} X$ as $n \to \infty$. □

Theorem 5.2 also implies by Theorem 4.8 that if $X_n \xrightarrow{r} X$ as $n \to \infty$ then $X_n \xrightarrow{d} X$ as $n \to \infty$. Convergence in probability and convergence in r^{th} mean are not equivalent as there are sequences that converge in probability that do not converge in r^{th} mean.

Example 5.5. Let $\{X_n\}_{n=1}^\infty$ be a sequence of independent random variables where X_n has probability distribution function

$$
f_n(x) = \begin{cases} 1 - n^{-\alpha} & x = 0 \\ n^{-\alpha} & x = n \\ 0 & \text{otherwise,} \end{cases}
$$

where $\alpha > 0$. Let $\varepsilon > 0$, then it follows that

$$\lim_{n \to \infty} P(|X_n - 0| > \varepsilon) \le n^{-\alpha} = 0,$$

as long as $\alpha > 0$. Therefore, $X_n \xrightarrow{p} 0$ as $n \to \infty$. However, it does not necessarily follow that $X_n \xrightarrow{r} 0$ as $n \to \infty$. To see this note that

$$E(|X_n - 0|^r) = E(|X_n|^r) = (0)(1 - n^{-\alpha}) + (n^r)(n^{-\alpha}) = n^{r-\alpha}.$$

If $\alpha > r$ then

$$\lim_{n \to \infty} E(|X_n|^r) = 0,$$

and it follows from Definition 5.1 that $X_n \xrightarrow{r} 0$ as $n \to \infty$. However, if $\alpha \le r$ then

$$E(|X_n|^r) = \begin{cases} 1 & \alpha = r, \\ \infty & \alpha < r. \end{cases}$$

In either case X_n does not converge in r^{th} mean to 0 as $n \to \infty$. ∎

Since convergence in probability does not always imply convergence in r^{th} mean, it is of interest to determine under what conditions such an implication might follow. The following result gives one such condition.

Theorem 5.3. *Suppose that $\{X_n\}_{n=1}^\infty$ is a sequence of random variables that converge in probability to a random variable X as $n \to \infty$. Suppose that $P(|X_n| \le |Y|) = 1$ for all $n \in \mathbb{N}$ and that $E(|Y|^r) < \infty$. Then $X_n \xrightarrow{r} X$ as $n \to \infty$.*

Proof. We begin proving this result by first showing that under the stated conditions it follows that $P(|X| \le |Y| = 1)$. To show this let $\delta > 0$. Since $P(|X_n| \le |Y|) = 1$ for all $n \in \mathbb{N}$ it follows that $P(|X| > |Y| + \delta) \le P(|X| > |X_n| + \delta) = P(|X| - |X_n| > \delta)$ for all $n \in \mathbb{N}$. Now Theorem A.18 implies that $|X| - |X_n| \le |X_n - X|$ which implies that $P(|X| > |Y| + \delta) \le P(|X_n - X| > \delta)$, for all $n \in \mathbb{N}$. Therefore,

$$P(|X| > |Y| + \delta) \le \lim_{n \to \infty} P(|X_n - X| > \delta) = 0,$$

where the limiting value follows from Definition 3.1 because we have assumed that $X_n \xrightarrow{p} X$ as $n \to \infty$. Therefore, we can conclude that $P(|X| > |Y| + \delta) = 0$ for all $\delta > 0$ and hence it follows that $P(|X| > |Y|) = 0$. Therefore, $P(|X| \le |Y|) = 1$. Now we can begin to work on the problem of interest. Theorem A.18 implies that $P(|X_n - X| \le |X_n| + |X|) = 1$ for all $n \in \mathbb{N}$. We have assumed that $P(|X_n| \le |Y|) = 1$ and have established in the arguments above that $P(|X| \le |Y|) = 1$, hence it follows that $P(|X_n - X| \le |2Y|) = 1$. The assumption that $E(|Y|^r) < \infty$ implies that

$$\lim_{b \to \infty} E(|Y|^r \delta\{Y; (b, \infty)\}) = \lim_{b \to \infty} \int_b^\infty |Y|^r dF = 0,$$

where F is taken to be the distribution of Y. Therefore, there exists a constant

$A_\varepsilon \in \mathbb{R}$ such that $A_\varepsilon > \varepsilon$ and

$$E(|Y|^r \delta\{|2Y|; (A_\varepsilon, \infty)\}) \le \varepsilon.$$

Hence,

$$
\begin{aligned}
E(|X_n - X|^r) \;=\; & E(|X_n - X|^r \delta\{|X_n - X|; (A_\varepsilon, \infty)\}) \\
& + E(|X_n - X|^r \delta\{|X_n - X|; [0, \varepsilon]\}) \\
& + E(|X_n - X|^r \delta\{|X_n - X|; (\varepsilon, A_\varepsilon]\}).
\end{aligned}
$$

We have established earlier that $P(|X_n - X| \le |2Y|) = 1$ so that it follows that

$$E(|X_n - X|^r \delta\{|X_n - X|; (A_\varepsilon, \infty)\}) \le E(|2Y|^r \delta\{|2Y|; (A_\varepsilon, \infty)\}).$$

Also,

$$
\begin{aligned}
E(|X_n - X|^r \delta\{|X_n - X|; [0, \varepsilon]\}) \le & \\
E(\varepsilon^r \delta\{|X_n - X|; [0, \varepsilon]\}) = \varepsilon^r & P(|X_n - X| \le \varepsilon) \le \varepsilon^r.
\end{aligned}
$$

For the remaining term in the sum we have that

$$
\begin{aligned}
E(|X_n - X|^r \delta\{|X_n - X|; (\varepsilon, A_\varepsilon]\}) \le & \\
E(A_\varepsilon^r \delta\{|X_n - X|; (\varepsilon, A_\varepsilon]\}) = A_\varepsilon^r & P[|X_n - X| \in (\varepsilon, A_\varepsilon]].
\end{aligned}
$$

Because $(\varepsilon, A_\varepsilon] \subset (\varepsilon, \infty)$, Theorem 2.3 implies that $A_\varepsilon^r P[|X_n - X| \in (\varepsilon, A_\varepsilon]] \le A_\varepsilon^r P(|X_n - X| > \varepsilon)$. Combining these results implies that

$$E(|X_n - X|^r) \le E(|2Y|^r \delta\{|2Y|; (A_\varepsilon, \infty)\}) + \varepsilon^r + A_\varepsilon^r P(|X_n - X| > \varepsilon).$$

Now,

$$E(|2Y|^r \delta\{|2Y|; (A_\varepsilon, \infty)\}) = 2^r E\{|Y|^r \delta\{|2Y|; (A_\varepsilon, \infty)\}\} \le 2^r \varepsilon.$$

Therefore $E(|X_n - X|^r) \le 2^r \varepsilon + \varepsilon^r + A_\varepsilon^r P(|X_n - X| > \varepsilon)$. Since $X_n \xrightarrow{p} X$ as $n \to \infty$ it follows that

$$\lim_{n \to \infty} P(|X_n - X| > \varepsilon) = 0.$$

Hence

$$\lim_{n \to \infty} E(|X_n - X|^r) \le (2^r + 1)\varepsilon^r,$$

for all $\varepsilon > 0$. Since ε is arbitrary, it follows that

$$\lim_{n \to \infty} E(|X_n - X|^r) = 0,$$

and hence we have proven that $X_n \xrightarrow{r} X$ as $n \to \infty$.   □

The relationship between convergence in r^{th} mean, complete convergence, and almost certain convergence is more difficult to obtain as one mode does not necessarily imply the other. For example, to ensure that the sequence converges almost certainly, additional assumptions are required.

Theorem 5.4. *Let $\{X_n\}_{n=1}^{\infty}$ be a sequence of random variables that converge in r^{th} mean to a random variable X as $n \to \infty$ for some $r > 0$. If*

$$\sum_{n=1}^{\infty} E(|X_n - X|^r) < \infty,$$

then $X_n \xrightarrow{a.c.} X$ as $n \to \infty$.

The proof of Theorem 5.4 is given in Exercise 6. The result of Theorem 5.4 can actually be strengthened to the conclusion that $X_n \xrightarrow{c} X$ as $n \to \infty$ without any change to the assumptions. Theorem 5.4 provides us with a condition under which convergence in r^{th} mean implies complete convergence and almost certain convergence. Could almost certain convergence imply convergence in r^{th} mean? The following example of Serfling (1980) shows that this is not always the case.

Example 5.6. Let $\{X_n\}_{n=1}^{\infty}$ be a sequence of independent and identically distributed random variables such that $E(X_n) = 0$ and $V(X_n) = 1$. Consider a new sequence of random variables defined by

$$Y_n = \{n \log[\log(n)]\}^{-1/2} \sum_{k=1}^{n} X_k,$$

for all $n \in \mathbb{N}$. In Example 5.3 we showed that $Y_n \xrightarrow{qm} 0$ as $n \to \infty$. Does it also follow that $Y_n \xrightarrow{a.c.} 0$? It does not follow since Theorem 3.15 (Hartman and Wintner) and Definition 1.3 imply that

$$P\left(\lim_{n \to \infty} \{n \log[\log(n)]\}^{-1/2} \sum_{k=1}^{n} X_k = 0 \right) = 0.$$

Hence, Y_n does not converge almost certainly to 0 as $n \to \infty$. ∎

5.2 Uniform Integrability

In the previous section we considered defining the convergence of a sequence of random variables in terms of expectation. We were able to derive a few results which established the relationship between this type of convergence and the types of convergence studied previously. In this section we consider the converse problem. That is, suppose that $\{X_n\}_{n=1}^{\infty}$ is a sequence of random variables that converge in some mode to a random variable X. Does it necessarily follow that the moments of the random variable converge as well?

It is not surprising that when we consider the case where X_n converges in r^{th} mean to X, the answer is relatively straightforward. For example, when $X_n \xrightarrow{am} X$ as $n \to \infty$ we know that

$$\lim_{n \to \infty} E(|X_n - X|) = 0.$$

Now, noting that $|E(X_n) - E(X)| = |E(X_n - X)| \le E(|X_n - X|)$ it follows that

$$\lim_{n\to\infty} E(X_n) = E(X), \tag{5.4}$$

and the corresponding sequence of first moments converge. In fact, Theorem 5.1 implies that if $X_n \xrightarrow{r} X$ as $n \to \infty$ for any $r > 1$, the result in Equation (5.4) follows. Can we find similar results for other modes of convergence? For example, suppose that $X_n \xrightarrow{p} X$ as $n \to \infty$. Under what conditions can we conclude that at least some of the corresponding moments of Y_n converge to the moments of Y? Example 5.5 and Theorem 5.3 provide important clues to solving this problem.

In Example 5.5 we consider a sequence of random variables that converge in probability to a degenerate random variable at zero. Hence, the expectation of the limiting distribution is also zero. However, the moments of the random variables in the sequence do not always converge to zero. One example where the convergence does not occur is when $\alpha = 1$, in which case the distribution of X_n is given by

$$f_n(x) = \begin{cases} 1 - n^{-1} & x = 0 \\ n^{-1} & x = n \\ 0 & \text{otherwise.} \end{cases}$$

What occurs with this distribution is that, as $n \to \infty$, the random variable can take on an arbitrarily large value (n). The probability associated with this value converges to zero so that it follows that $X_n \xrightarrow{p} 0$ as $n \to \infty$. However, the probability associated with this value is not small enough for the expectation of the X_n to converge to zero as well. In fact, in the case where $\alpha = 1$, the probability associated with the point at n is such that $E(X_n) = 1$ for all $n \in \mathbb{N}$. Hence the expectations do not converge to zero as $n \to \infty$.

Considering Theorem 5.3 with $r = 1$ provides a demonstration of a condition under which the expectations do converge to the proper value when the sequence converges in probability. The key idea in this result is that all of the random variables in the sequence must be absolutely bounded by an integrable random variable Y. This condition does not allow the type of behavior that is observed in Example 5.5. See Exercise 7. Hence, Theorem 5.3 provides a sufficient condition under which convergence in probability implies that the corresponding expectations of the sequence converge to the expectation of the limiting random variable. The main question of interest is then in developing a set of minimal conditions under which the result remains valid. Such development requires a more refined view of the integrability of sequences, known as *uniform integrability*.

Definition 5.2. *Let* $\{X_n\}_{n=1}^{\infty}$ *be a sequence of random variables. The sequence is uniformly integrable if* $E(|X_n|\delta\{|X_n|; (a, \infty)\})$ *converges to zero uniformly in* n *as* $a \to \infty$.

Note that Definition 5.2 requires more than just that each random variable in the sequence be integrable. Indeed, if this was true, then $E(|X_n|\delta\{|X_n|;(a,\infty)\})$ would converge to zero as $a \to \infty$ for all $n \in \mathbb{N}$. Rather, Definition 5.2 requires that this convergence be uniform in n, meaning that the rate at which $E(|X_n|\delta\{|X_n|;(a,\infty)\})$ convergences to zero as $a \to \infty$ *must be the same for all $n \in \mathbb{N}$, or that the rate of convergence cannot depend on n.*

Example 5.7. Let $\{X_n\}_{n=1}^{\infty}$ be a sequence of independent and identically distributed random variables from a distribution F with finite mean μ. That is $E(X) = \mu < \infty$. The fact that the mean is finite implies that the expectation $E(|X_n|\delta\{|X_n|;(a,\infty)\})$ converges to zero as $a \to \infty$, and since the random variables have identical distributions, the convergence rate for each expectation is exactly the same. Therefore, by Definition 5.2, the sequence $\{X_n\}_{n=1}^{\infty}$ is uniformly integrable. ∎

Example 5.8. Let $\{X_n\}_{n=1}^{\infty}$ be a sequence of independent $N(0,\sigma_n^2)$ random variables where $\{\sigma_n\}_{n=1}^{\infty}$ is a sequence of real numbers such that $0 < \sigma_n \le \sigma < \infty$ for all $n \in \mathbb{N}$. For this sequence of random variables it is true that $E(|X_n|) < \infty$ for all $n \in \mathbb{N}$, which ensures that $E(|X_n|\delta\{|X_n|;(a,\infty)\})$ converges to zero as $a \to \infty$. To have uniform integrability we must further establish that this convergence is uniform in $n \in \mathbb{R}$. To see this note that for every $n \in \mathbb{N}$, we have that

$$
\begin{aligned}
E(|X_n|\delta\{|X_n|;(a,\infty)\}) &= \int_{|x|>a} (2\pi\sigma_n^2)^{-1/2}|x|\exp(-\tfrac{1}{2}\sigma_n^2 x^2)dx \\
&= 2\int_a^{\infty} (2\pi\sigma_n^2)^{-1/2} x\exp(-\tfrac{1}{2}\sigma_n^2 x^2)dx \\
&= 2\sigma_n \int_{a\sigma_n^{-1}}^{\infty} (2\pi)^{-1/2} v\exp(-\tfrac{1}{2}v^2)dv.
\end{aligned}
$$

If $\sigma_n \le \sigma$ then $\sigma_n^{-1} \ge \sigma^{-1}$ and hence $a\sigma_n^{-1} \ge a\sigma^{-1}$, so that

$$
E(|X_n|\delta\{|X_n|;(a,\infty)\}) \le 2\sigma \int_{a\sigma^{-1}}^{\infty} (2\pi)^{-1/2} v\exp(-\tfrac{1}{2}v^2)dv < \infty,
$$

for all $n \in \mathbb{N}$. Therefore, if we choose $0 < a_\varepsilon < \infty$ so that

$$
2\sigma \int_{a_\varepsilon\sigma^{-1}}^{\infty} (2\pi)^{-1/2} v\exp(-\tfrac{1}{2}v^2)dv < \varepsilon,
$$

then $E(|X_n|\delta\{|X_n|;(a_\varepsilon,\infty)\}) < \varepsilon$ for all $n \in \mathbb{N}$. Since a_ε does not depend on n, the convergence of $E(|X_n|\delta\{|X_n|;(a,\infty)\})$ to zero is uniform in n, and hence Definition 5.2 implies that the sequence $\{X_n\}_{n=1}^{\infty}$ is uniformly integrable. ∎

Example 5.9. Let $\{X_n\}_{n=1}^{\infty}$ be a sequence of independent random variables where X_n has probability distribution function

$$
f_n(x) = \begin{cases} 1 - n^{-\alpha} & x = 0 \\ n^{-\alpha} & x = n \\ 0 & \text{otherwise,} \end{cases}
$$

where $\alpha > 1$. Note that the expected value of X_n is given by $E(X_n) = n^{1-\alpha}$, which is finite. Now, for fixed values of $n \in \mathbb{N}$ we have that for $a > 0$,

$$E(|X_n|\delta\{|X_n|; (a_\varepsilon, \infty)\}) = \begin{cases} n^{1-\alpha} & a \leq n, \\ 0 & a > n. \end{cases}$$

Let $\varepsilon > 0$ be given. Then $E(|X_n|\delta\{|X_n|; (a_\varepsilon, \infty)\}) < \varepsilon$ as long as $a > n$. This makes it clear that the convergence of $E(|X_n|\delta\{|X_n|; (a_\varepsilon, \infty)\})$ is not uniform in this case since the value of a required to obtain $E(|X_n|\delta\{|X_n|; (a_\varepsilon, \infty)\}) < \varepsilon$ is directly related to n. The only way to obtain a value of a that would ensure $E(|X_n|\delta\{|X_n|; (a_\varepsilon, \infty)\}) < \varepsilon$ for all $n \in \mathbb{N}$, would be to let $a \to \infty$. ∎

One fact that follows from the uniform integrability of a sequence of random variables is that the sequence expectations, and its least upper bound, must stay finite.

Theorem 5.5. *Let $\{X_n\}_{n=1}^{\infty}$ be a sequence of uniformly integrable random variables. Then,*

$$\sup_{n \in \mathbb{N}} E(|X_n|) < \infty.$$

Proof. Note that $E(|X_n|) = E(|X_n|\delta\{|X_n|; [0, a]\}) + E(|X_n|\delta\{|X_n|; (a, \infty)\})$. We can bound the first term as

$$E(|X_n|\delta\{|X_n|; [0, a]\}) = \int_0^a |X_n| dF_n \leq \int_0^a a\, dF_n = a \int_0^a dF_n \leq a,$$

where F_n is the distribution function of X_n for all $n \in \mathbb{N}$. For the second term we note that from Definition 5.2 we can choose $a < \infty$ large enough so that $E(|X_n|\delta\{|X_n|; (a, \infty)\}) \leq 1$. The uniformity of the convergence implies that a single choice of a will suffice for all $n \in \mathbb{N}$. Hence, it follows that for a large enough that $E(|X_n|) \leq a + 1$. Therefore,

$$\sup_{n \in \mathbb{N}} E(|X_n|) \leq a + 1 < \infty.$$

□

Uniform integrability can be somewhat difficult to verify in practice as the uniformity of the convergence of $E(|X_n|\delta\{|X_n|; (a_\varepsilon, \infty)\})$ to zero is not always easy to verify. There are several equivalent ways to characterize uniform integrability, though many of these conditions are technical in nature and are not always easier to apply than Definition 5.2. Therefore we will leave these equivalent conditions to the reader for further study. See Gut (2005). There are several sufficient conditions for uniform integrability which we will present as the conditions on these results are relatively simple.

Theorem 5.6. *Let $\{X_n\}_{n=1}^{\infty}$ be a sequence of random variables such that*

$$\sup_{n \in \mathbb{N}} E(|X_n|^{1+\varepsilon}) < \infty,$$

for some $\varepsilon > 0$. Then $\{X_n\}_{n=1}^{\infty}$ is uniformly integrable.

Proof. Let F_n be the distribution function of X_n for all $n \in \mathbb{N}$ and let $\varepsilon > 0$. Then

$$
\begin{aligned}
E(|X_n|\delta\{|X_n|; (a, \infty)\}) &= E(|X_n|^{1+\varepsilon}|X_n|^{-\varepsilon}\delta\{|X_n|; (a, \infty)\}) \\
&\leq E(|X_n|^{1+\varepsilon}a^{-\varepsilon}\delta\{|X_n|; (a, \infty)\}) \\
&= a^{-\varepsilon}E(|X_n|^{1+\varepsilon}\delta\{|X_n|; (a, \infty)\}) \\
&\leq a^{-\varepsilon}E(|X_n|^{1+\varepsilon}) \\
&\leq a^{-\varepsilon}\sup_{n \in \mathbb{N}} E(|X_n|^{1+\varepsilon}).
\end{aligned}
\tag{5.5}
$$

If $\varepsilon > 0$ and

$$
\sup_{n \in \mathbb{N}} E(|X_n|^{1+\varepsilon}) < \infty,
$$

then the upper bound in Equation (5.5) will converge to zero as $a \to \infty$. The convergence is guaranteed to be uniform because the upper bound does not depend on n. $\qquad\square$

Example 5.10. Suppose that $\{X_n\}_{n=1}^{\infty}$ is a sequence of independent random variables where X_n has a GAMMA(α_n, β_n) distribution for all $n \in \mathbb{N}$ where $\{\alpha_n\}_{n=1}^{\infty}$ and $\{\beta_n\}_{n=1}^{\infty}$ are real sequences. Suppose that $\alpha_n \leq \alpha < \infty$ and $\beta_n \leq \beta < \infty$ for all $n \in \mathbb{N}$. Then $E(|X_n|^2) = \alpha_n\beta_n^2 + \alpha_n^2\beta_n^2 \leq \alpha\beta^2 + \alpha^2\beta^2 < \infty$ for all $n \in \mathbb{N}$. Then,

$$
\sup_{n \in \mathbb{N}} E(|X_n|^2) \leq \alpha\beta^2 + \alpha^2\beta^2 < \infty,
$$

so that Theorem 5.6 implies that the sequence $\{X_n\}_{n=1}^{\infty}$ is uniformly integrable. $\qquad\blacksquare$

A similar result requires that the sequence of random variables be bounded almost certainly by an integrable random variable.

Theorem 5.7. *Let $\{X_n\}_{n=1}^{\infty}$ be a sequence of random variables such that $P(|X_n| \leq Y) = 1$ for all $n \in \mathbb{N}$ where Y is a positive integrable random variable. Then the sequence $\{X_n\}_{n=1}^{\infty}$ is uniformly integrable.*

Theorem 5.7 is proven in Exercise 8. The need for having a integrable random that bounds the sequence can be eliminated by replacing the random variable Y in Theorem 5.7 with the supremum of the sequence $\{X_n\}_{n=1}^{\infty}$.

Corollary 5.1. *Let $\{X_n\}_{n=1}^{\infty}$ be a sequence of random variables such that*

$$
E\left(\sup_{n \in \mathbb{N}} |X_n|\right) < \infty.
$$

Then the sequence $\{X_n\}_{n=1}^{\infty}$ is uniformly integrable.

Corollary 5.1 is proven in Exercise 9. The final result we highlight in this section shows that it is also sufficient for a sequence to be bounded by a uniformly integrable sequence to conclude that a sequence is uniformly integrable.

Theorem 5.8. *Let $\{X_n\}_{n=1}^{\infty}$ be a sequence of random variables and $\{Y_n\}_{n=1}^{\infty}$ be a sequence of positive integrable random variables such that $P(|X_n| \leq Y_n) = 1$ for all $n \in \mathbb{N}$. If the sequence $\{Y_n\}_{n=1}^{\infty}$ is uniformly integrable then the sequence $\{X_n\}_{n=1}^{\infty}$ is uniformly integrable.*

Proof. Note that because $P(|X_n| \leq Y_n) = 1$ for all $n \in \mathbb{N}$, it follows that $E(|X_n|\delta\{|X_n|; (a,\infty)\}) \leq E(Y_n\delta\{Y_n; (a,\infty)\})$. For every $\varepsilon > 0$ there exists an a_ε such that $E(Y_n\delta\{Y_n; (a_\varepsilon,\infty)\}) < \varepsilon$, uniformly in n since the sequence $\{Y_n\}_{n=1}^{\infty}$ is uniformly integrable. This value of a also ensures that $E(|X_n|\delta\{|X_n|; (a_\varepsilon,\infty)\}) < \varepsilon$ uniformly in $n \in \mathbb{N}$. Therefore, Definition 5.2 implies that the sequence $\{X_n\}_{n=1}^{\infty}$ is uniformly integrable. $\square$

Example 5.11. Let $\{Y_n\}_{n=1}^{\infty}$ be a sequence of independent random variables where Y_n has a $N(0,1)$ distribution for all $n \in \mathbb{N}$. Define a new sequence $\{X_n\}_{n=1}^{\infty}$ where $X_n = \min\{|Y_1|, \ldots, |Y_n|\}$ for all $n \in \mathbb{N}$. The uniform integrability of the sequence $\{X_n\}_{n=1}^{\infty}$ then follows from the uniform integrability of the sequence $\{Y_n\}_{n=1}^{\infty}$ by Theorem 5.8. $\blacksquare$

The final result of this section links the uniform integrability of the sum of sequences of random variables to the uniform integrability of the individual sequences.

Theorem 5.9. *Let $\{X_n\}_{n=1}^{\infty}$ and $\{Y_n\}_{n=1}^{\infty}$ be sequences of uniformly integrable random variables. Then the sequence $\{X_n + Y_n\}_{n=1}^{\infty}$ is uniformly integrable.*

Proof. From Definition 5.2 we know that to show that the sequence $\{X_n + Y_n\}_{n=1}^{\infty}$ is uniformly integrable, we must show that

$$\lim_{a \to \infty} E(|X_n + Y_n|\delta\{|X_n + Y_n|; (a,\infty)\}) = 0,$$

uniformly in n. To accomplish this we will bound the expectation by a sum of expectations depending on the individual sequences $\{X_n\}_{n=1}^{\infty}$ and $\{Y_n\}_{n=1}^{\infty}$. Let a be a positive real number. Theorem A.19 implies that $|X_n + Y_n| \leq 2\max\{|X_n|, |Y_n|\}$, and hence

$$\delta\{|X_n + Y_n|; (a,\infty)\} \leq \delta\{2\max\{|X_n|, |Y_n|\}; (a,\infty)\},$$

with probability one. Therefore,

$$|X_n + Y_n|\delta\{|X_n + Y_n|; (a,\infty)\} \leq$$
$$2\max\{|X_n|, |Y_n|\}\delta\{2\max\{|X_n|, |Y_n|\}; (a,\infty)\},$$

with probability one. Now, it can be shown that

$$2\max\{|X_n|, |Y_n|\}\delta\{2\max\{|X_n|, |Y_n|\}; (a,\infty)\} \leq$$
$$2|X_n|\delta\{|X_n|; (\tfrac{1}{2}a,\infty)\} + 2|Y_n|\delta\{|Y_n|; (\tfrac{1}{2}a,\infty)\}.$$

See Exercise 10. Therefore, we have proven that

$$|X_n + Y_n|\delta\{|X_n + Y_n|; (a, \infty)\} \le$$
$$2|X_n|\delta\{|X_n|; (\tfrac{1}{2}a, \infty)\} + 2|Y_n|\delta\{|Y_n|; (\tfrac{1}{2}a, \infty)\},$$

with probability one. Theorem A.16 then implies that

$$E(|X_n + Y_n|\delta\{|X_n + Y_n|; (a, \infty)\}) \le$$
$$2E(|X_n|\delta\{|X_n|; (\tfrac{1}{2}a, \infty)\}) + 2E(|Y_n|\delta\{|Y_n|; (\tfrac{1}{2}a, \infty)\}),$$

and hence,

$$\lim_{a\to\infty} E(|X_n + Y_n|\delta\{|X_n + Y_n|; (a, \infty)\}) \le$$
$$\lim_{a\to\infty} 2E(|X_n|\delta\{|X_n|; (\tfrac{1}{2}a, \infty)\}) + \lim_{a\to\infty} 2E(|Y_n|\delta\{|Y_n|; (\tfrac{1}{2}a, \infty)\}). \quad (5.6)$$

Now we use the fact that the individual sequences $\{X_n\}_{n=1}^\infty$ and $\{Y_n\}_{n=1}^\infty$ are uniformly integrable. This fact implies that

$$\lim_{a\to\infty} 2E(|X_n|\delta\{|X_n|; (\tfrac{1}{2}a, \infty)\}) = 0,$$

and

$$\lim_{a\to\infty} 2E(|Y_n|\delta\{|Y_n|; (\tfrac{1}{2}a, \infty)\}) = 0,$$

and that in both cases the convergence is uniform. This means that for any $\varepsilon > 0$ there exist b_ε and c_ε that do not depend on n such that

$$E(|X_n|\delta\{|X_n|; (\tfrac{1}{2}a, \infty)\}) < \tfrac{1}{2}\varepsilon,$$

for all $a > b_\varepsilon$ and

$$E(|Y_n|\delta\{|Y_n|; (\tfrac{1}{2}a, \infty)\}) < \tfrac{1}{2}\varepsilon,$$

for all $a > c_\varepsilon$. Let $a_\varepsilon = \max\{b_\varepsilon, c_\varepsilon\}$ and note that a_ε does not depend on n. It then follows from Equation (5.6) that

$$\lim_{a\to\infty} E(|X_n + Y_n|\delta\{|X_n + Y_n|; (a, \infty)\}) \le \varepsilon,$$

for all $a > a_\varepsilon$. Definition 5.2 then implies that the sequence $\{X_n + Y_n\}_{n=1}^\infty$ is uniformly integrable. $\qquad\square$

5.3 Convergence of Moments

In this section we will explore the relationship between the convergence of a sequence of random variables and the convergence of the corresponding moments of a random variable. As we shall observe, the uniform integrability of the sequence is the key idea in establishing this correspondence. We will first develop the results for almost certain convergence in some detail, and then point out parallel results to convergence in probability, whose development is quite similar. We will end the section by addressing weak convergence. To develop the results for the case of almost certain convergence we need a preliminary result that is another version of *Fatou's Lemma*.

Theorem 5.10 (Fatou). *Let $\{X_n\}_{n=1}^{\infty}$ be a sequence of random variables that converge almost certainly to a random variable X as $n \to \infty$. Then,*

$$E(|X|) \leq \liminf_{n \to \infty} E(|X_n|).$$

Proof. Consider the sequence of sets given by

$$\left\{ \inf_{k \geq n} |X_k| \right\}_{n=1}^{\infty},$$

and note that this sequence is monotonically increasing. Therefore, it follows from Theorem 1.12 (Lebesgue), that

$$\lim_{n \to \infty} E\left(\inf_{k \geq n} |X_k| \right) = E\left(\liminf_{n \to \infty} |X_n| \right) = E(|X|),$$

where the second equality follows from the fact that $X_n \xrightarrow{a.c.} X$ as $n \to \infty$. Note further that

$$\inf_{k \geq n} |X_k| \leq |X_n|,$$

for all $n \in \mathbb{N}$. Therefore, it follows from Theorem A.16 that

$$E\left(\inf_{k \geq n} |X_k| \right) \leq E(|X_n|),$$

for all $n \in \mathbb{N}$, and hence

$$E(|X|) = \lim_{n \to \infty} E\left(\inf_{k \geq n} |X_k| \right) = \liminf_{n \to \infty} E\left(\inf_{k \geq n} |X_k| \right) \leq \liminf_{n \to \infty} E(|X_n|).$$

Note that the second equality follows from Definition 1.3 because we have proven above that the limit in the second term exists and equals $E(|X|)$. □

We now have developed sufficient theory to prove the main result which equates uniform integrability with the convergence of moments.

Theorem 5.11. *Let $\{X_n\}_{n=1}^{\infty}$ be a sequence of random variables that converge almost certainly to a random variable X as $n \to \infty$. Let $r > 0$, then*

$$\lim_{n \to \infty} E(|X_n|^r) = E(|X|^r)$$

if and only if $\{|X_n|^r\}_{n=1}^{\infty}$ is uniformly integrable.

Proof. We begin by assuming that the sequence $\{|X_n|^r\}_{n=1}^{\infty}$ is uniformly integrable. Then, Theorem 5.5 implies that

$$\sup_{n \in \mathbb{N}} E(|X_n|^r) < \infty,$$

and Theorem 5.10 implies that

$$E(|X|^r) \leq \liminf_{n \to \infty} E(|X_n|^r).$$

From Theorem 1.5 we further have that

$$\liminf_{n\to\infty} E(|X_n|^r) \le \sup_{n\in\mathbb{N}} E(|X_n|^r),$$

so that it follows that $E(|X|^r) < \infty$. We will now establish the uniform integrability of the sequence $\{|X_n - X|^r\}_{n=1}^{\infty}$. Theorem A.18 implies that $|X_n - X|^r \le (|X_n| + |X|)^r$ and Theorem A.20 implies that $(|X_n| + |X|)^r \le 2^r(|X_n|^r + |X|^r)$ for $r > 0$. Therefore $|X_n - X|^r \le 2^r(|X_n|^r + |X|^r)$ and Theorem 5.9 then implies that since both $\{|X_n|^r\}_{n=1}^{\infty}$ and $|X|^r$ are uniformly integrable, it follows that the sequence $\{|X_n - X|^r\}_{n=1}^{\infty}$ is uniformly integrable. Let $\varepsilon > 0$ and note that

$$
\begin{aligned}
E(|X_n - X|^r) &= E(|X_n - X|^r \delta\{|X_n - X|; [0, \varepsilon]\}) \\
&\quad + E(|X_n - X|^r \delta\{|X_n - X|; (\varepsilon, \infty)\}) \\
&\le \varepsilon^r + E(|X_n - X|^r \delta\{|X_n - X|; (\varepsilon, \infty)\}).
\end{aligned}
$$

Therefore, Theorem 1.6 implies that

$$\limsup_{n\to\infty} E(|X_n - X|^r) \le \varepsilon^r + \limsup_{n\to\infty} E(|X_n - X|^r \delta\{|X_n - X|; (\varepsilon, \infty)\}).$$

For the second term on the right hand side we use Theorem 2.14 (Fatou) to conclude that

$$\limsup_{n\to\infty} E(|X_n - X|^r \delta\{|X_n - X|; (\varepsilon, \infty)\}) \le$$

$$E\left(\limsup_{n\to\infty} |X_n - X|^r \delta\{|X_n - X|; (\varepsilon, \infty)\}\right).$$

The fact that $X_n \xrightarrow{a.c.} X$ as $n \to \infty$ implies that

$$P\left(\limsup_{n\to\infty} |X_n - X| \le \varepsilon\right) = P\left(\limsup_{n\to\infty} \delta\{|X_n - X|; (\varepsilon, \infty)\} = 0\right) = 1.$$

Therefore, Theorem A.14 implies that

$$E\left(\limsup_{n\to\infty} |X_n - X|^r \delta\{|X_n - X|; (\varepsilon, \infty)\}\right) = 0,$$

and we have shown that

$$\limsup_{n\to\infty} E(|X_n - X|^r) \le \varepsilon^r.$$

Since ε is arbitrary, and $E(|X_n - X|^r) \ge 0$ it follows that we have proven that

$$\lim_{n\to\infty} E(|X_n - X|^r) = 0,$$

or equivalently that $X_n \xrightarrow{r} X$ as $n \to \infty$. We now intend to use this result to show that the corresponding expectations converge. To do this we need to consider two cases. In the first case we assume that $0 < r \le 1$, for which Theorem 2.8 implies that

$$E(|X_n|^r) = E[|(X_n - X) + X|^r] \le E(|X_n - X|^r) + E(|X|^r)$$

or equivalently that $E(|X_n|^r) - E(|X|^r) \le E(|X_n - X|^r)$. Therefore, the fact that $X_n \xrightarrow{r} X$ as $n \to \infty$ implies that

$$\lim_{n \to \infty} E(|X_n|^r) - E(|X|^r) \le \lim_{n \to \infty} E(|X_n - X|^r) = 0,$$

and hence we have proven that

$$\lim_{n \to \infty} E(|X_n|^r) = E(|X|^r).$$

Similar arguments, based on Theorem 2.9 in place of Theorem 2.8, are used for the case where $r > 1$. See Exercise 11. For a proof on the converse see Section 5.5 of Gut (2005).   □

The result of Theorem 5.11 also holds for convergence in probability. The proof in this case in nearly the same, except one needs to prove Theorem 5.10 for the case when the random variables converge in probability.

Theorem 5.12. *Let $\{X_n\}_{n=1}^{\infty}$ be a sequence of random variables that converge in probability to a random variable X as $n \to \infty$. Let $r > 0$, then*

$$\lim_{n \to \infty} E(|X_n|^r) = E(|X|^r)$$

if and only if $\{|X_n|^r\}_{n=1}^{\infty}$ is uniformly integrable.

The results of Theorems 5.11 and 5.12 also hold for convergence in distribution, though the proof is slightly different.

Theorem 5.13. *Let $\{X_n\}_{n=1}^{\infty}$ be a sequence of random variables that converge in distribution to a random variable X as $n \to \infty$. Let $r > 0$, then*

$$\lim_{n \to \infty} E(|X_n|^r) = E(|X|^r)$$

if and only if $\{|X_n|^r\}_{n=1}^{\infty}$ is uniformly integrable.

Proof. We will prove the sufficiency of the uniform integrability of the sequence following the method of proof used by Serfling (1980). For a proof of the necessity see Section 5.5 of Gut (2005). Suppose that the limiting random variable X has a distribution function F. Let $\varepsilon > 0$ and choose a positive real value a such that both a and $-a$ are continuity points of F and that

$$\sup_{n \in \mathbb{N}} E(|X_n|^r \delta\{|X_n|; [a, \infty)\}) < \varepsilon.$$

This is possible because we have assumed that the sequence $\{|X_n|^r\}_{n=1}^{\infty}$ is uniformly integrable and we can therefore find a real value a that does not depend on n such that $E(|X_n|^r \delta\{|X_n|; [a, \infty)\}) < \varepsilon$ for all $n \in \mathbb{N}$. Now choose a real number b such that $b > a$ and that b and $-b$ are also continuity points of F. Consider the function $|x|^r \delta\{|x|; [a, b]\}$, which is a continuous function on the $[a, b]$. It therefore follows from Theorem 4.3 (Helly and Bray) that

$$\lim_{n \to \infty} E(|X_n|^r \delta\{|X_n|; [a, b]\}) = E(|X|^r \delta\{|X|; [a, b]\}).$$

Now, note that for every $n \in \mathbb{N}$ we have that

$$|X_n|^r \delta\{|X_n|; [a, b]\} \leq |X_n|^r \delta\{|X_n|; [a, \infty)\},$$

with probability one. Theorem A.16 then implies that

$$E(|X_n|^r \delta\{|X_n|; [a, b]\}) \leq E(|X_n|^r \delta\{|X_n|; [a, \infty)\}) < \varepsilon,$$

for every $n \in \mathbb{N}$. Therefore it follows that

$$\lim_{n \to \infty} E(|X_n|^r \delta\{|X_n|; [a, b]\}) = E(|X|^r \delta\{|X|; [a, b]\}) < \varepsilon.$$

This result holds for every $b > a$. Hence, it further follows from Theorem 1.12 (Lebesgue) that

$$\lim_{b \to \infty} E(|X|^r \delta\{|X|; [a, b]\}) = E(|X|^r \delta\{|X|; [a, \infty)\}) < \varepsilon.$$

This also in turn implies that $E(|X|^r) < \infty$. Keeping the value of a as specified above, we have that

$$|E(|X_n|^r) - E(|X|^r)| = |E(|X_n|^r \delta\{|X_n|; [0, a]\}) + E(|X_n|^r \delta\{|X_n|; (a, \infty)\}) \\ - E(|X|^r \delta\{|X|; [0, a]\}) - E(|X|^r \delta\{|X|; (a, \infty)\})|.$$

Theorem A.18 implies that

$$|E(|X_n|^r) - E(|X|^r)| \leq |E(|X_n|^r \delta\{|X_n|; [0, a]\}) - E(|X|^r \delta\{|X|; [0, a]\})| + \\ |E(|X_n|^r \delta\{|X_n|; (a, \infty)\}) - E(|X|^r \delta\{|X|; (a, \infty)\})|.$$

The expectations in the second term are each less that ε so that

$$|E(|X_n|^r) - E(|X|^r)| \leq \\ |E(|X_n|^r \delta\{|X_n|; [0, a]\}) - E(|X|^r \delta\{|X|; [0, a]\})| + 2\varepsilon.$$

Noting once again that the function $|x|^r \delta\{|x|; [0, a]\}$ is continuous on $[0, a]$, we apply Theorem 4.3 (Helly and Bray) to find that

$$\lim_{n \to \infty} |E(|X_n|^r) - E(|X|^r)| \leq \\ \lim_{n \to \infty} |E(|X_n|^r \delta\{|X_n|; [0, a]\}) + E(|X|^r \delta\{|X|; [0, a]\})| + 2\varepsilon = 2\varepsilon.$$

Because ε is arbitrary, it follows that

$$\lim_{n \to \infty} E(|X_n|^r) = E(|X|^r).$$

$\square$

Example 5.12. Let $\{Y_n\}_{n=1}^\infty$ be a sequence of independent random variables where Y_n has a N(0,1) distribution for all $n \in \mathbb{N}$. Define a new sequence $\{X_n\}_{n=1}^\infty$ where $X_n = \min\{|Y_1|, \ldots, |Y_n|\}$ for all $n \in \mathbb{N}$. In Example 5.11 it was shown that the sequence $\{X_n\}_{n=1}^\infty$ is uniformly integrable. We can also note that for every $\varepsilon > 0$

$$P(|X_n - 0| > \varepsilon) = P(\min\{|Y_1|, \ldots, |Y_n|\} > \varepsilon) =$$

$$P\left(\bigcap_{i=1}^{n}\{|Y_i| > \varepsilon\}\right) = \prod_{i=1}^{n} P(|Y_i| > \varepsilon) = \{2[1 - \Phi(\varepsilon)]\}^n,$$

so that

$$\lim_{n \to \infty} P(|X_n| > \varepsilon) = 0.$$

Hence $X_n \xrightarrow{p} 0$ as $n \to \infty$. Theorem 5.12 further implies that

$$\lim_{n \to \infty} E(X_n) = E(0) = 0.$$

∎

Example 5.13. Let $\{U_n\}_{n=1}^{\infty}$ be a sequence of random variables where U_n has a UNIFORM$(0, n^{-1})$ distribution for all $n \in \mathbb{N}$. It can be shown that $U_n \xrightarrow{a.c.} 0$ as $n \to \infty$. Let $f_n(u) = \delta\{u; (0, n^{-1})\}$ denote the density of U_n for all $n \in \mathbb{N}$, and let F_n denote the corresponding distribution function. Note that $E(|X_n|\delta\{|X_n|; (a, \infty)\}) = 0$ for all $a > 1$. This proves that the sequence is uniformly integrable, and hence it follows from Theorem 5.11 that

$$\lim_{n \to \infty} E(X_n) = 0.$$

Note that this property could have been addressed directly by noting that $E(X_n) = (2n)^{-1}$ for all $n \in \mathbb{N}$, and therefore

$$\lim_{n \to \infty} E(X_n) = \lim_{n \to \infty} (2n)^{-1} = 0.$$

∎

Example 5.14. Suppose that $\{X_n\}_{n=1}^{\infty}$ is a sequence of independent random variables. Suppose that X_n has a GAMMA(α_n, β_n) distribution for all $n \in \mathbb{N}$ where $\{\alpha_n\}_{n=1}^{\infty}$ and $\{\beta_n\}_{n=1}^{\infty}$ are real sequences. Suppose that $\alpha_n \leq \alpha < \infty$ and $\beta_n \leq \beta < \infty$ for all $n \in \mathbb{N}$ and that $\alpha_n \to \alpha$ and $\beta_n \to \beta$ as $n \to \infty$. It was shown in Example 5.10 that the sequence $\{X_n\}_{n=1}^{\infty}$ is uniformly integrable. It also follows that $X_n \xrightarrow{d} X$ as $n \to \infty$ where X is a Gamma(α, β) random variable. Theorem 5.13 implies that $E(X_n) \to E(X) = \alpha\beta$ as $n \to \infty$. ∎

5.4 Exercises and Experiments

5.4.1 Exercises

1. Consider an arbitrary probability measure space $(\Omega, \mathcal{F}, P)$ and let $\mathcal{X}_r$ be the collection of all possible random variables X that map Ω to $\mathbb{R}$ subject to the restriction that $E(|X|^r) < \infty$. Define the operators $\oplus$ and $\otimes$ such that $X \oplus Y$ is the function $X(\omega) + Y(\omega)$ for all $\omega \in \Omega$ when $X \in \mathcal{X}_r$ and $Y \in \mathcal{X}_r$, and for a scalar $a \in \mathbb{R}$, $a \otimes X$ is the function $aX(\omega)$ for all $\omega \in \Omega$. Prove that $\mathcal{X}_r$ is a vector space by showing the following properties:

a. For each $X \in \mathcal{X}_r$ and $Y \in \mathcal{X}_r$, $X \oplus Y \in \mathcal{X}_r$.

b. The operation $\oplus$ is commutative, that is $X \oplus Y = Y \oplus X$ for each $X \in \mathcal{X}_r$ and $Y \in \mathcal{X}_r$.

c. The operation $\oplus$ is associative, that is $X \oplus (Y \oplus Z) = (X \oplus Y) \oplus Z$ for each $X \in \mathcal{X}_r$, $Y \in \mathcal{X}_r$ and $Z \in \mathcal{X}_r$.

d. There exists a random variable $0 \in \mathcal{X}_r$, called the origin, such that $0 \oplus X = X$ for all $X \in \mathcal{X}_r$.

e. For every $X \in \mathcal{X}_r$ there exists a unique random variable $-X$ such that $X + (-X) = 0$.

f. Multiplication by scalars is associative, that is $a \otimes (b \otimes X) = (ab) \otimes X$ for every $a \in \mathbb{R}$, $b \in \mathbb{R}$ and $X \in \mathcal{X}_r$.

g. For every $X \in \mathcal{X}_r$, $1 \otimes X = X$.

h. Multiplication by scalars is distributive, that is $a \otimes (X \oplus Y) = a \otimes X + a \otimes Y$ for all $a \in \mathbb{R}$, $X \in \mathcal{X}_r$ and $Y \in \mathcal{X}_r$.

i. Multiplication by random variables is distributive, that is $(a + b) \otimes X = (a \otimes X) \oplus (b \otimes Y)$ for all $a \in \mathbb{R}$, $b \in \mathbb{R}$ and $X \in \mathcal{X}_r$.

2. Within the context of Exercise 1, let $||X||_r$ be defined for $X \in \mathcal{X}_r$ as

$$||X||_r = \left[\int_\Omega |X(\omega)|^r dP(\omega) \right]^{1/r}.$$

Prove that $||X||_r$ is a norm. That is, show that $||X||_r$ has the following properties:

a. $||X||_r \geq 0$ for all $X \in \mathcal{X}_r$.

b. $||X||_r = 0$ if and only if $X = 0$.

c. $||a \otimes X||_r = |a| \cdot ||X||_r$

d. Prove the triangle inequality, that is $||X \oplus Y|| \leq ||X|| + ||Y||$ for all $X \in \mathcal{X}_r$ and $Y \in \mathcal{X}_r$.

3. Let $X_1, \ldots, X_n$ be a set of independent and identically distributed random variables from a distribution F that has parameter θ. Let $\hat{\theta}_n$ be an unbiased estimator of θ based on the observed sample where $V(\hat{\theta}_n) = \tau_n^2$ where

$$\lim_{n \to \infty} \tau_n^2 = 0.$$

Prove that $\hat{\theta}_n \xrightarrow{qm} \theta$ as $n \to \infty$.

4. Consider a sequence of random variables $\{X_n\}_{n=1}^\infty$ where X_n has probability distribution function

$$f_n(x) = \begin{cases} [\log(n+1)]^{-1} & x = n \\ 1 - [\log(n+1)]^{-1} & x = 0 \\ 0 & \text{elsewhere}, \end{cases}$$

for all $n \in \mathbb{N}$.

a. Prove that $X_n \xrightarrow{p} 0$ as $n \to \infty$.

b. Let $r > 0$. Determine whether $X_n \xrightarrow{r} 0$ as $n \to \infty$.

5. Suppose that $\{X_n\}_{n=1}^{\infty}$ is a sequence of independent random variables from a common distribution that has mean μ and variance σ^2, such that $E(|X_n|^4) < \infty$. Prove that

$$
\begin{aligned}
E(\bar{X}_n^4) &= n^{-4}[n\lambda + 4n(n-1)\mu\gamma + 6n(n-1)(n-2)\mu^2(\mu^2+\sigma^2) \\
&\quad + n(n-1)(n-2)(n-3)\mu^4] \\
&= B(n) + n^{-3}(n-1)(n-2)(n-3)\mu^4,
\end{aligned}
$$

where $B(n) = O(n^{-1})$ as $n \to \infty$, $\gamma = E(X_n^3)$, and $\lambda = E(X_n^4)$.

6. Let $\{X_n\}_{n=1}^{\infty}$ be a sequence of random variables that converge in r^{th} mean to a random variable X as $n \to \infty$ for some $r > 0$. Prove that if

$$
\sum_{n=1}^{\infty} E(|X_n - X|^r) < \infty,
$$

then $X_n \xrightarrow{a.c.} X$ as $n \to \infty$.

7. Let $\{X_n\}_{n=1}^{\infty}$ be a sequence of independent random variables where X_n has probability distribution function

$$
f_n(x) = \begin{cases} 1 - n^{-\alpha} & x = 0 \\ n^{-\alpha} & x = n \\ 0 & \text{otherwise,} \end{cases}
$$

and $0 < \alpha < 1$. Prove that there does not exist a random variable Y such that $P(|X_n| \leq |Y|) = 1$ for all $n \in \mathbb{N}$ and $E(|Y|) < \infty$.

8. Let $\{X_n\}_{n=1}^{\infty}$ be a sequence of random variables such that $P(|X_n| \leq Y) = 1$ for all $n \in \mathbb{N}$ where Y is a positive integrable random variable. Prove that the sequence $\{X_n\}_{n=1}^{\infty}$ is uniformly integrable.

9. Let $\{X_n\}_{n=1}^{\infty}$ be a sequence of random variables such that

$$
E\left(\sup_{n \in \mathbb{N}} |X_n|\right) < \infty.
$$

Prove that the sequence $\{X_n\}_{n=1}^{\infty}$ is uniformly integrable.

10. Prove that if a, x, and y are positive real numbers then

$$
2\max\{x,y\}\delta\{2\max\{x,y\}; (a,\infty)\} \leq 2x\delta\{x; (\tfrac{1}{2}a,\infty)\} + 2y\delta\{y; (\tfrac{1}{2}a,\infty)\}.
$$

11. Suppose that $\{X_n\}_{n=1}^{\infty}$ is a sequence of random variables such that $X_n \xrightarrow{r} X$ as $n \to \infty$ for some random variable X. Prove that for $r > 1$ that

$$
\lim_{n \to \infty} E(|X_n|^r) = E(|X|^r).
$$

12. Let $\{X_n\}_{n=1}^{\infty}$ be a sequence of independent random variables where X_n has an EXPONENTIAL(θ_n) distribution for all $n \in \mathbb{N}$, and $\{\theta_n\}_{n=1}^{\infty}$ is a sequence

of real numbers such that $\theta_n > 0$ for all $n \in \mathbb{N}$. Find the necessary properties for the sequence $\{\theta_n\}_{n=1}^{\infty}$ that will ensure that $\{X_n\}_{n=1}^{\infty}$ is uniformly
integrable.

13. Let $\{X_n\}_{n=1}^{\infty}$ be a sequence of independent random variables where X_n
has a TRIANGULAR$(\alpha_n, \beta_n, \gamma_n)$ distribution for all $n \in \mathbb{N}$, where $\{\alpha_n\}_{n=1}^{\infty}$,
$\{\beta_n\}_{n=1}^{\infty}$, and $\{\gamma_n\}_{n=1}^{\infty}$ are sequences of real numbers such that $\alpha_n < \gamma_n <$
β_n for all $n \in \mathbb{N}$. Find the necessary properties for the sequences $\{\alpha_n\}_{n=1}^{\infty}$,
$\{\beta_n\}_{n=1}^{\infty}$, and $\{\gamma_n\}_{n=1}^{\infty}$ that will ensure that $\{X_n\}_{n=1}^{\infty}$ is uniformly integrable.

14. Let $\{X_n\}_{n=1}^{\infty}$ be a sequence of independent random variables where X_n has
a BETA(α_n, β_n) distribution for all $n \in \mathbb{N}$, where $\{\alpha_n\}_{n=1}^{\infty}$ and $\{\beta_n\}_{n=1}^{\infty}$ are
sequences of real numbers such that $\alpha_n > 0$ and $\beta_n > 0$ for all $n \in \mathbb{N}$. Find
the necessary properties for the sequences $\{\alpha_n\}_{n=1}^{\infty}$, and $\{\beta_n\}_{n=1}^{\infty}$ that will
ensure that $\{X_n\}_{n=1}^{\infty}$ is uniformly integrable.

15. Let $\{X_n\}_{n=1}^{\infty}$ be a sequence of random variables where X_n has distribution
F_n which has mean θ_n for all $n \in \mathbb{N}$. Suppose that

$$\lim_{n \to \infty} \theta_n = \theta,$$

for some $\theta \in (0, \infty)$. Hence it follows that

$$\lim_{n \to \infty} E(X_n) = E(X),$$

where X has distribution F with mean θ. Does it necessarily follow that
the sequence $\{X_n\}_{n=1}^{\infty}$ is uniformly integrable?

16. Let $\{X_n\}_{n=1}^{\infty}$ be a sequence of random variables such that X_n has distribution F_n for all $n \in \mathbb{N}$. Suppose that $X_n \xrightarrow{a.c.} X$ as $n \to \infty$ for some random
variable X. Suppose that there exists a subset $A \subset \mathbb{R}$ such that

$$\int_A dF_n(t) = 1,$$

for all $n \in \mathbb{N}$, and

$$\int_A dt < \infty.$$

Do these conditions imply that $E(X_n) \to E(X)$ as $n \to \infty$?

17. Let $\{X_n\}_{n=1}^{\infty}$ be a sequence of random variables such that X_n has a $N(0, \sigma_n^2)$
distribution, conditional on σ_n. For each case detailed below, determine if
the sequence $\{X_n\}_{n=1}^{\infty}$ is uniformly integrable and determine if X_n converges weakly to a random variable X as $n \to \infty$.

a. $\sigma_n = n^{-1}$ for all $n \in \mathbb{N}$.
b. $\sigma_n = n$ for all $n \in \mathbb{N}$.
c. $\sigma_n = 10 + (-1)^n$ for all $n \in \mathbb{N}$.
d. $\{\sigma_n\}_{n=1}^{\infty}$ is a sequence of independent random variables where σ_n has an
EXPONENTIAL(θ) distribution for each $n \in \mathbb{N}$.

18. Let $\{X_n\}_{n=1}^\infty$ be a sequence of random variables such that X_n has a $N(\mu_n, \sigma_n^2)$ distribution, conditional on μ_n and σ_n. For each case detailed below determine if the sequence $\{X_n\}_{n=1}^\infty$ is uniformly integrable and determine if X_n converges weakly to a random variable X as $n \to \infty$.

a. $\mu_n = n$ and $\sigma_n = n^{-1}$ for all $n \in \mathbb{N}$.

b. $\mu_n = n^{-1}$ and $\sigma_n = n$ for all $n \in \mathbb{N}$.

c. $\mu_n = n^{-1}$ and $\sigma_n = n^{-1}$ for all $n \in \mathbb{N}$.

d. $\mu_n = n$ and $\sigma_n = n$ for all $n \in \mathbb{N}$.

e. $\mu_n = (-1)^n$ and $\sigma_n = 10 + (-1)^n$ for all $n \in \mathbb{N}$.

f. $\{\mu_n\}_{n=1}^\infty$ is a sequence of independent random variables where μ_n has a $N(0,1)$ distribution for each $n \in \mathbb{N}$, and $\{\sigma_n\}_{n=1}^\infty$ is a sequence of random variables where σ_n has an EXPONENTIAL(θ) distribution for each $n \in \mathbb{N}$.

5.4.2 Experiments

1. Write a program in R that simulates a sequence of independent and identically distributed random variables $X_1, \ldots, X_{100}$ where X_n follows a distribution F that is specified below. For each $n = 1, \ldots, 100$ compute $\bar{X}_n$ on $X_1, \ldots, X_n$ along with $|\bar{X}_n - \mu|^2$ where μ is the mean of the distribution F. Repeat the experiment five times and plot each sequence $\{|\bar{X}_n - \mu|^2\}_{n=1}^\infty$ against n on the same set of axes. Describe the behavior observed in each case and compare it to whether $X_n \xrightarrow{qm} \mu$ as $n \to \infty$.

a. F is a $N(0,1)$ distribution.

b. F is a CAUCHY$(0,1)$ distribution, where μ is taken to be the median of the distribution.

c. F is a EXPONENTIAL(1) distribution.

d. F is a T(2) distribution.

e. F is a T(3) distribution.

2. Write a program in R that simulates a sequence of independent and identically distributed random variables $B_1, \ldots, B_{500}$ where B_n is a BERNOULLI(θ) random variable where θ is specified below. For each $n = 3, \ldots, 500$, compute
$$n^{1/2}\{\log[\log(n)]\}^{-1/2}(\bar{B}_n - \theta),$$
where $\bar{B}_n$ is the sample mean computed on $B_1, \ldots, B_n$. Repeat the experiment five times and plot each sequence
$$\{n^{1/2}\{\log[\log(n)]\}^{-1/2}(\bar{B}_n - \theta)\}_{n=3}^{500}$$
against n on the same set of axes. Describe the behavior observed for the sequence in each case in terms of the result of Example 5.3. Repeat the experiment for each $\theta \in \{0.01, 0.10, 0.50, 0.75\}$.

3. Write a program in R that simulates a sequence of independent random variables $X_1, \ldots, X_{100}$ where X_n as probability distribution function

$$f_n(x) = \begin{cases} 1 - n^{-\alpha} & x = 0 \\ n^{-\alpha} & x = n \\ 0 & \text{elsewhere.} \end{cases}$$

Repeat the experiment five times and plot each sequence X_n against n on the same set of axes. Describe the behavior observed for the sequence in each case. Repeat the experiment for each $\alpha \in \{0.5, 1.0, 1.5, 2.0\}$.

4. Write a program in R that simulates a sequence of independent random variables $X_1, \ldots, X_{100}$ where X_n is a $N(0, \sigma_n^2)$ random variable where the sequence $\{\sigma_n\}_{n=1}^{\infty}$ is specified below. Repeat the experiment five times and plot each sequence X_n against n on the same set of axes. Describe the behavior observed for the sequence in each case, and relate the behavior to the results of Exercise 17.

a. $\sigma_n = n^{-1}$ for all $n \in \mathbb{N}$.

b. $\sigma_n = n$ for all $n \in \mathbb{N}$.

c. $\sigma_n = 10 + (-1)^n$ for all $n \in \mathbb{N}$.

d. $\{\sigma_n\}_{n=1}^{\infty}$ is a sequence of independent random variables where σ_n has an EXPONENTIAL(θ) distribution for each $n \in \mathbb{N}$.

5. Write a program in R that simulates a sequence of independent random variables $X_1, \ldots, X_{100}$ where X_n is a $N(\mu_n, \sigma_n^2)$ random variable where the sequences $\{\mu_n\}_{n=1}^{\infty}$ and $\{\sigma_n\}_{n=1}^{\infty}$ are specified below. Repeat the experiment five times and plot each sequence X_n against n on the same set of axes. Describe the behavior observed for the sequence in each case, and relate the behavior to the results of Exercise 18.

a. $\mu_n = n$ and $\sigma_n = n^{-1}$ for all $n \in \mathbb{N}$.

b. $\mu_n = n^{-1}$ and $\sigma_n = n$ for all $n \in \mathbb{N}$.

c. $\mu_n = n^{-1}$ and $\sigma_n = n^{-1}$ for all $n \in \mathbb{N}$.

d. $\mu_n = n$ and $\sigma_n = n$ for all $n \in \mathbb{N}$.

e. $\mu_n = (-1)^n$ and $\sigma_n = 10 + (-1)^n$ for all $n \in \mathbb{N}$.

f. $\{\mu_n\}_{n=1}^{\infty}$ is a sequence of independent random variables where μ_n has a $N(0, 1)$ distribution for each $n \in \mathbb{N}$, and $\{\sigma_n\}_{n=1}^{\infty}$ is a sequence of random variables where σ_n has an EXPONENTIAL(θ) distribution for each $n \in \mathbb{N}$.

CHAPTER 6

Central Limit Theorems

They formed a unit of the sort that normally can be formed only by matter that is lifeless.

The Trial by Franz Kafka

6.1 Introduction

One of the important and interesting features of the Central Limit Theorem is that the weak convergence of the mean holds under many situations beyond the simple situation where we observe a sequence of independent and identically distributed random variables. In this chapter we will explore some of these extensions. The two main direct extensions of the Central Limit Theorem we will consider are to non-identically distributed random variables and to triangular arrays. Of course, other generalizations are possible, and we only present some of the simpler cases. For a more general presentation of this subject see Gnedenko and Kolmogorov (1968). We will also consider transformations of asymptotically normal statistics that either result in asymptotically NORMAL statistics, or statistics that follow a CHISQUARED distribution. As we will show, the difference between these two outcomes depends on the smoothness of the transformation.

6.2 Non-Identically Distributed Random Variables

The Lindeberg, Lévy, and Feller version of the Central Limit Theorem relaxes the assumption that the random variables in the sequence need to be identically distributed, but still retains the assumption of independence. The result originates from the work of Lindeberg (1922), Lévy (1925), and Feller (1935), who each proved various parts of the final result.

Theorem 6.1 (Lindeberg, Lévy, and Feller). *Let $\{X_n\}_{n=1}^{\infty}$ be a sequence of independent random variables where $E(X_n) = \mu_n$ and $V(X_n) = \sigma_n^2 < \infty$ for all $n \in \mathbb{N}$ where $\{\mu_n\}_{n=1}^{\infty}$ and $\{\sigma_n^2\}_{n=1}^{\infty}$ are sequences of real numbers. Let*

$$\bar{\mu}_n = n^{-1} \sum_{k=1}^{n} \mu_k,$$

$$\tau_n^2 = \sum_{k=1}^{n} \sigma_k^2, \tag{6.1}$$

and suppose that

$$\lim_{n\to\infty} \max_{k\in\{1,2,\ldots,n\}} \tau_n^{-2}\sigma_k^2 = 0, \tag{6.2}$$

then $Z_n = n\tau_n^{-1}(\bar{X}_n - \bar{\mu}_n) \xrightarrow{d} Z$, *as* $n \to \infty$ *where* Z *has a* $N(0,1)$ *distribution if and only if*

$$\lim_{n\to\infty} \tau_n^{-2} \sum_{k=1}^{n} E(|X_k - \mu_k|^2 \delta\{|X_k - \mu_k|; (\varepsilon\tau_n, \infty)\}) = 0, \tag{6.3}$$

for every $\varepsilon > 0$.

Proof. We will prove the sufficiency of the condition given in Equation (6.3), but not its necessity. See Section 7.2 of Gut (2005) for details on the necessity part of the proof. The main argument of this proof is based on the same idea that we used in proving Theorem 4.20 (Lindeberg and Lévy) in that we will show that the characteristic function of Z_n converges to the characteristic function of a $N(0,1)$ random variable. As in the proof of Theorem 4.20 we begin by assuming that $\mu_n = 0$ for all $n \in \mathbb{N}$ which does not reduce the generality of the proof since the numerator of Z_n has the form

$$\bar{X}_n - \bar{\mu}_n = n^{-1} \sum_{k=1}^{n} (X_k - \mu_k) = n^{-1} \sum_{k=1}^{n} X_k^*,$$

where $E(X_k^*) = 0$ for all $k \in \{1,\ldots,n\}$. Let ψ_k be the characteristic function of X_k for all $k \in \mathbb{N}$. Theorem 2.33 implies that the characteristic function of the sum of $X_1,\ldots,X_n$ is

$$\prod_{k=1}^{n} \psi_k(t).$$

Note that

$$Z_n = n\tau_n^{-1}(\bar{X}_n - \bar{\mu}_n) = \tau_n^{-1} \sum_{k=1}^{n} X_k,$$

under the assumption that $\mu_k = 0$ for all $k \in \mathbb{N}$. Therefore, it follows from Theorem 2.32 that the characteristic function of Z_n, which we will denote as ψ, is

$$\psi(t) = \prod_{k=1}^{n} \psi_k(\tau_n^{-1}t) = \exp\left\{\sum_{k=1}^{n} \log\left[\psi_k(\tau_n^{-1}t)\right]\right\}.$$

Some algebra in the exponent can be used to show that

$$\exp\left\{\sum_{k=1}^{n}\log\left[\psi_k(\tau_n^{-1}t)\right]\right\} =$$

$$\exp\left\{\sum_{k=1}^{n}\log[\psi_k(\tau_n^{-1}t)] + \sum_{k=1}^{n}\left[1 - \psi_k(\tau_n^{-1}t)\right]\right\} \times$$

$$\exp\left\{\sum_{k=1}^{n}\psi_k(\tau_n^{-1}t) - \sum_{k=1}^{n}(1 - \tfrac{1}{2}\tau_n^{-2}\sigma_k^2 t^2)\right\}\exp\left\{\sum_{k=1}^{n}-\tfrac{1}{2}\tau_n^{-2}\sigma_k^2 t^2\right\}.$$

Note that the final term in the product has the form

$$\exp\left\{\sum_{k=1}^{n}-\tfrac{1}{2}\tau_n^{-2}\sigma_k^2 t^2\right\} = \exp\left\{-\tfrac{1}{2}t^2\tau_n^{-2}\sum_{k=1}^{n}\sigma_k^2\right\} = \exp(-\tfrac{1}{2}t^2),$$

which is the characteristic function of a standard normal random variable. Therefore, the proof now depends on demonstrating that

$$\lim_{n\to\infty}\exp\left\{\sum_{k=1}^{n}\log[\psi_k(\tau_n^{-1}t)] + \sum_{k=1}^{n}\left[1 - \psi_k(\tau_n^{-1}t)\right]\right\} = 1,$$

and

$$\lim_{n\to\infty}\exp\left\{\sum_{k=1}^{n}\psi_k(\tau_n^{-1}t) - \sum_{k=1}^{n}(1 - \tfrac{1}{2}\tau_n^{-2}\sigma_k^2 t^2)\right\} = 1,$$

which is equivalent to showing that

$$\lim_{n\to\infty}\left|\sum_{k=1}^{n}\log[\psi_k(\tau_n^{-1}t)] + \sum_{k=1}^{n}\left[1 - \psi_k(\tau_n^{-1}t)\right]\right| = 0, \tag{6.4}$$

and

$$\lim_{n\to\infty}\left|\sum_{k=1}^{n}\psi_k(\tau_n^{-1}t) - \sum_{k=1}^{n}(1 - \tfrac{1}{2}\tau_n^{-2}\sigma_k^2 t^2)\right| = 0. \tag{6.5}$$

We work on showing the property in Equation (6.4) first. Theorem 2.30 implies that $|\psi_k(t) - 1 - itE(X_k)| \le E(\tfrac{1}{2}t^2 X_k^2)$. In our case we are evaluating the characteristic function at $\tau_n^{-1}t$ and we have assumed that $E(X_k) = 0$. Therefore, we have that

$$|\psi_k(\tau_n^{-1}t) - 1| \le E\left(\frac{t^2 X_k^2}{2\tau_n^2}\right) = \frac{t^2\sigma_k^2}{2\tau_n^2} \le \tfrac{1}{2}t^2\tau_n^{-2}\max_{1\le k\le n}\sigma_k^2. \tag{6.6}$$

The assumption given in Equation (6.2) then implies that

$$\lim_{n\to\infty}|\psi(\tau_n^{-1}t) - 1| \le \lim_{n\to\infty}\tfrac{1}{2}t^2\tau_n^{-2}\max_{1\le k\le n}\sigma_k^2 = 0. \tag{6.7}$$

It is also noteworthy that the convergence is uniform in k, due to the bound provided in the assumption of Equation (6.2). Now, Theorem A.8 implies that for $z \in \mathbb{C}$ and $|z| \le \tfrac{1}{2}$ we have that $|\log(1-z)+z| \le |z|^2$. Equation (6.7) implies that there exists an integer n' such that $|\psi_k(\tau_n^{-1}t) - 1| \le \frac{1}{2}$. The uniformity of the convergence implies that n' does not depend on k. Therefore, Theorem A.8 implies that for all $n \ge n'$ we have that

$$|\log\{1 - [1 - \psi_k(\tau_n^{-1}t)]\} + 1 - \psi_k(\tau_n^{-1}t)| \le |\psi_k(\tau_n^{-1}t) - 1|^2.$$

Hence, for all $n \ge n'$,

$$\left| \sum_{k=1}^n \log[\psi_k(\tau_n^{-1}t)] + \sum_{k=1}^n [1 - \psi_k(\tau_n^{-1}t)] \right| =$$

$$\left| \sum_{k=1}^n \{\log\{1 - [1 - \psi_k(\tau_n^{-1}t)]\} + [1 - \psi_k(\tau_n^{-1}t)]\} \right| \le$$

$$\sum_{k=1}^n |\{\log\{1 - [1 - \psi_k(\tau_n^{-1}t)]\} + [1 - \psi_k(\tau_n^{-1}t)]\}| \le \sum_{k=1}^n |\psi_k(\tau_n^{-1}t) - 1|^2.$$

Now we use the fact that

$$|\psi_k(\tau_n^{-1}t) - 1| \le \max_{1 \le k \le n} |\psi_k(\tau_n^{-1}t) - 1|,$$

and Equation (6.6) to show that

$$\begin{aligned}
\sum_{k=1}^n |\psi_k(\tau_n^{-1}t) - 1|^2 &\le \max_{1 \le k \le n} |\psi_k(\tau_n^{-1}t) - 1| \sum_{k=1}^n |\psi_k(\tau_n^{-1}t) - 1| \\
&\le \tfrac{1}{2}\tau_n^{-2}t^2 \left(\max_{1 \le k \le n} \sigma_k^2 \right) \sum_{k=1}^n E(\tfrac{1}{2}\tau_n^{-2}t^2 X_k^2) \\
&= \tfrac{1}{4}t^4\tau_n^{-2} \left(\max_{1 \le k \le n} \sigma_k^2 \right) \tau_n^{-2} \sum_{k=1}^n \sigma_k^2 \\
&= \tfrac{1}{4}t^4\tau_n^{-2} \left(\max_{1 \le k \le n} \sigma_k^2 \right).
\end{aligned}$$

Therefore, Equation (6.2) implies that

$$\lim_{n\to\infty} \left| \sum_{k=1}^n \log[\psi_k(\tau_n^{-1}t)] + \sum_{k=1}^n [1 - \psi_k(\tau_n^{-1}t)] \right| \le$$

$$\lim_{n\to\infty} \tfrac{1}{4}t^4\tau_n^{-2} \left(\max_{1 \le k \le n} \sigma_k^2 \right) = 0.$$

This completes our first task in this proof since we have proven that Equation (6.4) is true. We now take up the task of proving that Equation (6.5) is true. Theorem 2.30 implies that

$$|\psi_k(t\tau_n^{-1}) - 1 - it\tau_n^{-1}E(X_k) + \tfrac{1}{2}t^2\tau_n^{-2}E(X_k^2)| \le E\left(\frac{t^2 X_k^2}{\tau_n^2} \right), \qquad (6.8)$$

which, due to the assumption that $E(X_k) = 0$ simplifies to

$$\left| \psi_k(t\tau_n^{-1}) - 1 + \frac{t^2\sigma_k^2}{2\tau_n^2} \right| \leq E\left(\frac{t^2 X_k^2}{\tau_n^2} \right). \tag{6.9}$$

Theorem 2.30 and similar calculations to those used earlier can also be used to establish that

$$\left| \psi_k(t\tau_n^{-1}) - 1 + \frac{t^2\sigma_k^2}{2\tau_n^2} \right| \leq E\left(\frac{|t|^3 |X_k|^3}{6\tau_n^3} \right).$$

Now, Theorem A.18 and Equations (6.8) and (6.9) imply that

$$\left| \sum_{k=1}^{n} \left[\psi_k(t\tau_n^{-1}) - \left(1 - \frac{t^2\sigma_k^2}{2\tau_n^2}\right) \right] \right| \leq \sum_{k=1}^{n} \left| \psi_k(t\tau_n^{-1}) - \left(1 - \frac{t^2\sigma_k^2}{2\tau_n^2}\right) \right|$$

$$\leq \sum_{k=1}^{n} \min\left\{ E\left(\frac{t^2 X_k^2}{\tau_n^2} \right), E\left(\frac{|t|^3 |X_k|^3}{6\tau_n^3} \right) \right\}.$$

We now split each of the expectations across small and large values of $|X_k|$. Let $\varepsilon > 0$, then

$$E\left(\frac{t^2 X_k^2}{\tau_n^2} \right) = E\left(\frac{t^2 X_k^2}{\tau_n^2} \delta\{|X_k|; [0, \varepsilon\tau_n]\} \right) + E\left(\frac{t^2 X_k^2}{\tau_n^2} \delta\{|X_k|; (\varepsilon\tau_n, \infty)\} \right),$$

and

$$E\left(\frac{|t|^3 |X_k|^3}{6\tau_n^3} \right) = E\left(\frac{|t|^3 |X_k|^3}{6\tau_n^3} \delta\{|X_k|; [0, \varepsilon\tau_n]\} \right) +$$

$$E\left(\frac{|t|^3 |X_k|^3}{6\tau_n^3} \delta\{|X_k|; (\varepsilon\tau_n, \infty)\} \right),$$

which implies that

$$\min\left\{ E\left(\frac{t^2 X_k^2}{\tau_n^2} \right), E\left(\frac{|t|^3 |X_k|^3}{6\tau_n^3} \right) \right\} \leq E\left(\frac{|t|^3 |X_k|^3}{6\tau_n^3} \delta\{|X_k|; [0, \varepsilon\tau_n]\} \right) +$$

$$E\left(\frac{t^2 X_k^2}{\tau_n^2} \delta\{|X_k|; (\varepsilon\tau_n, \infty)\} \right).$$

Therefore,

$$\left| \sum_{k=1}^{n} \left[\psi_k(t\tau_n^{-1}) - \left(1 - \frac{t^2\sigma_k^2}{2\tau_n^2}\right) \right] \right| \leq \sum_{k=1}^{n} E\left(\frac{|t|^3 |X_k|^3}{6\tau_n^3} \delta\{|X_k|; [0, \varepsilon\tau_n]\} \right) +$$

$$\sum_{k=1}^{n} E\left(\frac{t^2 X_k^2}{\tau_n^2} \delta\{|X_k|; (\varepsilon\tau_n, \infty)\} \right).$$

The first term on the right hand side can be simplified by bounding $|X_k|^3 \leq$

$\varepsilon\tau_n|X_k|^2$ due to the condition imposed by the indicator function. That is,

$$\sum_{k=1}^{n} E\left(\frac{|t|^3|X_k|^3}{6\tau_n^3}\delta\{|X_k|;[0,\varepsilon\tau_n]\}\right) \leq \sum_{k=1}^{n}\frac{|t|^3\varepsilon\tau_n}{6\tau_n^3}E(|X_k|^2\delta\{|X_k|;[0,\varepsilon\tau_n]\})$$

$$\leq \tfrac{1}{6}|t|^3\varepsilon\tau_n^{-2}\sum_{k=1}^{n}E(|X_k|^2)$$

$$= \tfrac{1}{6}|t|^3\varepsilon\tau_n^{-2}\sum_{k=1}^{n}\sigma_k^2$$

$$= \tfrac{1}{6}|t|^3\varepsilon.$$

The second inequality follows from the fact that

$$E(|X_k|^2\delta\{|X_k|;[0,\varepsilon\tau_n]\}) \leq E(|X_k|^2), \tag{6.10}$$

where we note that expectation on the right hand side of Equation (6.10) is finite by assumption. This was the reason for bounding one of the $|X_k|$ in the term $|X_k|^3$ by $\varepsilon\tau_n$. Therefore,

$$\left|\sum_{k=1}^{n}\left[\psi_k(t\tau_n^{-1})-\left(1-\frac{t^2\sigma_k^2}{2\tau_n^2}\right)\right]\right| \leq \tfrac{1}{6}|t|^3\varepsilon+$$

$$t^2\tau_n^{-2}\sum_{k=1}^{n}E(|X_k|^2\delta\{|X_k|;(\varepsilon\tau_n,\infty)\}). \tag{6.11}$$

Equation (6.3) implies that the second term on the right hand side of Equation (6.11) converges to zero as $n \to \infty$, therefore

$$\limsup_{n\to\infty}\left|\sum_{k=1}^{n}\left[\psi_k(t\tau_n^{-1})-\left(1-\frac{t^2\sigma_k^2}{2\tau_n^2}\right)\right]\right| \leq \tfrac{1}{6}|t|^3\varepsilon,$$

for every $\varepsilon > 0$. Since ε is arbitrary, it follows that

$$\lim_{n\to\infty}\left|\sum_{k=1}^{n}\left[\psi_k(t\tau_n^{-1})-\left(1-\frac{t^2\sigma_k^2}{2\tau_n^2}\right)\right]\right| = 0,$$

which verifies Equation (6.5), and the result follows. ☐

The condition given in Equation (6.3) is known as the *Lindeberg Condition*. As Serfling (1980) points out, the Lindeberg Condition actually implies the condition given in Equation (6.2), so that the main focus in applying this result is on the verification of Equation (6.3). This condition regulates the tail behavior of the sequence of distribution functions that correspond to the sequence of random variables $\{X_n\}_{n=1}^{\infty}$. Without becoming too technical about the meaning of this condition, it is clear in the proof of Theorem 6.1 where the condition arises. Indeed, this condition is exactly what is required to complete the final step of the proof. Unfortunately, the Lindeberg Condition can be difficult to verify in practice, and hence we will explore a simpler sufficient condition.

Corollary 6.1. *Let $\{X_n\}_{n=1}^{\infty}$ be a sequence of independent random variables where $E(X_n) = \mu_n$ and $V(X_n) = \sigma_n^2 < \infty$ for all $n \in \mathbb{N}$ where $\{\mu_n\}_{n=1}^{\infty}$ and $\{\sigma_n^2\}_{n=1}^{\infty}$ are sequences of real numbers. Let*

$$\bar{\mu}_n = n^{-1}\sum_{k=1}^{n}\mu_k,$$

$$\tau_n^2 = \sum_{k=1}^{n}\sigma_k^2,$$

and suppose that for some $\eta > 2$ that

$$\sum_{k=1}^{n}E(|X_k - \mu_k|^{\eta}) = o(\tau_n^{\eta}), \tag{6.12}$$

as $n \to \infty$, then $Z_n = n\tau_n^{-1}(\bar{X}_n - \bar{\mu}_n) \xrightarrow{d} Z$, as $n \to \infty$, where Z has a $N(0,1)$ distribution.

Proof. We will follow the method of proof of Serfling (1980). We will show that the conditions of Equations (6.2) and (6.3) follow from the condition given in Equation (6.12). Let $\varepsilon > 0$ and focus on the term inside the summation of Equation (6.3). We note that

$$E(|X_k - \mu_k|^2\delta\{|X_k - \mu_k|; (\varepsilon\tau_n, \infty)\}) =$$
$$E(|X_k - \mu_k|^{2-\eta}|X_k - \mu_k|^{\eta}\delta\{|X_k - \mu_k|; (\varepsilon\tau_n, \infty)\}) \leq$$
$$E(|\varepsilon\tau_n|^{2-\eta}|X_k - \mu_k|^{\eta}\delta\{|X_k - \mu_k|; (\varepsilon\tau_n, \infty)\}),$$

where we note that the inequality comes from the fact that $\eta > 2$ so that the exponent $2 - \eta < 0$, and hence

$$|X_k - \mu_k|^{2-\eta}\delta\{|X_k - \mu_k|; (\varepsilon\tau_n, \infty)\} \leq |\varepsilon\tau_n|^{2-\eta}\delta\{|X_k - \mu_k|; (\varepsilon\tau_n, \infty)\}.$$

The inequality follows from an application of Theorem A.16. It, then, further follows that

$$E(|\varepsilon\tau_n|^{2-\eta}|X_k - \mu_k|^{\eta}\delta\{|X_k - \mu_k|; (\varepsilon\tau_n, \infty)\}) =$$
$$|\varepsilon\tau_n|^{2-\eta}E(|X_k - \mu_k|^{\eta}\delta\{|X_k - \mu_k|; (\varepsilon\tau_n, \infty)\}) \leq$$
$$|\varepsilon\tau_n|^{2-\eta}E(|X_k - \mu_k|^{\eta}).$$

Now,

$$\limsup_{n\to\infty} \tau_n^{-2} \sum_{k=1}^{n} E(|X_k - \mu_k|^2 \delta\{|X_k - \mu_k|; (\varepsilon\tau_n, \infty)\}) \le$$

$$\limsup_{n\to\infty} \tau_n^{-2} |\varepsilon\tau_n|^{2-\eta} \sum_{k=1}^{n} E(|X_k - \mu_k|^{\eta}) =$$

$$\limsup_{n\to\infty} \varepsilon\tau_n^{-\eta} \sum_{k=1}^{n} E(|X_k - \mu_k|^{\eta}) = 0,$$

by Equation (6.12). A similar result follows for limit infimum, so that it follows that

$$\lim_{n\to\infty} \tau_n^{-2} \sum_{k=1}^{n} E(|X_k - \mu_k|^2 \delta\{|X_k - \mu_k|; (\varepsilon\tau_n, \infty)\}) = 0,$$

and therefore the condition of Equation (6.3) is satisfied. We now show that the condition of Equation (6.2) is also satisfied. To do this, we note that for all $k \in \{1, \ldots, n\}$, we have that

$$\sigma_k^2 = E[(X_k - \mu_k)^2] = E[(X_k - \mu_k)^2 \delta\{|X_k - \mu_k|; [0, \varepsilon\tau_n]\}] +$$
$$E[(X_k - \mu_k)^2 \delta\{|X_k - \mu_k|; (\varepsilon\tau_n, \infty)\}].$$

The first term can be bounded as

$$\begin{aligned} E[(X_k - \mu_k)^2 \delta\{|X_k - \mu_k|; [0, \varepsilon\tau_n]\}] &\le E[(\varepsilon\tau_n)^2 \delta\{|X_k - \mu_k|; [0, \varepsilon\tau_n]\}] \\ &= (\varepsilon\tau_n)^2 E[\delta\{|X_k - \mu_k|; [0, \varepsilon\tau_n]\}] \\ &= (\varepsilon\tau_n)^2 P(|X_k - \mu_k| \le \varepsilon\tau_n) \\ &\le (\varepsilon\tau_n)^2, \end{aligned}$$

where the final inequality follows because the probability is bounded above by one. Therefore, it follows that

$$\sigma_k^2 \le (\varepsilon\tau_n)^2 + E[(X_k - \mu_k)^2 \delta\{|X_k - \mu_k|; (\varepsilon\tau_n, \infty)\}],$$

for all $k \in \{1, \ldots, n\}$. Hence, it follows that

$$\max_{k\in\{1,\ldots,n\}} \sigma_k^2 \le (\varepsilon\tau_n)^2 + \sum_{k=1}^{n} E[(X_k - \mu_k)^2 \delta\{|X_k - \mu_k|; (\varepsilon\tau_n, \infty)\}].$$

Therefore,

$$\lim_{n\to\infty} \max_{k\in\{1,\ldots,n\}} \tau_n^{-2} \sigma_k^2 \le$$

$$\lim_{n\to\infty} \varepsilon^2 + \lim_{n\to\infty} \tau_n^{-2} \sum_{k=1}^{n} E[(X_k - \mu_k)^2 \delta\{|X_k - \mu_k|; (\varepsilon\tau_n, \infty)\}]. \quad (6.13)$$

The condition given in Equation (6.3) implies that the second term on the right hand side of Equation (6.13) converges to zero as $n \to \infty$. Therefore

$$\lim_{n \to \infty} \max_{k \in \{1,\dots,n\}} \tau_n^{-2} \sigma_k^2 \le \varepsilon^2,$$

for all $\varepsilon > 0$, so that

$$\lim_{n \to \infty} \max_{k \in \{1,\dots,n\}} \tau_n^{-2} \sigma_k^2 = 0,$$

and the result follows. □

Example 6.1. Let $\{X_n\}_{n=1}^{\infty}$ be a sequence of independent random variables where X_n has an EXPONENTIAL(θ_n) distribution. In this case $\mu_n = \theta_n$ and $\sigma_n^2 = \theta_n^2$ for all $n \in \mathbb{N}$. Let us consider the case where $\theta_n = n^{-1/2}$ for all $n \in \mathbb{N}$ and take $\eta = 4$. In this case $E(|X_n - \theta_n|^4) = 9\theta_n^4 = 9n^{-2}$ for all $n \in \mathbb{N}$. Hence,

$$\lim_{n \to \infty} \frac{\sum_{k=1}^{n} E(|X_n - \theta_n|^4)}{\left(\sum_{k=1}^{n} \sigma_k^2\right)^2} = \lim_{n \to \infty} \frac{9 \sum_{k=1}^{n} n^{-2}}{\left(\sum_{k=1}^{n} n^{-1}\right)^2} = 0,$$

since the Harmonic series diverges, but the series in the numerator converges to $\frac{1}{6}\pi^2$. Therefore, it follows from Definition 1.7 that

$$\sum_{k=1}^{n} E(|X_n - \theta_n|^4) = o(\tau_n^4),$$

as $n \to \infty$. Therefore, Corollary 6.1 implies that $Z_n = n\tau_n^{-1}(\bar{X}_n - \bar{\mu}_n) \xrightarrow{d} Z$, as $n \to \infty$ where Z has a $N(0,1)$ distribution. ∎

6.3 Triangular Arrays

Triangular arrays generalize the sequences of non-identically distributed random variables studied in Section 6.2. When the value of n changes in a triangular array, the distribution of the entire sequence of random variables up to the nth random variable may change as well. Such sequences can be represented as doubly indexed sequences of random variables.

Definition 6.1. Let $\{\{X_{nm}\}_{m=1}^{u_n}\}_{n=1}^{\infty}$ be a doubly indexed sequence of random variables where $\{u_n\}_{n=1}^{\infty}$ is a sequence of increasing integers such that $u_n \to \infty$ as $n \to \infty$. Then $\{\{X_{nm}\}_{m=1}^{u_n}\}_{n=1}^{\infty}$ is called a double array of random variables. In the special case that $u_n = n$ for all $n \in \mathbb{N}$, then the sequence is called a triangular array.

Under certain conditions the result of Theorem 4.20 can be extended to double arrays as well. For simplicity we give the result for triangular arrays.

Theorem 6.2. Let $\{\{X_{nk}\}_{k=1}^{n}\}_{n=1}^{\infty}$ be a triangular array where $X_{n1}, \dots, X_{nn}$ are mutually independent random variables for each $n \in \mathbb{N}$. Suppose that X_{nk} has mean μ_{nk} and variance $\sigma_{nk}^2 < \infty$ for all $k \in \{1, \dots, n\}$ and $n \in \mathbb{N}$. Let

$$\mu_n. = \sum_{k=1}^{n} \mu_{nk},$$

and

$$\sigma_{n\cdot}^2 = \sum_{k=1}^{n} \sigma_{nk}^2.$$

Then

$$\lim_{n \to \infty} \max_{k \in \{1,\dots,n\}} P(|X_{nk} - \mu_{nk}| > \varepsilon \sigma_{n\cdot}) = 0, \qquad (6.14)$$

for each $\varepsilon > 0$ and

$$Z_n = \sigma_{n\cdot}^{-1} \left(\sum_{k=1}^{n} X_{nk} - \mu_{n\cdot} \right) \xrightarrow{d} Z,$$

as $n \to \infty$, where Z is a $N(0,1)$ random variable, together hold if and only if

$$\lim_{n \to \infty} \sigma_{n\cdot}^{-2} \sum_{k=1}^{n} E[(X_{nk} - \mu_{nk})^2 \delta\{|X_{nk} - \mu_{nk}|; (\varepsilon \sigma_{n\cdot}, \infty)\}] = 0, \qquad (6.15)$$

for each $\varepsilon > 0$.

Theorem 6.2 can be generalized to double arrays without much modification to the result above. See Theorem 1.9.3 of Serfling (1980). The condition in Equation (6.14) is called the *uniform asymptotic negligibility* condition which essentially establishes bounds on the amount of probability in the tails of the distribution of X_{nk} uniformly within each row. Double arrays of random variables that have this property are said to be *holospoudic*. See Section 7.1 of Chung (1974). The condition given in Equation (6.15) is the same Lindeberg Condition used in Theorem 6.1. One can also note that in fact Theorem 6.1 is a special case of Theorem 6.2 if we assume that distribution of the random variable X_{nk} does not change with k. Since the Lindeberg Condition also shows up in this result, we are once again confronted with the fact the Equation (6.15) can be difficult to apply in practice. However, there is an analog of Corollary 6.1 which can be applied to the case of triangular arrays as well.

Corollary 6.2. *Let $\{\{X_{nk}\}_{k=1}^{n}\}_{n=1}^{\infty}$ be a triangular array where $X_{n1}, \dots, X_{nn}$ are mutually independent random variables for each $n \in \mathbb{N}$. Suppose that X_{nk} has mean μ_{nk} and variance $\sigma_{nk}^2 < \infty$ for all $k \in \{1, \dots, n\}$ and $n \in \mathbb{N}$. Let*

$$\mu_{n\cdot} = \sum_{k=1}^{n} \mu_{nk},$$

and

$$\sigma_{n\cdot}^2 = \sum_{k=1}^{n} \sigma_{nk}^2.$$

Suppose that for some $\eta > 2$

$$\sum_{k=1}^{n} E(|X_{nk} - \mu_{nk}|^{\eta}) = o(\sigma_{n\cdot}^{\eta}),$$

as $n \to \infty$, then

$$Z_n = \sigma_{n\cdot}^{-1} \left(\sum_{k=1}^{n} X_{nk} - \mu_{n\cdot} \right) \xrightarrow{d} Z,$$

as $n \to \infty$ where Z is a N(0,1) random variable.

For the proof of Corollary 6.2, see Exercise 5.

Example 6.2. Consider a triangular array $\{\{X_{nk}\}_{k=1}^{n}\}_{n=1}^{\infty}$ where the sequence $X_{n1}, \ldots, X_{nn}$ is a set of independent and identically distributed random variables from an EXPONENTIAL(θ_n) distribution where $\{\theta_n\}_{n=1}^{\infty}$ is a sequence of positive real numbers that converges to a real value θ as $n \to \infty$. In this case $\mu_{nk} = \theta_n$ and $\sigma_{nk}^2 = \theta_n^2$ for all $k \in \{1, \ldots, n\}$ and $n \in \mathbb{N}$ so that

$$\mu_{n\cdot} = \sum_{k=1}^{n} \mu_{nk} = \sum_{k=1}^{n} \theta_n = n\theta_n,$$

and

$$\sigma_{n\cdot}^2 = \sum_{k=1}^{n} \sigma_{nk}^2 = \sum_{k=1}^{n} \theta_n^2 = n\theta_n^2.$$

We will use Corollary 6.2 with $\eta = 4$, so that we have that

$$\sum_{k=1}^{n} E(|X_{nk} - \mu_{nk}|^4) = \sum_{k=1}^{n} E(|X_{nk} - \theta_n|^4) = \sum_{k=1}^{n} 9\theta_n^4 = 9n\theta_n^4.$$

Therefore, it follows that

$$\lim_{n\to\infty} \frac{\sum_{k=1}^{n} E(|X_{nk} - \mu_{nk}|^4)}{\sigma_{n\cdot}^4} = \lim_{n\to\infty} \frac{9n\theta_n^4}{(n\theta_n^2)^2} = \lim_{n\to\infty} 9n^{-1} = 0,$$

and hence,

$$\sum_{k=1}^{n} E(|X_{nk} - \mu_{nk}|^4) = o(\sigma_{n\cdot}^4),$$

as $n \to \infty$. Therefore, Corollary 6.2 implies that

$$Z_n = n^{1/2}\theta_n^{-1} \left(\sum_{k=1}^{n} X_{nk} - n\theta_n \right) \xrightarrow{d} Z,$$

where Z is a N(0,1) random variable. ∎

One application of triangular arrays occurs when we consider estimates based on the empirical distribution function such as bootstrap estimates as described in Efron (1979). See Beran and Ducharme (1991) for further details on this type of application.

6.4 Transformed Random Variables

Consider a problem in statistical inference where we have an asymptotically NORMAL estimate $\hat{\theta}_n$ for a parameter θ which is computed from a sequence of independent and identically distributed random variables $X_1, \ldots, X_n$. Suppose that the real interest is not in the parameter θ itself, but in a function of the parameter given by $g(\theta)$. An obvious estimate of $g(\theta)$ is $g(\hat{\theta}_n)$. The properties of this estimate depend on the properties of both the estimator $\hat{\theta}_n$ and the function g. For example, if $\hat{\theta}_n$ is a consistent estimator of θ, and g is continuous at θ, then $g(\hat{\theta}_n)$ is a consistent estimator of $g(\theta)$ by Theorem 3.7. On the other hand, if $\hat{\theta}_n$ is an unbiased estimator of θ, then $g(\hat{\theta}_n)$ will typically not be an unbiased estimator of $g(\theta)$ unless g is a linear function. However, it could be the case that $g(\hat{\theta}_n)$ is asymptotically unbiased as $n \to \infty$. See Chapter 10. In this chapter we have been examining conditions under which the sample mean is asymptotically NORMAL. We now examine under what conditions a function of an asymptotically NORMAL sequence of random variables remains asymptotically NORMAL. For example, suppose that $\sigma_n^{-1}(\hat{\theta}_n - \theta) \xrightarrow{d} Z$ as $n \to \infty$ where σ_n is a sequence of positive real numbers such that $\sigma_n \to 0$ as $n \to \infty$. Under what conditions can we find a function of $g(\hat{\theta}_n)$ of the form $\tau_n^{-1}[g(\hat{\theta}_n) - g(\theta)]$ that converges in distribution to a $N(0,1)$ random variable as $n \to \infty$ where τ_n is a sequence of positive constants that converge to zero as $n \to \infty$? That answer turns out to depend on the properties of the function g and finding the correct sequence $\{\tau_n\}_{n=1}^\infty$ to properly scale the sequence. To find the sequence $\{\tau_n\}_{n=1}^\infty$ we must account for how the transformation of $\hat{\theta}_n$ changes the variability of the sequence.

Theorem 6.3. *Let $\{X_n\}_{n=1}^\infty$ be a sequence of random variables such that $\sigma_n^{-1}(X_n - \mu) \xrightarrow{d} Z$ as $n \to \infty$ where Z is a $N(0,1)$ random variable and $\{\sigma_n\}_{n=1}^\infty$ is a real sequence such that*

$$\lim_{n\to\infty} \sigma_n = 0.$$

Let g be a real function that has a non-zero derivative at μ. Then,

$$\frac{g(X_n) - g(\mu)}{\sigma_n g'(\mu)} \xrightarrow{d} Z,$$

as $n \to \infty$.

Proof. The method of proof used here is to use Theorem 4.11 (Slutsky) to show that $\sigma_n^{-1}(X_n - \mu)$ and $[\sigma_n g'(\mu)]^{-1}[g(X_n) - g(\mu)]$ have the same limiting distribution. To this end, define a function h as $h(x) = (x - \mu)^{-1}[g(x) - g(\mu)] - g'(\mu)$ for all $x \neq \mu$. The definition of derivative motivates us to define $h(\mu) = 0$. Now, Theorem 4.21 implies than $X_n \xrightarrow{p} \mu$ as $n \to \infty$, and therefore Theorem 3.7 implies that $h(X_n) \xrightarrow{p} h(\mu) = 0$ as $n \to \infty$. Theorem 4.11 then implies that $h(X_n)\sigma_n^{-1}(X_n - \mu) \xrightarrow{p} 0$ as $n \to \infty$. But, this implies that

$$\left(\frac{g(X_n) - g(\mu)}{X_n - \mu} - g'(\mu) \right) \left(\frac{X_n - \mu}{\sigma_n} \right) =$$

$$\frac{g(X_n) - g(\mu)}{\sigma_n} - g'(\mu)\frac{X_n - \mu}{\sigma_n} \xrightarrow{p} 0,$$

as $n \to \infty$, which in turn implies that

$$\frac{g(X_n) - g(\mu)}{\sigma_n g'(\mu)} - \frac{X_n - \mu}{\sigma_n} \xrightarrow{p} 0,$$

as $n \to \infty$. Therefore, Theorem 4.11 implies that

$$\frac{g(X_n) - g(\mu)}{\sigma_n g'(\mu)} - \frac{X_n - \mu}{\sigma_n} + \frac{X_n - \mu}{\sigma_n} \xrightarrow{d} Z,$$

where Z is a N$(0,1)$ random variable. Hence

$$\frac{g(X_n) - g(\mu)}{\sigma_n g'(\mu)} \xrightarrow{d} Z,$$

as $n \to \infty$, and the result is proven. $\qquad\qquad\qquad\square$

Theorem 6.3 indicates that the change in variation required to maintain the asymptotic normality of a transformation of an asymptotically normal random variable is related to the derivative of the transformation. This is because the asymptotic variation in $g(X_n)$ is related to the local change in g near μ due to the fact that $X_n \xrightarrow{p} \mu$ as $n \to \infty$. To visualize this consider Figures 6.1 and 6.2. In Figure 6.1 the variation around μ decreases through the function g due to small derivative of g in a neighborhood of μ, while in Figure 6.2 the variation around μ increases through the function g due to large derivative of g in a neighborhood of μ. If the derivative of g is zero at μ, then there will be no variation in g as X_n approaches μ as $n \to \infty$. This fact does not allow us to obtain an asymptotic NORMAL result for the transformed sequence of random variables, though other types of weak convergence are possible.

Example 6.3. Let $\{X_n\}_{n=1}^{\infty}$ be a sequence of independent and identically distributed random variables from a distribution with variance $\sigma^2 < \infty$. Let S_n^2 be the sample variance. Then, under some minor conditions, it can be shown that $n^{1/2}(\mu_4 - \sigma^4)^{-1/2}(S_n^2 - \sigma^2) \xrightarrow{d} Z$ as $n \to \infty$ where Z is a N$(0,1)$ random variable and μ_4 is the fourth moment of X_n, which is assumed to be finite. Asymptotic normality can also be obtained for the sample standard deviation by considering the function $g(x) = x^{1/2}$ where

$$\left.\frac{d}{dx}g(x)\right|_{x=\sigma^2} = \left.\frac{d}{dx}x^{1/2}\right|_{x=\sigma^2} = \tfrac{1}{2}\sigma^{-1}.$$

Theorem 6.3 then implies that

$$\frac{n^{1/2}[g(S_n^2) - g(\sigma^2)]}{(\mu_4 - \sigma^4)^{1/2}g'(\sigma^2)} = \frac{2n^{1/2}\sigma(S_n - \sigma)}{\mu_4 - \sigma^4} \xrightarrow{d} Z,$$

as $n \to \infty$. $\qquad\qquad\qquad\blacksquare$

Example 6.4. Let $\{X_n\}_{n=1}^{\infty}$ be a sequence of independent and identically distributed random variables from a distribution with $E(X_n) = \mu$ and $V(X_n) = \sigma^2 < \infty$. Theorem 4.20 (Lindeberg and Lévy) implies that $n^{1/2}\sigma^{-1}(\bar{X}_n - \mu) \xrightarrow{d} Z$ as $n \to \infty$ where Z is a N$(0,1)$ random variable. Suppose that we wish

Figure 6.1 *When the derivative of g is small in a neighborhood of μ, the variation in g, represented by the vertical grey band about μ, decreases. This is indicated by the horizontal grey band around g(μ).*

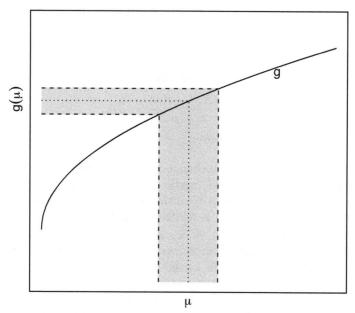

to estimate $\theta = \exp(\mu)$ with the plug-in estimator $\hat{\theta} = \exp(\bar{X}_n)$. Letting $g(x) = \exp(x)$ implies that $g'(\mu) = \exp(\mu)$, so that Theorem 6.3 then implies that $n^{1/2} \exp(-\mu)\sigma^{-1}[\exp(\bar{X}_n) - \exp(\mu)] \xrightarrow{d} Z$ as $n \to \infty$.   ∎

In the case where the derivative of g is zero at μ, the limiting distribution is no longer NORMAL. In this case the asymptotic distribution depends how how many derivatives at μ are equal to zero. For example, if $g'(\mu) = 0$ but $g''(\mu) \neq 0$ it follows that there is a function of the sequence that converges in distribution to Z^2 as $n \to \infty$ where Z has a $N(0,1)$ distribution and hence Z^2 is a CHISQUARED(1) distribution.

Theorem 6.4. *Let $\{X_n\}_{n=1}^{\infty}$ be a sequence of random variables such that $\sigma_n^{-1}(X_n - \mu) \xrightarrow{d} Z$ as $n \to \infty$ where Z is a $N(0,1)$ random variable and $\{\sigma_n\}_{n=1}^{\infty}$ is a real sequence such that*

$$\lim_{n \to \infty} \sigma_n = 0.$$

Let g be a real function such that

$$\left.\frac{d^m}{dx^m}g(x)\right|_{x=\mu} \neq 0,$$

Figure 6.2 *When the derivative of g is large in a neighborhood of μ, the variation in g, represented by the vertical grey band about μ, increases. This is indicated by the horizontal grey band around g(μ).*

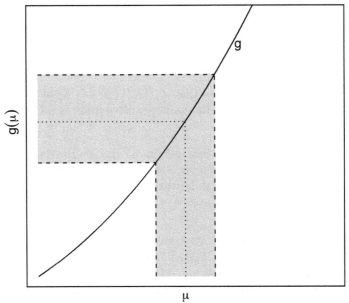

for some $m \in \mathbb{N}$ such that

$$\frac{d^k}{dx^k}g(x)\bigg|_{x=\mu} = 0,$$

for all $k \in \{1, \ldots, m-1\}$. Then,

$$\frac{m![g(X_n) - g(\mu)]}{\sigma_n^m g^{(m)}(\mu)} \xrightarrow{d} Z^m,$$

as $n \to \infty$.

The proof of Theorem 6.4 is the subject of Exercise 8.

Example 6.5. Let $\{X_n\}_{n=1}^{\infty}$ be a sequence of random variables such that $\sigma_n^{-1}(X_n - \mu) \xrightarrow{d} Z$ as $n \to \infty$ where Z is a $\mathrm{N}(0,1)$ random variable and $\{\sigma_n\}_{n=1}^{\infty}$ is a real sequence such that

$$\lim_{n \to \infty} \sigma_n = 0.$$

Consider the transformation $g(x) = x^2$. If $\mu \neq 0$ then,

$$\frac{d}{dx}g(x)\bigg|_{x=\mu} = 2\mu,$$

and Theorem 6.3 implies that $(2\mu\sigma_n)^{-1}(X_n^2 - \mu^2) \xrightarrow{d} Z$ as $n \to \infty$. On the other hand, if $\mu = 0$ then,

$$\frac{d}{dx}g(x)\bigg|_{x=\mu} = 0,$$

but

$$\frac{d^2}{dx^2}g(x)\bigg|_{x=0} = 2.$$

In this case Theorem 6.4 implies that $\sigma_n^{-2}(X_n^2 - \mu^2) \xrightarrow{d} Z^2$ as $n \to \infty$. Therefore, in the case where $\mu = 0$, the limiting distribution is a CHISQUARED(1) distribution. ∎

It is also possible to extend these results to the multivariate case, though the result gets slightly more complicated, as the proper transformation now depends on the matrix of partial derivatives. We will begin by first extending the result of Theorem 6.3 to the case where g is a function that maps $\mathbb{R}^d$ to $\mathbb{R}$.

Theorem 6.5. *Let $\{\mathbf{X}_n\}_{n=1}^{\infty}$ be a sequence of random vectors from a d-dimensional distribution such that $n^{1/2}(\mathbf{X}_n - \boldsymbol{\theta}) \xrightarrow{d} \mathbf{Z}$ as $n \to \infty$ where $\mathbf{Z}$ has a $\mathbf{N}(\mathbf{0}, \boldsymbol{\Sigma})$ distribution, $\boldsymbol{\theta}$ is a $d \times 1$ constant vector and $\boldsymbol{\Sigma}$ is a positive definite covariance matrix. Let g be a real function that maps $\mathbb{R}^d$ to $\mathbb{R}$ and let*

$$\mathbf{d}(\boldsymbol{\theta}) = \frac{\partial}{\partial \mathbf{x}}g(\mathbf{x})\bigg|_{\mathbf{x}=\boldsymbol{\theta}},$$

be the vector of partial derivatives of g evaluated at $\boldsymbol{\theta}$. If $\mathbf{d}(\boldsymbol{\theta})$ is not equal to the zero vector and $\mathbf{d}(\mathbf{x})$ is continuous in a neighborhood of $\boldsymbol{\theta}$, then $n^{1/2}[g(\mathbf{X}_n) - g(\boldsymbol{\theta})] \xrightarrow{d} Z$ as $n \to \infty$ where Z is a $N[0, \mathbf{d}'(\boldsymbol{\theta})\boldsymbol{\Sigma}\mathbf{d}(\boldsymbol{\theta})]$ random variable.

Proof. We generalize the argument used to prove Theorem 6.3. Define a function h that maps $\mathbb{R}^d$ to $\mathbb{R}$ as

$$h(\mathbf{x}) = \|\mathbf{x} - \boldsymbol{\theta}\|^{-1}[g(\mathbf{x}) - g(\boldsymbol{\theta}) - \mathbf{d}'(\boldsymbol{\theta})(\mathbf{x} - \boldsymbol{\theta})],$$

where we define $h(\boldsymbol{\theta}) = 0$. Therefore,

$$n^{1/2}[g(\mathbf{X}_n) - g(\boldsymbol{\theta})] = n^{1/2}\|\mathbf{X}_n - \boldsymbol{\theta}\|h(\mathbf{X}_n) + n^{1/2}\mathbf{d}'(\boldsymbol{\theta})(\mathbf{X}_n - \boldsymbol{\theta}).$$

By assumption we know that $n^{1/2}(\mathbf{X}_n - \boldsymbol{\theta}) \xrightarrow{d} \mathbf{Z}$ as $n \to \infty$ where $\mathbf{Z}$ has a $\mathbf{N}(\mathbf{0}, \boldsymbol{\Sigma})$ distribution. Therefore Theorem 4.17 (Cramér and Wold) implies that $n^{1/2}\mathbf{d}'(\boldsymbol{\theta})(\mathbf{X}_n - \boldsymbol{\theta}) \xrightarrow{d} Z_1$ as $n \to \infty$ where Z_1 is a $N[0, \mathbf{d}'(\boldsymbol{\theta})\boldsymbol{\Sigma}\mathbf{d}(\boldsymbol{\theta})]$ random variable. Since $\|\cdot\|$ denotes a vector norm on $\mathbb{R}^d$ we have that $n^{1/2}\|\mathbf{X}_n - \boldsymbol{\theta}\| = \|n^{1/2}(\mathbf{X}_n - \boldsymbol{\theta})\|$. Assuming that the norm is continuous, Theorem 4.18 implies that $\|n^{1/2}(\mathbf{X}_n - \boldsymbol{\theta})\| \xrightarrow{d} \|\mathbf{Z}\|$ as $n \to \infty$. Serfling (1980) argues that the function h is continuous, and therefore since $\mathbf{X}_n \xrightarrow{p} \boldsymbol{\theta}$ as $n \to \infty$ it follows that $h(\mathbf{X}_n) \xrightarrow{p} h(\boldsymbol{\theta}) = 0$ as $n \to \infty$. Therefore, Theorem 4.11 (Slutsky) implies that $n^{1/2}\|\mathbf{X}_n - \boldsymbol{\theta}\|h(\mathbf{X}_n) \xrightarrow{p} 0$ as $n \to \infty$, and another application of Theorem 4.11 implies that

$$n^{1/2}\|\mathbf{X}_n - \boldsymbol{\theta}\|h(\mathbf{X}_n) + n^{1/2}\mathbf{d}'(\boldsymbol{\theta})(\mathbf{X}_n - \boldsymbol{\theta}) \xrightarrow{d} \mathbf{Z},$$

as $n \to \infty$ and the result is proven. $\qquad\square$

Example 6.6. Let $\{\mathbf{X}_n\}_{n=1}^{\infty}$ be a sequence of independent and identically distributed bivariate random variables from a distribution with mean vector $\boldsymbol{\mu}$ and positive definite covariance matrix $\boldsymbol{\Sigma}$ where we will assume that $\mathbf{X}'_n = (X_{1n}, X_{2n})$ for all $n \in \mathbb{N}$, $\boldsymbol{\mu}' = (\mu_1, \mu_2)$, and that

$$\boldsymbol{\Sigma} = \begin{bmatrix} \sigma_1^2 & \tau \\ \tau & \sigma_2^2 \end{bmatrix},$$

where all of the elements of $\boldsymbol{\mu}$ and $\boldsymbol{\Sigma}$ will be assumed to be finite. Suppose we are interested in the correlation coefficient given by $\rho = \tau \sigma_x^{-1} \sigma_y^{-1}$, which can be estimated with $\hat{\rho}_n = S_{12} S_1^{-1} S_2^{-1}$, where

$$S_{12} = n^{-1} \sum_{k=1}^{n} (X_{1k} - \bar{X}_1)(X_{2k} - \bar{X}_2),$$

$$S_i^2 = n^{-1} \sum_{k=1}^{n} (X_{ik} - \bar{X}_i)^2,$$

and

$$\bar{X}_i = n^{-1} \sum_{k=1}^{n} X_{ik},$$

for $i = 1, 2$. We will first show that a properly normalized function of the random vector (S_1^2, S_2^2, S_{12}) converges in distribution to a NORMAL distribution. Following the arguments of Sen and Singer (1993) we first note that

$$
\begin{aligned}
S_{12} &= n^{-1} \sum_{k=1}^{n} (X_{1k} - \bar{X}_1)(X_{2k} - \bar{X}_2) \\
&= n^{-1} \sum_{k=1}^{n} [(X_{1k} - \mu_1) + (\mu_1 - \bar{X}_1)][(X_{2k} - \mu_2) + (\mu_2 - \bar{X}_2)] \\
&= n^{-1} \sum_{k=1}^{n} (X_{1k} - \mu_1)(X_{2k} - \mu_2) + n^{-1} \sum_{k=1}^{n} (X_{1k} - \mu_1)(\mu_2 - \bar{X}_2) + \\
&\quad n^{-1} \sum_{k=1}^{n} (\mu_1 - \bar{X}_1)(X_{2k} - \mu_2) + n^{-1} \sum_{k=1}^{n} (\mu_1 - \bar{X}_1)(\mu_2 - \bar{X}_2). \quad (6.16)
\end{aligned}
$$

The two middle terms in Equation (6.16) can be simplified as

$$n^{-1}\sum_{k=1}^{n}(X_{1k} - \mu_1)(\mu_2 - \bar{X}_2) =$$

$$n^{-1}(\mu_2 - \bar{X}_2)\sum_{k=1}^{n}(X_{1k} - \mu_1) = (\mu_2 - \bar{X}_2)(\bar{X}_1 - \mu_1),$$

and

$$n^{-1}\sum_{k=1}^{n}(\mu_1 - \bar{X}_1)(X_{2k} - \mu_2) = (\bar{X}_1 - \mu_1)(\mu_2 - \bar{X}_2).$$

Therefore, it follows that

$$
\begin{aligned}
S_{12} &= n^{-1}\sum_{k=1}^{n}(X_{1k} - \mu_1)(X_{2k} - \mu_2) - (\mu_1 - \bar{X}_1)(\mu_2 - \bar{X}_2) \\
&= n^{-1}\sum_{k=1}^{n}(X_{1k} - \mu_1)(X_{2k} - \mu_2) + R_{12}.
\end{aligned}
$$

Theorem 3.10 (Weak Law of Large Numbers) implies that $\bar{X}_1 \xrightarrow{p} \mu_1$ and $\bar{X}_2 \xrightarrow{p} \mu_2$ as $n \to \infty$, so that Theorem 3.9 implies that $R_{12} \xrightarrow{p} 0$ as $n \to \infty$. It can similarly be shown that

$$S_i^2 = n^{-1}\sum_{k=1}^{n}(X_{ik} - \mu_i)^2 + R_i,$$

where $R_i \xrightarrow{p} 0$ as $n \to \infty$ for $i = 1, 2$. See Exercise 11. Now define a random vector $\mathbf{U}_n' = (S_1^2 - \sigma_1^2, S_2^2 - \sigma_2^2, S_{12} - \tau)$ for all $n \in \mathbb{N}$. Let $\boldsymbol{\lambda} \in \mathbb{R}^3$ where $\boldsymbol{\lambda}' = (\lambda_1, \lambda_2, \lambda_3)$ and observe that

$$
\begin{aligned}
n^{1/2}\boldsymbol{\lambda}'\mathbf{U}_n &= n^{1/2}\left[\lambda_1(S_1^2 - \sigma_1^2) + \lambda_2(S_2^2 - \sigma_2^2) + \lambda_3(S_{12} - \tau)\right] \\
&= n^{1/2}\left\{\lambda_1\left[n^{-1}\sum_{k=1}^{n}(X_{1k} - \mu_1)^2 + R_1 - \sigma_1^2\right]\right. \\
&\quad + \lambda_2\left[n^{-1}\sum_{k=1}^{n}(X_{2k} - \mu_2)^2 + R_2 - \sigma_2^2\right] \\
&\quad \left. + \lambda_3\left[n^{-1}\sum_{k=1}^{n}(X_{1k} - \mu_1)(X_{2k} - \mu_2) + R_{12} - \tau\right]\right\}.
\end{aligned}
$$

Combine the remainder terms into one term and combine the remaining terms into one sum to find that

$$
\begin{aligned}
n^{1/2}\boldsymbol{\lambda}'\mathbf{U}_n = n^{-1/2}\sum_{k=1}^{n}\{&\lambda_1[(X_{1k} - \mu_1)^2 - \sigma_1^2] + \lambda_2[(X_{2k} - \mu_2)^2 - \sigma_2^2] \\
&+ \lambda_3[(X_{1k} - \mu_1)(X_{2k} - \mu_2) - \tau]\} + R,
\end{aligned}
$$

where $R = n^{1/2}(R_1 + R_2 + R_{12})$. Note that even though each individual remainder term converges to zero in probability, it may not necessarily follow that $n^{1/2}$ times the remainder also converges to zero. However, we will show in Chapter 8 that this does follow in this case. Now, define a sequence of random variables $\{\mathbf{V}_n\}_{n=1}^{\infty}$ as

$$V_k = \lambda_1[(X_{1k} - \mu_1)^2 - \sigma_1^2] + \lambda_2[(X_{2k} - \mu_2)^2 - \sigma_2^2] + \lambda_3[(X_{1k} - \mu_1)(X_{2k} - \mu_2) - \tau],$$

so that it follows that

$$n^{1/2}\boldsymbol{\lambda}'\mathbf{U}_n = n^{-1/2}\sum_{k=1}^{n} V_k + R,$$

where $R \xrightarrow{p} 0$ as $n \to \infty$. The random variables $V_1, \ldots, V_n$ are a set of independent and identically distributed random variables. The expectation of V_k is given by

$$
\begin{aligned}
E(V_k) &= E\{\lambda_1[(X_{1k} - \mu_1)^2 - \sigma_1^2] + \lambda_2[(X_{2k} - \mu_2)^2 - \sigma_2^2] \\
&\quad + \lambda_3[(X_{1k} - \mu_1)(X_{2k} - \mu_2) - \tau]\} \\
&= \lambda_1\{E[(X_{1k} - \mu_1)^2] - \sigma_1^2\} + \lambda_2\{E[(X_{2k} - \mu_2)^2] - \sigma_2^2\} \\
&\quad + \lambda_3\{E[(X_{1k} - \mu_1)(X_{2k} - \mu_2)] - \tau]\} \\
&= 0,
\end{aligned}
$$

where we have used the fact that $E[(X_{1k} - \mu_1)^2] = \sigma_1^2$, $E[(X_{2k} - \mu_2)^2] = \sigma_2^2$, and $E[(X_{1k} - \mu_1)(X_{2k} - \mu_2)] = \tau$. The variance of V_k need not be found explicitly, but is equal to $\boldsymbol{\lambda}'\boldsymbol{\Lambda}\boldsymbol{\lambda}$ where $\boldsymbol{\Lambda} = V(\mathbf{U}_n)$. Therefore, Theorem 4.20 (Lindeberg and Lévy) implies that

$$n^{-1/2}\sum_{k=1}^{n} V_k \xrightarrow{d} Z,$$

as $n \to \infty$ where Z has a $\mathrm{N}(0, \boldsymbol{\lambda}'\boldsymbol{\Lambda}\boldsymbol{\lambda})$ distribution. Theorem 4.11 (Slutsky) then further implies that

$$n^{1/2}\boldsymbol{\lambda}'\mathbf{U}_n = n^{-1/2}\sum_{k=1}^{n} V_k + R \xrightarrow{d} Z,$$

as $n \to \infty$, since $R \xrightarrow{p} 0$ as $n \to \infty$. Because $\boldsymbol{\lambda}$ is arbitrary, it follows from Theorem 4.17 (Cramér and Wold) that $n^{1/2}\mathbf{U}_n \xrightarrow{d} \mathbf{Z}$ as $n \to \infty$, where $\mathbf{Z}$ has a $\mathbf{N}(\mathbf{0}, \boldsymbol{\Lambda})$ distribution. Using this result, we shall prove that there is a function of the sample correlation that converges in distribution to a NORMAL distribution. Let $\boldsymbol{\theta}' = (\sigma_1^2, \sigma_2^2, \tau)$ and consider the function $g(\mathbf{x}) = g(x_1, x_2, x_3) = x_3(x_1 x_2)^{-1/2}$. Then, it follows that

$$\mathbf{d}'(\mathbf{x}) = [-\tfrac{1}{2}x_3 x_1^{-3/2} x_2^{-1/2}, -\tfrac{1}{2}x_3 x_1^{-1/2} x_2^{-3/2}, (x_1 x_2)^{-1/2}],$$

so that

$$\mathbf{d}'(\boldsymbol{\theta}) = [-\tfrac{1}{2}\rho\sigma_1^{-2}, -\tfrac{1}{2}\rho\sigma_2^{-2}, (\sigma_1\sigma_2)^{-1}],$$

which has been written in terms of the correlation $\rho = \sigma_1 \sigma_2 \tau$. Theorem 6.5 then implies that $n^{1/2}(\hat{\rho} - \rho) \xrightarrow{d} Z$ as $n \to \infty$ where Z is a $N[0, \mathbf{d}'(\boldsymbol{\theta}) \boldsymbol{\Lambda} \mathbf{d}(\boldsymbol{\theta})]$ random variable. ■

The result of Theorem 6.5 extends to the more general case where g is a function that maps $\mathbb{R}^d$ to $\mathbb{R}^m$. The main requirement for the transformed sequence to remain asymptotically NORMAL is that the matrix of partial derivatives of g must have elements that exist and are non-zero at $\boldsymbol{\theta}$.

Theorem 6.6. *Let* $\{\mathbf{X}_n\}_{n=1}^{\infty}$ *be a sequence of random vectors from a d-dimensional distribution such that* $n^{1/2}(\mathbf{X}_n - \boldsymbol{\theta}) \xrightarrow{d} \mathbf{Z}$ *as $n \to \infty$ where $\mathbf{Z}$ is a d-dimensional $\mathbf{N}(\mathbf{0}, \boldsymbol{\Sigma})$ random vector, $\boldsymbol{\theta}$ is a $d \times 1$ constant vector, and $\boldsymbol{\Sigma}$ is a $d \times d$ covariance matrix. Let g be a real function that maps $\mathbb{R}^d$ to $\mathbb{R}^m$ such that $g(\mathbf{x}) = [g_1(\mathbf{x}), \dots, g_m(\mathbf{x})]'$ for all $x \in \mathbb{R}^d$ where $g_k(\mathbf{x})$ is a real function that maps $\mathbb{R}^d$ to $\mathbb{R}$. Let $\mathbf{D}(\boldsymbol{\theta})$ be the $m \times d$ matrix of partial derivatives of g whose $(i,j)^{th}$ element is given by*

$$D_{ij}(\boldsymbol{\theta}) = \left. \frac{\partial}{\partial x_j} g_i(\mathbf{x}) \right|_{\mathbf{x}=\boldsymbol{\theta}}$$

for $i = 1, \dots, m$ and $j = 1, \dots, d$ where $\mathbf{x}' = (x_1, \dots, x_d)$. If $\mathbf{D}(\boldsymbol{\theta})$ exists and $D_{ij}(\boldsymbol{\theta}) \neq 0$ for all $i = 1, \dots, m$ and $j = 1, \dots, d$ then $n^{1/2}[g(\mathbf{X}_n - g(\boldsymbol{\mu})] \xrightarrow{d} \mathbf{Z}$ as $n \to \infty$ where $\mathbf{Z}$ is an m-dimensional random variable that has a $\mathbf{N}[\mathbf{0}, \mathbf{D}(\boldsymbol{\theta}) \boldsymbol{\Sigma} \mathbf{D}'(\boldsymbol{\theta})]$ distribution.

Proof. We follow the method of Serfling (1980) which is based on Theorem 4.17 (Cramér and Wold) and on generalizing the argument used to prove Theorem 6.5. Define functions $h_1, \dots, h_m$ where

$$h_i(\mathbf{x}) = \|\mathbf{x} - \boldsymbol{\theta}\|^{-1}[g_i(\mathbf{x}) - g_i(\boldsymbol{\theta}) - \mathbf{d}_i'(\boldsymbol{\theta})(\mathbf{x} - \boldsymbol{\theta})],$$

for $i = 1, \dots, m$ where $\mathbf{d}_i'(\boldsymbol{\theta})$ is the ith row of the matrix $\mathbf{D}(\boldsymbol{\theta})$, and we note once again that we define $h_i(\boldsymbol{\theta}) = 0$ and that $h_i(\mathbf{x})$ is continuous for all $i = 1, \dots, m$. Suppose that $\mathbf{v} \in \mathbb{R}^m$ and note that

$$\mathbf{v}'[g(\mathbf{X}_n) - g(\boldsymbol{\theta})] = \sum_{i=1}^{m} v_i[g_i(\mathbf{X}_n) - g_i(\boldsymbol{\theta})] =$$

$$\sum_{i=1}^{m} v_i[\|\mathbf{X}_n - \boldsymbol{\theta}\| h_i(\mathbf{X}_n) + \mathbf{d}_i'(\boldsymbol{\theta})(\mathbf{X}_n - \boldsymbol{\theta})] =$$

$$\sum_{i=1}^{m} v_i \|\mathbf{X}_n - \boldsymbol{\theta}\| h_i(\mathbf{X}_n) + \sum_{i=1}^{m} v_i \mathbf{d}_i'(\boldsymbol{\theta})(\mathbf{X}_n - \boldsymbol{\theta}). \quad (6.17)$$

The second term of the right hand side of Equation (6.17) can be written as

$$\sum_{i=1}^{m} v_i \mathbf{d}_i'(\boldsymbol{\theta})(\mathbf{X}_n - \boldsymbol{\theta}) = \mathbf{v}' \mathbf{D}(\boldsymbol{\theta})(\mathbf{X}_n - \boldsymbol{\theta}).$$

By assumption we have that $n^{1/2}(\mathbf{X}_n - \boldsymbol{\theta}) \xrightarrow{d} \mathbf{Z}$ as $n \to \infty$ where $\mathbf{Z}$ is a $\mathbf{N}(\mathbf{0}, \boldsymbol{\Sigma})$ random vector. Therefore, Theorem 4.17 implies that $n^{1/2}\mathbf{v}'\mathbf{D}(\boldsymbol{\theta})(\mathbf{X}_n - \boldsymbol{\theta}) \xrightarrow{d} \mathbf{v}'\mathbf{D}(\boldsymbol{\theta})\mathbf{Z} = W$ where W has a $\mathrm{N}[0, \mathbf{v}'\mathbf{D}(\boldsymbol{\theta})\boldsymbol{\Sigma}\mathbf{D}(\boldsymbol{\theta})\mathbf{v}]$ random vector. For the first term in Equation (6.17), we note that

$$\sum_{i=1}^{m} v_i \|\mathbf{X} - \boldsymbol{\theta}\| h_i(\mathbf{X}_n) = \mathbf{v}'\mathbf{h}(\mathbf{X}_n)\|\mathbf{X}_n - \boldsymbol{\theta}\|,$$

where $\mathbf{h}'(\mathbf{x}) = [h_1(\mathbf{x}), \ldots, h_m(\mathbf{x})]$. The fact that $n^{1/2}\|\mathbf{X}_n - \boldsymbol{\theta}\| = \|n^{1/2}(\mathbf{X}_n - \boldsymbol{\theta})\|$ and that $n^{1/2}(\mathbf{X}_n - \boldsymbol{\theta}) \xrightarrow{d} \mathbf{Z}$ implies once again that $\|n^{1/2}(\mathbf{X}_n - \boldsymbol{\theta})\| \xrightarrow{d} \|\mathbf{Z}\|$ as $n \to \infty$, as in the proof of Theorem 6.5. It also follows that $\mathbf{h}(\mathbf{X}_n) \xrightarrow{p} \mathbf{h}(\boldsymbol{\theta}) = \mathbf{0}$ as $n \to \infty$ so that Theorem 4.11 (Slutsky) and Theorem 3.9 imply that $n^{1/2}\mathbf{v}'\mathbf{h}(\mathbf{X}_n)\|\mathbf{X}_n - \boldsymbol{\theta}\| \xrightarrow{p} \mathbf{v}'\mathbf{0}\|\mathbf{Z}\| = 0$, as $n \to \infty$ and another application of Theorem 4.11 (Slutsky) implies that

$$n^{1/2}\mathbf{v}'[g(\mathbf{X}_n) - g(\boldsymbol{\theta})] =$$
$$n^{1/2}\mathbf{v}'\mathbf{h}(\mathbf{X}_n)\|\mathbf{X}_n - \boldsymbol{\theta}\| + n^{1/2}\mathbf{v}'\mathbf{D}(\boldsymbol{\theta})(\mathbf{X}_n - \boldsymbol{\theta}) \xrightarrow{d} \mathbf{v}'\mathbf{D}(\boldsymbol{\theta})\mathbf{Z},$$

as $n \to \infty$. Theorem 4.17 (Cramér and Wold) implies then that $n^{1/2}[g(\mathbf{X}_n) - g(\boldsymbol{\theta})] \xrightarrow{d} \mathbf{D}(\boldsymbol{\theta})\mathbf{Z}$ as $n \to \infty$ due to the fact that $\mathbf{v}$ is arbitrary. Therefore, the result is proven. $\square$

Example 6.7. Let $\{X_n\}_{n=1}^{\infty}$ be a sequence of independent and identically distributed random variables from a distribution where $E(|X_n|^6) < \infty$. Consider the sequence of three-dimensional random vectors defined by

$$\mathbf{Y}_n = \begin{bmatrix} n^{-1}\sum_{k=1}^{n} X_k \\ n^{-1}\sum_{k=1}^{n} X_k^2 \\ n^{-1}\sum_{k=1}^{n} X_k^3 \end{bmatrix} = \begin{bmatrix} \hat{\mu}'_1 \\ \hat{\mu}'_2 \\ \hat{\mu}'_3 \end{bmatrix},$$

for all $n \in \mathbb{N}$. Let $\boldsymbol{\mu}' = (\mu'_1, \mu'_2, \mu'_3)$, then Theorem 4.25 implies that $n^{1/2}(\mathbf{Y}_n - \boldsymbol{\mu}) \xrightarrow{d} \mathbf{Z}$ as $n \to \infty$, where $\mathbf{Z}$ has a $\mathbf{N}(\mathbf{0}, \boldsymbol{\Sigma})$ distribution with

$$\boldsymbol{\Sigma} = \begin{bmatrix} \mu'_2 - (\mu'_1)^2 & \mu'_3 - \mu'_1\mu'_2 & \mu'_4 - \mu'_1\mu'_3 \\ \mu'_3 - \mu'_1\mu'_2 & \mu'_4 - (\mu'_2)^2 & \mu'_5 - \mu'_2\mu'_3 \\ \mu'_4 - \mu'_1\mu'_3 & \mu'_5 - \mu'_2\mu'_3 & \mu'_6 - (\mu'_3)^2 \end{bmatrix}.$$

We will consider a function that maps the first three moments to the first three moments about the mean. That is, consider a function

$$g(\mathbf{x}) = \begin{bmatrix} g_1(\mathbf{x}) \\ g_2(\mathbf{x}) \\ g_3(\mathbf{x}) \end{bmatrix} = \begin{bmatrix} x_1 \\ x_2 - x_1^2 \\ x_3 - 3x_1x_2 + 2x_1^3 \end{bmatrix},$$

where $\mathbf{x}' = (x_1, x_2, x_3) \in \mathbb{R}^3$. In this case the matrix $D(\boldsymbol{\theta})$ is given by

$$\mathbf{D}(\boldsymbol{\theta}) = \begin{bmatrix} 1 & 0 & 0 \\ -2\mu'_1 & 1 & 0 \\ -3\mu'_2 + 6(\mu'_1)^2 & -3\mu'_1 & 1 \end{bmatrix},$$

and Theorem 6.6 implies that $n^{1/2}(\mathbf{Y}_n - \boldsymbol{\mu}) \overset{d}{\to} \mathbf{Z}$ as $n \to \infty$ where $\mathbf{Z}$ is a random vector in $\mathbb{R}^3$ that has a $\mathbf{N}[\mathbf{0}, \mathbf{D}(\boldsymbol{\theta})\boldsymbol{\Sigma}\mathbf{D}'(\boldsymbol{\theta})]$ distribution.   ∎

Theorem 6.4 showed how it was possible in some cases to obtain limiting distributions that were related to the CHISQUARED distribution. In these cases the result depended on the fact that the function of the asymptotically normal random variable had a first derivative that vanished at the limiting value, but whose second derivative did not. This is not the only way to find sequences of random variables that have an asymptotic CHISQUARED distribution. For example, suppose that $\{X_n\}_{n=1}^\infty$ is a sequence of random variables such that $X_n \overset{d}{\to} Z$ as $n \to \infty$ where Z is a $N(0,1)$ random variable. Then, Theorem 4.12 implies that $X_n^2 \overset{d}{\to} Z^2$ as $n \to \infty$ where Z^2 has a CHISQUARED(1) distribution. In the multivariate case we extend this result to *quadratic forms* of sequences of random vectors that have a limiting normal distribution. This development is very important to the development of the asymptotic theory of regression analysis and linear models.

Definition 6.2. *Let* $\mathbf{X}$ *be a* d-*dimensional random variable and let* $\mathbf{C}$ *be a* $d \times d$ *symmetric matrix of real values, then* $\mathbf{X}'\mathbf{C}\mathbf{X}$ *is a quadratic form of* $\mathbf{X}$.

It is clear that a quadratic form is a polynomial function of the elements of $\mathbf{X}$, which is a function that is continuous everywhere. Theorem 4.18 then implies that a quadratic form of any sequence of random vectors that converge in distribution to a random vector, converges in distribution to the quadratic form of the limiting random vector.

Theorem 6.7. *Let* $\{\mathbf{X}_n\}_{n=1}^\infty$ *be a sequence of* d-*dimensional random vectors that converge in distribution to a random vector* $\mathbf{X}$ *as* $n \to \infty$. *Let* $\mathbf{C}$ *be a* $d \times d$ *symmetric matrix of real values, then* $\mathbf{X}_n'\mathbf{C}\mathbf{X}_n \overset{d}{\to} \mathbf{X}'\mathbf{C}\mathbf{X}$ *as* $n \to \infty$.

When a random vector $\mathbf{X}$ has a normal distribution, under certain conditions on the covariance matrix of $\mathbf{X}$ and the form of the matrix $\mathbf{C}$ it is possible to obtain a quadratic form that has a non-central CHISQUARED distribution. There are many conditions under which this type of result can be obtained. We will consider one such condition, following along the general development of Serfling (1980). Additional results on quadratic forms can be found in Chapter 1 of Christensen (1996) and Chapter 4 of Graybill (1976).

Theorem 6.8. *Suppose that* $\mathbf{X}$ *has a* d-*dimensional* $\mathbf{N}(\boldsymbol{\mu}, \boldsymbol{\Sigma})$ *distribution and let* $\mathbf{C}$ *be a* $d \times d$ *symmetric matrix. Assume that* $\boldsymbol{\eta}'\boldsymbol{\Sigma} = \mathbf{0}$ *implies that* $\boldsymbol{\eta}'\boldsymbol{\mu} = 0$, *then* $\mathbf{X}'\mathbf{C}\mathbf{X}$ *has a non-central* CHISQUARED *distribution with* trace$(\mathbf{C}\boldsymbol{\Sigma})$ *degrees of freedom and non-centrality parameter equal to* $\boldsymbol{\mu}'\mathbf{C}\boldsymbol{\mu}$ *if and only if* $\boldsymbol{\Sigma}\mathbf{C}\boldsymbol{\Sigma}\mathbf{C}\boldsymbol{\Sigma} = \boldsymbol{\Sigma}\mathbf{C}\boldsymbol{\Sigma}$.

Theorem 6.8 implies that if $\{\mathbf{X}_n\}_{n=1}^\infty$ is a sequence of d-dimensional random vectors that converge to a $\mathbf{N}(\boldsymbol{\mu}, \boldsymbol{\Sigma})$ distribution, then a quadratic form of this sequence will converge to a non-central CHISQUARED distribution as long as the limiting covariance matrix and the form of the quadratic form follow the assumptions outlined in the result.

Example 6.8. Suppose that $\{\mathbf{X}_n\}_{n=1}^{\infty}$ is a sequence of random vectors such that $\mathbf{X}_n$ has a MULTINOMIAL$(n, d, \mathbf{p})$ distribution for all $n \in \mathbb{N}$ where $\mathbf{p}' = (p_1, p_2, \ldots, p_d)$ and $p_k > 0$ for all $k \in \{1, \ldots, d\}$. Note that $\mathbf{X}_n$ can be generated by summing n independent MULTINOMIAL$(1, d, \mathbf{p})$ random variables. That is, we can take

$$\mathbf{X}_n = \sum_{k=1}^{n} \mathbf{D}_k,$$

where $\mathbf{D}_k$ has a MULTINOMIAL$(1, d, \mathbf{p})$ distribution for $k = 1, \ldots, n$ and $\mathbf{D}_1$, $\ldots$, $\mathbf{D}_n$ are mutually independent. Therefore, noting that $E(\mathbf{X}_n) = n\mathbf{p}$, Theorem 4.22 implies that $n^{1/2}\boldsymbol{\Sigma}^{-1/2}(n^{-1}\mathbf{X}_n - \mathbf{p}) \xrightarrow{d} \mathbf{Z}$, as $n \to \infty$ where $\mathbf{Z}$ is a $\mathbf{N}(\mathbf{0}, \mathbf{I})$ random vector, and $\boldsymbol{\Sigma}$ is the covariance matrix of $\mathbf{D}_n$ which has $(i, j)^{\text{th}}$ element given by $p_i(\delta_{ij} - p_j)$ for $i = 1, \ldots, d$ and $j = 1, \ldots, d$, where

$$\delta_{ij} = \begin{cases} 1 & \text{when } i = j, \\ 0 & \text{when } i \neq j. \end{cases}$$

A popular test statistic for testing the null hypothesis that the probability vector is equal to a proposed model $\mathbf{p}$ is given by

$$T_n = \sum_{k=1}^{d} np_k^{-1}(X_{nk} - np_k)^2,$$

where we assume that $\mathbf{X}_n' = (X_{n1}, \ldots, X_{nd})$ for all $n \in \mathbb{N}$. Defining $\mathbf{Y}_n = n^{1/2}(n^{-1}\mathbf{X}_n - \mathbf{p})$ we note that T_n can be written as a quadratic form in terms of $\mathbf{Y}_n$ as $T_n = \mathbf{Y}_n' \mathbf{C} \mathbf{Y}_n$ where

$$\mathbf{C} = \begin{bmatrix} p_1^{-1} & 0 & \cdots & 0 \\ 0 & p_2^{-1} & \cdots & 0 \\ \vdots & \vdots & \ddots & \vdots \\ 0 & 0 & \cdots & p_d^{-1} \end{bmatrix}.$$

Therefore, C has $(i, j)^{\text{th}}$ element $C_{ij} = \delta_{ij} p_i^{-1}$. In order to verify that T_n has an asymptotic non-central CHISQUARED distribution, we first need to verify that for every vector $\boldsymbol{\eta}$ such that $\boldsymbol{\eta}'\boldsymbol{\Sigma} = \mathbf{0}$ then it follows that $\boldsymbol{\eta}'\boldsymbol{\mu} = 0$. In this case the limiting distribution of $\mathbf{Y}_n$ is a $\mathbf{N}(\mathbf{0}, \mathbf{I})$ distribution so that $\boldsymbol{\mu} = \mathbf{0}$. Hence $\boldsymbol{\eta}'\boldsymbol{\mu} = 0$ for all $\boldsymbol{\eta}$ and the property follows. We now need to verify that $\boldsymbol{\Sigma}\mathbf{C}\boldsymbol{\Sigma}\mathbf{C}\boldsymbol{\Sigma} = \boldsymbol{\Sigma}\mathbf{C}\boldsymbol{\Sigma}$. To verify this property we first note that the $(i, j)^{\text{th}}$ element of the product $\mathbf{C}\boldsymbol{\Sigma}$ has the form

$$(\mathbf{C}\boldsymbol{\Sigma})_{ij} = \sum_{k=1}^{d} \delta_{ik} p_i^{-1} p_k (\delta_{kj} - p_j) = \delta_{ii}(\delta_{ij} - p_j) = \delta_{ij} - p_j.$$

This implies that the $(i, j)^{\text{th}}$ element of $\mathbf{C\Sigma C\Sigma}$ is

$$
\begin{aligned}
(\mathbf{C\Sigma C\Sigma})_{ij} &= \sum_{k=1}^{d}(\delta_{ik} - p_k)(\delta_{kj} - p_j) \\
&= \sum_{k=1}^{d}(\delta_{ik}\delta_{kj} - p_k\delta_{kj} - \delta_{ik}p_j + p_kp_j) \\
&= \delta_{ij} - p_j - p_j + p_j\sum_{k=1}^{d}p_k \\
&= \delta_{ij} - p_j - p_j + p_j \\
&= \delta_{ij} - p_j.
\end{aligned}
$$

Thus, it follows that $\mathbf{C\Sigma C\Sigma} = \mathbf{C\Sigma}$ and hence $\mathbf{\Sigma C\Sigma C\Sigma} = \mathbf{\Sigma C\Sigma}$. It then follows from Theorem 6.8 that $T_n \xrightarrow{d} Q$ as $n \to \infty$ where Q is a CHISQUARED (trace($\mathbf{C\Sigma}$), $\boldsymbol{\mu'}\mathbf{C}\boldsymbol{\mu}$) random variable. Noting that

$$
\text{trace}(\mathbf{C\Sigma}) = \sum_{i=1}^{d}(\delta_{ii} - p_i) = \sum_{i=1}^{d}(1 - p_i) = d - 1,
$$

and that $\boldsymbol{\mu'}\mathbf{C}\boldsymbol{\mu} = 0$ implies that the random variable Q has a CHISQUARED($d-1$) distribution. ∎

6.5 Exercises and Experiments

6.5.1 Exercises

1. Prove that Theorem 6.1 (Lindeberg, Lévy, and Feller) reduces to Theorem 4.20 (Lindeberg and Lévy) when $\{X_n\}_{n=1}^{\infty}$ is a sequence of independent and identically distributed random variables.

2. Let $\{X_n\}_{n=1}^{\infty}$ be a sequence of independent random variables where X_n has a GAMMA($\theta_n, 2$) distribution where $\{\theta_n\}_{n=1}^{\infty}$ is a sequence of positive real numbers.

   a. Find a non-trivial sequence $\{\theta_n\}_{n=1}^{\infty}$ such that the assumptions of Theorem 6.1 (Lindeberg, Lévy, and Feller) hold, and describe the resulting conclusion for the weak convergence of $\bar{X}_n$.

   b. Find a non-trivial sequence $\{\theta_n\}_{n=1}^{\infty}$ such that the assumptions of Theorem 6.1 (Lindeberg, Lévy, and Feller) do not hold.

3. Let $\{X_n\}_{n=1}^{\infty}$ be a sequence of independent random variables where X_n has a BERNOULLI(θ_n) distribution where $\{\theta_n\}_{n=1}^{\infty}$ is a sequence of real numbers. Find a non-trivial sequence $\{\theta_n\}_{n=1}^{\infty}$ such that the assumptions of Theorem 6.1 (Lindeberg, Lévy, and Feller) hold, and describe the resulting conclusion for the weak convergence of $\bar{X}_n$.

4. In the context of Theorem 6.1 (Lindeberg, Lévy, and Feller), prove that Equation (6.3) implies Equation (6.2).

5. Prove Corollary 6.2. That is, let $\{\{X_{nk}\}_{k=1}^{n}\}_{n=1}^{\infty}$ be a triangular array where $X_{11}, \ldots, X_{n1}$ are mutually independent random variables for each $n \in \mathbb{N}$. Suppose that X_{nk} has mean μ_{nk} and variance σ_{nk}^2 for all $k \in \{1, \ldots, n\}$ and $n \in \mathbb{N}$. Let

$$\mu_{n\cdot} = \sum_{k=1}^{n} \mu_{nk},$$

and

$$\sigma_{n\cdot}^2 = \sum_{k=1}^{n} \sigma_{nk}^2.$$

Suppose that for some $\eta > 2$

$$\sum_{k=1}^{n} E(|X_{nk} - \mu_{nk}|^{\eta}) = o(\sigma_{n\cdot}^{\eta}),$$

as $n \to \infty$, then

$$Z_n = \sigma_{n\cdot}^{-1} \left(\sum_{k=1}^{n} X_{nk} - \mu_{n\cdot} \right) \xrightarrow{d} Z,$$

as $n \to \infty$, where Z is a $N(0,1)$ random variable.

6. Let $\{\{X_{n,k}\}_{k=1}^{n}\}_{n=1}^{\infty}$ be a triangular array of random variables where $X_{n,k}$ has a GAMMA$(\theta_{n,k}, 2)$ distribution where $\{\{\theta_{n,k}\}_{k=1}^{n}\}_{n=1}^{\infty}$ is a triangular array of positive real numbers. Find a non-trivial triangular array of the form $\{\{\theta_{n,k}\}_{k=1}^{n}\}_{n=1}^{\infty}$ such that the assumptions of Theorem 6.2 hold and describe the resulting conclusion for the weak convergence of

$$\sum_{k=1}^{n} X_{nk}.$$

7. Let $\{\{X_{n,k}\}_{k=1}^{n}\}_{n=1}^{\infty}$ be a triangular array of random variables where $X_{n,k}$ has a BERNOULLI$(\theta_{n,k})$ distribution where $\{\{\theta_{n,k}\}_{k=1}^{n}\}_{n=1}^{\infty}$ is a triangular array of real numbers that are between zero and one. Find a non-trivial triangular array $\{\{\theta_{n,k}\}_{k=1}^{n}\}_{n=1}^{\infty}$ such that the assumptions of Theorem 6.2 hold and describe the resulting conclusion for the weak convergence of

$$\sum_{k=1}^{n} X_{nk}.$$

8. Prove Theorem 6.4. That is, suppose that $\{X_n\}_{n=1}^{\infty}$ be a sequence of random variables such that $\sigma_n^{-1}(X_n - \mu) \xrightarrow{d} Z$ as $n \to \infty$ where Z is a $N(0,1)$ random variable and $\{\sigma_n\}_{n=1}^{\infty}$ is a real sequence such that

$$\lim_{n \to \infty} \sigma_n = 0.$$

Let g be a real function such that

$$\left. \frac{d^m}{dx^m} g(x) \right|_{x=\mu} \neq 0,$$

for some $m \in \mathbb{N}$ where

$$\left. \frac{d^k}{dx^k} g(x) \right|_{x=\mu} = 0,$$

for all $k \in \{1, \ldots, m-1\}$. Then prove that

$$\frac{m![g(X_n) - g(\mu)]}{\sigma_n^m g'(\mu)} \xrightarrow{d} Z^m$$

as $n \to \infty$.

9. Let $\{X_n\}_{n=1}^{\infty}$ be a sequence of random variables such that $\sigma_n^{-1}(X_n - \mu) \xrightarrow{d} Z$ as $n \to \infty$ where Z is a $\mathrm{N}(0,1)$ random variable and $\{\sigma_n\}_{n=1}^{\infty}$ is a real sequence such that

$$\lim_{n \to \infty} \sigma_n = 0.$$

Consider the transformation $g(x) = ax + b$ where a and b are known real constants, and $a \neq 0$. Derive the asymptotic behavior, normal or otherwise, of $g(X_n)$ as $n \to \infty$.

10. Let $\{X_n\}_{n=1}^{\infty}$ be a sequence of random variables such that $\sigma_n^{-1}(X_n - \mu) \xrightarrow{d} Z$ as $n \to \infty$ where Z is a $\mathrm{N}(0,1)$ random variable and $\{\sigma_n\}_{n=1}^{\infty}$ is a real sequence such that

$$\lim_{n \to \infty} \sigma_n = 0.$$

Consider the transformation $g(x) = x^3$.

   a. Suppose that $\mu > 0$. Derive the asymptotic behavior, normal or otherwise, of $g(X_n)$ as $n \to \infty$.

   b. Suppose that $\mu = 0$. Derive the asymptotic behavior, normal or otherwise, of $g(X_n)$ as $n \to \infty$.

11. Let $\{X_n\}_{n=1}$ be a set of independent and identically distributed random variables from a distribution with me μ and finite variance σ^2. Show that

$$S^2 = n^{-1} \sum_{k=1}^{n} (X_k - \bar{X})^2 = n^{-1} \sum_{k=1}^{n} (X_k - \mu)^2 + R,$$

where $R \xrightarrow{p} 0$ as $n \to \infty$.

12. In Example 6.6, find Λ and $\mathbf{d}'(\theta)\Lambda\mathbf{d}(\theta)$.

13. Let $\{\mathbf{X}_n\}$ be a sequence of d-dimensional random vectors where $\mathbf{X}_n \xrightarrow{d} \mathbf{Z}$ as $n \to \infty$ where $\mathbf{Z}$ has a $\mathbf{N}(\mathbf{0}, \mathbf{I})$ distribution. Let $\mathbf{A}$ be a $m \times d$ matrix and find the asymptotic distribution of the sequence $\{\mathbf{A}\mathbf{X}_n\}_{n=1}^{\infty}$ as $n \to \infty$. Describe any additional assumptions that may need to be made for the matrix $\mathbf{A}$.

14. Let $\{\mathbf{X}_n\}$ be a sequence of d-dimensional random vectors where $\mathbf{X}_n \xrightarrow{d} \mathbf{Z}$ as $n \to \infty$ where $\mathbf{Z}$ has a $\mathbf{N}(\mathbf{0}, \mathbf{I})$ distribution. Let $\mathbf{A}$ be a $m \times d$ matrix and let $\mathbf{b}$ be a $m \times 1$ real valued vector. Fnd the asymptotic distribution of the sequence $\{\mathbf{A}\mathbf{X}_n + \mathbf{b}\}_{n=1}^{\infty}$ as $n \to \infty$. Describe any additional assumptions that may need to be made for the matrix $\mathbf{A}$ and the vector $\mathbf{b}$. What effect does adding the vector $\mathbf{b}$ have on the asymptotic result?

15. Let $\{\mathbf{X}_n\}$ be a sequence of d-dimensional random vectors where $\mathbf{X}_n \xrightarrow{d} \mathbf{Z}$ as $n \to \infty$ where $\mathbf{Z}$ has a $\mathbf{N}(\mathbf{0}, \mathbf{I})$ distribution. Let $\mathbf{A}$ be a symmetric $d \times d$ matrix and find the asymptotic distribution of the sequence $\{\mathbf{X}_n' \mathbf{A} \mathbf{X}_n\}_{n=1}^{\infty}$ as $n \to \infty$. Describe any additional assumptions that need to be made for the matrix $\mathbf{A}$.

16. Let $\{\mathbf{X}_n\}_{n=1}^{\infty}$ be a sequence of two-dimensional random vectors where $\mathbf{X}_n \xrightarrow{d} \mathbf{Z}$ as $n \to \infty$ where $\mathbf{Z}$ has a $\mathbf{N}(\mathbf{0}, \mathbf{I})$ distribution. Consider the transformation $g(\mathbf{x}) = x_1 + x_2 + x_1 x_2$ where $\mathbf{x}' = (x_1, x_2)$. Find the asymptotic distribution of $g(\mathbf{X}_n)$ as $n \to \infty$.

17. Let $\{\mathbf{X}_n\}_{n=1}^{\infty}$ be a sequence of three-dimensional random vectors where $\mathbf{X}_n \xrightarrow{d} \mathbf{Z}$ as $n \to \infty$ where $\mathbf{Z}$ has a $\mathbf{N}(\mathbf{0}, \mathbf{I})$ distribution. Consider the transformation $\mathbf{g}(\mathbf{x}) = [x_1 x_2 + x_3, x_1 x_3 + x_2, x_2 x_3 + x_1]$ where $\mathbf{x}' = (x_1, x_2, x_3)$. Find the asymptotic distribution of $g(\mathbf{X}_n)$ as $n \to \infty$.

6.5.2 Experiments

1. Write a program in R that simulates 1000 samples of size n from an EX-PONENTIAL(1) distribution. On each sample compute $n^{1/2}(\bar{X}_n - 1)$ and $n^{1/2}(\bar{X}_n^2 - 1)$. Make a density histogram of the 1000 values of $n^{1/2}(\bar{X}_n - 1)$, and on a separate plot make a histogram with the same scale of the 1000 values of $n^{1/2}(\bar{X}_n^2 - 1)$. Compare the variability of the two histograms with what is predicted by theory, and describe how the transformation changes the variability of the sequences. Repeat the experiment for $n = 5, 10, 25, 100$ and 500 and describe how both sequences converge to a NORMAL distribution.

2. Write a program in R that simulates 1000 observations from a MULTINO-MIAL$(n, 3, \mathbf{p})$ distribution where $\mathbf{p}' = (\frac{1}{4}, \frac{1}{4}, \frac{1}{2})$. On each observation compute $T_n = \sum_{k=1}^{3} n p_k^{-1}(X_{nk} - n p_k)^2$ where $\mathbf{X}' = (X_{n1}, X_{n2}, X_{n3})$. Make a density histogram of the 1000 values of T_n and overlay the plot with a plot of CHISQUARED(2) distribution for comparison. Repeat the experiment for $n = 5, 10, 25, 100$ and 500 and describe how both sequences converge to a CHISQUARED(2) distribution.

3. Write a program in R that simulates 1000 sequences of independent random variables of length n where the kth variable in the sequence has an EXPONENTIAL(θ_k) distribution where $\theta_k = k^{-1/2}$ for all $k \in \mathbb{N}$. For each

simulated sequence, compute $Z_n = n^{1/2}\tau_n^{-1}(\bar{X}_n - \bar{\mu}_n)$, where

$$\tau_k^2 = \sum_{i=1}^{k} k^{-1}.$$

Plot the 1000 values of Z_n on a density histogram and overlay the histogram with a plot of a $N(0,1)$ density. Repeat the experiment for $n = 5$, 10, 25, 100 and 500 and describe how the distribution converges.

4. Write a program in R that simulates 1000 samples of size n from a UNI-FORM(θ_1, θ_2) distribution where n, θ_1, and θ_2 are specified below. For each sample compute $Z_n = n^{1/2}\sigma^{-1}(\bar{X}_n^2 - \mu^2)$ where $\bar{X}_n$ is the mean of the observed sample and μ and σ correspond to the mean and standard deviation of a UNIFORM(θ_1, θ_2) distribution. Plot a histogram of the 1000 observed values of Z_n for each case listed below and compare the shape of the histograms to what would be expected.

a. $\theta_1 = -1$, $\theta_2 = 1$.
b. $\theta_1 = 0$, $\theta_2 = 1$.

CHAPTER 7

Asymptotic Expansions for Distributions

But when he talks about them and compares them with himself and his colleagues there's a small error running through what he says, and, just for your interest, I'll tell you about it.

The Trial by Franz Kafka

7.1 Approximating a Distribution

Let us consider the case of a sequence of independent and identically distributed random variables $\{X_n\}_{n=1}^{\infty}$ from a distribution F whose mean is μ and variance is σ^2. Assume that $E(|X_n|^3) < \infty$. Then Theorem 4.20 (Lindeberg and Lévy) implies that $Z_n = n^{1/2}\sigma^{-1}(\bar{X}_n - \mu) \xrightarrow{d} Z$ as $n \to \infty$ where Z has a $N(0,1)$ distribution. This implies that

$$\lim_{n\to\infty} P(Z_n \le t) = \Phi(t),$$

for all $t \in \mathbb{R}$. Define $R_n(t) = \Phi(t) - P(Z_n \le t)$ for all $t \in \mathbb{R}$ and $n \in \mathbb{N}$. This in turn implies that $P(Z_n \le t) = \Phi(t) + R_n(t)$. Theorem 4.24 (Berry and Esseen) implies that $|R_n(t)| \le n^{-1/2}B\rho$ where B is a finite constant that does not depend on n or t and ρ is the third absolute moment about the mean of F. Noting that ρ also does not depend on n and t, we have that

$$\lim_{n\to\infty} n^{1/2}|R_n(t)| \le B\rho,$$

uniformly in t. Therefore, Definition 1.7 implies that $R_n(t) = O(n^{-1/2})$ and we obtain the asymptotic expansion $P(Z_n \le t) = \Phi(t) + O(n^{-1/2})$ as $n \to \infty$. Similar arguments also lead to the alternate expansion $P(Z_n \le t) = \Phi(t) + o(1)$ as $n \to \infty$.

The purpose of this chapter is to extend this idea to higher order expansions. Our focus will be on distributions that are asymptotically normal via Theorem 4.20 (Lindeberg and Lévy). In the case of distributions of a sample mean, it is possible to obtain an asymptotic expansion for the density or distribution function that has an error term that is $O(n^{-(p+1)/2})$ or $o(n^{-p/2})$ as $n \to \infty$ with the addition of several assumptions.

In this chapter we will specifically consider the case of obtaining an asymptotic expansion that has an error term that is $o(n^{-1/2})$ as $n \to \infty$. We will also consider inverting this expansion to obtain an asymptotic expansion for the quantile function whose error term is also $o(n^{-1/2})$ as $n \to \infty$. These results can be extended beyond the sample mean to smooth functions of vector means through what is usually known as the smooth function model and we will explore both this model and the corresponding expansion theory. This chapter concludes with a brief description of saddlepoint expansions which are designed to provide more accuracy to these approximations under certain conditions.

7.2 Edgeworth Expansions

An *Edgeworth expansion* is an asymptotic expansion for the standardized distribution of a sample mean. In essence, the expansion improves the accuracy of Theorem 4.20 (Lindeberg and Lévy) under additional assumptions on the finiteness of the moments and the smoothness of the underlying distribution of the population. Some key elements of Edgeworth expansions are that the error term is uniform over the real line and that the terms of the expansion depend on the moments of the underlying distribution of the data. The fact that the terms of the expansion depend on the moments of the population implies that we are able to observe what properties of a population affect the accuracy of Theorem 4.20.

The historical development of the Edgeworth expansions begins with the work of P. Tchebycheff and F. Y. Edgeworth. Specifically, one can refer to Tchebycheff (1890) and Edgeworth (1896, 1905). In both cases the distribution function was the focus of the work, and both worked with sums of independent random variables as we do in this section. A rigorous treatment of these expansions was considered by Cramér (1928). Chapter VII of Cramér (1946) also addresses this subject. Similar types of expansions, known as Gram–Charlier and Brun–Charlier expansions, can be seen as the basis of the work of Tchebycheff and Edgeworth, though these expansions have less developed convergence properties. See Cramér (1946,1970), Edgeworth (1907), and Johnson, Kotz, and Balakrishnan (1994) for further details.

In this section we will explicitly derive a one-term Edgeworth expansion, which has the form $\phi(t) + \frac{1}{6}\sigma^{-3}n^{-1/2}\mu_3(t^3 - 3t)\phi(t)$, and we will demonstrate that the error associated with this expansion is $o(n^{-1/2})$ as $n \to \infty$. This result provides an approximation for the standardized distribution of a sample mean that provides a more accurate approximation than what is given by Theorem 4.20 (Lindeberg and Lévy). We provide a detailed proof of this result, relying on the method of Feller (1971) for our arguments. The main idea of the proof is based on the inversion formula for characteristic functions, given by Theorem 2.28. Let $\{\psi_n(t)\}_{n=1}^{\infty}$ be a sequence of integrable characteristic functions and let $\psi(t)$ be another integrable characteristic function such that

$$\lim_{n\to\infty} \int_{-\infty}^{\infty} |\psi_n(t) - \psi(t)|dt = 0.$$

If F_n is the distribution associated with ψ_n for all $n \in \mathbb{N}$ and F is the distribution function associate with ψ then Theorem 2.28 implies that F_n has a bounded and continuous density f_n for all $n \in \mathbb{N}$ and that F has a bounded and continuous density f. Further, Theorem 2.28 implies that

$$
\begin{aligned}
|f_n(x) - f(x)| &= \left| (2\pi)^{-1} \int_{-\infty}^{\infty} \exp(-itx)\psi_n(t)dt - \right.\\
&\qquad \left. (2\pi)^{-1} \int_{-\infty}^{\infty} \exp(-itx)\psi(t)dt \right|\\
&= \left| (2\pi)^{-1} \int_{-\infty}^{\infty} \exp(-itx)[\psi_n(t) - \psi(t)]dt \right|\\
&\leq (2\pi)^{-1} \int_{-\infty}^{\infty} |\exp(-itx)||\psi_n(t) - \psi(t)|dt,
\end{aligned}
$$

where the inequality comes from an application of Theorem A.6. Noting that $|\exp(-itx)| \leq 1$, it then follows that

$$|f_n(x) - f(x)| \leq (2\pi)^{-1} \int_{-\infty}^{\infty} |\psi_n(t) - \psi(t)|dt. \qquad (7.1)$$

For further details see Section XV.3 of Feller (1971). The general method of proof for determining the error associated with the one-term Edgeworth expansion is based on computing the integral of the distance between the characteristic function of the standardized density of the sample mean and the characteristic function of the approximating expansion, which then bounds the difference between the corresponding densities. It is important to note that the bound given in Equation (7.1) is uniform in x, a property that will translate to the error term of the Edgeworth expansions.

A slight technicality should be addressed at this point. The expansion $\phi(t) + \frac{1}{6}\sigma^{-3}n^{-1/2}\mu_3'(t^3 - 3t)\phi(t)$ is typically not a valid density function as it usually does not integrate to one and is not always non-negative. In this case we are really computing a Fourier transformation on the expansion. Under the assumptions we shall impose it follows that the Fourier inversion theorem still works as detailed in Theorem 2.28 with the exception that the Fourier transformation of the expansion may not strictly be a valid characteristic function. See Theorem 4.1 of Bhattacharya and Rao (1976). The arguments producing the bound in Equation (7.1) also follow, with the right hand side being called the *Fourier norm* by Feller (1971). See Exercise 1.

Another technical matter arises when taking the Fourier transformation of the expansion. The second term in the expansion is given by a constant multiplied by $H_3(x)\phi(x)$. It is therefore convenient to have a mechanism for computing the Fourier transformation of a function of this type.

Theorem 7.1. *The Fourier transformation of* $H_k(x)\phi(x)$ *is* $(it)^k \exp(-\frac{1}{2}t^2)$.

Proof. We will prove one specific case of this result. For a proof of the general result see Exercise 2. In this particular case we will evaluate the integral

$$\int_{-\infty}^{\infty} \phi^{(3)}(x)\exp(-itx)dx,$$

which from Definition 1.6 is the Fourier transformation of $-H_3(x)\phi(x)$. Using Theorem A.4 with $u = \exp(-itx)$, $v = \phi^{(2)}(x)$, $du = -it\exp(-itx)$, and $dv = \phi^{(3)}(t)$, we have that

$$\int_{-\infty}^{\infty} \phi^{(3)}(x)\exp(-itx)dx =$$

$$\exp(-itx)\phi^{(2)}(x)\Big|_{-\infty}^{\infty} + (it)\int_{-\infty}^{\infty} \phi^{(2)}(x)\exp(-itx)dx. \quad (7.2)$$

Since $u = \exp(-itx)$ is a bounded function it follows that by taking the appropriate limits of $\phi^{(2)}(x)$ that the first term of Equation (7.2) is zero. To evaluate the second term in Equation (7.2) we use another application of Theorem A.4 to show that

$$(it)\int_{-\infty}^{\infty} \phi^{(2)}(x)\exp(-itx)dx = (it)^2 \int_{-\infty}^{\infty} \phi^{(1)}(x)\exp(-itx)dx,$$

for which yet another application of Theorem A.4 implies that

$$(it)^2 \int_{-\infty}^{\infty} \phi^{(1)}(x)\exp(-itx)dx = (it)^3 \exp(-\tfrac{1}{2}t^2).$$

This proves the result for the special case when $k = 3$. $\qquad\qquad\square$

The development of the Edgeworth expansion is also heavily dependent on the properties of characteristic functions. We first require a bound for the characteristic function for continuous random variables. The result given below provides a somewhat more general result that characterizes the behavior of characteristic functions in relation to the properties of F.

Theorem 7.2. *Let X be a random variable with distribution F and characteristic function ψ. Then either*

1. $|\psi(t)| < 1$ *for all $t \neq 0$, or*
2. $|\psi(u)| = 1$ *and $|\psi(t)| < 1$ for $t \in (0, u)$ for which the values of X are concentrated on a regular grid, or*
3. $|\psi(t)| = 1$ *for all $t \in \mathbb{R}$, for which X is concentrated at a single point.*

A more specific result concerning the nature of the random variable and the corresponding characteristic function for the second condition is addressed in Theorem 7.6, which is presented later in this section. Further discussion about Theorem 7.2 and its proof can be found in Section XV.1 of Feller (1971). We will also require some characterization about the asymptotic tail behavior of characteristic functions.

Theorem 7.3 (Riemann and Lebesgue). *Suppose that X is an absolutely continuous random variable with characteristic function ψ. Then,*

$$\lim_{t \to \pm\infty} |\psi(t)| = 0.$$

A proof of Theorem 7.3 can be found in Section 4.1 of Gut (2005). We now have a collection of theory that is suitable for determining the asymptotic characteristics of the error term of the Edgeworth expansion.

Theorem 7.4. *Let $\{X_n\}_{n=1}^{\infty}$ be a sequence of independent and identically distributed random variables where X_n has characteristic function ψ for all $n \in \mathbb{N}$. Let $F_n(x) = P[n^{1/2}\sigma^{-1}(\bar{X}_n - \mu) \le x]$, with density $f_n(x)$ for all $x \in \mathbb{R}$ where $\mu = E(X_n)$ and $\sigma^2 = V(X_n)$. Assume that $E(X_n^3) < \infty$, $|\psi|^{\nu}$ is integrable for some $\nu \ge 1$, and that $f_n(x)$ exists for $n \ge \nu$. Then,*

$$f_n(x) - \phi(x) - \tfrac{1}{6}\sigma^{-3}n^{-1/2}\mu_3(x^3 - 3x)\phi(x) = o(n^{-1/2}), \qquad (7.3)$$

as $n \to \infty$.

Proof. As usual, we will consider without loss of generality the case where $\mu = 0$. Let us first consider computing the Fourier transform of the function

$$f_n(x) - \phi(x) - \tfrac{1}{6}\sigma^{-3}n^{-1/2}\mu_3(x^3 - 3x)\phi(x). \qquad (7.4)$$

Because the Fourier transform is an integral, we can accomplish this with term-by-term integration. Previous calculations using Theorems 2.32 and 2.33 imply that the Fourier transform of $f_n(x)$ can be written as $\psi^n(t\sigma^{-1}n^{-1/2})$. Similarly, it is also known that the characteristic function of $\phi(x)$ is $\exp(-\tfrac{1}{2}t^2)$. Finally, we require the Fourier transform of $-\tfrac{1}{6}\sigma^{-3}n^{-1/2}\mu_3(x^3 - 3x)\phi(x)$. To simplify this matter we note that by Definition 1.6, the third Hermite polynomial is given by $H_3(x) = x^3 - 3x$ and, therefore, Definition 1.6 implies that

$$-(x^3 - 3x)\phi(t) = (-1)^3 H_3(x)\phi(x) = \phi^{(3)}(x).$$

Theorem 7.1 therefore implies that the Fourier transformation of $-H_3(x)\phi(x) = \phi^{(3)}(x)$ is $(it)^3 \exp(-\tfrac{1}{2}t^2)$. Therefore, the Fourier transformation of Equation (7.4) is given by

$$\psi^n(n^{-1/2}\sigma^{-1}t) - \exp(-\tfrac{1}{2}t^2) - \tfrac{1}{6}\mu_3\sigma^{-3}n^{-1/2}(it)^3 \exp(-\tfrac{1}{2}t^2).$$

Now, using the Fourier norm of Equation (7.1), we have that

$$|f_n(x) - \phi(x) - \tfrac{1}{6}\sigma^{-3}n^{-1/2}\mu_3(x^3 - 3x)\phi(x)| \le$$

$$\int_{-\infty}^{\infty} |\psi^n(n^{-1/2}\sigma^{-1}t) - \exp(-\tfrac{1}{2}t^2) - \tfrac{1}{6}\mu_3\sigma^{-3}n^{-1/2}(it)^3 \exp(-\tfrac{1}{2}t^2)|dt. \quad (7.5)$$

Our task now is to show that the integral of the right hand side of Equation (7.5) is $o(n^{-1/2})$ as $n \to \infty$. Let $\delta > 0$ and note that since ψ is a characteristic function of a density we have from Theorem 7.2 that $|\psi(t)| \le 1$ for all $t \in \mathbb{R}$ and that $|\psi(t)| < 1$ for all $t \ne 0$. This result, combined with the result of

Theorem 7.3 (Riemann and Lebesgue) implies that there is a real number q_δ such that $|\psi(t)| \le q_\delta < 1$ for all $|t| \ge \delta$. We now begin approximating the Fourier norm in Equation (7.5). We begin by breaking up the Fourier norm in Equation (7.5) into two integrals: the first over an interval near the origin and the second over the remaining tails. That is,

$$\int_{-\infty}^{\infty} |\psi^n(n^{-1/2}\sigma^{-1}t) - \exp(-\tfrac{1}{2}t^2) - \tfrac{1}{6}\mu_3\sigma^{-3}n^{-1/2}(it)^3 \exp(-\tfrac{1}{2}t^2)|dt =$$

$$\int_{|t|>\delta\sigma n^{1/2}} |\psi^n(n^{-1/2}\sigma^{-1}t) - \exp(-\tfrac{1}{2}t^2) - \tfrac{1}{6}\mu_3\sigma^{-3}n^{-1/2}(it)^3 \exp(-\tfrac{1}{2}t^2)|dt +$$

$$\int_{|t|<\delta\sigma n^{1/2}} |\psi^n(n^{-1/2}\sigma^{-1}t) - \exp(-\tfrac{1}{2}t^2) - \tfrac{1}{6}\mu_3\sigma^{-3}n^{-1/2}(it)^3 \exp(-\tfrac{1}{2}t^2)|dt.$$

$$(7.6)$$

Working with the first term on the right hand side of Equation (7.6), we note that Theorem A.18 implies that

$$\int_{|t|>\delta\sigma n^{1/2}} |\psi^n(n^{-1/2}\sigma^{-1}t) - \exp(-\tfrac{1}{2}t^2) - \tfrac{1}{6}\mu_3\sigma^{-3}n^{-1/2}(it)^3 \exp(-\tfrac{1}{2}t^2)|dt \le$$

$$\int_{|t|>\delta\sigma n^{1/2}} |\psi^n(n^{-1/2}\sigma^{-1}t)| + \exp(-\tfrac{1}{2}t^2)(1 + |\tfrac{1}{6}\mu_3\sigma^{-3}n^{-1/2}(it)^3|)dt =$$

$$\int_{|t|>\delta\sigma n^{1/2}} |\psi^n(n^{-1/2}\sigma^{-1}t)|dt +$$

$$\int_{|t|>\delta\sigma n^{1/2}} \exp(-\tfrac{1}{2}t^2)(1 + |\tfrac{1}{6}\mu_3\sigma^{-3}n^{-1/2}(it)^3|)dt. \quad (7.7)$$

For the first integral on the right hand side of Equation (7.7), we note that

$$\psi^n(n^{-1/2}\sigma^{-1}t) = \psi^\nu(n^{-1/2}\sigma^{-1}t)\psi^{n-\nu}(n^{-1/2}\sigma^{-1}t).$$

Now $|t| > \delta\sigma n^{1/2}$ implies that $n^{-1/2}\sigma^{-1}|t| > \delta$ so the fact that $|\psi(t)| \le q_\delta$ for all $|t| \ge \delta$ implies that $\psi^{n-\nu}(n^{-1/2}\sigma^{-1}t) \le q_\delta^{n-\nu}$. Hence, it follows from Theorem A.7 that

$$\int_{|t|>\delta\sigma n^{1/2}} |\psi^n(n^{-1/2}\sigma^{-1}t)|dt \le q_\delta^{n-\nu} \int_{|t|>\delta\sigma n^{1/2}} |\psi^\nu(n^{-1/2}\sigma^{-1}t)|dt$$

$$\le q_\delta^{n-\nu} \int_{-\infty}^{\infty} |\psi^\nu(n^{-1/2}\sigma^{-1}t)|dt \quad (7.8)$$

$$= n^{1/2}\sigma q_\delta^{n-\nu} \int_{-\infty}^{\infty} |\psi^\nu(u)|du. \quad (7.9)$$

Equation (7.9) follows from a change of variable in the integral in Equation (7.8). It is worthwhile to note the importance of the method used in establishing Equation (7.8). We cannot assume that $|\psi|^n$ is integrable for every $n \in \mathbb{N}$ since we have only assumed that $|\psi|^\nu$ is integrable. By bounding $|\psi|^{n-\nu}$ separately we are able then to take advantage of the integrability of $|\psi|^\nu$. Now, using the integrability of $|\psi|^\nu$, we note that

$$nq_\delta^{n-\nu}\sigma n^{1/2}\int_{-\infty}^{\infty}|\psi^\nu(u)|du = n^{3/2}q_\delta^{n-\nu}C,$$

where C is a constant that does not depend on n. Noting that $|q_\delta| < 1$, we have that

$$\lim_{n\to\infty} n^{3/2}q_\delta^{n-\nu}C = 0.$$

Therefore, using Definition 1.7, we have that

$$q_\delta^{n-\nu}\int_{-\infty}^{\infty}|\psi^\nu(n^{-1/2}\sigma^{-1}t)|dt = o(n^{-1}),$$

as $n \to \infty$. For the second integral on the right hand side of Equation (7.7) we have that

$$\int_{|t|>\delta\sigma n^{1/2}}\exp(-\tfrac{1}{2}t^2)(1+|\tfrac{1}{6}\mu_3\sigma^{-3}n^{-1/2}(it)^3|)dt \le$$

$$\int_{|t|>\delta\sigma n^{1/2}}\exp(-\tfrac{1}{2}t^2)(1+|\mu_3\sigma^{-3}t^3|)dt =$$

$$\int_{|t|>\delta\sigma n^{1/2}}\exp(-\tfrac{1}{2}t^2)dt + \int_{|t|>\delta\sigma n^{1/2}}\exp(-\tfrac{1}{2}t^2)|\mu_3\sigma^{-3}t^3|dt, \quad (7.10)$$

since $|\tfrac{1}{6}n^{-1/2}i^3| < 1$ for all $n \in \mathbb{N}$. Now

$$\int_{|t|>\delta\sigma n^{1/2}}\exp(-\tfrac{1}{2}t^2)dt = (2\pi)^{1/2}\int_{|t|>\delta\sigma n^{1/2}}(2\pi)^{-1/2}\exp(-\tfrac{1}{2}t^2)dt =$$

$$(2\pi)^{1/2}P(|Z| > \delta\sigma n^{1/2}),$$

where Z is a N$(0,1)$ random variable. Theorem 2.6 (Markov) implies that $P(|Z| > \delta\sigma n^{1/2}) \le (\delta\sigma n^{1/2})^{-3}E(|Z|^3)$. Therefore, we have that

$$\int_{|t|>\delta\sigma n^{1/2}}\exp(-\tfrac{1}{2}t^2)dt \le (2\pi)^{1/2}(\delta\sigma n^{1/2})^{-3}E(|Z|^3),$$

and hence,

$$\lim_{n\to\infty} n\int_{|t|>\delta\sigma n^{1/2}}\exp(-\tfrac{1}{2}t^2)dt \le \lim_{n\to\infty} n^{-1/2}(2\pi)^{1/2}(\delta\sigma)^{-3}E(|Z|^3) = 0.$$

Therefore, using Definition 1.7, we have proven that

$$\int_{|t|>\delta\sigma n^{1/2}}\exp(-\tfrac{1}{2}t^2)dt = o(n^{-1}),$$

as $n \to \infty$. For the second integral on the right hand side of Equation (7.10) we note that

$$\int_{|t|>\delta\sigma n^{1/2}} \exp(-\tfrac{1}{2}t^2)|\mu_3\sigma^{-3}t^3|dt = |\mu_3\sigma^{-3}|\int_{|t|>\delta\sigma n^{1/2}} |t|^3\exp(-\tfrac{1}{2}t^2)dt$$

$$= 2|\mu_3\sigma^{-3}|\int_{t>\delta\sigma n^{1/2}} t^3\exp(-\tfrac{1}{2}t^2)dt. \quad (7.11)$$

Note that the final equality in Equation (7.11) follows from the fact that the integrand is an even function. Now consider a change of variable where $u = \tfrac{1}{2}t^2$ so that $du = tdt$ and the lower limit of the integral becomes $\tfrac{1}{2}\delta^2\sigma^2 n$. Hence, we have that

$$\int_{|t|>\delta\sigma n^{1/2}} \exp(-\tfrac{1}{2}t^2)|\mu_3\sigma^{-3}t^3|dt = 4|\mu_3\sigma^{-3}|\int_{\frac{1}{2}\delta^2\sigma^2 n}^{\infty} u\exp(-u)du.$$

Now use Theorem A.4 with $dv = \exp(-u)$ so that $v = -\exp(-u)$ to find that

$$4|\mu_3\sigma^{-3}|\int_{\frac{1}{2}\delta^2\sigma^2 n}^{\infty} u\exp(-u)du =$$

$$-4|\mu_3\sigma^{-3}|u\exp(-u)\Big|_{\frac{1}{2}\delta^2\sigma^2 n}^{\infty} + 4|\mu_3\sigma^{-3}|\int_{\frac{1}{2}\delta^2\sigma^2 n}^{\infty} \exp(-u)du =$$

$$2|\mu_3\sigma^{-3}|\delta^2\sigma^2 n\exp(-\tfrac{1}{2}\delta^2\sigma^2 n) + 4|\mu_3\sigma^{-3}|\exp(-\tfrac{1}{2}\delta^2\sigma^2 n), \quad (7.12)$$

where we have used the fact that

$$\lim_{n\to\infty} n^k\exp(-n) = 0,$$

for all fixed $k \in \mathbb{N}$. Therefore, using the same property, we can also conclude using Definition 1.7 that

$$\int_{|t|>\delta\sigma n^{1/2}} \exp(-\tfrac{1}{2}t^2)|\mu_3\sigma^{-3}t^3|dt = o(n^{-1}),$$

as $n \to \infty$. Therefore, it follows that

$$(2\pi)^{-1}\int_{-\infty}^{\infty} |\psi^n(t\sigma^{-1}n^{-1/2}) - \exp(-\tfrac{1}{2}t^2) - \tfrac{1}{6}n^{-1/2}\sigma^{-3}\mu_3(it)^3\exp(-\tfrac{1}{2}t^2)|dt =$$

$$(2\pi)^{-1}\int_{|t|<\delta\sigma n^{1/2}} |\psi^n(t\sigma^{-1}n^{-1/2}) - \exp(-\tfrac{1}{2}t^2) -$$

$$\tfrac{1}{6}n^{-1/2}\sigma^{-3}\mu_3(it)^3\exp(-\tfrac{1}{2}t^2)|dt + o(n^{-1}),$$

as $n \to \infty$. We will now follow Feller (1971) in simplifying the problem through the introduction of the function $\Lambda(t) = \log[\psi(t)] + \tfrac{1}{2}\sigma^2 t^2$. To see why this function is useful, note that

$$\exp[n\Lambda(t\sigma^{-1}n^{-1/2})] = \exp\{n\log[\psi(t\sigma^{-1}n^{-1/2})] + \tfrac{1}{2}n\sigma^2 t^2\sigma^{-2}n^{-1}\}$$

$$= \exp\{\log[\psi^n(t\sigma^{-1}n^{-1/2})]\}\exp(\tfrac{1}{2}t^2)$$

$$= \psi^n(t\sigma^{-1}n^{-1/2})\exp(\tfrac{1}{2}t^2).$$

Therefore, it follows that

$$(2\pi)^{-1} \int_{|t|<\delta\sigma n^{1/2}} |\psi^n(t\sigma^{-1}n^{-1/2}) - \exp(-\tfrac{1}{2}t^2) -$$

$$\tfrac{1}{6}n^{-1/2}\sigma^{-3}\mu_3(it)^3 \exp(-\tfrac{1}{2}t^2)|dt =$$

$$(2\pi)^{-1} \int_{|t|<\delta\sigma n^{1/2}} \exp(-\tfrac{1}{2}t^2)| \exp[n\Lambda(t\sigma^{-1}n^{-1/2})] - 1 - \tfrac{1}{6}n^{-1/2}\mu_3\sigma^{-3}(it)^3|dt.$$

$$(7.13)$$

To place a bound on the integral in Equation (7.13), we will use the inequality from Theorem A.9 which states that $|\exp(a) - 1 - b| \le (|a - b| + \tfrac{1}{2}|b|^2)\exp(\gamma)$ where $\gamma \ge \max\{|a|, |b|\}$. Note that this inequality is valid for a and b whether they are real or complex valued. In this instance we will take $a = n\Lambda(t\sigma^{-1}n^{-1/2})$ and $b = \tfrac{1}{6}n^{-1/2}\mu_3\sigma^{-3}(it)^3$. Therefore, it follows that

$$|\exp[n\Lambda(t\sigma^{-1}n^{-1/2})] - 1 - \tfrac{1}{6}n^{-1/2}\mu_3\sigma^{-3}(it)^3| \le$$

$$[|n\Lambda(t\sigma^{-1}n^{-1/2}) - \tfrac{1}{6}n^{-1/2}\mu_3\sigma^{-3}(it)^3| + \tfrac{1}{2}|\tfrac{1}{6}n^{-1/2}\mu_3\sigma^{-3}(it)^3|^2]\exp(\gamma) =$$

$$[|n\Lambda(t\sigma^{-1}n^{-1/2}) - \tfrac{1}{6}n^{-1/2}\mu_3\sigma^{-3}(it)^3| + \tfrac{1}{72}n^{-1}\mu_3^2\sigma^{-6}|it|^6]\exp(\gamma).$$

We now develop some properties for the function Λ. Recalling that $\Lambda(t) = \log[\psi(t)] + \tfrac{1}{2}\sigma^2t^2$ we have that $\Lambda(0) = \log[\psi(0)] = \log(1) = 0$. We will also need to evaluate the first three derivatives of Λ at zero so that we may approximate $\Lambda(t)$ with a Taylor expansion. The first derivative is $\Lambda'(t) = \psi'(t)/\psi(t) + \sigma^2t$ so that Theorem 2.31 implies that $\Lambda'(0) = \psi'(0)/\psi(0) = i\mu = 0$ since we have assumed that $\mu = 0$. Similarly, the second derivative is given by $\Lambda''(t) = [\psi'(t)/\psi(t)]^2 - \psi''(t)/\psi(t) + \sigma^2$ so that Theorem 2.31 implies that

$$\Lambda''(0) = -[\psi'(0)/\psi(0)]^2 + \psi''(0)/\psi(0) + \sigma^2 = -(i\mu)^2 + (i\sigma)^2 + \sigma^2 = 0.$$

For the third derivative we have

$$\Lambda'''(t) = 2\left(\frac{\psi'(t)}{\psi(t)}\right)^3 - \frac{3\psi'(t)\psi''(t)}{\psi^2(t)} + \frac{\psi'''(t)}{\psi(t)},$$

so that $\Lambda'''(0) = i^3\mu_3$. Theorem 1.13 implies that

$$\Lambda(t + \delta) = \Lambda(t) + \delta\Lambda'(t) + \tfrac{1}{2}\delta^2\Lambda''(t) + E_2(t, \delta),$$

so that

$$\Lambda(\delta) = \Lambda(0) + \delta\Lambda'(0) + \tfrac{1}{2}\delta^2\Lambda''(0) + E_2(\delta) = E_2(\delta), \qquad (7.14)$$

where $E_2(\delta) = \tfrac{1}{6}\Lambda'''(\xi)\delta^3$ for some $\xi \in (0, \delta)$. Let $\varepsilon > 0$ and note that since $\Lambda'''(t)$ is a continuous function, there is a neighborhood $|t| < \delta$ such that $\Lambda'''(t)$ varies less than $6\varepsilon\sigma^3$ from $\Lambda'''(0) = i^3\mu_3$. Therefore $|\Lambda'''(\xi) - i^3\mu_3| \le 6\varepsilon\sigma^3$ and hence $|\tfrac{1}{6}\Lambda'''(\xi)t^3 - \tfrac{1}{6}i^3\mu_3t^3| \le \varepsilon\sigma^3t^3$. Since $\Lambda(t) = \tfrac{1}{6}\Lambda'''(\xi)t^3$ we have that

$$|\Lambda(t\sigma^{-1}n^{-1/2}) - \tfrac{1}{6}\mu_3 i^3(t\sigma^{-1}n^{-1/2})^3| \le |\varepsilon t^3 n^{-3/2}|. \qquad (7.15)$$

Multiplying both sides of Equation (7.15) by n implies that

$$|n\Lambda(t\sigma^{-1}n^{-1/2}) - \tfrac{1}{6}\mu_3(it)^3\sigma^{-3}n^{-1/2}| \le |\varepsilon t^3 n^{-1/2}|$$

for $|t| < \delta$. Therefore, using the bound from Theorem A.9, we have that

$$|\exp[n\Lambda(t\sigma^{-1}n^{-1/2})] - 1 - \tfrac{1}{6}n^{-1/2}\mu_3\sigma^{-3}(it)^3| \le$$
$$[|\varepsilon t^3 n^{-1/2}| + \tfrac{1}{72}n^{-1}\mu_3^2\sigma^{-6}(it)^6]\exp(\gamma) =$$
$$[\varepsilon n^{-1/2}|t|^3 + \tfrac{1}{72}n^{-1}\mu_3^2\sigma^{-6}t^6|i^6|]\exp(\gamma).$$

To find an appropriate value for γ we note that

$$\gamma \ge \max\{|n\Lambda(t\sigma^{-1}n^{-1/2})|, |\tfrac{1}{6}n^{-1/2}\mu_3\sigma^{-3}(it)^3|\}.$$

To find this bound we first note that

$$|\Lambda(t)| = |\tfrac{1}{6}\Lambda'''(\xi)\delta^3| \le |i^3\mu_3 + \sigma^3\varepsilon||t^3|.$$

Suppose that δ is chosen small enough so that $|\delta| < (\tfrac{1}{4}\sigma^2)|i^3\mu_3 + \sigma^3\varepsilon|$. Then

$$|\Lambda(t)| \le |i^3\mu_3 + \sigma^3\varepsilon||t^3| \le |i^3\mu_3 + \sigma^3\varepsilon||\delta||t^2| = \tfrac{1}{4}\sigma^2 t^2,$$

since $|t| < \delta$. Therefore,

$$|\Lambda(t\sigma^{-1}n^{-1/2})| \le \tfrac{1}{4}\sigma^2 t^2 (\sigma^{-2}n^{-1}) = \tfrac{1}{4}n^{-1}t^2,$$

and hence $|n\Lambda(t\sigma^{-1}n^{-1/2})| \le \tfrac{1}{4}t^2$. Similarly, we note that

$$|\tfrac{1}{6}\mu_3(it)^3| = |\tfrac{1}{6}\mu_3 t^3| = |\tfrac{1}{6}||t^2||t|,$$

and we additionally choose δ small enough so that $\delta < \tfrac{1}{4}|\tfrac{1}{6}\mu_3|^{-1}\sigma^2$. Therefore $|\tfrac{1}{6}\mu_3(it)^3| < \tfrac{1}{4}\sigma^2 t^2$. Evaluating this bound at $t\sigma^{-1}n^{-1/2}$ implies that

$$|\tfrac{1}{6}\mu_3 i^3 t^3 \sigma^{-3}n^{-3/2}| = |\tfrac{1}{6}\mu_3(it)^3\sigma^{-3}n^{-1/2}| \le \tfrac{1}{4}t^2.$$

Therefore, we can take $\gamma = \tfrac{1}{4}t^2$ and it follows that

$$|\exp[n\Lambda(t\sigma^{-1}n^{-1/2})] - 1 - \tfrac{1}{6}n^{-1/2}\mu_3\sigma^{-3}(it)^3| \le$$
$$[|\varepsilon t^3 n^{-1/2}| + \tfrac{1}{72}n^{-1}\mu_3^2\sigma^6|it|^6]\exp(\tfrac{1}{4}t^2).$$

It, then, follows that the integral in Equation (7.13) can be bounded as

$$(2\pi)^{-1}\int_{|t|<\delta\sigma n^{1/2}}\exp(-\tfrac{1}{2}t^2)|\exp[n\Lambda(t\sigma^{-1}n^{-1/2})] -$$
$$1 - \tfrac{1}{6}n^{-1/2}\mu_3\sigma^{-3}(it)^3|dt \le$$
$$(2\pi)^{-1}\int_{|t|<\delta\sigma n^{1/2}}[|\varepsilon t^3 n^{-1/2}| + \tfrac{1}{72}n^{-1}\mu_3^2\sigma^{-6}|it|^6]\exp(-\tfrac{1}{4}t^2)dt,$$

where

$$\lim_{n\to\infty}n^{1/2}(2\pi)^{-1}\int_{|t|<\delta\sigma n^{1/2}}\varepsilon|t|^3 n^{-1/2}\exp(-\tfrac{1}{4}t^2)dt =$$
$$\lim_{n\to\infty}\varepsilon\int_{|t|<\delta\sigma n^{1/2}}|t|^3\exp(-\tfrac{1}{4}t^2)dt = \varepsilon\int_{-\infty}^{\infty}|t|^3\exp(-\tfrac{1}{4}t^2)dt. \quad (7.16)$$

Note that the integral on the right hand side of Equation (7.16) is finite. Let B_1 be a finite bound that is larger than this integral, then it follows that

$$\lim_{n\to\infty} n^{1/2}(2\pi)^{-1} \int_{|t|<\delta\sigma n^{1/2}} \varepsilon|t|^3 n^{-1/2} \exp(-\tfrac{1}{4}t^2)dt \le \varepsilon B_1,$$

for every $\varepsilon > 0$, and hence the limit is zero. Similarly,

$$\lim_{n\to\infty} n^{1/2}(2\pi)^{-1} \int_{|t|<\delta\sigma n^{1/2}} \tfrac{1}{72}n^{-1}\mu_3^2\sigma^{-6}|it|^6 \exp(-\tfrac{1}{4}t^2)dt =$$

$$\lim_{n\to\infty} n^{-1/2}(2\pi)^{-1} \int_{|t|<\delta\sigma n^{1/2}} \tfrac{1}{72}\mu_3^2\sigma^{-6}|it|^6 \exp(-\tfrac{1}{4}t^2)dt = 0,$$

since

$$\int_{-\infty}^{\infty} |it|^6 \exp(-\tfrac{1}{4}t^2)dt < \infty.$$

It, then, follows that

$$(2\pi)^{-1} \int_{|t|<\delta\sigma n^{1/2}} \exp(-\tfrac{1}{2}t^2)|\exp[n\Lambda(t\sigma^{-1}n^{-1/2})]-$$

$$1 - \tfrac{1}{6}n^{-1/2}\mu_3\sigma^{-3}(it)^3|dt = o(n^{-1/2}),$$

as $n \to \infty$, and the result follows. $\qquad\square$

The error in Equation 7.3 can also be characterized as having the property $O(n^{-1})$ as $n \to \infty$. Theorem 7.4 offers a potential improvement in the accuracy of approximating the distribution of the standardized sample mean over Theorem 4.20 (Lindeberg and Lévy) which approximates $f_n(t)$ with a normal density. In that case, we have that $f_n(t) - \phi(t) = o(1) = O(n^{-1/2})$ as $n \to \infty$. Further reductions in the order of the error are possible if terms are added to the expansion and further conditions on the existence of moments and the smoothness of f can be assumed.

Theorem 7.5. *Let $\{X_n\}_{n=1}^{\infty}$ be a sequence of independent and identically distributed random variables. Let $F_n(x) = P[n^{1/2}\sigma^{-1}(\bar{X}_n - \mu) \le x]$, with density $f_n(x)$ for all $x \in \mathbb{R}$. Assume that $E(|X_n|^p) < \infty$, $|\psi|^\nu$ is integrable for some $\nu \ge 1$, and that $f_n(x)$ exists for $n \ge \nu$. Then,*

$$f_n(x) - \phi(x) - \phi(x)\sum_{k=1}^{p} n^{-k/2}p_k(x) = o(n^{-p/2}), \qquad (7.17)$$

uniformly in x as $n \to \infty$, where $p_k(x)$ is a real polynomial whose coefficients depend only on $\mu_1, \ldots, \mu_k$. In particular, $p_k(x)$ does not depend on n, p or on other properties of F.

A proof of Theorem 7.5 can be found in Section XVI.2 of Feller (1971). The polynomial $p_1(x)$ has already been identified in Theorem 7.4 as being equal to $p_1(x) = \tfrac{1}{6}\mu_3\sigma^{-3}H_3(x)$. It can be similarly shown that the polynomial $p_2(x)$ is given by $p_2(x) = \tfrac{1}{72}\mu_3^2\sigma^{-6}H_6(x) + \tfrac{1}{24}\sigma^{-4}(\mu_4 - 3\sigma^4)H_4(x)$. It turns out that these polynomials are often easier to write in terms of cumulants. In particular, noting that $\kappa_4 = \mu_4 - 3\sigma^4$ we have that $p_2(x) = \frac{1}{72}\mu_3^2\sigma^{-6}H_6(x) + \frac{1}{24}\sigma^{-4}\kappa_4 H_4(x)$. Even further simplification is possible if we define the *standardized cumulants* as $\rho_k = \sigma^{-k}\kappa_k$ for all $k \in \mathbb{N}$. In this case we have $p_1(x) = \frac{1}{6}\rho_3 H_3(x)$ and $p_2(x) = \frac{1}{72}\rho_3^2 H_6(x) + \frac{1}{24}\rho_4 H_4(x)$. The form of these, and additional terms, can be motivated by considering the form of the cumulant generating function for a standardized mean. For simplicity, suppose that $X_1, \dots, X_n$ are a set of independent and identically distributed random variables from a distribution F that has mean equal to zero, unit variance, and cumulant generating function $c(t)$. Theorem 2.24 and Definition 2.13 implies that the cumulant generating function of $n^{1/2}\bar{X}_n$, the standardized version of the mean, is $nc(n^{-1/2}t)$. Equation (2.23) gives the form of the cumulant generating function for the special case where $\kappa_1 = 0$ and $\kappa_2 = 1$ as $c(t) = \frac{1}{2}t^2 + \frac{1}{6}\kappa_3 t^3 + \frac{1}{24}\kappa_4 t^4 + o(t^4)$, as $t \to 0$. Therefore,

$$nc(n^{-1/2}t) = \tfrac{1}{2}t^2 + \tfrac{1}{6}n^{-1/2}\kappa_3 t^3 + \tfrac{1}{24}n^{-1}\kappa_4 t^4 + o(n^{-1}),$$

as $n \to \infty$. We now convert this result to an equivalent expression for the moment generating function of the standardized mean. Definition 2.13 implies that the moment generating function of $n^{1/2}\bar{X}_n$ is

$$m(t) = \exp[\tfrac{1}{2}t^2 + \tfrac{1}{6}n^{-1/2}\kappa_3 t^3 + \tfrac{1}{24}n^{-1}\kappa_4 t^4 + o(n^{-1})].$$

Using Theorem 1.13 on the exponential function implies that

$$m(t) = \exp(\tfrac{1}{2}t^2) + [\tfrac{1}{6}n^{-1/2}\kappa_3 t^3 + \tfrac{1}{24}n^{-1}\kappa_4 t^4 + o(n^{-1})]\exp(\tfrac{1}{2}t^2) +$$
$$\tfrac{1}{2}[\tfrac{1}{6}n^{-1/2}\kappa_3 t^3 + \tfrac{1}{24}n^{-1}\kappa_4 t^4 + o(n^{-1})]\exp(\tfrac{1}{2}t^2) +$$
$$o\{[\tfrac{1}{6}n^{-1/2}\kappa_3 t^3 + \tfrac{1}{24}n^{-1}\kappa_4 t^4 + o(n^{-1})]^2\},$$

as $n \to \infty$. Collecting terms of order $o(n^{-1})$ and higher into the error term yields

$$m(t) = \exp(\tfrac{1}{2}t^2) + \tfrac{1}{6}n^{-1/2}\kappa_3 t^3 \exp(\tfrac{1}{2}t^2) + \tfrac{1}{24}n^{-1}\kappa_4 t^4 \exp(\tfrac{1}{2}t^2) +$$
$$\tfrac{1}{72}n^{-1}\kappa_3^2 t^6 \exp(\tfrac{1}{2}t^2) + o(n^{-1}), \quad (7.18)$$

as $n \to \infty$. To obtain an expansion for the density of the standardized mean we must now invert the expansion for the moment generating function term by term. While we have not explicitly discussed inversion of moment generating functions, we need only one specialized result in this case. It can be shown that

$$\int_{-\infty}^{\infty} \exp(tx)\phi(x)H_k(x)dx = t^k \exp(\tfrac{1}{2}t^2). \quad (7.19)$$

See Exercise 10. We again find ourselves dealing with functions that are not strictly densities, but nonetheless we can view the left hand side of Equation (7.19) as the moment generating function of $\phi(x)H_k(x)$. Therefore, if we invert the term on the right hand side of Equation (7.19) we should get the function $\phi(x)H_k(x)$. Therefore, inverting $\frac{1}{6}n^{-1/2}\kappa_3 t^2 \exp(\tfrac{1}{2}t^2)$ results in

$\frac{1}{6}n^{-1/2}\kappa_3\phi(x)H_3(x)$. The remaining terms can be inverted using the same method to obtain $\frac{1}{24}n^{-1}\kappa_4 H_4(x)\phi(x)$ and $\frac{1}{72}n^{-1}\kappa_3^2 H_6(x)\phi(x)$, respectively. Therefore, the expansion for the density of $n^{1/2}\bar{X}_n$ that results from inverting the expansion for the moment generating function given in Equation (7.18), is given by

$$\phi(x) + \frac{1}{6}n^{-1/2}\kappa_3\phi(x)H_3(x) + \frac{1}{24}n^{-1}\kappa_4 H_4(x)\phi(x) + \frac{1}{72}n^{-1}\kappa_3^2 H_6(x)\phi(x) + E_n,$$

where E_n is the error term, whose asymptotic properties are not considered at this time. Note that we did not end up with the standardized version of the cumulants in our expansion because we have assumed that the variance is one. Additional terms of the expansion are available, though the computation of these terms does become rather tedious. These terms are based on an asymptotic expansion of the cumulants of Z_n. See Chapter 2 of Hall (1992) and Exercises 11–12 for further details.

Note that the error term of the pth-order Edgeworth expansion given in Equation (7.17) is $o(n^{-p/2})$, as $n \to \infty$. We can also show that this error term is $O(n^{-(p+1)/2})$ as $n \to \infty$. To see this, note that Equation (7.17) implies that a $(p+1)^{\text{st}}$-order Edgeworth expansion has the form

$$f_n(x) - \phi(x) - \phi(x)\sum_{k=1}^{p} n^{-k/2}p_k(x) =$$

$$n^{-p/2-1/2}\phi(x)p_{p+1}(x) + o(n^{-(p+1)/2}), \quad (7.20)$$

as $n \to \infty$, where we have moved the last term to the right hand side of the equation. Note that

$$\frac{n^{-(p+1)/2}\phi(x)p_{p+1}(x)}{n^{-(p+1)/2}} = \phi(x)p_{p+1}(x).$$

Since $p_{p+1}(x)$ is a polynomial that does not depend on n, $|\phi(x)p_{p+1}(x)|$ remains bounded as $n \to \infty$ for each fixed value of x. Therefore, it follows that $n^{-(p+1)/2}\phi(x)p_{p+1}(x) = O(n^{-(p+1)/2})$ as $n \to \infty$. The second term on the right hand side of Equation (7.20) converges to zero as $n \to \infty$ when divided by $n^{-(p+1)/2}$. Hence, it follows that

$$f_n(x) - \phi(x) - \phi(x)\sum_{k=1}^{p} n^{-k/2}p_k(x) = O(n^{-(p+1)/2}),$$

as $n \to \infty$.

Example 7.1. Let $\{X_n\}_{n=1}^{\infty}$ be a sequence of independent and identically distributed random variables where X_n has an EXPONENTIAL(θ) distribution for all $n \in \mathbb{N}$. In this case the third and fourth standardized cumulants are given by $\rho_3 = 2$ and $\rho_4 = 6$. Therefore we have that $p_1(x) = \frac{1}{3}H_3(x)$ and $p_2(x) = \frac{1}{18}H_6(x) + \frac{1}{4}H_4(x)$, with the two-term Edgeworth expansion for the density of $Z_n = n^{1/2}\sigma^{-1}(\bar{X}_n - \mu)$ being given by

$$\phi(x) + \frac{1}{3}n^{-1/2}\phi(x)H_3(x) + n^{-1}\phi(x)[\tfrac{1}{18}H_6(x) + \tfrac{1}{4}H_4(x)] + o(n^{-1}),$$

Figure 7.1 *The true density of $Z_n = n^{1/2}\sigma^{-1}(\bar{X}_n - \mu)$ (solid line), the standard normal density (dashed line) and the one-term Edgeworth expansion (dotted line) when $n = 5$ and $X_1, \ldots, X_n$ is a set of independent and identically distributed random variables following an* EXPONENTIAL(θ) *distribution.*

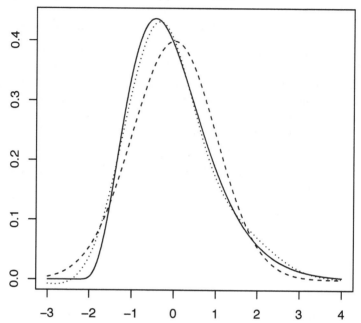

as $n \to \infty$. To illustrate the correction that the Edgeworth expansion makes in this case, we note that from Example 4.36 that Z_n has a location translated GAMMA$(n, n^{-1/2})$ density whose form is given in Equation (4.3). Figure 7.1 compares the true density of Z_n with the standard normal density and the one-term Edgeworth expansion when $n = 5$. One can observe from Figure 7.1 that the Edgeworth expansion does a much better job of capturing the true density of Z_n than the normal approximation does. Figure 7.2 provides the same comparison when $n = 10$. Both approximations are better, but one can observe that the one-term Edgeworth expansion is still more accurate.

Example 7.2. Let $\{X_n\}_{n=1}^{\infty}$ be a sequence of independent and identically distributed random variables where X_n has a linear type density of the form

$$f(x) = 2[\theta + x(1 - 2\theta)]\delta\{x; (0,1)\}, \tag{7.21}$$

where $\theta \in [0, 1]$. The kth moment about the origin for this density is given by

$$\mu_k' = E(X^k) = \frac{2}{k+2} - \frac{2k\theta}{(k+1)(k+2)}.$$

Therefore, the first four moments of this density are given by $\mu = \mu_1 =$

Figure 7.2 *The true density of $Z_n = n^{1/2}\sigma^{-1}(\bar{X}_n - \mu)$ (solid line), the standard normal density (dashed line) and the one-term Edgeworth expansion (dotted line) when $n = 10$ and $X_1, \ldots, X_n$ is a set of independent and identically distributed random variables following an* EXPONENTIAL(θ) *distribution.*

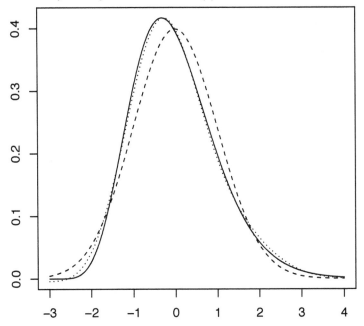

$\frac{2}{3} - \frac{1}{3}\theta$, $\sigma^2 = \mu_2 = \frac{1}{18} + \frac{1}{9}\theta - \frac{1}{9}\theta^2$, $\mu_3 = -\frac{1}{135} - \frac{1}{45}\theta + \frac{1}{9}\theta^2 - \frac{2}{27}\theta^3$, and $\mu_4 = \frac{1}{135} + \frac{4}{135}\theta - \frac{1}{15}\theta^2 + \frac{2}{27}\theta^3 - \frac{1}{27}\theta^4$. It then follows that the standardized cumulants are given by

$$\rho_3 = \frac{-\frac{1}{135} - \frac{1}{45}\theta + \frac{1}{9}\theta^2 - \frac{2}{27}\theta^3}{\left(\frac{1}{18} + \frac{1}{9}\theta - \frac{1}{9}\theta^2\right)^{3/2}},$$

and

$$\rho_4 = \frac{\mu_4 - 3\sigma^4}{\sigma^4}$$

$$= \frac{\frac{1}{135} + \frac{4}{135}\theta - \frac{1}{15}\theta^2 + \frac{2}{27}\theta^3 - \frac{1}{27}\theta^4}{\left(\frac{1}{18} + \frac{1}{9}\theta - \frac{1}{9}\theta^2\right)^2} - 3.$$

The standardized cumulants are plotted as a function of $\theta \in [0, 1]$ in Figure 7.3. Of interest here is the fact that the third standardized cumulant equals zero when $\theta = \frac{1}{2}$, which corresponds to the linear distribution being equal to a UNIFORM$(0, 1)$ distribution, which is symmetric. When the third standardized cumulant is zero it follows that the second term of the Edgeworth expansion, which has the form $\frac{1}{6}n^{-1/2}\phi(x)\rho_3 H_3(x)$, is also zero and the term

Figure 7.3 *The third (solid line) and fourth (dashed line) standardized cumulants of the linear density given in Equation (7.21) as a function of θ.*

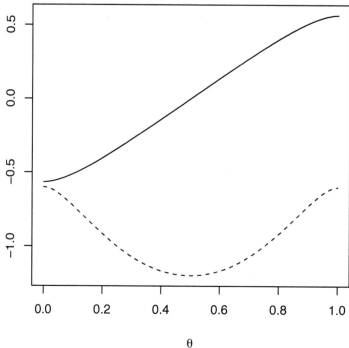

is eliminated. This means that we now have an expansion of the form

$$\phi(x) + n^{-1}\phi(x)[\tfrac{1}{72}\rho_3^2 H_6(x) + \tfrac{1}{24}\rho_4 H_4(x)] + o(n^{-1}) = \phi(x) + o(n^{-1/2}),$$

as $n \to \infty$. Hence, the normal approximation provided by Theorem 4.20 (Lindeberg and Lévy) which usually has an error term that is $o(1)$ as $n \to \infty$, is now $o(n^{-1/2})$ as $n \to \infty$. Therefore, in the case where the population is symmetric, Theorem 4.20 is more accurate than when the population is not symmetric. Is it possible to choose θ such that the second term of the expansion, which has the form $n^{-1}\phi(x)[\tfrac{1}{72}\rho_3^2 H_6(x) + \tfrac{1}{24}\rho_4 H_4(x)]$, would also be eliminated? That is, can we find a value of θ for which $\tfrac{1}{72}\rho_3^2 H_3(x) + \tfrac{1}{24}\rho_4 H_4(x) = 0$ for all $x \in \mathbb{R}$? This would require both the third and fourth standardized cumulants to be equal to zero, which does not occur for any value of $\theta \in [0, 1]$. Therefore, it is clear that there are not any situations where we can gain any further accuracy past the case from where the distribution is symmetric and the first term of the expansion is eliminated. ∎

Example 7.3. Let $\{X_n\}_{n=1}^{\infty}$ be a sequence of independent and identically distributed random variables where X_n has a density that is a mixture of two NORMAL densities given by $f(x) = \tfrac{1}{2}\phi(x) + \tfrac{1}{2}\phi(x - \theta)$ for all $x \in \mathbb{R}$ where

$\theta \in \mathbb{R}$. When considering the moments of a normal mixture such as this, it is worthwhile to note that

$$E(X_n^k) = \int_{-\infty}^{\infty} x^k [\tfrac{1}{2}\phi(x) + \tfrac{1}{2}\phi(x-\theta)]dx$$

$$= \tfrac{1}{2}\int_{-\infty}^{\infty} x^k \phi(x)dx + \tfrac{1}{2}\int_{-\infty}^{\infty} x^k \phi(x-\theta)dx.$$

Therefore, if we suppose that Y is a $N(0,1)$ random variable and W is a $N(\theta,1)$ random variable, then $E(X_n^k) = \tfrac{1}{2}E(Y^k) + \tfrac{1}{2}E(W^k)$. Therefore, we have that $E(X_n) = \tfrac{1}{2}\theta$, $E(X_n^2) = 1 + \tfrac{1}{2}\theta^2$, $E(X_n^3) = \tfrac{1}{2}\theta^3 + \tfrac{3}{2}\theta$, and $E(X_n^4) = -\tfrac{1}{2}\theta^4 + \tfrac{3}{2}\theta^2 + 3$. From these calculations we can find that the third and fourth moments of the normal mixture are $\mu_3 = 0$ and $\mu_4 = -\tfrac{3}{16}\theta^4 + \tfrac{3}{2}\theta^2 + 3$. The third cumulant is also zero, and the fourth cumulant is $\kappa_4 = -\tfrac{15}{16}\theta^4 - \tfrac{3}{2}\theta^2$. There are two key elements to the Edgeworth expansion in this case. First, note that since the third cumulant is always zero, as this particular normal mixture is always symmetric, the first term of the Edgeworth expansion is always zero, and hence the error term for this expansion is always $o(n^{-1/2})$, as $n \to \infty$, no matter what the value of θ is. Secondly, note that the fourth cumulant is zero when $\theta = 0$, which means that the error term of the Edgeworth expansion is no larger than $o(n^{-1})$ as $n \to \infty$ in this case. In fact, all of the cumulants except the second are zero when $\theta = 0$ since the normal mixture in this case coincides with a $N(0,1)$ distribution, where the Edgeworth expansion has no terms beyond the $\phi(x)$ term and the normal approximation of Theorem 4.20 (Lindeberg and Lévy) is exact. ∎

The development of the Edgeworth expansion in Theorems 7.4 and 7.5 focuses on the density of the random variable $Z_n = n^{1/2}\sigma^{-1}(\bar{X}_n - \mu)$. Similar expansions can also be developed for the distribution function of the standardized random variable Z_n. From one viewpoint such an expansion can be obtained through term-by-term integration of the expansion given in Equation (7.3), for which we would obtain the result

$$F_n(x) - \Phi(x) - \tfrac{1}{6}\sigma^{-3}n^{-1/2}\mu_3(1-x^2)\phi(t) = o(n^{-1/2}), \tag{7.22}$$

under the conditions of Theorem 7.4. However, such an expansion is still valid under even weaker conditions. In fact, it is not necessary that the distribution F have a density, but that certain distributions known as lattice distributions must be avoided.

Definition 7.1. *A distribution F is a lattice distribution if the jump points of F are all on set of the form $\{a + bz : z \in \mathbb{Z}\}$ where $a \in \mathbb{R}$ and b is a real number called the span of the lattice.*

We have encountered lattice distribution previously when we considered the inversion of the characteristic function, where we pointed out that examples of distributions that are lattice distributions include the BINOMIAL and Poisson distributions, which both have span equal to one. See Example 2.20. The reason that lattice distributions are important is that their characteristic functions have a particular form that make the derivation of Edgeworth expansions of the form given in Theorem 7.5 impossible without changing the form of the expansion.

Theorem 7.6. *Let X be a random variable with distribution F and characteristic function ψ.*

1. *Then F is a lattice distribution on the set $\{a+bz : z \in \mathbb{Z}\}$ then the function $\zeta(t) = \exp(-ita)\psi(t)$ is a periodic function with period $2\pi b^{-1}$.*
2. *If $\psi(t)$ is a periodic function, then X is a lattice random variable.*

Proof. If F is a lattice distribution on the set $\{a + bz : z \in \mathbb{Z}\}$, then the characteristic function of X is given by

$$
\begin{aligned}
\psi(t) &= E[\exp(itX)] \\
&= \sum_{k \in \mathbb{Z}} p_k \exp[it(a + bk)] \\
&= \exp(ita) \sum_{k \in \mathbb{Z}} p_k \exp(itbk),
\end{aligned}
$$

where $p_k = P(X = a + bk)$ for all $k \in \mathbb{Z}$. Therefore,

$$
\zeta(t) = \sum_{k \in \mathbb{Z}} p_k \exp(itbk),
$$

and

$$
\begin{aligned}
\zeta(t + 2\pi b^{-1}) &= \sum_{k \in \mathbb{Z}} p_k [ibk(t + 2\pi b^{-1})] \\
&= \sum_{k \in \mathbb{Z}} p_k \exp(itbk) \exp(2\pi ik) \\
&= \sum_{k \in \mathbb{Z}} p_k \exp(itbk) \\
&= \zeta(t),
\end{aligned}
$$

since $\exp(2\pi ik) = 1$ when $k \in \mathbb{Z}$.

To prove the second result, we now assume that X is a random variable with a periodic characteristic function that has period a. We wish now to show that X is a lattice random variable. Noting that $\psi(0) = E[\exp(i0X)] = E(1) = 1$ it follows from the periodicity of ψ that $\psi(0) = \psi(a) = 1$. Now, Definition A.6 implies that

$$
\psi(a) = E[\exp(aiX)] = E[\cos(aX) + i\sin(aX)] =
$$
$$
E[\cos(aX)] + iE[\sin(aX)] = 1. \quad (7.23)
$$

Note that the right hand side of Equation (7.23) is a real number. Therefore, $iE[\sin(aX)] = 0$ and hence $E[\cos(aX)] = 1$, or equivalently $E[1 - \cos(aX)] = 0$. Since $P[-1 \leq \cos(aX) \leq 1] = 1$ it follows that $P[1 - \cos(aX) \geq 0] = 1$.

Therefore Theorem A.15 implies that since $E[1 - \cos(aX)] = 0$ it follows that $P[1 - \cos(aX) = 0] = 1$. This implies that the only values of X that have non-zero probabilities are $x \in \mathbb{R}$ such that $\cos(ax) = 1$. The periodicity of the cosine function implies that X must be a lattice random variable. $\qquad\square$

Further discussion on this topic can be found in Chapter XV of Feller (1971) and Chapter 8 of Chow and Teicher (2003). See also Lukacs (1956). From Theorem 7.6 we can observe that when F corresponds to a lattice distribution, then the characteristic function of F has periodic components that do not decay as $|t| \to \infty$. This means that the truncation argument used to prove Theorem 7.4 can no longer be used, and in fact the form of the expansion must be changed. For further information on Edgeworth type expansions for lattice distributions see Esseen (1945) and Chapter XVI of Feller (1971).

The proof of the validity of the expansion in Equation (7.22) for non-lattice distributions is based on Theorem 4.23 (Smoothing Theorem). However, as with the expansion for densities, Theorem 4.23 is not directly applicable because the expansion does not yield a valid distribution function. In particular, $\Phi(x) + \frac{1}{6}\sigma^{-3}n^{-1/2}\mu_3(1 - x^2)\phi(t)$ may not be non-decreasing, may go outside the range $[0, 1]$, and may not have the proper limits as $x \to \pm\infty$. To address this problem we require a more general version of Theorem 4.23. For our particular case we will present a specific version of this theorem that focuses on the case of bounding the difference between the normal distribution function and another function.

Theorem 7.7. *Let G be a function such that*

$$\lim_{x \to \infty} |\Phi(x) - G(x)| = 0,$$

and

$$\lim_{x \to -\infty} |\Phi(x) - G(x)| = 0.$$

Suppose that G has bounded derivative that has a continuously differentiable Fourier transformation γ such that $\gamma(0) = 1$ and

$$\left. \frac{d}{dt}\gamma(t) \right|_{t=0} = 0.$$

If F is a distribution function with zero expectation, then

$$|F(x) - G(x)| \leq \int_{-t}^{t} \frac{|\psi(x) - \zeta(x)|}{\pi|x|}dx + 24(\pi t)^{-1}\sup_{x \in \mathbb{R}}|G'(x)|,$$

for all $x \in \mathbb{R}$ and $t > 0$.

A proof of Theorem 7.7 can be found in Section XVI.3 of Feller (1971). We now have the required results to justify the Edgeworth expansion for non-lattice distributions.

Theorem 7.8. *Let $\{X_n\}_{n=1}^{\infty}$ be a sequence of independent and identically distributed random variables from a distribution F. Let $F_n(x) = P[n^{1/2}\sigma^{-1}(\bar{X}_n -$*

$\mu) \le x]$ and assume that $E(X_n^3) < \infty$. If F is a non-lattice distribution then

$$F_n(x) - \Phi(x) - \tfrac{1}{6}\sigma^{-3}n^{-1/2}\mu_3(1 - x^2)\phi(x) = o(n^{-1/2}), \qquad (7.24)$$

as $n \to \infty$ uniformly in x.

Proof. We shall give a partial proof of the result, leaving the details for Exercise 6. Let

$$G(x) = \Phi(x) + \tfrac{1}{6}\sigma^{-3}n^{-1/2}\mu_3(1 - x^2)\phi(x).$$

We must first prove that our choice for $G(x)$ satisfies the assumptions of Theorem 7.7. We begin by noting that since $\phi(x) \to 0$ as $x \to \pm\infty$ at a faster rate than $(1 - x^2)$ diverges to $-\infty$ it follows that

$$\lim_{x \to \infty} (1 - x^2)\phi(x) = 0,$$

and

$$\lim_{x \to -\infty} (1 - x^2)\phi(x) = 0.$$

Therefore,

$$\lim_{x \to \infty} G(x) = \lim_{x \to \infty} \Phi(x) + \lim_{x \to \infty} \tfrac{1}{6}\sigma^{-3}n^{-1/2}\mu_3(1 - x^2)\phi(x) = 1,$$

and

$$\lim_{x \to -\infty} G(x) = \lim_{x \to -\infty} \Phi(x) + \lim_{x \to -\infty} \tfrac{1}{6}\sigma^{-3}n^{-1/2}\mu_3(1 - x^2)\phi(x) = 0.$$

Similarly, under the assumptions a similar conclusion can be obtained for $F_n(x)$. Therefore, it follows that

$$\lim_{x \to \infty} |F_n(x) - G(x)| = 0,$$

and

$$\lim_{x \to \infty} |F_n(x) - G(x)| = 0.$$

The derivative of $G(x)$ is given by

$$\begin{aligned}
\frac{d}{dx}G(x) &= \frac{d}{dx}\left[\Phi(x) + \tfrac{1}{6}\sigma^{-3}n^{-1/2}\mu_3(1 - x^2)\phi(x)\right] \\
&= \phi(x) - \tfrac{1}{6}\sigma^{-3}n^{-1/2}\mu_3(3x - x^3)\phi(x) \\
&= \phi(x) + \tfrac{1}{6}\sigma^{-1}n^{-1/2}\mu_3 H_3(x)\phi(x),
\end{aligned}$$

which can be shown to be bounded. We now compute the Fourier transformation of $G'(x)$ as

$$\begin{aligned}
\gamma(t) &= \int_{-\infty}^{\infty} \exp(itx)[\phi(x) + \tfrac{1}{6}\sigma^{-1}n^{-1/2}\mu_3 H_3(x)\phi(x)]dx \\
&= \int_{-\infty}^{\infty} \exp(itx)\phi(x)dx + \tfrac{1}{6}\sigma^{-1}n^{-1/2}\mu_3 \int_{-\infty}^{\infty} \exp(itx)H_3(x)\phi(x)dx \\
&= \exp(-\tfrac{1}{2}t^2) + (it)^3 \exp(-\tfrac{1}{2}t^2) \\
&= \exp(-\tfrac{1}{2}t^2)[1 + \tfrac{1}{6}\mu_3\sigma^{-3}n^{-1/2}(it)^3]. \qquad (7.25)
\end{aligned}$$

The first term in Equation (7.25) comes from the fact that the characteristic function of a $N(0,1)$ distribution is $\exp(-\frac{1}{2}t^2)$. The second term in Equation (7.25) is the result of an application of Theorem 7.1. Note that $\gamma(0) = 1$. The derivative of the Fourier transformation is given by

$$\frac{d}{dt}\exp(-\tfrac{1}{2}t^2)[1 + \tfrac{1}{6}\mu_3\sigma^{-3}n^{-1/2}(it)^3]\Big|_{t=0} =$$

$$-t\exp(-\tfrac{1}{2}t^2)[1 + \tfrac{1}{6}\mu_3\sigma^{-3}n^{-1/2}(it)^3]\Big|_{t=0} +$$

$$\exp(-\tfrac{1}{2}t^2)[\tfrac{1}{3}\mu_3\sigma^{-3}n^{-1/2}i^3t^2]\Big|_{t=0} = 0.$$

Therefore, we can apply Theorem 7.7 to find that

$$|F_n(x) - G(x)| \le$$

$$\pi^{-1}\int_{-T}^{T}|t^{-1}[\psi^n(t\sigma^{-1}n^{-1/2}) - \gamma(t)]dt + 24(T\pi)^{-1}\sup_{x\in\mathbb{R}}|G'(x)|.$$

Let $\varepsilon > 0$ and choose α to be large enough so that $24|G'(t)| < \alpha\varepsilon$ for all $x \in \mathbb{R}$. Then let $T = \alpha n^{1/2}$ and we have that

$$|F_n(x) - G(x)| \le \pi^{-1}\int_{-T}^{T}|t^{-1}[\psi^n(t\sigma^{-1}n^{-1/2}) - \gamma(t)]dt + \varepsilon n^{-1/2}.$$

The remainder of the proof proceeds much along the same path as the proof of Theorem 7.4. See Exercise 6. □

As in the case of expansions for densities, the result of Theorem 7.8 can be expanded under some additional assumptions so that the error is $o(n^{-p/2})$ as $n \to \infty$.

Theorem 7.9. *Let $\{X_n\}_{n=1}^{\infty}$ be a sequence of independent and identically distributed random variables from a distribution F that has characteristic function ψ. Let $F_n(t) = P[n^{1/2}\sigma^{-1}(\bar{X}_n - \mu) \le x]$ and assume that $E(X_n^p) < \infty$. If*

$$\limsup_{|t|\to\infty}|\psi(t)| < 1, \tag{7.26}$$

then

$$F_n(x) - \Phi(x) - \phi(x)\sum_{k=1}^{p}n^{-k/2}r_k(x) = o(n^{-p/2}), \tag{7.27}$$

uniformly in x as $n \to \infty$, where $r_k(x)$ is a real polynomial whose coefficients depend only on $\mu_1, \ldots, \mu_k$. In particular, $r_k(x)$ does not depend on n, p or on other properties of F.

To find the polynomials $r_1, \ldots, r_k$ we note that the Edgeworth expansion for the distribution function is the integrated version of the expansion for the density. Therefore, it follows that

$$\phi(x)p_k(x) = \frac{d}{dx}\phi(x)r_k(x), \tag{7.28}$$

which yields

$$-r_1(x) = \tfrac{1}{6}\rho_3 H_2(x) \tag{7.29}$$

and

$$-r_2(x) = \tfrac{1}{24}\rho_4 H_3(x) + \tfrac{1}{72}\rho_3^2 H_5(x). \tag{7.30}$$

See Exercise 13.

Example 7.4. Let $\{X_n\}_{n=1}^\infty$ be a sequence of independent and identically distributed random variables where X_n has an EXPONENTIAL(θ) distribution for all $n \in \mathbb{N}$, where the third and fourth standardized cumulants are given by $\rho_3 = 2$ and $\rho_4 = 6$. Therefore we have that $-r_1(x) = \tfrac{1}{3}H_2(x)$ and $-r_2(x) = \tfrac{1}{4}H_3(x) + \tfrac{1}{18}H_5(x)$, with the two-term Edgeworth expansion for the distribution function of $Z_n = n^{1/2}\sigma^{-1}(\bar{X}_n - \mu)$ being given by

$$\Phi(x) - \tfrac{1}{3}n^{-1/2}\phi(x)H_2(x) - n^{-1}\phi(x)[\tfrac{1}{4}H_3(x) + \tfrac{1}{18}H_5(x)] + o(n^{-1}),$$

as $n \to \infty$. To illustrate the correction that the Edgeworth expansion makes in this case, we note that from Example 4.36 that Z_n has a location translated GAMMA($n, n^{-1/2}$) density whose form is given in Equation (4.3). Figure 7.4 compares the true distribution function of Z_n with the standard normal distribution function and the one-term Edgeworth expansion when $n = 5$. One can observe from Figure 7.4 that the Edgeworth expansion does a much better job of capturing the true distribution function of Z_n than the normal approximation does. Figure 7.5 provides the same comparison when $n = 10$. Both approximations are better, but one can observe that the one-term Edgeworth expansion is still more accurate.   ■

7.3 The Cornish–Fisher Expansion

The Edgeworth expansion given in Equation (7.27) provides a more accurate approximation to the distribution function of $Z_n = n^{1/2}\sigma^{-1}(\bar{X}_n - \mu)$ than the normal approximation given by Theorem 4.20 (Lindeberg and Lévy). While it is useful to see how the distribution of Z_n changes with respect to the moments of F, it might also be useful to use these results to provide new methods of statistical inference based on these corrections. For example, if $X_1, \ldots, X_n$ is a set of independent and identically distributed random variables from a N(θ, σ^2) distribution, then a $100\alpha\%$ confidence interval for θ is given by $[\bar{X}_n - n^{-1/2}\sigma z_{(1+\alpha)/2}, \bar{X}_n - n^{-1/2}\sigma z_{(1-\alpha)/2}]$ when σ is known. The form of this interval is based on the fact that when F has a N(θ, σ^2) distribution, Z_n has a N($0, 1$) distribution. When F does not have a N(θ, σ^2) distribution, then the confidence interval can still be justified asymptotically in the sense that $Z_n \xrightarrow{d} Z$ as $n \to \infty$ where Z is a N($0, 1$) random variable. This interval is approximate for finite sample sizes in that the actual coverage probability may not be exactly α. Of interest, then, is the possibility that the performance of this interval could be improved using the information that we obtain about the distribution of Z_n from the Edgeworth expansion given

Figure 7.4 *The true distribution function of* $Z_n = n^{1/2}\sigma^{-1}(\bar{X}_n - \mu)$ *(solid line), the standard normal density (dashed line) and the one-term Edgeworth expansion (dotted line) when* $n = 5$ *and* $X_1, \ldots, X_n$ *is a set of independent and identically distributed random variables following an* EXPONENTIAL(θ) *distribution.*

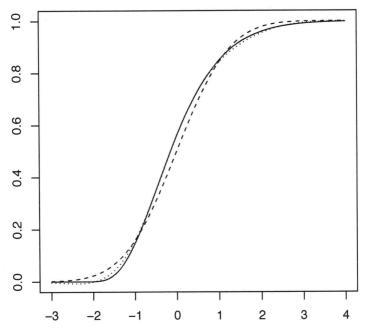

in Equation (7.27). However, in order to improve the confidence interval we need to have corrections to the normal approximation for the *quantiles* of Z_n, not the distribution function of Z_n. Therefore, we would like to develop an Edgeworth-type expansion for the quantiles of Z_n.

Expansions for the quantiles of the random variable Z_n are called *Cornish–Fisher* expansions and were first developed by Cornish and Fisher (1937) and Fisher and Cornish (1960). See also Hall (1983a). We will develop these expansions using the inversion method described in Section 1.6. In particular we will begin with the two-term Edgeworth expansion for the distribution function of Z_n given by

$$F_n(x) = \Phi(x) - \tfrac{1}{6}n^{-1/2}\phi(x)\rho_3 H_2(x) -$$
$$n^{-1}\phi(x)[\tfrac{1}{24}\rho_4 H_3(x) + \tfrac{1}{72}\rho_3^2 H_5(x)] + o(n^{-1}), \quad (7.31)$$

where we would like to find an asymptotic expansion for a value $g_{\alpha,n}$ where $F_n(g_{\alpha,n}) = \alpha + o(n^{-1})$, as $n \to \infty$. To begin the process we will assume that the asymptotic expansion for $g_{\alpha,n}$ has the form $g_{\alpha,n} = v_0(\alpha) + n^{-1/2}v_1(\alpha) + n^{-1}v_2(\alpha) + o(n^{-1})$ as $n \to \infty$. We now substitute this expansion for x in

Figure 7.5 *The true distribution function of $Z_n = n^{1/2}\sigma^{-1}(\bar{X}_n - \mu)$ (solid line), the standard normal density (dashed line) and the one-term Edgeworth expansion (dotted line) when $n = 10$ and $X_1, \ldots, X_n$ is a set of independent and identically distributed random variables following an* EXPONENTIAL(θ) *distribution.*

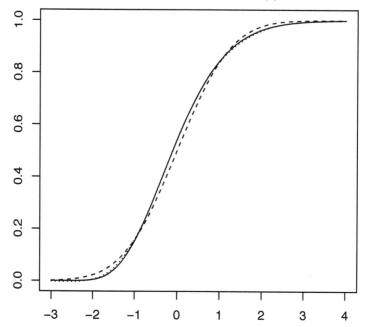

Equation (7.31) and use Theorem 1.15 to find an asymptotic expansion in terms of the functions v_0, v_1, and v_2. We first recall from Theorem 1.15 that $\Phi(x+\delta) = \Phi(x) + \delta\phi(x) - \frac{1}{2}\delta^2 H_1(x)\phi(x) + o(\delta^2)$, as $\delta \to 0$ for any fixed $x \in \mathbb{R}$. Therefore, taking $x = v_0(\alpha)$ and $\delta = n^{-1/2}v_1(\alpha) + n^{-1}v_2(\alpha) + o(n^{-1})$ we have that

$$\Phi(g_{\alpha,n}) = \Phi[v_0(\alpha)] + [n^{-1/2}v_1(\alpha) + n^{-1}v_2(\alpha) + o(n^{-1})]\phi[v_0(\alpha)] -$$
$$\frac{1}{2}[n^{-1/2}v_1(\alpha) + n^{-1}v_2(\alpha) + o(n^{-1})]^2 H_1[v_0(\alpha)]\phi[v_0(\alpha)] +$$
$$o\{[n^{-1/2}v_1(\alpha) + n^{-1}v_2(\alpha) + o(n^{-1})]^2\}, \quad (7.32)$$

as $n \to \infty$. We will simplify each term and consolidate all terms that are of order $o(n^{-1})$ or smaller into the error term. The first term on the right hand side of Equation (7.32) needs no simplification, so we move to the second term where we have that

$$[n^{-1/2}v_1(\alpha) + n^{-1}v_2(\alpha) + o(n^{-1})]\phi[v_0(\alpha)] =$$
$$n^{-1/2}v_1(\alpha)\phi[v_0(\alpha)] + n^{-1}v_2(\alpha)\phi[v_0(\alpha)] + o(n^{-1}),$$

as $n \to \infty$ where we conclude that the term that has the form $o(n^{-1})\phi[v_0(\alpha)] =$

$o(n^{-1})$ because $\phi[v_0(\alpha)]$ is constant with respect to n. The third term is simplified by noting that

$$[n^{-1/2}v_1(\alpha) + n^{-1}v_2(\alpha) + o(n^{-1})]^2 = n^{-1}v_1^2(\alpha) + o(n^{-1}), \qquad (7.33)$$

as $n \to \infty$. To see why this is true, let $\{R_n\}_{n=1}^{\infty}$ be a sequence of real numbers such that $R_n = o(n^{-1})$ as $n \to \infty$. Then, note that

$$[n^{-1/2}v_1(\alpha) + n^{-1}v_2(\alpha) + R_n]^2 =$$
$$n^{-1}v_1^2(\alpha) + 2n^{-3/2}v_1(\alpha)v_2(\alpha) + 2n^{-1/2}v_1(\alpha)R_n +$$
$$n^{-2}v_2^2(\alpha) + 2n^{-1}v_2(\alpha)R_n + R_n^2.$$

Now, $2n^{-3/2}v_1(\alpha)v_2(\alpha) = o(n^{-1})$ as $n \to \infty$ since

$$\lim_{n\to\infty} n[2n^{-3/2}v_1(\alpha)v_2(\alpha)] = \lim_{n\to\infty} 2n^{-1/2}v_1(\alpha)v_2(\alpha) = 0,$$

based once again on the fact that $v_1(\alpha)$ and $v_2(\alpha)$ do not depend on n. Similarly, it can be shown that $n^{-2}v_2^2(\alpha) = o(n^{-1})$ as $n \to \infty$. For the term $2n^{-1/2}v_1(\alpha)R_n$ we note that $R_n = o(n^{-1})$ as $n \to \infty$, which implies that

$$\lim_{n\to\infty} n[2n^{-1/2}v_1(\alpha)R_n] = \lim_{n\to\infty} 2n^{-1/2}v_1(\alpha)(nR_n) = 0,$$

so that $2n^{-1/2}v_1(\alpha)R_n = o(n^{-1})$ as $n \to \infty$. For the remaining term we also have that $R_n^2 = o(n^{-1})$ as $n \to \infty$. See Exercise 7. Consolidating these results leads to the result given in Equation (7.33). We finally note that $H_1[v_0(\alpha)] = v_0(\alpha)$, so that $H_1[v_0(\alpha)]\phi[v_0(\alpha)] = v_0(\alpha)\phi[v_0(\alpha)]$ which does not depend on n. Therefore, it follows that

$$H_1[v_0(\alpha)][n^{-1/2}v_1(\alpha) + n^{-1}v_2(\alpha) + o(n^{-1})]^2\phi[v_0(\alpha)] =$$
$$n^{-1}v_0(\alpha)v_1^2(\alpha)\phi[v_0(\alpha)].$$

Similar arguments can also be used to show that

$$o\{[n^{-1/2}v_1(\alpha) + n^{-1}v_2(\alpha) + o(n^{-1})]^2\} = o(n^{-1}),$$

as $n \to \infty$. See Exercise 8. Therefore, it follows that

$$\Phi(g_{\alpha,n}) = \Phi[v_0(\alpha)] + n^{-1/2}v_1(\alpha)\phi[v_0(\alpha)] +$$
$$n^{-1}\phi[v_0(\alpha)][v_2(\alpha) - \tfrac{1}{2}v_0(\alpha)v_1^2(\alpha)] + o(n^{-1}), \quad (7.34)$$

as $n \to \infty$.

We now consider the second term in Equation (7.31). Note that the leading coefficient of this term is $n^{-1/2}$. Because we are only keeping terms that are of lower order than $o(n^{-1})$, we only need to keep track of the terms of $\tfrac{1}{6}\rho_3\phi(g_{\alpha,n})H_2(g_{\alpha,n})$ that are of smaller order than $o(n^{-1})$ as $n \to \infty$. We again rely on Theorem 1.15 to find that

$$\phi(x + \delta) = \phi(x) - \delta H_1(x)\phi(x) + o(\delta^2) = \phi(x) - \delta x\phi(x) + o(\delta^2),$$

as $\delta \to 0$. Therefore,

$$\phi(g_{\alpha,n}) = \phi[v_0(\alpha)] - v_0(\alpha)\phi[v_0(\alpha)][n^{-1/2}v_1(\alpha) + n^{-1}v_2(\alpha) + o(n^{-1})] +$$
$$o\{[n^{-1/2}v_1(\alpha) + n^{-1}v_2(\alpha)]^2\},$$

as $n \to \infty$. Keeping only terms of order larger than $o(n^{-1/2})$ yields

$$\phi(g_{\alpha,n}) = \phi[v_0(\alpha)] - n^{-1/2}v_0(\alpha)v_1(\alpha)\phi[v_0(\alpha)] + o(n^{-1/2}),$$

as $n \to \infty$. Now $H_2(x) = x^2 - 1$ so that using calculations similar to those detailed above

$$
\begin{aligned}
H_2(g_{\alpha,n}) &= [v_0(\alpha) + n^{-1/2}v_1(\alpha) + n^{-1}v_2(\alpha) + o(n^{-1})]^2 - 1 \\
&= v_0^2(\alpha) + n^{-1}v_1^2(\alpha) + n^{-2}v_2^2(\alpha) + 2n^{-1/2}v_0(\alpha)v_1(\alpha) \\
&\quad + 2n^{-1}v_0(\alpha)v_2(\alpha) + 2n^{-3/2}v_1(\alpha)v_2(\alpha) - 1 + o(n^{-1}),
\end{aligned}
$$

as $n \to \infty$. But, noting that

$$n^{-1}v_1^2(\alpha) + n^{-2}v_2^2(\alpha) + 2n^{-1}v_0(\alpha)v_2(\alpha) + 2n^{-3/2}v_1(\alpha)v_2(\alpha) = o(n^{-1/2}),$$

implies that

$$
\begin{aligned}
H_2(g_{\alpha,n}) &= v_0^2(\alpha) + 2n^{-1/2}v_0(\alpha)v_1(\alpha) - 1 + o(n^{-1/2}) \\
&= H_2[v_0(\alpha)] + 2n^{-1/2}v_0(\alpha)v_1(\alpha) + o(n^{-1/2}),
\end{aligned}
$$

as $n \to \infty$. Therefore,

$$
\begin{aligned}
\phi(g_{\alpha,n})H_2(g_{\alpha,n}) &= \{\phi[v_0(\alpha)] - n^{-1/2}v_0(\alpha)v_1(\alpha)\phi[v_0(\alpha)] + o(n^{-1/2})\} \times \\
&\quad \{H_2[v_0(\alpha)] + 2n^{-1/2}v_0(\alpha)v_1(\alpha) + o(n^{-1/2})\} \\
&= \phi[v_0(\alpha)]H_2[v_0(\alpha)] + 2n^{-1/2}v_0(\alpha)v_1(\alpha)\phi[v_0(\alpha)] \\
&\quad - n^{-1/2}v_0(\alpha)v_1(\alpha)H_2[v_0(\alpha)]\phi[v_0(\alpha)] \\
&\quad - 2n^{-1}v_0^2(\alpha)v_1^2(\alpha)\phi[v_0(\alpha)] + o(n^{-1/2}),
\end{aligned}
$$

as $n \to \infty$. However $2n^{-1}v_0^2(\alpha)v_1^2(\alpha)\phi[v_0(\alpha)] = o(n^{-1/2})$ so that

$$\phi(g_{\alpha,n})H_2(g_{\alpha,n}) = \phi[v_0(\alpha)]H_2[v_0(\alpha)] +$$
$$n^{-1/2}v_0(\alpha)v_1(\alpha)\{2 - H_2[v_0(\alpha)]\}\phi[v_0(\alpha)] + o(n^{-1/2}),$$

as $n \to \infty$. This, therefore, implies that

$$\tfrac{1}{6}n^{-1/2}\rho_3\phi(g_{\alpha,n})H_2(g_{\alpha,n}) = \tfrac{1}{6}n^{-1/2}\rho_3\phi[v_0(\alpha)]H_2[v_0(\alpha)] +$$
$$\tfrac{1}{6}n^{-1}\rho_3v_0(\alpha)v_1(\alpha)\{2 - H_2[v_0(\alpha)]\}\phi[v_0(\alpha)] + o(n^{-1}), \quad (7.35)$$

as $n \to \infty$. For the final term in Equation (7.31) we require an asymptotic expansion for $\phi(g_{\alpha,n})[\tfrac{1}{24}\rho_4 H_3(g_{\alpha,n}) + \tfrac{1}{72}\rho_3^2 H_5(g_{\alpha,n})]$ whose coefficient, which is n^{-1}, implies that our error term for this expansion may be $o(1)$ as $n \to \infty$. We begin by noting that $H_3(g_{\alpha,n}) = g_{\alpha,n}^3 - 3g_{\alpha,n}$, where it can be shown that

$$g_{\alpha,n}^3 = [v_0(\alpha) + n^{-1/2}v_1(\alpha) + n^{-1}v_2(\alpha) + o(n^{-1})]^3 = v_0^3(\alpha) + o(n^{-1}),$$

as $n \to \infty$. Along with the direct result that $3g_{\alpha,n} = 3v_0(\alpha) + o(1)$ as $n \to \infty$ implies that $H_3(g_{\alpha,n}) = H_3[v_0(\alpha)] + o(1)$ as $n \to \infty$. Similarly, it can be shown that $H_4(g_{\alpha,n}) = H_4[v_0(\alpha)] + o(1)$ as $n \to \infty$. See Exercise 9. Previous calculations imply that $\phi(g_{\alpha,n}) = \phi[v_0(\alpha)] + o(1)$ as $n \to \infty$ so that

$$n^{-1}\phi(g_{\alpha,n})[\tfrac{1}{24}\rho_4 H_3(g_{\alpha,n}) + \tfrac{1}{72}\rho_3^2 H_5(g_{\alpha,n})] =$$
$$n^{-1}\phi[v_0(\alpha)]\{\tfrac{1}{24}\rho_4 H_3[v_0(\alpha)] + \tfrac{1}{72}\rho_3^2 H_5[v_0(\alpha)]\} + o(n^{-1}), \quad (7.36)$$

as $n \to \infty$. Combining the results of Equations (7.34), (7.35), and (7.36) implies that

$$F_n(g_{\alpha,n}) = \Phi[v_0(\alpha)] + n^{-1/2}\phi[v_0(\alpha)]\{v_1(\alpha) - \tfrac{1}{6}\rho_3 H_2[v_0(\alpha)]\} +$$
$$n^{-1}\phi[v_0(\alpha)]\{v_2(\alpha) - \tfrac{1}{2}v_0(\alpha)v_1^2(\alpha) - \tfrac{1}{6}\rho_3 v_0(\alpha)v_1(\alpha)(2 - H_2[v_0(\alpha)]) -$$
$$\tfrac{1}{24}\rho_4 H_3[v_0(\alpha)] - \tfrac{1}{72}\rho_3^2 H_5[v_0(\alpha)]\} + o(n^{-1}), \quad (7.37)$$

as $n \to \infty$. To obtain expressions for $v_0(\alpha)$, $v_1(\alpha)$, and $v_2(\alpha)$ we set $F(g_{\alpha,n}) = \alpha + o(n^{-1})$ and match the coefficients of the successive powers of $n^{-1/2}$. Matching the constant terms with respect to n in Equation (7.37) with the constant term α implies that $\Phi[v_0(\alpha)] = \alpha$ or equivalently that $v_0(\alpha) = z_\alpha$. We similarly match coefficient of $n^{-1/2}$ in Equation (7.37) with zero which implies that $\phi(z_\alpha)[v_1(\alpha) - \tfrac{1}{6}\rho_3 H_2(z_\alpha)] = 0$, where we have already made the substitution $v_0(\alpha) = z_\alpha$. This implies that $v_1(\alpha) = \tfrac{1}{6}\rho_3 H_2(z_\alpha)$. We finally match the coefficients of n^{-1} in Equation (7.37) also to zero. This implies

$$v_2(\alpha) + \tfrac{1}{72}z_\alpha \rho_3^2 H_2^2(z_\alpha) - \tfrac{1}{18}\rho_3^2 z_\alpha H_2(z_\alpha) - \tfrac{1}{24}\rho_4 H_3(z_\alpha) - \tfrac{1}{72}\rho_3^2 H_5(z_\alpha) = 0,$$

so that

$$v_2(\alpha) = \tfrac{1}{18}\rho_3^2 z_\alpha H_2(z_\alpha) - \tfrac{1}{72}z_\alpha \rho_3^2 H_2^2(z_\alpha) + \tfrac{1}{24}\rho_4 H_3(z_\alpha) + \tfrac{1}{72}\rho_3^2 H_5(z_\alpha).$$

Therefore, it follows that

$$\begin{aligned}
g_{\alpha,n} &= v_0(\alpha) + n^{-1/2}v_1(\alpha) + n^{-1}v_2(\alpha) + o(n^{-1}) \\
&= z_\alpha + \tfrac{1}{6}n^{-1/2}\rho_3 H_2(z_\alpha) + n^{-1}[\tfrac{1}{18}\rho_3^2 z_\alpha H_2(z_\alpha) - \tfrac{1}{72}z_\alpha \rho_3^2 H_2^2(z_\alpha) + \\
&\quad \tfrac{1}{24}\rho_4 H_3(z_\alpha) + \tfrac{1}{72}\rho_3^2 H_5(z_\alpha)] + o(n^{-1}), \quad (7.38)
\end{aligned}$$

as $n \to \infty$. The asymptotic expansion given in Equation (7.38) is called a second-order Cornish–Fisher expansion. The expansion can also be terminated after the first order term to obtain an expansion of the form $z_\alpha + \tfrac{1}{6}n^{-1/2}\rho_3 H_2(z_\alpha) + o(n^{-1/2})$ as $n \to \infty$, which is called a first-order Cornish–Fisher expansion. Note that the constant term of the expansion is z_α, so as in the case of the Edgeworth expansions, Cornish–Fisher expansions can be interpreted as corrections to the approximation given by Theorem 4.20 (Lindeberg and Lévy). The result can be expanded to form a pth-order Cornish–Fisher expansion that has an error term of order $o(n^{-p/2})$ as $n \to \infty$.

Theorem 7.10. *Let $\{X_n\}_{n=1}^{\infty}$ be a sequence of independent and identically distributed random variables from a distribution F that has characteristic function ψ. Let $F_n(t) = P[n^{1/2}\sigma^{-1}(\bar{X}_n - \mu) \le t]$, $g_{\alpha,n} = \inf\{x \in \mathbb{R} : F_n(x) \ge$*

$\alpha\}$, *and assume that* $E(X_n^p) < \infty$. *If*

$$\limsup_{|t| \to \infty} |\psi(t)| < 1,$$

then

$$g_{n,\alpha} - z_\alpha - \sum_{k=1}^{p} n^{-k/2} q_k(z_\alpha) = o(n^{-p/2}), \tag{7.39}$$

as $n \to \infty$ *uniformly in* $\varepsilon < \alpha < 1 - \varepsilon$ *for each* $\varepsilon > 0$ *where* $q_3, \dots, q_p$ *are polynomials that depend on the moments of* $\mathbf{X}_n$ *and not on* n.

We shall not endeavor to present a more formal proof of Theorem 7.10 other than what is presented above. The polynomials $q_1, \dots, q_p$ are related to the polynomials $r_1, \dots, r_p$, though the relationship can become quite complicated as more terms are added to the expansion. In paticular $q_1(x) = -r_1(x)$ and $q_2(x) = r_1(x)r_1'(x) - \frac{1}{2}xr_1^2(x) - r_2(x)$. These relationships can be determined by directly inverting the asymptotic expansion in Equation (7.27). See Exercise 14. As with the Edgeworth expansion, the expansion given by Theorem 7.10 can also be written so that the error term has order $O(n^{-(p+1)/2})$ as $n \to \infty$.

Because of the relationship between the two methods, many similar conclusions about the accuracy of the normal approximation for the distribution of $n^{1/2}\sigma^{-1}(\bar{X}_n - \mu)$ also hold for the quantiles of $n^{1/2}\sigma^{-1}(\bar{X}_n - \mu)$. In general, we can observe from Theorem 7.10 that the normal approximation provides an approximation for the quantiles of $n^{1/2}\sigma^{-1}(\bar{X}_n - \mu)$ that has error that is $o(1)$, or $O(n^{-1/2})$, as $n \to \infty$. However, if the third standardized cumulant ρ_3 is zero, corresponding to the case where the population is symmetric, the normal approximation provides an approximation for the quantiles of $n^{1/2}\sigma^{-1}(\bar{X}_n - \mu)$, where the error is $o(n^{-1/2})$, or $O(n^{-1})$, as $n \to \infty$. Similarly, if the fourth standardized cumulant ρ_4 is zero, corresponding to the case where the population has the same kurtosis as a NORMAL population, the normal approximation provides an approximation for the quantiles of $n^{1/2}\sigma^{-1}(\bar{X}_n - \mu)$, where the error is $o(n^{-1})$, or $O(n^{-3/2})$, as $n \to \infty$.

Example 7.5. Let $\{X_n\}_{n=1}^\infty$ be a sequence of independent and identically distributed random variables where X_n has an EXPONENTIAL(θ) distribution for all $n \in \mathbb{N}$. In Example 7.4 it was shown that the distribution of $n^{1/2}\sigma^{-1}(\bar{X}_n - \theta)$ has an Edgeworth expansion with $-r_1(x) = \frac{1}{3}H_2(x)$ and $-r_2(x) = \frac{3}{8}H_3(x) + \frac{1}{18}H_5(x)$. Theorem 7.10 then implies that the αth quantile of the distribution of $n^{1/2}\sigma^{-1}(\bar{X}_n - \theta)$ has Cornish–Fisher expansion

$$g_{\alpha,n} = z_\alpha + n^{-1/2}q_1(z_\alpha) + n^{-1}q_2(z_\alpha) + o_p(n^{-1}),$$

as $n \to \infty$, where $q_1(z_\alpha) = -r_1(z_\alpha) = \frac{1}{3}H_3(z_\alpha)$ and $q_2(z_\alpha) = r_1(z_\alpha)r_1'(z_\alpha) - \frac{1}{2}z_\alpha r_1^2(z_\alpha) - r_2(z_\alpha)$. Evaluating the derivative of r_1 we find that

$$r_1'(z_\alpha) = \frac{d}{dx}r_1(z_\alpha)\Big|_{x=z_\alpha} = -\frac{2}{3}H_1(z_\alpha) = -\frac{2}{3}z_\alpha.$$

Therefore

$$q_2(z_\alpha) = \tfrac{2}{9} z_\alpha H_2(z_\alpha) - \tfrac{1}{18} H_2^2(z_\alpha) + \tfrac{3}{8} H_3(z_\alpha) + \tfrac{1}{18} H_5(z_\alpha).$$

∎

7.4 The Smooth Function Model

The sample mean is not the only function of a set of independent and identically distributed random variables that has an asymptotic NORMAL distribution. For example, in Section 6.4, we studied several conditions under which transformations of an asymptotically NORMAL random vector also have an asymptotic NORMAL distribution. The key condition for such transformations to be asymptotically NORMAL was based on the smoothness of the transformation. One particular application of this theory is based on looking at smooth functions of sample mean vectors. Edgeworth and Cornish–Fisher expansions can also be applied to these problems as well, though the results are slightly more complicated and the function of the sample mean must have a certain type of smooth representation. This section will focus on the model for which these expansions are valid. Section 7.5 will provide details about the expansions themselves.

The specification of the smooth function model begins with a sequence of d-dimensional random vectors $\{\mathbf{X}_n\}_{n=1}^\infty$ following a d-dimensional distribution F. It is assumed that these random vectors are independent and identically distributed. Let $\boldsymbol{\mu} = E(\mathbf{X}_n)$ and assume that the components of $\boldsymbol{\mu}$ are finite. The parameter of interest is assumed to be a smooth function of the vector mean $\boldsymbol{\mu}$. That is, we assume that there exists a smooth function $g : \mathbb{R}^d \to \mathbb{R}$ such that $\theta = g(\boldsymbol{\mu})$. The exact requirements on the smoothness of g will be detailed in Section 7.5.

The parameter of interest will be estimated using a plug-in estimate based on the sample mean. That is, we will assume that the mean $\boldsymbol{\mu}$ is estimated by the sample mean

$$\hat{\boldsymbol{\mu}} = \bar{\mathbf{X}}_n = n^{-1} \sum_{k=1}^{n} \mathbf{X}_k.$$

An estimate of θ can then be computed by substituting the sample mean into the function g. That is, $\hat{\theta}_n = g(\hat{\boldsymbol{\mu}})$. Note that under the conditions we have stated thus far $\hat{\theta}_n$ is a consistent estimator of θ owing to the consistency of the sample mean from Theorem 3.10 (Weak Law of Large Numbers) and Theorem 3.9.

In order to correctly standardize the distribution of $\hat{\theta}_n$ we also require the standard error of $\hat{\theta}_n$. In particular we will assume that the asymptotic variance of $n^{1/2}\hat{\theta}_n$ is a constant σ that is also a smooth function of $\boldsymbol{\mu}$. That is, we assume that there is a smooth function $h : \mathbb{R}^d \to \mathbb{R}$ such that

$$\sigma^2 = h^2(\mu) = \lim_{n \to \infty} V(n^{1/2} \hat{\theta}_n).$$

The smooth function model can be summarized as a model for parameters that are smooth functions of a vector mean where the standard error of the estimated parameter is also a smooth function of the vector mean. The smooth function model allows one to study statistics well beyond just the mean. Parameters that have representations within the smooth function model include the variance, the correlation between two random variables, a ratio of means, a ratio of variances, the skewness, the kurtosis, and many more. The key to setting up the model correctly comes from identifying the sequence of random vectors $\{\mathbf{X}_n\}_{n=1}^{\infty}$, their dimension d, and the functions g and h. Note that the specification of the function h further requires one to evaluate the asymptotic variance of $\hat{\theta}_n$. It is important to note that when specifying a smooth function model, the dimension d must be chosen large enough so that both θ and σ can be specified. In a typical specification of the smooth function model, the sequence $\{\mathbf{X}_n\}_{n=1}^{\infty}$ will often consist the random vectors whose components are powers of another random variable or vector. Therefore, the vector mean $\boldsymbol{\mu}$ will be a vector that contains various moments of the original random variable or vector. Therefore both $\{\mathbf{X}_n\}_{n=1}^{\infty}$ and the dimension d must be specified so that there are sufficient moments to specify both θ and σ in the model. Some example smooth function model specifications are given below.

Example 7.6. Let $\{W_n\}_{n=1}^{\infty}$ be a sequence of independent and identically distributed random variables from a distribution F with mean θ. We wish to represent the parameter θ in a smooth function model. Assuming that μ'_k is the kth moment of W_n, we note that $\theta = \mu'_1 = E(W_n)$. While we have not specified the sequence $\{\mathbf{X}_n\}_{n=1}^{\infty}$, the dimension d, or the function h as of yet, the estimate of θ will consist of the component of the sample mean of the vectors $\mathbf{X}_1, \ldots, \mathbf{X}_n$ corresponding to W_n. Therefore, the estimate of θ will be given by $\hat{\theta}_n = \bar{W}_n$. Hence, the asymptotic variance of $\hat{\theta}_n$ is given by

$$\sigma^2 = \lim_{n \to \infty} V(n^{1/2} \hat{\theta}_n) = \lim_{n \to \infty} V(n^{1/2} \bar{W}_n) = \mu'_2 - (\mu'_1)^2.$$

Since the parameter and the asymptotic variance rely on the first two moments of W_n we can specify a smooth function model for θ with $d = 2$ and $\mathbf{X}_n = (W_n, W_n^2)$ so that the vector mean is given by $\boldsymbol{\mu} = E(\mathbf{X}'_n) = (\mu'_1, \mu'_2)$ for all $n \in \mathbb{N}$. The functions g and h are given by $g(\mathbf{x}) = x_1$ and $h(\mathbf{x}) = x_2 - x_1^2$ where $\mathbf{x}' = (x_1, x_2)$. ∎

Example 7.7. Let $\{W_n\}_{n=1}^{\infty}$ be a sequence of independent and identically distributed random variables from a distribution F with mean η and variance θ. We wish to represent the parameter θ in a smooth function model. Assuming that μ'_k is the kth moment of W_n, we note that $\theta = \mu'_2 - (\mu'_1)^2$ so that we will need at least the first two moments of W_n to be represented in our $\mathbf{X}_n$ vector. Theorem 3.21 implies that the asymptotic variance of the sample variance is

$$\mu_4 - \mu_2^2 = \mu'_4 - 4\mu'_1 \mu'_3 + 6(\mu'_1)^2 \mu_2 - 3(\mu'_1)^4 - [\mu'_2 - (\mu'_1)^2]^2,$$

so that to represent the variance in the smooth function model we require $d = 4$ with $\mathbf{X}'_n = (W_n, W_n^2, W_n^3, W_n^4)$ for all $n \in \mathbb{N}$. With this representation we have $g(\mathbf{x}) = x_2 - x_1^2$ and

$$h(\mathbf{x}) = x_4 - 4x_1 x_3 + 6x_1^2 x_2 - 3x_1^4 - (x_2 - x_1^2)^2,$$

where $\mathbf{x}' = (x_1, x_2, x_3, x_4)$. See Exercise 15. ∎

Example 7.8. Let $\{\mathbf{W}_n\}_{n=1}^\infty$ be a sequence of independent and identically distributed bivariate random vectors from a distribution F having mean vector $\boldsymbol{\eta}$ and covariance matrix $\boldsymbol{\Sigma}$. Let $\mathbf{W}'_n = (W_{n1}, W_{n2})$ for all $n \in \mathbb{N}$ and define $\mu_{ij} = E\{[W_{n1} - E(W_{n2})]^i [W_{n2}^j - E(W_{n2})]^j\}$ and $\mu'_{ij} = E(W_{n1}^j W_{n2}^j)$. The parameter of interest in this example is the correlation between W_{n1} and W_{n2}. That is, $\theta = \mu_{11} \mu_{20}^{-1/2} \mu_{02}^{-1/2}$. The estimate of θ, based on replacing the population moments with the sample moments as described above, has the form

$$\hat{\theta}_n = \frac{\sum_{k=1}^n (W_{k1} - \bar{W}_1)(W_{k2} - \bar{W}_2)}{[\sum_{k=1}^n (W_{k1} - \bar{W}_1)^2]^{1/2} [\sum_{k=1}^n (W_{k2} - \bar{W}_2)^2]^{1/2}}.$$

When constructing the smooth function model representation for this parameter, we will need to specify the sequence $\{\mathbf{X}\}_{n=1}^\infty$ to include moments of both W_{n1} and W_{n2}, but also various products of these random variables as well. The correlation parameter itself can be specified readily enough as a function of these moments as shown above, but the asymptotic variance is more challenging. For our current application we will use the result from Section 27.8 of Cramér (1946) which gives the asymptotic variance of $n^{1/2}\hat{\theta}_n$ as

$$\sigma^2 = \tfrac{1}{4}\theta(\mu_{40}\mu_{20}^{-2} + \mu_{04}\mu_{02}^{-2} + 2\mu_{22}\mu_{20}^{-1}\mu_{02}^{-1} +$$
$$4\mu_{22}\mu_{11}^{-2} - 4\mu_{31}\mu_{11}^{-1}\mu_{20}^{-1} - 4\mu_{13}\mu_{11}^{-1}\mu_{02}^{-1}),$$

which makes it apparent that we require moments up to order four from each random variable plus several products of powers of these random variables. It then suffices to define the sequence $\{\mathbf{X}\}_{n=1}^\infty$ as

$$\mathbf{X}'_n = (W_{n1}, W_{n1}^2, W_{n1}^3, W_{n1}^4, W_{n2}, W_{n2}^2, W_{n2}^3, W_{n2}^4,$$
$$W_{n1}W_{n2}, W_{n1}^2 W_{n2}, W_{n1}W_{n2}^2, W_{n1}^2 W_{n2}^2, W_{n1}^3 W_{n2}, W_{n1}W_{n2}^3),$$

for all $n \in \mathbb{N}$, where we have $d = 14$ and

$$\boldsymbol{\mu}' = (\mu'_{10}, \mu'_{20}, \mu'_{30}, \mu'_{40}, \mu'_{01}, \mu'_{02}, \mu'_{03}, \mu'_{04}, \mu'_{11}, \mu'_{21}, \mu'_{12}, \mu'_{22}, \mu'_{31}, \mu'_{13}).$$

From this definition we can define the function

$$g(\mathbf{x}) = \frac{x_9 - x_1 x_5}{(x_2 - x_1^2)^{1/2}(x_6 - x_5^2)^{1/2}},$$

for which $\theta = g(\boldsymbol{\mu})$ and $\hat{\theta}_n = g(\bar{\mathbf{X}}_n)$. The asymptotic variance calculation is somewhat complicated, so we will introduce some helpful functions to make the resulting expressions more compact. We begin by noting that

$$\mu_{40} = \mu'_{40} - 4\mu'_{10}\mu'_{30} + 6\mu'_{20}(\mu'_{10})^2 - 3(\mu'_{10})^4,$$

we can define a function $h_{40}(\mathbf{x})$ as

$$h_{40}(\mathbf{x}) = x_4 - 4x_1 x_3 + 6x_2 x_1^2 - 3x_1^4,$$

so that $h_{40}(\boldsymbol{\mu}) = \mu_{40}$. Similarly, we can define

$$h_{04}(\mathbf{x}) = x_8 - 4x_5 x_7 + 6x_6 x_5^2 - 3x_5^4,$$

so that $h_{40}(\boldsymbol{\mu}) = \mu_{04}$. Extending this idea we define the functions $h_{20}(\mathbf{x}) = x_2 - x_1^2$, $h_{02}(\mathbf{x}) = x_6 - x_5^2$, $h_{11}(\mathbf{x}) = x_9 - x_1 x_5$,

$$h_{22}(\mathbf{x}) = x_{12} - 2x_5 x_{10} + x_5^2 x_2 - 2x_1 x_{11} + 4x_1 x_5 x_9 - 3x_1^2 x_5^2 + x_1^2 x_6,$$

$$h_{31}(\mathbf{x}) = x_{13} - 3x_1 x_{10} + 3x_1^2 x_9 - x_3 x_5 - 3x_1 x_5 x_2 - 3x_1^3 x_5,$$

and

$$h_{13}(\mathbf{x}) = x_{14} - 3x_5 x_{11} + 3x_5^2 x_9 - x_7 x_1 + 3x_1 x_5 x_6 - 3x_1 x_5^3,$$

where it follows that $h_{20}(\boldsymbol{\mu}) = \mu_{20}$, $h_{02}(\boldsymbol{\mu}) = \mu_{02}$, $h_{11}(\boldsymbol{\mu}) = \mu_{11}$, $h_{22}(\boldsymbol{\mu}) = \mu_{22}$, $h_{31}(\boldsymbol{\mu}) = \mu_{31}$, and $h_{13}(\boldsymbol{\mu}) = \mu_{13}$. See Exercise 16. In terms of these functions the asymptotic variance can be written as

$$\sigma^2 = h^2(\boldsymbol{\mu}) = \tfrac{1}{4}g(\boldsymbol{\mu})\{h_{40}(\boldsymbol{\mu})[h_{20}(\boldsymbol{\mu})]^{-2} + h_{04}(\boldsymbol{\mu})[h_{02}(\boldsymbol{\mu})]^{-2} +$$
$$2h_{22}(\boldsymbol{\mu})[h_{20}(\boldsymbol{\mu})h_{02}(\boldsymbol{\mu})]^{-1} + 4h_{22}(\boldsymbol{\mu})[h_{11}(\boldsymbol{\mu})]^{-1} -$$
$$4h_{31}(\boldsymbol{\mu})[h_{11}(\boldsymbol{\mu})h_{20}(\boldsymbol{\mu})]^{-1} - 4h_{13}(\boldsymbol{\mu})[h_{11}(\boldsymbol{\mu})h_{02}(\boldsymbol{\mu})]^{-1}\}.$$

Therefore, the correlation between two random variables is a parameter that can be represented in the smooth function model with $d = 14$.   ∎

The previous examples demonstrate that reasonably smooth functions of moments can always be represented in the smooth function model, though expressions for the h function can become quite complicated. It is worthwhile to note that there are also many parameters which cannot be represented in the smooth function model. For example, a median, or any other quantile for that matter, cannot be written in terms of the smooth function model as there is not any general method for representing the median in terms of the moments of a distribution. Another example of such a parameter is the mode of a density. Further, non-smooth functions of the moments of a distribution will also not fit within the smooth function model.

7.5 General Edgeworth and Cornish–Fisher Expansions

The results in this section rely heavily on the smooth function model presented in Section 7.4. In particular, we will assume that $\{\mathbf{X}_n\}_{n=1}^{\infty}$ is a sequence of independent and identically distributed d-dimensional random vectors from a d-dimensional distribution F. Let $\boldsymbol{\mu} = E(\mathbf{X}_n)$ and assume that the parameter of interest θ is related to $\boldsymbol{\mu}$ through a smooth function g. The parameter θ is estimated with the plug-in estimate $g(\bar{\mathbf{X}}_n)$. Finally, we assume that

$$\sigma^2 = \lim_{n \to \infty} V[n^{1/2} g(\bar{\mathbf{X}}_n)],$$

is also related to μ through a smooth function h.

It is worth noting the effect that transformations have on the normal approximation. We know from Theorem 4.22 that $n^{1/2}\Sigma^{-1/2}(\bar{\mathbf{X}}_n - \mu) \xrightarrow{d} \mathbf{Z}$ as $n \to \infty$ where $\mathbf{Z}$ is a d-dimensional $\mathbf{N}(\mathbf{0}, \mathbf{I})$ random variable and Σ is the covariance matrix of $\mathbf{X}_n$ as long as the elements of Σ are all finite. Given this result, we can then apply Theorem 6.5 to find that $n^{1/2}[g(\mathbf{X}_n) - g(\mu)] \xrightarrow{d} Z$ as $n \to \infty$ where Z is a $N[0, \mathbf{d}'(\mu)\Sigma\mathbf{d}(\mu)]$ random variable and $\mathbf{d}(\mu)$ is the vector of partial derivatives of g evaluated at μ. To simplify the notation in this section we will let d_i represent the ith element of $\mathbf{d}(\mu)$. We now encounter some smoothness conditions required in our smooth function model. In particular, we must now assume that $\mathbf{d}(\mu)$ is not equal to the zero vector and $\mathbf{d}(\mathbf{x})$ is continuous in a neighborhood of μ.

An important issue in developing Edgeworth expansions for under the smooth function model is related to finding expressions for the cumulants of $\hat{\theta}_n$ in terms of the moments of the distribution F. This is important because, as with the usual Edgeworth expansion, the coefficients of the terms of the expansion will be related to the cumulants of $\hat{\theta}_n$, and therefore the specification of the cumulants is required to specify the exact form for the expansion. As an example we will consider the case of the specification of σ^2. Suppose that $\mathbf{X}' = (X_1, \ldots, X_d)$ where $\mathbf{X}$ has distribution F and define $\mu_{ij} = E\{[X_i - E(X_i)][X_j - E(X_j)]\}$ so that the $(i, j)^{\text{th}}$ element of Σ is given by μ_{ij}. The quadratic form representing the asymptotic variance of $n^{1/2}(\hat{\theta}_n - \theta)$, which is equivalently the asymptotic variance of $n^{1/2}\hat{\theta}_n$, is then given by

$$\sigma^2 = \mathbf{d}'(\mu)\Sigma\mathbf{d}(\mu) = \sum_{i=1}^{d}\sum_{j=1}^{d} d_i d_j \mu_{ij}.$$

It turns out that all of the cumulants of $n^{1/2}(\hat{\theta}_n - \theta)$ can be written using similar expressions to these forms, though the proof becomes rather complicated for each additional cumulant. We shall present only the results. A detailed argument supporting these results can be found in Chapter 2 of Hall (1992). Define $A(\mathbf{x}) = [g(\mathbf{x}) - g(\mu)]/h(\mu)$ where $\mathbf{x}' = (x_1, \ldots, x_d)$. Let

$$a_{i_1 \cdots i_k} = \left.\frac{\partial^k}{\partial x_{i_1} \cdots \partial x_{i_k}} A(\mathbf{x})\right|_{\mathbf{x}=\mu}.$$

It then follows that the first cumulant $n^{1/2}A(\bar{X}_n) = n^{1/2}\sigma^{-1}(\hat{\theta}_n - \theta)$ equals $n^{-1/2}A_1 + O(n^{-1})$ as $n \to \infty$ where

$$A_1 = \tfrac{1}{2}\sum_{i=1}^{d}\sum_{j=1}^{d} a_{ij}\mu_{ij}. \qquad (7.40)$$

If we have chosen our h function correctly, the second cumulant of $n^{1/2}A(\bar{X}_n)$ should be one. See Page 55 of Hall (1992) for further details. The third cumulant of $n^{1/2}A(\bar{X}_n)$ is given by $n^{-1/2}A_2 + O(n^{-1})$ where

$$A_2 = \sum_{i=1}^{d}\sum_{j=1}^{d}\sum_{k=1}^{d} a_i a_j a_k \mu_{ijk} + 3\sum_{i=1}^{d}\sum_{j=1}^{d}\sum_{k=1}^{d}\sum_{l=1}^{d} a_i a_j a_{kl}\mu_{ik}\mu_{jl}, \qquad (7.41)$$

where $\mu_{ijk} = E\{[X_i - E(X_i)][X_j - E(X_j)][X_k - E(X_k)]\}$.

The main result of this section is that the form of the Edgeworth expansion for the distribution function of $n^{1/2}\sigma^{-1}(\hat{\theta}_n - \theta)$ has the same form as that for the standardized sample mean, with the exception that the cumulants are replaced by those above. This result was established by Bhattacharya and Ghosh (1978).

Theorem 7.11 (Bhattacharya and Ghosh). *Let $\{\mathbf{X}_n\}_{n=1}^{\infty}$ be a sequence of independent and identically distributed d-dimensional random vectors from a distribution F with mean vector $\boldsymbol{\mu}$. Let $\theta = g(\boldsymbol{\mu})$ for some function g and suppose that $\hat{\theta}_n = g(\bar{\mathbf{X}}_n)$. Let σ^2 be the asymptotic variance of $n^{1/2}\hat{\theta}_n$ and assume that $\sigma = h(\boldsymbol{\mu})$ for some function h. Define $A(\mathbf{x}) = \sigma^{-1}[g(\mathbf{x}) - g(\boldsymbol{\mu})]$ and assume that A has $p+2$ continuous derivatives in a neighborhood of $\boldsymbol{\mu}$ and that $E(||\mathbf{X}||^{p+2}) < \infty$. Let ψ be the characteristic function of F and assume that*

$$\limsup_{||\mathbf{t}||\to\infty} |\psi(\mathbf{t})| < \infty. \qquad (7.42)$$

Let $G_n(x) = P[n^{1/2}A(\bar{\mathbf{X}}_n) \le x]$, then

$$G_n(x) = \Phi(x) + \sum_{k=1}^{p} n^{-k/2} r_k(x)\phi(x) + o(n^{-p/2}), \qquad (7.43)$$

as $n \to \infty$, uniformly in x where $r_1, \ldots, r_p$ are polynomials that depend on the moments of $\mathbf{X}$ up to order $p + 2$.

As indicated previously, the form of the polynomials is the same as in Theorem 7.9. In particular we have that

$$r_1(x) = -[\tilde{\sigma}^{-1}A_1 + \tfrac{1}{6}\tilde{\sigma}^{-3}A_2 H_2(x)], \qquad (7.44)$$

where A_1 and A_2 are defined in Equations (7.40) and (7.41) and

$$\tilde{\sigma} = \sum_{i=1}^{d}\sum_{j=1}^{d} a_i a_j \mu_{ij}. \qquad (7.45)$$

In most applications we will have $\tilde{\sigma} = 1$. One may note that the extra term involving the cumulant A_1 in Equation (7.44) that does not appear in the polynomial defined in Theorem 7.9. This is due to the fact that the first cumulant of $\bar{X}_n - \mu$ is zero. Further polynomials can be obtained using the methodology of Withers (1983, 1984). For example, Polansky (1995) obtains

$$r_2(x) = \tfrac{1}{2}\tilde{\sigma}^{-2}(A_1^2 + A_{22})H_1(x) +$$
$$\tfrac{1}{24}(4\tilde{\sigma}^{-2}A_1 A_2 + \tilde{\sigma}^{-4}A_{43})H_3(x) + \tfrac{1}{72}\tilde{\sigma}^{-2}A_2^2 H_5(x), \quad (7.46)$$

where

$$A_{22} = \sum_{i=1}^{d}\sum_{j=1}^{d}\sum_{k=1}^{d} a_i a_{jk} \mu_{ijk} + \frac{1}{2}\sum_{i=1}^{d}\sum_{j=1}^{d}\sum_{k=1}^{d}\sum_{l=1}^{d} a_{ij} a_{kl} \mu_{ik} \mu_{jl} +$$

$$\sum_{i=1}^{d}\sum_{j=1}^{d}\sum_{k=1}^{d}\sum_{l=1}^{d} a_i a_{jkl} \mu_{ij} \mu_{kl}, \quad (7.47)$$

and

$$A_{43} = \frac{1}{2}\sum_{i=1}^{d}\sum_{j=1}^{d}\sum_{k=1}^{d}\sum_{l=1}^{d} a_i a_j a_k a_l \mu_{ijkl} - 3\left(\frac{1}{2}\sum_{i=1}^{d}\sum_{j=1}^{d} a_i a_j \mu_{ij}\right)^2 +$$

$$12\sum_{i=1}^{d}\sum_{j=1}^{d}\sum_{k=1}^{d}\sum_{l=1}^{d}\sum_{m=1}^{d} a_i a_j a_k a_{lm} \mu_{il} \mu_{jkm} +$$

$$12\sum_{i=1}^{d}\sum_{j=1}^{d}\sum_{k=1}^{d}\sum_{l=1}^{d}\sum_{m=1}^{d}\sum_{u=1}^{d} a_i a_j a_{kl} a_{mu} \mu_{ik} \mu_{jm} \mu_{lu} +$$

$$4\sum_{i=1}^{d}\sum_{j=1}^{d}\sum_{k=1}^{d}\sum_{l=1}^{d}\sum_{m=1}^{d}\sum_{u=1}^{d} a_i a_j a_k a_{lmu} \mu_{il} \mu_{jm} \mu_{ku}. \quad (7.48)$$

Inversion of the asymptotic expansion given in Equation (7.43) will result in a Cornish–Fisher expansion for the quantile function of $n^{1/2} A(\bar{\mathbf{X}}_n)$, which we will denote as $g_{\alpha,n}(\alpha) = G_n^{-1}(\alpha)$. The form of this expansion and the involved polynomials are the same as is detailed in Theorem 7.10, with the exception that the cumulants are defined as above. We will again call these polynomials q_k. Edgeworth expansions also exist for the density of $n^{1/2} A(\bar{\mathbf{X}}_n)$ under additional assumptions. These expansions have the same form as given in Theorem 7.5 with the exception of the form of the cumulants. See Theorem 2.5 of Hall (1992) for further details.

An important assumption in Theorem 7.11 is the multivariate form of Cramér's continuity condition given by Equation (7.49), which is equivalent to the random vector $\mathbf{X}_n$ having a non-degenerate absolutely continuous component. The importance of this condition is that it implies that all the points in the distribution of $\bar{\mathbf{X}}_n$ have exponentially small probabilties so that the distribution of $n^{1/2} A(\bar{\mathbf{X}}_n)$ is virtually continuous. We will not attempt to verify this condition directly, but will use a result from Hall (1992) which verifies the condition in the cases we will encounter in this chapter.

Theorem 7.12 (Hall). *Let W be a random variable that has a nonsingular distribution and define $\mathbf{X}' = (W, W^2, \ldots, W^d)$. If ψ is the characteristic function of $\mathbf{X}$, then*

$$\limsup_{||\mathbf{t}||\to\infty} |\psi(\mathbf{t})| < \infty.$$

A proof of Theorem 7.12 is given in Section Chapter 2 of Hall (1992). Note that the form of the random vector used in Theorem 7.12 is of the same form we used in our specification of the smooth function models in Examples 7.6–7.8. Therefore, as long as the distribution in those examples is nonsingular, the required result for Cramér's continuity condition should follow.

Example 7.9. Let $\{W_n\}_{n=1}^{\infty}$ be a sequence of independent and identically distributed random variables from a distribution F with mean η and variance θ. It was shown in Example 7.7 that θ can be represented using a smooth function model with $d = 4$ with $\mathbf{X}_n' = (W_n, W_n^2, W_n^3, W_n^4)$ for all $n \in \mathbb{N}$. In this representation we specified $g(\mathbf{x}) = x_2 - (x_1)^2$ and

$$h(\mathbf{x}) = x_4 - 4x_1x_3 + 6x_1^2x_2 - 3x_1^4 - (x_2 - x_1^2)^2,$$

where $\mathbf{x}' = (x_1, x_2, x_3, x_4)$. We will now demonstrate how this model can be used to find the form of the Edgeworth expansion for the standardized distribution of $\hat{\theta}_n$. For simplicity we will examine the special case where $\eta = 0$ and $\theta = 1$. In this special case direct calculations can be used to show that $\boldsymbol{\mu}' = (0, 1, \gamma, \kappa)$ where we have defined $\gamma = E(W_n^3)$ and $\kappa = E(W_n^4)$. It then follows that $\sigma^2 = h(\boldsymbol{\mu}) = \kappa - \theta^2 = \kappa - 1$. To obtain the polynomial r_1 we need to find expressions for the constants a_i, a_{ij}, and a_{ijk} for $i = 1, \ldots, 4$, $j = 1, \ldots, 4$, and $k = 1, \ldots, 4$. In this case $A(\mathbf{x}) = \sigma^{-1}[g(\mathbf{x}) - g(\boldsymbol{\mu})]$ where $\sigma^2 = h(\boldsymbol{\mu})$. Therefore,

$$a_1 = \left. \frac{\partial}{\partial x_1} \sigma^{-1}[x_2 - x_1^2 - \theta] \right|_{\mathbf{x}=\boldsymbol{\mu}} = -2\sigma^{-1}\eta = 0,$$

since we have assumed that $\eta = 0$. Similarly,

$$a_2 = \left. \frac{\partial}{\partial x_2} \sigma^{-1}[x_2 - x_1^2 - \theta] \right|_{\mathbf{x}=\boldsymbol{\mu}} = \sigma^{-1},$$

and $a_3 = a_4 = 0$. Similarly,

$$a_{11} = \left. \frac{\partial^2}{\partial x_1^2} \sigma^{-1}[x_2 - x_1^2 - \theta] \right|_{\mathbf{x}=\boldsymbol{\mu}} = -2\sigma^{-1},$$

$a_{12} = a_{21} = a_{22} = 0$, $a_{3i} = a_{i3} = 0$, and $a_{4i} = a_{i4} = 0$ for $i \in \{1, 2, 3, 4\}$. Therefore, it follows from Equation (7.40) that

$$A_1 = \frac{1}{2} \sum_{i=1}^{d} \sum_{j=1}^{d} a_{ij}\mu_{ij} = \frac{1}{2}a_{11}\mu_{11} = -\sigma^{-1}\mu_{11},$$

where $\mu_{11} = E(W_n^2) = \theta = 1$. Therefore $A_1 = -\sigma^{-1}$. Similarly, Equation (7.41) implies that

$$A_2 = \sum_{i=1}^{d} \sum_{j=1}^{d} \sum_{k=1}^{d} a_i a_j a_k \mu_{ijk} + 3 \sum_{i=1}^{d} \sum_{j=1}^{d} \sum_{k=1}^{d} \sum_{l=1}^{d} a_i a_j a_{kl} \mu_{ik}\mu_{jl} =$$

$$a_2^3\mu_{222} + 3a_{11}a_2^2\mu_{21}^2.$$

Let

$$\mu_{222} = E[(W_n^2 - \theta)^3] = E[(W_n^2 - 1)^3] = E[W_n^6 - 3W_n^4 + 3W_n^2 - 1] = \zeta,$$

and

$$\mu_{21} = E[(W_n^2 - \theta)W_n] = E[W_n^3 - W_n] = \gamma.$$

Hence

$$A_2 = \sigma^{-3}\zeta - 6\sigma^{-3}\gamma^2.$$

Finally, Equation (7.45) implies that

$$\tilde{\sigma} = \sum_{i=1}^{d}\sum_{j=1}^{d} a_i a_j \mu_{ij} = a_2^2 \mu_{22} = \sigma^{-2}\mu_{22}$$

where $\mu_{22} = E[(W_n^2 - 1)^2] = E(W_n^4) - 2E(W_n^2) + 1 = \kappa - 1 = \sigma^2$. Therefore, $\tilde{\sigma} = 1$ and it follows that the polynomial r_1 is given by

$$r_1(x) = \sigma^{-1} - \sigma^{-3}[\tfrac{1}{6}\zeta - \gamma^2]H_2(x).$$

∎

7.6 Studentized Statistics

In the theory of Edgeworth and Cornish–Fisher expansions developed so far we have dealt with standardized statistics of the form $n^{1/2}\sigma^{-1}(\hat{\theta}_n - \theta)$ where θ is a parameter that fits within the smooth function model. Note that this setup requires that the asymptotic variance of $n^{1/2}\hat{\theta}_n$ be known. In most practical cases the variance is unknown and therefore must be estimated from the observed sample. If we note that within the smooth function model we have $\sigma = h(\boldsymbol{\mu})$ and that $\bar{\mathbf{X}}_n$ is a consistent estimate of $\boldsymbol{\mu}$, then it follows from Theorem 3.9 that we can obtain a consistent estimate of σ as $\hat{\sigma}_n = h(\bar{\mathbf{X}}_n)$. We can now apply Theorem 4.11 (Slutsky) as detailed in Example 4.13 and conclude that $n^{1/2}\hat{\sigma}_n^{-1}(\hat{\theta}_n - \theta)$ is still asymptotically NORMAL. The most well known example of this type of application occurs when $\hat{\theta}_n$ is the sample mean. Even when the population is NORMAL, $n^{1/2}\hat{\sigma}_n^{-1}(\hat{\theta}_n - \theta)$ does not have a N$(0,1)$ distribution but instead has a T$(n-1)$ distribution. However, $n^{1/2}\hat{\sigma}_n^{-1}(\hat{\theta}_n - \theta)$ still has an asymptotic NORMAL distribution. The case of the sample mean under a normal population is interesting because the distribution of $n^{1/2}\hat{\sigma}_n^{-1}(\hat{\theta}_n - \theta)$ is known for finite values of n. The T$(n-1)$ distribution is similar in shape to a N$(0,1)$ distribution in that both distributions are symmetric with a single peak at the origin. However, the T$(n-1)$ distribution has heavier tails, with the difference between the two distributions vanishing as $n \to \infty$. This suggests that an Edgeworth type correction to the N$(0,1)$ distribution might be possible for this case. It turns out that this type of correction is possible even when the population is not normal.

We will need to specify some additional notation before presenting the result.

Define $B(\mathbf{x}) = [g(\mathbf{x}) - g(\boldsymbol{\mu})]/h(\mathbf{x})$ where $\mathbf{x} \in \mathbb{R}^d$ and we note that the $B(\bar{\mathbf{X}}_n) = \hat{\sigma}_n^{-1}(\hat{\theta}_n - \theta)$. While $n^{1/2}A(\bar{\mathbf{X}}_n)$ was called the standardized version of $\hat{\theta}_n$, the function $n^{1/2}B(\bar{\mathbf{X}}_n)$ is usually called the *studentized* version of $\hat{\theta}_n$, alluding to the distribution of $n^{1/2}B(\bar{\mathbf{X}}_n)$ when the population is NORMAL. Let

$$b_{i_1 \cdots i_k} = \frac{\partial^k}{\partial x_{i_1} \cdots \partial x_{i_k}} B(\mathbf{x}) \bigg|_{\mathbf{x} = \boldsymbol{\mu}}.$$

With this definition, the Edgeworth type correction for the distribution function of $B(\bar{\mathbf{X}}_n)$ then has the same form as that of $A(\bar{\mathbf{X}}_n)$, with the exception that the constants $a_{i_1 \cdots i_k}$ are replaced by $b_{i_1 \cdots i_k}$.

Theorem 7.13. *Let $\{\mathbf{X}_n\}_{n=1}^{\infty}$ be a sequence of independent and identically distributed d-dimensional random vectors from a distribution F with mean vector $\boldsymbol{\mu}$. Let $\theta = g(\boldsymbol{\mu})$ for some function g and suppose that $\hat{\theta}_n = g(\bar{\mathbf{X}}_n)$. Let σ^2 be the asymptotic variance of $n^{1/2}\hat{\theta}_n$ and assume that $\sigma = h(\boldsymbol{\mu})$ for some function h. Define $B(\mathbf{x}) = [g(\mathbf{x}) - g(\boldsymbol{\mu})]/h(\mathbf{x})$ and assume that B has $p + 2$ continuous derivatives in a neighborhood of $\boldsymbol{\mu}$ and that $E(\|\mathbf{X}\|^{p+2}) < \infty$. Let ψ be the characteristic function of F and assume that*

$$\limsup_{\|\mathbf{t}\| \to \infty} |\psi(\mathbf{t})| < \infty. \tag{7.49}$$

Let $H_n(x) = P[n^{1/2}B(\bar{\mathbf{X}}_n) \le x]$, then

$$H_n(x) = \Phi(x) + \sum_{k=1}^{p} n^{-k/2} v_k(x)\phi(x) + o(n^{-p/2}), \tag{7.50}$$

as $n \to \infty$, uniformly in x where $v_1, \ldots, v_p$ are polynomials that depend on the moments of $\mathbf{X}$ up to order $p + 2$.

The polynomial q_1 is given by

$$v_1(x) = -[B_1 + \tfrac{1}{6} B_2 H_2(x)] \tag{7.51}$$

where

$$B_1 = \tfrac{1}{2} \sum_{i=1}^{d} \sum_{j=1}^{d} b_{ij} \mu_{ij}, \tag{7.52}$$

$$B_2 = \sum_{i=1}^{d} \sum_{j=1}^{d} \sum_{k=1}^{d} b_i b_j b_k \mu_{ijk} + 3 \sum_{i=1}^{d} \sum_{j=1}^{d} \sum_{k=1}^{d} \sum_{l=1}^{d} b_i b_j b_{kl} \mu_{ik} \mu_{jl}, \tag{7.53}$$

and μ_{ij} and μ_{ijk} are as defined previously. Even though r_1 and q_1 have the same form, the two polynomials are not equal as $a_{i_1 \cdots i_k} \ne b_{i_1 \cdots i_k}$. In fact, Hall (1988a) points out that

$$r_1(x) - v_1(x) = -\tfrac{1}{2}\sigma^{-3} \sum_{i=1}^{d} \sum_{j=1}^{d} a_i c_j \mu_{ij} x^2,$$

where

$$c_k = \frac{\partial}{\partial x_k} g(\mathbf{x}) \Big|_{\mathbf{x}=\boldsymbol{\mu}}.$$

See Exercise 18. Further polynomials may be obtained using similar methods.

The expansion in Equation (7.50) can also be inverted to obtain a Cornish–Fisher type expansion for the quantiles of the distribution of $B(\bar{\mathbf{X}}_n)$. This expansion has the form

$$h_{\alpha,n} = z_\alpha + \sum_{k=1}^{p} n^{-k/2} s_k(z_\alpha) + o(n^{-p/2}),$$

as $n \to \infty$ where $s_1(x) = -v_1(x)$ and $s_2(x) = v_1(x)v_1'(x) - \frac{1}{2}xv_1^2(x) - v_2(x)$.

Example 7.10. Let $\{W_n\}_{n=1}^{\infty}$ be a sequence of independent and identically distributed random variables from a distribution F with mean θ and variance σ^2. It was shown in Example 7.6 that this parameter can be represented in the smooth function model with $\mathbf{X}_n' = (W_n, W_n^2)$, $g(\mathbf{x}) = x_1$, and $h(\mathbf{x}) = x_2 - x_1^2$, where we would have $\hat{\theta}_n = \bar{W}_n$ and

$$\hat{\sigma}_n^2 = n^{-1} \sum_{k=1}^{n} (W_k - \bar{W}_n)^2.$$

In this example, we will derive the one-term Edgeworth expansion for the studentized distribution of $\hat{\theta}_n$ and compare it to the Equation (7.24). For simplicity, we will consider the case where $\theta = 0$ and $\sigma = 1$. To obtain this expansion, we must first find the constants b_1, b_2, b_{11}, b_{12}, b_{21} and b_{22}, where $B(\mathbf{x}) = (x_1 - \theta)(x_2 - x_1^2)^{-1/2}$. We, first, note that

$$
\begin{aligned}
b_1 &= \frac{\partial}{\partial x_1} B(\mathbf{x}) \Big|_{\mathbf{x}=\boldsymbol{\mu}} \\
&= x_1(x_1 - \theta)(x_2 - x_1^2)^{-3/2} + (x_2 - x_1^2)^{-1/2} \Big|_{\mathbf{x}=\boldsymbol{\mu}} \\
&= \theta(\theta - \theta)(\beta - \theta^2)^{-3/2} + (\beta - \theta^2)^{-1/2} \\
&= (\beta - \theta^2)^{-1/2} \\
&= 1,
\end{aligned}
$$

where $\beta = E(W_n^2)$ so that $\beta - \theta^2 = \sigma^2 = 1$. Similarly,

$$
\begin{aligned}
b_2 &= \frac{\partial}{\partial x_2} B(\mathbf{x}) \Big|_{\mathbf{x}=\boldsymbol{\mu}} \\
&= -\tfrac{1}{2}(x_1 - \theta)(x_2 - x_1^2)^{-3/2} \Big|_{\mathbf{x}=\boldsymbol{\mu}} \\
&= 0.
\end{aligned}
$$

The second partial derivatives are given by

$$
\begin{aligned}
b_{11} &= \left.\frac{\partial^2}{\partial x_1^2}B(\mathbf{x})\right|_{\mathbf{x}=\mu} \\
&= 2x_1(x_2-x_1^2)^{-3/2} + 3(x_1-\theta)x_1^2(x_2-x_1^2)^{-5/2} + \\
&\quad \left.(x_1-\theta)(x_2-x_1^2)^{-3/2}\right|_{\mathbf{x}=\mu} \\
&= 0,
\end{aligned}
$$

$$
\begin{aligned}
b_{12}=b_{21} &= \left.\frac{\partial^2}{\partial x_1 x_2}B(\mathbf{x})\right|_{\mathbf{x}=\mu} \\
&= \left.-\tfrac{3}{2}x_1(x_1-\theta)(x_2-x_1^2)^{-5/2} - \tfrac{1}{2}(x_2-x_1^2)^{-3/2}\right|_{\mathbf{x}=\mu} \\
&= -\tfrac{1}{2}(\beta-\theta^2)^{-3/2} \\
&= -\tfrac{1}{2},
\end{aligned}
$$

and

$$
\begin{aligned}
b_{22} &= \left.\frac{\partial^2}{\partial x_2^2}B(x)\right|_{\mathbf{x}=\mu} \\
&= \left.\tfrac{3}{4}(x_1-\theta)(x_2-x_1^2)^{-5/2}\right|_{\mathbf{x}=\mu} \\
&= 0.
\end{aligned}
$$

The expressions for B_1 and B_2 also require us to find the moments of the form μ_{ij} and μ_{ijk} for $i=1,2$, $j=1,2$, and $k=1,2$. However, due to the fact that $b_2=b_{11}=b_{22}=0$ we only need to find μ_{11}, μ_{12}, and μ_{111}. Letting κ_3 denote the third cumulant of W_n we have in this case that $\mu_{11}=E[(W_n-\theta)^2]=\beta-\theta^2=1$, $\mu_{12}=\mu_{21}=E[W_n^3]-\theta\beta=\kappa_3$, $\mu_{111}=E[(W_n-\theta)^3]=\kappa_3$, We now have enough information to compute B_1 and B_2. From Equation (7.52), we have that

$$
B_1 = \tfrac{1}{2}\sum_{i=1}^2\sum_{j=1}^2 b_{ij}\mu_{ij} = \tfrac{1}{2}(b_{12}\mu_{12}+b_{21}\mu_{21}) = -\tfrac{1}{2}\kappa_3.
$$

To find B_2 we first note that

$$
\sum_{i=1}^2\sum_{j=1}^2\sum_{k=1}^2 b_ib_jb_k\mu_{ijk} = b_1^3\mu_{111} = \kappa_3.
$$

Similarly,

$$
\sum_{i=1}^2\sum_{j=1}^2\sum_{k=1}^2\sum_{l=1}^2 b_ib_jb_{kl}\mu_{ik}\mu_{jl} = b_1^2b_{12}\mu_{11}\mu_{12}+b_1^2b_{21}\mu_{12}\mu_{11} = -\kappa_3.
$$

Hence, it follows from Equation (7.53) that $B_2=-2\kappa_3$, so that Equation

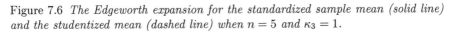

Figure 7.6 *The Edgeworth expansion for the standardized sample mean (solid line) and the studentized mean (dashed line) when $n = 5$ and $\kappa_3 = 1$.*

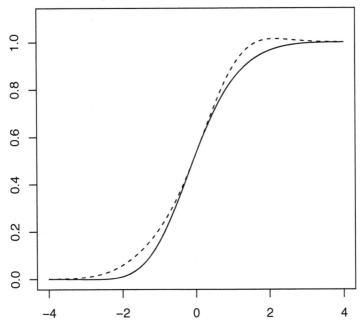

(7.51) implies that

$$v_1(x) = [\tfrac{1}{2}\kappa_3 + \tfrac{1}{3}\kappa_3 H_2(x)] = \tfrac{1}{6}\kappa_3(2x^2 + 1).$$

Hence, under the assumptions of Theorem 7.13 we have that

$$P[n^{1/2}\hat{\sigma}^{-1}(\bar{X}_n - \theta) \leq x] = \Phi(x) + \tfrac{1}{6}n^{-1/2}\phi(x)\kappa_3(2x^2 + 1) + o(n^{-1/2}), \quad (7.54)$$

as $n \to \infty$. For comparison, the one-term Edgeworth expansion from Equation (7.24) and the expansion from Equation (7.54) are plotted in Figure 7.6 for the case when $\kappa_3 = 1$ and $n = 5$. One can observe from these plots the heavier lower tail associated with the studentized distribution of the sample mean. This heavier tail accounts for the extra variability that is introduced into the distribution because the parameter σ has been replaced by the random variable $\hat{\sigma}_n$, along with the fact that the distribution is positively skewed. For the viewpoint of statistical inference, this heavier tail can be interpreted as having less precise information about the underlying population which results in a wider upper confidence limit and possibly less powerful hypothesis tests for the mean.   ∎

Further terms in the Edgeworth expansion for studentized statistics can be determined using calculations of the same form used to find r_2 in the case of standardized statistics. See, for example, Exercise 19. Cornish–Fisher type expansions can also be obtained for the quantile functions of studentized statistics within the smooth function model. The form of these expansions are the same as given in Section 7.3.

7.7 Saddlepoint Expansions

Exponential tilting is a methodology that has been developed to increase the accuracy of Edgeworth expansion in the tails of the distributions of the standardized sample mean. Consider the case where $\{X_n\}_{n=1}^{\infty}$ is a sequence of independent and identically distributed random variables from a distribution F. Under the assumptions of Theorem 7.4 we have that

$$f_n(t) = \phi(t) + \tfrac{1}{6}n^{-1/2}\rho_3 H_3(t)\phi(t) + O(n^{-1}),$$

as $n \to \infty$ where $f_n(t)$ is the density of $n^{1/2}\sigma^{-1}(\bar{X}_n - \mu)$ and ρ_3 is the third standardized cumulant of F. The error term of this expansion is $O(n^{-1})$ as $n \to \infty$ which provides a description of the asymptotic error of the expansion. This error is uniform in t, meaning that it applies to any value of $t \in \mathbb{R}$. However, there are points where the error of the expansion may be quite a bit less than is indicated by the form of the error term. For example, if we consider the form of the expansion at the point $x = 0$ we note that $\phi(0)H_3(0) = 0$ and therefore we obtain the expansion $f_n(0) = \phi(0) + O(n^{-1})$ as $n \to \infty$. Hence, the normal approximation provided by Theorem 4.20 (Lindeberg-Lévy) is actually more accurate near zero, or near the center of the distribution of the standardized mean.

An idea of how this term affects the expansion can be obtained by observing the behavior in Figure 7.7. The function is zero at the origin but quickly increases as we move away from the origin. Asymptotically, however, we also observe that the exponential function in the Normal density eventually dominates the function $H_3(t)$ and the term returns *near* zero once again. Hence, the point $t = 0$ is the only point where the normal approximation attains the smaller asymptotic error term. In practical terms this analysis may provide a somewhat distorted view of how well the Edgeworth expansion performs in the tails of the distribution. While the actual error of the first term of the Edgeworth expansion may become smaller as $t \to \pm\infty$, the error may be large compared to the actual density we are attempting to approximate. A more practical viewpoint looks at this error relative to the actual density of the standardized sample mean. In general this density is unknown without further specification of F, but we can obtain an idea of the relative error by comparing it to the normal density. That is, we would just look at the function $H_3(x)$, which is plotted in Figure 7.8. It is clear from this plot that the relative error of the one-term Edgeworth expansion becomes quite large as $t \to \pm\infty$.

The idea behind exponential tilting is to take advantage of the error term in the Edgeworth expansion when $t = 0$. The consequence of using this methodology is that the resulting approximation will have a relative error of $O(n^{-1})$

Figure 7.7 *The function $\phi(t)H_3(t)$ from the first term of an Edgeworth expansion for the standardized sample mean.*

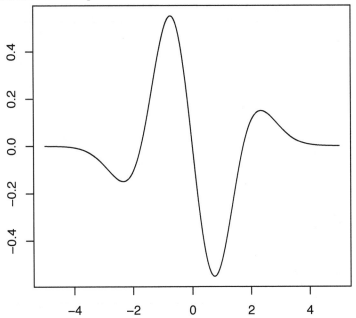

as $n \to \infty$ over a large interval of t values in $\mathbb{R}$ whereas the usual Edgeworth expansion has an absolute error of $O(n^{-1})$ as $n \to \infty$ only at the origin.

To begin our development, let f be the density associated with F and assume that f has moment generating function $m(u)$ and cumulant generating function $c(u)$, both of which are assumed to exist. Define $f_\lambda(t) = \exp(\lambda t)f(t)/m(\lambda)$ for some real parameter λ. It then follows that $f_\lambda(t)$ is a density since $f_\lambda(t) \geq 0$ and

$$\int_{-\infty}^{\infty} \exp(\lambda t)f(t)dt = m(\lambda),$$

which implies

$$\int_{-\infty}^{\infty} f_\lambda(t)dt = [m(\lambda)]^{-1} \int_{-\infty}^{\infty} \exp(\lambda t)f(t)dt = 1.$$

Several properties of this density can be obtained. Let E_λ denote the expectation with respect to f_λ and suppose that Y is a random variable following the density f_λ. That is,

$$E_\lambda(Y) = \int_{-\infty}^{\infty} tf_\lambda(t)dt = \int_{-\infty}^{\infty} [m(\lambda)]^{-1}t\exp(\lambda t)f(t)dt.$$

Figure 7.8 *The function $H_3(t)$ from the first term of an Edgeworth expansion for the standardized sample mean which provides an approximation to the relative error of the Edgeworth expansion as a function of t.*

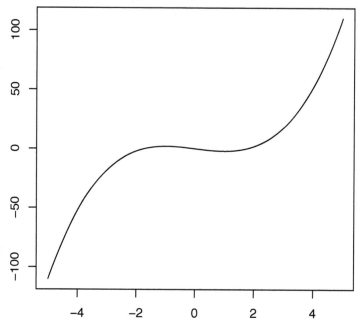

Now the moment generating function of $f_\lambda(t)$ is given by

$$m_\lambda(u) = E_\lambda[\exp(uY)] = \int_{-\infty}^{\infty} [m(\lambda)]^{-1} \exp(ut) \exp(\lambda t) f(t) dt =$$

$$[m(\lambda)]^{-1} \int_{-\infty}^{\infty} \exp[(u + \lambda)t] f(t) dt = [m(\lambda)]^{-1} m(u + \lambda).$$

Now suppose that $\{Y_n\}_{n=1}^{\infty}$ is a sequence of independent and identically distributed random variables following the density f_λ. Then Theorem 2.25 implies that the moment generating function of $n\bar{Y}_n$ can be found to be

$$E_\lambda[\exp(un\bar{Y}_n)] = E_\lambda^n[\exp(uY_n)] = \left[\frac{m(u + \lambda)}{m(\lambda)}\right]^n.$$

The central idea to exponential tilting is to develop a connecting between the distribution of $n\bar{X}_n$ and $n\bar{Y}_n$. The density of $n\bar{X}_n$ can then be written in terms of the density of $n\bar{Y}_n$ where we are free to choose the parameter λ as we please. By choosing a specific choice for λ, the density of $n\bar{X}_n$ can be written in terms of the center of the density $n\bar{Y}_n$. An Edgeworth expansion is then applied to the center of the density of $n\bar{Y}_n$ to obtain the expansion we want.

We must first find the connection between the density of $n\bar{X}_n$ and $n\bar{Y}_n$. Let f_n denote the density of $n\bar{X}_n$ and $f_{n,\lambda}$ denote the density of $n\bar{Y}_n$. We have obtained the fact that the moment generating functions of f_n and $f_{n,\lambda}$ are $m^n(u)$ and $m^n(u+\lambda)/m^n(\lambda)$ respectively. We now informally follow the argument in Section 4.3 of Barndorff-Nielsen and Cox (1989). Note that since

$$m(u) = \int_{-\infty}^{\infty} \exp(ut)f(t)dt,$$

it follows that if we invert $m(u)$ we get the density $f(t)$. Similarly, we note that

$$m(u+\lambda) = \int_{-\infty}^{\infty} \exp[t(u+\lambda)]f(t)dt = \int_{-\infty}^{\infty} \exp(ut)[\exp(\lambda t)f(t)]dt.$$

Therefore, if we invert the moment generating function $m(u+\lambda)$ we should get the function $\exp(\lambda t)f(t)$. Noting that $m(\lambda)$ is a constant with respect to u, it follows that if we invert $m(u+\lambda)/m(\lambda)$ we obtain $\exp(\lambda t)f(t)/m(\lambda)$. We now shift this same argument to the moment generating functions of $n\bar{X}_n$ and $n\bar{Y}_n$. We know that $m^n(u)$ is the moment generating function of $n\bar{X}_n$ so that $m^n(u)$ should invert to $f_n(t)$. Similarly, following the above pattern, $m^n(u+\lambda)$ should invert to $\exp(\lambda t)f_{n,\lambda}(t)$, and hence the density associated with $m^n(u+\lambda)/m^n(\lambda)$ is $\exp(\lambda t)f_n(t)/m^n(\lambda)$. This establishes that $f_{n,\lambda}(t) = \exp(\lambda t)f_n(t)/m^n(\lambda)$. Note that from Definition 2.13 we have that $m^n(\lambda) = \exp\{n\log[m(\lambda)]\} = \exp[nc(\lambda)]$ so that we have established the identity

$$f_{n,\lambda}(t) = \exp[\lambda t - nc(\lambda)]f_n(t), \tag{7.55}$$

which connects the density of $n\bar{X}_n$ given by $f_n(t)$ to the density of $n\bar{Y}_n$ given by $f_{n,\lambda}(t)$. The informality in this argument comes from the fact that $\exp(\lambda t)f(t)$ may not be a density, but we may think of the inversion of the moment generating functions as a type of LaPlace transformation. Alternately, we could work with characteristic functions. See Section 4.3 of Barndorff-Nielsen and Cox (1989) for further details.

Suppose that our aim is to obtain an approximation of $f_n(t)$ at a point $t \in \mathbb{R}$ and we wish to take advantage of the identity in Equation (7.55) so that we can choose λ to give us an accurate Edgeworth expansion. To begin, we note that Equation (7.55) implies that $f_n(t) = \exp[nc(\lambda) - \lambda t]f_{n,\lambda}(t)$. We will take an Edgeworth expansion for $f_{n,\lambda}$ but we wish to do so where the expansion is most accurate, which is in the center. The key to this is to choose λ so that t corresponds to the expected value of $n\bar{Y}_n$. That is, we choose $\lambda = \tilde{\lambda}$ so that $E_\lambda(n\bar{Y}_n) = t$. Now $E_\lambda(n\bar{Y}_n) = nE_\lambda(Y)$ where we can find the expectation of Y using Theorem 2.21 along with the moment generating function of Y. That is

$$E_\lambda(Y) = \frac{d}{du}\left.\frac{m(u+\lambda)}{m(\lambda)}\right|_{u=0} = \frac{m'(\lambda)}{m(\lambda)} = \frac{d}{du}\log[m(u)]\Big|_{u=\lambda} = c'(\lambda).$$

Therefore, it follows that we choose $\lambda = \tilde{\lambda}$ such that $nc'(\tilde{\lambda}) = t$.

To find the approximation we must now find an Edgeworth expansion for $f_{n,\lambda}(t)$ evaluated at the mean of $n\bar{Y}_n$. Theorem 7.4 provides an expansion for the density of $n^{1/2}\sigma^{-1}(\bar{Y}_n - \mu)$. To connect these two densities, we first note that

$$F_{n,\lambda}(t) = P(n\bar{Y}_n \le t) =$$
$$P[n^{1/2}\tilde{\sigma}^{-1}(\bar{Y}_n - \tilde{\mu}) \le n^{-1/2}\tilde{\sigma}^{-1}(t - n\tilde{\mu})] = G_n[n^{-1/2}\tilde{\sigma}^{-1}(t - n\tilde{\mu})],$$

where G_n is the distribution function of $n^{1/2}\sigma^{-1}(\bar{Y}_n - \mu)$, $\tilde{\mu} = E(Y)$, and $\tilde{\sigma}^2 = V(Y)$. Therefore, it follows that the density of $n\bar{Y}_n$ can be written as

$$f_{n,\lambda}(t) = \frac{d}{dt}G_n[n^{-1/2}\tilde{\sigma}^{-1}(t - n\tilde{\mu})] = n^{-1/2}\tilde{\sigma}^{-1}g_n[n^{-1/2}\tilde{\sigma}^{-1}(t - n\tilde{\mu})].$$

Now, Theorem 7.4 implies that

$$g_n(t) = \phi(t) + \tfrac{1}{6}n^{-1/2}\phi(t)\rho_3 H_3(t) + O(n^{-1}), \qquad (7.56)$$

as $n \to \infty$. Evaluating $f_{n,\lambda}$ at the mean of $n\bar{Y}_n$ implies that $f_{n,\lambda}(n\tilde{\mu}) = n^{-1/2}\tilde{\sigma}^{-1}g_n(0)$. Equation (7.56) then implies that

$$g_n(0) = \phi(0) + \tfrac{1}{6}n^{-1/2}\phi(0)\rho_3 H_3(0) + O(n^{-1}) = (2\pi)^{-1/2}[1 + O(n^{-1})].$$

Therefore, choosing $\lambda = \tilde{\lambda}$ so that $t = n\tilde{\mu}$ implies that

$$\begin{aligned} f_n(t) &= \exp[nc(\tilde{\lambda}) - \tilde{\lambda}t]f_{n,\lambda}(n\tilde{\mu}) \\ &= \exp[nc(\tilde{\lambda}) - \tilde{\lambda}t]n^{-1/2}\tilde{\sigma}^{-1}(2\pi)^{-1/2}[1 + O(n^{-1})]. \end{aligned}$$

To complete the expansion we must find an expression for $\tilde{\sigma}$. It can be shown that $\tilde{\sigma}^2 = nc''(\tilde{\lambda})$. See Exercise 21. Therefore, the final form of the expansion is given by

$$f_n(t) = [2\pi nc''(\tilde{\lambda})]^{-1/2}\exp[nc(\tilde{\lambda}) - \tilde{\lambda}t][1 + O(n^{-1})], \qquad (7.57)$$

as $n \to \infty$. The expansion in Equation (7.57) is called a *tilted* Edgeworth expansion, *indirect* Edgeworth expansion, or a *saddlepoint* expansion. The name tilted Edgeworth expansion comes from the fact that forming the density f_λ is called exponential tilting. The name saddlepoint expansion has its basis in an alternate method for deriving the approximation using a contour integral. See Daniels (1954). The derivation used in this section is based on the derivation from Section 4.3 of Barndorff-Nielsen and Cox (1989).

There are several relevant issues about the expansion in Equation (7.57) that warrant further discussion. The first concerns the parameter value $\tilde{\lambda}$ which is defined explicitly through the equation $nc'(\tilde{\lambda}) = t$, and whether one should expect that the equation can always be solved. This will not present a problem for the cases studied in this book, but for a more general discussion see Chapter 6 of Barndorff-Nielsen and Cox (1989). The second issue concerns the applicability of the method for various values of t. The expansion in Equation (7.57) guarantees a *relative* error of $O(n^{-1})$, as long as we are able to find the value $\tilde{\lambda}$ that corresponds to the t value we wish to create the expansion around. Section 4.3 of Barndorff-Nielsen and Cox (1989) points out that the expansion should be valid for any value of t for which $|t - nE(X_n)| < bn$ for a fixed value of b. There are even cases where the expansion is valid for all $t \in \mathbb{R}$. For further information see Daniels (1954) and Jensen (1988). For a detailed book-length exposition of saddlepoint methods, with many useful applications, see Butler (2007).

Example 7.11. Let $\{X_n\}_{n=1}^{\infty}$ be a sequence of independent and identically distributed random variables following a $N(\mu, \sigma^2)$ distribution. We wish to derive a saddlepoint expansion that approximates the density of $n\bar{X}_n$ at a point x with relative error $O(n^{-1})$ as $n \to \infty$. From Example 2.22 we know that the cumulant generating function of X_n is $c(t) = t\mu + \frac{1}{2}t^2\sigma^2$ for all $t \in \mathbb{R}$. Therefore, setting $x = nc'(\tilde{\lambda})$ implies that $x = n\mu + n\tilde{\lambda}\sigma^2$ so that $\tilde{\lambda} = \sigma^{-2}(n^{-1}x - \mu)$. Further we have that $c''(t) = \sigma^2$ for all $t \in \mathbb{R}$ so that $c''(\tilde{\lambda}) = \sigma^2$. We also need an expression for $nc(\tilde{\lambda})$ which is given by

$$
\begin{aligned}
nc(\tilde{\lambda}) &= n\mu[\sigma^{-2}(n^{-1}x - \mu)] + \tfrac{1}{2}n\sigma^2[\sigma^{-4}(n^{-1}x - \mu)^2] \\
&= n\sigma^{-2}(n^{-1}\mu x - \mu^2 + \tfrac{1}{2}n^{-2}x^2 - \mu n^{-1}x + \tfrac{1}{2}\mu^2) \\
&= \tfrac{1}{2}\sigma^{-2}(-n\mu^2 + n^{-1}x^2).
\end{aligned}
$$

It then follows that the saddlepoint expansion given in Equation (7.57) has the form

$$
\begin{aligned}
f_n(x) &= [2\pi nc''(\tilde{\lambda})]^{-1/2} \exp[nc(\tilde{\lambda}) - \tilde{\lambda}x][1 + O(n^{-1})] \\
&= (2\pi n\sigma^2)^{-1/2} \exp[\tfrac{1}{2}\sigma^{-2}(-n\mu^2 + n^{-1}x^2) - x\sigma^{-2}(n^{-1}x - \mu)] \\
&\quad \times [1 + O(n^{-1})] \\
&= (2\pi n\sigma^2)^{-1/2} \exp[-\tfrac{1}{2}n^{-1}\sigma^{-2}(x - n\mu)^2][1 + O(n^{-1})],
\end{aligned}
$$

which has a leading term equal to a $N(n\mu, n\sigma^2)$ density, which matches the exact density of $n\bar{X}_n$. Therefore, the saddlepoint expansion provides the exact answer in the leading term. ∎

Example 7.12. Let $\{X_n\}_{n=1}^{\infty}$ be a sequence of independent and identically distributed random variables following a CHISQUARED(θ) distribution. We wish to derive a saddlepoint expansion that approximates the density of $n\bar{X}_n$ at a point x with relative error $O(n^{-1})$ as $n \to \infty$. The cumulant generating function of a CHISQUARED(θ) distribution is given by $c(t) = -\frac{1}{2}\theta \log(1 - 2t)$ for $t < \frac{1}{2}$. Therefore it follows that $c'(t) = \theta(1 - 2t)^{-1}$ and solving $nc'(\tilde{\lambda}) = x$ for $\tilde{\lambda}$ yields $\tilde{\lambda} = \frac{1}{2} - \frac{1}{2}nx^{-1}\theta$. Note that since $c(t)$ only exists for $t < \frac{1}{2}$ we can only find the correct value of $\tilde{\lambda}$ for values of x such that $\tilde{\lambda} < \frac{1}{2}$. This turns out not to be a problem, since $\tilde{\lambda} < \frac{1}{2}$ for all $x > 0$. Further, it follows that $c''(t) = 2\theta(1 - 2t)^{-2}$ so that $c''(\tilde{\lambda}) = 2x^2n^{-2}\theta$. Finally, we have that $c(\tilde{\lambda}) = -\frac{1}{2}\theta \log[n\theta x^{-1}]$. Therefore, the saddlepoint expansion has the form

$$
f_n(x) = [4\pi n^{-1}x^2\theta]^{-1/2} \exp[-\tfrac{1}{2}n\theta \log(n\theta x^{-1}) - \tfrac{1}{2}(1 - nx^{-1}\theta)][1 + O(n^{-1})].
$$

Barndorff-Nielsen and Cox (1989) point out that an application of Theorem

1.8 (Stirling) implies that the leading term of this expansion has the same form as a GAMMA distribution. See Exercise 25. ∎

7.8 Exercises and Experiments

7.8.1 Exercises

1. Let f be a real function and define the Fourier norm as Feller (1971) does as

$$(2\pi)^{-1} \int_{-\infty}^{\infty} |f(x)| dx.$$

For a fixed value of x, is this function a norm?

2. Prove that the Fourier transformation of $H_k(x)\phi(x)$ is $(it)^k \exp(-\frac{1}{2}t^2)$. Hint: Use induction and integration by parts as in the partial proof of Theorem 7.1.

3. Let $\{X_n\}_{n=1}^{\infty}$ be a sequence of independent and identically distributed random variables where X_n has a GAMMA(α, β) distribution for all $n \in \mathbb{N}$.

   a. Compute one- and two- term Edgeworth expansions for the density of $n^{1/2}\sigma^{-1}(\bar{X}_n - \mu)$ where in this case $\mu = \alpha\beta$ and $\sigma^2 = \alpha\beta^2$. What effect do the values of α and β have on the accuracy of the expansion? Is it possible to eliminate either the first or second term through a specific choice of α and β?

   b. Compute one- and two-term Edgeworth expansions for the distribution function of $n^{1/2}\sigma^{-1}(\bar{X}_n - \mu)$.

   c. Compute one- and two-term Cornish–Fisher expansions for the quantile function of $n^{1/2}\sigma^{-1}(\bar{X}_n - \mu)$.

4. Let $\{X_n\}_{n=1}^{\infty}$ be a sequence of independent and identically distributed random variables where X_n has a BETA(α, β) distribution for all $n \in \mathbb{N}$.

   a. Compute one- and two-term Edgeworth expansions for the density of $n^{1/2}\sigma^{-1}(\bar{X}_n - \mu)$ where in this case $\mu = \alpha/(\alpha + \beta)$ and

   $$\sigma^2 = \frac{\alpha\beta}{(\alpha + \beta)^2(\alpha + \beta + 1)}.$$

   What effect do the values of α and β have on the accuracy of the expansion? Is it possible to eliminate either the first or second term through a specific choice of α and β?

   b. Compute one- and two-term Edgeworth expansions for the distribution function of $n^{1/2}\sigma^{-1}(\bar{X}_n - \mu)$.

   c. Compute one- and two-term Cornish–Fisher expansions for the quantile function of $n^{1/2}\sigma^{-1}(\bar{X}_n - \mu)$.

5. Let $\{X_n\}_{n=1}^\infty$ be a sequence of independent and identically distributed random variables where X_n has a density that is a mixture of two NORMAL densities of the form $f(x) = \theta\phi(x) + (1-\theta)\phi(x-1)$ for all $x \in \mathbb{R}$, where $\theta \in [0,1]$.

   a. Compute one- and two-term Edgeworth expansions for the density of $n^{1/2}\sigma^{-1}(\bar{X}_n - \mu)$ where in this case $\mu = E(X_n)$ and $\sigma^2 = V(X_n)$. What effect does the value of θ have on the accuracy of the expansion? Is it possible to eliminate either the first or second term through a specific choice of θ?

   b. Compute one- and two-term Edgeworth expansions for the distribution function of $n^{1/2}\sigma^{-1}(\bar{X}_n - \mu)$.

   c. Compute one- and two-term Cornish–Fisher expansions for the quantile function of $n^{1/2}\sigma^{-1}(\bar{X}_n - \mu)$.

6. Prove Theorem 7.8. That is, let $\{X_n\}_{n=1}^\infty$ be a sequence of independent and identically distributed random variables from a distribution F. Let $F_n(t) = P[n^{1/2}\sigma^{-1}(\bar{X}_n - \mu) \le t]$ and assume that $E(X_n^3) < \infty$. If F is a non-lattice distribution then prove that

$$F_n(x) - \Phi(x) - \tfrac{1}{6}\sigma^{-3}n^{-1/2}\mu_3(1-x^2)\phi(t) = o(n^{-1/2}),$$

as $n \to \infty$ uniformly in x. The first part of this proof is provided after Theorem 7.4. At what point is it important that the distribution be non-lattice?

7. Let $\{R_n\}_{n=1}^\infty$ be a sequence of real numbers such that $R_n = o(n^{-1})$ as $n \to \infty$. Prove that $R_n^2 = o(n^{-1})$ as $n \to \infty$.

8. Suppose that $v_1(\alpha)$ and $v_2(\alpha)$ are constant with respect to n. Prove that if $R_n = [n^{-1/2}v_1(\alpha) + n^{-1}v_2(\alpha) + o(n^{-1})]^2$ then a sequence that is $o(R_n)$ as $n \to \infty$ is also $o(n^{-1})$ as $n \to \infty$.

9. Suppose that $g_{\alpha,n} = v_0(\alpha) + n^{-1/2}v_1(\alpha) + n^{-1}v_2(\alpha) + o(n^{-1})$ as $n \to \infty$ where $v_0(\alpha)$, $v_1(\alpha)$, and $v_2(\alpha)$ are constant with respect to n. Prove that $H_3(g_{\alpha,n}) = H_3[v_0(\alpha)] + o(1)$ and $H_4(g_{\alpha,n}) = H_4[v_0(\alpha)] + o(1)$ as $n \to \infty$.

10. Prove that

$$\int_{-\infty}^\infty \exp(tx)\phi(x)H_k(x)dx = t^k \exp(\tfrac{1}{2}t^2).$$

11. Suppose that $X_1, \ldots, X_n$ are a set of independent and identically distributed random variables from a distribution F that has mean equal to zero, unit variance, and cumulant generating function $c(t)$. Find the form of the polynomial $p_3(x)$ from Theorem 7.5 by considering the form of an expansion for the cumulant generating function of $n^{1/2}\bar{X}_n$ that has an error term equal to $o(n^{-3/2})$ as $n \to \infty$. What assumptions must be made in order to apply Theorem 7.5 to this problem?

12. Suppose that $X_1, \ldots, X_n$ are a set of independent and identically distributed random variables from a distribution F that has mean equal to

θ, variance equal to σ^2, and cumulant generating function $c(t)$. Find the form of the polynomials $p_1(x)$, $p_2(x)$ and $p_3(x)$ from Theorem 7.5 by considering the form of an expansion for the cumulant generating function of $n^{1/2}\sigma^{-1}(\bar{X}_n - \theta)$ that has an error term equal to $o(n^{-3/2})$ as $n \to \infty$.

13. Using Equation (7.28), prove that $-r_1(x) = \frac{1}{6}\rho_3 H_2(x)$ and

$$-r_2(x) = \frac{1}{24}\rho_4 H_3(x) + \frac{1}{72}\rho_3^2 H_5(x).$$

14. Let $\{X_n\}_{n=1}^{\infty}$ be a sequence of independent and identically distributed random variables from a distribution F. Let $F_n(t) = P[n^{1/2}\sigma^{-1}(\bar{X}_n - \mu) \le t]$ and assume that $E(X_n^3) < \infty$. Suppose that F is a non-lattice distribution, then

$$F_n(x) = \Phi(x) + n^{-1/2}r_1(x)\phi(x) + n^{-1}r_2(x)\phi(x) + o(n^{-1}), \qquad (7.58)$$

as $n \to \infty$ uniformly in x. Prove that $q_1(x) = -r_1(x)$ and

$$q_2(x) = r_1(x)r_1'(x) - \tfrac{1}{2}xr_1^2(x) - r_2(x)$$

by inverting the expansion given in Equation (7.58).

15. Let $\{W_n\}_{n=1}^{\infty}$ be a sequence of independent and identically distributed random variables from a distribution F with mean η and variance θ. Prove that the parameter θ can be represented in the smooth function model with $d = 4$, $\mathbf{X}_n' = (W_n, W_n^2, W_n^3, W_n^4)$ for all $n \in \mathbb{N}$, $g(\mathbf{x}) = x_2 - (x_1)^2$, and

$$h(\mathbf{x}) = x_4 - 4x_1 x_3 + 6x_1^2 x_2 - 3x_1^4 - (x_2 - x_1^2)^2,$$

where $\mathbf{x}' = (x_1, x_2, x_3, x_4)$. This will verify the results of Example 7.7.

16. In the context of Example 7.8, let $\{W_n\}_{n=1}^{\infty}$ be a sequence of independent and identically distributed bivariate random vectors from a distribution F having mean vector $\boldsymbol{\eta}$ and covariance matrix $\boldsymbol{\Sigma}$. Let $\mathbf{W}_n' = (W_{n1}, W_{n2})$ for all $n \in \mathbb{N}$ and define $\mu_{ij} = E\{[W_{n1} - E(W_{n2})]^i [W_{n2} - E(W_{n2})]^j\}$ and $\mu_{ij}' = E(W_{n1}^j W_{n2}^j)$. Let

$$\boldsymbol{\mu}' = (\mu_{10}', \mu_{20}', \mu_{30}', \mu_{40}', \mu_{01}', \mu_{02}', \mu_{03}', \mu_{04}', \mu_{11}', \mu_{21}', \mu_{12}', \mu_{22}', \mu_{31}', \mu_{13}'),$$

and prove that if we define $h_{20}(\mathbf{x}) = x_2 - x_1^2$, $h_{02}(\mathbf{x}) = x_6 - x_5^2$, $h_{11}(\mathbf{x}) = x_9 - x_1 x_5$,

$$h_{22}(\mathbf{x}) = x_{12} - 2x_5 x_{10} + x_5^2 x_2 - 2x_1 x_{11} + 4x_1 x_5 x_9 - 3x_1^2 x_5^2 + x_1^2 x_6,$$

$$h_{31}(\mathbf{x}) = x_{13} - 3x_1 x_{10} + 3x_1^2 x_9 - x_3 x_5 - 3x_1 x_5 x_2 - 3x_1^3 x_5,$$

and

$$h_{13}(\mathbf{x}) = x_{14} - 3x_5 x_{11} + 3x_5^2 x_9 - x_7 x_1 + 3x_1 x_5 x_6 - 3x_1 x_5^3,$$

then it follows that $h_{20}(\boldsymbol{\mu}) = \mu_{20}$, $h_{02}(\boldsymbol{\mu}) = \mu_{02}$, $h_{11}(\boldsymbol{\mu}) = \mu_{11}$, $h_{22}(\boldsymbol{\mu}) = \mu_{22}$, $h_{31}(\boldsymbol{\mu}) = \mu_{31}$, and $h_{13}(\boldsymbol{\mu}) = \mu_{13}$.

17. Prove that the polynomials given in Equations (7.44) and (7.46) reduce to those given in Equations (7.29) and (7.30), when θ is taken to be the univariate mean.

18. Let $\{\mathbf{X}_n\}_{n=1}^\infty$ be a sequence of independent and identically distributed d-dimensional random vectors from a distribution F with mean vector $\boldsymbol{\mu}$. Let $\theta = g(\boldsymbol{\mu})$ for some function g and suppose that $\hat{\theta}_n = g(\bar{\mathbf{X}}_n)$. Let σ^2 be the asymptotic variance of $n^{1/2}\hat{\theta}_n$ and assume that $\sigma = h(\boldsymbol{\mu})$ for some function h. Define $A(\mathbf{x}) = [g(\mathbf{x}) - g(\boldsymbol{\mu})]/h(\boldsymbol{\mu})$, $B(\mathbf{x}) = [g(\mathbf{x}) - g(\boldsymbol{\mu})]/h(\mathbf{x})$ and assume that B has $p+2$ continuous derivatives in a neighborhood of $\boldsymbol{\mu}$ and that $E(||\mathbf{X}||^{p+2}) < \infty$. Finally, define the constants $a_{i_1\cdots i_k}$ and $b_{i_1\cdots i_k}$ by

$$a_{i_1\cdots i_k} = \left.\frac{\partial^k}{\partial x_{i_1}\cdots\partial x_{i_k}}A(\mathbf{x})\right|_{\mathbf{x}=\boldsymbol{\mu}},$$

and

$$b_{i_1\cdots i_k} = \left.\frac{\partial^k}{\partial x_{i_1}\cdots\partial x_{i_k}}B(\mathbf{x})\right|_{\mathbf{x}=\boldsymbol{\mu}}.$$

a. Prove that $b_i = a_i\sigma^{-1}$.

b. Prove that $b_{ij} = a_{ij}\sigma^{-1} - \frac{1}{2}(a_ic_j + a_jc_i)\sigma^{-3}$, where

$$c_k = \left.\frac{\partial}{\partial x_k}g(\mathbf{x})\right|_{\mathbf{x}=\boldsymbol{\mu}}.$$

c. Use the first two parts of this problem to prove that

$$r_1(x) - q_1(x) = -\frac{1}{2}\sigma^{-3}\sum_{i=1}^d\sum_{j=1}^d a_ic_j\mu_{ij}x^2,$$

where $r_1(x) = -[A_1 + \frac{1}{6}A_2H_2(x)]$, $q_1(x) = -[B_1 + \frac{1}{6}B_2H_2(x)]$ and A_1, A_2, B_1 and B_2 are defined in Equations (7.40), (7.41), (7.52), and (7.53).

19. Let $\{W_n\}_{n=1}^\infty$ be a sequence of independent and identically distributed random variables from a distribution F with mean θ and variance σ^2. It was shown in Example 7.6 that this parameter can be represented in the smooth function model with $\mathbf{X}'_n = (W_n, W_n^2)$, $g(\mathbf{x}) = x_1$, and $h(\mathbf{x}) = x_2 - x_1^2$ where we have $\hat{\theta}_n = \bar{W}_n$ and

$$\hat{\sigma}^2 = n^{-1}\sum_{k=1}^n(W_k - \bar{W}_n)^2.$$

Assuming that $\theta = 0$ and $\sigma = 1$ and using the form from Equations (7.44)–(7.48), derive a two-term Edgeworth expansion for the studentized distribution of $\hat{\theta}$ and compare it to Equation (7.54). In particular, show that

$$q_2(x) = x[\tfrac{1}{12}\kappa_4(x^2 - 3) - \tfrac{1}{18}\kappa_3^2(x^4 + 2x^2 - 3) - \tfrac{1}{4}(x^2 + 3)],$$

which can be found in Section 2.6 of Hall (1992), where κ_3 and κ_4 are the third and fourth cumulants of F, respectively.

20. Let $\{X_n\}_{n=1}^\infty$ be a sequence of independent and identically distributed random variables from a distribution F that has density f, characteristic func-

tion $\psi(u)$, and cumulant generating function $c(u)$. Assume that the characteristic function is real valued in this case and use the alternate definition of the cumulant generating function given by $c(u) = \log[\psi(u)]$. Define the density $f_\lambda(t) = \exp(\lambda t)f(t)/\psi(\lambda)$ and let $\{Y_n\}_{n=1}^\infty$ be a sequence of independent and identically distributed random variables following the density f_λ. Let f_n denote the density of $n\bar{X}_n$ and $f_{n,\lambda}$ denote the density of $n\bar{Y}_n$. Using characteristic functions, prove that $f_{n,\lambda}(t) = \exp[\lambda t - nc(\lambda)]f_n(t)$.

21. Let X be a random variable with moment generating function $m(u)$ and cumulant generating function $c(u)$. Assuming that both functions exist, prove that

$$\frac{d^2}{du^2}\frac{m(u+\lambda)}{m(\lambda)}\bigg|_{t=0} = c''(\lambda).$$

22. Let $\{X_n\}_{n=1}^\infty$ be a sequence of independent and identically distributed random variables following an EXPONENTIAL(1) density.

a. Prove that if $\tilde{\lambda} = x^{-1}(x-n)$ then $nc'(\tilde{\lambda}) = x$.

b. Prove that $c(\tilde{\lambda}) = \log(n^{-1}x)$ and that $c''(\tilde{\lambda}) = n^{-2}x^2$.

c. Prove that the leading term of the saddlepoint expansion for $f_n(x)$ is given by $(2\pi)^{-1/2}n^{-n+1/2}x^{n-1}\exp(n-x)$.

d. The exact form for $f_n(x)$ in this case is $x^{n-1}\exp(-x)/\Gamma(n)$. Approximate $\Gamma(n)$ using Theorem 1.8 (Stirling), and show that the resulting approximation matches the saddlepoint approximation.

23. Let $\{X_n\}_{n=1}^\infty$ be a sequence of independent and identically distributed random variables following an GAMMA(α, β) density.

a. Find the value of $\tilde{\lambda}$ that is the solution to $nc'(\tilde{\lambda}) = x$.

b. Find $c(\tilde{\lambda})$ and $c''(\tilde{\lambda})$.

c. Derive the saddlepoint expansion for $f_n(x)$.

24. Let $\{X_n\}_{n=1}^\infty$ be a sequence of independent and identically distributed random variables following an WALD(α, β) density.

a. Find the value of $\tilde{\lambda}$ that is the solution to $nc'(\tilde{\lambda}) = x$.

b. Find $c(\tilde{\lambda})$ and $c''(\tilde{\lambda})$.

c. Derive the saddlepoint expansion for $f_n(x)$.

25. Let $\{X_n\}_{n=1}^\infty$ be a sequence of independent and identically distributed random variables following a CHISQUARED(θ) distribution. In Example 7.12 we derived a saddlepoint expansion that approximates the density of $n\bar{X}_n$ at a point x with relative error $O(n^{-1})$ as $n \to \infty$, given by

$$f_n(x) = [4\pi n^{-1}x^2\theta]^{-1/2}\exp[-\tfrac{1}{2}n\theta\log(n\theta x^{-1}) - \tfrac{1}{2}(1-nx^{-1}\theta)][1+O(n^{-1})].$$

Prove that an application of Theorem 1.8 (Stirling) implies that the leading term of this expansion has the same form as a GAMMA distribution.

7.8.2 Experiments

1. Write a program in R that generates b samples of size n from a specified distribution F (specified below). For each sample compute the statistic $Z_n = n^{1/2}\sigma^{-1}(\bar{X}_n - \mu)$ where μ and σ correspond to the mean and standard deviation of the specified distribution F. Produce a *density* histogram of the b values of Z_n. Run this simulation for $n = 10, 25, 50$, and 100 for each of the distributions listed below. On each histogram overlay a plot of the standard normal density and the function given by the one-term Edgeworth expansion for each case studied. Discuss how these histograms compare to what would be expected for large n, as regulated by the underlying theory given by Theorems 4.20 and 7.4.

   a. F corresponds to a $N(0, 1)$ distribution.
   b. F corresponds to an EXPONENTIAL(1) distribution.
   c. F corresponds to a GAMMA(2, 2) distribution.
   d. F corresponds to a UNIFORM(0, 1) distribution.

2. Write a program in R that generates b samples of size n from a specified distribution F (specified below). For each sample compute the statistic $Z_n = n^{1/2}\sigma^{-1}(\bar{X}_n - \mu)$ where μ and σ correspond to the mean and standard deviation of the specified distribution F. Produce a plot of the empirical distribution function of the b values of Z_n. On each plot overlay a plot of the standard normal distribution function and the function given by the one-term Edgeworth expansion for each case studied. Discuss how these functions compare to what would be expected for large n as regulated by the underlying theory given by Theorems 4.20 and 7.4. Run this simulation for $n = 10, 25, 50$, and 100 for each of the distributions listed below.

   a. F corresponds to a $N(0, 1)$ distribution.
   b. F corresponds to an EXPONENTIAL(1) distribution.
   c. F corresponds to a GAMMA(2, 2) distribution.
   d. F corresponds to a UNIFORM(0, 1) distribution.

3. Write a program in R that generates b samples of size n from the linear density $f(x) = 2[\theta + x(1 - 2\theta)]\delta\{x; (0, 1)\}$ studied in Example 7.2. Recall that the first four moments of this density are given by $\mu = \mu_1 = \frac{2}{3} - \frac{1}{3}\theta$, $\sigma^2 = \mu_2 = \frac{1}{18} + \frac{1}{9}\theta - \frac{1}{9}\theta^2$, $\mu_3 = -\frac{1}{135} - \frac{1}{45}\theta + \frac{1}{9}\theta^2 - \frac{2}{27}\theta^3$, and $\mu_4 = \frac{1}{135} + \frac{4}{135}\theta - \frac{1}{15}\theta^2 + \frac{2}{27}\theta^3 - \frac{1}{27}\theta^4$. For each sample compute the statistic $Z_n = n^{1/2}\sigma^{-1}(\bar{X}_n - \mu)$ where μ and σ correspond to the mean and standard deviation specified above. Produce a plot of the empirical distribution function of the b values of Z_n. On each plot overlay a plot of the standard normal distribution function and the function given by the one-term Edgeworth expansion. Discuss how these functions compare to what would be expected for large n as regulated by the underlying theory given by Theorems 4.20 and 7.4. Run this simulation for $n = 10, 25, 50$, and 100 each with $\theta = 0.10, 0.25$, and 0.50.

4. Write a program in R that generates b samples of size n from a density that is a mixture of two NORMAL densities given by $f(x) = \frac{1}{2}\phi(x) + \frac{1}{2}\phi(x-\theta)$ for all $x \in \mathbb{R}$ where $\theta \in \mathbb{R}$. In Example 7.3 we found that $E(X_n) = \frac{1}{2}\theta$, $E(X_n^2) = 1 + \frac{1}{2}\theta^2$, $E(X_n^3) = \frac{1}{2}\theta^3 + \frac{3}{2}\theta$, and $E(X_n^4) = -\frac{1}{2}\theta^4 + \frac{3}{2}\theta^2 + 3$. The third and fourth moments of the normal mixture are $\mu_3 = 0$ and $\mu_4 = -\frac{3}{16}\theta^4 + \frac{3}{2}\theta^2 + 3$. The third cumulant is also zero, and the fourth cumulant is $\kappa_4 = -\frac{15}{16}\theta^4 - \frac{3}{2}\theta^2$. For each sample compute the statistic $Z_n = n^{1/2}\sigma^{-1}(\bar{X}_n - \mu)$ where μ and σ correspond to the mean and standard deviation specified above. Produce a plot of the empirical distribution function of the b values of Z_n. On each plot overlay a plot of the standard normal distribution function and the function given by the one-term Edgeworth expansion. Discuss how these functions compare to what would be expected for large n as regulated by the underlying theory given by Theorems 4.20 and 7.4. Run this simulation for $n = 10$, 25, 50, and 100 each with $\theta = 0$, 0.50, 1.00, and 3.00.

5. Write a program in R that generates b samples of size n from a specified distribution F (specified below). For each sample compute the approximate $100\alpha\%$ confidence interval for the mean given by $[\bar{X}_n - n^{-1/2}\sigma z_{(1+\alpha)/2}, \bar{X}_n - n^{-1/2}\sigma z_{(1-\alpha)/2}]$ using the known value of the population variance from the distributions specified below. Also compute two additional confidence intervals for the mean where the $N(0,1)$ quantiles are replaced by ones based on one- and two-term Cornish–Fisher expansions for $g_{(1-\alpha)/2,n}$ and $g_{(1+\alpha)/2,n}$. Compute the percentage of time each method for computing the confidence interval contained the true value of the mean over the b simulated samples. Use $\alpha = 0.10$, $b = 1000$ and repeat the simulation for the sample sizes $n = 5$, 10, 25, 50, 100, 250, and 500.

   a. F corresponds to a $N(0,1)$ distribution where $\mu = 0$, σ is known to be one, $\rho_3 = 0$ and $\rho_4 = 0$. Note: In this case one needs only to compute the confidence interval for the mean given by the normal approximation.

   b. F corresponds to an EXPONENTIAL(1) distribution.

   c. F corresponds to a GAMMA(2, 2) distribution.

   d. F corresponds to a UNIFORM(0, 1) distribution.

6. Let $\{X_n\}_{n=1}^\infty$ be a sequence of independent and identically distributed random variables following a $N(\mu, \sigma^2)$ distribution. A saddlepoint expansion that approximates the density of $n\bar{X}_n$ at a point x with relative error $O(n^{-1})$ as $n \to \infty$ was shown in Example 7.11 to have the form

$$
\begin{aligned}
f_n(x) &= [2\pi n c''(\tilde{\lambda})]^{-1/2} \exp[nc(\tilde{\lambda}) - \tilde{\lambda}x][1 + O(n^{-1})] \\
&= (2\pi n\sigma^2)^{-1/2} \exp[\tfrac{1}{2}\sigma^{-2}(-n\mu^2 + n^{-1}x^2) - x\sigma^{-2}(n^{-1}x - \mu)] \\
&\quad \times [1 + O(n^{-1})] \\
&= (2\pi n\sigma^2)^{-1/2} \exp[-\tfrac{1}{2}n^{-1}\sigma^{-2}(x - n\mu)^2][1 + O(n^{-1})],
\end{aligned}
$$

which has a leading term equal to a $N(n\mu, n\sigma^2)$ density, which matches the exact density of $n\bar{X}_n$. Plot a $N(n\mu, n\sigma^2)$ density and the saddlepoint approximation for $\mu = 1$, $\sigma^2 = 1$, and $n = 2, 5, 10, 25$ and 50. Discuss how well the saddlepoint approximation appears to be doing in each case.

7. Let $\{X_n\}_{n=1}^{\infty}$ be a sequence of independent and identically distributed random variables following a CHISQUARED(θ) distribution. A saddlepoint expansion that approximates the density of $n\bar{X}_n$ at a point x with relative error $O(n^{-1})$ as $n \to \infty$ was shown in Example 7.12 to have the form

$$f_n(x) = [4\pi n^{-1}x^2\theta]^{-1/2} \exp[-\tfrac{1}{2}n\theta \log(n\theta x^{-1}) - \tfrac{1}{2}(1 - nx^{-1}\theta)][1 + O(n^{-1})].$$

Plot the correct GAMMA density of $n\bar{X}_n$ and the saddlepoint approximation for $\theta = 1$ and $n = 2, 5, 10, 25$ and 50. Discuss how well the saddlepoint approximation appears to be doing in each case.

CHAPTER 8

Asymptotic Expansions for Random Variables

I trust them far more than what Barnabas says. Even if they are old, worthless letters picked at random out of a pile of equally worthless letters, with no more understanding than the canaries at fairs have, pecking out people's fortunes at random, well, even if that is so, at least these letters bear some relation to my work.

The Castle by Franz Kafka

8.1 Approximating Random Variables

So far in this book we have relied at many times on the ability to approximate a function $f(x + \delta)$ for a sequence of constants δ that converge to zero based on the value $f(x)$. That is, we are able to approximate values of $f(x + \delta)$ for small values of δ as long as we know $f(x)$. The main tool for developing these approximations was based on Theorem 1.13 (Taylor), though we also briefly talked about other methods as well. The main strength of the theory is based on the idea that the accuracy of these approximations are well known and have properties that are easily represented using the asymptotic order notation introduced in Section 1.5. For instance, in Example 1.23 we found that the distribution function of a $N(0, 1)$ random variable could be approximated near zero with $\frac{1}{2} + \delta(2\pi)^{-1/2} + \frac{1}{6}\delta^3(2\pi)^{-1/2}$. The error of this approximation can be represented as $o(\delta^3)$, which means that the error, when divided by δ^3, converges to zero as δ converges to zero. We also saw that this error can be represented as $O(\delta^4)$, which means that the error, when divided by δ^4, remains bounded as δ converges to zero.

In some cases it would be useful to develop methods for approximating random variables. As an example, consider the situation where we have observed $X_1, \ldots, X_n$, a set of independent and identically distributed random variables from a distribution F with mean θ and variance σ^2. If the distribution F is not known, then an approximate $100\alpha\%$ upper confidence limit for θ is given by $\hat{U}_n(X_1, \ldots, X_n) = \bar{X}_n - n^{-1/2}\hat{\sigma}z_{1-\alpha}$, where $\hat{\sigma}$ is the sample standard deviation. This confidence limit is approximate in the sense that $P[\theta \leq \hat{U}_n(X_1, \ldots, X_n)] \simeq \alpha$. Let us assume that there is an *exact* upper confidence limit $U_n(X_1, \ldots, X_n)$ such that $P[\theta \leq U_n(X_1, \ldots, X_n)] = \alpha$. In this case it would be useful to be able to compare $\hat{U}_n(X_1, \ldots, X_n)$ and $U_n(X_1, \ldots, X_n)$ to determine the quality of the approximation. That is, we are interested in the behavior of $R_n = |\hat{U}_n(X_1, \ldots, X_n) - U_n(X_1, \ldots, X_n)|$ as $n \to \infty$. For example, we would like to be able to conclude that $R_n \to 0$ as $n \to \infty$. However, there is a small problem in this case in that both $U_n(X_1, \ldots, X_n)$ and $\hat{U}_n(X_1, \ldots, X_n)$ are random variables due to their dependence on the sample $X_1, \ldots, X_n$. Therefore, we cannot characterize the asymptotic behavior of R_n using a limit for real sequences, we must characterize this behavior in terms of one of the modes of convergence for random variables discussed in Chapter 3. For example, we can use Definition 3.1 and determine whether $R_n \xrightarrow{p} 0$ as $n \to \infty$.

An alternative method for approximating the upper confidence limit would be to use the correction suggested by the Cornish-Fisher expansion of Theorem 7.10. That is, replace $z_{1-\alpha}$ with $h_{1-\alpha,n}$ to obtain an approximate upper confidence limit given by $\tilde{U}_n(X_1, \ldots, X_n) = \bar{X}_n - n^{-1/2}\hat{\sigma} h_{1-\alpha,n}$. By defining $R_n = |\tilde{U}_n(X_1, \ldots, X_n) - U_n(X_1, \ldots, X_n)|$ we could again ascertain whether $R_n \xrightarrow{p} 0$ as $n \to \infty$. However, a more useful analysis would consider whether $\tilde{U}_n(X_1, \ldots, X_n)$ is an asymptotically more accurate approximate upper confidence limit than $\hat{U}_n(X_1, \ldots, X_n)$. If all we know is that both methods converge to zero in probability then we cannot compare the two methods directly. In this case we require some information about the *rate* at which the two methods converge in probability to zero.

The goal of this chapter is to develop methods for approximating a random variable, or a sequence of random variables, with an asymptotic expansion whose terms may be random variables. The error terms of these sequences will also necessarily be random variables as well. That is, let $\{X_n\}_{n=1}^{\infty}$ be a sequence of random variables. Then we would like to find random variables $Y_0, Y_1, \ldots, Y_n$ such that $X_n = Y_0 + n^{-1/2}Y_1 + n^{-1}Y_2 + \cdots + n^{-p/2}Y_p + R(n, p)$, where $R(n, p)$ is a random variable that serves as a remainder term that depends on both n and p. Note that the random variables $Y_0, \ldots, Y_p$ themselves do not depend on n. As with asymptotic expansions for sequences of real numbers, the form of the expansion and the rate at which the error term converges to zero are both important properties of the expansion. Therefore, another focus of this chapter is on defining rates of convergence for sequences of random variables. In particular, we will develop stochastic analogs of the asymptotic order notation from Definition 1.7. We will then apply these methods to applications, such as the delta method and the asymptotic distribution of the sample central moments.

8.2 Stochastic Order Notation

Extending the asymptotic order notation from sequences of real numbers to sequences of random variables in principle involves converting the limits in the asymptotic order notation to limits of random variables based on one of the modes of convergence studied in Chapter 3. For sequences of real numbers $\{x_n\}_{n=1}^\infty$ and $\{y_n\}_{n=1}^\infty$ we conclude that $x_n = o(y_n)$ as $n \to \infty$ if

$$\lim_{n \to \infty} \frac{x_n}{y_n} = 0. \tag{8.1}$$

Therefore, to define a stochastic version of this notation, we replace the limit in Equation (8.1) with convergence in probability.

Definition 8.1. *Let $\{X_n\}_{n=1}^\infty$ and $\{Y_n\}_{n=1}^\infty$ be sequences of random variables. The notation $X_n = o_p(Y_n)$ as $n \to \infty$ means that $X_n Y_n^{-1} \xrightarrow{p} 0$ as $n \to \infty$.*

Example 8.1. Let $\{X_n\}_{n=1}^\infty$ be a sequence of independent discrete random variables where X_n has probability distribution function

$$f_n(x) = \begin{cases} \frac{1}{2} & x \in \{0, n^{-1}\} \\ 0 & \text{otherwise,} \end{cases}$$

for all $n \in \mathbb{N}$. Note that $n^{1/2} X_n$ therefore has probability distribution function

$$g_n(x) = \begin{cases} \frac{1}{2} & x \in \{0, n^{-1/2}\} \\ 0 & \text{otherwise,} \end{cases}$$

for all $n \in \mathbb{N}$, where it can then be shown that for any $\varepsilon > 0$,

$$\lim_{n \to \infty} P(n^{1/2} X_n > \varepsilon) \le \lim_{n \to \infty} n^{-1/2} = 0,$$

and therefore Definition 3.1 implies that $n^{1/2} X_n \xrightarrow{p} 0$ as $n \to \infty$. Therefore, it follows from Definition 8.1 that $X_n = o_p(n^{-1/2})$ as $n \to \infty$. Note that X_n is *not* $o_p(n^{-1})$ since nX_n has a BERNOULLI($\frac{1}{2}$) distribution for all $n \in \mathbb{N}$, which is a sequence of random variables that does not depend on n. ∎

Example 8.2. Let $\{X_n\}_{n=1}^\infty$ be a sequence of independent discrete random variables where X_n has probability distribution function

$$f_n(x) = \begin{cases} \frac{1}{2} & x \in \{0, n^{-1}\} \\ 0 & \text{otherwise,} \end{cases}$$

for all $n \in \mathbb{N}$ and let $\{Y_n\}_{n=1}^\infty$ be a sequence of independent random variables where Y_n has probability distribution function

$$g_n(y) = \begin{cases} \frac{1}{2} & y \in \{n^{-1/2}, 1\} \\ 0 & \text{otherwise,} \end{cases}$$

for all $n \in \mathbb{N}$. We will assume that X_n is independent of Y_n for all $n \in \mathbb{N}$. Consider the sequence of random variables given by $\{Z_n\}_{n=1}^\infty$ where $Z_n = X_n Y_n^{-1}$ for all $n \in \mathbb{N}$. Because of the independence between X_n and Y_n, the distribution of Z_n can be found to be

$$h_n(z) = \begin{cases} \frac{1}{2} & z = 0 \\ \frac{1}{4} & z \in \{n^{-1/2}, n^{-1}\} \\ 0 & \text{otherwise,} \end{cases}$$

for all $n \in \mathbb{N}$. Now, it can then be shown that for any $\varepsilon > 0$,

$$\lim_{n \to \infty} P(Z_n > \varepsilon) \leq \lim_{n \to \infty} n^{-1/2} = 0,$$

and therefore Definition 3.1 implies that $Z_n \overset{p}{\to} 0$ as $n \to \infty$. Therefore, it follows from Definition 8.1 that $X_n = o_p(Y_n)$ as $n \to \infty$. ∎

Example 8.3. Let $\{X_n\}_{n=1}^{\infty}$ be a sequence of independent random variables from a distribution F with mean μ. From Theorem 3.10 (Weak Law of Large Numbers) we know that the sample mean converges in probability to μ as $n \to \infty$. Further, Theorem 4.11 (Slutsky) implies that $\bar{X}_n - \mu \overset{p}{\to} 0$ as $n \to \infty$. Therefore, it follows from Definition 8.1 that $\bar{X}_n - \mu = o_p(1)$ as $n \to \infty$. ∎

For sequences of real numbers $\{x_n\}_{n=1}^{\infty}$ and $\{y_n\}_{n=1}^{\infty}$ we conclude that $x_n = O(y_n)$ as $n \to \infty$ if $|x_n y_n^{-1}|$ remains bounded as $n \to \infty$. To develop a stochastic version of this operator we first note that a sequence of real numbers $\{z_n\}_{n=1}^{\infty}$ is bounded if there exists a positive real number b such that $|z_n| < b$ for all $n \in \mathbb{N}$. The same sequence remains bounded as $n \to \infty$ if there exists a positive real number m and integer n_b such that $|z_n| < b$ for all $n > n_b$. For a stochastic version of this property we require that behavior to apply with at least probability $1 - \varepsilon$ for every $\varepsilon > 0$. This matches the concept that a sequence is bounded in probability from Definition 4.3. Using this definition, we can now define a stochastic version of the O operator.

Definition 8.2. *Let $\{X_n\}_{n=1}^{\infty}$ and $\{Y_n\}_{n=1}^{\infty}$ be sequences of random variables. The notation $X_n = O_p(Y_n)$ as $n \to \infty$ means that $|X_n Y_n^{-1}|$ is bounded in probability.*

Example 8.4. Let $\{X_n\}_{n=1}^{\infty}$ and $\{Y_n\}_{n=1}^{\infty}$ be sequences of independent random variables. Suppose that Y_n is a POISSON(θ) random variable for all $n \in \mathbb{N}$, and that, conditional on Y_n, the random variable X_n has a UNIFORM$\{1, 2, \ldots, Y_n\}$ distribution. Therefore, it follows that $X_n Y_n^{-1}$ has a distribution that guarantees that $P(0 \leq X_n Y_n^{-1} \leq 1) = 1$ for all $n \in \mathbb{N}$. Let $\varepsilon > 0$ and choose $b_\varepsilon = 1$. Then it follows that $P(|X_n Y_n^{-1}| \leq b_\varepsilon) = 1$ for all $n \in \mathbb{N}$ and therefore Theorem 4.3 implies that $|X_n Y_n^{-1}|$ is bounded in probability. Definition 8.2 then implies that $X_n Y_n^{-1} = O_p(1)$, or equivalently that $X_n = O_p(Y_n)$ as $n \to \infty$. ∎

Example 8.5. Let $\{X_n\}_{n=1}^{\infty}$ be a sequence of independent random variables where X_n has a UNIFORM$(-n^{-1}, 1+n^{-1})$ distribution for all $n \in \mathbb{N}$. Let $\varepsilon > 0$ and consider the postive real number $b_\varepsilon = 2$. Then it follows that $P(|X_n| \leq b_\varepsilon) = 1$ for all $n \in \mathbb{N}$ and therefore by Definition 4.3 the sequence $\{X_n\}_{n=1}^{\infty}$ is bounded in probability. Hence, Definition 8.2 implies that $X_n = O_p(1)$ as $n \to \infty$. ∎

Example 8.6. Let $\{X_n\}_{n=1}^{\infty}$ be a sequence of independent random variables where X_n has a N$(0, n^{-1})$ distribution for all $n \in \mathbb{N}$. Let $\varepsilon > 0$ and let b be any positive real number. Then

$$\lim_{n \to \infty} P(|X_n| < b) = \lim_{n \to \infty} P(|Z| \leq n^{1/2} b) = 1,$$

where Z is a $N(0,1)$ random variable. Therefore, there exists a positive real number b_ε and positive integer n_ε such that $P(|X_n| \le b_\varepsilon) > 1 - \varepsilon$ for all $n > n_\varepsilon$, and hence $X_n = O_p(1)$, as $n \to \infty$. ∎

One can observe from Example 8.5 that a very useful special case of sequences that are bounded in probability are those that converge in distribution. As discussed in Section 4.2, we will always assume that sequences that converge in distribution do so to distributions that have valid distribution functions. It is this property that assures that such sequences are bounded in probability.

Theorem 8.1. *Let $\{X_n\}_{n=1}^\infty$ be a sequence of random variables that converge in distribution to a random variable X as $n \to \infty$, then $X_n = O_p(1)$ as $n \to \infty$.*

Proof. Let F_n denote the distribution function of X_n for all $n \in \mathbb{N}$ and let F denote the distribution of X. Since $X_n \xrightarrow{d} X$ as $n \to \infty$ it follows from Definition 4.1 that

$$\lim_{n \to \infty} F_n(x) = F(x),$$

for all $x \in C(F)$. By assumption F is a distribution function such that

$$\lim_{x \to \infty} F(x) = 1,$$

and

$$\lim_{x \to -\infty} F(x) = 0.$$

Therefore Theorem 4.1 implies that $\{X_n\}_{n=1}^\infty$ is bounded in probability Definition 8.2 implies then that $X_n = O_p(1)$ as $n \to \infty$. □

Example 8.7. Let $\{X_n\}_{n=1}^\infty$ be a sequence of independent random variables from a distribution F with mean μ and variance σ^2. Theorem 4.20 (Lindeberg and Lévy) implies that $n^{1/2}\sigma^{-1}(\bar{X}_n - \mu) \xrightarrow{d} Z$ where Z has a $N(0,1)$ distribution. Therefore Theorem 8.1 implies that $n^{1/2}\sigma^{-1}(\bar{X}_n - \mu) = O_p(1)$ as $n \to \infty$ and Definition 8.2 implies that $\bar{X}_n - \mu = O_p(n^{-1/2})$ or equivalently that $\bar{X}_n = \mu + O_p(n^{-1/2})$ as $n \to \infty$. ∎

Example 8.8. Let $Z_1, \ldots, Z_m$ be a set of independent and identically distributed $N(0,1)$ random variables and let $\delta > 0$. Define

$$X_{m,\delta} = (Z_1 + \delta)^2 + \sum_{k=2}^m Z_k^2.$$

Note that if $\delta = 0$ then $X_{m,\delta}$ has a CHISQUARED(m) distribution. If $\delta > 0$ then $X_{m,\delta}$ has a non-central CHISQUARED distribution. Following Section 2.3 of Barndorff-Nielsen and Cox (1989), we will investigate the asymptotic distribution of $(2\delta)^{-1}(X_{m,\delta} - \delta^2)$ as $\delta \to \infty$. We first note that

$$X_{m,\delta} = \delta^2 + 2\delta Z_1 + \sum_{k=1}^m Z_k^2,$$

so that

$$(2\delta)^{-1}(X_{m,\delta} - \delta^2) = Z_1 + (2\delta)^{-1}\sum_{k=1}^{m} Z_k^2 = Z_1 + R_\delta.$$

Now $Z_1^2 + \cdots + Z_m^2$ has a CHISQUARED(m) distribution and does not depend on δ. Hence it follows that δR_δ is a random variable equal to half of a CHISQUARED(m) random variable and hence $\delta R_\delta \xrightarrow{d} R$ as $\delta \to \infty$ where $2R$ has a CHISQUARED(m) distribution. Therefore Theorem 8.1 implies that $\delta R_\delta = O_p(1)$ or equivalently that $R_\delta = O_p(\delta^{-1})$ as $\delta \to \infty$. It then follows that $(2\delta)^{-1}(X_{m,\delta} - \delta^2) = Z_1 + O_p(\delta^{-1})$ as $\delta \to \infty$. Note that this expansion provides a normal approximation for the non-central CHISQUARED distribution when the non-centrality parameter is large. ∎

Example 8.9. Let $\{X_n\}_{n=1}^{\infty}$ and $\{Y_n\}_{n=1}^{\infty}$ be sequences of independent and identically distributed random variables from distributions F and G respectively. Suppose that the two sequences are mutually independent of one another and that the distributions F and G both have means equal to zero and variances equal to one. Theorem 4.22 implies that $n^{1/2}\bar{X}_n \xrightarrow{d} Z_1$ and $n^{1/2}\bar{Y}_n \xrightarrow{d} Z_2$ where Z_1 and Z_2 are independent $N(0,1)$ random variables. Theorem 4.18 implies that $\bar{X}_n\bar{Y}_n^{-1} \xrightarrow{d} W$ as $n \to \infty$ where W is a CAUCHY$(0,1)$ random variable. Therefore, Theorem 8.1 implies that $\bar{X}_n\bar{Y}_n^{-1} = O_p(1)$ as $n \to \infty$, or equivalently $X_n = O_p(Y_n)$ as $n \to \infty$. ∎

It is important to note that a sequence being bounded in probability is not equivalent to the sequence converging in distribution, or in any other mode of convergence. A sequence that is bounded in probability need not even be convergent, it only needs to stay within some bounds as $n \to \infty$.

Example 8.10. Let $\{X_n\}_{n=1}^{\infty}$ be a sequence of independent random variables such that $X_n = (-1)^n X$ where X is a BERNOULLI$(\frac{1}{2})$ random variable. This sequence does not converge in distribution to any random variable, but the sequence is bounded in probability. To see this let $\varepsilon > 0$ and define $b_\varepsilon = \frac{3}{2}$ and note that $P(|X_n| \leq b_\varepsilon) = 1 > 1 - \varepsilon$ for all $n \in \mathbb{N}$. ∎

When using the order notation for real valued sequences, we found that if a sequence was $o(y_n)$ as $n \to \infty$ for some real valued sequence $\{y_n\}_{n=1}^{\infty}$ then the sequence was also $O(y_n)$ as $n \to \infty$. This same type of relationship holds for sequences of random variables using the stochastic order notation.

Theorem 8.2. Let $\{X_n\}_{n=1}^{\infty}$ and $\{Y_n\}_{n=1}^{\infty}$ be sequences of random variables. Suppose that $X_n = o_p(Y_n)$ as $n \to \infty$, then $X_n = O_p(Y_n)$ as $n \to \infty$.

For a proof of Theorem 8.2 see Exercise 2. To effectively work with stochastic order notation we must also establish how the sequences of each order interact with each other. The result below also provides results as to how real sequences and sequences of random variables interact with one another.

Theorem 8.3. Let $\{X_n\}_{n=1}^{\infty}$ and $\{Y_n\}_{n=1}^{\infty}$ be sequences of random variables and let $\{y_n\}_{n=1}^{\infty}$ be a sequence of real numbers.

1. *If $X_n = O_p(n^{-a})$ and $Y_n = O_p(n^{-b})$ as $n \to \infty$, then $X_n Y_n = O_p(n^{-(a+b)})$ as $n \to \infty$.*

2. *If $X_n = O_p(n^{-a})$ and $y_n = o(n^{-b})$ as $n \to \infty$, then $X_n y_n = o_p(n^{-(a+b)})$ as $n \to \infty$.*

3. *If $X_n = O_p(n^{-a})$ and $Y_n = o_p(n^{-b})$ as $n \to \infty$, then $X_n Y_n = o_p(n^{-(a+b)})$ as $n \to \infty$.*

4. *If $X_n = o_p(n^{-a})$ and $y_n = o(n^{-b})$ as $n \to \infty$, then $X_n y_n = o_p(n^{-(a+b)})$ as $n \to \infty$.*

5. *If $X_n = O_p(n^{-a})$ and $y_n = O(n^{-b})$ as $n \to \infty$, then $X_n y_n = O_p(n^{-(a+b)})$ as $n \to \infty$.*

6. *If $X_n = o_p(n^{-a})$ and $y_n = O(n^{-b})$ as $n \to \infty$, then $X_n Y_n = o_p(n^{-(a+b)})$ as $n \to \infty$.*

7. *If $X_n = o_p(n^{-a})$ and $Y_n = o_p(n^{-b})$ as $n \to \infty$, then $X_n Y_n = o_p(n^{-(a+b)})$ as $n \to \infty$.*

Proof. We will prove the first two parts of this theorem, leaving the remaining parts for Exercise 9. To prove the first result, suppose that $\{X_n\}_{n=1}^{\infty}$ and $\{Y_n\}_{n=1}^{\infty}$ are sequences of random variables such that $X_n = O_p(n^{-a})$ and $Y_n = O_p(n^{-b})$ as $n \to \infty$. Therefore, it follows from Definition 8.2 that the sequences $\{n^a X_n\}_{n=1}^{\infty}$ and $\{n^b Y_n\}_{n=1}^{\infty}$ are both bounded in probability. Therefore, for every $\varepsilon > 0$ there exist bounds x_ε and y_ε and positive integers $n_{x,\varepsilon}$ and $n_{y,\varepsilon}$ such that $P(|n^a X_n| \leq x_\varepsilon) > 1 - \varepsilon$ and $P(|m^b Y_m| \leq y_\varepsilon) > 1 - \varepsilon$ for all $n > n_{x,\varepsilon}$ and $m > n_{y,\varepsilon}$. Define $b_\varepsilon = \max\{x_\varepsilon, y_\varepsilon\}$ and $n_\varepsilon = \max\{n_{x,\varepsilon}, n_{y,\varepsilon}\}$. Therefore, it follows that $P(|n^a X_n| \leq b_\varepsilon) > 1 - \varepsilon$ and $P(|n^b Y_n| \leq b_\varepsilon) > 1 - \varepsilon$ for all $n > n_\varepsilon$. In accordance with Definition 8.2, we must now prove that the sequence $\{n^{a+b} X_n Y_n\}_{n=1}^{\infty}$ is bounded in probability. Let $\varepsilon > 0$ and note that

$$
\begin{aligned}
P(|n^{a+b} X_n Y_n| \leq b_\varepsilon^2) &= P\left(|n^{a+b} X_n Y_n| \leq b_\varepsilon^2 \big| |n^a X_n| > b_\varepsilon\right) P(|n^a X_n| > b_\varepsilon) + \\
&\quad\; P\left(|n^{a+b} X_n Y_n| \leq b_\varepsilon^2 \big| |n^a X_n| \leq b_\varepsilon\right) P(|n^a X_n| \leq b_\varepsilon) \\
&\geq P\left(|n^{a+b} X_n Y_n| \leq b_\varepsilon^2 \big| |n^a X_n| \leq b_\varepsilon\right) P(|n^a X_n| \leq b_\varepsilon) \\
&\geq P(|n^b Y_n| \leq b_\varepsilon) P(|n^a X_n| \leq b_\varepsilon) \\
&\geq (1 - \varepsilon)^2 \\
&= 1 - 2\varepsilon + \varepsilon^2 \\
&> 1 - 2\varepsilon.
\end{aligned}
$$

Therefore, Definition 4.3 implies that the sequence $\{n^{a+b} X_n Y_n\}_{n=1}^{\infty}$ is bounded in probability and Definition 8.2 implies that $X_n Y_n = O_p(n^{-(a+b)})$ as $n \to \infty$.

To prove the second result suppose that $\{X_n\}_{n=1}^{\infty}$ is a sequence of random variables and that $\{y_n\}_{n=1}^{\infty}$ is a sequence of real numbers such that $X_n = O_p(n^{-a})$ and $y_n = o(n^{-b})$ as $n \to \infty$. Since $X_n = O_p(n^{-a})$ as $n \to \infty$ it follows from Definitions 4.3 and 8.2 that for every $\varepsilon > 0$ there exists a positive real number b_ε and a positive integer n_ε such that $P(|n^a X_n| \leq b_\varepsilon) \geq 1 - \varepsilon$ for all $n > n_\varepsilon$. Since $y_n = o(n^{-b})$ as $n \to \infty$ it follows from Definition 1.7 that for every $\delta > 0$ there exists a positive integer n_δ such that $|n^b Y_n| < \delta$ for all $n > n_\delta$. Now let $\varepsilon > 0$ and $\xi > 0$ be given, and choose δ so that $b_\varepsilon \delta < \xi$. Then it follows that for all $n > \max\{n_\varepsilon, n_\delta\}$ we have that

$$P(|n^b y_n||n^a X_n| \leq b_\varepsilon \delta) = P(|n^{a+b} X_n y_n| \leq \xi) \geq 1 - \varepsilon.$$

Since ε is arbitrary it follows from Definition 1.1 that

$$\lim_{n \to \infty} P(|n^{a+b} X_n y_n| \leq \xi) = 1,$$

and therefore Definition 3.1 implies that $n^{a+b} X_n y_n \xrightarrow{p} 0$ as $n \to \infty$. Therefore, Definition 8.1 implies that $X_n y_n = o_p(n^{-(a+b)})$ as $n \to \infty$.   □

With the introduction of the stochastic order notation in Definitions 8.1 and 8.2, we can now define an asymptotic expansion for a sequence of random variables $\{X_n\}_{n=1}^\infty$ as an expansion of the form

$$X_n = Y_0 + n^{-1/2} Y_1 + n^{-1} Y_2 + \cdots + Y_p n^{-p/2} + O_p(n^{-(p+1)/2}), \qquad (8.2)$$

or of the form

$$X_n = Y_0 + n^{-1/2} Y_1 + n^{-1} Y_2 + \cdots + Y_p n^{-p/2} + o_p(n^{-p/2}), \qquad (8.3)$$

as $n \to \infty$. Such expansions are often called *stochastic asymptotic expansions*. We can also define these expansion in terms of the powers of n^{-1}, or any other sequence in n that converges to zero as $n \to \infty$. We have already seen some expansions of this form in Examples 8.7 and 8.8. If we consider a stochastic expansion of the form given in Equation (8.2) or (8.3) with $p = 1$ then we have that $X_n = Y_0 + o_p(1)$ as $n \to \infty$, so that the error in this expansion converges in probability to zero. Therefore it follows that $X_n - Y_0 \xrightarrow{p} 0$, or equivalently that $X_n \xrightarrow{p} Y_0$. In this sense, such an expansion can be seen as having a leading term equal to a limiting random variable plus some random error that converges to zero in probability as $n \to \infty$. We now return to our motivating example from Section 8.1.

Example 8.11. Let $\{X_n\}_{n=1}^\infty$ be a sequence of independent and identically distributed random variables from a distribution F with mean μ and variance σ^2. Earlier in this section we considered an approximate $100\alpha\%$ upper confidence limit for θ given by $\hat{U}_n(X_1, \ldots, X_n) = \bar{X}_n - n^{-1/2} \hat{\sigma} z_{1-\alpha}$. Now let us consider the accuracy of this confidence limit. Define the studentized distribution of $\bar{X}_n$ as $H_n(x) = P[n^{1/2} \hat{\sigma}_n^{-1}(\bar{X}_n - \theta) \leq x]$ and let $h_{\alpha,n}$ be the αth quantile of this distribution, where $\hat{\sigma}_n$ is the sample standard deviation. Note that an exact $100\alpha\%$ upper confidence limit for θ is given by

$U_n(X_1, \ldots, X_n) = \bar{X}_n - n^{-1/2}\hat{\sigma}_n h_{1-\alpha,n}$ since

$$
\begin{aligned}
P[\theta < U_n(X_1, \ldots, X_n)] &= P(\theta < \bar{X}_n - n^{-1/2}\hat{\sigma}_n h_{1-\alpha,n}) \\
&= P[n^{1/2}\hat{\sigma}_n^{-1}(\bar{X}_n - \theta) > h_{1-\alpha,n}] \\
&= 1 - H_n(g_{1-\alpha,n}) \\
&= \alpha.
\end{aligned}
$$

Now suppose that F follows the assumptions of Theorem 7.13, then $h_{\alpha,n}$ has an asymptotic expansion of the form

$$
h_{1-\alpha,n} = z_{1-\alpha} + n^{-1/2}s_1(z_{1-\alpha}) + n^{-1}s_2(z_{1-\alpha}) + o(n^{-1}),
$$

as $n \to \infty$. Therefore

$$
\begin{aligned}
U_n(X_1, \ldots, X_n) = \\
\bar{X}_n - n^{-1/2}\hat{\sigma}_n z_{1-\alpha} - n^{-1}\hat{\sigma}_n s_1(z_{1-\alpha}) - n^{-3/2}\hat{\sigma}_n s_2(z_{1-\alpha}) + o(n^{-3/2}),
\end{aligned}
$$

as $n \to \infty$. Hence, if we define $R_n = |\hat{U}_n(X_1, \ldots, X_n) - U_n(X_1, \ldots, X_n)|$, then it follows that

$$
R_n = |n^{-1}\hat{\sigma}_n s_1(z_{1-\alpha}) + n^{-3/2}\hat{\sigma}_n s_2(z_{1-\alpha}) + o(n^{-3/2})|,
$$

as $n \to \infty$. Assuming that $\hat{\sigma}_n$ is a consistent estimator of σ, which can be verified using Theorem 3.19, it follows that $\hat{\sigma}_n = \sigma + o_p(1)$, as $n \to \infty$. Therefore, it follows that $n^{-1}\hat{\sigma}_n s_1(z_{1-\alpha}) = o_p(n^{-1/2})$ and hence $R_n = o_p(n^{-1/2})$, as $n \to \infty$. Hence we have shown that $\hat{U}_n(X_1, \ldots, X_n) = U_n(X_1, \ldots, X_n) + o_p(n^{-1/2})$ as $n \to \infty$. Now recall that $s_1(z_{1-\alpha})$ and $s_2(z_{1-\alpha})$ are polynomials whose coefficients are functions of the moments of F. Suppose that we can estimate $s_1(z_{1-\alpha})$ and $s_2(z_{1-\alpha})$ with consistent estimators $\hat{s}_1(z_{1-\alpha})$ and $\hat{s}_2(z_{1-\alpha})$ so that $\hat{s}_1(z_{1-\alpha}) = s_1(z_{1-\alpha}) + o_p(1)$ and $\hat{s}_2(z_{1-\alpha}) = s_2(z_{1-\alpha}) + o_p(1)$, as $n \to \infty$. We can then approximate the true upper confidence limit with

$$
\tilde{U}_n(X_1, \ldots, X_n) = \bar{X}_n - n^{-1/2}\hat{\sigma}_n z_\alpha - n^{-1}\hat{\sigma}_n \hat{s}_1(z_{1-\alpha}) - n^{-3/2}\hat{\sigma}_n \hat{s}_2(z_{1-\alpha}).
$$

To find the accuracy of this approximate note that

$$
n^{-1/2}\hat{\sigma}_n \hat{s}_1(z_{1-\alpha}) = n^{-1/2}\sigma s_1(z_{1-\alpha}) + o_p(n^{-1/2}),
$$

and

$$
n^{-1}\hat{\sigma}_n \hat{s}_2(z_{1-\alpha}) = n^{-1}\sigma s_2(z_{1-\alpha}) + o_p(n^{-1}),
$$

as $n \to \infty$. Therefore,

$$
\tilde{U}_n(X_1, \ldots, X_n) = \bar{X}_n - n^{-1/2}\hat{\sigma}_n z_{1-\alpha} - n^{-1}\hat{\sigma}_n s_1(z_{1-\alpha}) + o_p(n^{-1}),
$$

as $n \to \infty$ and it follows that $|\tilde{U}_n(X_1, \ldots, X_n) - U_n(X_1, \ldots, X_n)| = o_p(n^{-1})$, as $n \to \infty$, which is more accurate than the normal approximation given by $\hat{U}_n(X_1, \ldots, X_n)$. Note that estimating $s_2(z_{1-\alpha})$ in this case makes no difference from an asymptotic viewpoint, because the error from estimating $s_1(z_{1-\alpha})$ is as large, asymptotically, as this term. Therefore, an asymptotically equivalent substitute for $\tilde{U}_n(X_1, \ldots, X_n)$ is the approximation $\bar{X}_n - n^{-1/2}\hat{\sigma}_n z_{1-\alpha} - n^{-1}\hat{\sigma}_n \hat{s}_1(z_{1-\alpha})$.

8.3 The Delta Method

In several instances we have considered how taking a function of a sequence of convergent random variables affects the convergence properties of the sequence. For example, in certain cases the function of the sequence of random variables converges to the same function of the limit random variable. That is, if $\{X_n\}_{n=1}^{\infty}$ is a sequence of random variables that converges in some mode to a random variable X as $n \to \infty$, and g is a smooth function, then we observed in Theorems 3.8 and 4.12 that $g(X_n)$ will converge in the same mode to $g(X)$ as $n \to \infty$. In Section 6.4 we observed that certain functions of asymptotically NORMAL sequences are also asymptotically NORMAL. In this section we use the asymptotic properties of a sequence of random variables to develop a stochastic asymptotic expansion for a function of the sequence of random variables. In this development we will require both that the sequence converges in probability to a constant and that a monotonically increasing function of n multiplied by the sequence converges in distribution. Our results will include a stochastic asymptotic expansion for the function of the sequence along with a conclusion about the weak convergence of the transformed sequence. This method is often called the *delta method* or *approximate local linearization*.

To begin this development, let $\{X_n\}_{n=1}^{\infty}$ be a sequence of random variables such that $X_n \xrightarrow{p} \theta$ as $n \to \infty$ for some real number θ and assume that there is a random variable Z such that

$$n^{1/2}(X_n - \theta) = Z + o_p(1), \tag{8.4}$$

as $n \to \infty$. Therefore, it follows from Theorem 4.11 (Slutsky) that $n^{1/2}(X_n - \theta) \xrightarrow{d} Z$ as $n \to \infty$ and hence X_n has the stochastic asymptotic expansion $X_n = \theta + n^{-1/2}Z + o_p(n^{-1/2})$ as $n \to \infty$.

Now consider transforming this sequence with a function g, where we will assume that g has two continuous derivatives on the real line such that $g'(\theta) \neq 0$ and g'' is bounded. The transformed sequence is $\{g(X_n)\}_{n=1}^{\infty}$. Theorem 1.13 (Taylor) implies that $g(x) = g(\theta) + (x - \theta)g'(\theta) + R_x$ where $R_x = \frac{1}{2}g''(\xi_x)(x - \theta)^2$ and ξ_x is real number between x and θ that depends on x. Therefore, substituting the random variable X_n for x we have that $g(X_n) = g(\theta) + (X_n - \theta)g'(\theta) + R_n$ where $R_n = \frac{1}{2}g''(\xi_n)(X_n - \theta)^2$ and ξ_n is a random variable that is between X_n and θ with probability one, and therefore depends on n. Now, it follows from Equation (8.4) that

$$
\begin{aligned}
n^{1/2}[g(X_n) - g(\theta)] &= n^{1/2}[g(\theta) + (X_n - \theta)g'(\theta) + R_n - g(\theta)] \\
&= n^{1/2}(X_n - \theta)g'(\theta) + n^{1/2}R_n \\
&= [Z + o_p(1)]g'(\theta) + n^{1/2}R_n \\
&= Zg'(\theta) + n^{1/2}R_n + o_p(1),
\end{aligned}
$$

as $n \to \infty$. To ascertain the asymptotic behavior of the term $n^{1/2}R_n$ we first note that

$$
\begin{aligned}
n^{1/2}(X_n - \theta)^2 &= n^{1/2}(X_n - \theta)(X_n - \theta) \\
&= [Z + o_p(1)][n^{-1/2}Z + o_p(n^{-1/2})] \\
&= n^{-1/2}Z^2 + (n^{-1/2}Z)o_p(1) + (Z)o_p(n^{-1/2}) \\
&\quad + o_p(1)o_p(n^{-1/2}),
\end{aligned}
$$

as $n \to \infty$. Considering the first term of this expression, we observe that $n^{-1/2} = o(1)$ and assuming that Z has a valid distribution function, Definition 8.2 implies that $Z = O_p(1)$ as $n \to \infty$ and hence $Z^2 = O_p(1)O_p(1) = O_p(1)$ as $n \to \infty$ by Theorem 8.3. Therefore, Theorem 8.3 also implies that $n^{-1/2}Z^2 = o_p(1)$ as $n \to \infty$. Similar arguments using Theorem 8.3 can be used to show that $(n^{-1/2}Z)o_p(1) = o(1)O_p(1)o_p(1) = o_p(1)$, $(Z)o_p(n^{-1/2}) = O_p(1)o_p(n^{-1/2}) = o_p(n^{-1/2}) = o_p(1)$, and finally $o_p(1)o_p(n^{-1/2}) = o_p(n^{-1/2}) = o_p(1)$, as $n \to \infty$. Because we have assumed that g'' is uniformly bounded in the range of X_n it follows that $g''(\xi_n)$ is uniformly bounded with probability one and therefore $g''(\xi_n) = O_p(1)$ as $n \to \infty$. Therefore, it follows that

$$
n^{1/2}R_n = \tfrac{1}{2}n^{1/2}g''(\xi_n)(X_n - \theta)^2 = o_p(1)O_p(1) = o_p(1),
$$

as $n \to \infty$, and hence $n^{1/2}[g(X_n) - g(\theta)] = Zg'(\theta) + o_p(1)$ as $n \to \infty$. Applying Theorem 4.11 to this stochastic asymptotic expansion leads us to the conclusion that $n^{1/2}[g(X_n) - g(\theta)] \xrightarrow{d} Zg'(\theta)$ as $n \to \infty$.

A common application of this theory is when Z is a $N(0, \sigma^2)$ random variable for which the conclusion would be, under the assumptions of this section on g, that $n^{1/2}[g(X_n) - g(\theta)]$ converges in distribution to a $g'(\theta)Z$ random variable which has a $N(0, [g'(\theta)\sigma]^2)$ distribution. This is the same conclusion we encountered in Theorem 6.3. As this section shows, the conclusion is more general and can be motivated through the use of stochastic asymptotic expansions. If the condition that $g'(\theta) \neq 0$ is violated, then additional terms in the Taylor expansion must be used and we obtain a result that has the same form as in Theorem 6.4.

Example 8.12. Suppose that $\{W_n\}_{n=1}^\infty$ is a sequence of independent random variables such that W_n has a $N(\theta, \sigma^2)$ distribution for all $n \in \mathbb{N}$ where $\theta \neq 0$. Define a sequence of random variables $\{X_n\}_{n=1}^\infty$ where $X_n = \bar{W}_n$ for all $n \in \mathbb{N}$ so that X_n has a $N(\theta, n^{-1}\sigma^2)$ distribution for all $n \in \mathbb{N}$ and $X_n \xrightarrow{p} \theta$ as $n \to \infty$. Let $g(x) = x^2$ so that $g'(x) = 2x$ and $g''(x) = 2$, which is bounded on the real line. The earlier conclusions imply that $n^{1/2}(X_n^2 - \theta^2) = 2\theta Z + o_p(1)$ for a random variable Z that has a $N(0, \sigma^2)$ distribution and that $n^{1/2}(X_n^2 - \theta^2)$ converges in distribution to a random variable that has a $N(0, 4\theta^2\sigma^2)$ distribution. ∎

Example 8.13. One problem in statistical inference is that the standard error of an estimator may depend on the unknown parameter of interest. In this case the standard error can only be estimated using the estimate of the unknown parameter. *Variance stabilization* is a method that can sometimes be used to circumvent this problem by transforming the estimator so that the standard error does not depend on the unknown parameter. As an example, consider $\{X_n\}_{n=1}^{\infty}$, a sequence of random variables where we will assume that $n^{1/2}(X_n - \theta) \overset{d}{\to} Z$ and that $X_n \overset{p}{\to} \theta$ as $n \to \infty$. Hence, we will consider X_n to be a consistent estimator of θ. The distribution of Z is not overly important for this process, but for simplicity we will assume that Z has a $N[0, h(\theta)]$ where h is some known function of θ. We will further assume, consistent with the assumptions above, that the variance of X_n is $n^{-1}h(\theta)$. We would now like to transform this sequence so that the variance of X_n does not depend on θ. Let g be a function that follows the assumptions of this section. Then $n^{1/2}[g(X_n) - g(\theta)] \overset{d}{\to} Y$ where Y has a $N\{0, [g'(\theta)]^2 h(\theta)\}$ distribution. Therefore, to eliminate θ from the variance we need to choose g such that $[g'(\theta)]^2 h(\theta) = 1$, or equivalently that $g'(\theta) = [h(\theta)]^{-1/2}$. ■

8.4 The Sample Moments

Let $\{X_n\}_{n=1}^{\infty}$ be a sequence of independent and identically distributed random variables. In Section 4.6 we established that the sample moments $\hat{\mu}'_k$ are asymptotically normal. In this section we demonstrate how to use stochastic asymptotic expansions to prove that the sample central moments are also asymptotically normal. Therefore, let μ_k be the kth central moment defined in Definition 2.9 and let $\hat{\mu}_k$ be the kth sample central moment defined in Section 3.8 as

$$\hat{\mu}_k = n^{-1} \sum_{i=1}^{n} (X_i - \bar{X}_n)^k.$$

The presence of the sample mean in the expression for $\hat{\mu}_k$ complicates the discussion of the limiting distribution of these statistics. If the sample mean was replaced by the population mean μ'_1 then the asymptotic NORMALITY of $\hat{\mu}_k$ could be established directly through the use of Theorem 4.20. One approach to simplifying this problem is based on finding an approximation to $\hat{\mu}_k$ that has this simpler form.

Theorem 8.4. *Let $\{X_n\}_{n=1}^{\infty}$ be a sequence of independent and identically distributed random variables from a distribution F with mean μ and whose first k moments exist. Define*

$$\tilde{\mu}_k = n^{-1} \sum_{i=1}^{n} (X_i - \mu)^k,$$

then,

$$n^{1/2}(\hat{\mu}_k - \mu_k) = n^{1/2}(\tilde{\mu}_k - \mu_k - k\mu_{k-1}\tilde{\mu}_1) + o_p(1), \qquad (8.5)$$

as $n \to \infty$.

Proof. To begin this argument, first note that using Theorem A.22 we have

that

$$n^{-1}\sum_{i=1}^{n}(X_i - \bar{X}_n)^k = n^{-1}\sum_{i=1}^{n}[(X_i - \mu) - (\bar{X}_n - \mu)]^k$$

$$= n^{-1}\sum_{i=1}^{n}\sum_{j=0}^{k}(-1)^{k-j}\binom{k}{j}(X_i - \mu)^j(\bar{X}_n - \mu)^{k-j}$$

$$= \sum_{j=0}^{k}(-1)^{k-j}\binom{k}{j}n^{-1}(\bar{X}_n - \mu)^{k-j}\left[\sum_{i=1}^{n}(X_i - \mu)^j\right]$$

$$= \sum_{j=0}^{k}(-1)^{k-j}\binom{k}{j}(\bar{X}_n - \mu)^{k-j}\tilde{\mu}_j.$$

Now note that

$$\tilde{\mu}_1 = n^{-1}\sum_{i=1}^{n}(X_i - \mu) = \bar{X}_n - \mu,$$

so that

$$n^{-1}\sum_{i=1}^{n}(X_i - \bar{X}_n)^k = \sum_{j=0}^{k}(-1)^{k-j}\binom{k}{j}\tilde{\mu}_1^{k-j}\tilde{\mu}_j$$

$$= \sum_{j=0}^{k-2}(-1)^{k-j}\binom{k}{j}\tilde{\mu}_1^{k-j}\tilde{\mu}_j + (-1)^1\binom{k}{k-1}\tilde{\mu}_1\tilde{\mu}_{k-1}$$

$$+(-1)^0\binom{k}{k}\tilde{\mu}_k$$

$$= \sum_{j=0}^{k-2}(-1)^{k-j}\binom{k}{j}\tilde{\mu}_1^{k-j}\tilde{\mu}_j - k\tilde{\mu}_1\tilde{\mu}_{k-1} + \tilde{\mu}_k.$$

Therefore, it follows that

$$n^{1/2}(\hat{\mu}_k - \mu_k) = n^{1/2}\left[\tilde{\mu}_k - \mu_k - k\tilde{\mu}_1\tilde{\mu}_{k-1} + \sum_{j=0}^{k-2}(-1)^{k-j}\binom{k}{j}\tilde{\mu}_1^{k-j}\tilde{\mu}_j\right]$$

$$= n^{1/2}\left[\tilde{\mu}_k - \mu_k - k\tilde{\mu}_1\mu_{k-1} + k\tilde{\mu}_1\mu_{k-1} - k\tilde{\mu}_1\tilde{\mu}_{k-1}+\right.$$

$$\left.\sum_{j=0}^{k-2}(-1)^{k-j}\binom{k}{j}\tilde{\mu}_1^{k-j}\tilde{\mu}_j\right]$$

$$= n^{1/2}(\tilde{\mu}_k - \mu_k - k\tilde{\mu}_1\mu_{k-1}) + n^{1/2}\tilde{\mu}_1\left[k\mu_{k-1} - k\tilde{\mu}_{k-1}+\right.$$

$$\left.\sum_{j=0}^{k-2}(-1)^{k-j}\binom{k}{j}\tilde{\mu}_1^{k-j-1}\tilde{\mu}_j\right].$$

Now Theorem 4.20 implies that $n^{1/2}\tilde{\mu}_1 = n^{1/2}(\bar{X}_n - \mu) \xrightarrow{d} Z$ as $n \to \infty$

where Z is a $N(0, \mu_2)$ random variable. Therefore, Theorem 8.1 implies that $n^{1/2}\tilde{\mu}_1 = O_p(1)$ as $n \to \infty$. Next we note that Theorem 3.10 implies that

$$\tilde{\mu}_{k-1} = n^{-1}\sum_{i=1}^{n}(X_i - \mu)^{k-1} \xrightarrow{p} E[(X_i - \mu)^{k-1}] = \mu_{k-1}, \qquad (8.6)$$

as $n \to \infty$ so that Theorem 3.8 implies that $k\mu_{k-1} - k\tilde{\mu}_{k-1} \xrightarrow{p} 0$ as $n \to \infty$. Finally, we note that in the special case when $k = 2$ in Equation (8.6) that

$$\tilde{\mu}_1 = n^{-1}\sum_{i=1}^{n}(X_i - \mu) \xrightarrow{p} E(X_i - \mu) = \mu_1 = 0,$$

so that $\tilde{\mu}_1 \xrightarrow{p} 0$ as $n \to \infty$. Therefore, Theorem 3.8 implies that $\tilde{\mu}_1^{k-j-1}\tilde{\mu}_j \xrightarrow{p} 0$ as $n \to \infty$ for $j = 0, \ldots, k - 2$, and hence

$$\sum_{j=0}^{k-2}(-1)^{k-j}\binom{k}{j}\tilde{\mu}_1^{k-j-1}\tilde{\mu}_j = o_p(1),$$

as $n \to \infty$. Therefore,

$$n^{1/2}(\hat{\mu}_k - \mu_k) = n^{1/2}(\tilde{\mu}_k - \mu_k - k\mu_{k-1}\tilde{\mu}_1) + o_p(1),$$

as $n \to \infty$. $\qquad\qquad\qquad\qquad\qquad\qquad\qquad\qquad\qquad\qquad\square$

The key idea in developing the stochastic approximation in Theorem 8.4 is that the statistics on the right hand side of Equation (8.5) can be written as the sum of independent and identically distributed random variables, whose asymptotic behavior can the be linked directly to Theorem 4.20 (Lindeberg and Lévy). In the following result we use Theorem 8.4 along with Theorem 4.22 to develop a multivariate NORMAL result for a vector of estimated central moments.

Theorem 8.5. *Let $\{X_n\}_{n=1}^{\infty}$ be a sequence of independent and identically distributed random variables from a distribution F with mean μ and k^{th} central moment equal to μ_k. If $\mu_{2k} < \infty$ then*

$$n^{1/2}(\hat{\mu}_2 - \mu_2, \ldots, \hat{\mu}_k - \mu_k)' \xrightarrow{d} \mathbf{Z},$$

as $n \to \infty$ where $\mathbf{Z}$ is a $N(\mathbf{0}, \boldsymbol{\Sigma})$ random vector. The covariance matrix $\boldsymbol{\Sigma}$ has $(i, j)^{th}$ element equal to

$$\mu_{i+j+2} - \mu_{i+1}\mu_{j+1} - (i+1)\mu_i\mu_{j+2} - (j+1)\mu_{i+2}\mu_j + (i+1)(j+1)\mu_i\mu_j\mu_2. \quad (8.7)$$

Proof. Let $\mathbf{Z}_n = n^{1/2}(\hat{\mu}_2 - \mu_2, \ldots, \hat{\mu}_k - \mu_k)'$ and $\mathbf{W}_n = n^{1/2}(\tilde{\mu}_2 - \mu_2 - 2\mu_1\tilde{\mu}_1, \ldots, \tilde{\mu}_k - \mu_k - k\mu_{k-1}\tilde{\mu}_1)'$. Let us assume for the moment that $\mathbf{Z}_n \xrightarrow{d} \mathbf{Z}$ for some random vector $\mathbf{Z}$ whose distribution we will specify later in the proof. Theorem 8.4 tells us that $\mathbf{W}_n - \mathbf{Z}_n \xrightarrow{p} 0$ as $n \to \infty$. Therefore, Theorem 4.19 (Multivariate Slutsky Theorem) implies that $\mathbf{W}_n = \mathbf{Z}_n + (\mathbf{W}_n - \mathbf{Z}_n) \xrightarrow{d} \mathbf{Z}$ as $n \to \infty$. Hence $\mathbf{Z}_n$ and $\mathbf{W}_n$ converge in distribution to the same random variable. Therefore, we need only find the limit distribution of $\mathbf{W}_n$. To find this limiting distribution, we first note that

$$\tilde{\mu}_k - \mu_k - k\mu_{k-1}\tilde{\mu}_1 = n^{-1}\sum_{i=1}^{n}(X_i - \mu)^k - \mu_k - k\mu_{k-1}n^{-1}\sum_{i=1}^{n}(X_i - \mu) =$$

$$n^{-1}\sum_{i=1}^{n}\left[(X_i - \mu)^k - \mu_k - k\mu_{k-1}(X_i - \mu)\right].$$

Define a sequence of random vectors $\{\mathbf{Y}_n\}_{n=1}^{\infty}$ as

$$\mathbf{Y}_n' = [(X_n - \mu)^2 - \mu_2 - 2\mu_1(X_n - \mu), \ldots, (X_n - \mu)^k - \mu_k - k\mu_{k-1}(X_n - \mu)],$$

for all $n \in \mathbb{N}$. Note that

$$E[(X_n - \mu)^k - \mu_k - k\mu_{k-1}(X_n - \mu)] = \mu_k - \mu_k - k\mu_{k-1}\mu_1 = 0,$$

since $\mu_1 = 0$. Then it follows that $\mathbf{W}_n = n^{1/2}\bar{\mathbf{Y}}_n$ and Theorem 4.22 implies that $\mathbf{W}_n \overset{d}{\to} \mathbf{Z}$ as $n \to \infty$ where $\mathbf{Z}$ is a $\mathbf{N}(\mathbf{0}, \boldsymbol{\Sigma})$ random vector. As discussed above, the random vector $\mathbf{Z}_n$ has this same limit distribution, and therefore all there is left to do is verify the form of the covariance matrix. The form of the covariance matrix can be determined from the covariance matrix of $\mathbf{Y}_n$. The $(i,j)^{\text{th}}$ element of this covariance matrix is given by the covariance between $(X_n-\mu)^{i+1}-\mu_{i+1}-(i+1)\mu_i(X_n-\mu)$ and $(X_n-\mu)^{j+1}-\mu_{j+1}-(j+1)\mu_j(X_n-\mu)$. Since both random variables have expectation equal to zero, this covariance is equal to the expectation of the product

$$E\{[(X_n - \mu)^{i+1} - \mu_{i+1} - (i+1)\mu_i(X_n - \mu)]\times$$
$$[(X_n - \mu)^{j+1} - \mu_{j+1} - (j+1)\mu_j(X_n - \mu)]\} =$$
$$E[(X_n - \mu)^{i+1}(X_n - \mu)^{j+1}] - E[(X_n - \mu)^{i+1}\mu_{j+1}]$$
$$- E[(X_n - \mu)^{i+1}(j+1)\mu_j(X_n - \mu)] - E[\mu_{i+1}(X_n - \mu)^{j+1}] + E[\mu_{i+1}\mu_{j+1}]$$
$$+ E[\mu_{i+1}(j+1)\mu_j(X_n - \mu)] - E[(i+1)\mu_i(X_n - \mu)^{j+1}]$$
$$+ E[(i+1)\mu_i(X_n - \mu)\mu_{j+1}] + E[(i+1)(j+1)(X_n - \mu)^2\mu_i\mu_j] =$$
$$\mu_{i+j+2} - \mu_{i+1}\mu_{j+1} - (j+1)\mu_j\mu_{i+2} - \mu_{i+1}\mu_{j+1} + \mu_{i+1}\mu_{j+1}$$
$$+ (j+1)\mu_{i+1}\mu_j\mu_1 - (i+1)\mu_i\mu_{j+2} + (i+1)\mu_i\mu_{j+1}\mu_1 + (i+1)(j+1)\mu_i\mu_j\mu_2 =$$
$$\mu_{i+j+2} - \mu_{i+1}\mu_{j+1} - (j+1)\mu_j\mu_{i+2} - (i+1)\mu_i\mu_{j+2} + (i+1)(j+1)\mu_i\mu_j\mu_2,$$

where we note that $(i+1)\mu_i\mu_{j+1}\mu_1 = (j+1)\mu_{i+1}\mu_j\mu_1 = 0$. This expression matches the covariance given in Equation (8.7). $\qquad\square$

Example 8.14. Let $\{X_n\}_{n=1}^{\infty}$ be a sequence of independent and identically distributed random variables from a distribution F with k^{th} central moment given by μ_k. Suppose that $\mu_6 < \infty$, then Theorem 8.5 implies that $n^{1/2}(\hat{\mu}_2 - \mu_2, \hat{\mu}_3 - \mu_3)' \overset{d}{\to} \mathbf{Z}$ as $n \to \infty$ where $\mathbf{Z}$ is a $\mathbf{N}(\mathbf{0}, \boldsymbol{\Sigma})$ random vector where

$$\boldsymbol{\Sigma} = \begin{bmatrix} \mu_4 - \mu_2^2 & \mu_5 - 4\mu_2\mu_3 \\ \mu_5 - 4\mu_2\mu_3 & \mu_6 - \mu_3^2 - 6\mu_3\mu_4 + 9\mu_2^3 \end{bmatrix},$$

where we note that $\mu_1 = 0$.   ∎

8.5 Exercises and Experiments

8.5.1 Exercises

1. Let $\{X_n\}_{n=1}^{\infty}$ be a sequence of independent random variables where X_n has a $N(0, n^{-1})$ distribution for all $n \in \mathbb{N}$. Prove that $X_n = O_p(n^{-1/2})$ as $n \to \infty$.

2. Let $\{X_n\}_{n=1}^{\infty}$ and $\{Y_n\}_{n=1}^{\infty}$ be sequences of random variables. Suppose that $X_n = o_p(Y_n)$ as $n \to \infty$. Prove that $X_n = O_p(Y_n)$ as $n \to \infty$.

3. Let $\{X_n\}_{n=1}^{\infty}$ be a sequence of independent and identically distributed random variables where $X_n = n^{-1}U_n$ where U_n has a UNIFORM$(0, 1)$ distribution for all $n \in \mathbb{N}$. Prove that $X_n = o_p(n^{-1/2})$ and that $X_n = O_p(n^{-1})$ as $n \to \infty$.

4. Let $\{X_n\}_{n=1}^{\infty}$ and $\{Y_n\}_{n=1}^{\infty}$ be sequences of independent random variables. Suppose that Y_n is a BETA(α_n, β_n) random variable where $\{\alpha_n\}_{n=1}^{\infty}$ and $\{\beta_n\}_{n=1}^{\infty}$ are sequences of positive real numbers that converge to α and β, respectively. Suppose further that, conditional on Y_n, the random variable X_n has a BINOMIAL(m, Y_n) distribution where m is a fixed positive integer for all $n \in \mathbb{N}$. Prove that $X_n = O_p(1)$ as $n \to \infty$.

5. Let $\{X_n\}_{n=1}^{\infty}$ and $\{Y_n\}_{n=1}^{\infty}$ be sequences of independent random variables. Suppose that Y_n is a POISSON(θ) random variable where θ is a positive real number. Suppose further that, conditional on Y_n, the random variable X_n has a BINOMIAL(Y_n, τ) distribution for all $n \in \mathbb{N}$ where τ is a fixed real number in the interval $[0, 1]$. Prove that $X_n = O_p(Y_n)$ as $n \to \infty$.

6. Let $\{X_n\}_{n=1}^{\infty}$ be a sequence of independent random variables where X_n has a GAMMA(α_n, β_n) distribution for all $n \in \mathbb{N}$ and $\{\alpha_n\}_{n=1}^{\infty}$ and $\{\beta_n\}_{n=1}^{\infty}$ are bounded sequences of positive real numbers. That is, there exist real numbers α and β such that $0 < \alpha_n \leq \alpha$ and $0 < \beta_n \leq \beta$ for all $n \in \mathbb{N}$. Prove that $X_n = O_p(1)$ as $n \to \infty$.

7. Let $\{X_n\}_{n=1}^{\infty}$ be a sequence of independent random variables where X_n has a GEOMETRIC(θ_n) distribution where $\{\theta_n\}_{n=1}^{\infty}$ is described below. For each sequence determine whether $X_n = O_p(1)$ as $n \to \infty$.

   a. $\theta_n = n(n + 10)^{-1}$ for all $n \in \mathbb{N}$.
   b. $\theta_n = n^{-1}$ for all $n \in \mathbb{N}$.
   c. $\theta_n = n^{-2}$ for all $n \in \mathbb{N}$.
   d. $\theta_n = \frac{1}{2}$ for all $n \in \mathbb{N}$.

8. Let $\{X_n\}_{n=1}^{\infty}$ and $\{Y_n\}_{n=1}^{\infty}$ be sequences of independent random variables, where X_n has a UNIFORM$(0, n)$ distribution and Y_n has a UNIFORM$(0, n^2)$ distribution for all $n \in \mathbb{N}$. Prove that $X_n = o_p(Y_n)$ as $n \to \infty$.

9. Let $\{X_n\}_{n=1}^{\infty}$ and $\{Y_n\}_{n=1}^{\infty}$ be sequences of random variables and let $\{y_n\}_{n=1}^{\infty}$ be a sequence of real numbers.

   a. Prove that if $X_n = O_p(n^{-a})$ and $Y_n = o_p(n^{-b})$ as $n \to \infty$, then $X_n Y_n = o_p(n^{-(a+b)})$ as $n \to \infty$.

   b. Prove that if $X_n = o_p(n^{-a})$ and $y_n = o(n^{-b})$ as $n \to \infty$, then $X_n y_n = o_p(n^{-(a+b)})$ as $n \to \infty$.

   c. Prove that if $X_n = O_p(n^{-a})$ and $y_n = O(n^{-b})$ as $n \to \infty$, then $X_n y_n = O_p(n^{-(a+b)})$ as $n \to \infty$.

   d. Prove that if $X_n = o_p(n^{-a})$ and $y_n = O(n^{-b})$ as $n \to \infty$, then $X_n y_n = o_p(n^{-(a+b)})$ as $n \to \infty$.

   e. Prove that if $X_n = o_p(n^{-a})$ and $Y_n = o_p(n^{-b})$ as $n \to \infty$, then $X_n Y_n = o_p(n^{-(a+b)})$ as $n \to \infty$.

10. Suppose that $\{W_n\}_{n=1}^{\infty}$ is a sequence of independent random variables such that W_n has a $N(\theta, \sigma^2)$ distribution for all $n \in \mathbb{N}$ where $\theta \neq 0$. Define a sequence of random variables $\{X_n\}_{n=1}^{\infty}$ where $X_n = \bar{W}_n$ for all $n \in \mathbb{N}$ so that $N(\theta, n^{-1}\sigma^2)$ distribution for all $n \in \mathbb{N}$ and $X_n \xrightarrow{p} \theta$ as $n \to \infty$. Find the asymptotic distribution of $n^{1/2}[\exp(-X_n^2) - \exp(-\theta^2)]$.

11. Let $\{B_n\}_{n=1}^{\infty}$ be a sequence of independent random variables where B_n has a BERNOULLI(θ) distribution for all $n \in \mathbb{N}$. Define a sequence of random variables $\{X_n\}_{n=1}^{\infty}$ where

$$X_n = n^{-1} \sum_{k=1}^{n} B_k,$$

which is the proportion of the first n BERNOULLI random variables that equal one. Prove that $n^{1/2}(X_n - \theta) \xrightarrow{d} Z$ as $n \to \infty$ where Z has a $N[0, n^{-1}\theta(1 - \theta)]$ distribution and that $X_n \xrightarrow{p} \theta$ as $n \to \infty$. Using these conclusions, find the asymptotic distribution of $n^{1/2}[X_n(1 - X_n) - \theta(1 - \theta)]$.

12. Let $\{X_n\}_{n=1}^{\infty}$ be a sequence of independent and identically distributed random variables from a distribution F with k^{th} central moment given by μ_k. Suppose that $\mu_{10} < \infty$. Use Theorem 8.5 to find the asymptotic distribution of $n^{1/2}(\hat{\mu}_3 - \mu_3, \hat{\mu}_4 - \mu_4, \hat{\mu}_5 - \mu_5)'$.

8.5.2 Experiments

1. Write a program in R that first simulates 1000 observations from a POISSON(10) distribution. For each observation, simulate a BINOMIAL($n, \frac{1}{2}$) observation where n is equal to the corresponding observation from the POISSON(10) distribution. Repeat this experiment five times and plot the resulting sequences of ratios of the BINOMIAL observations to the POISSON observations. Describe the plots and address whether the behavior in the plots appears to indicate that the theory given in Exercise 5 has been observed.

2. Write a program in R that simulates two sequences of random variables. The first sequence is given by $X_1, \ldots, X_{100}$ where X_n has a UNIFORM$(0, n)$ distribution for $n = 1, \ldots, 100$. The second sequence is given by $Y_1, \ldots, Y_{100}$ where Y_n has a UNIFORM$(0, n^2)$ distribution for $n = 1, \ldots, 100$. Given the two sequences, compute the sequence $X_1 Y_1^{-1}, \ldots, X_{100} Y_{100}^{-1}$. Repeat the experiment five times and plot the resulting sequences of ratios. Describe the plots and address whether the behavior in the plots appears to indicate that the theory given in Exercise 6 has been observed.

3. Write a program in R that simulates 1000 samples of size 100 from a distribution F, where F is specified below. For each sample compute the second and third sample central moments. Use the 1000 simulated values of the two moments to estimate the variance of each sample moment, along with the covariance between the moments. Note that to estimate the covariance it is important to keep the sample central moments from each sample paired with one another. Compare these estimates to what would be expected given the theory in Example 8.14, noting that the assumptions of the example are not met for all of the specified distributions given below.

   a. N$(0, 1)$
   b. CAUCHY$(0, 1)$
   c. T(3)
   d. EXPONENTIAL(1)

4. Write a program in R that simulates samples of size $n = 1, \ldots, 100$ from a distribution F, where F is specified below. For each sample compute the fourth sample central moment, as well as $\tilde{\mu}_1$ and $\tilde{\mu}_4$ where

$$\tilde{\mu}_k = n^{-1} \sum_{i=1}^{n} (X_i - \mu)^k,$$

and $X_1, \ldots, X_n$ denotes the simulated sample. For each value of n, compute $n^{1/2}(\hat{\mu}_4 - \mu_4)$, $n^{1/2}(\tilde{\mu}_4 - \mu_4 - 4\mu_3\tilde{\mu}_1)$, and the absolute difference between the two expressions. Repeat the experiment five times and plot the resulting sequences of differences against n. Compare the limiting behavior of the absolute differences to what would be expected given the theory in Theorem 8.4, noting that the assumptions of the example are not met for all of the specified distributions given below.

   a. N$(0, 1)$
   b. CAUCHY$(0, 1)$
   c. T(3)
   d. EXPONENTIAL(1)

5. Write a program in R that simulates the sequence $\{X_n\}_{n=1}^{100}$ where X_n has a GEOMETRIC(θ_n) distribution where the sequence θ_n is specified below. Repeat the experiment five times and plot the five realizations against n on

the same set of axes. Describe the behavior of each sequence and compare this to the theoretical results of Exercise 7.

a. $\theta_n = n(n + 10)^{-1}$ for all $n \in \mathbb{N}$.

b. $\theta_n = n^{-1}$ for all $n \in \mathbb{N}$.

c. $\theta_n = n^{-2}$ for all $n \in \mathbb{N}$.

d. $\theta_n = \frac{1}{2}$ for all $n \in \mathbb{N}$.

CHAPTER 9

Differentiable Statistical Functionals

K. stopped talking with them; do I, he thought to himself, do I really have to carry on getting tangled up with the chattering of base functionaries like this?

The Trial by Franz Kafka

9.1 Introduction

This chapter will introduce a class of parameters that are known as *functional parameters*. These types of parameters are functions of a distribution function and can therefore be estimated by taking the same function of the empirical distribution function computed on a sample from the distribution. A novel approach to finding the asymptotic distribution of statistics of this type was developed by von Mises (1947). In essence von Mises (1947) showed that statistics of this type could be approximated based on a type of Taylor expansion. Under some regularity conditions the asymptotic distribution of the approximation can be found using standard methods such as Theorem 4.20 (Linbdeberg and Lévy). In this chapter we will first introduce functional parameters and statistics. We will then develop the Taylor type expansion by first introducing a differential for functional statistics, and then introducing the expansion itself. We will then proceed to develop the asymptotic theory.

There have been many advances in this theory since its inception, mostly in regard to developing more useful differentials. Much of the mathematical theory required to study these advances is beyond the mathematical level of this book. The purpose of this chapter is to provide a general overview of the subject at a mathematical level that is consistent with the rest of our presentation so far. Those who find an interest in this topic should consult Fernholz (1983), Chapter 6 of Serfling (1980), and Chapter 20 of van der Vart (1998).

9.2 Functional Parameters and Statistics

In many cases in statistical inference the parameter of interest can be written as a function of the underlying distribution. That is, suppose that we have observed a set of independent and identically distributed random variables from a distribution F, and we are interested in a certain characteristic θ of this distribution. Usually, θ can be written as a function of F. That is, we can take $\theta = T(F)$ for some function T. It is important to note that the domain of this function is not the real line. Rather, T takes a distribution function and maps it to the real line. Therefore the domain of this function is a space containing distribution functions. We will work with a few spaces of distribution functions. In particular, we will consider the set of all continuous distribution functions on $\mathbb{R}$, the set of all discrete distribution functions on $\mathbb{R}$, and the set of all distribution functions on $\mathbb{R}$. Some examples of functional parameters are given below.

Example 9.1. Let $F \in \mathcal{F}$, the collection of all distribution functions on the real line, and let θ be the kth central moment of F. Then θ is a functional parameter that can be written as

$$\theta = T(F) = \int_{-\infty}^{\infty} t^k dF(t).$$

Noting that this parameter may not exist for all $F \in \mathcal{F}$, we may consider taking F from $\mathcal{F}_k$, the collection of all distribution functions that have at least k finite moments. ∎

Example 9.2. Let $F \in \mathcal{F}_k$, where $\mathcal{F}_k$ is defined in Example 9.1, and let θ be the kth moment of F about the mean. Then θ is a functional parameter that can be written as

$$\theta = T(F) = \int_{-\infty}^{\infty} \left[t - \int_{-\infty}^{\infty} u dF(u) \right]^k dF(t).$$

The variance parameter is a special case of this functional parameter with $k = 2$. ∎

Example 9.3. Let F be a distribution on $\mathbb{R}$ and let $\{R_i\}_{i=1}^{k}$ be a sequence of subsets of $\mathbb{R}$ that form a partition of $\mathbb{R}$. That is, $R_i \cap R_j = \emptyset$ for all $i \neq j$, and

$$\bigcup_{i=1}^{k} R_i = \mathbb{R}.$$

For simplicity we will assume that k is finite. Let $\{p_i\}_{i=1}^{k}$ be a hypothesized model for the probabilities associated with the subsets in the sequence $\{R_i\}_{i=1}^{\infty}$. That is, if the model is correct, then

$$p_i = \int_{R_i} dF,$$

for all $i \in \{1, \ldots, k\}$. The hypothesized model can then be compared to the true model, using the functional

$$\theta = T(F) = \sum_{i=1}^{k} p_i^{-1} \left(\int_{R_i} dF - p_i \right)^2,$$

which is the sum of the square differences of the probabilities relative to the model probabilities. Note that when the proposed model is correct then $\theta = 0$, and that when the proposed model is incorrect then $\theta > 0$. ∎

Estimation of functional parameters can be based on finding an estimator of the distribution function $\hat{F}$ based on the observed sample $X_1, \ldots, X_n$. An estimator of $\theta = T(F)$ is then developed by computing the functional of the estimator of the distribution function. That is, $\hat{\theta}_n = T(\hat{F})$. Such estimates are often called *plug-in* or *substitution* estimators. The most common estimator of F is the empirical distribution function defined in Definition 3.5, though other estimates can be used as well. See Putter and Van Zwet (1996) for general conditions about the consistency of such estimators.

A key property that applies to many plug-in estimators based on the empirical distribution function is that, conditional on the observed sample $X_1, \ldots, X_n$, the empirical distribution function is a discrete distribution that associates a probability of n^{-1} with each value in the sample. Therefore, integrals with respect to the empirical distribution function simplify according to Definition 2.10 as a sum. That is, if g is a real valued function we have that

$$\int_{-\infty}^{\infty} g(x) d\hat{F}_n(x) = n^{-1} \sum_{k=1}^{n} g(X_i).$$

Example 9.4. Let F be a distribution with at least k finite moments, and let θ be the kth central moment of F which is a functional parameter that can be written as

$$\theta = T(F) = \int_{-\infty}^{\infty} t^k dF(t).$$

Suppose that a sample $X_1, \ldots, X_n$ is observed from F and the empirical distribution function $\hat{F}_n$ is used to estimate F. Then a plug-in estimator for θ is given by

$$\hat{\theta}_n = T(\hat{F}_n) = \int_{-\infty}^{\infty} t^k d\hat{F}_n(t) = n^{-1} \sum_{i=1}^{n} X_i^k,$$

which is the kth sample moment $\hat{\mu}'_k$. ∎

Example 9.5. Let F be a distribution with finite k^{th} moment and let θ be the kth moment of F about the mean. Then θ is a functional parameter that can be written as

$$\theta = T(F) = \int_{-\infty}^{\infty} \left[t - \int_{-\infty}^{\infty} u \, dF(u) \right]^k dF(t).$$

Suppose that a sample $X_1, \ldots, X_n$ is observed from F and the empirical distribution function $\hat{F}_n$ is used to estimate F. Then a plug-in estimator for θ is given by

$$
\begin{aligned}
\hat{\theta}_n = T(\hat{F}_n) &= \int_{-\infty}^{\infty} \left[t - \int_{-\infty}^{\infty} u \, d\hat{F}_n(u) \right]^k d\hat{F}_n(t) \\
&= \int_{-\infty}^{\infty} \left[t - n^{-1} \sum_{i=1}^{n} X_i \right]^k d\hat{F}_n(t) \\
&= n^{-1} \sum_{j=1}^{n} \left[X_j - n^{-1} \sum_{i=1}^{n} X_i \right]^k,
\end{aligned}
$$

which is the kth sample central moment $\hat{\mu}_k$. ∎

Example 9.6. Let F be a distribution on $\mathbb{R}$ and let $\{R_i\}_{i=1}^{k}$ be a sequence of subsets of $\mathbb{R}$ that form a partition of $\mathbb{R}$ where we will assume that k is finite. Let $\{p_i\}_{i=1}^{k}$ be a hypothesized model for the probabilities associated with the subsets in the sequence $\{R_i\}_{i=1}^{\infty}$. In Example 9.3 we considered the functional parameter

$$
\theta = T(F) = \sum_{i=1}^{k} p_i^{-1} \left(\int_{R_i} dF - p_i \right)^2,
$$

which compares the hypothesized model to the true model. Suppose that a sample $X_1, \ldots, X_n$ is observed from F and the empirical distribution function $\hat{F}_n$ is used to estimate F. Then a plug-in estimator for θ is given by

$$
\begin{aligned}
\hat{\theta}_n = T(\hat{F}_n) &= \sum_{i=1}^{k} p_i^{-1} \left(\int_{R_i} d\hat{F}_n - p_i \right)^2 \\
&= \sum_{i=1}^{k} p_i^{-1} \left[n^{-1} \sum_{j=1}^{n} \delta(X_j; R_i) - p_i \right]^2 \\
&= \sum_{i=1}^{k} p_i^{-1} (\hat{p}_i - p_i)^2,
\end{aligned}
$$

where $\hat{p}_i$ is the proportion of the sample that was observed in subset R_i. Suppose that we alternatively considered estimating the distribution function F with a NORMAL distribution whose parameters were estimated from the observed sample. That is, we could estimate F with a $N(\bar{X}_n, S)$ distribution, conditional on the observed sample $X_1, \ldots, X_n$, where S is the sample standard deviation. In this case, the plug-in estimator for θ is given by

$$
\hat{\theta}_n = T(\hat{F}_n) = \sum_{i=1}^{k} p_i^{-1} (\tilde{p}_i - p_i)^2,
$$

where $\tilde{p}_i$ is the probability that a $N(\bar{X}_n, S)$ random variable is in the region R_i for all $i \in \{1, \ldots, k\}$. ∎

9.3 Differentiation of Statistical Functionals

The development of a Taylor type expansion for functional parameters and statistics requires us first to develop a derivative, or differential, for functional parameters. To motivate defining such a derivative, we will first briefly recall how the derivative of a real valued function is computed. Let g be a real valued function and suppose that we wish to compute the derivative of g at a point $x \in \mathbb{R}$. The derivative of g at x is the instantaneous slope of the function g at the point x. This instantaneous slope can be defined by taking the slope of a line that connects $g(x)$ with the point $g(x + \delta)$ as δ approaches zero. That is,

$$g'(x) = \lim_{\delta \to 0} \frac{g(x + \delta) - g(x)}{\delta}.$$

Of course we could approach the point x in the opposite direction and define the derivative to be

$$g'(x) = \lim_{\delta \to 0} \frac{g(x) - g(x - \delta)}{\delta}.$$

The derivative in his case is said to exist if both definitions agree. If the two definitions do not agree then the function is not differentiable at x.

Developing a derivative for functionals is more complicated since the domain of a functional is a space of functions. Let T be a functional that maps a function space $\mathcal{F}$ to the real line. In order to compute a derivative of the functional, we need to find the instantaneous change in the functional at a point in the space $\mathcal{F}$. As with the derivative of a real function, we will define this instantaneous change by comparing the difference between $T(F)$ and $T(F_\delta)$, where F_δ is a function such that $F_\delta \in \mathcal{F}$ for all $\delta \in \mathbb{R}$, and F_δ converges to F as $\delta \to \infty$. It is important to note that there are many potential paths through the space $\mathcal{F}$ that F_δ may take to reach F as $\delta \to 0$, and that, just as with the derivative of a real function, this path may have an effect of the derivative at F. Another problem that arises when we attempt to build such a derivative is that we also need a measure of the amount of change between F_δ and F for the denominator of the limit.

The *Gâteaux differential* of T uses the path in $\mathcal{F}$ defined by $F_\delta = F + \delta(G - F)$ for some function $G \in \mathcal{F}$ and $\delta \in [0, 1]$. The amount of change between F_δ and F is measured by δ.

Definition 9.1 (Gâteaux). *Let T be a functional that maps a space of functions $\mathcal{F}$ to $\mathbb{R}$ and let F and G be members of $\mathcal{F}$. The Gâteaux differential of T at F in the direction of G is defined to be*

$$\Delta_1 T(F; G - F) = \lim_{\delta \downarrow 0} \delta^{-1}[T(F_\delta) - T(F)],$$

provided the limit exists.

The function $\Delta_1 T(F; G - F)$ is usually called a *differential* and not a *derivative* due to the fact that the function may not always have the same properties as a derivative for real functions. However, if we define a real valued function $h(\delta) = T[F + \delta(G - F)]$ then we have that $h(0) = T(F)$ and hence

$$\Delta_1 T(F; G - F) = \lim_{\delta \downarrow 0} \delta^{-1}[h(\delta) - h(0)] = \frac{d}{d\delta} h(\delta) \Big|_{\delta \downarrow 0},$$

which is the usual derivative, from the right, of the real function h evaluated at zero. This result allows us to easily define higher order Gâteaux differentials.

Definition 9.2 (Gâteaux). *Let T be a functional that maps a space of functions $\mathcal{F}$ to $\mathbb{R}$ and let F and G be members of $\mathcal{F}$. The kth order Gâteaux differential of T at F in the direction of G is defined to be*

$$\Delta_k T(F; G - F) = \frac{d^k}{d\delta^k} h(\delta) \Big|_{\delta \downarrow 0},$$

where $h(\delta) = T[F + \delta(G - F)]$, provided the derivative exists.

Many of the functionals studied in this chapter will all have a relatively simple form that can be written as a multiple integral of a symmetric function where each integral is integrated with respect to dF. In this case the Gâteaux differential has a simple form.

Theorem 9.1. *Consider a functional of the form*

$$T(F) = \int_{-\infty}^{\infty} \cdots \int_{-\infty}^{\infty} h(x_1, \ldots, x_r) \prod_{i=1}^{r} dF(x_i),$$

where $F \in \mathcal{F}$, and $\mathcal{F}$ is a collection of distribution functions. Then if $k \leq r$,

$$\Delta_k T(F; G - F) = \left[\prod_{i=1}^{k} (r - i + 1) \right] \times$$

$$\int_{-\infty}^{\infty} \cdots \int_{-\infty}^{\infty} h(x_1, \ldots, x_k, y_1, \ldots, y_{r-k}) \left[\prod_{i=1}^{r-k} dF(y_i) \right] \left\{ \prod_{i=1}^{k} d[G(x_i) - F(x_i)] \right\}.$$

If $k > r$ then $\Delta_k T(F; G - F) = 0$.

Proof. We will prove this result for $r = 1$ and $r = 2$. For a general proof see Exercise 9. We first note that when $r = 1$ we have that

$$T[F + \delta(G - F)] = \int_{-\infty}^{\infty} h(x_1) d\{F(x_1) + \delta[G(x_1) - F(x_1)]\} =$$

$$\int_{-\infty}^{\infty} h(x_1) dF(x_1) + \delta \int_{-\infty}^{\infty} h(x_1) d[G(x_1) - F(x_1)].$$

Therefore, Definition 9.2 implies that

$$
\begin{aligned}
\Delta_1 T(F; G - F) &= \frac{d}{d\delta}\left\{\int_{-\infty}^{\infty} h(x_1)dF(x_1)+\right. \\
&\qquad \left. \delta \int_{-\infty}^{\infty} h(x_1)d[G(x_1) - F(x_1)]\right\}\bigg|_{\delta \downarrow 0} \\
&= \int_{-\infty}^{\infty} h(x_1)d[G(x_1) - F(x_1)] \\
&= \int_{-\infty}^{\infty} h(x_1)dG(x_1) - \int_{-\infty}^{\infty} h(x_1)dF(x_1) \\
&= T(G) - T(F).
\end{aligned}
$$

We also note that when $k > 1$ it follows that $\Delta_k T(F; G - F) = 0$. When $r = 2$, we have that

$$
T[F + \delta(G - F)] =
$$
$$
\int_{-\infty}^{\infty}\int_{-\infty}^{\infty} h(x_1, x_2)d\{F(x_1)+\delta[G(x_1)-F(x_1)]\}d\{F(x_2)+\delta[G(x_2)-F(x_2)]\}.
$$

To simplify this expression we first work with the differential in the double integral. Note that

$$
\begin{aligned}
d\{F(x_1) + \delta[G(x_1) - F(x_1)]\}&d\{F(x_2) + \delta[G(x_2) - F(x_2)]\} = \\
\{dF(x_1) + \delta d[G(x_1) - F(x_1)]\}&\{dF(x_2) + \delta d[G(x_2) - F(x_2)]\} = \\
dF(x_1)dF(x_2) &+ \delta d[G(x_1) - F(x_1)]dF(x_2)+ \\
\delta dF(x_1)d[G(x_2) - F(x_2)] &+ \delta^2 d[G(x_1) - F(x_1)]d[G(x_2) - F(x_2)].
\end{aligned}
$$

Therefore,

$$
\begin{aligned}
T[F + \delta(G - F)] &= \int_{-\infty}^{\infty}\int_{-\infty}^{\infty} h(x_1, x_2)dF(x_1)dF(x_2)+ \\
&\delta \int_{-\infty}^{\infty}\int_{-\infty}^{\infty} h(x_1, x_2)d[G(x_1) - F(x_1)]dF(x_2)+ \\
&\delta \int_{-\infty}^{\infty}\int_{-\infty}^{\infty} h(x_1, x_2)dF(x_1)d[G(x_2) - F(x_2)]+ \\
&\delta^2 \int_{-\infty}^{\infty}\int_{-\infty}^{\infty} h(x_1, x_2)d[G(x_1) - F(x_1)]d[G(x_2) - F(x_2)].
\end{aligned}
$$

We now take advantage of the fact that $h(x_1, x_2)$ is a symmetric function in its arguments so that $h(x_1, x_2) = h(x_2, x_1)$. With this assumption it follows

that

$$\int_{-\infty}^{\infty} \int_{-\infty}^{\infty} h(x_1, x_2) d[G(x_1) - F(x_1)] dF(x_2) =$$

$$\int_{-\infty}^{\infty} \int_{-\infty}^{\infty} h(x_1, x_2) dF(x_1) d[G(x_2) - F(x_2)],$$

and, therefore,

$$T[F + \delta(G - F)] = \int_{-\infty}^{\infty} \int_{-\infty}^{\infty} h(x_1, x_2) dF(x_1) dF(x_2) +$$

$$2\delta \int_{-\infty}^{\infty} \int_{-\infty}^{\infty} h(x_1, x_2) d[G(x_1) - F(x_1)] dF(x_2) +$$

$$\delta^2 \int_{-\infty}^{\infty} \int_{-\infty}^{\infty} h(x_1, x_2) d[G(x_1) - F(x_1)] d[G(x_2) - F(x_2)].$$

Therefore, the first two Gâteaux differentials follow from Definition 9.2 as

$$\Delta_1 T(F; G - F) = \frac{d}{d\delta} \left[\int_{-\infty}^{\infty} \int_{-\infty}^{\infty} h(x_1, x_2) dF(x_1) dF(x_2) + \right.$$

$$2\delta \int_{-\infty}^{\infty} \int_{-\infty}^{\infty} h(x_1, x_2) d[G(x_1) - F(x_1)] dF(x_2) +$$

$$\left. \delta^2 \int_{-\infty}^{\infty} \int_{-\infty}^{\infty} h(x_1, x_2) d[G(x_1) - F(x_1)] d[G(x_2) - F(x_2)] \right] \Bigg|_{\delta \downarrow 0}$$

$$= 2 \int_{-\infty}^{\infty} \int_{-\infty}^{\infty} h(x_1, x_2) d[G(x_1) - F(x_1)] dF(x_2) +$$

$$2\delta \int_{-\infty}^{\infty} \int_{-\infty}^{\infty} h(x_1, x_2) d[G(x_1) - F(x_1)] d[G(x_2) - F(x_2)] \Bigg|_{\delta \downarrow 0}$$

$$= 2 \int_{-\infty}^{\infty} \int_{-\infty}^{\infty} h(x_1, x_2) d[G(x_1) - F(x_1)] dF(x_2),$$

and

$$\Delta_2 T(F; G - F) = \frac{d}{d\delta} \left[2 \int_{-\infty}^{\infty} \int_{-\infty}^{\infty} h(x_1, x_2) d[G(x_1) - F(x_1)] dF(x_2) + \right.$$

$$\left. 2\delta \int_{-\infty}^{\infty} \int_{-\infty}^{\infty} h(x_1, x_2) d[G(x_1) - F(x_1)] d[G(x_2) - F(x_2)] \right] \Bigg|_{\delta \downarrow 0}$$

$$= 2 \int_{-\infty}^{\infty} \int_{-\infty}^{\infty} h(x_1, x_2) d[G(x_1) - F(x_1)] d[G(x_2) - F(x_2)]$$

$$= 2T(G - F).$$

We also note that when $k > 2$ it follows that $\Delta_k T(F; G - F) = 0$. $\qquad \square$

Example 9.7. Consider the mean functional given by

$$T(F) = \int_{-\infty}^{\infty} x \, dF(x).$$

This functional has the form given in Theorem 9.1 with $r = 1$ and $h(x) = x$. Therefore, Theorem 9.1 implies that the first Gâteaux differential has the form

$$\Delta_1 T(F; G - F) = T(G) - T(F) = \int_{-\infty}^{\infty} x \, dG(x) - \int_{-\infty}^{\infty} x \, dF(x).$$

Higher order Gâteaux differentials are equal to zero. ∎

Example 9.8. Consider the variance functional given by

$$T(F) = \int_{-\infty}^{\infty} \left[t - \int_{-\infty}^{\infty} t \, dF(t) \right]^2 dF(t),$$

where $F \in \mathcal{F}$, a collection of distribution functions with mean μ and finite variance σ^2. In order to take advantage of the result given in Theorem 9.1 we must first write this functional as a multiple integral of a symmetric function. To find such a function we can note that if X_1 and X_2 are independent random variables both following the distribution F, then

$$E[(X_1 - X_2)^2] = E(X_1^2 + X_2^2 - 2X_1 X_2) = 2\mu^2 + 2\sigma^2 - 2\mu^2 = 2\sigma^2,$$

so that $\frac{1}{2} E[(X_1 - X_2)^2] = \sigma^2$. Therefore, it follows that $T(F)$ can equivalently be written as

$$T(F) = \int_{-\infty}^{\infty} \int_{-\infty}^{\infty} h(x_1, x_2) \, dF(x_1) \, dF(x_2),$$

where $h(x_1, x_2) = \frac{1}{2}(x_1^2 + x_2^2 - 2x_1 x_2)$, which is a symmetric function in x_1 and x_2. We can now apply Theorem 9.1 to this functional to find the Gâteaux differentials for this functional. We first find that

$$
\begin{aligned}
\Delta_1 T(F, G - F) &= 2 \int_{-\infty}^{\infty} \int_{-\infty}^{\infty} \tfrac{1}{2}(x_1^2 + x_2^2 - 2x_1 x_2) \, d[G(x_1) - F(x_1)] \, dF(x_2) \\
&= \int_{-\infty}^{\infty} \int_{-\infty}^{\infty} x_1^2 \, d[G(x_1) - F(x_1)] \, dF(x_2) \\
&\quad + \int_{-\infty}^{\infty} \int_{-\infty}^{\infty} x_2^2 \, d[G(x_1) - F(x_1)] \, dF(x_2) \\
&\quad - 2 \int_{-\infty}^{\infty} \int_{-\infty}^{\infty} x_1 x_2 \, d[G(x_1) - F(x_1)] \, dF(x_2).
\end{aligned}
$$

Now

$$\int_{-\infty}^{\infty}\int_{-\infty}^{\infty} x_1^2 d[G(x_1) - F(x_1)]dF(x_2) =$$

$$\int_{-\infty}^{\infty}\int_{-\infty}^{\infty} x_1^2 dG(x_1)dF(x_2) - \int_{-\infty}^{\infty}\int_{-\infty}^{\infty} x_1^2 dF(x_1)dF(x_2) =$$

$$\left[\int_{-\infty}^{\infty} x_1^2 dG(x_1)\right]\left[\int_{-\infty}^{\infty} dF(x_2)\right] - \left[\int_{-\infty}^{\infty} x_1^2 dF(x_1)\right]\left[\int_{-\infty}^{\infty} dF(x_2)\right] =$$

$$\nu_2' - \mu_2',$$

where ν_2' and μ_2' are the second moments of G and F, respectively. Similarly, the second term is given by

$$\int_{-\infty}^{\infty}\int_{-\infty}^{\infty} x_2^2 d[G(x_1) - F(x_1)]dF(x_2) =$$

$$\left[\int_{-\infty}^{\infty} dG(x_1)\right]\left[\int_{-\infty}^{\infty} x_2^2 dF(x_2)\right] - \left[\int_{-\infty}^{\infty} dF(x_1)\right]\left[\int_{-\infty}^{\infty} x_2^2 dF(x_2)\right] = 0.$$

Finally, the third term is given by

$$2\int_{-\infty}^{\infty}\int_{-\infty}^{\infty} x_1 x_2 d[G(x_1) - F(x_1)]dF(x_2) =$$

$$2\left[\int_{-\infty}^{\infty} x_1 dG(x_1)\right]\left[\int_{-\infty}^{\infty} x_2 dF(x_2)\right] - 2\left[\int_{-\infty}^{\infty} x_1 dF(x_1)\right]\left[\int_{-\infty}^{\infty} x_2 dF(x_2)\right]$$

$$= 2\nu_1'\mu_1' - 2(\mu_1')^2,$$

where μ_1 and ν_1 are the first moments of F and G, respectively. Therefore, $\Delta_1 T(F, G - F) = \nu_2' - \mu_2' - 2\nu_1'\mu_1' + 2(\mu_1')^2$. The second derivative has the form

$$2T(G - F) = \int_{-\infty}^{\infty}\int_{-\infty}^{\infty} x_1^2 d[G(x_1) - F(x_1)]d[G(x_2) - F(x_2)]+$$

$$\int_{-\infty}^{\infty}\int_{-\infty}^{\infty} x_2^2 d[G(x_1) - F(x_1)]d[G(x_2) - F(x_2)]-$$

$$2\int_{-\infty}^{\infty}\int_{-\infty}^{\infty} x_1 x_2 d[G(x_1) - F(x_1)]d[G(x_2) - F(x_2)].$$

Simplifying the first term we observe that

$$\int_{-\infty}^{\infty}\int_{-\infty}^{\infty} x_1^2 d[G(x_1) - F(x_1)]d[G(x_2) - F(x_2)] =$$

$$\left[\int_{-\infty}^{\infty} x_1^2 d[G(x_1) - F(x_1)]\right]\left[\int_{-\infty}^{\infty} d[G(x_2) - F(x_2)]\right] = 0.$$

Similarly,

$$\int_{-\infty}^{\infty}\int_{-\infty}^{\infty} x_2^2 d[G(x_1) - F(x_1)]d[G(x_2) - F(x_2)] = 0.$$

Therefore,

$$\Delta_2 T(F, G - F) = -2 \int_{-\infty}^{\infty} \int_{-\infty}^{\infty} x_1 x_2 d[G(x_1) - F(x_1)] d[G(x_2) - F(x_2)] =$$

$$-2 \left\{ \int_{-\infty}^{\infty} x_1 d[G(x_1) - F(x_1)] \right\} \left\{ \int_{-\infty}^{\infty} x_2 d[G(x_2) - F(x_2)] \right\} =$$

$$-2 \left[\int_{-\infty}^{\infty} x_1 dG(x_1) - \int_{-\infty}^{\infty} x_1 dF(x_1) \right]^2 = -2(\nu_1' - \mu_1')^2.$$

∎

The Gâteaux differential is not the only approach to defining a derivative type methodology for functionals, and in some sense this differential does not have sufficient properties for a full development of the type of asymptotic theory that we wish to seek in general settings. The original approach to using differentials of this type was developed by von Mises (1947), who used a differential similar to the Gâteaux differential. If we are working in a linear space that is equipped with a norm then one can define what is commonly known as the *Fréchet derivative*. See Dieudonné (1960), Section 2.3 of Fernholz (1983), Fréchet (1925), Nashed (1971), and Chapter 6 of Serfling (1980), for further details on this type of derivative. However, Fernholz (1983) points out that few statistical functions are Fréchet differentiable. The *Hadamard differential*, first used in this context by Reeds (1976), exists under weaker conditions than the Fréchet derivative, but still exhibits the required properties. The Hadamard differential is preferred by Fernholz (1983) as a compromise. For further details on the Hadamard differential see Averbukh and Smolyanov (1968), Fernholz (1983), Keller (1974), and Yamamuro (1974). In our presentation we will continue to use the Gâteaux differential and will limit ourselves to problems where this differential is useful. This generally follows the development of Serfling (1980), without addressing the issues related to the Fréchet derivative. This provides a reasonable and practical overview of this topic without becoming too deep with mathematical technicalities.

9.4 Expansion Theory for Statistical Functionals

The concept that a differential can be developed for a functional motivates a Taylor type approximation for functionals. Recall from Theorem 1.13 that a function f can be approximated at points near $x \in \mathbb{R}$ by $f(x) + \delta f'(x)$ where the error between $f(x + \delta)$ and the approximation converges to zero as $\delta \to 0$. In principle a similar approximation can be obtained for functionals evaluated at distributions near F. That is, we can approximate $T(G)$ with $T(F) + \Delta_1 T(F, G - F)$ where the distribution G is near F. As Serfling (1980) points out, no further theory is required to use this approximation. However, if we wish for this approximation to have similar properties to the Taylor approximation, then we need to consider how the error of the approximation behaves. That is, define $E_1(F, G)$ such that $T(G) = T(F) + \Delta_1 T(F, G - F) + E_1(F, G)$ for all F, G and T where $\Delta_1 T(F, G - F)$ exists. Then we must show that $E_1(F, G) \to 0$ at some rate as $G \to F$.

One case that is of particular interest comes from the fact that the statistical properties of plug-in estimators of T can be studied by replacing G with the empirical distribution function of Definition 3.5. That is, we can consider the expansion

$$T(\hat{F}_n) = T(F) + \Delta_1 T(F, \hat{F}_n - F) + E_1(\hat{F}_n, F). \qquad (9.1)$$

We know from Theorem 3.18 (Glivenko and Cantelli) that $\hat{F}_n \overset{u}{\to} F$ with probability one as $n \to \infty$ since $||\hat{F}_n - F||_\infty \overset{a.c.}{\longrightarrow} 0$ as $n \to \infty$. We would like to use this fact to prove that $E_1(\hat{F}_n, F) \overset{p}{\to} 0$ as $n \to \infty$. We now begin developing a theory that will lead to the asymptotic behavior of this error term. The following result helps to simplify the form of the error term.

Theorem 9.2. *Let $G \in \mathcal{F}$ and let F be a fixed distribution from $\mathcal{F}$. Consider a functional of the form*

$$T(G) = \int_{-\infty}^{\infty} \cdots \int_{-\infty}^{\infty} t(x_1, \ldots, x_r) \prod_{k=1}^{r} d[G(x_k) - F(x_k)].$$

Then, there exists a function $\tilde{t}(x_1, \ldots, x_r | F)$ such that

$$T(G) = \int_{-\infty}^{\infty} \cdots \int_{-\infty}^{\infty} \tilde{t}(x_1, \ldots, x_r | F) \prod_{k=1}^{r} dG(x_k),$$

for all $G \in \mathcal{F}$.

Proof. We prove this result for the special cases when $r = 1$ and $r = 2$. For the general case see Exercise 10. For the case when $r = 1$, we follow the constructive proof of Serfling (1980) and consider the function

$$\tilde{t}(x_1 | F) = t(x_1) - \int_{-\infty}^{\infty} t(x_2) dF(x_2).$$

Using this function, we observe that

$$\int_{-\infty}^{\infty} \tilde{t}(x_1 | F) dG(x_1) = \int_{-\infty}^{\infty} \left[t(x_1) - \int_{-\infty}^{\infty} t(x_2) dF(x_2) \right] dG(x_1) =$$

$$\int_{-\infty}^{\infty} t(x_1) dG(x_1) - \int_{-\infty}^{\infty} \int_{-\infty}^{\infty} t(x_2) dF(x_2) dG(x_1) =$$

$$\int_{-\infty}^{\infty} t(x_1) [dG(x_1) - dF(x_1)],$$

which proves the result when $r = 1$. For $r = 2$ we use the function

$$\tilde{t}_2(x_1, x_2|F) = t(x_1, x_2) - \int_{-\infty}^{\infty} t(t_1, x_2)dF(t_1) - \int_{-\infty}^{\infty} t(x_1, t_2)dF(t_2) +$$
$$\int_{-\infty}^{\infty} \int_{-\infty}^{\infty} t(t_1, t_2)dF(t_1)dF(t_2),$$

so that

$$\int_{-\infty}^{\infty} \int_{-\infty}^{\infty} \tilde{t}_2(x_1, x_2|F)dG(x_1)dG(x_2) = \int_{-\infty}^{\infty} \int_{-\infty}^{\infty} t(x_1, x_2)dG(x_1)dG(x_2) -$$
$$\int_{-\infty}^{\infty} \int_{-\infty}^{\infty} \int_{-\infty}^{\infty} t(t_1, x_2)dF(t_1)dG(x_1)dG(x_2) -$$
$$\int_{-\infty}^{\infty} \int_{-\infty}^{\infty} \int_{-\infty}^{\infty} t(x_1, t_2)dF(t_2)dG(x_1)dG(x_2) +$$
$$\int_{-\infty}^{\infty} \int_{-\infty}^{\infty} \int_{-\infty}^{\infty} \int_{-\infty}^{\infty} t(t_1, t_2)dF(t_1)dF(t_2)dG(x_1)dG(x_2) =$$
$$\int_{-\infty}^{\infty} \int_{-\infty}^{\infty} t(x_1, x_2)dG(x_1)dG(x_2) - \int_{-\infty}^{\infty} \int_{-\infty}^{\infty} t(t_1, x_2)dF(t_1)dG(x_2) -$$
$$\int_{-\infty}^{\infty} \int_{-\infty}^{\infty} t(x_1, t_2)dG(x_1)dF(t_2) + \int_{-\infty}^{\infty} \int_{-\infty}^{\infty} t(t_1, t_2)dF(t_1)dF(t_2) =$$
$$\int_{-\infty}^{\infty} \int_{-\infty}^{\infty} t(x_1, x_2)d[G(x_1) - F(x_1)]d[G(x_2) - F(x_2)],$$

which completes the proof for $k = 2$. $\qquad\qquad\square$

The result of Theorem 9.2 can have a profound impact on the form of a Gâteaux differential. For example, Theorem 9.1 implies that if T is a functional of the form

$$T(F) = \int_{-\infty}^{\infty} \cdots \int_{-\infty}^{\infty} h(x_1, \ldots, x_r) \prod_{i=1}^{r} dF(x_i),$$

then

$$\Delta_k T(F, \hat{F}_n - F) = \int_{-\infty}^{\infty} \cdots \int_{-\infty}^{\infty} t(x_1, \ldots, x_k|F) \prod_{i=1}^{k} d[\hat{F}_n(x_i) - F(x_i)],$$

where

$$t(x_1, \ldots, x_k|F) = \int_{-\infty}^{\infty} \cdots \int_{-\infty}^{\infty} h(x_1, \ldots, x_k, y_1, \ldots, y_{r-k}) \prod_{i=1}^{r-k} dF(y_i).$$

Theorem 9.2 then implies that

$$\Delta_k T(F, \hat{F}_n - F) = \int_{-\infty}^{\infty} \cdots \int_{-\infty}^{\infty} \tilde{t}(x_1, \ldots, x_k|F) \prod_{i=1}^{k} d\hat{F}_n(x_i),$$

for some function $\tilde{t}(x_1, \ldots, x_k | F)$. Applying Definition 2.10 to the integrals with respect to the empirical distribution function then yields

$$\Delta_k T(F, \hat{F}_n - F) = n^{-k} \sum_{i_1=1}^{n} \cdots \sum_{i_k=1}^{n} \tilde{t}(X_{i_1}, \ldots, X_{i_k} | F).$$

This form of the differential can often be used to establish the weak convergence properties of the differential. For example, if $k = 1$, we have that

$$\Delta_1 T(F, \hat{F}_n - F) = n^{-1} \sum_{i=1}^{n} \tilde{t}(X_i | F),$$

which is asymptotically NORMAL by Theorem 4.20 (Lindeberg and Lévy). Hence, if it can also be shown that $E_1(F, \hat{F}_n) \xrightarrow{p} 0$ as $n \to \infty$, the Theorem 4.11 can be used to show that $T(\hat{F}_n)$ is asymptotically NORMAL. To establish this type of property we use the result given below.

Theorem 9.3. *Let $\{X_n\}_{n=1}^{\infty}$ be a sequence of independent and identically distributed random variables from a distribution F and let $t(x_1, \ldots, x_r)$ be a function such that $E[t^2(X_{i_1}, \ldots, X_{i_r})] < \infty$ for all sets of indices $i_1, \ldots, i_r$ where $i_j \in \{1, \ldots, r\}$ for all $j \in \{1, \ldots, r\}$. Then*

$$E\left\{ \left[\int_{-\infty}^{\infty} \cdots \int_{-\infty}^{\infty} t(x_1, \ldots, x_r) \prod_{i=1}^{r} d[\hat{F}_n(x_i) - F(x_i)] \right]^2 \right\} = O(n^{-r}),$$

as $n \to \infty$.

Proof. Theorem 9.2 implies that there exists a function $\tilde{t}(x_1, \ldots, x_r | F)$ such that

$$\int_{-\infty}^{\infty} \cdots \int_{-\infty}^{\infty} t(x_1, \ldots, x_r) \prod_{i=1}^{r} d[\hat{F}_n(x_i) - F(x_i)] =$$

$$\int_{-\infty}^{\infty} \cdots \int_{-\infty}^{\infty} \tilde{t}(x_1, \ldots, x_r | F) d\hat{F}_n(x_1) \cdots d\hat{F}_n(x_r).$$

Repeated application of Definition 2.10 then implies that

$$\int_{-\infty}^{\infty} \cdots \int_{-\infty}^{\infty} \tilde{t}(x_1, \ldots, x_r | F) d\hat{F}_n(x_1) \cdots d\hat{F}_n(x_r) =$$

$$n^{-r} \sum_{i_1=1}^{n} \cdots \sum_{i_r=1}^{n} \tilde{t}(x_{i_1}, \ldots, x_{i_r} | F).$$

Therefore,

$$
E\left\{\left[\int_{-\infty}^{\infty}\cdots\int_{-\infty}^{\infty}t(x_1,\ldots,x_r)\prod_{i=1}^{r}d[\hat{F}_n(x_i)-F(x_i)]\right]^2\right\}=
$$

$$
E\left\{\left[n^{-r}\sum_{i_1=1}^{n}\cdots\sum_{i_r=1}^{n}\tilde{t}(X_{i_1},\ldots,X_{i_r}|F)\right]^2\right\}=
$$

$$
n^{-2r}\sum_{i_1=1}^{n}\cdots\sum_{i_r=1}^{n}\sum_{j_1=1}^{n}\cdots\sum_{j_r=1}^{n}E[\tilde{t}(X_{i_1},\ldots,X_{i_r}|F)\tilde{t}(X_{j_1},\ldots,X_{j_r}|F)].
$$

(9.2)

Now we take advantage of the fact that

$$
\int_{-\infty}^{\infty}\tilde{t}(x_{i_1},\ldots,x_{i_r}|F)dF(x_{i_k})=0,
$$

for all $i_k \in \{1,\ldots,r\}$ and $k \in \{1,\ldots,r\}$. See Exercise 8. This implies that the expectation in Equation (9.2) will be zero unless each index in the function occurs at least twice. Serfling (1980) concludes that the number of non-zero terms is $O(n^r)$, as $n \to \infty$. Assuming that the remaining absolute expectations are bounded by some real value, as indicated by the assumptions, it follows that

$$
\left|E\left\{\left[\int_{-\infty}^{\infty}\cdots\int_{-\infty}^{\infty}t(x_1,\ldots,x_r)\prod_{i=1}^{r}d[\hat{F}_n(x_i)-F(x_i)]\right]^2\right\}\right|\leq
$$

$$
E\left\{\left|\int_{-\infty}^{\infty}\cdots\int_{-\infty}^{\infty}t(x_1,\ldots,x_r)\prod_{i=1}^{r}d[\hat{F}_n(x_i)-F(x_i)]\right|^2\right\}=
$$

$$
n^{-2r}\sum_{i_1=1}^{n}\cdots\sum_{i_r=1}^{n}\sum_{j_1=1}^{n}\cdots\sum_{j_r=1}^{n}E[\tilde{t}(X_{i_1},\ldots,X_{i_r})\tilde{t}(X_{j_1},\ldots,X_{j_r})]=O(n^{-r}),
$$

as $n \to \infty$. $\qquad\square$

We now combine these results in order to prove that the error term in the expansion in Equation (9.1) has the desired asymptotic properties. We begin by noting the Definition 9.2 implies that

$$
\begin{aligned}
T(\hat{F}_n) &= T(F)+\Delta_1 T(F,\hat{F}_n-F)+E_1(F,\hat{F}_n) \\
&= T(F)+\frac{d}{d\delta}T[F+\delta(\hat{F}_n-F)]\Big|_{\delta\downarrow 0}+E_1(F,\hat{F}_n). \quad (9.3)
\end{aligned}
$$

Let us first consider the case when $r=1$. Noting that $\Delta_1 T(F,\hat{F}_n-F)=T(\hat{F}_n)-T(F)$ implies that $T(F)+\Delta_1 T(F,\hat{F}_n-F)=T(F)+T(\hat{F}_n)-T(F)=T(\hat{F}_n)$. Therefore, $E_1(F,\hat{F}_n)$ is identically zero for all $n \in \mathbb{N}$.

For the case when $r = 2$ we define $v(\delta) = T[F + \delta(\hat{F}_n - F)]$ as a function of δ so that $v(0) = T(F)$ and $v(1) = T(\hat{F}_n)$. With this notation, Equation (9.3) can be written as

$$v(1) = v(0) + \frac{d}{d\delta}v(\delta)\Big|_{\delta=0} + E_1(\delta) = v(0) + v'(0) + E_1(\delta), \qquad (9.4)$$

which has the same form of a Taylor expansion for the function v provided by Theorem 1.13, and hence $E_1(\delta) = \frac{1}{2}\delta^2 v''(\xi)$, for some $\xi \in (0, 1)$.

Let δ be an arbitrary member of the unit interval. Following the arguments of the proof of Theorem 9.1, if the functional T has the form

$$T(F) = \int_{-\infty}^{\infty} h(x_1, x_2) dF(x_1) dF(x_2),$$

then

$$v''(\xi) = 2 \int_{-\infty}^{\infty} \int_{-\infty}^{\infty} h(x_1, x_2) d[\hat{F}_n(x_1) - F(x_1)] d[\hat{F}_n(x_1) - F(x_2)],$$

where we note that the derivative does not depend on ξ. Theorem 9.2 implies that there exists a function $\tilde{h}(x_1, x_2 | F)$ such that

$$v''(\xi) = \int_{-\infty}^{\infty} \int_{-\infty}^{\infty} \tilde{h}(x_1, x_2) d\hat{F}_n(x_1) d\hat{F}_n(x_2) = n^{-2} \sum_{i=1}^{n} \sum_{j=1}^{n} \tilde{h}(X_1, X_j | F).$$

Suppose that we can assume that $E[\tilde{h}^2(X_i, X_j)] < \infty$ for all i and j from the index set $\{1, \ldots, n\}$. Then Theorem 9.3 implies that

$$n^{1/2} E\left\{ \left| n^{-2} \sum_{i=1}^{n} \sum_{j=1}^{n} \tilde{h}(X_i, X_j | F) \right|^2 \right\} = O(n^{-3/2}),$$

as $n \to \infty$. Therefore, it follows that

$$\lim_{n \to \infty} n^{1/2} E\left\{ \left| n^{-2} \sum_{i=1}^{n} \sum_{j=1}^{n} \tilde{h}(X_i, X_j | F) \right|^2 \right\} = 0.$$

Definition 5.1 implies that $n^{1/2}|v''(\xi)| \xrightarrow{qm} 0$ as $n \to \infty$ and hence Theorem 5.2 implies that $n^{1/2}|v''(\xi)| \xrightarrow{p} 0$ as $n \to \infty$. This in turn implies that $n^{1/2}E_1(\delta) \xrightarrow{p} 0$ as $n \to \infty$ for all $\delta \in (0, 1)$. This type of argument can be generalized to the following result.

Theorem 9.4. *Let $X_1, \ldots, X_n$ be a set of independent and identically distributed random variables from a distribution F and let θ be a functional parameter of the form*

$$\theta = T(F) = \int_{-\infty}^{\infty} \cdots \int_{-\infty}^{\infty} h(x_1, \ldots, x_r) \prod_{i=1}^{r} dF(x_i).$$

Then,

$$n^{1/2} \sup_{\lambda \in [0,1]} \left| \frac{d^2}{d\lambda^2} T[F + \lambda(\hat{F}_n - F)] \right| \xrightarrow{p} 0,$$

as $n \to \infty$, and therefore $n^{1/2} E_1(F, \hat{F}_n) \xrightarrow{p} 0$, as $n \to \infty$.

Discussion about some general conditions under which the result of Theorem 9.4 holds for other types of functional parameters and differentials can be found in Chapter 6 of Serfling (1980).

Example 9.9. Consider the variance functional of Example 9.8 which has the form

$$T(F) = \int_{-\infty}^{\infty} \int_{-\infty}^{\infty} h(x_1, x_2) dF(x_1) dF(x_2),$$

where $h(x_1, x_2) = \frac{1}{2}(x_1^2 + x_2^2 - 2x_1 x_2)$. In Example 9.8 is was shown that $\Delta_1 T(F, \hat{F}_n - F) = \hat{\mu}_2' - \mu_2' - 2\hat{\mu}_1' \mu_1' + 2(\mu_1')^2$. Therefore, it follows that $T(F) + \Delta_1 T(F, \hat{F}_n - F) = \hat{\mu}_2' + (\mu_1')^2 - 2\hat{\mu}_1' \mu_1'$, and hence in this case we have that

$$E_1(F, \hat{F}_n) = \hat{\mu}_2' - (\hat{\mu}_1')^2 - \mu_2' - (\mu_1')^2 + 2\hat{\mu}_1' \mu_1' = -(\hat{\mu}_1' - \mu_1')^2.$$

Theorems 8.1 and 8.5 then imply that $E_1(F, \hat{F}_n) = o_p(n^{-1/2})$ as $n \to \infty$, which verifies the arguments given earlier for this example. ∎

9.5 Asymptotic Distribution

In this section we will use the results of the previous section, along with the expansion given in Equation (9.1), to find the asymptotic distribution of the sample functional $T(\hat{F}_n)$. From an initial view, the use of the expansion given in Equation (9.1) may not seem as if it would be necessarily useful as it is not readily apparent that the asymptotic properties of $\Delta_1 T(F, \hat{F}_n - F)$ would be easier to establish than that of $T(\hat{F}_n)$ directly. Indeed, it is the case is some problems that either approach may have the same level of difficulty. However, we have established that in some cases $\Delta_1 T(F, \hat{F}_n - F)$ can be written as a sum of random variables whose asymptotic properties follow from established results such as Theorem 4.20 (Lindeberge-Lévy). Another important key ingredient in establishing these results is the asymptotic behavior of the error for the expansion given in Equation (9.1). In the previous section we observed that in certain cases this error term can be guaranteed to converge to zero at a certain rate. For the development of the result below, the error term will need to converge in probability to zero at least as fast as $n^{-1/2}$ as $n \to \infty$.

Theorem 9.5. *Let $\{X_n\}_{n=1}^{\infty}$ be a sequence of independent and identically distributed random variables from a distribution F. Let $\theta = T(F)$ be a functional with estimator $\hat{\theta}_n = T(\hat{F}_n)$, where $\hat{F}_n$ is the empirical distribution function of $X_1, \ldots, X_n$. Suppose that*

$$\Delta_1 T(F, \hat{F}_n - F) = n^{-1} \sum_{i=1}^{n} \tilde{t}(X_i | F),$$

for some function $\tilde{t}(X_i|F)$. *Let* $\tilde{\mu} = E[\tilde{t}(X_i|F)]$ *and* $\tilde{\sigma}^2 = V[\tilde{t}(X_i|F)]$ *and assume that* $0 < \tilde{\sigma}^2 < \infty$. *If*

$$n^{1/2}E_1(F, \hat{F}_n) = n^{1/2}[T(\hat{F}_n) - T(F) - \Delta_1(F, \hat{F}_n - F)] \xrightarrow{p} 0,$$

as $n \to \infty$, *then* $n^{1/2}(\hat{\theta}_n - \theta - \tilde{\mu}) \xrightarrow{d} Z$ *as* $n \to \infty$ *where* Z *has a* $N(0, \tilde{\sigma}^2)$ *distribution.*

Proof. We begin proving this result by first finding the asymptotic distribution of $\Delta_1 T(F, \hat{F}_n - F)$. First, note that $\{\tilde{t}(X_n|F)\}_{n=1}^{\infty}$ is a sequence of independent and identically distributed random variables with mean $\tilde{\mu}$ and variance $0 < \tilde{\sigma}^2 < \infty$. Therefore, Theorem 4.20 (Lindeberge-Lévy) implies that

$$n^{1/2}\left[n^{-1}\sum_{i=1}^{n}\tilde{t}(X_i|F) - \tilde{\mu}\right] \xrightarrow{d} Z, \tag{9.5}$$

as $n \to \infty$ where Z is a $N(0, \tilde{\sigma}^2)$ random variable. Now we use the fact that we have defined $E_1(F, \hat{F}_n)$ so that $T(\hat{F}_n) = T(F) + \Delta_1(F, \hat{F}_n - F) + E_1(F, \hat{F}_n)$ so that it follows that

$$n^{1/2}(\hat{\theta}_n - \theta - \tilde{\mu}) = n^{1/2}[\Delta_1 T(F, \hat{F}_n - F) - \tilde{\mu}] + n^{1/2}E_1(F, \hat{F}_n). \tag{9.6}$$

It follows that the first term on the right hand side of Equation (9.6) converges in distribution to Z, while the second term converges in probability to zero as $n \to \infty$. Therefore, Theorem 4.11 (Slutsky) implies that the sum of these two terms converges to Z, and the result is proven. □

When the functional parameter has the form studied in the previous section, the conditions under which we obtain the result given in Theorem 9.5 greatly simplify.

Corollary 9.1. *Let* $\{X_n\}_{n=1}^{\infty}$ *be a sequence of independent and identically distributed random variables from a distribution* F. *Let* $\theta = T(F)$ *be a functional parameter of the form*

$$T(F) = \int_{-\infty}^{\infty} \cdots \int_{-\infty}^{\infty} h(x_1, \ldots, x_r) \prod_{i=1}^{r} dF(x_i), \tag{9.7}$$

with estimator $\hat{\theta}_n = T(\hat{F}_n)$, *where* $\hat{F}_n$ *is the empirical distribution function of* $X_1, \ldots, X_n$. *Define* $\tilde{\mu} = E[\tilde{t}(X_i|F)]$, $\tilde{\sigma}^2 = V[\tilde{t}(X_i|F)]$, *where*

$$\Delta_1 T(F, \hat{F}_n - F) = n^{-1}\sum_{i=1}^{n}\tilde{t}(X_i|F),$$

for some function $\tilde{t}(X_i|F)$, *assume that* $0 < \tilde{\sigma}^2 < \infty$, *and that* $\Delta_1 T(F, \hat{F}_n - F)$ *is not functionally equal to zero, then* $n^{1/2}(\hat{\theta}_n - \theta - \tilde{\mu}) \xrightarrow{d} Z$ *as* $n \to \infty$ *where* Z *has a* $N(0, \tilde{\sigma}^2)$ *distribution.*

Proof. We begin by noting that if the functional T has the form indicated in

Equation (9.7) then Theorem 9.1 implies that

$$\Delta_1 T(F; \hat{F}_n - F) = T(\hat{F}_n) - T(F) =$$

$$r \int_{-\infty}^{\infty} \cdots \int_{-\infty}^{\infty} h(x_1, y_1, \ldots, y_{r-1}) \left(\prod_{i=1}^{r-1} dF(y_i) \right) d[\hat{F}_n(x_1) - F(x_1)] =$$

$$\int_{-\infty}^{\infty} \tilde{h}(x_1|F) d[\hat{F}_n(x_1) - F(x_1)].$$

Now, Theorem 9.2 implies that there exists a function $\tilde{t}(x_1|F)$ such that

$$\Delta_1 T(F; \hat{F}_n - F) = \int_{-\infty}^{\infty} \tilde{t}(x_1|F) d\hat{F}_n(x_1) = n^{-1} \sum_{i=1}^{n} \tilde{t}(X_i|F),$$

which yields the form required by Theorem 9.5. Theorem 9.4 then provides the required behavior of the error term, which then proves the result. □

Hence, the form of the differential is key in establishing asymptotic NORMAL-ITY in this case, in that $\Delta_1 T(F; \hat{F}_n - F)$ can be written as a sum of independent and identically distributed random variables. If $\Delta_1 T(F; \hat{F}_n - F) = 0$, then the first term of the approximation is zero and the asymptotic behavior changes. In particular, Theorem 6.4.B of Serfling (1980) demonstrates conditions under which $\Delta_1 T(F; \hat{F}_n - F) = 0$ and $\Delta_2 T(F; \hat{F}_n - F) \neq 0$, and the resulting asymptotic distribution is a weighted sum of independent CHISQUARED random variables.

Example 9.10. Let $\{X_n\}_{n=1}^{\infty}$ be a sequence of independent and identically distributed random variables from a distribution F and consider once again the variance functional $\theta = T(F)$ from Example 9.8. In Example 9.9 it was shown that $T(\hat{F}_n) = T(F) + \Delta_1 T(F, \hat{F}_n - F) + E_1(F, \hat{F}_n - F)$ where

$$\begin{aligned}
\Delta_1 T(F, \hat{F}_n - F) &= n^{-1} \sum_{i=1}^{n} X_i^2 - \mu_2' - 2\bar{X}_n \mu_1' + 2(\mu_1')^2 \\
&= n^{-1} \sum_{i=1}^{n} [X_i^2 - 2X_i \mu_1' + (\mu_1')^2 + (\mu_1')^2 - \mu_2'] \\
&= n^{-1} \sum_{i=1}^{n} [(X_i - \mu_1')^2 - \theta] \\
&= n^{-1} \sum_{i=1}^{n} \tilde{t}(X_i|F),
\end{aligned}$$

where $\tilde{t}(X_i|F) = (X_i - \mu_1')^2 - \theta$. Example 9.9 shows that $E_1(F, \hat{F}_n) = o_p(n^{-1/2})$ as $n \to \infty$ so that Theorem 9.5 implies that if $\tilde{\sigma}$ is finite, then $n^{1/2}(\hat{\theta}_n - \theta - \tilde{\mu}) \xrightarrow{d} Z$ as $n \to \infty$ where Z is a $N(0, \tilde{\sigma}^2)$ random variable. In this case $\hat{\theta}_n$ is the sample variance given by

$$\hat{\theta}_n = n^{-1} \sum_{i=1}^{n} (X_i - \bar{X}_n)^2.$$

Direct calculations can be used to show that $\tilde{\mu} = E[\tilde{t}(X_i|F)] = 0$ and $\tilde{\sigma}^2 = V[\tilde{t}(X_i|F)] = V[(X_i - \mu)^2 - \theta] = \mu_4 - \theta^2$. Therefore we have shown that $n^{1/2}(\hat{\theta}_n - \theta) \xrightarrow{d} Z$ as $n \to \infty$ where Z is a $N(0, \mu_4 - \theta^2)$ random variable. ∎

9.6 Exercises and Experiments

9.6.1 Exercises

1. Let $\{X_n\}_{n=1}^{\infty}$ be a sequence of independent and identically distributed random variables from a distribution F. Let θ be defined as the pth quantile of F.

   a. Write θ as a functional parameter $T(F)$ of F.

   b. Develop a plug-in estimator for θ based on using the empirical distribution function to estimate F.

   c. Consider estimating F with a $N(\bar{X}_n, S)$ distribution. Write this estimator in terms of z_p, the pth quantile of a $N(0, 1)$ distribution.

2. Consider a functional of the form

$$T(F) = \int_{-\infty}^{\infty} \int_{-\infty}^{\infty} h(x_1, x_2) dF(x_1) dF(X_2),$$

   where $F \in \mathcal{F}$, a collection of distribution functions. Find an expression for the kth Gâteaux differential of $T(F)$ without using the assumption that h is a symmetric function of its arguments. Compare this result to that of Theorem 9.1.

3. Consider the skewness functional given by

$$T(F) = \int_{-\infty}^{\infty} \left[t - \int_{-\infty}^{\infty} t dF(t) \right]^3 dF(t),$$

   where $F \in \mathcal{F}$, a collection of distribution functions with mean μ and variance σ^2 and third central moment γ.

   a. Using the method demonstrated in Example 9.8, write this functional in a form suitable for the application of Theorem 9.1.

   b. Using Theorem 9.1, find the first two Gâteaux differentials of $T(F)$.

   c. Determine whether Corollary 9.1 applies to this functional and derive the asymptotic normality of $T(\hat{F}_n)$ if it does.

4. Consider the kth moment functional given by

$$T(F) = \int_{-\infty}^{\infty} t^k dF(t),$$

where $F \in \mathcal{F}$, a collection of distribution functions where $\mu'_k < \infty$.

a. Using the method demonstrated in Example 9.8, write this functional in a form suitable for the application of Theorem 9.1.

b. Using Theorem 9.1, find the first two Gâteaux differentials of $T(F)$.

c. Determine whether Corollary 9.1 applies to this functional and derive the asymptotic normality of $T(\hat{F}_n)$ if it does.

5. Consider the kth central moment functional given by

$$T(F) = \int_{-\infty}^{\infty} \left[t - \int_{-\infty}^{\infty} t\, dF(t) \right]^k dF(t),$$

where $F \in \mathcal{F}$, a collection of distribution functions where $\mu_k < \infty$.

a. Using the method demonstrated in Example 9.8, write this functional in a form suitable for the application of Theorem 9.1.

b. Using Theorem 9.1, find the first two Gâteaux differentials of $T(F)$.

c. Determine whether Corollary 9.1 applies to this functional and derive the asymptotic normality of $T(\hat{F}_n)$ if it does.

6. Let F be a symmetric and continuous distribution function and consider the functional parameter θ that corresponds to the expectation

$$(1 - 2\alpha)^{-1} E(X\delta\{X; [\xi_\alpha, \xi_{1-\alpha}]\}),$$

where $\alpha \in [0, \frac{1}{2})$ is a fixed constant, $\xi_\alpha = F^{-1}(\alpha)$, and $\xi_{1-\alpha} = F^{-1}(1 - \alpha)$.

a. Prove that this functional parameter can be written as

$$T(F) = (1 - 2\alpha)^{-1} \int_{\alpha}^{1-\alpha} \xi_u\, du.$$

b. Let G denote a degenerate distribution at a real constant δ. Prove that

$$\Delta_1 T(F; G - F) = (1 - 2\alpha)^{-1} \int_{\alpha}^{1-\alpha} \frac{u - \delta\{F(\delta); (-\infty, u]\}}{f(\xi_u)} du.$$

c. Prove that the differential given above can be written as

$$\Delta_1 T(F; G - F) = \begin{cases} (1 - 2\alpha)^{-1}[F^{-1}(\alpha) - \xi_{1/2}] & \delta \in (-\infty, \xi_\alpha) \\ (1 - 2\alpha)^{-1}(\delta - \xi_{1/2}) & \delta \in [\xi_\alpha, \xi_{1-\alpha}] \\ (1 - 2\alpha)^{-1}[F^{-1}(\alpha) - \xi_{1/2}] & \delta \in (\xi_{1-\alpha}, \infty). \end{cases}$$

7. Let X be a random variable following a distribution $F(x|\theta)$ where $\theta \in \Omega$. Assume that $F(x|\theta)$ has continuous density $f(x|\theta)$. The *maximum likelihood estimator* of θ is given by the value of θ that maximizes $F(x|\theta)$ with respect to θ. Assuming that $F(x|\theta)$ is maximized at a unique interior point of Ω, argue that the functional corresponding to the maximum likelihood estimator can be written implicitly as

$$\int_{-\infty}^{\infty} \frac{f'(x|\theta)}{f(x|\theta)} dF(x|\theta) = 0,$$

where the derivative in the integral is taken with respect to θ. Discuss any additional assumptions that you need to make.

8. Let $G \in \mathcal{F}$ and let F be a fixed distribution from $\mathcal{F}$. Consider a functional of the form

$$I(G) = \int_{-\infty}^{\infty} \cdots \int_{-\infty}^{\infty} t(x_1, \ldots, x_r) \prod_{k=1}^{r} d[G(x_k) - F(x_k)],$$

where there exists a function $\tilde{t}(x_1, \ldots, x_r | F)$ such that

$$I(G) = \int_{-\infty}^{\infty} \cdots \int_{-\infty}^{\infty} \tilde{t}(x_1, \ldots, x_r | F) \prod_{k=1}^{r} dG(x_k),$$

for all $G \in \mathcal{F}$ as given by Theorem 9.2. Prove that for the functions $\tilde{t}(x_1, \ldots, x_r | F)$ specified in the proof of Theorem 9.2 that

$$\int_{-\infty}^{\infty} \tilde{t}(x_{i_1}, \ldots, x_{i_r} | F) dF(x_{i_k}) = 0.$$

9. Consider a functional of the form

$$T(F) = \int_{-\infty}^{\infty} \cdots \int_{-\infty}^{\infty} h(x_1, \ldots, x_r) \prod_{i=1}^{r} dF(x_i),$$

where $F \in \mathcal{F}$, a collection of distribution functions. Prove that if $k \leq r$ then

$$\Delta_k T(F; G - F) = \left[\prod_{i=1}^{k}(r - i + 1)\right] \times$$

$$\int_{-\infty}^{\infty} \cdots \int_{-\infty}^{\infty} h(x_1, \ldots, x_k, y_1, \ldots, y_{r-k}) \left[\prod_{i=1}^{r-k} dF(y_i)\right] \left[\prod_{i=1}^{k} d[G(x_i) - F(x_i)]\right],$$

and if $k > r$ then $\Delta_k T(F; G - F) = 0$. This result has been proven for $r = 1$ and $r = 2$ in Section 9.3. Establish this result for a general value of r.

10. Let $t(x_1, \ldots, x_r)$ be a real valued function. Prove that

$$I(G) = \int_{-\infty}^{\infty} \cdots \int_{-\infty}^{\infty} t(x_1, \ldots, x_r) \prod_{i=1}^{r} d[G(x_i) - F(x_i)],$$

where F and G are members of a collection of distribution functions given by $\mathcal{F}$ and F is fixed, then there exists a function $\tilde{t}(x_1,\ldots,x_r|F)$ such that

$$I(G) = \int_{-\infty}^{\infty} \cdots \int_{-\infty}^{\infty} \tilde{t}(x_1,\ldots,x_r|F)dG(x_1)\cdots dG(x_r).$$

Use a constructive proof using the function

$$\tilde{t}(x_1,\ldots,x_r|F) = t(x_1,\ldots,x_r) - \sum_{i=1}^{r} \int_{-\infty}^{\infty} t(x_1,\ldots,x_r)dF(x_i)+$$

$$\sum_{i=1}^{r}\sum_{\substack{j=1\\j>i}}^{r} \int_{-\infty}^{\infty}\int_{-\infty}^{\infty} t(x_1,\ldots,x_r)dF(x_i)dF(x_j) - \cdots$$

$$+ (-1)^r \int_{-\infty}^{\infty} \cdots \int_{-\infty}^{\infty} t(x_1,\ldots,x_r)dF(x_1)\cdots dF(x_r).$$

9.6.2 Experiments

1. Write a program in R that simulates 100 samples of size n from distributions that are specified below. Let $T(F)$ be the variance functional described in Example 9.8. For each sample compute $T(\hat{F}_n) - T(F)$ and $\Delta_1 T(F, \hat{F}_n - F)$ where the differential was found in Example 9.8. For each sample size and distribution, construct a scatterplot the 100 values $T(\hat{F}_n) - T(F)$ and $\Delta_1 T(F, \hat{F}_n - F)$. Describe the behavior observed in the plots with relation to the expansion given in Equation (9.1). Use $n = 5, 10, 25, 50$, and 100.

   a. N$(0,1)$
   b. EXPONENTIAL(1)
   c. UNIFORM$(0,1)$
   d. CAUCHY$(0,1)$

2. Write a program in R that simulates 100 samples of size n from distributions that are specified below. Let $T(F)$ correspond to the quantile functional so that $T(F) = F^{-1}(\alpha)$ where $\alpha \in (0,1)$ is a specified constant. For each sample compute $\hat{\theta}_n = T(\hat{F})$ where $\hat{F}$ corresponds to the empirical distribution function, and again where $\hat{F}$ corresponds to a NORMAL distribution with mean equal to the sample mean, and variance equal to the sample variance. Construct a scatterplot of the pairs of estimated quantiles and overlay a line on the plot that corresponds to the true value of the population quantile in each case. Repeat these calculations for $\alpha = 0.05, 0.10, 0.25, 0.50, 0.75, 0.90$, and 0.95, and for $n = 10, 25, 50$ and 100. Describe the behavior found in each of these plots and discuss how the assumption of NORMALITY affects the performance of the estimator based on the NORMAL distribution.

a. $N(0,1)$
b. EXPONENTIAL(1)
c. UNIFORM$(0,1)$
d. CAUCHY$(0,1)$

CHAPTER 10

Parametric Inference

But sitting in front of him and taken by surprise by his dismissal, K. would be able easily to infer everything he wanted from the lawyer's face and behaviour, even if he could not be induced to say very much.

The Trial by Franz Kafka

10.1 Introduction

Classical statistical inference is usually concerned with estimation and hypothesis testing for a unknown parameter θ within a parametric framework. This means that we will assume that the unknown population follows a known parametric family and that only the parameter is unknown. Furthermore, the parameter space is assumed to have a finite dimension. In this chapter we will be interested in developing asymptotic methods for studying how these methods perform. For point estimation, asymptotic methods will be developed to obtain approximate expressions for the bias and variance of estimators that are functions of the sample mean. We will also be interested in establishing if asymptotically unbiased estimators have a variance that approaches optimality as the sample size increases to infinity. The optimality properties of maximum likelihood estimators will also be studied. In the case of confidence intervals we will consider how the confidence coefficient behaves as $n \to \infty$. In statistical hypothesis testing we will use asymptotic comparisons to compare the power functions of tests. We will also consider the asymptotic properties of observed confidence levels, a method for solving multiple testing problems. We will conclude the chapter with a look at conditions under which Bayes estimators are asymptotically optimal within the frequentist framework.

10.2 Point Estimation

Let $X_1, \ldots, X_n$ be a set of real valued independent and identically distributed random variables from a distribution $F(x|\theta)$ where θ is an unknown parameter with parameter space Ω. In point estimation we are concerned with estimating θ based on $X_1, \ldots, X_n$. That is, we would like to find a plausible value for θ based on an observed sample from $F(x|\theta)$.

Definition 10.1. *Any function T that maps a sample $X_1, \ldots, X_n$ to a parameter space Ω is a point estimator of θ.*

We will usually denote an estimator of a parameter θ as $\hat{\theta}_n = T(X_1, \ldots, X_n)$, or simply as $\hat{\theta}_n$. Note that there is nothing special about a point estimator, it is simply a function of the observed data that does not depend on θ, or any other unknown quantities. The search for a *good* point estimator requires us to define the types of properties that a good estimator should have. Usually we would like our estimator to be close to θ in some respect. Let ρ be a metric on Ω. Then we can measure the distance between $\hat{\theta}_n$ and θ as $\rho(\hat{\theta}_n, \theta)$. But $\rho(\hat{\theta}_n, \theta)$ is a random variable and hence we need some way of summarizing the behavior of $\rho(\hat{\theta}_n, \theta)$. This is usually accomplished by taking the expectation of $\rho(\hat{\theta}_n, \theta)$. In decision theory the distance $\rho(\hat{\theta}_n, \theta)$ is usually called the *loss* associated with estimating θ with $\hat{\theta}_n$ and the function ρ is called the *loss function*. The expected loss, given by $R(\hat{\theta}, \theta) = E[\rho(\hat{\theta}_n, \theta)]$ is called the *risk* associated with estimating θ with the estimator $\hat{\theta}_n$.

A common loss function that is often used in practice is $\rho(\hat{\theta}, \theta) = (\hat{\theta}_n - \theta)^2$, which is called *squared error loss*. The associated risk, given by $\text{MSE}(\hat{\theta}_n) = E[(\hat{\theta}_n - \theta)^2]$ is called the *mean squared error*, which measures the expected square distance between $\hat{\theta}_n$ and θ. It can be shown that $\text{MSE}(\hat{\theta}_n)$ can be decomposed into two parts given by $\text{MSE}(\hat{\theta}_n) = \text{Bias}^2(\hat{\theta}_n) + V(\hat{\theta}_n)$, where $\text{Bias}(\hat{\theta}_n) = E(\hat{\theta}_n) - \theta$ is called the *bias* of the estimator $\hat{\theta}_n$. See Exercise 1. The bias of an estimator $\hat{\theta}_n$ measures the expected systematic departure of $\hat{\theta}_n$ from θ. A special case occurs when the bias always equals zero.

Definition 10.2. *An estimator $\hat{\theta}$ of θ is unbiased if $\text{Bias}(\hat{\theta}_n) = 0$ for all $\theta \in \Omega$.*

If an estimator is unbiased then the mean squared error and the variance of the estimator coincide. In this case the variance of the estimator can be used alone as a measure of the quality of the estimator. Usually the standard deviation of the estimator, called the *standard error* of $\hat{\theta}_n$ is often reported as a measure of the quality of the estimator, since it is in the same units as the parameter space, whereas the variance is in square units.

An important special case in estimation theory is the case of estimating the mean of a population that has a finite variance. Let $\{X_n\}_{n=1}^{\infty}$ be a sequence of independent and identically distributed random variables from a distribution F with mean θ and finite variance σ^2. It is well known that if we estimate θ with the sample mean $\bar{X}_n$ then $\text{Bias}(\bar{X}_n) = 0$ and $V(\bar{X}_n) = n^{-1}\sigma^2$ for all $\theta \in \Omega$. Suppose now that we are interested in estimating $g(\theta)$ for some real function g. An obvious estimator of $g(\theta)$ is $g(\bar{X}_n)$. If g is a linear function of the form $g(x) = a+bx$ then the bias and variance of $g(\bar{X}_n)$ can be found by directly using the properties of $\bar{X}_n$. That is $E[g(\bar{X}_n)] = a + bE(\bar{X}_n) = a + b\theta = g(\theta)$ so that $g(\bar{X}_n)$ is an unbiased estimator of $g(\theta)$. The variance of $g(\bar{X}_n)$ can be found to be $V[g(\bar{X}_n)] = b^2 V(\bar{X}_n) = n^{-1}b^2\sigma^2$.

If g is not a linear function then the bias and variance of $g(\bar{X}_n)$ cannot be found using only the properties of $\bar{X}_n$. In this case more information about the population F must be known.

Example 10.1. Let $\{X_n\}_{n=1}^{\infty}$ be a sequence of independent and identically distributed random variables from a distribution F with mean θ and finite variance σ^2. Suppose now that we are interested in estimating $g(\theta) = \theta^2$ using the estimator $g(\bar{X}_n) = \bar{X}_n^2$. We first find the bias for this estimator. Taking the expectation we have that

$$
\begin{aligned}
E(\bar{X}_n^2) &= E\left(n^{-2}\sum_{i=1}^{n}\sum_{j=1}^{n}X_iX_j\right) \\
&= n^{-2}\sum_{i=1}^{n}E(X_i^2) + n^{-2}\sum_{i=1}^{n}\sum_{\substack{j=1 \\ j\neq i}}^{n}E(X_i)E(X_j) \\
&= n^{-1}\theta^2 + n^{-1}\sigma^2 + n^{-1}(n-1)\theta^2 \\
&= \theta^2 + n^{-1}\sigma^2.
\end{aligned}
$$

Hence the bias is given by $\text{Bias}[g(\bar{X}_n)] = -n^{-1}\sigma^2 = O(n^{-1})$, as $n \to \infty$. To find the variance we have that $E[(\bar{X}_n^2)^2] = E(\bar{X}_n^4)$. Therefore we must find an expression for $E(\bar{X}_n^4)$. Direct calculations yield

$$
E(\bar{X}_n^4) = E\left(n^{-4}\sum_{i=1}^{n}\sum_{j=1}^{n}\sum_{k=1}^{n}\sum_{l=1}^{n}X_iX_jX_kX_l\right) =
$$

$$
n^{-4}\sum_{i=1}^{n}E(X_i^4) + 4n^{-4}\sum_{i=1}^{n}\sum_{\substack{j=1 \\ j\neq i}}^{n}E(X_i^3)E(X_j) + 3n^{-4}\sum_{i=1}^{n}\sum_{\substack{j=1 \\ j\neq i}}^{n}E(X_i^2)E(X_j^2)+
$$

$$
6n^{-4}\sum_{i=1}^{n}\sum_{\substack{j=1 \\ j\neq i}}^{n}\sum_{\substack{k=1 \\ k\neq j \\ k\neq i}}^{n}E(X_i^2)E(X_j)X(X_k)+
$$

$$
n^{-4}\sum_{i=1}^{n}\sum_{\substack{j=1 \\ j\neq i}}^{n}\sum_{\substack{k=1 \\ k\neq j \\ k\neq i}}^{n}\sum_{\substack{l=1 \\ l\neq k \\ l\neq j \\ l\neq i}}^{n}E(X_i)E(X_j)E(X_k)E(X_l) =
$$

$$
n^{-3}\mu_4' + 4n^{-3}(n-1)\theta\mu_3' + 3n^{-3}(n-1)(\mu_2')^2+
$$
$$
6n^{-3}(n-1)(n-2)\theta^2\mu_2' + n^{-3}(n-1)(n-2)(n-3)\theta^4.
$$

A great deal of algebraic perseverance could be used on these expressions to obtain an exact, but quite complicated, expression for the variance. To simplify matters, we will only keep terms that are larger than $O(n^{-2})$ as $n \to \infty$ for this analysis. Therefore,

$$E(\bar{X}_n^4) = 6n^{-3}(n-1)(n-2)\theta^2(\theta^2+\sigma^2)+$$
$$n^{-3}(n-1)(n-2)(n-3)\theta^4 + O(n^{-2}), \quad (10.1)$$

as $n \to \infty$. However, this expression does not eliminate all of the terms that are $O(n^{-2})$ or smaller as $n \to \infty$. Expanding the first term on the right hand side of Equation (10.1) yields

$$6n^{-3}(n-1)(n-2)\theta^2(\theta^2+\sigma^2) = 6n^{-3}(n^2-3n+2)\theta^2(\theta^2+\sigma^2) =$$
$$6n^{-1}\theta^2(\theta^2+\sigma^2) + O(n^{-2}),$$

as $n \to \infty$. Similarly, expanding the second term on the right hand side of Equation (10.1) yields

$$n^{-3}(n-1)(n-2)(n-3)\theta^4 = n^{-3}(n^3-6n^2+11n-6)\theta^4 = \theta^4 - 6n^{-1}\theta^4 + O(n^{-2}),$$

as $n \to \infty$. Therefore, it follows that

$$E[\bar{X}_n^4] = 6n^{-1}\theta^2\sigma^2 + \theta^4 + O(n^{-2}),$$

as $n \to \infty$. To complete finding the variance we note that

$$E^2(\bar{X}_n^2) = (\theta^2 + n^{-1}\sigma^2)^2 = \theta^4 + 2\theta^2 n^{-1}\sigma^2 + O(n^{-2}),$$

as $n \to \infty$. This yields a variance of the form $V(\bar{X}_n^2) = 4n^{-1}\theta^2\sigma^2 + O(n^{-2})$, as $n \to \infty$. ∎

Example 10.1 highlights the increased complexity one finds when working with estimating a non-linear function of θ. The fact that we are able to compute the bias and variance in the closed forms indicated are a result of the function being a sum of powers of $\bar{X}_n$. If we alternatively considered functions such as $\sin(\bar{X}_n)$ or $\exp(-\bar{X}_n^2)$ such a direct approach would no longer be possible. However, an approximate approach can be developed for certain functions by approximating the function of interest with a linear expression obtained using a Taylor expansion from Theorem 1.13.

Example 10.2. Let $\{X_n\}_{n=1}^{\infty}$ be a sequence of independent and identically distributed random variables from a distribution F with mean θ and finite variance σ^2. Suppose now that we are interested in estimating $g(\theta) = \exp(-\theta^2)$ using the estimator $g(\bar{X}_n) = \exp(-\bar{X}_n^2)$. Without further specific information of the form of F, simple closed form expressions for the mean and variance of $\exp(\bar{X}_n^2)$ are difficult to obtain. However, note that Theorem 1.13 implies that we can find a Taylor expansion for the exponential function at $-\bar{X}_n^2$ around the point $-\theta^2$. That is,

$$\exp(-\bar{X}_n^2) = \exp(-\theta^2) - 2(\bar{X}_n - \theta)\theta \exp(-\theta^2) + E_2(\bar{X}_n, \theta),$$

where $E_2(\bar{X}_n, \theta) = (\bar{X}_n - \theta)^2(2\xi^2 - 1)\exp(-\xi^2)$, where ξ is a random variable that is always between $\bar{X}_n$ and θ with probability one. Taking expectations, we find that

$$\begin{aligned} E[\exp(-\bar{X}_n^2)] &= \exp(-\theta^2) - 2\theta E(\bar{X}_n - \theta) \exp(-\theta^2) + E[E_2(\bar{X}_n, \theta)] \\ &= \exp(-\theta^2) + E[(\bar{X}_n - \theta)^2 (2\xi^2 - 1) \exp(-\xi^2)], \end{aligned}$$

since $E(\bar{X}_n - \theta) = 0$. Therefore, the bias of $\exp(-\bar{X}_n^2)$ as an estimator of $\exp(-\theta^2)$ is given by $E[(\bar{X}_n - \theta)^2 (2\xi^2 - 1) \exp(-\xi^2)]$. The expectation of the error can be troublesome to compute due to the random variable ξ unless we happen to note in this case that the function $(2\xi^2 - 1) \exp(-\xi^2)$ is a bounded function so that there exists a finite real value m such that $|(2\xi^2 - 1) \exp(-\xi^2)| < m$ for all $\xi \in \mathbb{R}$. Therefore, it follows that

$$\begin{aligned} |E[(\bar{X}_n - \theta)^2 (2\xi^2 - 1) \exp(-\xi^2)]| &\leq \\ E[|(\bar{X}_n - \theta)^2 (2\xi^2 - 1) \exp(-\xi^2)|] &\leq mE[(\bar{X}_n - \theta)^2] = n^{-1} m \sigma^2, \end{aligned}$$

and hence Definition 1.7 implies that $E[E_2(\bar{X}_n, \theta)] = O(n^{-1})$, as $n \to \infty$. Therefore, we have proven that $\mathrm{Bias}[\exp(-\bar{X}_n)^2] = O(n^{-1})$, as $n \to \infty$. To find the approximate variance we note $[\exp(-\bar{X}_n^2)]^2 = \exp(-2\bar{X}_n^2)$, and use a Taylor expansion for this function to find

$$\exp(-2\bar{X}_n^2) = \exp(-2\theta^2) - 4\theta(\bar{X}_n - \theta) \exp(-2\theta^2) + \tilde{E}_2(\bar{X}_n, \theta), \quad (10.2)$$

where $\tilde{E}_2(\bar{X}_n, \theta) = 2(\bar{X}_n - \theta)^2 (4\xi^2 - 1) \exp(-2\xi^2)$ for some random variable ξ that is between θ and $\bar{X}_n$ with probability one. Taking the expectation of both sides of Equation (10.2), we note that the second term is zero since $E(\bar{X}_n - \theta) = 0$, and therefore we need only find the rate of convergence for the error term. Using the same reasoning as above, we note that the function $(4\xi^2 - 1) \exp(-2\xi^2)$ is bounded for all $\xi \in \mathbb{R}$ and therefore it follows that

$$\left| E[2(\bar{X}_n - \theta)^2 (4\xi^2 - 1) \exp(-2\xi^2)] \right| \leq 2mE[(\bar{X}_n - \theta)^2] = 2mn^{-1}\sigma^2,$$

for some real valued bound m. Hence, it follows that

$$E[\exp(-2\bar{X}_n^2)] = \exp(-2\theta^2) + O(n^{-1}), \quad (10.3)$$

as $n \to \infty$. To obtain an expression for the variance we also need to find an expansion for the square expectation of $\exp(-\bar{X}_n^2)$. From the previous result, we note that

$$E^2[\exp(-\bar{X}_n^2)] = [\exp(-\theta^2) + O(n^{-1})]^2 = \exp(-2\theta^2) + O(n^{-1}), \quad (10.4)$$

as $n \to \infty$. Combining the results of Equations (10.3) and (10.4) implies that $V[\exp(-\bar{X}_n^2)] = O(n^{-1})$ as $n \to \infty$. ∎

The general methodology of Example 10.2, based on taking expectations of Taylor expansions, can be generalized to a wider class of functions. However, it is worth noting that the key property of this development is the boundedness of certain derivatives of the function of $\bar{X}_n$. In this context, assumptions of this form cannot be relaxed without additional assumptions on the distribution F.

Theorem 10.1. *Let $\{X_n\}_{n=1}^{\infty}$ be a sequence of independent and identically distributed random variables from a distribution F with mean θ and variance*

σ^2. Suppose that f has a finite fourth moment and let g be a function that has at least four derivatives.

1. If the fourth derivative of g is bounded, then

$$E[g(\bar{X}_n)] = g(\theta) + \tfrac{1}{2}n^{-1}\sigma^2 g''(\theta) + O(n^{-2}),$$

as $n \to \infty$.

2. If the fourth derivative of g^2 is also bounded, then

$$V[g(\bar{X}_n)] = n^{-1}\sigma^2[g'(\theta)]^2 + O(n^{-2}),$$

as $n \to \infty$.

Proof. We prove Part 1, leaving the proof of Part 2 as Exercise 2. We begin by noting that Theorem 1.13 (Taylor) implies that

$$g(\bar{X}_n) = g(\theta) + (\bar{X}_n - \theta)g'(\theta)+$$
$$\tfrac{1}{2}(\bar{X}_n - \theta)^2 g''(\theta) + \tfrac{1}{6}(\bar{X}_n - \theta)^3 g'''(\theta) + \tfrac{1}{24}(\bar{X}_n - \theta)^4 g''''(\xi), \quad (10.5)$$

for some ξ that is between θ and $\bar{X}_n$ with probability one. Taking the expectation of both sides of Equation (10.5) yields

$$E[g(\bar{X}_n)] = g(\theta)+$$
$$\tfrac{1}{2}n^{-1}\sigma^2 g''(\theta) + \tfrac{1}{6}E[(\bar{X}_n - \theta)^3]g'''(\theta) + \tfrac{1}{24}E[(\bar{X}_n - \theta)^4 g''''(\xi)], \quad (10.6)$$

since $E(\bar{X}_n - \theta) = 0$ and $E[(\bar{X}_n - \theta)^2] = n^{-1}\sigma^2$. Note that we cannot factor $g''''(\xi)$ out of the expectation in Equation (10.6) because ξ is a random variable due to the fact that ξ is always between θ and $\bar{X}_n$. For the third term in Equation (10.6), we note that

$$E[(\bar{X}_n - \theta)^3] = n^{-3}E\left\{\left[\sum_{i=1}^n (X_i - \theta)\right]^3\right\}$$
$$= n^{-3}\sum_{i=1}^n E[(X_i - \theta)^3] + 3n^{-3}\sum_{i=1}^n\sum_{\substack{j=1\\j\neq i}}^n E[(X_i - \theta)^2(X_j - \theta)]$$
$$+n^{-3}\sum_{i=1}^n\sum_{\substack{j=1\\j\neq i}}^n\sum_{\substack{k=1\\k\neq i\\k\neq j}}^n E[(X_i - \theta)(X_j - \theta)(X_k - \theta)]. \quad (10.7)$$

The second and third term on the right hand side of Equation (10.7) are zero due to independence and the fact that $E(X_i - \theta) = 0$. Therefore, it follows that $E[(\bar{X}_n - \theta)^3] = n^{-2}E[(X_i - \theta)^3] = O(n^{-2})$, as $n \to \infty$. By assumption, there exists a bound $m \in \mathbb{R}$ such that $g''''(t) \leq m$ for all $t \in \mathbb{R}$. Therefore, it follows that $g''''(\xi)(\bar{X}_n - \theta)^4 \leq m(\bar{X}_n - \theta)^4$, with probability one. Hence Theorem A.16 implies that $E[\tfrac{1}{24}g''''(\xi)(\bar{X}_n - \theta)^4] \leq \tfrac{1}{24}mE[(\bar{X}_n - \theta)^4] < \infty$, since we have assumed that the fourth moment is finite. To obtain the rate of convergence for the error term we note that

$$E[(\bar{X}_n - \theta)^4] = E\left\{ n^{-4} \left[\sum_{i=1}^{n} (X_i - \theta) \right]^4 \right\}$$

$$= n^{-4} \sum_{i=1}^{n} E[(X_i - \theta)^4] + 4n^{-4} \sum_{i=1}^{n} \sum_{\substack{j=1 \\ j \neq i}}^{n} E[(X_i - \theta)^3 (X_j - \theta)]$$

$$+ 3n^{-4} \sum_{i=1}^{n} \sum_{\substack{j=1 \\ j \neq i}}^{n} E[(X_i - \theta)^2 (X_j - \theta)^2]$$

$$+ 6n^{-4} \sum_{i=1}^{n} \sum_{\substack{j=1 \\ j \neq i}}^{n} \sum_{\substack{k=1 \\ k \neq i \\ k \neq j}}^{n} E[(X_i - \theta)^2 (X_j - \theta)(X_k - \theta)]$$

$$+ n^{-4} \sum_{i=1}^{n} \sum_{\substack{j=1 \\ j \neq i}}^{n} \sum_{\substack{k=1 \\ k \neq i \\ k \neq j}}^{n} \sum_{\substack{l=1 \\ l \neq i \\ l \neq j \\ l \neq k}}^{n} E[(X_i - \theta)(X_j - \theta)(X_k - \theta)(X_l - \theta)].$$

Using similar arguments to those used for the third moment we find that $E[(\bar{X}_n - \theta)^4] = n^{-3} E[(X_n - \theta)^4] + 3n^{-3}(n-1) E[(X_n - \theta)^2 (X_m - \theta)^2]$ so that it follows that $E[(\bar{X}_n - \theta)^4] = O(n^{-2})$ as $n \to \infty$. Therefore, it follows that

$$E[g(\bar{X}_n)] = g(\theta) + \tfrac{1}{2} n^{-1} \sigma^2 g''(\theta) + O(n^{-2}),$$

as $n \to \infty$. $\quad\square$

Example 10.3. Let $X_1, \ldots, X_n$ be a set of independent and identically distributed random variables from a distribution F with mean θ and finite variance σ^2. In Example 10.1 we considered estimating θ^2 using the estimator $\bar{X}_n^2$. Taking $g(t) = t^2$ we have that $g'(t) = 2t$, $g''(t) = 2$, and that $g^k(t) = 0$ for all $k \geq 3$. Therefore, we can apply Theorem 10.1, assuming that the fourth moment of F is finite, to find $E(\bar{X}_n^2) = \theta^2 + n^{-1}\sigma^2 + O(n^{-2})$, as $n \to \infty$. This compares exactly with the result of Example 10.1 with the exception of the error term. The error term in this case is identically zero in this case because the derivatives of order higher than two are zero for the function $g(t) = t^2$. For the variance, we consider derivatives of the function $h(t) = g^2(t) = t^4$, where we have that $h'(t) = 4t^3$, $h''(t) = 12t^2$, $h'''(t) = 24t$ and $h''''(t) = 24$. The fourth derivative of h is bounded and therefore we can apply Theorem 10.1 to find that $V(\bar{X}_n) = 4n^{-1}\theta^2\sigma^2 + O(n^{-2})$, as $n \to \infty$, which matches the variance expansion found in Example 10.1. The error term in this case is not identically zero as we encountered non-zero terms of order $O(n^{-2})$ in Example 10.1. This is also indicated by the fact that the fourth derivative of h is not zero. $\quad\blacksquare$

Example 10.4. Consider the framework presented in Example 10.2 where $\{X_n\}_{n=1}^{\infty}$ is a sequence of independent and identically distributed random variables from a distribution F with mean θ and finite variance σ^2. We are interested in estimating $g(\theta) = \exp(-\theta^2)$ using the estimator $g(\bar{X}_n) = \exp(-\bar{X}_n^2)$. If we assume that F has a finite fourth moment then the assumptions of Theorem 10.1 hold and we have that

$$E[\exp(-\bar{X}_n^2)] = \exp(-\theta^2) - n^{-1}\sigma^2(1 - 2\theta^2)\exp(-\theta^2) + O(n^{-2}),$$

and

$$V[\exp(-\bar{X}_n^2)] = 4n^{-1}\sigma^2\theta^2\exp(-2\theta^2) + O(n^{-2}),$$

as $n \to \infty$. We can compare this result to the result supplied by Theorem 6.3, which implies that $n^{1/2}(2\theta\sigma)^{-1}\exp(\theta^2)[\exp(-\bar{X}_n^2) - \exp(-\theta^2)] \xrightarrow{d} Z$, as $n \to \infty$ where Z is a $N(0,1)$ random variable. Note that the asymptotic variance given by Theorem 6.3 matches the first term of the asymptotic expansion for the variance given in Theorem 10.1.   ∎

While the result of Theorem 10.1 is general in the sense that the distribution F need not be specified, the assumption on the boundedness of the fourth derivative of F will often be violated in practice. This does not mean that no asymptotic result of this kind can be obtained, but that such results will probably rely on methods more specific to the problem considered.

Example 10.5. Let $\{X_n\}_{n=1}^{\infty}$ be a sequence of independent and identically distributed random variables following a $N(\theta, \sigma^2)$ distribution and consider estimating $\exp(\theta)$ with $\exp(\bar{X}_n)$. We are not able to apply Theorem 10.1 directly to this case because the fourth derivative of $\exp(t)$, which is still $\exp(t)$, is not a bounded function. If we attempt to apply Theorem 1.13 to this problem we will end up with an error term of the form $\frac{1}{24}(\bar{X}_n - \theta)^2\exp(\xi)$, where ξ is a random variable that is always between θ and $\bar{X}_n$. This type of error term is difficult to deal with in this case because the exponential function is not bounded and hence we cannot directly bound the expectation as we did in Example 10.2 and the proof of Theorem 10.1. Instead we follow the approach of Lehmann (1999) and use the convergent Taylor series for the exponential function instead. That is,

$$\exp(\bar{X}_n) = \sum_{i=0}^{\infty} \frac{\exp(\theta)(\bar{X}_n - \theta)^i}{i!}. \tag{10.8}$$

We will assume that it is permissible in this case to exchange the expectation and the infinite sum, so that taking the expectation of both sides of Equation (10.8) yields

$$E[\exp(\bar{X}_n)] = \sum_{i=0}^{\infty} \frac{\exp(\theta)E[(\bar{X}_n - \theta)^i]}{i!} = \exp(\theta)\sum_{i=0}^{\infty} \frac{E[(\bar{X}_n - \theta)^{2i}]}{(2i)!}. \tag{10.9}$$

The second equality in Equation (10.9) is due to the fact that $\bar{X}_n - \theta$ has a $N(0, n^{-1}\sigma^2)$ distribution, whose odd moments are zero. Hence it follows that we must evaluate

$$E[(\bar{X}_n - \theta)^{2i}] = \left(\frac{n}{2\pi\sigma^2}\right)^{1/2} \int_{-\infty}^{\infty} t^{2i} \exp(-\tfrac{1}{2}n\sigma^{-2}t^2)dt.$$

To evaluate this integral consider the change of variable $u = \tfrac{1}{2}n\sigma^{-2}t^2$ so that $du = n\sigma^{-2}tdt$, where one must be careful to note that the transformation in the change of variable is not one-to-one. It follows that

$$\begin{aligned}
E[(\bar{X}_n - \theta)^{2i}] &= \pi^{-1/2}\sigma^{2i}n^{-i}2^i \int_0^{\infty} u^{i-1/2} \exp(-u)du \\
&= \pi^{-1/2}\sigma^{2i}n^{-i}2^i \Gamma(i + \tfrac{1}{2}).
\end{aligned}$$

Now, noting that $\Gamma(\tfrac{1}{2}) = \pi^{1/2}$ we have that

$$\begin{aligned}
\Gamma(i + \tfrac{1}{2}) &= (i - \tfrac{1}{2})(i - \tfrac{3}{2})\cdots\tfrac{3}{2}\tfrac{1}{2}\Gamma(\tfrac{1}{2}) \\
&= 2^{-i}(2i - 1)(2i - 3)\cdots(3)(1)\pi^{1/2} \\
&= \frac{(2i)!\pi^{1/2}}{2^i(2i)(2i - 1)\cdots(4)(2)} \\
&= (2i)!\pi^{1/2}2^{-2i}(i!)^{-1}.
\end{aligned}$$

Therefore, it follows that

$$E[(\bar{X}_n - \theta)^{2i}] = \frac{\sigma^{2i}(2i)!}{n^i 2^i i!}. \tag{10.10}$$

Putting the result of Equation (10.10) into the sum in Equation (10.9) yields

$$\begin{aligned}
E[\exp(\bar{X}_n)] &= \exp(\theta) \sum_{i=0}^{\infty} \frac{E[(\bar{X}_n - \theta)^{2i}]}{(2i)!} \\
&= \exp(\theta) \sum_{i=0}^{\infty} \frac{\sigma^{2i}(2i)!}{n^i 2^i i!(2i)!} \\
&= \exp(\theta) \sum_{i=0}^{\infty} \frac{\sigma^{2i}}{n^i 2^i i!} \\
&= \exp(\theta) \sum_{i=0}^{\infty} \frac{1}{i!}\left(\frac{\sigma^2}{2n}\right)^i \\
&= \exp(\theta)\exp(\tfrac{1}{2}n^{-1}\sigma^2).
\end{aligned}$$

Now, Theorem 1.13 (Taylor) implies that

$$\exp(\tfrac{1}{2}n^{-1}\sigma^2) = \exp(0) + \tfrac{1}{2}n^{-1}\sigma^2 \exp(0) + O(n^{-2}) =$$
$$1 + \tfrac{1}{2}n^{-1}\sigma^2 + O(n^{-2}),$$

as $n \to \infty$. Therefore, it follows that

$$E[\exp(\bar{X}_n)] = \exp(\theta)(1 + \tfrac{1}{2}n^{-1}\sigma^2) + O(n^{-2}) =$$
$$\exp(\theta) + \tfrac{1}{2}n^{-1}\sigma^2\exp(\theta) + O(n^{-2}),$$

as $n \to \infty$. Hence, it follows that the bias is $O(n^{-1})$ as $n \to \infty$. The variance can be found using a similar argument. See Exercise 7. ∎

It is noteworthy that not all functions of the sample mean have nice properties, even when the population is normal. For example, when $X_1, \ldots, X_n$ are independent and identically distributed $N(\theta, 1)$ random variables, $n^{1/2}(\bar{X}_n^{-1} - \theta^{-1}) \overset{d}{\to} Z$ as $n \to \infty$ where Z is a $N(0, \theta^{-4})$ random variable, but $E(\bar{X}_n^{-1})$ does not exist for any $n \in \mathbb{N}$. See Example 4.2.4 of Lehmann (1999) and Lehmann and Shaffer (1988) for further details.

If we consider the case where $\hat{\theta}_n$ is an estimator of a parameter θ with an expansion for its expectation of the form $E(\hat{\theta}_n) = \theta + n^{-1}b + O(n^{-2})$ as $n \to \infty$ where b is a real constant, then it follows that the bias of $\hat{\theta}_n$ is $O(n^{-1})$, and hence the square bias of $\hat{\theta}_n$ is $O(n^{-2})$, as $n \to \infty$. If the variance of $\hat{\theta}_n$ has an expansion of the form $V(\hat{\theta}_n) = n^{-1}v + O(n^{-2})$ as $n \to \infty$ where v is a real constant, then it follows that the mean squared error of $\hat{\theta}_n$ has the expansion $\text{MSE}(\hat{\theta}_n) = n^{-1}v + O(n^{-2})$ as $n \to \infty$. Therefore, it is the constant v that is the important factor in determining the asymptotic performance of $\hat{\theta}_n$, under these assumptions on the form of the bias and variance.

Now suppose that $\tilde{\theta}_n$ is another estimator of θ, and that $\tilde{\theta}_n$ has similar properties to $\hat{\theta}_n$ in the sense that the mean squared error for $\tilde{\theta}_n$ has asymptotic expansion $\text{MSE}(\tilde{\theta}_n) = n^{-1}w + O(n^{-2})$ as $n \to \infty$, where w is a real constant. If we wish to compare the performance of these two estimators from an asymptotic viewpoint it follows that we need only compare the constants v and w.

Definition 10.3. *Let $\hat{\theta}_n$ and $\tilde{\theta}_n$ be two estimators of a parameter θ such that $\text{MSE}(\hat{\theta}_n) = n^{-1}v + O(n^{-2})$ and $\text{MSE}(\tilde{\theta}_n) = n^{-1}w + O(n^{-2})$ as $n \to \infty$ where v and w are real constants. Then the asymptotic relative efficiency of $\hat{\theta}_n$ relative to $\tilde{\theta}_n$ is given by $\text{ARE}(\hat{\theta}_n, \tilde{\theta}_n) = wv^{-1}$.*

The original motivation for the form of the asymptotic relative efficiency comes from comparing the sample sizes required for each estimator to have the same mean squared error. From the asymptotic viewpoint we would require sample sizes n and m so that $n^{-1}v = m^{-1}w$, with the asymptotic relative efficiency of $\hat{\theta}_n$ relative to $\tilde{\theta}_n$ is given by $\text{ARE}(\hat{\theta}_n, \tilde{\theta}_n) = mn^{-1}$. However, note that if $n^{-1}v = m^{-1}w$ then it follows that $wv^{-1} = mn^{-1}$, yielding the form of Definition 10.3.

Example 10.6. Let $\{X_n\}_{n=1}^{\infty}$ be a sequence of independent and identically distributed random variables from a distribution F with finite variance σ^2 and continuous density f. We will assume that F has a unique median θ

such that $f(\theta) > 0$. In the case where f is symmetric about θ we have two immediate possible choices for estimating θ given by the sample mean $\hat{\theta}_n = \bar{X}_n$ and the sample median $\tilde{\theta}_n$. It is known that $V(\hat{\theta}_n) = n^{-1}\sigma^2$, and it follows from Corollary 4.4 that the leading term for the variance of $\tilde{\theta}_n$ is $\frac{1}{4}[f(\theta)]^{-2}$. Therefore, Definition 10.3 implies that $\mathrm{ARE}(\hat{\theta}_n, \tilde{\theta}_n) = \frac{1}{4}[\sigma f(\theta)]^{-2}$. If F corresponds to $N(\theta, \sigma^2)$ random variable then $f(\theta) = (2\pi\sigma^2)^{-1/2}$ and we have that $\mathrm{ARE}(\hat{\theta}_n, \tilde{\theta}_n) = \frac{1}{2}\pi \simeq 1.5708$, which indicates that the sample mean is about one and one half times more efficient than the sample median. If F corresponds to a $T(\nu)$ distribution where $\nu > 2$ we have that $\theta = 0$, $\sigma^2 = v/(v-2)$, and

$$f(\theta) = \frac{\Gamma(\frac{\nu+1}{2})}{(\pi v)^{1/2}\Gamma(\frac{\nu}{2})}.$$

Therefore, it follows that

$$\mathrm{ARE}(\hat{\theta}_n, \tilde{\theta}_n) = \frac{\pi(\nu-2)\Gamma^2(\frac{\nu}{2})}{4\Gamma^2(\frac{\nu+1}{2})}.$$

The values of $\mathrm{ARE}(\hat{\theta}_n, \tilde{\theta}_n)$ for $\nu = 3, 4, 5, 10, 25, 50$, and 100 are given in Table 10.1. From Table 10.1 one can observe that for heavy tailed $T(\nu)$ distributions, the sample median is more efficient than the sample mean. In these cases the variance of the sample mean is increased by the high likelihood of observing outliers in samples. However, as the degrees of freedom increase, the median becomes less efficient so that when $\nu = 5$ the sample median and the sample mean are almost equally efficient from an asymptotic viewpoint. From this point on, the sample mean becomes more efficient. In the limit we find that $\mathrm{ARE}(\hat{\theta}_n, \tilde{\theta}_n)$ approaches the value associated with the normal distribution given by $\frac{1}{2}\pi$. ∎

Example 10.7. Let $B_1, \ldots, B_n$ be a set of independent and identically distributed BERNOULLI(θ) random variables, where the success probability θ is the parameter of interest. We will also assume that the parameter space is $\Omega = (0, 1)$. The usual unbiased estimator of θ is the sample mean $\hat{\theta}_n = \bar{B}_n$, which corresponds to the proportion of successes observed in the sample. The properties of the sample mean imply that this estimator is unbiased with variance $n^{-1}\theta(1-\theta)$. When θ is small the estimator $\hat{\theta}_n$ is often considered unsatisfactory because $\hat{\theta}_n$ will be equal to zero with a large probability. For example, calculations based on the BINOMIAL(n, θ) distribution can be used to show that if $n = 100$ and $\theta = 0.001$ then $P(\hat{\theta}_n = 0) = 0.9048$. Since zero is not in the parameter space of θ, this may not be considered a reasonable estimator of θ. An alternative approach to estimating θ in this case is based on adding in one success and one failure to the observed sample. That is, we consider the alternative estimator

$$\tilde{\theta}_n = (n+2)^{-1}\left(\sum_{i=1}^{n} B_i + 1\right) = (1 + 2n^{-1})^{-1}(\bar{B}_n + n^{-1}).$$

Table 10.1 *The asymptotic relative efficiency of the sample mean relative to the sample median when the population follows a* T(ν) *distribution.*

ν	3	4	5	10	25	50	100
ARE($\hat{\theta}_n, \tilde{\theta}_n$)	0.617	0.889	1.041	1.321	1.4743	1.5231	1.5471

This estimator will never equal zero, but is not unbiased as

$$E(\tilde{\theta}_n) = (1 + 2n^{-1})^{-1}(\theta + n^{-1}) = \theta + O(n^{-1}),$$

as $n \to \infty$, so that the bias of $\tilde{\theta}_n$ is $O(n^{-1})$ as $n \to \infty$. Focusing on the variance we have that

$$V(\tilde{\theta}_n) = (1 + 2n^{-1})^{-2}[n^{-1}\theta(1-\theta)] = (n+2)^{-2}n\theta(1-\theta).$$

Hence, the efficiency of $\hat{\theta}_n$, relative to $\tilde{\theta}_n$ is given by

$$\text{ARE}(\hat{\theta}_n, \tilde{\theta}_n) = \lim_{n\to\infty} \frac{n\theta(1-\theta)}{(n+2)^2} \frac{n}{\theta(1-\theta)} = \lim_{n\to\infty} \frac{n^2}{(n+2)^2} = 1.$$

Therefore, from an asymptotic viewpoint, the estimators have the same efficiency. ∎

Asymptotic optimality refers to the condition that an estimator achieves the best possible performance as $n \to \infty$. For unbiased estimators, optimality is defined in terms of the Cramér–Rao bound.

Theorem 10.2 (Cramér and Rao). *Let* $X_1, \ldots, X_n$ *be a set of independent and identically distributed random variables from a distribution* $F(x|\theta)$ *with parameter* θ, *parameter space* Ω, *and density* $f(x|\theta)$. *Let* $\hat{\theta}_n$ *be any estimator of* θ *computed on* $X_1, \ldots, X_n$ *such that* $V(\hat{\theta}_n) < \infty$. *Assume that the following regularity conditions hold.*

1. *The parameter space* Ω *is an open interval which can be finite, semi-infinite, or infinite.*

2. *The set* $\{x : f(x|\theta) > 0\}$ *does not depend on* θ.

3. *For any* $x \in A$ *and* $\theta \in \Omega$ *the derivative of* $f(x|\theta)$ *with respect to* θ *exists and is finite.*

4. *The first two derivatives of*

$$\int_{-\infty}^{\infty} f(x|\theta)dx,$$

   with respect to θ *can be obtained by exchanging the derivative and the integral.*

5. *The first two derivatives of* $\log[f(x|\theta)]$ *with respect to* θ *exist for all* $x \in \mathbb{R}$ *and* $\theta \in \Omega$.

Then $V(\hat{\theta}_n) \geq [nI(\theta)]^{-1}$ where

$$I(\theta) = V\left\{\frac{\partial}{\partial\theta}\log[f(X_n|\theta)]\right\}.$$

The development of this result can be found in Section 2.6 of Lehmann and Casella (1998). The value $I(\theta)$ is called the *Fisher information number*. Noting that

$$\frac{\partial}{\partial\theta}\log[f(x|\theta)] = \frac{f'(x|\theta)}{f(x|\theta)},$$

it then follows that the random variable within the expectation measures the relative rate of change of $f(x|\theta)$ with respect to changes in θ. If this rate of change is large, then samples with various values of θ will be easily distinguished from one another and θ will be easier to estimate. In this case the bound on the variance will be small. If this rate of change is small then the parameter is more difficult to estimate and the variance bound will be larger. Several alternate expressions are available for $I(\theta)$ under various assumptions. To develop some of these, let X be a random variable that follows the distribution $F(x|\theta)$ and note that

$$V\left\{\frac{\partial}{\partial\theta}\log[f(X|\theta)]\right\} = V\left[\frac{f'(X|\theta)}{f(X|\theta)}\right] =$$

$$E\left\{\left[\frac{f'(X|\theta)}{f(X|\theta)}\right]^2\right\} - E^2\left[\frac{f'(X|\theta)}{f(X|\theta)}\right]. \quad (10.11)$$

Evaluating the second term on the right hand side of Equation (10.11) yields

$$E\left[\frac{f'(X|\theta)}{f(X|\theta)}\right] = \int_{-\infty}^{\infty}\frac{f'(x|\theta)}{f(x|\theta)}f(x|\theta)dx = \int_{-\infty}^{\infty}f'(x|\theta)dx. \quad (10.12)$$

Because $f(x|\theta)$ is a density we know that

$$\int_{-\infty}^{\infty}f(x|\theta)dx = 1,$$

and hence,

$$\frac{\partial}{\partial\theta}\int_{-\infty}^{\infty}f(x|\theta)dx = 0.$$

If we can exchange the partial derivative and the integral, then it follows that

$$\int_{-\infty}^{\infty}f'(x|\theta)dx = 0. \quad (10.13)$$

Therefore, under this condition,

$$I(\theta) = E\left\{\left[\frac{f'(X|\theta)}{f(X|\theta)}\right]^2\right\}. \quad (10.14)$$

Note further that

$$\frac{\partial^2}{\partial\theta^2}\log[f(x|\theta)] = \frac{\partial}{\partial\theta}\frac{f'(x|\theta)}{f(x|\theta)} = \frac{f''(x|\theta)}{f(x|\theta)} - \frac{[f'(x|\theta)]^2}{f^2(x|\theta)}.$$

Therefore,

$$E\left[\frac{\partial^2}{\partial\theta^2}\log[f(X|\theta)]\right] = E\left[\frac{f''(x|\theta)}{f(x|\theta)}\right] - E\left\{\left[\frac{f'(x|\theta)}{f(x|\theta)}\right]^2\right\}.$$

Under the assumption that the second partial derivative with respect to θ and the integral in the expectation can be exchanged we have that

$$E\left[\frac{f''(x|\theta)}{f(x|\theta)}\right] = \int_{-\infty}^{\infty} f''(x|\theta)dx = 0. \qquad (10.15)$$

Therefore, under this assumption

$$I(\theta) = E\left[\frac{\partial^2}{\partial\theta^2}\log[f(X|\theta)]\right],$$

which is usually the simplest form useful for computing $I(\theta)$.

Classical estimation theory strives to find unbiased estimators that are optimal in the sense that they have a mean squared error that attains the lower bound specified by Theorem 10.2. This can be a somewhat restrictive approach due to the fact that the lower bound is not sharp, and hence is not always attainable.

Example 10.8. Let $X_1, \ldots, X_n$ be a set of independent and identically distributed random variables from a $N(\mu, \theta)$ distribution where θ is finite. Theorem 10.2 (Cramér and Rao) implies that the optimal mean squared error for unbiased estimators of θ is $2n^{-1}\theta^2$. The usual unbiased estimator of θ is given by

$$\hat{\theta}_n = (n-1)^{-1}\sum_{i=1}^{n}(X_i - \bar{X}_n)^2.$$

The mean squared error of this estimator is given by $2(n-1)^{-1}\theta^2$, which is strictly larger than the bound given in Theorem 10.2. As pointed out in Example 7.3.16 of Casella and Berger (2002), the bound is not attainable in this case because the optimal estimator of θ is given by

$$\tilde{\theta}_n = n^{-1}\sum_{i=1}^{n}(X_i - \mu)^2,$$

which depends on the unknown parameter μ. Note however that the bound is attained asymptotically since

$$\lim_{n\to\infty}\mathrm{ARE}(\hat{\theta}_n, \tilde{\theta}_n) = \lim_{n\to\infty}\frac{2n^{-1}\theta^2}{2(n-1)^{-1}\theta^2} = 1.$$

∎

A less restrictive approach is to consider estimators that attain the lower bound asymptotically as the sample size increases to ∞. Within this approach we will consider consistent estimators of θ that have an asymptotic NORMAL distribution. For these estimators, mild regularity conditions exist for which there are estimators that attain the bound given in Theorem 10.2. To develop this approach let $X_1, \ldots, X_n$ be a set of independent and identically distributed random variables from a distribution F with parameter θ. Let $\hat{\theta}_n$ be an estimator of θ based on $X_1, \ldots, X_n$ such that $n^{1/2}(\hat{\theta}_n - \theta) \xrightarrow{d} Z$ as $n \to \infty$ where Z is a $N[0, \sigma^2(\theta)]$ random variable. Note that in this setup $\hat{\theta}_n$ is consistent and that $\sigma^2(\theta)$ is the asymptotic variance of $n^{1/2}\hat{\theta}_n$, which does not depend on n. The purpose of this approach is then to determine under what conditions $\sigma^2(\theta) = [I(\theta)]^{-1}$.

Definition 10.4. *Let $X_1, \ldots, X_n$ be a set of independent and identically distributed random variables from a distribution F with parameter θ. Let $\hat{\theta}_n$ be an estimator of θ such that $n^{1/2}(\hat{\theta}_n - \theta) \xrightarrow{d} Z$ as $n \to \infty$ where Z is a $N[0, \sigma^2(\theta)]$ random variable. If $\sigma^2(\theta) = [I(\theta)]^{-1}$ for all $\theta \in \Omega$, then $\hat{\theta}_n$ is an asymptotically efficient estimator of θ.*

There are several differences between finding an efficient estimator, that is one that attains the bound given in Theorem 10.2 for every $n \in \mathbb{N}$, and one that is asymptotically efficient, which attains this bound only in the limit as $n \to \infty$. We first note that an asymptotically efficient estimator is not unique, and may not even be unbiased, even asymptotically. However, some regularity conditions on the type of estimator considered are generally necessary as demonstrated by the famous example given below, which is due to Hodges. See Le Cam (1953) for further details.

Example 10.9. Let $X_1, \ldots, X_n$ be a set of independent and identically distributed random variables following a $N(\theta, 1)$ distribution. In this case $I(\theta) = 1$ so that an asymptotically efficient estimator is one with $\sigma^2(\theta) = 1$. Consider the estimator given by $\hat{\theta}_n = \bar{X}_n + (a - 1)\bar{X}_n \delta_n$, where $\delta_n = \delta\{|\bar{X}_n|; [0, n^{-1/4})\}$ for all $n \in \mathbb{N}$. We know from Theorem 4.20 (Lindeberg and Lévy) that $n^{1/2}(\bar{X}_n - \theta) \xrightarrow{d} Z$ as $n \to \infty$ where Z is a $N(0, 1)$ random variable. Hence $\bar{X}_n$ is asymptotically efficient by Definition 10.4. To establish the asymptotic behavior of $\hat{\theta}_n$ we consider two distinct cases. When $\theta \neq 0$ we have that $\bar{X}_n \xrightarrow{p} \theta \neq 0$ as $n \to \infty$ by Theorem 3.10 and hence it follows that $\delta\{|\bar{X}_n| : [0, n^{-1/4})\} \xrightarrow{p} 0$ as $n \to \infty$. See Exercise 12. Therefore, Theorem 3.9 implies that $\hat{\theta}_n \xrightarrow{p} \theta$ as $n \to \infty$ when $\theta \neq 0$. However, we can demonstrate an even stronger result which will help us establish the weak convergence of $\hat{\theta}_n$. Consider the sequence of random variables given by $n^{1/2}(a - 1)\bar{X}_n \delta_n$. Note that Theorem 4.20 (Lindeberg and Lévy) implies that $n^{1/2}\bar{X}_n \xrightarrow{d} Z$ as $n \to \infty$ where Z is a $N(0, 1)$ random variable. Therefore, Theorem 4.11 (Slutsky) implies that $n^{1/2}(a - 1)\bar{X}_n \delta_n \xrightarrow{p} 0$ as $n \to \infty$. Now note that Theorem 4.11 implies that

$$
\begin{aligned}
n^{1/2}(\hat{\theta}_n - \theta) &= n^{1/2}[\bar{X}_n + (a - 1)\delta_n \bar{X}_n - \theta] \\
&= n^{1/2}(\bar{X}_n - \theta) + n^{1/2}(a - 1)\delta_n \bar{X}_n,
\end{aligned}
$$

converges in distribution to a $N(0,1)$ distribution as $n \to \infty$. Therefore, we have established that $\sigma^2(\theta) = 1$ when $\theta \neq 0$ and hence $\hat{\theta}_n$ is asymptotically efficient in this case.

For the case when $\theta = 0$ we have that $\delta_n \xrightarrow{p} 1$ as $n \to \infty$. See Exercise 12. Noting that $n^{1/2}\bar{X}_n \xrightarrow{d} Z$ as $n \to \infty$ where Z is a $N(0,1)$ random variable, it then follows from Theorem 4.11 (Slutsky) that $n^{1/2}\hat{\theta}_n = n^{1/2}\bar{X}_n[1 + (a - 1)\delta_n] \xrightarrow{d} aZ$ as $n \to \infty$, so that $\sigma^2(\theta) = a^2$. Note then that if $|a| < 1$ then it follows that $\sigma^2(\theta) < 1$, and therefore in this case the estimator is *more efficient* that the best estimator.   ∎

Example 10.9 demonstrates the somewhat disturbing result that there are conditions under which we might have what are known as *super-efficient* estimators, whose variance is less than the minimum given by the bound of Theorem 10.2. However, it is noteworthy that we only obtain such an estimator at a single point in the parameter space Ω. In fact, Le Cam (1953) shows that under some regularity conditions similar to those given in Theorem 10.2, the set of points in Ω for which there are super-efficient estimators always has a Lebesgue measure equal to zero. In particular, there are no uniformly super-efficient estimators over Ω. For further details on this result see Bahadur (1964) and Section 6.1 of Lehmann and Casella (1998). However, it has been pointed out by Le Cam (1953), Huber (1966), and Hájek (1972) that the violation of the inequality in Theorem 10.2 even at a single point can produce certain unfavorable properties of the risk of the estimators in a neighborhood of that point. See Example 1.1 in Section 6.2 of Lehmann and Casella (1998) for further details.

Lehmann and Casella (1998) also point out that there are no additional assumptions on the distribution of F that can be made which will avoid this difficulty. However, restricting the types of estimators considered is possible. For example, one could require both that

$$\lim_{n \to \infty} V[n^{1/2}(\hat{\theta}_n - \theta)] = v(\theta),$$

and

$$\lim_{n \to \infty} \frac{\partial}{\partial \theta} \text{Bias}(\hat{\theta}_n) = 0,$$

which would avoid super-efficient estimators. Another assumption that can avoid this difficulty is to require $v(\theta)$ to be a continuous function in θ. Still another possibility suggested by Rao (1963) and Wolfowitz (1965) is to require the weak convergence on $n^{1/2}(\hat{\theta}_n - \theta)$ to Z to be uniform in θ. Further results are proven by Phanzagl (1970).

We now consider a specific methodology for obtaining estimators that is applicable to many types of problems: *maximum likelihood estimation*. Under specific conditions, we will be able to show that this method provides asymptotically optimal estimators that are also asymptotically NORMAL. If we observe $X_1, \ldots, X_n$, a set of independent and identically distributed random variables from a distribution $F(x|\theta)$, then the joint density of the observed sample if given by

$$f(x_1,\ldots,x_n|\theta) = \prod_{k=1}^{n} f(x_i|\theta),$$

where $\theta \in \Omega$. We have assumed that $F(x|\theta)$ has density $f(x|\theta)$ and that the observed random variables are continuous. In the discrete case the density $f(x|\theta)$ is replaced by the probability distribution function associated with $F(x|\theta)$. The *likelihood function* considers the joint density $f(x_1,\ldots,x_n|\theta)$ as a function of θ where the observed sample is taken to be fixed. That is

$$L(\theta|x_1,\ldots,x_n) = \prod_{k=1}^{n} f(x_i|\theta).$$

For simpler notation we will often use $L(\theta|\mathbf{x})$ in place of $L(\theta|x_1,\ldots,x_n)$ where $\mathbf{x}' = (x_1,\ldots,x_n)$. The *maximum likelihood estimators* of θ are taken to be the points that maximize the function $L(\theta|x_1,\ldots,x_n)$ with respect to θ. That is, $\hat{\theta}_n$ is a maximum likelihood estimator of θ if

$$L(\hat{\theta}_n|x_1,\ldots,x_n) = \sup_{\theta \in \Omega} L(\theta|x_1,\ldots,x_n).$$

Assuming that $L(\theta|x_1,\ldots,x_n)$ has at least two derivatives and that the parameter space of θ is Ω, candidates for the maximum likelihood estimator of θ have the properties

$$\left.\frac{d}{d\theta}L(\theta|x_1,\ldots,x_n)\right|_{\theta=\hat{\theta}_n} = 0,$$

and

$$\left.\frac{d^2}{d\theta^2}L(\theta|x_1,\ldots,x_n)\right|_{\theta=\hat{\theta}_n} < 0.$$

Other candidates for a maximum likelihood estimator are the points on the boundary of the parameter space. The maximum likelihood estimators are the candidates for which the likelihood is the largest. Therefore, maximum likelihood estimators may not be unique and hence there may be two or more values of $\hat{\theta}_n$ that all maximize the likelihood.

One must be careful when interpreting a maximum likelihood estimator. A maximum likelihood estimator is not the most likely value of θ given the observed data. Rather, a maximum likelihood estimator is a value of θ which has the largest probability of generating the data when the distribution is discrete. In the continuous case a maximum likelihood estimator is a value of θ for which the joint density of the sample is greatest.

In many cases the distribution of interest often contains forms that may not be simple to differentiate after the product is taken to form the likelihood function. In these cases the problem can be simplified by taking the natural logarithm of the likelihood function. The resulting function is often called the *log-likelihood* function. Note that because the natural logarithm function is monotonic, the points that maximize $L(\theta|x_1,\ldots,x_n)$ will also maximize $l(\theta) = \log[L(\theta|x_1,\ldots,x_n)]$. Therefore, a maximum likelihood estimator of θ can be defined as the value $\hat{\theta}_n$ such that

$$l(\hat{\theta}_n) = \sup_{\theta \in \Omega} l(\theta).$$

Example 10.10. Suppose that $X_1,\ldots,X_n$ is a set of independent and identically distributed random variables from an EXPONENTIAL(θ) distribution, where the parameter space for θ is $\Omega = (0,\infty)$. We will denote the corresponding observed sample as $x_1,\ldots,x_n$. The likelihood function in this case is given by

$$
\begin{aligned}
L(\theta|x_1,\ldots,x_n) &= \prod_{k=1}^{n} f(x_k|\theta) \\
&= \prod_{k=1}^{n} \theta^{-1} \exp(-\theta^{-1} x_k) \\
&= \theta^{-n} \exp\left(-\theta^{-1} \sum_{k=1}^{n} x_k\right).
\end{aligned}
$$

Therefore, the log-likelihood function is given by

$$l(\theta) = -n\log(\theta) - \theta^{-1} \sum_{k=1}^{n} x_k.$$

The first derivative is

$$\frac{d}{d\theta} l(\theta) = -n\theta^{-1} + \theta^{-2} \sum_{k=1}^{n} x_k.$$

Setting this derivative equal to zero and solving for θ gives

$$\hat{\theta}_n = \bar{x}_n = n^{-1} \sum_{k=1}^{n} x_k,$$

as a candidate for the maximum likelihood estimator. The second derivative of the log-likelihood is given by

$$\frac{d^2}{d\theta^2} l(\theta) = n\theta^{-2} - 2\theta^{-3} \sum_{k=1}^{n} x_k,$$

so that

$$\left.\frac{d^2}{d\theta^2} l(\theta)\right|_{\theta=\hat{\theta}_n} = n\bar{x}_n^{-2} - 2n\bar{x}_n^{-2} = -n\bar{x}_n^{-2} < 0,$$

since $\bar{x}_n$ will be positive with probability one. It follows that $\bar{x}_n$ is a local maximum. We need only now check the boundary points of $\Omega = (0,\infty)$. Noting

that

$$\lim_{\theta \to 0} \theta^{-n} \exp\left(-\theta^{-1} \sum_{k=1}^{n} x_k\right) = \lim_{\theta \to \infty} \theta^{-n} \exp\left(-\theta^{-1} \sum_{k=1}^{n} x_k\right) = 0,$$

we need only show that $L(\bar{x}_n | X_1, \ldots, X_n) > 0$ to conclude that $\bar{x}_n$ is the maximum likelihood estimator of θ. To see this, note that

$$L(\bar{x}_n | X_1, \ldots, X_n) = \bar{x}_n^{-n} \exp\left(-\bar{x}_n^{-1} \sum_{k=1}^{n} x_k\right) = \bar{x}_n^{-1} \exp(-n) > 0.$$

Therefore, it follows that $\bar{x}_n$ is the maximum likelihood estimator of θ. ∎

Example 10.11. Suppose that $X_1, \ldots, X_n$ is a set of independent and identically distributed random variables from a UNIFORM$(0, \theta)$ distribution where the parameter space for θ is $\Omega = (0, \infty)$. The likelihood function for this case is given by

$$L(\theta | X_1, \ldots, X_n) = \prod_{k=1}^{n} \theta^{-1} \delta\{x_k; (0, \theta)\} = \theta^{-n} \prod_{k=1}^{n} \delta\{x_k; (0, \theta)\}. \qquad (10.16)$$

When taken as a function of θ, note that $\delta\{x_k; (0, \theta)\}$ is zero unless $\theta > x_k$. Therefore, the product on the right hand side of Equation (10.16) is zero unless $\theta > x_k$ for all $k \in \{1, \ldots, n\}$, or equivalently if $\theta > x_{(n)}$, where $x_{(n)}$ is the largest value in the sample. Therefore, the likelihood function has the form

$$L(\theta | x_1, \ldots, x_n) = \begin{cases} 0 & \theta \leq x_{(n)} \\ \theta^{-n} & \theta > x_{(n)}. \end{cases}$$

It follows then that the likelihood function is then maximized at $\hat{\theta}_n = x_{(n)}$. See Figure 10.1. ∎

Maximum likelihood estimators have many useful properties. For example, maximum likelihood estimators have an invariance property that guarantees that the maximum likelihood estimator of a function of a parameter is the same function of the maximum likelihood estimator of the parameter. See Theorem 7.2.10 of Casella and Berger (2002). In this section we will establish some asymptotic properties of maximum likelihood estimators. In particular, we will establish conditions under which maximum likelihood estimators are consistent and asymptotically efficient.

The main impediment to establishing a coherent asymptotic theory for maximum likelihood estimators is that the derivative of the likelihood, or log-likelihood, function may have multiple roots. This can cause problems for consistency, for example, since the maximum likelihood estimator may jump from root to root as new observations are obtained from the population. Because we intend to provide the reader with an overview of this subject we will concentrate on problems that have a single unique root. A detailed account of the case where there are multiple roots can be found in Section 6.4 of Lehmann and Casella (1998).

Figure 10.1 *The likelihood function for θ where $X_1, \ldots, X_n$ is a set of independent and identically distributed random variables from a* UNIFORM$(0, \theta)$ *distribution. The horizontal axis corresponds to the parameter space, and the jump occurs when $\theta = x_{(n)}$.*

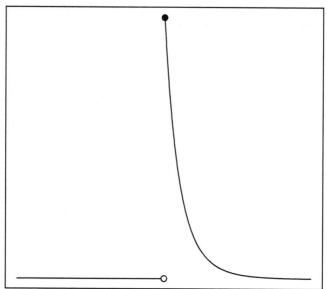

A first asymptotic motivation for maximum likelihood estimators comes from the fact that the likelihood function of θ computed on a set of independent and identically distributed random variables from a distribution $F(x|\theta)$ is asymptotically maximized at the true value of θ.

Theorem 10.3. *Let $X_1, \ldots, X_n$ be a set of independent and identically distributed random variables from a distribution $F(x|\theta)$ with density or probability distribution $f(x|\theta)$, where θ is the parameter of interest. Suppose that the following conditions hold.*

1. *The parameter θ is identifiable. That is, $f(x|\theta)$ is distinct for each value of θ in Ω.*

2. *The set $\{x \in \mathbb{R} : f(x|\theta) > 0\}$ is the same for all $\theta \in \Omega$.*

If θ_0 is the true value of θ, then

$$\lim_{n \to \infty} P[L(\theta_0|X_1, \ldots, X_n) > L(\theta|X_1, \ldots, X_n)] = 1,$$

for all $\theta \in \Omega \setminus \theta_0$.

Proof. We begin by noting that $L(\theta_0|X_1, \ldots, X_n) > L(\theta|X_1, \ldots, X_n)$ will occur if and only if

$$\prod_{i=1}^{n} f(X_i|\theta_0) > \prod_{i=1}^{n} f(X_i|\theta),$$

which is equivalent to the condition that

$$\left[\prod_{i=1}^{n} f(X_i|\theta_0)\right]^{-1} \prod_{i=1}^{n} f(X_i|\theta) < 1.$$

Taking the logarithm of this last expression and dividing by n implies that

$$n^{-1} \sum_{i=1}^{n} \log\{f(X_i|\theta)[f(X_i|\theta_0)]^{-1}\} < 0.$$

Since $X_1, \ldots, X_n$ are independent and identically distributed, it follows from Theorem 3.10 (Weak Law of Large Numbers) that

$$n^{-1} \sum_{i=1}^{n} \log[f(X_i|\theta)/f(X_i|\theta_0)] \overset{p}{\to} E\{\log[f(X_1|\theta)/f(X_1|\theta_0)]\}, \qquad (10.17)$$

as $n \to \infty$, where we assume that the expectation exists. Note that $-\log(t)$ is a strictly convex function on $\mathbb{R}$ so that Theorem 2.11 (Jensen) implies that

$$-E[\log\{f(X_1|\theta)[f(X_1|\theta_0)]^{-1}\}] > -\log[E\{f(X_1|\theta)[f(X_1|\theta_0)]^{-1}\}]. \quad (10.18)$$

We now use the fact that θ_0 is the true value of θ to find

$$E\{f(X_1|\theta)[f(X_1|\theta_0)]^{-1}\} = \int_{-\infty}^{\infty} \frac{f(x|\theta)}{f(x|\theta_0)} f(x|\theta_0) dx = \int_{-\infty}^{\infty} f(x|\theta) dx = 1.$$

Therefore, it follows that $\log[E\{f(X_1|\theta)[f(X_1|\theta_0)]^{-1}\}] = 0$, and hence Equation (10.18) implies that $E[\log\{f(X_1|\theta)[f(X_1|\theta_0)]^{-1}\}] < 0$. Therefore, we have from Equation (10.17) that

$$n^{-1} \sum_{i=1}^{n} \log\{f(X_i|\theta)[f(X_i|\theta_0)]^{-1}\} \overset{p}{\to} c,$$

where $c < 0$ is a real constant. Definition 3.1 then implies that

$$\lim_{n\to\infty} P\left(n^{-1} \sum_{i=1}^{n} \log\{f(X_i|\theta)[f(X_i|\theta_0)]^{-1}\} < 0\right) = 1,$$

which is equivalent to

$$\lim_{n\to\infty} P[L(\theta_0|X_1, \ldots, X_n) > L(\theta|X_1, \ldots, X_n)] = 1.$$

□

The assumptions of Theorem 10.3 are sufficient to ensure the consistency of the maximum likelihood estimator for the case when Ω has a finite number of elements. See Corollary 6.2 of Lehmann and Casella (1998). However, even when Ω is countable the result breaks down. See Bahadur (1958), Le Cam (1979), and Example 6.2.6 of Lehmann and Casella (1998), for further details. However, a few additional regularity conditions can provide a consistency result for the maximum likelihood estimator.

Theorem 10.4. *Let $X_1, \ldots, X_n$ be a set of independent and identically distributed random variables from a distribution $F(x|\theta)$ with density $f(x|\theta)$, where θ is the parameter of interest. Let $\mathbf{X} = (X_1, \ldots, X_n)'$ and suppose that the following conditions hold.*

1. *The parameter θ is identifiable. That is, $f(x|\theta)$ is distinct for each value of θ in Ω.*

2. *The set $\{x \in \mathbb{R} : f(x|\theta) > 0\}$ is the same for all $\theta \in \Omega$.*

3. *The parameter space Ω contains an open interval W where the true value of θ is an interior point.*

4. *For almost all $x \in \mathbb{R}$, $f(x|\theta)$ is differentiable with respect to θ in W.*

5. *The equation*

$$\frac{\partial}{\partial \theta} L(\theta|\mathbf{X}) = 0, \qquad (10.19)$$

has a single unique root for each $n \in \mathbb{N}$ and all $\mathbf{X} \in \mathbb{R}^n$.

Then, with probability one, if $\hat{\theta}_n$ is the root of Equation (10.19), then $\hat{\theta}_n \xrightarrow{p} \theta$, as $n \to \infty$.

Proof. Let $\theta_0 \in \Omega$ be the true value of the parameter. Condition 3 implies that we can select a positive real number δ such that $(\theta_0 - \delta, \theta_0 + \delta) \subset W$. Define the set

$$G_n(\delta) = \{\mathbf{X} \in \mathbb{R}^n : L(\theta_0|\mathbf{X}) > L(\theta_0 - \delta|\mathbf{X}) \text{ and } L(\theta_0|\mathbf{X}) > L(\theta_0 + \delta|\mathbf{X})\}.$$

Theorem 10.3 implies that

$$\lim_{n \to \infty} P[G_n(\delta)] = 1, \qquad (10.20)$$

since $\theta_0 - \delta$ and $\theta_0 + \delta$ are not the true values of θ. Note that if $\mathbf{x} \in G_n(\delta)$ then it follows that there must be a local maximum in the interval $(\theta_0 - \delta, \theta_0 + \delta)$ since $L(\theta_0|\mathbf{X}) > L(\theta_0 - \delta|\mathbf{x})$ and $L(\theta_0|\mathbf{X}) > L(\theta_0 + \delta|\mathbf{x})$. Condition 4 implies that there always exists $\hat{\theta}_n \in (\theta_0 - \delta, \theta_0 + \delta)$ such that $L'(\hat{\theta}_n|\mathbf{X}) = 0$. Hence Equation (10.20) implies that there is a sequence of roots $\{\hat{\theta}_n\}_{n=1}^{\infty}$ such that

$$\lim_{n \to \infty} P(|\hat{\theta}_n - \theta_0| < \delta) = 1,$$

and hence $\hat{\theta}_n \xrightarrow{p} \theta$, as $n \to \infty$. $\qquad \Box$

For further details on the result of Theorem 10.4 when there are multiple roots, see Section 6.4 of Lehmann and Casella (1998).

Example 10.12. Suppose that $X_1, \ldots, X_n$ is a set of independent and identically distributed random variables having an EXPONENTIAL(θ) distribution. In Example 10.10, the sample mean was shown to be the unique maximum likelihood estimator of θ. Assuming that θ is an interior point of the parameter space $\Omega = (0, \infty)$, the properties of Theorem 10.4 are satisfied and we can conclude that $\bar{X}_n \xrightarrow{p} \theta$ as $n \to \infty$. This result can also be found directly using Theorem 3.10. ∎

In Example 10.12 we can observe that if the maximum likelihood estimator of θ has a closed form, then it is often easier to establish the consistency directly using results like Theorem 3.10. The interest in results like Theorem 10.4 is that we can observe what types of assumptions are required to establish consistency. Note, however, that the assumptions of Theorem 10.4 are not necessary; there are examples of maximum likelihood estimators that do not follow all of the assumptions of Theorem 10.4 that are still consistent.

Example 10.13. Suppose that $X_1, \ldots, X_n$ is a set of independent and identically distributed random variables having an UNIFORM$(0, \theta)$ distribution, where the unique maximum likelihood estimator was found in Example 10.11 to be $\hat{\theta}_n = X_{(n)}$, the largest order statistic of the sample. This example violates the second assumption of Theorem 10.4, so that we cannot use that result to directly obtain the consistency of $\hat{\theta}_n$. However, noting that the parameter space in this case is $\Omega = (0, \infty)$, we can use a direct approach to find that if we let $0 < \varepsilon < \theta$, then

$$
\begin{aligned}
P(|\hat{\theta}_n - \theta| < \varepsilon) &= P(\theta - \varepsilon < \hat{\theta}_n < \theta) \\
&= 1 - P\left(\bigcap_{i=1}^{n} \{0 \leq X_i \leq \theta - \varepsilon\}\right) \\
&= 1 - \prod_{i=1}^{n} P(0 \leq X_i \leq \theta - \varepsilon) \\
&= 1 - [\theta^{-1}(\theta - \varepsilon)]^n.
\end{aligned}
$$

Therefore, it follows that

$$
\lim_{n \to \infty} P(|\hat{\theta}_n - \theta| < \varepsilon) = 1.
$$

When $\varepsilon > \theta$ we have that $P(|\hat{\theta}_n - \theta| < \varepsilon) = 1$ for all $n \in \mathbb{N}$. Therefore $\hat{\theta}_n \xrightarrow{p} \theta$ as $n \to \infty$. ∎

During the development of the fundamental aspects of estimation theory, there were several conjectures that maximum likelihood estimators had many globally applicable properties such as consistency. The following example demonstrates that there are, in fact, inconsistent maximum likelihood estimators.

Example 10.14. Consider a sequence of random variables $\{\{X_{ij}\}_{j=1}^{k}\}_{i=1}^{n}$ that are assumed to be mutually independent, each having a N(μ_i, θ) distribution for $i = 1, \ldots, n$. It can be shown that the maximum likelihood estimators of μ_i and θ are

$$\hat{\mu}_i = \bar{X}_i = k^{-1} \sum_{j=1}^{k} X_{ij},$$

for each $i = 1, \ldots, n$ and

$$\hat{\theta}_n = (nk)^{-1} \sum_{i=1}^{n} \sum_{j=1}^{k} (X_{ij} - \bar{X}_i)^2,$$

respectively. See Exercise 17. Note that

$$\hat{\theta}_n = (nk)^{-1} \sum_{i=1}^{n} \sum_{j=1}^{k} (X_{ij} - \bar{X}_i)^2 =$$

$$n^{-1} \sum_{i=1}^{n} \left[k^{-1} \sum_{j=1}^{k} (X_{ij} - \bar{X}_i)^2 \right] = n^{-1} \sum_{i=1}^{n} S_i^2,$$

where

$$S_i^2 = k^{-1} \sum_{j=1}^{k} (X_{ij} - \bar{X}_i)^2,$$

for $i = 1, \ldots, n$, which is the sample variance computed on $X_{i1}, \ldots, X_{ik}$. Define $C_i = k\theta^{-1} S_i^2$ for $i = 1, \ldots, n$ and note that C_i has a CHISQUARED$(k - 1)$ distribution for all $i = 1, \ldots, n$. Then

$$\hat{\theta}_n = n^{-1} \sum_{i=1}^{n} S_i^2 = (nk)^{-1} \sum_{i=1}^{n} \theta C_i.$$

Noting that $C_1, \ldots, C_n$ are mutually independent, if follows from Theorem 3.10 (Weak Law of Large Numbers) that $\hat{\theta}_n \xrightarrow{p} k^{-1} \theta E(C_1) = \theta k^{-1}(k - 1)$ as $n \to \infty$, so that the maximum likelihood estimator does not converge in probability to θ. The inconsistency in this example is due to the fact that the bias of the maximum likelihood estimator does not converge to zero as $n \to \infty$.   ∎

Example 10.14 was the first example of an inconsistent maximum likelihood estimator and is given by Neyman and Scott (1948). An interesting story regarding this example can be found in Barnard (1970). Other examples of inconsistent maximum likelihood estimators can be found in Bahadur (1958), Le Cam (1953), Basu (1955). Simple adjustments or a Bayesian approach can often be used to obtain consistent estimates in these cases. For instance, in Example 10.14, the estimator $k(k - 1)^{-1} \hat{\theta}_n$ will provide a consistent estimator of θ, though the adjusted estimator is not the maximum likelihood estimator. See Ghosh (1994) for further examples.

With a few more assumptions added to those of Theorem 10.4, we can establish conditions under which maximum likelihood estimators are both consistent and asymptotically efficient.

Theorem 10.5. *Suppose that $X_1, \ldots, X_n$ are a set of independent and identically distributed random variables from a distribution $F(x|\theta)$ with density $f(x|\theta)$, where θ has parameter space Ω. Let θ_0 denote the true value of θ. Suppose that*

1. *Ω is an open interval.*

2. *The set $\{x : f(x|\theta) > 0\}$ is the same for all $\theta \in \Omega$.*

3. *The density $f(x|\theta)$ has three continuous derivatives with respect to θ for each $x \in \{x : f(x|\theta) > 0\}$.*

4. *The integral*

$$\int_{-\infty}^{\infty} f(x|\theta)dx,$$

can be differentiated three times by exchanging the integral and the derivatives.

5. *The Fisher information number $I(\theta_0)$ is defined, positive, and finite.*

6. *For any $\theta_0 \in \Omega$ there exists a positive constant d and function $B(x)$ such that*

$$\left| \frac{\partial^3}{\partial \theta^3} \log[f(x|\theta)] \right| \leq B(x),$$

for all $x \in \{x : f(x|\theta) > 0\}$ and $\theta \in [\theta_0 - d, \theta_0 + d]$ such that $E[B(X_1)] < \infty$.

7. *There is a unique maximum likelihood estimator $\hat{\theta}_n$ for each $n \in \mathbb{N}$ and $\theta \in \Omega$.*

Then $n^{1/2}(\hat{\theta}_n - \theta) \xrightarrow{d} Z$ as $n \to \infty$ where Z has a $N[0, I^{-1}(\theta_0)]$ distribution.

Proof. Suppose that $X_1, \ldots, X_n$ is a set of independent and identically distributed random variables from a distribution $F(x|\theta)$ with density $f(x|\theta)$. Let $l(\theta|\mathbf{X})$ denote the log-likelihood function of θ given by

$$l(\theta|\mathbf{X}) = \log \left[\prod_{i=1}^{n} f(X_i|\theta) \right] = \sum_{i=1}^{n} \log[f(X_i|\theta)],$$

where $\mathbf{X} = (X_1, \ldots, X_n)'$. Assume that the maximum likelihood estimator, denoted by $\hat{\theta}_n$, is the unique solution to the equation

$$l'(\theta|\mathbf{X}) = \frac{d}{d\theta} \sum_{i=1}^{n} \log[f(X_i|\theta)] = \sum_{i=1}^{n} \frac{f'(X_i|\theta)}{f(X_i|\theta)} = 0, \qquad (10.21)$$

where the derivative indicated by $f'(X_i|\theta)$ is taken with respect to θ. Now apply Theorem 1.13 (Taylor) to $l'(\theta|\mathbf{X})$ to expand $l'(\hat{\theta}_n|\mathbf{X})$ about a point $\theta_0 \in \Omega$ as

$$l'(\hat{\theta}_n|\mathbf{X}) = l'(\theta_0|\mathbf{X}) + (\hat{\theta}_n - \theta_0)l''(\theta_0|\mathbf{X}) + \tfrac{1}{2}(\hat{\theta}_n - \theta_0)^2 l'''(\xi_n|\mathbf{X}),$$

where ξ_n is a random variable that is between θ_0 and $\hat{\theta}_n$ with probability one. Because $\hat{\theta}_n$ is the unique root of $l'(\theta|\mathbf{X})$, it follows that

$$l'(\theta_0|\mathbf{X}) + (\hat{\theta}_n - \theta_0)l''(\theta_0|\mathbf{X}) + \tfrac{1}{2}(\hat{\theta}_n - \theta_0)^2 l'''(\xi_n|\mathbf{X}) = 0,$$

or equivalently

$$n^{1/2}(\hat{\theta}_n - \theta_0) = \frac{n^{-1/2}l'(\theta_0|\mathbf{X})}{-n^{-1}l''(\theta_0|\mathbf{X}) - \tfrac{1}{2}n^{-1}(\hat{\theta}_n - \theta_0)l'''(\xi_n|\mathbf{X})}. \qquad (10.22)$$

The remainder of the proof is based on analyzing the asymptotic behavior of each of the terms on the right hand side of Equation (10.22). We first note that

$$n^{-1/2}l'(\theta_0|\mathbf{X}) = n^{-1/2}\sum_{i=1}^{n} \frac{f'(X_i|\theta_0)}{f(X_i|\theta_0)}.$$

Under the assumption that it is permissible to exchange the partial derivative and the integral, Equation (10.13) implies that

$$\int_{-\infty}^{\infty} f'(x|\theta)dx = 0,$$

and hence

$$E\left[\frac{f'(X_i|\theta_0)}{f(X_i|\theta_0)}\right] = 0.$$

Therefore,

$$n^{-1/2}l'(\theta_0|\mathbf{X}) = n^{1/2}\left\{n^{-1}\sum_{i=1}^{n} \frac{f'(X_i|\theta_0)}{f(X_i|\theta_0)} - E\left[\frac{f'(X_i|\theta_0)}{f(X_i|\theta_0)}\right]\right\}.$$

We can apply Theorem 4.20 (Lindeberg and Lévy) and Theorem 10.2 (Cramér and Rao) to this last expression to find that $n^{-1/2}l'(\theta_0|\mathbf{X}) \xrightarrow{d} Z$ as $n \to \infty$ where Z is a $\mathrm{N}[0, I(\theta_0)]$ random variable. We now consider the term $-n^{-1}l''(\theta_0)$. First note that

$$
\begin{aligned}
-n^{-1}l''(\theta_0) &= -n^{-1} \frac{\partial^2}{\partial\theta^2} \log[L(\theta|\mathbf{X})]\Big|_{\theta=\theta_0} \\
&= -n^{-1} \frac{\partial^2}{\partial\theta^2} \sum_{i=1}^{n} \log[f(X_i|\theta)]\Big|_{\theta=\theta_0} \\
&= -n^{-1} \frac{\partial}{\partial\theta} \sum_{i=1}^{n} \frac{f'(X_i|\theta)}{f(X_i|\theta)}\Big|_{\theta=\theta_0} \\
&= -n^{-1} \sum_{i=1}^{n} \frac{f''(X_i|\theta_0)}{f(X_i|\theta_0)} + n^{-1} \sum_{i=1}^{n} \frac{[f'(X_i|\theta_0)]^2}{f^2(X_i|\theta_0)} \\
&= n^{-1} \sum_{i=1}^{n} \frac{[f'(X_i|\theta_0)]^2 - f(X_i|\theta_0)f''(X_i|\theta_0)}{f^2(X_i|\theta_0)},
\end{aligned}
$$

which is the sum of a set of independent and identically distributed random
variables, each with expectation

$$E\left\{\frac{[f'(X_i|\theta_0)]^2 - f(X_i|\theta_0)f''(X_i|\theta_0)}{f^2(X_i|\theta_0)}\right\} =$$

$$E\left\{\left[\frac{f'(X_i|\theta_0)}{f(X_i|\theta_0)}\right]^2\right\} - E\left[\frac{f''(X_i|\theta_0)}{f(X_i|\theta_0)}\right], \quad (10.23)$$

where Equation (10.14) implies that

$$I(\theta_0) = E\left\{\left[\frac{f'(X_i|\theta_0)}{f(X_i|\theta_0)}\right]^2\right\}.$$

To evaluate the second term of the right hand side of Equation (10.23), we
note that Equation (10.15) implies that

$$E\left[\frac{f''(X_i|\theta_0)}{f(X_i|\theta_0)}\right] = 0,$$

and hence we have that

$$E\left\{\frac{[f'(X_i|\theta_0)]^2 - f(X_i|\theta_0)f''(X_i|\theta_0)}{f^2(X_i|\theta_0)}\right\} = I(\theta_0).$$

Therefore, Theorem 3.10 (Weak Law of Large Numbers) implies that

$$n^{-1}\sum_{i=1}^{n}\frac{[f'(X_i|\theta_0)]^2 - f(X_i|\theta_0)f''(X_i|\theta_0)}{f^2(X_i|\theta_0)} \xrightarrow{p} I(\theta_0),$$

as $n \to \infty$. For the last term we note that

$$n^{-1}l'''(\theta) = n^{-1}\frac{\partial^3}{\partial\theta^3}\log\left[\prod_{i=1}^{n}f(X_i|\theta)\right]$$

$$= n^{-1}\frac{\partial^3}{\partial\theta^3}\sum_{i=1}^{n}\log[f(X_i|\theta)]$$

$$= n^{-1}\sum_{i=1}^{n}\frac{\partial^3}{\partial\theta^3}\log[f(X_i|\theta)].$$

Therefore, Assumption 6 implies that

$$|n^{-1}l'''(\theta)| = \left|n^{-1}\sum_{i=1}^{n}\frac{\partial^3}{\partial\theta^3}\log[f(X_i|\theta)]\right| \leq n^{-1}\sum_{i=1}^{n}\left|\frac{\partial^3}{\partial\theta^3}\log[f(X_i|\theta)]\right|$$

$$\leq n^{-1}\sum_{i=1}^{n}B(X_i),$$

with probability one for all $\theta \in (\theta_0 - c, \theta_0 + c)$. Now ξ_n is a random variable
that is between θ_0 and $\hat{\theta}_n$ with probability one, and Theorem 10.4 implies

that $\hat{\theta}_n \xrightarrow{p} \theta$ as $n \to \infty$. Therefore, it follows that for any $c > 0$,

$$\lim_{n \to \infty} P[\xi_n \in (\theta_0 - c, \theta_0 + c)] = 1,$$

and hence

$$\lim_{n \to \infty} P\left[|n^{-1}l'''(\xi_n)| \leq n^{-1} \sum_{i=1}^{n} B(X_i)\right] = 1.$$

Now Theorem 3.10 (Weak Law of Large Numbers) implies that

$$n^{-1} \sum_{i=1}^{n} B(X_i) \xrightarrow{p} E[B(X_1)] < \infty,$$

as $n \to \infty$. Therefore, it follows from Definition 4.3 that $|n^{-1}l'''(\xi_n)|$ is bounded in probability as $n \to \infty$ and therefore $\frac{1}{2}n^{-1}(\hat{\theta}_n - \theta_0)l'''(\xi_n) \xrightarrow{p} 0$ as $n \to \infty$. Hence, Theorem 4.11 (Slutsky) implies that $n^{1/2}(\hat{\theta}_n - \theta_0) \xrightarrow{d} I^{-1}(\theta_0)Z_0$ as $n \to \infty$ where Z_0 has a $N[0, I(\theta_0)]$ distribution, and it follows that $n^{1/2}(\hat{\theta}_n - \theta_0) \xrightarrow{d} Z$ as $n \to \infty$ where Z is a random variable with a $N[0, I^{-1}(\theta_0)]$ distribution.   □

From the assumptions of Theorem 10.5 it is evident that the asymptotic efficiency of a maximum likelihood estimator depends heavily on f, including its support and smoothness properties. The main important assumption that may not always be obvious is that the integral of the density and the derivative of the density with respect to θ may be interchanged. In this context the following result is often useful.

Theorem 10.6. *Let $f(x|\theta)$ be a function that is differentiable with respect to $\theta \in \Omega$. Suppose there exists a function $g(x|\theta)$ and a real constant $\delta > 0$ such that*

$$\int_{-\infty}^{\infty} g(x|\theta)dx < \infty,$$

for all $\theta \in \Omega$ and

$$\left|\left.\frac{\partial}{\partial\theta}f(x,\theta)\right|_{\theta=\theta_0}\right| \leq g(x,\theta)$$

for all $\theta_0 \in \Omega$ such that $|\theta_0 - \theta| \leq \delta$. Then

$$\frac{\partial}{\partial\theta}\int_{-\infty}^{\infty} f(x,\theta)dx = \int_{-\infty}^{\infty} \frac{\partial}{\partial\theta}f(x,\theta)dx.$$

The proof of Theorem 10.6 follows from Theorem 1.11 (Lesbesgue). For further details on this result see Casella and Berger (2002) or Section 7.10 of Khuri (2003). The conditions of Theorem 10.6 holds for a wide range of problems, including those that fall within the *exponential family*.

Definition 10.5. *Let X be a continuous random variable with a density of the form $f(x|\theta) = \exp[\theta T(x) - A(\theta)]$ for all $x \in \mathbb{R}$ where θ is a parameter with parameter space Ω, T is a function that does not depend on θ, and A*

is a function that does not depend on x. Then X has a density from a one parameter exponential family.

Of importance for the exponential family in relation to our current discussion is the fact that derivatives and integrals of the density can be exchanged.

Theorem 10.7. *Let h be an integrable function and let θ be a interior point of Ω, then the integral*

$$\int_{-\infty}^{\infty} h(x) \exp[\theta T(x) - A(\theta)] dx,$$

is continuous and has derivatives of all orders with respect to θ, and these derivatives can be obtained by exchanging the derivative and the integral.

A proof of Theorem 10.7 can be found in Section 7.1 of Barndorff-Nielsen (1978) or Chapter 2 of Lehmann (1986).

Corollary 10.1. *Let $X_1, \ldots, X_n$ be a set of independent and identically distributed random variables from a distribution that has density*

$$f(x|\theta) = \exp[\theta T(x) - A(\theta)],$$

where θ is a parameter with parameter space Ω that is an open interval, T is a function that does not depend on θ, and A is a function that does not depend on x. Then the likelihood function of θ has a unique solution that is consistent and asymptotically normal and efficient.

Proof. The likelihood function of θ is given by

$$L(\theta|\mathbf{X}) = \prod_{i=1}^{n} \exp[\theta T(X_i) - A(\theta)] = \exp\left[\theta \sum_{i=1}^{n} T(X_i) - nA(\theta)\right],$$

so that the log-likelihood is given by

$$l(\theta|\mathbf{X}) = \theta \sum_{i=1}^{n} T(X_i) - nA(\theta).$$

Therefore, the maximum likelihood estimator of θ is the solution to the equation

$$\sum_{i=1}^{n} T(X_i) - nA'(\theta) = 0. \tag{10.24}$$

Now, noting that

$$\int_{-\infty}^{\infty} \exp[\theta T(x) - A(\theta)] dx = 1,$$

it follows from Theorem 10.7 that

$$\frac{d}{d\theta} \int_{-\infty}^{\infty} \exp[\theta T(x) - A(\theta)] dx = \int_{-\infty}^{\infty} \frac{d}{d\theta} \exp[\theta T(x) - A(\theta)] dx =$$

$$\int_{-\infty}^{\infty} [T(x) - A'(\theta)] \exp[\theta T(x) - A(\theta)] dx = 0.$$

This implies that $E[T(X_i)] = A'(\theta)$, and hence the likelihood function in Equation (10.24) is equivalent to

$$E[T(X_i)] = n^{-1} \sum_{i=1}^{n} T(X_i). \qquad (10.25)$$

Note further that

$$\frac{d}{d\theta^2} \int_{-\infty}^{\infty} \exp[\theta T(x) - A(\theta)] dx =$$

$$\frac{d}{d\theta} \int_{-\infty}^{\infty} [T(x) - A'(\theta)] \exp[\theta T(x) - A(\theta)] dx =$$

$$\int_{-\infty}^{\infty} [T(x) - A'(\theta)]^2 \exp[\theta T(x) - A(\theta)] dx -$$

$$\int_{-\infty}^{\infty} A''(\theta) \exp[\theta T(x) - A(\theta)] dx = 0. \qquad (10.26)$$

Noting that the first integral on the right hand side of Equation (10.26) is the expectation $E\{[T(X_i) - A'(\theta)]^2\}$ and that previous arguments have shown that $E[T(X_i)] = A'(\theta)$, it follows that $E\{[T(X_i) - A'(\theta)]^2\} = V[T(X_i)] = A''(\theta)$. Since the variance must be positive, we have that

$$A''(\theta) = \frac{d}{d\theta} A'(\theta) = \frac{d}{d\theta} E[T(X_i)] > 0.$$

Hence, the right hand side of Equation (10.25) is a strictly increasing function of θ, and therefore can have at most one solution. The remainder of the proof of this result is based on verifying the assumptions of Theorem 10.5 for this case. We have already assumed that Ω is an open interval. The form of the density from Definition 10.5, along with the fact that $T(x)$ is not a function of θ and that $A(\theta)$ is not a function of x, implies that the set $\{x : f(x|\theta) > 0\}$ does not depend on θ. The first three derivatives of $f(x|\theta)$, taken with respect to θ can be shown to be continuous in θ under the assumption that $A(\theta)$ has at least three continuous derivatives. The fact that the integral of $f(x|\theta)$ taken with respect to x can be differentiated three times with respect to θ by exchanging the integral and the derivative follows from Theorem 10.7. The Fisher information number for the parameter θ for this model is given by

$$\begin{aligned}
I(\theta) &= E\left\{\left[\frac{d}{d\theta} \log[f(X_i|\theta)]\right]^2\right\} \\
&= E\left[\left\{\frac{d}{d\theta}[\theta T(X_i) - A(\theta)]\right\}^2\right] \\
&= E\{[T(X_i) - A'(\theta)]^2\} \\
&\geq 0.
\end{aligned}$$

Under the assumption that $E[T^2(X_i)] < \infty$ we obtain the required behavior for Assumption 5. For Assumption 6 we note that the third derivative of $\log[f(x|\theta)]$, taken with respect to θ, is given by

$$\frac{\partial^3}{\partial\theta^3}\log[f(x|\theta)] = -A'''(\theta),$$

where we note that a suitable constant function is given by

$$B(x) = \sup_{\theta\in(\theta_0-c,\theta_0+c)} -A'''(\theta).$$

For further details see Lemma 2.5.3 of Lehmann and Casella (1998). It follows that the assumptions of Theorem 10.5 are satisfied and the result follows. □

Example 10.15. Let $X_1,\ldots,X_n$ be a set of independent and identically distributed random variables from a density of the form

$$f(x|\theta) = \begin{cases} \theta\exp(-\theta x) & x < 0 \\ 0 & x \leq 0, \end{cases} \tag{10.27}$$

where $\theta \in \Omega = (0,\infty)$. Calculations similar to those given in Example 10.10 can be used to show that the maximum likelihood estimator of θ is $\hat{\theta}_n = \bar{X}_n^{-1}$. Alternatively, one can use the fact that maximum likelihood estimators are transformation respecting. Because the density in Equation (10.27) has the form given in Definition 10.5 with $A(\theta) = \log(\theta)$ and $T(x) = -x$, Corollary 10.1 implies that $n^{1/2}(\hat{\theta}_n - \theta) \xrightarrow{d} Z$ as $n \to \infty$ where Z is a $N(0,\sigma^2)$ random variable. The asymptotic variance σ^2 is given by

$$\begin{aligned} \sigma^2 &= I(\theta) \\ &= E\left[-\frac{\partial^2}{\partial\theta^2}f(X_i|\theta)\right] \\ &= E\left\{-\frac{\partial^2}{\partial\theta^2}[\log(\theta) - \theta X_i]\right\} \\ &= \theta^{-2}. \end{aligned}$$

Therefore, it follows that $n^{1/2}\theta^{-1}(\hat{\theta}_n-\theta)$ converges in distribution to a $N(0,1)$ distribution. ∎

The definition of an exponential family given in Definition 10.5 is somewhat simplistic and has a restrictive form. We have used this form in order to keep the presentation simple. More general results are available that can be applied to many different distributions and parameters. As with the consistency of maximum likelihood estimators, it is sometimes easier to bypass Theorem 10.5 and try to obtain the asymptotic normality of a maximum likelihood estimator directly, especially if the estimator has a closed form. In many cases, results like Theorem 6.3, and other similar results, can be used to aid in this process. Additionally, there are cases of maximum likelihood estimators that do not follow all of the assumptions of Theorem 10.5 that are still asymptotically normal.

Efficiency results for maximum likelihood estimators can be extended to more general cases in several ways. A key assumption used in this section was that the equation $L'(\theta|\mathbf{X})$ has a unique root for all $n \in \mathbb{N}$. This need not be the case in general. When there are multiple roots, consistency and asymptotic efficiency results are obtainable, though the assumptions and arguments used for establishing the results are more involved. These results can also be extended to the case where there is more than one parameter. See Sections 6.5–6.7 of Lehmann and Casella (1998) for further details on these results.

10.3 Confidence Intervals

Confidence intervals specify an estimator for a parameter θ that accounts for the inherent random error associated with a point estimate of the parameter. This is achieved by identifying a function of the observed sample data that produces an interval, or region, that contains the true parameter value with a probability that is specified prior to sampling.

Definition 10.6. *Let $X_1,\ldots,X_n$ be a set of independent and identically distributed random variables following a distribution F with parameter θ that has parameter space $\Omega \subset \mathbb{R}$. Let $\hat{\theta}_{L,n}$ and $\hat{\theta}_{U,n}$ be functions of $X_1,\ldots,X_n$ and let $\alpha \in (0,1)$. Then $C_n(\alpha) = [\hat{\theta}_{L,n},\hat{\theta}_{U,n}]$ is a $100\alpha\%$ confidence interval for θ if $P[\theta \in C_n(\alpha)] = \alpha$ for all $\theta \in \Omega$ and $n \in \mathbb{N}$.*

The value α is called the *confidence coefficient* of the interval. It is important to remember that the confidence coefficient is the probability that the interval contains the parameter based on the random mechanism associated with taking the sample. After the sample is taken we can no longer say that the parameter is contained within the interval with probability α. Rather, in the post-sample interpretation α is usually interpreted as the expected number of intervals calculated using the same method that will contain the parameter. One-sided confidence intervals can also be defined by taking $\hat{\theta}_{L,n} = -\infty$ for an upper confidence interval or $\hat{\theta}_{U,n} = \infty$ for a lower confidence interval.

The development of a confidence interval usually results from inverting a statistical hypothesis test, or from a *pivotal quantity*, which is a function of the data and the unknown parameter θ, whose distribution does not depend on θ, or on any other unknown quantities. The typical example of a pivotal quantity comes from considering $X_1,\ldots,X_n$ to be a set of independent and identically distributed random variables from a $N(\theta,\sigma^2)$ distribution where σ is known. In this case the function $\sigma^{-1}n^{1/2}(\bar{X}_n - \theta)$ has a $N(0,1)$ distribution which does not depend on θ. It is from this fact that we are able to conclude that $C_n(\alpha) = [\bar{X}_n - n^{-1/2}\sigma z_{(1+\alpha)/2}, \bar{X}_n - n^{-1/2}\sigma z_{(1-\alpha)/2}]$ is a $100\alpha\%$ confidence interval for θ. Similarly, when σ is not known, the function $\hat{\sigma}_n^{-1}n^{1/2}(\bar{X}_n - \theta)$ has a $T(n-1)$ distribution which does not depend on θ or σ. Using this pivot we obtain the usual t-interval for the mean.

In many cases pivotal functions are difficult to obtain and we may then choose to use confidence intervals that may not have a specific confidence coefficient for finite sample sizes, but may have a confidence coefficient that converges to α as $n \to \infty$. That is, we wish to specify an approximate confidence interval $C_n(\alpha)$ such that

$$\lim_{n \to \infty} P[\theta \in C_n(\alpha)] = \alpha,$$

for all $\theta \in \Omega$ and $\alpha \in (0,1)$. We can further refine such a property to reflect both the *accuracy* and the *correctness* of the confidence interval.

Definition 10.7. *Suppose that $C_n(\alpha) = [\hat{\theta}_{L,n}, \hat{\theta}_{U,n}]$ is a $100\alpha\%$ confidence interval for a parameter θ such that $P[\theta \in C_n(\alpha)] = \alpha$ for all $\theta \in \Omega$ and $n \in \mathbb{N}$. Let $D_n(\alpha) = [\tilde{\theta}_{L,n}, \tilde{\theta}_{U,n}]$ be an approximate $100\alpha\%$ confidence interval for a parameter θ such that $P[\theta \in D_n(\alpha)] \to \alpha$ as $n \to \infty$ for all $\theta \in \Omega$.*

1. *The approximate confidence interval $D_n(\alpha)$ is accurate if $P[\theta \in D_n(\alpha)] = \alpha$ for all $n \in \mathbb{N}$, $\theta \in \Omega$ and $\alpha \in (0,1)$.*

2. *The approximate confidence interval $D_n(\alpha)$ is k^{th}-order accurate if $P[\theta \in D_n(\alpha)] = \alpha + O(n^{-k/2})$ as $n \to \infty$ for all $\theta \in \Omega$ and $\alpha \in (0,1)$.*

3. *The approximate confidence interval $D_n(\alpha)$ is correct if $\tilde{\theta}_{L,n} = \hat{\theta}_{L,n}$ and $\tilde{\theta}_{U,n} = \hat{\theta}_{U,n}$ for all $n \in \mathbb{N}$, $\theta \in \Omega$ and $\alpha \in (0,1)$.*

4. *The approximate confidence interval $D_n(\alpha)$ is k^{th}-order correct if $\tilde{\theta}_{L,n} = \hat{\theta}_{L,n} + O(n^{-(k+1)/2})$ and $\tilde{\theta}_{U,n} = \hat{\theta}_{U,n} + O(n^{-(k+1)/2})$ as $n \to \infty$.*

Similar definitions can be used to define the correctness and accuracy of one-sided confidence intervals.

Example 10.16. Let $X_1, \ldots, X_n$ be a set of independent and identically distributed random variables from a distribution F with mean θ and finite variance σ^2. Even when F does not have a normal distribution, Theorem 4.20 (Lindeberg and Lévy) implies that $n^{1/2}\sigma^{-1}(\bar{X}_n - \theta) \xrightarrow{d} Z$ as $n \to \infty$, where Z has a $N(0,1)$ distribution. Therefore, an asymptotically accurate $100\alpha\%$ confidence interval for θ is given by

$$C_n(\alpha) = [\bar{X}_n - n^{-1/2}\sigma z_{(1+\alpha)/2}, \bar{X}_n - n^{-1/2}\sigma z_{(1-\alpha)/2}],$$

under the condition that σ is known. If σ is unknown then a consistent estimator of σ is given by the usual sample standard deviation $\hat{\sigma}_n$. Therefore, in this case an asymptotically accurate $100\alpha\%$ confidence interval for θ is given by

$$C_n(\alpha) = [\bar{X}_n - n^{-1/2}\hat{\sigma}_n z_{(1+\alpha)/2}, \bar{X}_n - n^{-1/2}\hat{\sigma}_n z_{(1-\alpha)/2}]. \qquad \blacksquare$$

Moving beyond the case of constructing a confidence interval for a normal mean, we first consider the case where θ is a general parameter that can be estimated using an estimator $\hat{\theta}_n$ where $n^{1/2}\sigma^{-1}(\hat{\theta}_n - \theta) \xrightarrow{d} Z$ as $n \to \infty$, and Z has a $N(0,1)$ distribution. At this time we will assume that σ is known.

In this case we can consider the approximate upper confidence limit given by
$\tilde{\theta}_{U,n} = \hat{\theta}_n - n^{-1/2}\sigma z_{1-\alpha}$. Note that

$$
\begin{aligned}
\lim_{n\to\infty} P(\theta \le \tilde{\theta}_{U,n}) &= \lim_{n\to\infty} P(\theta \le \hat{\theta}_n - n^{-1/2}\sigma z_{1-\alpha}) \\
&= \lim_{n\to\infty} P[n^{1/2}\sigma^{-1}(\hat{\theta}_n - \theta) \ge z_{1-\alpha}] \\
&= 1 - \Phi(z_{1-\alpha}) = \alpha.
\end{aligned}
$$

Hence, the upper confidence limit $\tilde{\theta}_{U,n}$ is asymptotically accurate. The lower
confidence limit $\tilde{\theta}_{L,n} = \hat{\theta}_n - n^{-1/2}\sigma z_{\alpha}$ is also asymptotically accurate since

$$
\lim_{n\to\infty} P(\theta \ge \tilde{\theta}_{L,n}) = \lim_{n\to\infty} P[n^{1/2}\sigma^{-1}(\hat{\theta}_n - \theta) \le z_{\alpha}] = \Phi(z_{\alpha}) = \alpha.
$$

For two-sided confidence intervals we can use the interval $\tilde{C}_n(\alpha) = [\hat{\theta}_n - n^{-1/2}\sigma z_{(1+\alpha)/2}, \hat{\theta}_n - n^{-1/2}\sigma z_{(1-\alpha)/2}]$, so that the asymptotic probability that
the interval will contain the true parameter value is given by

$$
\begin{aligned}
\lim_{n\to\infty} P[\theta \in \tilde{C}_n(\alpha)] &= \lim_{n\to\infty} P[n^{-1/2}\sigma z_{(1-\alpha)/2} \le \hat{\theta}_n - \theta \le n^{-1/2}\sigma z_{(1+\alpha)/2}] \\
&= \lim_{n\to\infty} P[z_{(1-\alpha)/2} \le n^{1/2}\sigma^{-1}(\hat{\theta}_n - \theta) \le z_{(1+\alpha)/2}] \\
&= \Phi(z_{(1+\alpha)/2}) - \Phi(z_{(1-\alpha)/2}) = \alpha.
\end{aligned}
$$

Therefore the two-sided interval is also asymptotically accurate.

In the case where σ is unknown we consider the upper confidence limit given
by $\bar{\theta}_{U,n} = \hat{\theta}_n - n^{-1/2}\hat{\sigma}_n z_{1-\alpha}$, where we will assume that $\hat{\sigma}_n$ is a consistent
estimator of σ. That is, we assume that $\hat{\sigma}_n \xrightarrow{p} \sigma$ as $n \to \infty$. If $n^{1/2}\sigma^{-1}(\hat{\theta}_n - \theta) \xrightarrow{d} Z$ as $n \to \infty$, then Theorem 4.11 (Slutsky) can be used to show that
$n^{1/2}\hat{\sigma}_n^{-1}(\hat{\theta}_n - \theta) \xrightarrow{d} Z$ as $n \to \infty$ as well. Therefore,

$$
\begin{aligned}
\lim_{n\to\infty} P(\theta \le \bar{\theta}_{U,n}) &= \lim_{n\to\infty} P(\theta \le \hat{\theta}_n - n^{-1/2}\hat{\sigma}_n z_{1-\alpha}) \\
&= \lim_{n\to\infty} P[n^{1/2}\hat{\sigma}_n^{-1}(\hat{\theta}_n - \theta) \ge z_{1-\alpha}] \\
&= 1 - \Phi(z_{1-\alpha}) = \alpha.
\end{aligned}
$$

Hence, the upper confidence limit $\bar{\theta}_{U,n}$ is asymptotically accurate. Similar calculations to those used above can then be used to show that the corresponding
lower confidence limit and two-sided confidence interval are also asymptotically accurate.

Example 10.17. Let $X_1, \ldots, X_n$ be a set of independent and identically
distributed random variables from a BERNOULLI(θ) distribution. Theorem
4.20 (Lindeberg and Lévy) implies that $n^{1/2}[\theta(1-\theta)]^{-1/2}(\bar{X}_n - \theta) \xrightarrow{d} Z$ as
$n \to \infty$, where Z has a N$(0,1)$ distribution. Therefore, an asymptotically
accurate $100\alpha\%$ confidence interval for θ could be thought of as

$$
[\bar{X}_n - n^{-1/2}[\theta(1-\theta)]^{1/2}z_{(1+\alpha)/2}, \bar{X}_n - n^{-1/2}[\theta(1-\theta)]^{-1/2}z_{(1-\alpha)/2}],
$$

except for the fact that the standard deviation of $\bar{X}_n$ in this case depends on θ, the unknown parameter. However, Theorem 3.10 (Weak Law of Large Numbers) implies that $\bar{X}_n \overset{p}{\to} \theta$ as $n \to \infty$ and hence Theorem 3.7 implies that $[\bar{X}_n(1 - \bar{X}_n)]^{1/2} \overset{p}{\to} [\theta(1-\theta)]^{1/2}$ as $n \to \infty$, which provides a consistent estimator of $\theta(1-\theta)$. Therefore, it follows that

$$\hat{C}_n(\alpha) = [\bar{X}_n - n^{-1/2}[\bar{X}_n(1 - \bar{X}_n)]^{1/2}z_{(1+\alpha)/2},$$
$$\bar{X}_n - n^{-1/2}[\bar{X}_n(1 - \bar{X}_n)]^{-1/2}z_{(1-\alpha)/2}],$$

is an asymptotically accurate $100\alpha\%$ confidence interval for θ. ∎

Example 10.18. Let $X_1, \ldots, X_n$ be a set of independent and identically distributed random variables from a distribution F with mean η and variance $\theta < \infty$. If F corresponds to a $\mathrm{N}(\eta, \theta)$ distribution then an exact $100\alpha\%$ confidence interval for θ is given by

$$C_n(\alpha) = [(n-1)\hat{\theta}_n[\chi^2_{n-1;(1+\alpha)/2}]^{-1}, (n-1)\hat{\theta}_n[\chi^2_{n-1;(1-\alpha)/2}]^{-1}],$$

which uses the fact that $(n-1)\theta^{-1}\hat{\theta}_n$ is a pivotal quantity for θ that has a CHISQUARED$(n-1)$ distribution, and $\hat{\theta}_n$ is the unbiased version of the sample variance. This pivotal function is only valid for the normal distribution. If F is unknown then we can use the fact that Theorem 8.5 implies that $n^{1/2}(\mu_4 - \theta^2)^{-1/2}(\hat{\theta}_n - \theta) \overset{d}{\to} Z$ as $n \to \infty$ where Z has a $\mathrm{N}(0,1)$ distribution to construct an approximate confidence interval for θ. If $E(|X_1|^4) < \infty$ then Theorems 3.21 and 3.9 imply that $\hat{\mu}_4 - \hat{\theta}_n^2 \overset{p}{\to} \mu_4 - \theta^2$ as $n \to \infty$ and an asymptotically accurate confidence interval for θ is given by

$$\hat{C}_n(\alpha) = [\hat{\theta}_n - z_{(1+\alpha)/2}n^{-1/2}(\hat{\mu}_4 - \hat{\theta}_n^2)^{1/2}, \hat{\theta}_n - z_{(1-\alpha)/2}n^{-1/2}(\hat{\mu}_4 - \hat{\theta}_n^2)^{1/2}].$$

∎

Example 10.19. Suppose $X_1, \ldots, X_n$ is a set of independent and identically distributed random variables from a POISSON(θ) distribution. Garwood (1936) suggested a $100\alpha\%$ confidence interval for θ using the form

$$C_n(\alpha) = [\tfrac{1}{2}n^{-1}\chi^2_{2Y;(1-\alpha)/2}, \tfrac{1}{2}n^{-1}\chi^2_{2(Y+1);(1+\alpha)/2}],$$

where

$$Y = n\hat{\theta}_n = \sum_{i=1}^{n} X_i.$$

The coverage probability of this interval is at least α, but may also be quite conservative in some cases. See Figure 9.2.5 of Casella and Berger (2002). An asymptotically accurate confidence interval for θ based on Theorem 4.20 (Lindeberg and Lévy) and Theorem 3.10 (Weak Law of Large Numbers) is given by

$$\hat{C}_n(\alpha) = [\hat{\theta}_n - z_{(1+\alpha)/2}n^{-1/2}\hat{\theta}_n^{1/2}, \hat{\theta}_n - z_{(1-\alpha)/2}n^{-1/2}\hat{\theta}_n^{1/2}].$$

∎

More accurate asymptotic properties, such as the order of correctness and accuracy, can be obtained if we assume the framework of the smooth function model introduced in Section 7.4. That is, consider a sequence of independent and identically distributed d-dimensional random vectors $\{\mathbf{X}_n\}_{n=1}^{\infty}$ from a d-dimensional distribution F. Let $\boldsymbol{\mu} = E(\mathbf{X}_n)$ and assume that the components of $\boldsymbol{\mu}$ are finite and that there exists a smooth function $g : \mathbb{R}^d \to \mathbb{R}$ such that $\theta = g(\boldsymbol{\mu})$ and we estimate θ with $\hat{\theta}_n = g(\hat{\boldsymbol{\mu}})$. Finally, assume that there is a smooth function $h : \mathbb{R}^d \to \mathbb{R}$ such that

$$\sigma^2 = h^2(\boldsymbol{\mu}) = \lim_{n\to\infty} V(n^{1/2}\hat{\theta}_n).$$

If required, the asymptotic variance will be estimated with $\hat{\sigma}_n^2 = h^2(\bar{\mathbf{X}}_n)$. Let $G_n(t) = P[n^{1/2}\sigma^{-1}(\hat{\theta}_n - \theta) \leq t]$ and $H_n(t) = P[n^{1/2}\hat{\sigma}_n^{-1}(\hat{\theta}_n - \theta) \leq t]$ and define $g_{\alpha,n}$ and $h_{\alpha,n}$ to be the corresponding α quantiles of G_n and H_n so that $G_n(g_{\alpha,n}) = \alpha$ and $H_n(h_{\alpha,n}) = \alpha$.

In the same exact way that one would develop the confidence interval for a population mean we can develop a confidence interval for θ using the quantiles of G_n and H_n. In particular, if σ is known then it follows that a $100\alpha\%$ upper confidence limit for θ is given by $\hat{\theta}_{n,\text{ord}} = \hat{\theta}_n - n^{-1/2}\sigma g_{1-\alpha}$ and if σ is unknown then it follows that a $100\alpha\%$ upper confidence limit for θ is given by $\hat{\theta}_{n,\text{stud}} = \hat{\theta}_n - n^{-1/2}\hat{\sigma}_n h_{1-\alpha}$. In this case we are borrowing the notation and terminology of Hall (1988a) where $\hat{\theta}_{n,\text{ord}}$ is called the *ordinary* confidence limit and $\hat{\theta}_{n,\text{stud}}$ is called the *studentized* confidence limit, making reference to the t-interval for the mean where the population standard deviation is replaced by the sample standard deviation. In both cases these upper confidence limits are accurate and correct. See Exercise 18.

Note that in the case where F is a $N(\theta, \sigma^2)$ distribution, the distribution H_n is a $N(0,1)$ distribution and G_n is a $T(n-1)$ distribution as discussed above. When F is not a normal distribution, but θ still is contained within the smooth function model, it is often the case that the distributions G_n and H_n are unknown, complicated, or may depend on unknown parameters. In these cases the normal approximation, motivated by the fact that $G_n \rightsquigarrow \Phi$ and $H_n \rightsquigarrow \Phi$ as $n \to \infty$, is often used. The normal approximation replaces the quantiles g_α and h_α with z_α, obtaining an approximate upper $100\alpha\%$ confidence limits of the form $\tilde{\theta}_{n,\text{ord}} = \hat{\theta}_n - n^{-1/2}\sigma z_{1-\alpha}$ if σ is known, and $\tilde{\theta}_{n,\text{stud}} = \hat{\theta}_n - n^{-1/2}\hat{\sigma}_n z_{1-\alpha}$ if σ is unknown. The accuracy and correctness of these confidence limits can be studied with the aid of Edgeworth expansions.

Example 10.20. Suppose that $X_1, \ldots, X_n$ is a set of independent and identically distributed random variables from a distribution F with parameter θ. Suppose that F and θ fall within the smooth function model described above. A correct and exact $100\alpha\%$ upper confidence limit for θ is given by $\hat{\theta}_{n,\text{stud}} = \hat{\theta}_n - n^{-1/2}\hat{\sigma}_n h_{1-\alpha}$. According to Theorem 7.13, the quantile $h_{1-\alpha}$ has an asymptotic expansion of the form $h_{1-\alpha} = z_{1-\alpha} + n^{-1/2}s_1(z_{1-\alpha}) + n^{-1}s_2(z_{1-\alpha}) + O(n^{-3/2})$, as $n \to \infty$. Therefore, it follows that the exact

$100\alpha\%$ upper confidence limit for θ has asymptotic expansion

$$
\begin{aligned}
\hat{\theta}_{n,\text{stud}} &= \hat{\theta}_n - n^{-1/2}\hat{\sigma}_n h_{n,1-\alpha} \\
&= \hat{\theta}_n - n^{-1/2}\hat{\sigma}_n[z_{1-\alpha} + n^{-1/2}s_1(z_{1-\alpha}) + n^{-1}s_2(z_{1-\alpha})] + O(n^{-2}) \\
&= \hat{\theta}_n - n^{-1/2}\hat{\sigma}_n z_{1-\alpha} - n^{-1}\hat{\sigma}_n s_1(z_{1-\alpha}) - n^{-3/2}\hat{\sigma}_n s_2(z_{1-\alpha}) \\
&\quad + O_p(n^{-2}) \\
&= \tilde{\theta}_{n,\text{stud}} + O_p(n^{-1}),
\end{aligned}
$$

as $n \to \infty$, where we have used the fact that $\hat{\sigma}_n = \sigma + O_p(n^{-1/2})$ as $n \to \infty$. From this result we find that $|\hat{\theta}_{n,\text{stud}} - \tilde{\theta}_{n,\text{stud}}| = O_p(n^{-1})$ as $n \to \infty$, so that the normal approximation is first-order correct. ∎

Example 10.21. Consider the same setup as Example 10.20 where an exact and correct upper confidence limit for θ has asymptotic expansion

$$
\hat{\theta}_{n,\text{stud}} = \hat{\theta}_n - n^{-1/2}\hat{\sigma}_n z_{1-\alpha} - n^{-1}\hat{\sigma}_n s_1(z_{1-\alpha}) + O_p(n^{-3/2}),
$$

as $n \to \infty$. The polynomial s_1 depends on the moments of F that are usually unknown but can be estimated using the corresponding sample moments. The resulting estimate of $s_1(z_{1-\alpha})$, denoted by $\hat{s}_1(z_{1-\alpha})$, has the property that $\hat{s}_1(z_{1-\alpha}) = s_1(z_{1-\alpha}) + O_p(n^{-1/2})$, as $n \to \infty$. Therefore, we can consider the Edgeworth corrected version of the normal approximation given by the upper confidence limit

$$
\begin{aligned}
\bar{\theta}_{n,\text{stud}} &= \hat{\theta}_n - n^{-1/2}\hat{\sigma}_n z_{1-\alpha} - n^{-1}\hat{\sigma}_n \hat{s}_1(z_{1-\alpha}) \\
&= \hat{\theta}_n - n^{-1/2}\hat{\sigma}_n z_{1-\alpha} - n^{-1}\hat{\sigma}_n s_1(z_{1-\alpha}) + O_p(n^{-3/2}) \\
&= \hat{\theta}_{n,\text{stud}} + O_p(n^{-3/2}),
\end{aligned}
$$

as $n \to \infty$. Therefore, it follows that $\bar{\theta}_{n,\text{stud}}$ is second-order correct. ∎

In order to use Edgeworth expansions to study the accuracy of confidence intervals, let $\hat{\theta}_n(\alpha)$ denote a generic upper $100\alpha\%$ confidence limit for θ that has an asymptotic expansion of the form

$$
\hat{\theta}_n(\alpha) = \hat{\theta}_n + n^{-1/2}\hat{\sigma}_n z_\alpha + n^{-1}\hat{\sigma}_n \hat{u}_1(z_\alpha) + n^{-3/2}\hat{\sigma}_n \hat{u}_2(z_\alpha) + O_p(n^{-2}), \quad (10.28)
$$

as $n \to \infty$, where $\hat{u}_1(z_\alpha) = u_1(z_\alpha) + O_p(n^{-1/2})$ for an even polynomial u_1 and $\hat{u}_2(z_\alpha) = u_2(z_\alpha) + O_p(n^{-1/2})$ for an odd polynomial u_2. Following the development of Hall (1988a), the $100\alpha\%$ upper confidence interval for θ given by $C_n(\alpha) = (-\infty, \hat{\theta}_n(\alpha)]$ has coverage probability given by

$$
\begin{aligned}
\pi_n(\alpha) &= P[\theta \leq \hat{\theta}_n(\alpha)] \\
&= P[\theta \leq \hat{\theta}_n + n^{-1/2}\hat{\sigma}_n z_\alpha + n^{-1}\hat{\sigma}_n \hat{u}_1(z_\alpha) + n^{-3/2}\hat{\sigma}_n \hat{u}_2(z_\alpha) \\
&\quad + O_p[(n^{-2})] \\
&= P[n^{1/2}\hat{\sigma}_n^{-1}(\theta - \hat{\theta}_n) \leq z_\alpha + n^{-1/2}\hat{u}_1(z_\alpha) + n^{-1}\hat{u}_2(z_\alpha) \\
&\quad + O_p(n^{-3/2})] \\
&= P[n^{1/2}\hat{\sigma}_n^{-1}(\hat{\theta}_n - \theta) \geq -z_\alpha - n^{-1/2}\hat{u}_1(z_\alpha) - n^{-1}u_2(z_\alpha) \\
&\quad + O_p(n^{-3/2})]
\end{aligned}
\tag{10.29}
$$

where we have used the fact that $\hat{u}_2(z_\alpha) = u_2(z_\alpha) + O_p(n^{-1/2})$, as $n \to \infty$. We still have to contend with two random terms on the right hand side of the probability in Equation (10.29). Subtracting $n^{-1/2}u_1(z_\alpha)$ from both sides of the inequality yields

$$
\pi_n(\alpha) = P\{n^{1/2}\hat{\sigma}_n^{-1}(\hat{\theta}_n - \theta) + n^{-1/2}[\hat{u}_1(z_\alpha) - u_1(z_\alpha)] \geq \\
- z_\alpha - n^{-1/2}u_1(z_\alpha) - n^{-1}u_2(z_\alpha) + O_p(n^{-3/2})\}. \tag{10.30}
$$

The simplification of this probability is now taken in two steps. The first result accounts for how the error term of $O_p(n^{-3/2})$ in the event contributes to the corresponding probability.

Theorem 10.8 (Hall). *Under the assumptions of this section it follows that*

$$
\begin{aligned}
P\{n^{1/2}&\hat{\sigma}_n^{-1}(\hat{\theta}_n - \theta) + n^{-1/2}[\hat{u}_1(z_\alpha) - u_1(z_\alpha)] \geq \\
&- z_\alpha - n^{-1/2}u_1(z_\alpha) - n^{-1}u_2(z_\alpha) + O_p(n^{-3/2})\} = \\
&P\{n^{1/2}\hat{\sigma}_n^{-1}(\hat{\theta}_n - \theta) + n^{-1/2}[\hat{u}_1(z_\alpha) - u_1(z_\alpha)] \geq \\
&- z_\alpha - n^{-1/2}u_1(z_\alpha) - n^{-1}u_2(z_\alpha)\} + O_p(n^{-3/2}),
\end{aligned}
$$

as $n \to \infty$.

See Hall (1986a) for further details on the exact assumptions required for this result and its proof. The second step of the simplification of the probability involves relating the distribution function of $n^{1/2}\hat{\sigma}_n^{-1}(\hat{\theta}_n - \theta)$, whose expansion we know from Theorem 7.13, to the distribution function of $n^{1/2}\hat{\sigma}_n^{-1}(\hat{\theta}_n - \theta) + n^{-1/2}[\hat{u}_1(z_\alpha) - u_1(z_\alpha)]$, whose expansion we are not yet familiar with.

Theorem 10.9 (Hall). *Under the assumptions of this section it follows that for every $x \in \mathbb{R}$,*

$$
\begin{aligned}
P\{n^{1/2}\hat{\sigma}_n^{-1}(\hat{\theta}_n - \theta) &+ n^{-1/2}[\hat{u}_1(z_\alpha) - u_1(z_\alpha)] \leq x\} = \\
&P[n^{1/2}\hat{\sigma}_n^{-1}(\hat{\theta}_n - \theta) \leq x] - n^{-1}u_\alpha x\phi(x) + O(n^{-3/2}),
\end{aligned}
$$

as $n \to \infty$, where u_α is a constant satisfying

$$
E\{n\hat{\sigma}_n^{-1}(\hat{\theta}_n - \theta)[\hat{u}_1(z_\alpha) - u_1(z_\alpha)]\} = u_\alpha + O(n^{-1}), \tag{10.31}
$$

as $n \to \infty$.

See Hall (1986a) for additional details on the exact assumptions required for this result and its proof. The Edgeworth expansion from Theorem 7.13 can be used to find an expression for $P[n^{1/2}\hat{\sigma}_n^{-1}(\hat{\theta}_n - \theta) \leq x]$ and the result of Theorem 10.8 can then be used to relate this to the asymptotic coverage of the upper confidence limit. Picking up where we left off with Equation (10.30), Theorem 10.8 tells us that the error term of order $O_p(n^{-3/2})$ in the event contributes $O(n^{-3/2})$ to the corresponding probability. Therefore, the result of Theorem 10.9 yields

$$\pi(\alpha) = P\{n^{1/2}\hat{\sigma}_n^{-1}(\hat{\theta}_n - \theta) + n^{-1/2}[\hat{u}_1(z_\alpha) - u_1(z_\alpha)] \geq$$
$$- z_\alpha - n^{-1/2}u_1(z_\alpha) - n^{-1}u_2(z_\alpha)\} + O(n^{-3/2}) =$$
$$1 - P[n^{1/2}\hat{\sigma}_n^{-1}(\hat{\theta}_n - \theta) \leq -z_\alpha - n^{-1/2}u_1(z_\alpha) - n^{-1}u_2(z_\alpha)] +$$
$$n^{-1}u_\alpha[-z_\alpha - n^{-1/2}u_1(z_\alpha) - n^{-1}u_2(z_\alpha)] \times$$
$$\phi[-z_\alpha - n^{-1/2}u_1(z_\alpha) - n^{-1}u_2(z_\alpha)] + O(n^{-3/2}), \quad (10.32)$$

as $n \to \infty$ where u_α is defined in Equation (10.31). The remainder of this argument is now concerned with simplifying the expressions in Equation (10.32).

We begin by noting that Theorem 7.13 implies that

$$P[n^{1/2}\hat{\sigma}_n^{-1}(\hat{\theta}_n - \theta) \leq x] = \Phi(x) + n^{-1/2}v_1(x)\phi(x) + n^{-1}v_2(x)\phi(x) + O(n^{-3/2}),$$

as $n \to \infty$. Therefore, it follows that

$$P[n^{1/2}\hat{\sigma}_n^{-1}(\hat{\theta}_n - \theta) \leq -z_\alpha - n^{-1/2}u_1(z_\alpha) - n^{-1}u_2(z_\alpha)] =$$
$$\Phi[-z_\alpha - n^{-1/2}u_1(z_\alpha) - n^{-1}u_2(z_\alpha)] + n^{-1/2}v_1[-z_\alpha - n^{-1/2}u_1(z_\alpha) - n^{-1}u_2(z_\alpha)] \times$$
$$\phi[-z_\alpha - n^{-1/2}u_1(z_\alpha) - n^{-1}u_2(z_\alpha)] + n^{-1}v_2[-z_\alpha - n^{-1/2}u_1(z_\alpha) - n^{-1}u_2(z_\alpha)] \times$$
$$\phi[-z_\alpha - n^{-1/2}u_1(z_\alpha) - n^{-1}u_2(z_\alpha)] + O(n^{-3/2}), \quad (10.33)$$

as $n \to \infty$. We must now simplify the terms in Equation (10.33). From Example 1.18 we have that the first term in Equation (10.33) has the form

$$\Phi(-z_\alpha) - [n^{-1/2}u_1(z_\alpha) + n^{-1}u_2(z_\alpha)]\phi(-z_\alpha) +$$
$$\tfrac{1}{2}[n^{-1/2}u_1(z_\alpha) + n^{-1}u_2(z_\alpha)]^2\phi'(-z_\alpha) + O(n^{-3/2}) =$$
$$1 - \alpha - n^{-1/2}u_1(z_\alpha)\phi(z_\alpha) - n^{-1}u_2(z_\alpha)\phi(z_\alpha) +$$
$$\tfrac{1}{2}n^{-1}u_1^2(z_\alpha)\phi'(-z_\alpha) + O(n^{-3/2}), \quad (10.34)$$

as $n \to \infty$. Recalling that $\phi'(t) = -t\phi(t)$ it follows that the expression in Equation (10.34) is equivalent to

$$1 - \alpha - n^{-1/2}u_1(z_\alpha)\phi(z_\alpha) - n^{-1}u_2(z_\alpha)\phi(z_\alpha) +$$
$$\tfrac{1}{2}n^{-1}z_\alpha u_1^2(z_\alpha)\phi(z_\alpha) + O(n^{-3/2}), \quad (10.35)$$

as $n \to \infty$. For the second term in Equation (10.33) we can either use the exact form of the polynomial v_1 as specified in Section 7.6, or we can simply use Theorem 1.15 to conclude that $v_1(t+\delta) = v_1(t)+\delta v_1'(t)+O(\delta^2)$, as $\delta \to 0$. Therefore, keeping terms of order $O(n^{-1/2})$ or larger, yields

$$
\begin{aligned}
v_1[-z_\alpha - n^{-1/2}u_1(z_\alpha) - n^{-1}u_2(z_\alpha)] = \\
v_1(-z_\alpha) - [n^{-1/2}u_1(z_\alpha) + n^{-1}u_2(z_\alpha)]v_1'(-z_\alpha) + O(n^{-1}) = \\
v_1(-z_\alpha) - n^{-1/2}u_1(z_\alpha)v_1'(-z_\alpha) + O(n^{-1}), \quad (10.36)
\end{aligned}
$$

as $n \to \infty$. Now v_1 is an even function, and hence v_1' is an odd function. Therefore, the expression in Equation (10.36) equals

$$
v_1(z_\alpha) + n^{-1/2}u_1(z_\alpha)v_1'(z_\alpha) + O(n^{-1}), \qquad (10.37)
$$

as $n \to \infty$. Working further with the second term in Equation (10.33), we note that Theorem 1.15 implies that

$$
\begin{aligned}
\phi[-z_\alpha - n^{-1/2}u_1(z_\alpha) - n^{-1}u_2(z_\alpha)] = \\
\phi(-z_\alpha) - [n^{-1/2}u_1(z_\alpha) + n^{-1}u_2(z_\alpha)]\phi'(-z_\alpha) + O(n^{-1}) = \\
\phi(z_\alpha) - n^{-1/2}z_\alpha u_1(z_\alpha)\phi(z_\alpha) + O(n^{-1}), \quad (10.38)
\end{aligned}
$$

as $n \to \infty$. Combining the expressions of Equations (10.37) and (10.38) yields

$$
\begin{aligned}
n^{-1/2}v_1[-z_\alpha-n^{-1/2}u_1(z_\alpha)-n^{-1}u_2(z_\alpha)]\phi[-z_\alpha-n^{-1/2}u_1(z_\alpha)-n^{-1}u_2(z_\alpha)] = \\
n^{-1/2}v_1(z_\alpha)\phi(z_\alpha) + n^{-1}u_1(z_\alpha)v_1'(z_\alpha)\phi(z_\alpha)- \\
n^{-1}z_\alpha u_1(z_\alpha)v_1(z_\alpha)\phi(z_\alpha) + O(n^{-3/2}), \quad (10.39)
\end{aligned}
$$

as $n \to \infty$. The third term in Equation (10.33) requires less sophisticated arguments as we are only retaining terms of order $O(n^{-1})$ and larger, and the leading coefficient on this term is $O(n^{-1})$. Hence, similar calculations to those used above can be used to show that

$$
\begin{aligned}
n^{-1}v_2[-z_\alpha - n^{-1/2}u_1(z_\alpha) - n^{-1}u_2(z_\alpha)] = \\
-n^{-1}v_2(z_\alpha)\phi(z_\alpha) + O(n^{-3/2}), \quad (10.40)
\end{aligned}
$$

as $n \to \infty$, where we have used the fact that v_2 is an odd function. Combining the results of Equations (10.33), (10.35), (10.39), and (10.40) yields

$$
\begin{aligned}
P[n^{1/2}\hat{\sigma}_n^{-1}(\hat{\theta}_n - \theta) \le -z_\alpha - n^{-1/2}u_1(z_\alpha) - n^{-1}u_2(z_\alpha)] = \\
1 - \alpha - n^{-1/2}u_1(z_\alpha)\phi(z_\alpha) - n^{-1}u_2(z_\alpha)\phi(z_\alpha) + \tfrac{1}{2}n^{-1}z_\alpha u_1^2(z_\alpha)\phi(z_\alpha)+ \\
n^{-1/2}v_1(z_\alpha)\phi(z_\alpha) + n^{-1}u_1(z_\alpha)v_1'(z_\alpha)\phi(z_\alpha) - n^{-1}z_\alpha u_1(z_\alpha)v_1(z_\alpha)\phi(z_\alpha)- \\
n^{-1}v_2(z_\alpha)\phi(z_\alpha) + O(n^{-3/2}) = \\
1 - \alpha + n^{-1/2}[v_1(z_\alpha) - u_1(z_\alpha)]\phi(z_\alpha)+ \\
n^{-1}[\tfrac{1}{2}z_\alpha u_1^2(z_\alpha) - u_2(z_\alpha) + u_1(z_\alpha)v_1'(z_\alpha) - z_\alpha u_1(z_\alpha)v_1(z_\alpha) - v_2(z_\alpha)]\phi(z_\alpha) \\
+ O(n^{-3/2}),
\end{aligned}
$$

as $n \to \infty$. To complete simplifying the expression in Equation (10.32), we note that

$$n^{-1}u_\alpha[-z_\alpha-n^{-1/2}u_1(z_\alpha)-n^{-1}u_2(z_\alpha)]\phi[-z_\alpha-n^{-1/2}u_1(z_\alpha)-n^{-1}u_2(z_\alpha)] =$$
$$-n^{-1}u_\alpha z_\alpha\phi(z_\alpha)+O(n^{-3/2}),$$

as $n \to \infty$. Therefore, it follows that

$$\pi_n(\alpha) = \alpha + n^{-1/2}[u_1(z_\alpha) - v_1(z_\alpha)]\phi(z_\alpha) - n^{-1}[\tfrac{1}{2}z_\alpha u_1^2(z_\alpha) - u_2(z_\alpha)+$$
$$u_1(z_\alpha)v_1'(z_\alpha) - z_\alpha u_1(z_\alpha)v_1(z_\alpha) - v_2(z_\alpha) + u_\alpha z_\alpha]\phi(z_\alpha) + O(n^{-3/2}), \quad (10.41)$$

as $n \to \infty$. From Equation (10.41) it is clear that the determining factor in the accuracy of the one-sided confidence interval is the relationship between the polynomials u_1 and v_1. In particular, if $u_1(z_\alpha) = -s_1(z_\alpha) = v_1(z_\alpha)$ then the expansion for the coverage probability simplifies to

$$\pi_n(\alpha) = \alpha - n^{-1}[\tfrac{1}{2}z_\alpha v_1^2(z_\alpha) - u_2(z_\alpha) + v_1(z_\alpha)v_1'(z_\alpha)-$$
$$v_2(z_\alpha) + u_\alpha z_\alpha]\phi(z_\alpha) + O(n^{-3/2}),$$

as $n \to \infty$ and the resulting interval is second-order accurate. On the other hand, if $u_1(z_\alpha) \neq v_1(z_\alpha)$, then the interval is only first-order accurate.

Example 10.22. Consider the general setup of Example 10.20 with

$$\tilde{\theta}_{n,\text{stud}}(\alpha) = \hat{\theta}_n = n^{-1/2}\hat{\sigma}_n z_{1-\alpha},$$

denoting the upper confidence limit given by the normal approximation. In terms of the generic expansion given in Equation (10.28) we have that $u_1(z_\alpha) = u_2(z_\alpha) = 0$ for all $\alpha \in (0,1)$. Therefore, Equation (10.41) implies that the coverage probability of $\tilde{\theta}_{n,\text{stud}}(\alpha)$ has asymptotic expansion

$$\tilde{\pi}_n(\alpha) = \alpha - n^{-1/2}v_1(z_\alpha)\phi(z_\alpha) + n^{-1}[v_2(z_\alpha) - u_\alpha z_\alpha]\phi(z_\alpha) + O(n^{-3/2}),$$

as $n \to \infty$. Hence, the normal approximation is first-order accurate. ∎

Example 10.23. Consider the general setup of Example 10.20 with Edgeworth corrected upper confidence limit given by

$$\begin{aligned}\bar{\theta}_{n,\text{stud}}(\alpha) &= \hat{\theta}_n - n^{-1/2}\hat{\sigma}_n z_{1-\alpha} - n^{-1}\hat{\sigma}_n\hat{s}_1(z_{1-\alpha})\\ &= \hat{\theta}_n + n^{-1/2}\hat{\sigma}_n z_\alpha - n^{-1}\hat{\sigma}_n\hat{s}_1(z_\alpha).\end{aligned}$$

In terms of the generic expansion given in Equation (10.28) we have that $u_1(z_\alpha) = -\hat{s}_1(z_\alpha)$. Therefore, Equation (10.41) implies that the coverage probability of $\bar{\theta}_{n,\text{stud}}(\alpha)$ has asymptotic expansion $\alpha + O(n^{-1})$, as $n \to \infty$. Therefore, the Edgeworth corrected upper confidence limit for θ is second-order accurate. ∎

A two-sided $100\alpha\%$ confidence interval for θ is given by

$$[\hat{\theta}_n - n^{-1/2}\hat{\sigma}_n h_{(1+\alpha)/2}, \hat{\theta}_n - n^{-1/2}\hat{\sigma}_n h_{(1-\alpha)/2}], \quad (10.42)$$

which has coverage probability $\pi_n[\tfrac{1}{2}(1+\alpha)/2] - \pi_n[\tfrac{1}{2}(1-\alpha)/2]$ where we are using the definition of $\pi_n(\alpha)$ from earlier. Therefore, the coverage of the two-sided interval in Equation (10.42) is given by Equation (10.41) as

$$\tfrac{1}{2}(1+\alpha) - \tfrac{1}{2}(1-\alpha) + n^{-1/2}[u_1(z_{(1+\alpha)/2}) - v_1(z_{(1+\alpha)/2})] -$$
$$n^{-1/2}[u_1(z_{(1-\alpha)/2}) - v_1(z_{(1-\alpha)/2})] + O(n^{-1}),$$

as $n \to \infty$. Noting that $z_{(1+\alpha)/2} = -z_{(1-\alpha)/2}$ for all $\alpha \in (0,1)$ and that v_1 and u_1 are even functions, it follows that $u_1(z_{(1+\alpha)/2}) = u_1(z_{(1-\alpha)/2})$ and $v_1(z_{(1+\alpha)/2}) = v_1(z_{(1-\alpha)/2})$ so that the coverage of the two-sided interval is given by $\alpha + O(n^{-1})$, as $n \to \infty$ regardless of u_1. Therefore, within the smooth function model, second-order correctness has a major influence on the coverage probability of one-sided confidence intervals, but its influence on two-sided confidence intervals is greatly diminished.

The view of the accuracy and correctness of confidence intervals based on Edgeworth expansion theory is useful, but also somewhat restrictive. This is due to the fact that we require the added structure of the smooth function model to obtain the required asymptotic expansions. For example, these results cannot be used to find the order accuracy of population quantiles based on the quantiles of the empirical distribution function. This is because these quantile estimates do not fit within the framework of the smooth function model. The related expansion theory for sample quantiles turns out to be quite different from what is presented here. See Appendix IV of Hall (1992) for further details on this problem. Another limitation comes from the smoothness assumptions that must be made on the distribution F. Therefore, the theory presented above is not able to provide further details on the order of correctness or accuracy of the approximate confidence intervals studied in Examples 10.17 and 10.19. Esseen (1945) provides Edgeworth expansions for the case of lattice random variables, and again the resulting expansions are quite different from those presented here. Another interesting case which we have not studied in this section is the case where θ is a p-dimensional vector, and we seek a region $\mathbf{C}_n(\alpha)$ such that $P[\theta \in \mathbf{C}_n(\alpha)] = \alpha$. In this case there is a multivariate form of the smooth function model and well defined Edgeworth expansion theory that closely follows the general form of what is presented here, though there are some notable differences. See Section 4.2 of Hall (1992) and Chapter 3 of Polansky (2007) for further details. A very general view of coverage processes, well beyond that of just confidence intervals and regions, can be found in Hall (1988b).

10.4 Statistical Hypothesis Tests

Statistical hypothesis tests are procedures that decide the truth of a hypothesis about an unknown population parameter based on a sample from the population. The decision is usually made in accordance with a known rate of error specified by the researcher.

For this section we consider $X_1, \ldots, X_n$ to be a set of independent and identically distributed random variables from a distribution F with functional parameter θ that has parameter space Ω. A hypothesis test begins by dividing the parameter space into a partition of two regions called Ω_0 and Ω_1 for which there are associated hypotheses $H_0 : \theta \in \Omega_0$ and $H_1 : \theta \in \Omega_1$, respectively. The structure of the test is such that the hypothesis $H_0 : \theta \in \Omega_0$, called the *null hypothesis*, is initially assumed to be true. The data are observed and evidence that $H_0 : \theta \in \Omega_0$ is actually false is extracted from the data. If the amount of evidence against $H_0 : \theta \in \Omega_0$ is large enough, as specified by an acceptable error rate given by the researcher, then the null hypothesis $H_0 : \theta \in \Omega_0$ is rejected as being false, and the hypothesis $H_1 : \theta \in \Omega_1$, called the *alternative hypothesis*, is accepted as truth. Otherwise we fail to reject the null hypothesis and we conclude that there was not sufficient evidence in the observed data to conclude that the null hypothesis is false.

The measure of evidence in the observed data against the null hypothesis is measured by a statistic $T_n = T_n(X_1, \ldots, X_n)$, called a *test statistic*. Prior to observing the sample $X_1, \ldots, X_n$ the researcher specifies a set R that is a subset of the range of T_n. This set is constructed so that when $T_n \in R$, the researcher considers the evidence in the sample to be sufficient to warrant rejection of the null hypothesis. That is, if $T_n \in R$, then the null hypothesis is rejected and if $T_n \notin R$ then the null hypothesis is not rejected. The set R is called the *rejection region*.

The rejection region is usually constructed so that the probability of rejecting the null hypothesis when the null hypothesis is really true is set to a level α called the *significance level*. That is, $\alpha = P(T_n \in R | \theta \in \Omega_0)$. The error corresponding to this conclusion is called the *Type I error*, and it should be noted that the probability of this error may depend on the specific value of θ in Ω_0. In this case we control the largest probability of a Type I error for points in Ω_0 to be α with a slight difference in terminology separating the cases where the probability α can be achieved and where it cannot. Another terminology will be used when the error rate asymptotically achieves the probability α.

Definition 10.8. *Consider a test of a null hypothesis $H_0 : \theta \in \Omega_0$ with test statistic T_n and rejection region R.*

1. *The test is a size α test if*
$$\sup_{\theta_0 \in \Omega_0} P(T_n \in R | \theta = \theta_0) = \alpha.$$

2. *The test is a level α test if*
$$\sup_{\theta_0 \in \Omega_0} P(T_n \in R | \theta = \theta_0) \leq \alpha.$$

3. *The test has asymptotic size equal to α if*
$$\lim_{n \to \infty} P(T_n \in R | \theta = \theta_0) = \alpha.$$

4. *The test is k^{th}-order accurate if*

$$P(T_n \in R|\theta = \theta_0) = \alpha + O(n^{-k/2}),$$

as $n \to \infty$.

The other type of error that one can make is a *Type II error*, which occurs when the null hypothesis is not rejected but the alternative hypothesis is actually true. The probability of avoiding this error is called the *power* of the test. This probability, taken as a function of θ, is called the *power function* of the test and will be denoted as $\beta_n(\theta)$. That is, $\beta_n(\theta) = P(T_n \in R|\theta)$. The domain of the power function can be taken to be Ω so that the function $\beta_n(\theta)$ will also reflect the probability of a Type I error when $\theta \in \Omega_0$. In this context we would usually want $\beta_n(\theta)$ when $\theta \in \Omega_0$ to be smaller than $\beta_n(\theta)$ when $\theta \in \Omega_1$. That is, we should have a larger probability of rejecting the null hypothesis when the alternative is true than when the null is true. A test that has this property is called an *unbiased* test.

Definition 10.9. *A test with power function $\beta_n(\theta)$ is unbiased if $\beta_n(\theta_0) \leq \beta_n(\theta_1)$ for all $\theta_0 \in \Omega_0$ and $\theta_1 \in \Omega_1$.*

While unbiased tests are important, there are also asymptotic considerations that can be accounted for as well. The most common of these is that a test should be consistent against values of θ in the alternative hypothesis. This is an extension of the idea of consistency in the case of point estimation. In point estimation we like to have consistent estimators, that is ones that converge to the correct value of θ as $n \to \infty$. The justification for requiring this property is that if we could examine the entire population then we should know the parameter value exactly. The consistency concept is extended to statistical hypothesis tests by requiring that if we could examine the entire population, then we should be able to make a correct decision. That is, if the alternative hypothesis is true, then we should reject with probability one as $n \to \infty$. Note that this behavior is only specified for points in the alternative hypothesis since we always insist on erroneously rejecting the null hypothesis with probability at most α no matter what the sample size is.

Definition 10.10. *Consider a test of $H_0 : \theta \in \Omega_0$ against $H_1 : \theta \in \Omega_1$ that has power function $\beta_n(\theta)$. The test is consistent against the alternative $\theta \in \Omega_1$ if*

$$\lim_{n \to \infty} \beta_n(\theta) = 1.$$

If the test is consistent against all alternatives $\theta \in \Omega_1$, then the test is called a consistent test.

A consistent hypothesis test assures us that when the sample is large enough that we will reject the null hypothesis when the alternative hypothesis is true. For fixed sample sizes the probability of rejecting the null hypothesis when the alternative is true depends on the actual value of θ in the alternative hypothesis. Many tests will perform well when the actual value of θ is far away from the null hypothesis, and in the limiting case have a power equal to one in the limit. That is

$$\lim_{d(\theta,\Omega_0)\to\infty} \beta_n(\theta) = 1,$$

where d is a measure of distance between θ and the set Ω_0. Of greater interest then is how the tests will perform for values of θ that are in the alternative, but are close to the boundary between Ω_0 and Ω_1. One way to study this behavior is to consider a sequence of points in the alternative hypothesis that depend on n that converge to a point on the boundary of Ω_0. The rate at which the points converge to the boundary must be chosen carefully, otherwise the limit will be either α, if the sequence converges too quickly to the boundary, or will be one if the sequence does not converge fast enough. For many tests, choosing this sequence so that $d(\theta,\Omega_0) = O(n^{-1/2})$ as $n \to \infty$ will ensure that the resulting limit of the power function evaluated on this sequence will be between α and one. Such a limit is called the *asymptotic power* of the test against the specified sequence of points in the alternative hypothesis.

Definition 10.11. *Consider a test of $H_0 : \theta \in \Omega_0$ against $H_1 : \theta \in \Omega_1$ that has power function $\beta_n(\theta)$. Let $\{\theta_n\}_{n=1}^{\infty}$ be a sequence of points in Ω_1 such that*

$$\lim_{n\to\infty} d(\theta_n, \Omega_0) = 0,$$

at a specified rate. Then the asymptotic power of the test against the sequence of alternatives $\{\theta_n\}_{n=1}^{\infty}$ is given by

$$\lim_{n\to\infty} \beta_n(\theta_n).$$

In this section we will consider some asymptotic properties of hypothesis tests for the case when Ω is the real line and Ω_0 and Ω_1 are intervals. The null hypothesis will generally have the form $H_0 : \theta \in \Omega_0 = (-\infty, \theta_0]$ or $H_0 : \theta \in \Omega_0 = [\theta_0, \infty)$ with the corresponding alternative hypotheses given by $H_1 : \theta \in \Omega_1 = (\theta_0, \infty)$ and $H_1 : \theta \in \Omega_1 = (-\infty, \theta_0)$, respectively. We will first consider the case of asymptotically normal test statistics. In particular we will consider test statistics of the form $Z_n = n^{1/2}\sigma^{-1}(\hat{\theta}_n - \theta_0)$ where $Z_n \xrightarrow{d} Z$ as $n \to \infty$ where Z is a $N(0,1)$ random variable, and σ is a known constant that does not depend on n. Consider the problem of testing the null hypothesis $H_0 : \theta \le \theta_0$ against the alternative hypothesis $H_1 : \theta > \theta_0$. Because the alternative hypothesis specifies that θ is larger than θ_0 we will consider rejecting the null hypothesis when Z_n is too large, or when $Z_n > r_{\alpha,n}$ where $\{r_{\alpha,n}\}_{n=1}^{\infty}$ is a sequence or real numbers and α is the specified significance level of the test. If the distribution of Z_n under the null hypothesis is known then the sequence $\{r_{\alpha,n}\}_{n=1}^{\infty}$ can be specified so that the test has level or size α in accordance with Definition 10.8. If this distribution is unknown then the test can be justified from an asymptotic viewpoint by specifying that $r_{n,\alpha} \to z_{1-\alpha}$ as $n \to \infty$. In this case we have that

$$\lim_{n\to\infty} P(Z_n \ge r_{n,\alpha}|\theta = \theta_0) = \alpha.$$

Now consider this probability when $\theta = \theta_l$, a point that is away from the boundary of Ω_0, so that θ_l is strictly less than θ_0. In this case we have that

$$P(Z_n \geq r_{n,\alpha}|\theta = \theta_l) = P\left[\frac{n^{1/2}(\hat{\theta}_n - \theta_0)}{\sigma} \geq r_{n,\alpha} \middle| \theta = \theta_l\right]$$

$$= P\left[\frac{n^{1/2}(\hat{\theta}_n - \theta_l)}{\sigma} \geq r_{n,\alpha} + \frac{n^{1/2}(\theta_0 - \theta_l)}{\sigma} \middle| \theta = \theta_l\right]$$

$$\leq P\left[\frac{n^{1/2}(\hat{\theta}_n - \theta_l)}{\sigma} \geq r_{n,\alpha} \middle| \theta = \theta_l\right].$$

Therefore, it follows that

$$\lim_{n\to\infty} P(Z_n \geq r_{n,\alpha}|\theta = \theta_l) \leq \lim_{n\to\infty} P(Z_n \geq r_{n,\alpha}|\theta = \theta_0) = \alpha,$$

for all $\theta_l < \theta_0$. Hence, Definition 10.8 implies that this test has asymptotic size α. Similar arguments can be used to show that this test is also unbiased. See Exercise 26.

Example 10.24. Let $X_1, \ldots, X_n$ be a set of independent and identically distributed random variables following an EXPONENTIAL(θ) distribution where $\theta \in \Omega = (0, \infty)$. We will consider testing the null hypothesis $H_0 : \theta \leq \theta_0$ against the alternative hypothesis $H_1 : \theta > \theta_0$ for some specified $\theta_0 > 0$. Theorem 4.20 (Lindeberg-Lévy) implies that the test statistic $Z_n = n^{1/2}\theta_0^{-1}(\bar{X}_n - \theta_0)$ converges in distribution to a $N(0,1)$ distribution as $n \to \infty$ when $\theta = \theta_0$. Therefore, rejecting the null hypothesis when Z_n exceeds $z_{1-\alpha}$ will result in a test with asymptotic level equal to α. ∎

Example 10.25. Let $X_1, \ldots, X_n$ be a set of independent and identically distributed random variables following a POISSON(θ) distribution where $\theta \in \Omega = (0, \infty)$. We will consider testing the null hypothesis $H_0 : \theta \leq \theta_0$ against the alternative hypothesis $H_1 : \theta > \theta_0$ for some specified $\theta_0 > 0$. Once again, Theorem 4.20 implies that the test statistic $Z_n = n^{1/2}\theta_0^{-1/2}(\bar{X}_n - \theta_0)$ converges in distribution to a $N(0,1)$ distribution as $n \to \infty$ when $\theta = \theta_0$. Therefore, rejecting the null hypothesis when Z_n exceeds $z_{1-\alpha}$ will result in a test with asymptotic level equal to α. ∎

The variance σ^2 can either be known as a separate parameter, or can be known through the specification of the null hypothesis, as in the problem studied in Example 10.25. In some cases, however, σ will not be known. In these cases there is often a consistent estimator of σ that can be used. That is, we consider the case where σ can be estimated by $\hat{\sigma}_n$ where $\hat{\sigma}_n \xrightarrow{p} \sigma$ as $n \to \infty$. Theorem 4.11 (Slutsky) implies that $n^{1/2}\hat{\sigma}_n^{-1}(\hat{\theta}_n - \theta_0) \xrightarrow{d} Z$ as $n \to \infty$ and hence

$$\lim_{n\to\infty} P[n^{1/2}\hat{\sigma}_n^{-1}(\hat{\theta}_n - \theta_0) \geq r_{\alpha,n}] = \alpha,$$

as long as $r_{\alpha,n} \to Z_{1-\alpha}$ as $n \to \infty$.

Example 10.26. Let $X_1, \ldots, X_n$ be a set of independent and identically distributed random variables following a distribution F with mean θ and finite variance σ^2. Let $\hat{\sigma}_n$ be the usual sample variance, which is a consistent estimator of σ by Theorem 3.21 as long as $E(|X_n|^4) < \infty$. Therefore we have that $T_n = n^{1/2}\hat{\sigma}_n^{-1}(\bar{X}_n - \theta_0) \xrightarrow{d} Z$ as $n \to \infty$, and a test that rejects the null hypothesis $H_0 : \theta \le \theta_0$ in favor of the alternative hypothesis $H_1 : \theta > \theta_0$ when $T_n > z_{1-\alpha}$ is a test with asymptotic level equal to α. ∎

Example 10.27. Let $X_1, \ldots, X_n$ be a set of independent and identically distributed random variables from a $N(\eta, \theta)$ distribution. Consider the problem of testing the null hypothesis $H_0 : \theta \le \theta_0$ against the alternative hypothesis $H_1 : \theta > \theta_0$. Let $\hat{\theta}_n$ be the usual unbiased version of the sample variance. Under these assumptions it follows that $(n-1)\theta^{-1}\hat{\theta}_n$ has a CHISQUARED$(n-1)$ distribution, which motivates using the test statistic $X_n = (n-1)\theta_0^{-1}\hat{\theta}_n$, where the null hypothesis will be rejected when $X_n > \chi^2_{n-1;1-\alpha}$. This is an exact test under these assumptions. An approximate test can be motivated by Theorem 8.5, where we see that for any F that has a finite fourth moment it follows that

$$Z_n = \frac{n^{1/2}(\hat{\theta}_n - \theta_0)}{(\mu_4 - \theta^2)^{1/2}} \xrightarrow{d} Z$$

as $n \to \infty$ where Z has a $N(0,1)$ and μ_3 and μ_4 are the third and fourth moments of F, respectively. Of course the statistic Z_n could not be used to test the null hypothesis because the denominator is unknown, and is not fully specified by the null hypothesis. However, under the assumption that the fourth moment is finite, a consistent estimator of the denominator can be obtained by replacing the population moments with their sample counterparts. Therefore, it also follows that

$$T_n = \frac{n^{1/2}(\hat{\theta}_n - \theta_0)}{(\hat{\mu}_4 - \hat{\theta}_n^2)^{1/2}} \xrightarrow{d} Z$$

as $n \to \infty$, and hence a test that rejects the null hypothesis in favor of the alternative hypothesis when $T_n > z_{1-\alpha}$ is a test with asymptotic level equal to α. ∎

Under some additional assumptions we can address the issue of consistency. In particular we will assume that

$$P\left[\frac{n^{1/2}(\hat{\theta}_n - \theta_u)}{\sigma} \le t \,\Big|\, \theta = \theta_u\right] \rightsquigarrow \Phi(t),$$

as $n \to \infty$ for all $t \in \mathbb{R}$ and $\theta_u \in (\theta_0, \infty)$. It then follows that

$$
\begin{aligned}
\beta_n(\theta_u) &= P(Z_n \ge r_{n,\alpha} | \theta = \theta_u) \\
&= P\left[\frac{n^{1/2}(\hat{\theta}_n - \theta_0)}{\sigma} \ge r_{n,\alpha} \,\Big|\, \theta = \theta_u\right] \\
&= P\left[\frac{n^{1/2}(\hat{\theta}_n - \theta_u)}{\sigma} \ge r_{n,\alpha} - \frac{n^{1/2}(\theta_u - \theta_0)}{\sigma} \,\Big|\, \theta = \theta_u\right],
\end{aligned}
$$

where we note that $r_{n,\alpha} - n^{1/2}\sigma^{-1}(\theta_u - \theta_0) \to -\infty$ as $n \to \infty$, under the assumption that $r_{n,\alpha} \to z_{1-\alpha}$ as $n \to \infty$. It then follows that

$$\lim_{n\to\infty} \beta_n(\theta_u) = 1,$$

for all $\theta_u \in (\theta_0, \infty)$. Therefore, Definition 10.10 implies that the test is consistent against any alternative $\theta_u \in (\theta_0, \infty)$. See Exercise 27 for further details.

The asymptotic power of the test can be studied using the sequence of alternative hypotheses $\{\theta_{1,n}\}_{n=1}^\infty$ where $\theta_{1,n} = \theta_0 + n^{-1/2}\delta$ where δ is a positive real constant. Note that $\theta_{1,n} \to \theta_0$ as $n \to \infty$ and that $\theta_{1,n} = \theta_0 + O(n^{-1/2})$ as $n \to \infty$. For this sequence we have that

$$\begin{aligned}
\beta_n(\theta_{1,n}) &= P(Z_n \geq r_{\alpha,n} | \theta = \theta_{1,n}) \\
&= P\left[\frac{n^{1/2}(\hat{\theta}_n - \theta_{1,n})}{\sigma} \geq r_{\alpha,n} - \frac{n^{1/2}(\theta_{1,n} - \theta_0)}{\sigma}\middle| \theta = \theta_{1,n}\right] \\
&= P\left[\frac{n^{1/2}(\hat{\theta}_n - \theta_{1,n})}{\sigma} \geq r_{\alpha,n} - \sigma^{-1}\delta\middle| \theta = \theta_{1,n}\right].
\end{aligned}$$

As in the above case we must make some additional assumptions about the limiting distribution of the sequence $n^{1/2}\sigma^{-1}(\hat{\theta}_n - \theta_{1,n})$ when $\theta = \theta_{1,n}$. In this case we will assume that

$$P\left[\frac{n^{1/2}(\hat{\theta}_n - \theta_{1,n})}{\sigma} \leq t\middle| \theta = \theta_{1,n}\right] \rightsquigarrow \Phi(t),$$

as $n \to \infty$. Therefore, using the same arguments as above it follows that

$$\lim_{n\to\infty} \beta_n(\theta_{1,n}) = 1 - \Phi(z_{1-\alpha} - \sigma^{-1}\delta).$$

The symmetry of the NORMAL distribution can be used to conclude that

$$\lim_{n\to\infty} \beta_n(\theta_{1,n}) = \Phi(\sigma^{-1}\delta - z_{1-\alpha}). \tag{10.43}$$

For δ near zero, a further approximation based on the results of Theorem 1.15 can be used to find that

$$\lim_{n\to\infty} \beta_n(\theta_{1,n}) = \Phi(-z_{1-\alpha}) + \sigma^{-1}\delta\phi(z_\alpha) + O(\delta^2) = \alpha + \sigma^{-1}\delta\phi(z_\alpha) + O(\delta^2),$$

as $\delta \to 0$.

Note that in some cases, such as in Example 10.26, the standard deviation σ, depends on θ. That is, the standard deviation is $\sigma(\theta)$, for some function σ. Hence considering a sequence of alternatives $\{\theta_{1,n}\}_{n=1}^\infty$ will imply that we must also consider a sequence of standard deviations $\{\sigma_{1,n}\}_{n=1}^\infty$ where $\sigma_{1,n} = \sigma(\theta_{1,n})$ for all $n \in \mathbb{N}$. Therefore, if we can assume that $\sigma(\theta)$ is a continuous function of θ, it follows from Theorem 1.3 that $\sigma_{1,n} \to \sigma$ as $n \to \infty$ for some positive finite constant $\sigma = \sigma(\theta_0)$. In this case

$$
\begin{aligned}
\beta_n(\theta_{1,n}) &= P\left[\frac{n^{1/2}(\hat{\theta} - \theta_0)}{\sigma(\theta_0)} \geq r_{\alpha,n} \,\middle|\, \theta = \theta_{1,n}\right] \\
&= P\left[\frac{n^{1/2}(\hat{\theta} - \theta_{1,n})}{\sigma(\theta_0)} \geq r_{\alpha,n} - \frac{n^{1/2}(\theta_{1,n} - \theta_0)}{\sigma(\theta_0)} \,\middle|\, \theta = \theta_{1,n}\right].
\end{aligned}
$$

Now, assume that $n^{1/2}(\hat{\theta}_n - \theta)/\sigma(\theta) \xrightarrow{d} Z$ as $n \to \infty$ where Z is a $N(0,1)$ random variable, for all $\theta \in (\theta_0 - \varepsilon, \theta_0 + \varepsilon)$ for some $\varepsilon > 0$. Further, assume that

$$
P\left[\frac{n^{1/2}(\hat{\theta} - \theta_{1,n})}{\sigma(\theta_{1,n})} \leq t \,\middle|\, \theta = \theta_{1,n}\right] \rightsquigarrow \Phi(t),
$$

as $n \to \infty$. Then the fact that

$$
\lim_{n \to \infty} \frac{\sigma(\theta_{1,n})}{\sigma(\theta_0)} = 1,
$$

and Theorem 4.11 (Slutsky) implies that

$$
P\left[\frac{n^{1/2}(\hat{\theta} - \theta_{1,n})}{\sigma(\theta_0)} \leq t \,\middle|\, \theta = \theta_{1,n}\right] \rightsquigarrow \Phi(t),
$$

as $n \to \infty$, and hence the result of Equation (10.43) holds in this case as well.

These results can be summarized with the following result.

Theorem 10.10. *Let $X_1, \dots, X_n$ be a set of independent and identically distributed random variables from a distribution F with parameter θ. Consider testing the null hypothesis $H_0 : \theta \leq \theta_0$ against the alternative hypothesis $H_1 : \theta > \theta_0$ by rejecting the null hypothesis if $n^{1/2}(\hat{\theta}_n - \theta_0)/\sigma(\theta_0) > r_{\alpha,n}$. Assume that*

1. *$r_{\alpha,n} \to z_{1-\alpha}$ as $n \to \infty$.*

2. *$\sigma(\theta)$ is a continuous function of θ.*

3. *$n^{1/2}(\hat{\theta}_n - \theta)/\sigma(\theta) \xrightarrow{d} Z$ as $n \to \infty$ for all $\theta > \theta_0 - \varepsilon$ for some $\varepsilon > 0$ where Z is a $N(0,1)$ random variable.*

4. *$n^{1/2}(\hat{\theta}_n - \theta_{1,n})/\sigma(\theta_{1,n}) \xrightarrow{d} Z$ as $n \to \infty$ where $\theta = \theta_{1,n}$ and $\{\theta_{1,n}\}_{n=1}^{\infty}$ is a sequence such that $\theta_{1,n} \to \theta_0$ as $n \to \infty$.*

Then the test is consistent against all alternatives $\theta > \theta_0$ and

$$
\lim_{n \to \infty} \beta_n(\theta_{1,n}) = \Phi[\delta/\sigma(\theta_0) - z_{1-\alpha}].
$$

Example 10.28. Suppose that $X_1, \dots, X_n$ is a set of independent and identically distributed random variables from a POISSON(θ) distribution. In Example 10.25 we considered the test statistic $Z_n = n^{1/2}\theta_0^{-1/2}(\bar{X}_n - \theta_0)$. Theorem

4.20 implies that

$$P\left[\frac{n^{1/2}(\bar{X}_n - \theta_u)}{\theta_u^{1/2}} \leq t \,\middle|\, \theta = \theta_u\right] \rightsquigarrow \Phi(t),$$

as $n \to \infty$ for all $t \in \mathbb{R}$ and $\theta_u \in (\theta_0, \infty)$. Hence, it follows that the test statistic based on T_n is consistent. To obtain the asymptotic power of the test we need to show that

$$P\left[\frac{n^{1/2}(\bar{X}_n - \theta_{1,n})}{\theta_{n,1}^{1/2}} \leq t \,\middle|\, \theta = \theta_{1,n}\right] \rightsquigarrow \Phi(t),$$

as $n \to \infty$ for all $t \in \mathbb{R}$ and any sequence $\{\theta_{1,n}\}_{n=1}^{\infty}$ such that $\theta_{1,n} \to \theta_0$ as $n \to \infty$. To see why this holds we follow the approach of Lehmann (1999) and note that for each $n \in \mathbb{N}$, Theorem 4.24 (Berry and Esseen) implies that

$$\left|P\left[\frac{n^{1/2}(\bar{X}_n - \theta_{1,n})}{\theta_{n,1}^{1/2}} \leq t \,\middle|\, \theta = \theta_{1,n}\right] - \Phi(t)\right| \leq$$
$$n^{-1/2} B \theta_{1,n}^{-3/2} E(|X_1 - \theta_{1,n}|^3), \quad (10.44)$$

where B is a constant that does not depend on n. Therefore, it follows that for each $t \in \mathbb{R}$ that

$$\lim_{n \to \infty} P\left[\frac{n^{1/2}(\bar{X}_n - \theta_{1,n})}{\theta_{n,1}^{1/2}} \leq t \,\middle|\, \theta = \theta_{1,n}\right] = \Phi(t),$$

as long as the right hand side of Equation (10.44) converges to zero as $n \to \infty$. To show this we note that

$$\theta_n^{-3/2} E(|X_1 - \theta_{1,n}|^3) = \theta_n^{-3/2} E(|X_1^3 - 3X_1^2 \theta_{1,n} + 3X_1 \theta_{1,n}^2 - \theta_{1,n}^3|)$$
$$\leq \theta_n^{-3/2}[E(|X_1|^3) - 3E(X_1^2)\theta_{1,n} + 2\theta_{1,n}^3)].$$

Now $E(X_1^2) = \theta_{1,n} + \theta_{1,n}^2$ and $E(X_1^3) = \theta_{1,n}[(\theta_{1,n} + 1)^2 + \theta_{1,n}]$, so that

$$\theta_n^{-3/2} E(|X_1 - \theta_{1,n}|^3) \leq \theta_{1,n}^{-1/2}(7\theta_{1,n}^2 + 5\theta_{1,n} + 1).$$

Therefore, it follows that

$$\lim_{n \to \infty} \theta_n^{-3/2} E(|X_1 - \theta_{1,n}|^3) \leq \theta_0^{-1/2}(7\theta_0^2 + 5\theta_0 + 1).$$

Hence we have that the right hand side of Equation (10.44) converges to zero as $n \to \infty$. Therefore, it follows from Theorem 10.10 that the asymptotic power of the test is $\Phi(\theta_0^{-1/2} - Z_{1-\alpha})$ for the sequence of alternatives given by $\theta_{1,n} = \theta_0 + n^{-1/2}\delta$ where $\delta > 0$. This test is also consistent against all alternatives $\theta > \theta_0$. ∎

Example 10.29. Let $X_1, \ldots, X_n$ be a set of independent and identically distributed random variables from an EXPONENTIAL(θ) distribution. We wish to test the null hypothesis $H_0 : \theta \leq \theta_0$ against the alternative hypothesis $H_1 : \theta > \theta_0$ using the test statistic $n^{1/2}\theta_0^{-1}(\bar{X}_n - \theta_0)$. Theorem 4.20 (Lindeberg and Lévy) implies that $n^{1/2}\theta^{-1}(\bar{X}_n - \theta) \xrightarrow{d} Z$ as $n \to \infty$ for all $\theta > 0$ where Z is a $N(0,1)$ random variable. Hence, assumptions 1 and 2 of Theorem 10.10 have been satisfied and it only remains to show that $n^{1/2}\theta_{1,n}^{-1}(\bar{X}_n - \theta_{1,n}) \xrightarrow{d} Z$ as $n \to \infty$ when $\theta = \theta_{1,n}$ and $\theta_{1,n} \to \theta_0$ as $n \to \infty$. Using the same approach as shown in Example 10.28 that is based on Theorem 4.24 (Berry and Esseen) it follows that the assumptions of Theorem 10.10 hold and hence the test is consistent with asymptotic power function given by $\Phi(\theta_0^{-1}\delta - z_{1-\alpha})$ where the sequence of alternatives is given by $\theta_{1,n} = \theta_0 + n^{-1/2}\delta$. ∎

In the case where σ is unknown, but is replaced by a consistent estimator given by $\hat{\sigma}_n$, we have that the test statistic $T_n = n^{1/2}\hat{\sigma}_n^{-1}(\hat{\theta}_n - \theta) \xrightarrow{d} Z$ as $n \to \infty$ where Z is a $N(0,1)$ random variable by Theorem 4.11 (Slutsky). Under similar conditions to those given in Theorem 10.10, the consistency and asymptotic power of the test has the properties given in Theorem 10.10. See Section 3.3 of Lehmann (1999) for a complete development of this case.

More precise asymptotic behavior under the null hypothesis can be determined by adding further structure to our model. In particular we will now restrict out attention to the framework of the smooth function model described in Section 7.4. We will first consider the case where σ, which denotes the asymptotic variance of $n^{1/2}\hat{\theta}_n$, is known. In this case we will consider using the test statistic $Z_n = n^{1/2}\sigma^{-1}(\hat{\theta}_n - \theta_0)$ which follows the distribution G_n when θ_0 is the true value of θ. Calculations similar to those used above can be used to show that an unbiased test of size α of the null hypothesis $H_0 : \theta \le \theta_0$ against the alternative hypothesis $H_1 : \theta > \theta_0$ rejects the null hypothesis if $Z_n > g_{1-\alpha}$, where we recall that $g_{1-\alpha}$ is the $(1-\alpha)^{\text{th}}$ quantile of the distribution G_n. See Exercise 31.

From an asymptotic viewpoint it follows from the smooth function model that $Z_n \xrightarrow{d} Z$ as $n \to \infty$ where Z is a $N(0,1)$ random variable. Therefore, if the distribution G_n is unknown then one can develop a large sample test by rejecting $H_0 : \theta \le \theta_0$ if $Z_n \ge z_{1-\alpha}$. This test is similar to the approximate normal tests studied above, except with the additional framework of the smooth function model, we can study the effect of this approximation, as well as alternate approximations, more closely. Note that Theorem 7.11 (Bhattacharya and Ghosh) implies that the quantile $g_{1-\alpha}$ has a Cornish-Fisher expansion given by $g_{1-\alpha} = z_{1-\alpha} + n^{-1/2}q_1(z_{1-\alpha}) + O(n^{-1})$, as $n \to \infty$. Therefore, we have that $|g_{1-\alpha} - z_{1-\alpha}| = O(n^{-1/2})$, as $n \to \infty$. To see what effect this has on the size of the test we note that Theorem 7.11 implies that the Edgeworth expansion for the distribution Z_n is given by $P(Z_n \le t) = \Phi(t) + n^{-1/2}r_1(t)\phi(t) + O(n^{-1})$, as $n \to \infty$. Therefore, the probability of rejecting the null hypothesis when $\theta = \theta_0$ is given by

$$
\begin{aligned}
P(Z_n > z_{1-\alpha}|\theta = \theta_0) &= 1 - \Phi(z_{1-\alpha}) - n^{-1/2}r_1(z_{1-\alpha})\phi(z_{1-\alpha}) + O(n^{-1}) \\
&= \alpha - n^{-1/2}r_1(z_\alpha)\phi(z_\alpha) + O(n^{-1}), \quad (10.45)
\end{aligned}
$$

as $n \to \infty$, where we have used the fact that both r_1 and ϕ are even functions.

Therefore, Definition 10.8 implies that this test is first-order accurate. Note that the test may even be more accurate depending on the form of the term $r_1(z_\alpha)\phi(z_\alpha)$ in Equation (10.45). For example, if $r_1(z_\alpha) = 0$ then the test is at least second-order accurate.

Another strategy for obtaining a more accurate test is to use a quantile of the form $z_{1-\alpha} + n^{-1/2}q_1(z_{1-\alpha})$, under the assumption that the polynomial q_1 is known. Note that

$$P(Z_n > z_{1-\alpha} + n^{-1/2}q_1(z_{1-\alpha})|\theta = \theta_0) = 1 - \Phi[z_{1-\alpha} + n^{-1/2}q_1(z_{1-\alpha})] -$$
$$n^{-1/2}r_1[z_{1-\alpha} + n^{-1/2}q_1(z_{1-\alpha})]\phi[z_{1-\alpha} + n^{-1/2}q_1(z_{1-\alpha})] + O(n^{-1}),$$

as $n \to \infty$. Now, using Theorem 1.15 we have that

$$\Phi[z_{1-\alpha} + n^{-1/2}q_1(z_{1-\alpha})] = \Phi(z_{1-\alpha}) + n^{-1/2}q_1(z_{1-\alpha})\phi(z_{1-\alpha}) + O(n^{-1}),$$

$r_1[z_{1-\alpha}+n^{-1/2}q_1(z_{1-\alpha})] = r_1(z_{1-\alpha})+O(n^{-1/2})$, and $\phi[z_{1-\alpha}+n^{-1/2}q_1(z_{1-\alpha})] = \phi(z_{1-\alpha}) + O(n^{-1/2})$, as $n \to \infty$. Therefore,

$$P(Z_n > z_{1-\alpha} + n^{-1/2}q_1(z_{1-\alpha})|\theta = \theta_0) =$$
$$1 - \Phi(z_{1-\alpha}) - n^{-1/2}q_1(z_{1-\alpha})\phi(z_{1-\alpha}) - n^{-1/2}r_1(z_{1-\alpha})\phi(z_{1-\alpha}) + O(n^{-1}),$$

as $n \to \infty$. Noting that $q_1(z_{1-\alpha}) = -r_1(z_{1-\alpha})$ implies that

$$P(Z_n > z_{1-\alpha} + n^{-1/2}q_1(z_{1-\alpha})|\theta = \theta_0) = \alpha + O(n^{-1}), \tag{10.46}$$

as $n \to \infty$, and therefore the test is second-order accurate.

Unfortunately, because the coefficients of q_1 depend on the moments of the population, which are likely to be unknown, we cannot usually compute this rejection region in practice. On the other hand, if sample moments were substituted in place of the population moments, then we would have an estimator of q_1 with the property that $\hat{q}_1(z_{1-\alpha}) = q_1(z_{1-\alpha}) + O_p(n^{-1/2})$, as $n \to \infty$. Using this estimator we see that

$$z_{1-\alpha} + n^{-1/2}\hat{q}_1(z_{1-\alpha}) + O(n^{-1}) = z_{1-\alpha} + n^{-1/2}q_1(z_{1-\alpha}) + O_p(n^{-1}),$$

as $n \to \infty$. Using a result like Theorem 10.8, and the result of Equation (10.46) implies that

$$P(Z_n > z_{1-\alpha} + n^{-1/2}\hat{q}_1(z_{1-\alpha})|\theta = \theta_0) =$$
$$P(Z_n > z_{1-\alpha} + n^{-1/2}q_1(z_{1-\alpha}) + O_p(n^{-1})|\theta = \theta_0) =$$
$$P(Z_n > z_{1-\alpha} + n^{-1/2}q_1(z_{1-\alpha})|\theta = \theta_0) + O(n^{-1}) = \alpha + O(n^{-1}),$$

as $n \to \infty$. Therefore, this approximate Edgeworth-corrected test will also be second-order accurate.

For the case when σ is unknown we use the test statistic $T_n = n^{1/2}\hat{\sigma}_n^{-1}(\hat{\theta}_n - \theta_0)$ which has distribution H_n and associated quantile $h_{1-\alpha}$ when $\theta = \theta_0$. Calculations similar to those given above can be used to show that approximating

$h_{1-\alpha}$ with $z_{1-\alpha}$ results in a test that is first-order accurate while approximating $h_{1-\alpha}$ with $z_{1-\alpha} + n^{-1/2}\hat{s}_1(z_{1-\alpha})$ results in a test that is second-order accurate. See Exercise 32.

Example 10.30. Let $X_1, \ldots, X_n$ be a sequence of independent and identically distributed random variables that have a $N(\theta, \sigma^2)$ distribution. Consider testing the null hypothesis $H_0 : \theta \le \theta_0$ against the alternative hypothesis $H_1 : \theta > \theta_0$ for some $\theta \in \mathbb{R}$. This model falls within the smooth function model where G_n is a $N(0,1)$ distribution and H_n is a $T(n-1)$ distribution. Of course, rejecting the null hypothesis when $T_n > t_{n-1;1-\alpha}$ results in an exact test. However, when n is large it is often suggested that rejecting the null hypothesis when $T_n = n^{1/2}\hat{\sigma}_n^{-1}(\bar{X}_n - \theta_0) > z_{1-\alpha}$, where $\hat{\sigma}_n^2$ is the unbiased version of the sample variance, is a test that is approximately valid. Indeed, this is motivated by the fact that the $T(n-1)$ distribution converges in distribution to a $N(0,1)$ distribution as $n \to \infty$. For the test based on the normal approximation we have that

$$P(T_n > z_{1-\alpha}|\theta = \theta_0) = \alpha - n^{-1/2}v_1(z_{1-\alpha})\phi(z_{1-\alpha}) + O(n^{-1}),$$

as $n \to \infty$, so that the test is at least first-order accurate. In the case of the mean functional $v_1(t) = \frac{1}{6}\gamma(2t^2 + 1)$ where $\gamma = \sigma^{-3}\mu_3$, which is the standardized skewness of the population. For a normal population $\gamma = 0$ and hence $P(T_n > z_{1-\alpha}|\theta = \theta_0) = \alpha + O(n^{-1})$, as $n \to \infty$. We can also note that in this case

$$P(T_n > z_{1-\alpha}|\theta = \theta_0) = \alpha - n^{-1}v_2(z_{1-\alpha})\phi(z_{1-\alpha}) + O(n^{-3/2}),$$

as $n \to \infty$. For the case of the mean functional we have that

$$\begin{aligned} v_2(t) &= t[\tfrac{1}{12}\kappa(t^2 - 3) - \tfrac{1}{18}\gamma^2(t^4 + 2t^2 - 3) - \tfrac{1}{4}(t^2 + 3)] \\ &= t[\tfrac{1}{12}\kappa(t^2 - 3) - \tfrac{1}{4}(t^2 + 3)], \end{aligned}$$

since $\gamma = 0$ for the $N(\theta, \sigma^2)$ distribution, where $\kappa = \sigma^{-4}\mu_4 - 3$. The fourth moment of a $N(\theta, \sigma^2)$ distribution is $3\sigma^4$ so that $\kappa = 0$. Therefore $v_2(t) = -\frac{1}{4}t(t^2 + 3)$, and hence

$$P(T_n > z_{1-\alpha}|\theta = \theta_0) = \alpha + \tfrac{1}{4}n^{-1}z_{1-\alpha}(z_{1-\alpha}^2 + 3)\phi(z_{1-\alpha}) + O(n^{-3/2}),$$

as $n \to \infty$. Therefore, it follows that the normal approximation is second-order accurate for samples from a $N(\theta, \sigma^2)$ distribution. Note that the second-order accuracy will also hold for any symmetric population. If the population is not symmetric the test will only be first-order accurate. ∎

Example 10.31. Let $X_1, \ldots, X_n$ be a sequence of independent and identically distributed random variables that have a distribution F for all $n \in \mathbb{N}$. Let θ be the mean of F, and assume that F falls within the assumptions of the smooth function model. Consider testing the null hypothesis $H_0 : \theta \le \theta_0$ against the alternative hypothesis $H_1 : \theta > \theta_0$ for some $\theta \in \mathbb{R}$. Theorem 8.5 implies that we can estimate $\gamma = \sigma^{-3}\mu_3$ with $\hat{\gamma}_n = \hat{\sigma}_n^{-3}\hat{\mu}_3$ where $\hat{\gamma}_n = \gamma + O_p(n^{-1/2})$ as $n \to \infty$. In this case $\hat{\sigma}_n$ and $\hat{\mu}_3$ are the sample versions of the corresponding moments. Using this estimate, we can estimate

$s_1(z_{1-\alpha})$ with $\hat{s}_1(z_{1-\alpha}) = -\frac{1}{6}\hat{\gamma}_n(2z_{1-\alpha}^2 + 1)$, where it follows that $\hat{s}_1(z_{1-\alpha}) = s_1(z_{1-\alpha}) + O_p(n^{-1/2})$, as $n \to \infty$. Now consider the test that rejects the null hypothesis when $T_n = n^{1/2}\hat{\sigma}_n^{-1}(\bar{X}_n - \theta_0) > z_{1-\alpha} + \hat{s}_1(z_{1-\alpha})$. Then

$$P[T_n > z_{1-\alpha} + n^{-1/2}\hat{s}_1(z_{1-\alpha})|\theta = \theta_0] =$$

$$P[T_n > z_{1-\alpha} + n^{-1/2}s_1(z_{1-\alpha})|\theta = \theta_0] + O(n^{-1}) =$$

$$1 - \Phi[z_{1-\alpha} + n^{-1/2}s_1(z_{1-\alpha})] -$$

$$n^{-1/2}v_1[z_{1-\alpha} + n^{-1/2}s_1(z_{1-\alpha})]\phi[z_{1-\alpha} + n^{-1/2}s_1(z_{1-\alpha})] + O(n^{-1}) =$$

$$\alpha - n^{-1/2}s_1(z_{1-\alpha})\phi(z_{1-\alpha}) - n^{-1/2}v_1(z_{1-\alpha})\phi(z_{1-\alpha}) + O(n^{-1}),$$

as $n \to \infty$. Noting that $s_1(z_{1-\alpha}) = -v_1(z_{1-\alpha})$ implies that rejecting the null hypothesis when $T_n > z_{1-\alpha} + \hat{s}_1(z_{1-\alpha})$ yields a test that is second-order correct. ■

The normal distribution is not the only asymptotic distribution that is common for test statistics. Let $X_1, \ldots, X_n$ be a set of independent and identically distributed random variables from a distribution F with parameter θ. Let $L(\theta)$ be the likelihood function of θ and let $\hat{\theta}_n$ be the maximum likelihood estimator of θ, which we will assume is unique. We will consider Wilk's likelihood ratio test statistic for testing a null hypothesis of the form $H_0 : \theta = \theta_0$ against an alternative hypothesis of the form $H_1 : \theta \neq \theta_0$. The idea behind this test statistic is based on comparing the likelihood for the estimated value of θ, given by $L(\hat{\theta}_n)$, to the likelihood of the value of θ under the null hypothesis, given by $L(\theta_0)$. Since $\hat{\theta}_n$ is the maximum likelihood estimator of θ it follows that $L(\hat{\theta}_n) \geq L(\theta_0)$. If this difference is large then there is evidence that the likelihood for θ_0 is very low compared to the maximum likelihood and hence θ_0 is not a likely value of θ. In this case the null hypothesis would be rejected. One the other hand, if this difference is near zero, then the likelihoods are near one another, and there is little evidence that the null hypothesis is not true. The usual likelihood ratio test compares $L(\hat{\theta}_n)$ to $L(\theta_0)$ using the ratio $L(\theta_0)/L(\hat{\theta}_n)$, so that small values of this test statistic indicate that the null hypothesis should be rejected. Wilk's likelihood ratio test statistic is equal to -2 times the logarithm of this ratio. That is,

$$\Lambda_n = -2\log\left[\frac{L(\theta_0)}{L(\hat{\theta}_n)}\right] = 2\{\log[L(\hat{\theta}_n)] - \log[L(\theta_0)]\} = 2[l(\hat{\theta}_n) - l(\theta_0)].$$

Therefore, a test based on the statistic Λ_n rejects the null hypothesis when Λ_n is large. Under certain conditions the test statistic Λ_n has an asymptotic CHISQUARED(1) distribution.

Theorem 10.11. *Suppose that $X_1, \ldots, X_n$ is a set of independent and identically distributed random variables from a distribution $F(x|\theta)$ with density $f(x|\theta)$, where θ has parameter space Ω. Suppose that the conditions of Theorem 10.5 hold, then, under the null hypothesis $H_0 : \theta = \theta_0$, $\Lambda_n \xrightarrow{d} X$ as $n \to \infty$ where X has CHISQUARED(1) distribution.*

Proof. Theorem 1.13 (Taylor) implies that under the stated assumptions we have that

$$
\begin{aligned}
l(\hat{\theta}_n) &= l(\theta_0) + (\hat{\theta}_n - \theta_0)l'(\theta_0) + \tfrac{1}{2}(\hat{\theta}_n - \theta_0)^2 l''(\xi) \\
&= l(\theta_0) + n^{1/2}(\hat{\theta}_n - \theta_0)[n^{-1/2}l'(\theta_0)] \\
&\quad + \tfrac{1}{2}(\hat{\theta}_n - \theta_0)^2 l''(\xi),
\end{aligned}
\tag{10.47}
$$

where ξ is a random variable that is between θ_0 and $\hat{\theta}_n$ with probability one. Apply Theorem 1.13 to the second term on the right hand side of Equation (10.47) to obtain

$$
n^{-1/2}l'(\theta_0) = n^{-1/2}l'(\hat{\theta}_n) + n^{-1/2}(\theta_0 - \hat{\theta}_n)l''(\zeta),
\tag{10.48}
$$

where ζ is a random variable that is between $\hat{\theta}_n$ and θ_0 with probability one. To simplify the expression in Equation (10.48) we note that since $\hat{\theta}_n$ is the maximum likelihood estimator of θ it follows that $l'(\hat{\theta}_n) = 0$. Therefore

$$
n^{-1/2}l'(\theta_0) = n^{-1/2}l'(\hat{\theta}_n) + n^{-1/2}(\theta_0 - \hat{\theta}_n)l''(\zeta) = \\
- n^{-1}l''(\zeta)[n^{1/2}(\hat{\theta}_n - \theta_0)].
\tag{10.49}
$$

Substituting the expression from Equation (10.49) into Equation (10.47) yields

$$
\begin{aligned}
l(\hat{\theta}_n) &= l(\theta_0) + n^{1/2}(\hat{\theta}_n - \theta_0)\{-n^{-1}l''(\zeta)[n^{1/2}(\hat{\theta}_n - \theta_0)]\} \\
&\quad + \tfrac{1}{2}(\hat{\theta}_n - \theta_0)^2 l''(\xi) \\
&= l(\theta_0) - n(\hat{\theta}_n - \theta_0)^2[n^{-1}l''(\zeta)] + \tfrac{1}{2}(\hat{\theta}_n - \theta_0)^2 l''(\xi),
\end{aligned}
\tag{10.50}
$$

or equivalently that

$$
\Lambda_n = -2n(\hat{\theta}_n - \theta_0)^2[n^{-1}l''(\zeta)] + n(\hat{\theta}_n - \theta_0)^2[n^{-1}l''(\xi)].
$$

Using calculations similar to those used in the proof of Theorem 10.5, we have that Theorem 3.10 (Weak Law of Large Numbers) implies that $n^{-1}l''(\xi) \xrightarrow{p} -I(\theta_0)$ and $n^{-1}l''(\zeta) \xrightarrow{p} -I(\theta_0)$ as $n \to \infty$. Therefore it follows that $n^{-1}l''(\xi) = -I(\theta_0) + o_p(1)$ and $n^{-1}l''(\zeta) = -I(\theta_0) + o_p(1)$ as $n \to \infty$. Therefore, it follows that

$$
\Lambda_n = n(\hat{\theta}_n - \theta_0)^2 I(\theta_0) + o_p(1),
$$

as $n \to \infty$. Now, Theorem 10.5 implies that $n^{1/2}(\hat{\theta}_n - \theta)I^{1/2}(\theta_0) \xrightarrow{d} Z$ as $n \to \infty$ where Z has a $N(0, 1)$ distribution. Therefore, Theorem 4.12 implies that $n(\hat{\theta}_n - \theta)^2 I(\theta_0) \xrightarrow{d} X$ as $n \to \infty$, where X has a CHISQUARED(1) distribution, and the result is proven with an application of Theorem 4.11 (Slutsky). $\square$

Some other common test statistics that have asymtptotic CHISQUARED distributions include Wald's statistic and Rao's efficient score statistic. See Exercises 35 and 36. Under a sequence of alternative hypothesis $\{\theta_{1,n}\}_{n=1}^{\infty}$ where $\theta_{1,n} = \theta_0 + n^{-1/2}\delta$, it follows that Wilk's likelihood ratio test statistic has an asymptotic CHISQUARED statistic with a non-zero non-centrality parameter.

Theorem 10.12. *Suppose that $X_1, \ldots, X_n$ is a set of independent and identically distributed random variables from a distribution $F(x|\theta)$ with density $f(x|\theta)$, where θ has parameter space Ω. Suppose that the conditions of Theorem 10.5 hold, then, under the sequence of alternative hypotheses $\{\theta_{1,n}\}_{n=1}^{\infty}$ where $\theta_{1,n} = \theta_0 + n^{-1/2}\delta$, $\Lambda_n \overset{d}{\to} X$ as $n \to \infty$ where X has $\mathrm{CHISQUARED}[1, \delta^2 I(\theta_0)]$ distribution.*

Proof. We begin by noting that if $\theta_{1,n} = \theta_0 + n^{-1/2}\delta$, then retracing the steps in the proof of Theorem 10.11 implies that under the sequence of alternative hypotheses $\{\theta_{1,n}\}_{n=1}^{\infty}$ we have that $\Lambda_n = n(\hat{\theta}_n - \theta_0)^2 I(\theta_0) + o_p(n^{-1})$, as $n \to \infty$. See Exercise 34. Now, note that

$$n^{1/2}(\hat{\theta}_n - \theta_0) = n^{1/2}(\hat{\theta}_n - \theta_{1,n} + n^{-1/2}\delta) = n^{1/2}(\hat{\theta}_n - \theta_{1,n}) + \delta.$$

Therefore, Theorem 4.11 (Slutsky) implies that under the sequence of alternative hypotheses we have that $n^{1/2} I^{1/2}(\theta_0)(\hat{\theta}_n - \theta_0) \overset{d}{\to} Z$ as $n \to \infty$ where Z is a $N(\delta, 1)$ random variable. Therefore, Theorem 4.12 implies that $n(\hat{\theta}_n - \theta_0)^2 I(\theta_0) \overset{d}{\to} X$ as $n \to \infty$, where X has a $\mathrm{CHISQUARED}[1, \delta^2 I(\theta_0)]$ distribution, and the result is proven. $\qquad \square$

Similar results also hold for the asymptotic distributions of Wald's statistic and Rao's efficient score statistic under the same sequence of alternative hypotheses. See Exercises 37 and 38.

10.5 Observed Confidence Levels

Observed confidence levels provide useful information about the relative truth of hypotheses in multiple testing problems. In place of the repeated tests of hypothesis usually associated with multiple comparison techniques, observed confidence levels provide a level of confidence for each of the hypotheses. This level of confidence measures the amount of confidence there is, based on the observed data, that each of the hypotheses are true. This results in a relatively simple method for assessing the truth of a sequence of hypotheses.

The development of observed confidence levels begins by constructing a formal framework for the problem. This framework is based on the *problem of regions*, which was first formally proposed by Efron and Tibshirani (1998). The problem is constructed as follows. Let $\mathbf{X}$ be a d-dimensional random vector following a d-dimensional distribution function F. We will assume that F is a member of a collection, or family, of distributions given by $\mathcal{F}$. The collection $\mathcal{F}$ may correspond to a parametric family such as the collection of all d-variate normal distributions, or may be nonparametric, such as the collection of all continuous distributions with finite mean. Let $\boldsymbol{\theta}$ be the parameter of interest and assume that $\boldsymbol{\theta}$ is a functional parameter of F of the form $\boldsymbol{\theta} = T(F)$, with parameter space Ω. Let $\{\Omega_i\}_{i=1}^{\infty}$ be a countable sequence of subsets, or regions, of Ω such that

$$\bigcup_{i=1}^{\infty} \Omega_i = \Omega.$$

For simplicity each region will usually be assumed to be connected, though in many problems Ω_i is made up of disjoint regions. Further, in many practical examples the sequence is finite, but there are also practical examples that require countable sequences. There are also examples where the sequence $\{\Omega_i\}_{i=1}^{\infty}$ forms a partition of Ω in the sense that $\Omega_i \cap \Omega_j = \emptyset$ for all $i \neq j$. In the general case, the possibility that the regions can overlap will be allowed. In many practical problems the subsets technically overlap on their boundaries, but the sequence can often be thought of as a partition from a practical viewpoint. The statistical interest in such a sequence of regions arises from the structure of a specific inferential problem. Typically the regions correspond to competing models for the distribution of the random vector $\mathbf{X}$ and one is interested in determining which of the models is most reasonable based on the observed data vector $\mathbf{X}$. Therefore the problem of regions is concerned with determining which of the regions in the sequence $\{\Omega_i\}_{i=1}^{\infty}$ that $\boldsymbol{\theta}$ belongs to based on the observed data $\mathbf{X}$.

An obvious simple solution to this problem would be to estimate $\boldsymbol{\theta}$ based on $\mathbf{X}$ and conclude that $\boldsymbol{\theta}$ is in the region Ω_i whenever the estimate $\hat{\boldsymbol{\theta}}_n$ is in the region Ω_i. We will consider the estimator $\hat{\boldsymbol{\theta}}_n = T(\hat{F}_n)$ where $\hat{F}_n$ is the empirical distribution function defined in Definition 3.5. The problem with simply concluding that $\boldsymbol{\theta} \in \Omega_i$ whenever $\hat{\boldsymbol{\theta}}_n \in \Omega_i$ is that $\hat{\boldsymbol{\theta}}_n$ is subject to sample variability. Therefore, even though we may observe $\hat{\boldsymbol{\theta}}_n \in \Omega_i$, it may actually be true that $\hat{\boldsymbol{\theta}}_n \in \Omega_j$ for some $i \neq j$ where $\Omega_i \cap \Omega_j = \emptyset$, and that $\hat{\boldsymbol{\theta}} \in \Omega_i$ was observed simply due to chance. If such an outcome were rare, then the method may be acceptable. However, if such an outcome occurred relatively often, then the method would not be useful. Therefore, it is clear that the inherent variability in $\hat{\boldsymbol{\theta}}_n$ must be accounted for in order to develop a useful solution to the problem of regions.

Multiple comparison techniques solve the problem of regions using a sequence of hypothesis tests. Adjustments to the testing technique helps control the overall significance level of the sequence of tests. Modern techniques have been developed by Stefansson, Kim and Hsu (1988) and Finner and Strassburger (2002). Some general references that address issues concerned with multiple comparison techniques include Hochberg and Tamhane (1987), Miller (1981) and Westfall and Young (1993). Some practitioners find the results of these procedures difficult to interpret as the number of required tests can sometimes be quite large.

An alternate approach to multiple testing techniques was formally introduced by Efron and Tibshirani (1998). This approach computes a measure of confidence for each of the regions. This measure reflects the amount of confidence there is that $\boldsymbol{\theta}$ lies within the region based on the observed sample

X. The method used for computing the observed confidence levels studied here is based on the methodology of Polansky (2003a, 2003b, 2007). Let $C(\alpha, \omega) \subset \Theta$ be a $100\alpha\%$ confidence region for θ based on the sample **X**. That is, $C(\alpha, \omega) \subset \Omega$ is a function of the sample **X** with the property that

$$P[\theta \in C(\alpha, \omega)] = \alpha.$$

The vector $\omega \in W_\alpha \subset \mathbb{R}^q$ is called the *shape parameter vector* as it contains a set of parameters that control the shape and orientation of the confidence region, but do not have an effect on the confidence coefficient. Even though W_α is usually a function of α, the subscript α will often be omitted to simplify mathematical expressions. Now suppose that there exist sequences $\{\alpha_i\}_{i=1}^\infty \in [0,1]$ and $\{\omega_i\}_{i=1}^\infty \in W_\alpha$ such that $C(\alpha_i, \omega_i) = \Omega_i$ for $i = 1, 2, \ldots$, conditional on **X**. Then the sequence of confidence coefficients are defined to be the observed confidence levels for $\{\Omega_i\}_{i=1}^\infty$. In particular, α_i is defined to be the observed confidence level of the region Ω_i. That is, the region Ω_i corresponds to a $100\alpha_i\%$ confidence region for θ based on the observed data. This measure is similar to the measure suggested by Efron and Gong (1983), Felsenstein (1985) and Efron, Holloran, and Holmes (1996). It is also similar in application to the methods of Efron and Tibshirani (1998), though the formal definition of the measure differs slightly from the definition used above. See Efron and Tibshirani (1998) for further details on this definition.

Example 10.32. To demonstrate this idea, consider a simple example where $X_1, \ldots, X_n$ is a set of independent and identically distributed random variables from a $N(\theta, \sigma^2)$ distribution. Let $\hat{\theta}_n$ and $\hat{\sigma}_n$ be the usual sample mean and variance computed on $X_1, \ldots, X_n$. A confidence interval for the mean that is based on the assumption that the population is NORMAL is based on percentiles from the $T(n-1)$ distribution and has the form

$$C(\alpha, \omega) = (\hat{\theta}_n - t_{n-1;1-\omega_L} n^{-1/2} \hat{\sigma}_n, \hat{\theta}_n - t_{n-1;1-\omega_U} n^{-1/2} \hat{\sigma}_n), \qquad (10.51)$$

where $t_{\nu;\xi}$ is the ξth quantile of a $T(\nu)$ distribution. In order for the confidence interval in Equation (10.51) to have a confidence level equal to $100\alpha\%$ we take $\omega' = (\omega_L, \omega_U)$ to be the shape parameter vector where

$$W_\alpha = \{\omega : \omega_U - \omega_L = \alpha, \omega_L \in [0,1], \omega_U \in [0,1]\},$$

for $\alpha \in (0, 1)$. Note that selecting $\omega \in W_\alpha$ not only ensures that the confidence level is $100\alpha\%$, but also allows for several orientations and shapes of the interval. For example, a symmetric two-tailed interval can be constructed by selecting $\omega_L = (1-\alpha)/2$ and $\omega_U = (1+\alpha)/2$. An upper one-tailed interval is constructed by setting $\omega_L = 0$ and $\omega_U = \alpha$. A lower one-tailed interval uses $\omega_L = 1 - \alpha$ and $\omega_U = 1$.

Now consider the problem of computing observed confidence levels for the NORMAL mean for a sequence of interval regions of the form $\Omega_i = [t_i, t_{i+1}]$ where $-\infty < t_i < t_{i+1} < \infty$ for $i \in \mathbb{N}$. Setting $\Omega_i = C(\alpha, \omega)$ where the confidence interval used for this calculation is the one given in Equation (10.51) yields

$$\hat{\theta}_n - t_{n-1;1-\omega_L} n^{-1/2} \hat{\sigma}_n = t_i, \tag{10.52}$$

and

$$\hat{\theta}_n - t_{n-1;1-\omega_U} n^{-1/2} \hat{\sigma}_n = t_{i+1}. \tag{10.53}$$

Solving Equations (10.52) and (10.53) for ω_L and ω_U yields

$$\omega_L = 1 - T_{n-1}[n^{1/2} \hat{\sigma}_n^{-1}(\hat{\theta}_n - t_i)],$$

and

$$\omega_U = 1 - T_{n-1}[n^{1/2} \hat{\sigma}_n^{-1}(\hat{\theta}_n - t_{i+1})],$$

where T_{n-1} is the distribution function of a $T(n-1)$ distribution. Because $\boldsymbol{\omega} \in W_\alpha$ if and only if $\omega_U - \omega_L = \alpha$ it follows that the observed confidence level for the region Ω_i is given by

$$T_{n-1}[n^{1/2} \hat{\sigma}_n^{-1}(\hat{\theta}_n - t_i)] - T_{n-1}[n^{1/2} \hat{\sigma}_n^{-1}(\hat{\theta}_n - t_{i+1})].$$

∎

This section will consider the problem of developing the asymptotic theory of observed confidence levels for the case when there is a single parameter, that is, when $\Omega \subset \mathbb{R}$. To simplify the development in this case, it will be further assumed that Ω is an interval subset of $\mathbb{R}$. Most standard single parameter problems in statistical inference fall within these assumptions.

An observed confidence level is simply a function that takes a subset of Ω and maps it to a real number between 0 and 1. Formally, let α be a function and let $\mathcal{T}$ be a collection of subsets of Ω. Then an observed confidence level is a function $\alpha : \mathcal{T} \to [0, 1]$. Because confidence levels are closely related to probabilities, it is reasonable to assume that α has the axiomatic properties given in Definition 2.2 where we will suppose that $\mathcal{T}$ is a sigma-field of subsets of Ω. For most reasonable problems in statistical inference it should suffice to take $\mathcal{T}$ to be the Borel σ-field on Ω. Given this structure, it suffices to develop observed confidence levels for interval subsets of Θ. Observed confidence levels for other regions can be obtained through operations derived from the axioms in Definition 2.2.

To develop observed confidence levels for a general scalar parameter θ, consider a single interval region of the form $\Psi = (t_L, t_U) \in \mathcal{T}$. To compute the observed confidence level of Ψ, a confidence interval for θ based on the sample $\mathbf{X}$ is required. The general form of a confidence interval for θ based on $\mathbf{X}$ can usually be written as $C(\alpha, \boldsymbol{\omega}) = [L(\omega_L), U(\omega_U)]$, where ω_L and ω_U are shape parameters such that $(\omega_L, \omega_U) \in W_\alpha$, for some $W_\alpha \subset \mathbb{R}^2$. It can often be assumed that $L(\omega_L)$ and $U(\omega_U)$ are continuous monotonic functions of ω_L and ω_U onto Ω, respectively, conditional on the observed sample $\mathbf{X}$. See, for example, Section 9.2 of Casella and Berger (2002). If such an assumption is true, the observed confidence level of Ψ is computed by setting $\Psi = C(\alpha, \boldsymbol{\omega})$ and solving for $\boldsymbol{\omega}$. The value of α for which $\boldsymbol{\omega} \in W_\alpha$ is the observed confidence level of Ψ. For the form of the confidence interval given above, the solution is obtained by setting $\omega_L = L^{-1}(t_L)$ and $\omega_U = U^{-1}(t_U)$, conditional on $\mathbf{X}$. A unique solution will exist for both shape parameters given the assumptions on the functions L and U. Therefore, the observed confidence level of Ψ is the value of α such that $\boldsymbol{\omega} = (\omega_L, \omega_U) \in \Omega_\alpha$. Thus, the calculation of observed confidence levels in the single parameter case is equivalent to inverting the endpoints of a confidence interval for θ. Some simple examples illustrating this method is given below.

Example 10.33. Continuing with the setup of Example 10.32, suppose that $X_1, \ldots, X_n$ is a set of independent and identically distributed random variables from a $N(\mu, \theta)$ distribution where $\theta < \infty$. For the variance, the parameter space is $\Omega = (0, \infty)$, so that the region $\Psi = (t_L, t_U)$ is assumed to follow the restriction that $0 < t_L \leq t_U < \infty$. Let $\hat{\theta}_n$ be the unbiased version of the sample variance, then a $100\alpha\%$ confidence interval for θ is given by

$$C(\alpha, \boldsymbol{\omega}) = \left[\frac{(n-1)\hat{\theta}_n}{\chi^2_{n-1;1-\omega_L}}, \frac{(n-1)\hat{\theta}_n}{\chi^2_{n-1;1-\omega_U}} \right], \tag{10.54}$$

where $\boldsymbol{\omega} \in W_\alpha$ with

$$W_\alpha = \{\boldsymbol{\omega}' = (\omega_1, \omega_2) : \omega_L \in [0,1], \omega_U \in [0,1], \omega_U - \omega_L = \alpha\},$$

and $\chi^2_{\nu,\xi}$ is the ξth percentile of a CHISQUARED(ν) distribution. Therefore

$$L(\boldsymbol{\omega}) = \frac{(n-1)\hat{\theta}_n}{\chi^2_{n-1;1-\omega_L}},$$

and

$$U(\boldsymbol{\omega}) = \frac{(n-1)\hat{\theta}_n}{\chi^2_{n-1;1-\omega_U}}.$$

Setting $\Psi = C(\alpha, \boldsymbol{\omega})$ and solving for ω_L and ω_U yields

$$\omega_L = L^{-1}(t_L) = 1 - \chi^2_{n-1}[t_L^{-1}(n-1)\hat{\theta}_n],$$

and

$$\omega_U = U^{-1}(t_U) = 1 - \chi^2_{n-1}[t_U^{-1}(n-1)\hat{\theta}_n],$$

where χ^2_ν is the distribution function of a CHISQUARED(ν) distribution. This implies that the observed confidence limit for Ψ is given by

$$\alpha(\Psi) = \chi^2_{n-1}[t_L^{-1}(n-1)\hat{\theta}_n] - \chi^2_{n-1}[t_U^{-1}(n-1)\hat{\theta}_n].$$

∎

Example 10.34. Suppose $(X_1, Y_1), \ldots, (X_n, Y_n)$ is a set of independent and identically distributed bivariate random vectors from a bivariate normal distribution with mean vector $\boldsymbol{\mu}' = (\mu_X, \mu_Y)$ and covariance matrix

$$\boldsymbol{\Sigma} = \begin{bmatrix} \sigma_X^2 & \sigma_{XY} \\ \sigma_{XY} & \sigma_Y^2 \end{bmatrix},$$

where $V(X_i) = \sigma_X^2$, $V(Y_i) = \sigma_Y^2$ and the covariance between X and Y is

σ_{XY}. Let $\theta = \sigma_{XY}\sigma_X^{-1}\sigma_Y^{-1}$, the correlation coefficient between X and Y. The problem of constructing a reliable confidence interval for θ is usually simplified using Fisher's normalizing transformation. See Fisher (1915) and Winterbottom (1979). Using the fact that $\tanh^{-1}(\hat{\theta}_n)$ has an approximate $N[\tanh^{-1}(\theta), (n-3)^{-1/2}]$ distribution, the resulting approximate confidence interval for θ has the form

$$C(\alpha, \boldsymbol{\omega}) = [\tanh(\tanh^{-1}(\hat{\theta}_n) - z_{1-\omega_L}(n-3)^{-1/2}),$$
$$\tanh(\tanh^{-1}(\hat{\theta}_n) - z_{1-\omega_U}(n-3)^{-1/2})] \qquad (10.55)$$

where $\hat{\theta}_n$ is the sample correlation coefficient given by

$$\hat{\theta}_n = \frac{\sum_{i=1}^{n}(X_i - \bar{X})(Y_i - \bar{Y})}{\left[\sum_{i=1}^{n}(X_i - \bar{X})^2\right]^{1/2}\left[\sum_{i=1}^{n}(Y_i - \bar{Y})^2\right]^{1/2}}, \qquad (10.56)$$

and $\boldsymbol{\omega}' = (\omega_L, \omega_U) \in W_\alpha$ with

$$W_\alpha = \{\boldsymbol{\omega}' = (\omega_1, \omega_2) : \omega_L \in [0,1], \omega_U \in [0,1], \omega_U - \omega_L = \alpha\}.$$

Therefore

$$L(\omega_L) = \tanh(\tanh^{-1}(\hat{\theta}_n) - z_{1-\omega_L}(n-3)^{-1/2})$$

and

$$U(\omega_U) = \tanh(\tanh^{-1}(\hat{\theta}_n) - z_{1-\omega_U}(n-3)^{-1/2}).$$

Setting $\Psi = (t_L, t_U) = C(\alpha, \boldsymbol{\omega})$ yields

$$\omega_L = 1 - \Phi[(n-3)^{1/2}(\tanh^{-1}(\hat{\theta}_n) - \tanh^{-1}(t_L))],$$

and

$$\omega_U = 1 - \Phi[(n-3)^{1/2}(\tanh^{-1}(\hat{\theta}_n) - \tanh^{-1}(t_U))],$$

so that the observed confidence level for Ψ is given by

$$\alpha(\Psi) = \Phi[(n-3)^{1/2}(\tanh^{-1}(\hat{\theta}) - \tanh^{-1}(t_L))] -$$
$$\Phi[(n-3)^{1/2}(\tanh^{-1}(\hat{\theta}) - \tanh^{-1}(t_U))].$$

∎

For the asymptotic development in this section we will consider problems that occur within the smooth function model described in Section 7.4. As observed in Section 10.3, confidence regions for θ can be constructed using the ordinary upper confidence limit

$$\hat{\theta}_{\text{ord}}(\alpha) = \hat{\theta}_n - n^{-1/2}\sigma g_{1-\alpha}, \qquad (10.57)$$

for the case when σ is known, and the studentized critical point

$$\hat{\theta}_{\text{stud}}(\alpha) = \hat{\theta}_n - n^{-1/2}\hat{\sigma}_n h_{1-\alpha}, \qquad (10.58)$$

for the case when σ is unknown. Two-sided confidence intervals can be developed using each of these upper confidence limits. For example, if $\omega_L \in (0,1)$ and $\omega_U \in (0,1)$ are such that $\alpha = \omega_U - \omega_L \in (0,1)$ then $C_{\text{ord}}(\alpha, \boldsymbol{\omega}) =$

$[\hat{\theta}_{\mathrm{ord}}(\omega_L), \hat{\theta}_{\mathrm{ord}}(\omega_U)]$ is a $100\alpha\%$ confidence interval for θ when σ is known. Similarly

$$C_{\mathrm{stud}}(\alpha, \boldsymbol{\omega}) = [\hat{\theta}_{\mathrm{stud}}(\omega_L), \hat{\theta}_{\mathrm{stud}}(\omega_U)],$$

is a $100\alpha\%$ confidence interval for θ when σ is unknown.

For a region $\Psi = (t_L, t_U)$ the observed confidence level corresponding to each of the theoretical critical points can be computed by setting Ψ equal to the confidence interval and solving for ω_L and ω_U. For example, in the case of the ordinary theoretical critical point, setting $\Psi = C_{\mathrm{ord}}(\alpha, \boldsymbol{\omega}; \mathbf{X})$ yields the two equations

$$t_L = L(\boldsymbol{\omega}) = \hat{\theta}_n - n^{-1/2}\sigma g_{1-\omega_L}, \qquad (10.59)$$

and

$$t_U = U(\boldsymbol{\omega}) = \hat{\theta}_n - n^{-1/2}\sigma g_{1-\omega_U}. \qquad (10.60)$$

Let $G_n(x) = P[n^{1/2}\sigma^{-1}(\hat{\theta}_n - \theta)]$ and $H_n(x) = P[n^{1/2}\hat{\sigma}_n^{-1}(\hat{\theta}_n - \theta)]$. Solving for ω_L and ω_U in Equations (10.59) and (10.60) yields

$$\omega_L = L^{-1}(t_L) = 1 - G_n[n^{1/2}\sigma^{-1}(\hat{\theta}_n - t_L)],$$

and

$$\omega_U = U^{-1}(t_U) = 1 - G_n[n^{1/2}\sigma^{-1}(\hat{\theta}_n - t_U)],$$

so that the observed confidence level corresponding to the ordinary theoretical confidence limit is

$$\alpha_{\mathrm{ord}}(\Psi) = G_n[n^{1/2}\sigma^{-1}(\hat{\theta}_n - t_L)] - G_n[n^{1/2}\sigma^{-1}(\hat{\theta}_n - t_U)]. \qquad (10.61)$$

Similarly, the observed confidence levels corresponding to the studentized confidence interval is

$$\alpha_{\mathrm{stud}}(\Psi) = H_n[n^{1/2}\hat{\sigma}_n^{-1}(\hat{\theta}_n - t_L)] - H_n[n^{1/2}\hat{\sigma}_n^{-1}(\hat{\theta}_n - t_U)]. \qquad (10.62)$$

In the case where F is unknown, the asymptotic NORMAL behavior of the distributions G_n and H_n can be used to compute approximate observed confidence levels of the form

$$\hat{\alpha}_{\mathrm{ord}}(\Psi) = \Phi[n^{1/2}\sigma^{-1}(\hat{\theta}_n - t_L)] - \Phi[n^{1/2}\sigma^{-1}(\hat{\theta}_n - t_U)], \qquad (10.63)$$

and

$$\hat{\alpha}_{\mathrm{stud}}(\Psi) = \Phi[n^{1/2}\hat{\sigma}_n^{-1}(\hat{\theta}_n - t_L)] - \Phi[n^{1/2}\hat{\sigma}_n^{-1}(\hat{\theta}_n - t_U)], \qquad (10.64)$$

for the cases where σ is known, and unknown, respectively.

The observed confidence level based on the NORMAL approximation is just one of several alternative methods available for computing an observed confidence level for any given parameter. Indeed, as was pointed out earlier, any function that maps regions to the unit interval such that the three properties of a probability measure are satisfied is technically a method for computing an observed confidence level. Even if we focus on observed confidence levels that are derived from confidence intervals that at least guarantee their coverage level asymptotically, there may be many methods to choose from, and techniques for comparing the methods become paramount in importance.

This motivates the question as to what properties we would wish the observed confidence levels to possess. Certainly the issue of consistency would be relevant in that an observed confidence level computed on a region $\Omega_0 = (t_{0L}, t_{0U})$ that contains θ should converge to one as the sample size becomes large. Correspondingly, an observed confidence level computed on a region $\Omega_1 = (t_{1L}, t_{1U})$ that does not contain θ should converge to zero as the sample size becomes large. The issue of consistency is relatively simple to decide within the smooth function model studied. The normal approximation provides the simplest case and is a good starting point. Consider the ordinary observed confidence level given in Equation (10.63). Note that

$$\Phi[n^{1/2}\sigma^{-1}(\hat{\theta}_n - t_{0L})] = \Phi[n^{1/2}\sigma^{-1}(\theta - t_{0L}) + n^{1/2}\sigma^{-1}(\hat{\theta}_n - \theta)]$$

where $n^{1/2}(\theta - t_{0L})/\sigma \to \infty$ and $n^{1/2}(\hat{\theta}_n - \theta)/\sigma \xrightarrow{d} Z$ as $n \to \infty$, where Z is a $N(0,1)$ random variable. It is clear that the second sequence is bounded in probability, so that the first sequence dominates. It follows that

$$\Phi[n^{1/2}\sigma^{-1}(\hat{\theta}_n - t_{0L})] \xrightarrow{p} 1$$

as $n \to \infty$. Similarly, it can be shown that

$$\Phi[n^{1/2}\sigma^{-1}(\hat{\theta}_n - t_{0U})] \xrightarrow{p} 0$$

as $n \to \infty$, so that it follows that $\hat{\alpha}_{\text{ord}}(\Omega_0) \xrightarrow{p} 1$ as $n \to \infty$ when $\theta \in \Omega_0$. A similar argument, using the fact that $\hat{\sigma}_n \sigma^{-1} \xrightarrow{p} 1$ as $n \to \infty$ can be used to show that $\hat{\alpha}_{\text{stud}}(\Omega_0) \xrightarrow{p} 1$ as $n \to \infty$, as well. Arguments to show that $\alpha_{\text{ord}}(\Omega_0)$ and $\alpha_{\text{stud}}(\Omega_0)$ are also consistent follow in a similar manner, though one must use the fact that $G_n \rightsquigarrow \Phi$ and $H_n \rightsquigarrow \Phi$ as $n \to \infty$.

Beyond consistency, it is desirable for an observed confidence level to provide an accurate representation of the level of confidence there is that $\theta \in \Psi$, given the observed sample $\mathbf{X}_1, \ldots, \mathbf{X}_n$. Considering the definition of an observed confidence level, it is clear that if Ψ corresponds to a $100\alpha\%$ confidence interval for θ, conditional on $\mathbf{X}_1, \ldots, \mathbf{X}_n$, the observed confidence level for Ψ should be α. When σ is known the interval $C_{\text{ord}}(\alpha, \boldsymbol{\omega})$ will be used as the standard for a confidence interval for θ. Hence, a measure $\tilde{\alpha}$ of an observed confidence level is *accurate* if $\tilde{\alpha}[C_{\text{ord}}(\alpha, \boldsymbol{\omega})] = \alpha$. For the case when σ is unknown the interval $C_{\text{stud}}(\alpha, \boldsymbol{\omega})$ will be used as the standard for a confidence interval for θ, and an arbitrary measure $\tilde{\alpha}$ is defined to be accurate if $\tilde{\alpha}[C_{\text{stud}}(\alpha, \boldsymbol{\omega})] = \alpha$. Using this definition, it is clear that α_{ord} and α_{stud} are accurate when σ is known and unknown, respectively. When σ is known and $\tilde{\alpha}[C_{\text{ord}}(\alpha, \boldsymbol{\omega})] \neq \alpha$ or when σ is unknown and $\tilde{\alpha}[C_{\text{stud}}(\alpha, \boldsymbol{\omega})] \neq \alpha$ one can analyze how close $\tilde{\alpha}[C_{\text{ord}}(\alpha, \boldsymbol{\omega})]$ or $\tilde{\alpha}[C_{\text{stud}}(\alpha, \boldsymbol{\omega})]$ is to α using asymptotic expansion theory. In particular, if σ is known then a measure of confidence $\tilde{\alpha}$ is said to be k^{th}-*order accurate* if $\tilde{\alpha}[C_{\text{ord}}(\alpha, \boldsymbol{\omega})] = \alpha + O(n^{-k/2})$, as $n \to \infty$. Similarly, if σ is unknown a measure $\tilde{\alpha}$ is said to be k^{th}-order accurate if $\tilde{\alpha}[C_{\text{stud}}(\alpha, \boldsymbol{\omega})] = \alpha + O(n^{-k/2})$, as $n \to \infty$.

To analyze the normal approximations, let us first suppose that σ is known.

If $\alpha = \omega_U - \omega_L$ then

$$
\begin{aligned}
\hat{\alpha}_{\text{ord}}[C_{\text{ord}}(\alpha, \boldsymbol{\omega})] &= \Phi\{n^{1/2}\sigma^{-1}[\hat{\theta} - \hat{\theta}_{\text{ord}}(\omega_L)]\} - \\
&\quad \Phi\{n^{1/2}\sigma^{-1}[\hat{\theta} - \hat{\theta}_{\text{ord}}(\omega_U)]\} \\
&= \Phi(g_{1-\omega_L}) - \Phi(g_{1-\omega_U}).
\end{aligned}
\tag{10.65}
$$

If $G_n = \Phi$, then $\hat{\alpha}_{\text{ord}}[C_{\text{ord}}(\alpha, \boldsymbol{\omega})] = \alpha$, and the method is accurate. When $G_n \neq \Phi$ the Cornish-Fisher expansion for a quantile of G_n, along with an application of Theorem 1.13 (Taylor) to Φ yields

$$
\Phi(g_{1-\omega}) = 1 - \omega - n^{-1/2} r_1(z_\omega)\phi(z_\omega) + O(n^{-1}),
$$

as $n \to \infty$, for an arbitrary value of $\omega \in (0,1)$. It is then clear that

$$
\hat{\alpha}_{\text{ord}}[C_{\text{ord}}(\alpha, \boldsymbol{\omega})] = \alpha + n^{-1/2}\Delta(\omega_L, \omega_U) + O(n^{-1}),
\tag{10.66}
$$

as $n \to \infty$ where

$$
\Delta(\omega_L, \omega_U) = r_1(z_{\omega_U})\phi(z_{\omega_U}) - r_1(z_{\omega_L})\phi(z_{\omega_L}).
$$

One can observe that $\hat{\alpha}_{\text{ord}}$ is first-order accurate, unless the first-order term in Equation (10.66) is functionally zero. If it happens that $\omega_L = \omega_U$ or $\omega_L = 1 - \omega_U$, then it follows that $r_1(z_{\omega_L})\phi(z_{\omega_L}) = r_1(z_{\omega_U})\phi(z_{\omega_U})$ since r_1 is an even function and the first-order term vanishes. The first case corresponds to a degenerate interval with confidence measure equal to zero. The second case corresponds to the situation where $\hat{\theta}$ corresponds to the midpoint of the interval (t_L, t_U). Otherwise, the term is typically nonzero.

When σ is unknown and $\alpha = \omega_U - \omega_L$ we have that

$$
\begin{aligned}
\hat{\alpha}_{\text{stud}}[C_{\text{stud}}(\alpha, \boldsymbol{\omega})] &= \Phi\{n^{1/2}\hat{\sigma}_n^{-1}[\hat{\theta} - \hat{\theta}_{\text{stud}}(\omega_L)]\} - \\
&\quad \Phi\{n^{1/2}\hat{\sigma}_n^{-1}[\hat{\theta} - \hat{\theta}_{\text{stud}}(\omega_U)]\} \\
&= \Phi(h_{1-\omega_L}) - \Phi(h_{1-\omega_U}).
\end{aligned}
$$

A similar argument to the one given above yields

$$
\hat{\alpha}_{\text{stud}}[C_{\text{stud}}(\alpha, \boldsymbol{\omega})] = \alpha + n^{-1/2}\Lambda(\omega_L, \omega_U) + O(n^{-1}),
$$

as $n \to \infty$, where

$$
\Lambda(\omega_L, \omega_U) = v_1(z_{\omega_U})\phi(z_{\omega_U}) - v_1(z_{\omega_L})\phi(z_{\omega_L}).
$$

Therefore, the methods based on the normal approximation are first-order accurate. This accuracy can be improved using Edgeworth type corrections. In particular, by estimating the polynomials in the first term of the Edgeworth expansions for the distributions H_n and G_n, second-order correct observed confidence levels can be obtained. See Exercise 43. Observed confidence levels can also be applied to problems where the parameter is a vector, along with problems in regression, linear models, and density estimation. See Polansky (2007) for further details on these applications.

10.6 Bayesian Estimation

The statistical inference methods used so far in this book are classified as *frequentists* methods. In these methods the unknown parameter is considered to be a fixed constant that is an element of the parameter space. The random mechanism that produces the observed data is based on a distribution that depends on this unknown, but fixed, parameter. These methods are called frequentist methods because the results of the statistical analyses are interpreted using the frequency interpretation of probability. That is, the methods are justified by properties that hold under repeated sampling from the distribution of interest. For example, a $100\alpha\%$ confidence set is justified in that the probability that the set contains the true parameter value with a probability of α *before the sample is taken*, or that the expected proportion of the confidence sets that contain the parameter over repeated sampling from the same distribution is α.

An alternative view of statistical inference is based on *Bayesian* methods. In the Bayesian framework the unknown parameter is considered to be a random variable and the observed data is based on the joint distribution between the parameter and the observed random variables. In the usual formulation the distribution of the parameter, called the *prior* distribution, and the conditional distribution of the observed random variables given the parameter are specified. Inference is then carried out using the distribution of the parameter conditional on the sample that was observed. This distribution is called the *posterior* distribution. The computation of the posterior distribution is based on calculations justified by Bayes' theorem, from which the methods are named. The interpretation of using Bayes' theorem in this way is that the prior distribution can be interpreted as the knowledge of the parameter before the data was observed, while the posterior distribution is the knowledge of the parameter that has been updated based on the information from the observed sample. The advantage of this type of inference is that the theoretical properties of Bayesian methods are interpreted for the sample that was actually observed, and not over all possible samples. The interpretation of the results is also simplified due to the randomness of the parameter value. For example, while a confidence interval must be interpreted in view of all of the possible samples that could have been observed, a Bayesian confidence set produces a set that has a posterior probability of α, which can be interpreted solely on the basis of the observed sample and the prior distribution.

In some sense Bayesian methods do not need to rely on asymptotic properties for their justification because of their interpretability on the current sample. However, many Bayesian methods can be justified within the frequentist framework as well. In this section we will demonstrate that Bayes estimators can also be asymptotically efficient and have an asymptotic NORMAL distribution within the frequentist framework.

To formalize the framework for our study, consider a set of independent and identically distributed random variables from a distribution $F(x|\theta)$ which has either a density of probability distribution function given by $f(x|\theta)$ where we will assume for simplicity that $x \in \mathbb{R}$. The parameter θ will be assumed to follow the prior distribution $\pi(\theta)$ over some parameter space Ω, which again for simplicity we will often be taken to be $\mathbb{R}$. The object of a Bayesian analysis is then to obtain the posterior distribution $\pi(\theta|x_1, \ldots, x_n)$, which can be obtained using an argument based on Bayes' theorem of the form

$$\pi(\theta|x_1, \ldots, x_n) = \frac{f(x_1, \ldots, x_n, \theta)}{m(x_1, \ldots, x_n)},$$

where $f(x_1, \ldots, x_n, \theta)$ is the joint distribution of $X_1, \ldots, X_n$ and θ, and the marginal distribution of $X_1, \ldots, X_n$ is given by

$$m(x_1, \ldots, x_n) = \int_{\Omega} f(x_1, \ldots, x_n, \theta) d\theta.$$

Using the fact that the joint distribution of $X_1, \ldots, X_n$ and θ can be found as $f(x_1, \ldots, x_n, \theta) = f(x_1, \ldots, x_n|\theta)\pi(\theta)$ it follows that the posterior distribution can be found directly from $f(x_1, \ldots, x_n|\theta)$ and $\pi(\theta)$ as

$$\pi(\theta|x_1, \ldots, x_n) = \frac{f(x_1, \ldots, x_n|\theta)\pi(\theta)}{\int_{\Omega} f(x_1, \ldots, x_n|\theta)\pi(\theta) d\theta}. \tag{10.67}$$

Noting that the denominator of Equation (10.67) is a constant, it is often enough to conclude that

$$\pi(\theta|x_1, \ldots, x_n) \propto f(x_1, \ldots, x_n|\theta)\pi(\theta),$$

which eliminates the need to compute the integral, which can be difficult in some cases. Once the posterior distribution is computed then a Bayesian analysis will either use the distribution itself as the updated knowledge about the parameter. Alternately, point estimates, confidence regions and tests can be constructed, though their interpretation is necessarily different than the parallel frequentists methods. For an introduction to Bayesian methods see Bolstad (2007).

The derivation of a *Bayes estimator* of a parameter θ begins by specifying a loss function $L[\theta, \delta(x_1, \ldots, x_n)]$ where $\delta(x_1, \ldots, x_n)$ is a point estimator, called a *decision rule*. The *posterior expected loss*, or *Bayes risk*, is then given by

$$\int_{-\infty}^{\infty} L[\theta, \delta(x_1, \ldots, x_n)]\pi(\theta|x_1, \ldots, x_n) d\theta.$$

The Bayes estimator of θ is then taken to be the decision rule δ that minimizes the Bayes risk. The result given below, which is adapted from Lehmann and Casella (1998) provides conditions under which a Bayes estimator of θ can be found for two common loss functions.

Theorem 10.13. *Let θ have a prior distribution π over Ω and suppose that the density or probability distribution of $X_1, \ldots, X_n$, conditional on θ is given by $f_{\theta}(x_1, \ldots, x_n|\theta)$. If*

1. $L(\theta, \delta)$ is a non-negative loss function,
2. There exists a decision rule δ that has finite risk,
3. For almost all $(x_1, \ldots, x_n) \in \mathbb{R}^n$ there exists a decision rule $\delta(x_1, \ldots, x_n)$ that minimizes the Bayes risk,

then δ is the Bayes estimator and

1. If $L(\theta, \delta) = (\theta - \delta)^2$ then the Bayes estimator is $E(\theta | X_1, \ldots, X_n)$ and is unique.
2. If $L(\theta, \delta) = |\theta - \delta|$ then the Bayes estimator is any median of the posterior distribution.

For a proof of Theorem 10.13 see Section 4.1 of Lehmann and Casella (1998).

Example 10.35. Let X be a single observation from a discrete distribution with probability distribution function

$$f(x|\theta) = \begin{cases} \frac{1}{2}\theta & x \in \{-1, 1\}, \\ 1 - \theta & x = 0, \\ 0 & \text{elsewhere}, \end{cases}$$

where $\theta \in \Omega = \{\frac{1}{4}, \frac{1}{2}, \frac{3}{4}\}$. Suppose that we place a UNIFORM$\{\frac{1}{4}, \frac{1}{2}, \frac{3}{4}\}$ prior distribution on θ. The posterior distribution can then be found by direct calculation. For example if we observe $X = 0$, the posterior probability for $\theta = \frac{1}{4}$ is

$$P(\theta = \tfrac{1}{4} | X = 0) = \frac{P(X = 0 | \theta = \frac{1}{4})P(\theta = \frac{1}{4})}{P(X = 0)} = \frac{\frac{3}{4} \cdot \frac{1}{3}}{P(X = 0)}$$

where

$$P(X = 0) = P(X = 0 | \theta = \tfrac{1}{4})P(\theta = \tfrac{1}{4}) + P(X = 0 | \theta = \tfrac{1}{2})P(\theta = \tfrac{1}{2}) +$$
$$P(X = 0 | \theta = \tfrac{3}{4})P(\theta = \tfrac{3}{4}) = \tfrac{3}{4} \cdot \tfrac{1}{3} + \tfrac{1}{2} \cdot \tfrac{1}{3} + \tfrac{1}{4} \cdot \tfrac{1}{3} = \tfrac{1}{2}.$$

Therefore, the posterior probability is $P(\theta = \frac{1}{4} | X = 0) = \frac{3}{12} \cdot \frac{2}{1} = \frac{1}{2}$. Similar calculations can be used to show that $P(\theta = \frac{1}{2} | X = 0) = \frac{1}{3}$ and $P(\theta = \frac{3}{4} | X = 0) = \frac{1}{6}$. One can note that the lowest value of θ, which corresponds to the highest probability for $X = 0$ has the largest posterior probability. If we consider using squared error loss, then Theorem 10.13 implies that the Bayes estimator of θ is given by the mean of the posterior distribution, which is $\tilde{\theta} = \frac{5}{12}$. ∎

Example 10.36. Let $B_1, \ldots, B_n$ be a set of independent and identically distributed random variables each having a BERNOULLI(θ) distribution. Suppose that θ has a BETA(α, β) prior distribution where both α and β are specified and hence can be treated as constants. The conditional distribution of $B_1, \ldots, B_n$ given θ is given by

$$P(B_1 = b_1, \ldots, B_n = b_n | \theta) = \theta^{n\bar{B}_n}(1 - \theta)^{n - n\bar{B}_n},$$

where b_i can either be 0 or 1 for each $i \in \{1, \ldots, n\}$, and $\bar{B}_n$ is the sample mean of $b_1, \ldots, b_n$. Therefore it follows that the posterior distribution for θ given $B_1 = b_1, \ldots, B_n = b_n$ is proportional to

$$\left[\theta^{n\bar{B}_n} (1 - \theta)^{n - n\bar{B}_n} \right] \left[\theta^{\alpha - 1} (1 - \theta)^{\beta - 1} \right] = \theta^{\alpha - 1 + n\bar{B}_n} (1 - \theta)^{\beta + n - 1 - n\bar{B}_n},$$

which corresponds to a BETA$(\alpha + n\bar{B}_n, n + \beta - \bar{B}_n)$ distribution. Theorem 10.13 then implies that the Bayes estimator of θ, when the loss function is squared error loss, is given by the expectation of the posterior distribution which is $\tilde{\theta}_n = (\alpha + \beta + n)^{-1}(\alpha + n\bar{B}_n)$. From the frequentist perspective, we can first note that $\tilde{\theta}_n$ is a consistent estimator of θ using Theorem 3.9 since $\tilde{\theta}_n = (n^{-1}\alpha + n^{-1}\beta + 1)^{-1}(n^{-1}\alpha + \bar{B}_n)$ where $\bar{B}_n \xrightarrow{p} \theta$ by Theorem 3.10 (Weak Law of Large Numbers), $n^{-1}\alpha \to 0$, and $n^{-1}(n^{-1}\alpha + n^{-1}\beta + 1)^{-1} \to 1$, as $n \to \infty$. The efficiency of $\tilde{\theta}_n$ can then be studied by noting that the expectation and variance of $\tilde{\theta}_n$ are given by $E(\tilde{\theta}_n) = (\alpha + \beta + n)^{-1}(\alpha + \theta)$ and $V(\tilde{\theta}_n) = (\alpha + \beta + n)^{-2} n\theta(1 - \theta)$. Using these results we can compare the asymptotic relative efficiency of $\tilde{\theta}_n$ to the maximum likelihood estimator of θ which is given by $\hat{\theta}_n = \bar{B}_n$. Using the fact that the variance of $\hat{\theta}_n$ is given by $n^{-1}\theta(1 - \theta)$ we find that

$$\text{ARE}(\tilde{\theta}_n, \hat{\theta}_n) = \lim_{n \to \infty} \frac{\theta(1 - \theta)(\alpha + \beta + n)^2}{n^2 \theta(1 - \theta)} = 1.$$

It is known that $\hat{\theta}_n$ attains the lower bound for the variance given by Theorem 10.2 and hence it follows that $\tilde{\theta}_n$ is asymptotically efficient.   ∎

The asymptotic efficiency of the Bayes estimator in Example 10.36 is particularly intriguing because it demonstrates the possibility that there may be general conditions that would allow for Bayes estimators to have frequentists properties such as consistency and asymptotic efficiency. One necessary requirement for such properties is that the prior information must asymptotically have a negligible effect on the estimator as the sample size increases to infinity. The reason for this is that the prior information necessarily introduces a bias in the estimator, and this bias must be overcome by the sample information for the estimator to be consistent. From an intuitive standpoint we can argue that if we have full knowledge of the population, which is what the limiting sample size might represent, then any prior information should be essentially ignored. Note that this is a frequentist property and that the Bayesian viewpoint may not necessarily follow this intuition.

Example 10.37. Let $B_1, \ldots, B_n$ be a set of independent and identically distributed random variables, each having a BERNOULLI(θ) distribution. Suppose that θ has a BETA(α, β) prior distribution as in Example 10.36, where the Bayes estimator of θ was found to be $\tilde{\theta}_n = (\alpha + \beta + n)^{-1}(\alpha + n\bar{B}_n)$. We can observe the effect that the prior information has on $\tilde{\theta}_n$ by looking at some specific examples. In the first case consider taking $\alpha = 5$ and $\beta = 10$ which indicates that our prior information is that θ is very likely between zero and $\frac{4}{10}$. See Figure 10.2. If we observe $\bar{B}_n = \frac{1}{2}$ with $n = 10$ then the posterior distribution of θ is BETA$(10, 25)$ with $\tilde{\theta}_n = \frac{10}{35}$ while the sample proportion is $\hat{\theta}_n = \frac{1}{2}$. One can observe from Figure 10.2 that the peak of the posterior distribution has moved slightly toward the estimate $\hat{\theta}_n$, a result of accounting for both the prior information about θ and the information about θ from the observed sample. Alternatively, if $n = 10$ but the prior distribution of θ is taken to be a BETA$(5, 5)$ distribution, which emphasizes a wider range of values in our prior information with a preference for θ being near $\frac{1}{2}$, an observed value of $\bar{B}_n = \frac{1}{2}$ results in a BETA$(10, 10)$ distribution and a Bayes estimate equal to $\frac{1}{2}$. In this case our prior information and our observed information match very well, and the result is that the posterior distribution is still centered about $\frac{1}{2}$, but with a smaller variance indicating more posterior evidence that θ is near $\frac{1}{2}$. See Figure 10.3.

We can now consider the question as to how this framework behaves asymptotically as $n \to \infty$. Let us consider the first case where the prior distribution is BETA$(5, 10)$, and the sample size has been increased to 100. In this case the posterior distribution is BETA$(55, 60)$ with $\tilde{\theta}_n = \frac{55}{115}$. One can now observe a significant change in the posterior distribution when compared to the prior distribution as the large amount of information from the observed sample overwhelms the information contained in the prior distribution. See Figure 10.4. It is this type of behavior that must occur for Bayes estimators to be consistent and efficient. Note that this effect depends on the choice of the prior distribution. For example, Figure 10.5 compares a BETA$(100, 500)$ prior distribution with the BETA$(150, 550)$ posterior distribution under the same sample size of $n = 100$ and observed sample proportion of $\bar{B}_n = \frac{1}{2}$. There is little difference here because the variance of the prior distribution, which reflects the quality of our knowledge of θ, is quite a bit less and hence more observed information from the sample is required to overtake this prior information. However, as the sample size increases, the information from the observed sample will eventually overtake the prior information. The fact that this occurs can be seen from the fact that $\tilde{\theta}_n$ is a consistent and asymptotically efficient estimator of θ for all choices of α and β. See Example 10.36. ∎

Before proceeding to the main asymptotic results for this section, it is worthwhile to develop some connections between Bayesian estimation and the likelihood function. If we focus momentarily on the case where $\mathbf{X}' = (X_1, \dots, X_n)$ is a vector of independent and identically distributed random variables from a distribution $F(x|\theta)$ with density or probability distribution function $f(x|\theta)$, then

$$f(\mathbf{X}|\theta) = \prod_{i=1}^{n} f(X_i|\theta) = L(\theta|\mathbf{X}).$$

Therefore, the posterior distribution of θ can be written as

$$\pi(\theta|\mathbf{X}) = \frac{L(\theta|\mathbf{X})\pi(\theta)}{\int_\Omega L(\theta|\mathbf{X})\pi(\theta)d\theta}. \tag{10.68}$$

The assumptions required for the development of the asymptotic results rely

Figure 10.2 *The* BETA$(5, 10)$ *prior density (solid line) and the* BETA$(10, 25)$ *posterior density (dashed line) on θ from Example 10.37 when $n = 10$ and $\bar{B}_n = \frac{1}{2}$. The sample proportion is located at $\frac{1}{2}$ (dotted line) and the Bayes estimate is located at $\frac{10}{35}$ (dash and dot line).*

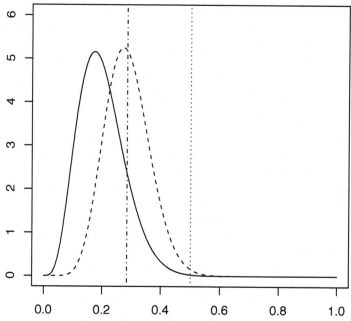

on properties of the error term of a Taylor expansion of the log-likelihood function. In particular, Theorem 1.13 implies that

$$l(\hat{\theta}_n|\mathbf{X}) = l(\theta_0|\mathbf{X}) + (\hat{\theta}_n - \theta_0)l'(\theta_0|\mathbf{X}) + \tfrac{1}{2}(\hat{\theta}_n - \theta_0)^2 l''(\xi_n|\mathbf{X}), \qquad (10.69)$$

where $\hat{\theta}_n$ is a sequence that converges in probability to θ_0 as $n \to \infty$, and ξ_n is a random variable that is between $\hat{\theta}_n$ and θ_0 with probability one. Note that this implies that $\xi_n \xrightarrow{p} \theta_0$ as $n \to \infty$. In the proof of Theorem 10.5 we show that $n^{-1}l''(\theta_0|\mathbf{X}) \xrightarrow{p} I(\theta_0)$ as $n \to \infty$, and hence the assumed continuity of $l''(\theta|\mathbf{X})$ with respect to θ implies that $n^{-1}l''(\xi_n|\mathbf{X}) \xrightarrow{p} I(\theta_0)$, as $n \to \infty$. Therefore, we have that $l''(\xi_n|\mathbf{X}) = -nI(\theta_0) - R_n(\hat{\theta}_n)$ where $n^{-1}R_n(\hat{\theta}_n) \xrightarrow{p} 0$ as $n \to \infty$. For the Bayesian framework we will require additional conditions on the convergence of $R_n(\hat{\theta}_n)$ as detailed in Theorem 10.14 below.

Theorem 10.14. *Suppose that $\mathbf{X}' = (X_1, \dots, X_n)$ is a vector of independent and identically distributed random variables from a distribution $f(x|\theta)$, conditional on θ, where the prior distribution of θ is $\pi(\theta)$ for $\theta \in \Omega$. Let $\tau(t|\mathbf{x})$ be the posterior density of $n^{1/2}(\theta - \tilde{\theta}_{0,n})$, where $\tilde{\theta}_{0,n} = \theta_0 + n^{-1}[I(\theta_0)]^{-1}l'(\theta_0)$ and $\theta_0 \in \Omega$ is the true value of θ. Suppose that*

Figure 10.3 *The* BETA$(5,5)$ *prior density (solid line) and the* BETA$(10,10)$ *posterior density (dashed line) on* θ *from Example 10.37 when* $n = 10$ *and* $\bar{B}_n = \frac{1}{2}$. *The sample proportion and Bayes estimate are both located at* $\frac{1}{2}$ *(dotted line).*

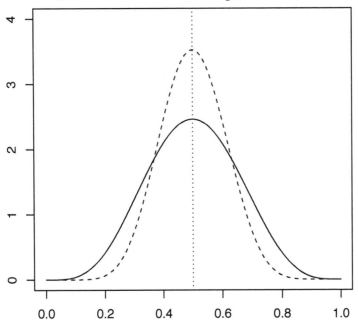

1. Ω *is an open interval.*

2. *The set* $\{x : f(x|\theta) > 0\}$ *is the same for all* $\theta \in \Omega$.

3. *The density* $f(x|\theta)$ *has two continuous derivatives with respect to* θ *for each* $x \in \{x : f(x|\theta) > 0\}$.

4. *The integral*

$$\int_{-\infty}^{\infty} f(x|\theta)dx,$$

can be twice differentiated by exchanging the integral and the derivative.

5. *The Fisher information number* $I(\theta_0)$ *is defined, positive, and finite.*

6. *For any* $\theta_0 \in \Omega$ *there exists a positive constant* d *and function* $B(x)$ *such that*

$$\left| \frac{\partial^2}{\partial \theta^2} \log[f(x|\theta)] \right| \leq B(x),$$

for all $x \in \{x : f(x|\theta) > 0\}$ *and* $\theta \in [\theta_0 - d, \theta_0 + d]$ *such that with* $E[B(X_1)] < \infty$.

Figure 10.4 *The* BETA$(5, 10)$ *prior density (solid line) and the* BETA$(55, 60)$ *posterior density (dashed line) on* θ *from Example 10.37 when* $n = 100$ *and* $\bar{B}_n = \frac{1}{2}$. *The sample proportion is located at* $\frac{1}{2}$ *(dotted line) and the Bayes estimate is located at* $\frac{55}{115}$ *(dash and dot line).*

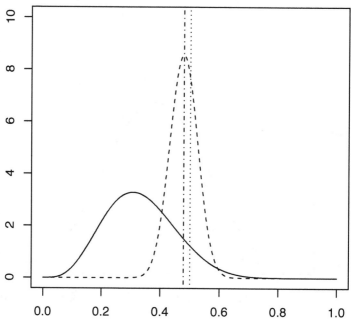

7. *For any* $\varepsilon > 0$ *there exists* $\delta > 0$ *such that*

$$\lim_{n \to \infty} P\left(\sup_{|\hat{\theta}_n - \theta_0| \le \delta} |n^{-1} R_n(\theta)| \ge \varepsilon \right) = 0.$$

8. *For any* $\delta > 0$ *there exists* $\varepsilon > 0$ *such that*

$$\lim_{n \to \infty} P\left(\sup_{|\hat{\theta}_n - \theta_0| \ge \delta} n^{-1}[l(\hat{\theta}_n) - l(\theta_0)] \le -\varepsilon \right) = 1.$$

9. *The prior density* π *on* θ *is continuous and positive for all* $\theta \in \Omega$.

Then,

$$\int_{-\infty}^{\infty} |\tau(t|\mathbf{x}) - [I(\theta_0)]^{1/2} \phi\{t[I(\theta_0)]^{1/2}\}| dt \xrightarrow{p} 0, \qquad (10.70)$$

as $n \to \infty$. *If we can additionally assume that the expectation*

$$\int_{\Omega} |\theta| \pi \theta d\theta,$$

Figure 10.5 *The* BETA(100, 500) *prior density (solid line) and the* BETA(150, 550) *posterior density (dashed line) on θ from Example 10.37 when n = 100 and* $\bar{B}_n = \frac{1}{2}$.

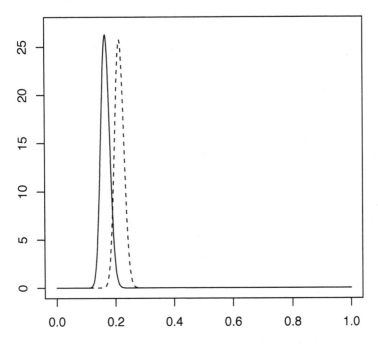

is finite, then

$$\int_{-\infty}^{\infty} (1 + |t|) |\tau(t|\mathbf{x}) - [I(\theta_0)]^{1/2} \phi\{t[I(\theta_0)]^{1/2}\}| dt \xrightarrow{p} 0, \qquad (10.71)$$

as $n \to \infty$.

The proof of Theorem 10.14 is rather complicated, and can be found in Section 6.8 of Lehmann and Casella (1998). Note that in Theorem 10.14 the integrals in Equations (10.70) and (10.71) are being used as a type of norm in the space of density functions, so that the results state that $\pi(t|\mathbf{x})$ and $I^{1/2}\phi\{t[I(\theta_0)]^{1/2}\}$ coincide as $n \to \infty$, since the integrals converge in probability to zero. Therefore, the conclusions of Theorem 10.14 state that the posterior density of $n^{1/2}[\theta - \theta_0 - n^{-1}l'(\theta_0)/I(\theta_0)]$ converges to that of a $N\{0, [I(\theta_0)]^{-1}\}$ density. Equivalently, one can conclude that when n is large, the posterior distribution of θ has an approximate $N\{[\theta_0 + n^{-1}l](\theta_0)[I(\theta_0)]^{-1}, [nI(\theta_0)]^{-1}\}$ distribution. The results of Equations (10.70) and (10.71) make the same type of conclusion, the difference being that the rate of convergence is faster in the tails of the density in Equation (10.71) due to the factor $(1 + |t|)$ in the integral.

Note that the assumptions required for Theorem 10.14 are quite a bit stronger than what is required to develop the asymptotic theory of maximum likelihood estimators. While Assumptions 1–6 are the same as in Theorem 10.5, the assumptions used for likelihood theory that imply that $n^{-1}R_n(\theta_0) \xrightarrow{p} 0$ as $n \to \infty$, are replaced by the stronger Assumption 7 in Theorem 10.14. Additionally, in the asymptotic theory of maximum likelihood estimation, the consistency and asymptotic efficiency of the maximum likelihood estimator requires us to only specify the behavior of the likelihood function in a neighborhood of the true parameter value. Because Bayes estimators involve integration of the likelihood over the entire range of the parameter space, Assumption 8 ensures that the likelihood function is well behaved away from θ_0 as well.

When the squared error loss function is used, the result of Theorem 10.14 is sufficient to conclude that the Bayes estimator is both consistent and efficient.

Theorem 10.15. *Suppose that $\mathbf{X}' = (X_1, \ldots, X_n)$ is a vector of independent and identically distributed random variables from a distribution $f(x|\theta)$, conditional on θ, where the prior distribution of θ is $\pi(\theta)$ for $\theta \in \Omega$. Let $\tau(t|\mathbf{x})$ be the posterior density of $n^{1/2}(\theta - \tilde{\theta}_{0,n})$, where $\tilde{\theta}_{0,n} = \theta_0 + n^{-1}[I(\theta_0)]^{-1}l'(\theta_0)$ and $\theta_0 \in \Omega$ is the true value of θ. Suppose that the conditions of Theorem 10.14 hold, then when the loss function is squared error loss, it follows that $n^{1/2}(\hat{\theta}_n - \theta_0) \xrightarrow{d} Z$ as $n \to \infty$ where Z is a $N\{0, [I(\theta_0)]^{-1}\}$ random variable.*

Proof. Note that

$$n^{1/2}(\hat{\theta}_n - \theta_0) = n^{1/2}(\hat{\theta}_n - \tilde{\theta}_{0,n}) + n^{1/2}(\tilde{\theta}_{0,n} - \theta_0). \tag{10.72}$$

The second term in Equation (10.72) has the form

$$
\begin{aligned}
n^{1/2}(\tilde{\theta}_{0,n} - \theta_0) &= n^{-1/2}[I(\theta_0)]^{-1}l'(\theta_0) \\
&= n^{-1/2}[I(\theta_0)]^{-1} \frac{\partial}{\partial \theta} \sum_{i=1}^{n} \log[f(X_i|\theta)]\Big|_{\theta=\theta_0} \\
&= n^{-1/2}[I(\theta_0)]^{-1} \sum_{i=1}^{n} \frac{f'(X_i|\theta_0)}{f(X_i|\theta_0)}.
\end{aligned}
$$

As noted in the proof of Theorem 10.5, Assumption 4 implies that

$$E\left[\frac{f'(X_i|\theta_0)}{f(X_i|\theta_0)}\right] = 0,$$

so that Theorem 4.20 (Lindeberg and Lévy) implies that $n^{1/2}(\tilde{\theta}_{0,n} - \theta_0) \xrightarrow{d} Z$ as $n \to \infty$ where Z is a $N\{0, [I(\theta_0)]^{-1}\}$ random variable. To study the first term in Equation (10.72) we begin by noting that

$$
\begin{aligned}
\tau(t|\mathbf{x}) &= \frac{\pi[n^{-1/2}(t + n^{1/2}\tilde{\theta}_{0,n})|\mathbf{x}]L(n^{-1/2}t + \tilde{\theta}_{0,n}|\mathbf{x})}{\int_\Omega \pi[n^{-1/2}(t + n^{1/2}\tilde{\theta}_{0,n}|\mathbf{x}]L(n^{-1/2}t + \tilde{\theta}_{0,n}|\mathbf{x})}dt \\
&= n^{-1/2}\pi(\tilde{\theta}_{0,n} + n^{-1/2}t|\mathbf{x}).
\end{aligned}
$$

Theorem 10.13 implies that the Bayes estimator of θ using squared error loss is given by

$$\tilde{\theta}_n = \int_\Omega \theta\pi(\theta|\mathbf{x})d\theta.$$

Consider the change of variable where we let $\theta = \tilde{\theta}_{0,n} + n^{-1/2}t$ so that $d\theta = n^{-1/2}dt$. Let $\tilde{\Omega}$ denote the corresponding transformation of Ω. Therefore

$$\tilde{\theta}_n = \int_\Omega \theta\pi(\theta|\mathbf{x})d\theta = \int_{\tilde{\Omega}} n^{-1/2}(\tilde{\theta}_{0,n} + n^{-1/2}t)\pi(\tilde{\theta}_{0,n} + n^{-1/2}t)dt.$$

Noting that $\tau(t|\mathbf{x}) = n^{-1/2}\pi(\tilde{\theta}_{0,n} + n^{-1/2}t|\mathbf{x})$ we have that

$$\tilde{\theta}_n = \int_{\tilde{\Omega}}(\tilde{\theta}_{0,n} + n^{-1/2}t)\tau(t|\mathbf{x})dt$$

$$= \int_{-\tilde{\Omega}}\tilde{\theta}_{0,n}\tau(t|\mathbf{x})dt + n^{-1/2}\int_{\tilde{\Omega}} t\tau(t|\mathbf{x})dt. \tag{10.73}$$

Note that $\tilde{\theta}_{0,n}$ does not depend on t so that the first term in Equation (10.73) is $\tilde{\theta}_{0,n}$. Therefore,

$$\tilde{\theta}_n = \tilde{\theta}_{0,n} + n^{-1/2}\int_{\tilde{\Omega}} t\tau(t|\mathbf{x})dt,$$

or that

$$n^{1/2}(\tilde{\theta}_n - \tilde{\theta}_{0,n}) = \int_{\tilde{\Omega}} t\tau(t|\mathbf{x})dt.$$

Now, note that

$$n^{1/2}|\tilde{\theta}_n - \tilde{\theta}_{0,n}| = \left|\int_{\tilde{\Omega}} t\tau(t|\mathbf{x})dt\right| = $$

$$\left|\int_{\tilde{\Omega}}\{t\tau(t|\mathbf{x}) - t[I(\theta_0)]^{1/2}\phi\{t[I(\theta_0)]^{1/2}\}\}dt\right|$$

which follows from the fact that

$$\int_{\tilde{\Omega}} t[I(\theta_0)]^{1/2}\phi\{t[I(\theta_0)]^{1/2}\}dt = 0,$$

since the integral represents the expectation of a $N\{0, [I(\theta_0)]^{-1}\}$ random variable. Theorem A.6 implies that

$$\left|\int_{\tilde{\Omega}}\{t\tau(t|\mathbf{x}) - tI^{1/2}(\theta_0)\phi[tI^{1/2}(\theta_0)]\}dt\right| \le$$

$$\int_{\tilde{\Omega}}\left|t\tau(t|\mathbf{x}) - tI^{1/2}(\theta_0)\phi[tI^{1/2}(\theta_0)]\right|dt =$$

$$\int_{\tilde{\Omega}}|t|\left|\tau(t|\mathbf{x}) - tI^{1/2}(\theta_0)\phi[tI^{1/2}(\theta_0)]\right|dt.$$

Therefore

$$n^{1/2}|\tilde{\theta}_n - \tilde{\theta}_{0,n}| \le \int_{\tilde{\Omega}}|t|\left|\tau(t|\mathbf{x}) - tI^{1/2}(\theta_0)\phi[tI^{1/2}(\theta_0)]\right|dt. \tag{10.74}$$

Theorem 10.14 implies that the integral on the right hand side of Equation (10.74) converges in probability to zero as $n \to \infty$. Therefore, it follows that $n^{1/2}|\tilde{\theta}_n - \tilde{\theta}_{0,n}| \xrightarrow{p} 0$ as $n \to \infty$. Combining this result with the fact that $n^{1/2}(\tilde{\theta}_{0,n} - \theta_0) \xrightarrow{d} Z$ as $n \to \infty$ and using Theorem 4.11 (Slutsky) implies that $n^{1/2}(\tilde{\theta}_n - \theta_0) \xrightarrow{d} Z$ as $n \to \infty$, which proves the result. $\square$

The conditions of Theorem 10.15 appear to be quite complicated, but in fact can be shown to hold for exponential families.

Theorem 10.16. *Consider an exponential family density of the form $f(x|\theta) = \exp[\theta T(x) - A(\theta)]$ where the parameter space Ω is an open interval, $T(x)$ is not a function of θ, and $A(\theta)$ is not a function of x. Then for this density the assumptions of Theorem 10.15 are satisifed.*

For a proof of Theorem 10.16 see Example 6.8.4 of Lehmann and Casella (1998).

Example 10.38. Let $X_1, \ldots, X_n$ be a set of independent and identically distributed random variables from a $N(\theta, \sigma^2)$ distribution, conditional on θ, where θ has a $N(\lambda, \tau^2)$ distribution where σ^2, λ and τ^2 are known. In this case it can be shown that the posterior distribution of θ is $N(\tilde{\theta}_n, \tilde{\sigma}_n^2)$ where

$$\tilde{\theta}_n = \frac{\tau^2 \bar{X}_n + n^{-1}\sigma^2 \lambda}{\tau^2 + n^{-1}\sigma^2},$$

and

$$\tilde{\sigma}_n^2 = \frac{\sigma^2 \tau^2}{n\tau^2 + \sigma^2}.$$

See Exercise 44. Under squared error loss, the Bayes estimator of θ is given by $E(\theta|X_1, \ldots, X_n) = \tilde{\theta}_n$. Treating $X_1, \ldots, X_n$ as a random sample from a $N(\theta, \sigma^2)$ where θ is fixed we have that $\bar{X}_n \xrightarrow{p} \theta$ as $n \to \infty$, by Theorem 3.10 (Weak Law of Large Numbers) and hence Theorem 3.7 implies that

$$\tilde{\theta}_n = \frac{\tau^2 \bar{X}_n + n^{-1}\sigma^2 \lambda}{\tau^2 + n^{-1}\sigma^2} \xrightarrow{p} \theta,$$

as $n \to \infty$. Therefore the Bayes estimator is consistent. Further, Theorem 4.20 (Lindeberg and Lévy) implies that $n^{1/2}\bar{X}_n \xrightarrow{d} Z$ as $n \to \infty$, where Z is a $N(\theta, \sigma^2)$ random variable. The result in this case is, in fact, exact for any sample size, but for the sake of argument here we will base our calculations on the asymptotic result. Note that

$$n^{1/2}\tilde{\theta}_n = \frac{n^{1/2}\tau^2 \bar{X}_n + n^{-1/2}\sigma^2 \lambda}{\tau^2 + n^{-1}\sigma^2} = n^{1/2}\bar{X}_n \frac{\tau^2}{\tau^2 + n^{-1}\sigma^2} + \frac{n^{-1/2}\sigma^2 \lambda}{\tau^2 + n^{-1}\sigma^2},$$

where

$$\lim_{n \to \infty} \frac{\tau^2}{\tau^2 + n^{-1}\sigma^2} = 1,$$

and

$$\lim_{n \to \infty} \frac{n^{-1/2}\sigma^2 \lambda}{\tau^2 + n^{-1}\sigma^2} = 0.$$

Therefore, Theorem 4.11 (Slutsky) implies that $n^{1/2}\tilde{\theta}_n \xrightarrow{d} Z$ as $n \to \infty$, which is the asymptotic behavior specified by Theorem 10.15. ∎

Example 10.39. Let $B_1, \ldots, B_n$ be a set of independent and identically distributed BERNOULLI(θ) random variables and suppose that θ has a BETA(α, β) prior distribution where both α and β are specified. In Example 10.36 it was shown that the Bayes estimator under squared error loss is given by

$$\tilde{\theta}_n = \frac{\alpha + n\bar{B}_n}{\alpha + \beta + n},$$

which was shown to be a consistent estimator of θ. Theorem 4.20 (Lindeberg and Lévy) implies that $n^{1/2}\bar{B}_n \xrightarrow{d} Z$ as $n \to \infty$ where Z is a $N[\theta, \theta(1-\theta)]$ random variable. Now, note that

$$n^{1/2}\tilde{\theta}_n = \frac{n^{1/2}\alpha + n^{3/2}\bar{B}_n}{\alpha + \beta + n} = \frac{n^{1/2}\alpha}{\alpha + \beta + n} + \frac{n^{1/2}\bar{B}_n}{n^{-1}\alpha + n^{-1}\beta + 1},$$

where

$$\lim_{n \to \infty} \frac{n^{1/2}\alpha}{\alpha + \beta + n} = 0,$$

and

$$\lim_{n \to \infty} n^{-1}\alpha + n^{-1}\beta + 1 = 1.$$

Therefore, Theorem 4.11 (Slutsky) implies that $n^{1/2}\tilde{\theta}_n \xrightarrow{d} Z$ as $n \to \infty$, which is the asymptotic behavior specified by Theorem 10.15. ∎

10.7 Exercises and Experiments

10.7.1 Exercises

1. Prove that $MSE(\hat{\theta}, \theta)$ can be decomposed into two parts given by

   $$\text{MSE}(\hat{\theta}, \theta) = \text{Bias}^2(\hat{\theta}, \theta) + V(\hat{\theta}),$$

   where $\text{Bias}(\hat{\theta}, \theta) = E(\hat{\theta}) - \theta$.

2. Let $\{X_n\}_{n=1}^\infty$ be a sequence of independent and identically distributed random variables from a distribution F with mean θ and variance σ^2. Suppose that f has a finite fourth moment and let g be a function that has at least four derivatives. If the fourth derivative of g^2 is bounded then prove that

   $$V[g(\bar{X}_n)] = n^{-1}\sigma^2[g'(\theta)]^2 + O(n^{-1}),$$

   as $n \to \infty$.

3. Let $\{X_n\}_{n=1}^\infty$ be a sequence of independent and identically distributed random variables from a distribution F with mean θ and finite variance σ^2. Suppose now that we are interested in estimating $g(\theta) = c\theta^3$ using the estimator $g(\bar{X}_n) = c\bar{X}_n^3$ where c is a known real constant.

   a. Find the bias and variance of $g(\bar{X}_n)$ as an estimator of $g(\theta)$ directly.

   b. Find asymptotic expressions for the bias and variance of $g(\bar{X}_n)$ as an estimator of $g(\theta)$ using Theorem 10.1. Compare this result to the exact expressions derived above.

4. Let $B_1, \ldots, B_n$ be a set of independent and identically distributed random variables from a BERNOULLI(θ) distribution. Suppose we are interested in estimating the variance $g(\theta) = \theta(1 - \theta)$ with the estimator $g(\bar{B}_n) = \bar{B}_n(1 - \bar{B}_n)$.

   a. Find the bias and variance of $g(\bar{X}_n)$ as an estimator of $g(\theta)$ directly.

   b. Find asymptotic expressions for the bias and variance of $g(\bar{X}_n)$ as an estimator of $g(\theta)$ using Theorem 10.1. Compare this result to the exact expressions derived above.

5. Let $X_1, \ldots, X_n$ be a set of independent and identically distributed random variables from an EXPONENTIAL(θ) distribution. Suppose that we are interested in estimating the variance $g(\theta) = \theta^2$ with the estimator $g(\bar{X}_n) = \bar{X}_n^2$.

   a. Find the bias and variance of $g(\bar{X}_n)$ as an estimator of $g(\theta)$ directly.

   b. Find asymptotic expressions for the bias and variance of $g(\bar{X}_n)$ as an estimator of $g(\theta)$ using Theorem 10.1. Compare this result to the exact expressions derived above.

6. Let $X_1, \ldots, X_n$ be a set of independent and identically distributed random variables from a POISSON(θ) distribution. Suppose that we are interested in estimating the variance $g(\theta) = \exp(-\theta)$ with the estimator $g(\bar{X}_n) = \exp(-\bar{X}_n)$. Find asymptotic expressions for the bias and variance of $g(\bar{X}_n)$ as an estimator of $g(\theta)$ using Theorem 10.1.

7. Let $\{X_n\}_{n=1}^{\infty}$ be a sequence of independent and identically distributed random variables following a N($0, \sigma^2$) distribution and consider estimating $\exp(\theta)$ with $\exp(\bar{X}_n)$. Find an asymptotic expression for the variance of $\exp(\bar{X}_n)$ using the methods detailed in Example 10.5.

8. Let $X_1, \ldots, X_n$ be a set of independent and identically distributed random variables from a N($0, \sigma^2$) distribution where σ^2 is finite. Let $k(n)$ be a function of $n \in \mathbb{N}$ that returns an integer between 1 and n, and consider the estimator of θ that returns the average of the first $k(n)$ observations. That is, define

$$\bar{X}_{k(n)} = [k(n)]^{-1} \sum_{i=1}^{k(n)} X_i.$$

   a. Prove that $\bar{X}_{k(n)}$ is an unbiased estimator of θ.

   b. Find the mean squared error of $\bar{X}_{k(n)}$.

   c. Assuming that the optimal mean squared error for unbiased estimators of θ is $n^{-1}\sigma^2$, under what conditions will $\bar{X}_{k(n)}$ be asymptotically optimal?

9. Let $\{X_n\}_{n=1}^{\infty}$ be a sequence of independent and identically distributed random variables from a LAPLACE$(\theta, 1)$ distribution. Let $\hat{\theta}_n$ denote the sample mean and $\tilde{\theta}_n$ denote the sample median computed on $X_1, \ldots, X_n$. Compute ARE$(\hat{\theta}_n, \tilde{\theta})$.

10. Let $X_1, \ldots, X_n$ be a set of independent and identically distributed random variables following a mixture of two NORMAL distributions, with a density given by $f(x) = \frac{1}{2}\phi(x - \zeta) + \frac{1}{2}\phi(x + \zeta)$ where θ is a positive real number. Compute the asymptotic relative efficiency of the sample mean relative to the sample median as an estimator of the mean of this density. Comment on the role that the parameter ζ has on the efficiency.

11. Let $X_1, \ldots, X_n$ be a set of independent and identically distributed random variables from a POISSON(θ) distribution. Consider two estimators of $P(X_n = 0) = \exp(-\theta)$ given by

$$\hat{\theta}_n = n^{-1} \sum_{i=1}^{n} \delta\{X_i; \{0\}\},$$

which is the proportion of values in the sample that are equal to zero, and $\tilde{\theta}_n = \exp(-\bar{X}_n)$. Compute the asymptotic relative efficiency of $\hat{\theta}_n$ relative to $\tilde{\theta}_n$ and comment on the results.

12. Let $X_1, \ldots, X_n$ be a set of independent and identically distributed random variables following a $N(\theta, 1)$ distribution.

   a. Prove that if $\theta \neq 0$ then $\delta\{|\bar{X}_n| : [0, n^{-1/4})\} \xrightarrow{P} 0$ as $n \to \infty$.
   b. Prove that if $\theta = 0$ then $\delta\{|\bar{X}_n| : [0, n^{-1/4})\} \xrightarrow{P} 1$ as $n \to \infty$.

13. Let $X_1, \ldots, X_n$ be a set of independent and identically distributed random variables from a GAMMA$(2, \theta)$ distribution.

   a. Find the maximum likelihood estimator for θ.
   b. Determine whether the maximum likelihood estimator is consistent and asymptotically efficient.

14. Let $X_1, \ldots, X_n$ be a set of independent and identically distributed random variables from a POISSON(θ) distribution.

   a. Find the maximum likelihood estimator for θ.
   b. Determine whether the maximum likelihood estimator is consistent and asymptotically efficient.

15. Let $X_1, \ldots, X_n$ be a set of independent and identically distributed random variables from a $N(\theta, 1)$ distribution.

   a. Find the maximum likelihood estimator for θ.
   b. Determine whether the maximum likelihood estimator is consistent and asymptotically efficient.

16. Let $X_1, \ldots, X_n$ be a set of independent and identically distributed random variables from a N$(0, \theta)$ distribution.

   a. Find the maximum likelihood estimator for θ.

   b. Determine whether the maximum likelihood estimator is consistent and asymptotically efficient.

17. Consider a sequence of random variables $\{\{X_{ij}\}_{j=1}^{k}\}_{i=1}^{n}$ that are assumed to be mutually independent, each having a N(μ_i, θ) distribution for $i = 1, \ldots, n$. Prove that the maximum likelihood estimators of μ_i and θ are

$$\hat{\mu}_i = \bar{X}_i = k^{-1} \sum_{j=1}^{k} X_{ij},$$

for each $i = 1, \ldots, n$ and

$$\hat{\theta}_n = (nk)^{-1} \sum_{i=1}^{n} \sum_{j=1}^{k} (X_{ij} - \bar{X}_i)^2.$$

18. Consider a sequence of independent and identically distributed d-dimensional random vectors $\{\mathbf{X}_n\}_{n=1}^{\infty}$ from a d-dimensional distribution F. Assume the structure smooth function model with $\boldsymbol{\mu} = E(\mathbf{X}_n)$, $\theta = g(\boldsymbol{\mu})$ with $\hat{\theta}_n = g(\bar{\mathbf{X}}_n)$. Further, assume that

$$\sigma^2 = h^2(\mu) = \lim_{n\to\infty} V(n^{1/2}\hat{\theta}),$$

with $\hat{\sigma}_n^2 = h^2(\bar{\mathbf{X}}_n)$. Let $G_n(t) = P[n^{1/2}\sigma^{-1}(\hat{\theta}_n - \theta) \le t]$ and $H_n(t) = P[n^{1/2}\hat{\sigma}_n^{-1}(\hat{\theta}_n - \theta) \le t]$ and define $g_{\alpha,n}$ and $h_{\alpha,n}$ to be the corresponding α quantiles of G_n and H_n. Define the ordinary and studentized $100\alpha\%$ upper confidence limits for θ as $\hat{\theta}_{n,\text{ord}} = \hat{\theta}_n - n^{-1/2}\sigma g_{1-\alpha}$ and $\hat{\theta}_{n,\text{stud}} = \hat{\theta}_n - n^{-1/2}\hat{\sigma}_n h_{1-\alpha}$. Prove that $\hat{\theta}_{n,\text{ord}}$ and $\hat{\theta}_{n,\text{stud}}$ are accurate upper confidence limits.

19. Suppose that $X_1, \ldots, X_n$ is a set of independent and identically distributed random variables from a distribution F with parameter θ. Suppose that F and θ fall within the smooth function model. For the case when σ is known, prove that $\tilde{\theta}_{n,\text{ord}}$ is a first-order correct and accurate $100\alpha\%$ confidence limit for θ.

20. Equation (10.41) provides an asymptotic expansion for the coverage probability of an upper confidence limit that has asymptotic expansion

$$\hat{\theta}_n(\alpha) = \hat{\theta}_n + n^{-1/2}\hat{\sigma}_n z_\alpha + n^{-1}\hat{\sigma}_n \hat{u}_1(z_\alpha) + n^{-3/2}\hat{\sigma}_n \hat{u}_2(z_\alpha) + O_p(n^{-2}),$$

as $n \to \infty$, as

$$\pi_n(\alpha) = \alpha + n^{-1/2}[u_1(z_\alpha) - v_1(z_\alpha)]\phi(z_\alpha) - n^{-1}[\tfrac{1}{2}z_\alpha u_1^2(z_\alpha) - u_2(z_\alpha) +$$
$$u_1(z_\alpha)v_1'(z_\alpha) - z_\alpha u_1(z_\alpha)v_1(z_\alpha) - v_2(z_\alpha) + u_\alpha z_\alpha]\phi(z_\alpha) + O(n^{-3/2}),$$

as $n \to \infty$. Prove that when $u_1(z_\alpha) = s_1(z_\alpha)$ and $u_2(z_\alpha) = s_2(z_\alpha)$ it follows that $\pi_n(\alpha) = n^{-1} u_\alpha z_\alpha \phi(z_\alpha) + O(n^{-3/2})$, as $n \to \infty$.

21. Let $X_1, \ldots, X_n$ be a sequence of independent and identically distributed random variables from a distribution F with mean θ and assume the framework of the smooth function model. Let $\sigma^2 = E[(X_1 - \theta)^2]$, $\gamma = \sigma^{-3} E[(X_1 - \theta)^3]$, and $\kappa = \sigma^{-4} E[(X_1 - \theta)^4] - 3$.

a. Prove that an exact $100\alpha\%$ upper confidence limit for θ has asymptotic expansion

$$\hat{\theta}_{n,\text{stud}}(\alpha) = \hat{\theta}_n + n^{-1/2} \hat{\sigma}_n \{ z_{1-\alpha} + \tfrac{1}{6} n^{-1/2} \gamma (2 z_{1-\alpha}^2 + 1) +$$
$$n^{-1} z_{1-\alpha} [-\tfrac{1}{12} \kappa (z_{1-\alpha}^2 - 3) + \tfrac{5}{72} \gamma^2 (4 z_{1-\alpha}^2 - 1) + \tfrac{1}{4} (z_{1-\alpha}^2 + 3)] \} + O_p(n^{-2}),$$

as $n \to \infty$.

b. Let $\hat{\gamma}_n$ and $\hat{\kappa}_n$ be the sample skewness and kurtosis, respectively, and assume that $\hat{\gamma}_n = \gamma + O_p(n^{-1/2})$ and $\hat{\kappa}_n = \kappa + O_p(n^{-1/2})$, as $n \to \infty$. Prove that the Edgeworth-corrected upper confidence limit for θ given by

$$\hat{\theta}_n + n^{-1/2} \hat{\sigma}_n \{ z_{1-\alpha} + \tfrac{1}{6} n^{-1/2} \hat{\gamma}_n (2 z_{1-\alpha}^2 + 1) +$$
$$n^{-1} z_{1-\alpha} [-\tfrac{1}{12} \hat{\kappa}_n (z_{1-\alpha}^2 - 3) + \tfrac{5}{72} \hat{\gamma}_n^2 (4 z_{1-\alpha}^2 - 1) + \tfrac{1}{4} (z_{1-\alpha}^2 + 3)] \},$$

has an asymptotic expansion for its coverage probability given by

$$\bar{\pi}_n(\alpha) = \alpha - \tfrac{1}{6} n^{-1} (\kappa - \tfrac{3}{2} \gamma^2) z_\alpha (2 z_\alpha^2 + 1) \phi(z_\alpha) + O(n^{-3/2}),$$

so that the confidence limit is second-order accurate. In this argument you will need to prove that $u_\alpha = \gamma^{-1}(\kappa - \tfrac{3}{2}\gamma^2) u_1(z_\alpha)$.

c. Let $\hat{\gamma}_n$ be the sample skewness and assume that $\hat{\gamma}_n = \gamma + O_p(n^{-1/2})$, as $n \to \infty$. Find an asymptotic expansion for the coverage probability of the Edgeworth-corrected upper confidence limit for θ given by

$$\hat{\theta}_n + n^{-1/2} \hat{\sigma}_n [z_{1-\alpha} + \tfrac{1}{6} n^{-1/2} \hat{\gamma}_n (2 z_{1-\alpha}^2 + 1)].$$

Does there appear to be any advantage from an asymptotic viewpoint to estimating the kurtosis?

22. Let $X_1, \ldots, X_n$ be a sequence of independent and identically distributed random variables from a distribution F with parameter θ and assume the framework of the smooth function model. Consider a general $100\alpha\%$ upper confidence limit for θ which has an asymptotic expansion of the form

$$\hat{\theta}_n(\alpha) = \hat{\theta}_n + n^{-1/2} \hat{\sigma}_n z_\alpha + n^{-1} \hat{\sigma}_n \hat{u}_1(z_\alpha) + n^{-3/2} \hat{\sigma}_n \hat{u}_2(z_\alpha) + O_p(n^{-2}),$$

as $n \to \infty$. Consider a two-sided confidence interval based on this confidence limit of the form $[\hat{\theta}_n[(1-\alpha)/2], \hat{\theta}_n[(1+\alpha)/2]]$. Prove that the length of this interval has an asymptotic expansion of the form

$$2 n^{-1/2} \hat{\sigma}_n z_{(1+\alpha)/2} + 2 n^{-1} \hat{\sigma}_n u_1(z_{(1+\alpha)/2}) + O_p(n^{-2}),$$

as $n \to \infty$.

23. Prove that the coverage probability of a $100\alpha\%$ upper confidence limit that has an asymptotic expansion of the form

$$\hat{\theta}_n(\alpha) = \hat{\theta}_n + n^{-1/2}\hat{\sigma}_n z_\alpha + n^{-1}\hat{\sigma}_n \hat{s}_1(z_\alpha) + n^{-3/2}\hat{\sigma}_n \hat{s}_2(z_\alpha) + O_p(n^{-2}),$$

has an asymptotic expansion of the form

$$\pi_n(\alpha) = n^{-1} u_\alpha z_\alpha \phi(z_\alpha) + O(n^{-3/2}),$$

as $n \to \infty$. Is there any case where $u_\alpha = 0$ so that the resulting confidence limit would be third-order accurate?

24. Let $X_1, \ldots, X_n$ be a sequence of independent and identically distributed random variables from a distribution F with parameter θ and assume the framework of the smooth function model. A common $100\alpha\%$ upper confidence limit proposed, using the very different motivation, for use with the bootstrap methodology of Efron (1979), is the *backwards* confidence limit given by $\hat{\theta}_n(\alpha) = \hat{\theta}_n + n^{-1/2}\hat{\sigma}_n g_\alpha$, where it is assumed that the standard deviation σ is unknown. Hall (1988a) named this confidence limit the backwards confidence limit because it is based on the upper quantile of G instead of the lower quantile of H, which is the correct quantile given in the form of the upper studentized confidence limit. Therefore, one justification of this method is based on assuming $g_\alpha \simeq h_{1-\alpha}$, which will be approximately true in the smooth function model when n is large.

   a. Find an asymptotic expansion for the confidence limit $\hat{\theta}_n(\alpha)$ and find the order of asymptotic correctness of the method.

   b. Find an asymptotic expansion for the coverage probability of the upper confidence $\hat{\theta}_n(\alpha)$ and find the order of asymptotic accuracy of the method.

   c. Find an asymptotic expansion for the coverage probability of a two-sided confidence interval based on this method.

25. Let $X_1, \ldots, X_n$ be a sequence of independent and identically distributed random variables from a distribution F with parameter θ and assume the framework of the smooth function model. Consider the *backwards* confidence limit given by $\hat{\theta}_n(\alpha) = \hat{\theta}_n + n^{-1/2}\hat{\sigma}_n g_\alpha$, where it is assumed that the standard deviation σ is unknown. The asymptotic correctness and accuracy is studied in Exercise 24. Efron (1981) attempted to improve the properties of the backwards method by adjusting the confidence coefficient to remove some of the bias from the method. The resulting method, called the *bias-corrected method*, uses an upper confidence limit equal to $\hat{\theta}_n[\beta(\alpha)] = \hat{\theta}_n + n^{-1/2}\hat{\sigma}_n g_{\beta(\alpha)}$, where $\beta(\alpha) = \Phi(z_\alpha + 2\tilde{\mu})$ and $\tilde{\mu} = \Phi^{-1}[G_n(0)]$.

   a. Prove that $\tilde{\mu} = n^{-1/2} r_1(0) + O(n^{-1})$, as $n \to \infty$.

   b. Prove that the bias-corrected critical point has the form $\hat{\theta}_n[\beta(\alpha)] = \hat{\theta}_n + n^{-1/2}\hat{\sigma}_n\{z_\alpha + n^{-1/2}[2r_1(0) - r_1(z_\alpha)] + O(n^{-1})\}$, as $n \to \infty$.

c. Prove that the coverage probability of the bias-corrected upper confidence limit is given by

$$\pi_{bc}(\alpha) = \alpha + n^{-1/2}[2r_1(0) - r_1(z_\alpha) - v_1(z_\alpha)]\phi(z_\alpha) + O(n^{-1}),$$

as $n \to \infty$.

d. Discuss the results given above in terms of the performance of this confidence interval.

26. Consider the problem of testing the null hypothesis $H_0 : \theta \leq \theta_0$ against the alternative hypothesis $H_1 : \theta > \theta_0$ using the test statistic $Z_n = n^{1/2}\sigma^{-1}(\hat{\theta}_n - \theta_0)$ where σ is known and the null hypothesis is rejected whenever $Z_n > r_{n,\alpha}$, a constant that depends on n and α. Prove that this test is unbiased.

27. Let $\{F_n\}_{n=1}^{\infty}$ be a sequence of distribution functions such that $F_n \rightsquigarrow F$ as $n \to \infty$ for some distribution function F. Let $\{t_n\}_{n=1}^{\infty}$ be a sequence of real numbers.

a. Prove that if $t_n \to \infty$ as $n \to \infty$ then

$$\lim_{n\to\infty} F_n(t_n) = 1.$$

b. Prove that if $t_n \to -\infty$ as $n \to \infty$ then

$$\lim_{n\to\infty} F_n(t_n) = 0.$$

c. Prove that if $t_n \to t$ where $t \in C(F)$ as $n \to \infty$ then

$$\lim_{n\to\infty} F_n(t_n) = F(t).$$

28. Let $B_1, \ldots, B_n$ be a sequence of independent and identically distributed random variables from a BERNOULLI(θ) distribution where the parameter space of θ is $\Omega = (0,1)$. Consider testing the null hypothesis $H_0 : \theta \leq \theta_0$ against the alternative hypothesis $H_1 : \theta > \theta_0$.

a. Describe an exact test of H_0 against H_1 whose rejection region is based on the BINOMIAL distribution.

b. Find an approximate test of H_0 against H_1 using Theorem 4.20 (Lindeberg and Lévy). Prove that this test is consistent and find an expression for the asymptotic power of this test for the sequence of alternatives given by $\theta_{1,n} = \theta_0 + n^{-1/2}\delta$ where $\delta > 0$.

29. Let $U_1, \ldots, U_n$ be a sequence of independent and identically distributed random variables from a UNIFORM($0, \theta$) distribution where the parameter space for θ is $\Omega = (0, \infty)$. Using the test statistic $U_{(n)}$, where $U_{(n)} = \max\{U_1, \ldots, U_n\}$, develop an unbiased test of the null hypothesis $H_0 : \theta \leq \theta_0$ against the alternative hypothesis $H_1 : \theta > \theta_0$ that is *not* based on an asymptotic NORMAL distribution.

30. Let $X_1, \ldots, X_n$ be a set of independent and identically distributed random variables from a distribution F with parameter θ. Consider testing the null hypothesis $H_0 : \theta \geq \theta_0$ against the alternative hypothesis $H_1 : \theta < \theta_0$. Assume that the test statistic is of the form $Z_n = n^{1/2}[\sigma(\theta_0)]^{-1}(\hat{\theta}_n - \theta_0)$ where $Z_n \xrightarrow{d} Z$ as $n \to \infty$ and the null hypothesis is rejected when $Z_n < r_{\alpha,n}$ where $r_{\alpha,n} \to z_\alpha$ as $n \to \infty$.

    a. Prove that the test is consistent against all alternatives $\theta < \theta_0$. State any additional assumptions that must be made in order for this result to be true.

    b. Develop an expression similar to that given in Theorem 10.10 for the asymptotic power of this test for a sequence of alternatives given by $\theta_{1,n} = \theta_0 - n^{-1/2}\delta$ where $\delta > 0$ is a constant. State any additional assumptions that must be made in order for this result to be true.

31. Consider the framework of the smooth function model where σ, which denotes the asymptotic variance of $n^{1/2}\hat{\theta}_n$, is known. Consider using the test statistic $Z_n = n^{1/2}\sigma^{-1}(\hat{\theta}_n - \theta_0)$ which follows the distribution G_n when θ_0 is the true value of θ. Prove that an unbiased test of size α of the null hypothesis $H_0 : \theta \leq \theta_0$ against the alternative hypothesis $H_1 : \theta > \theta_0$ rejects the null hypothesis if $Z_n > g_{1-\alpha}$, where we recall that $g_{1-\alpha}$ is the $(1-\alpha)^{\text{th}}$ quantile of the distribution G_n.

32. Consider the framework of the smooth function model where σ, which denotes the asymptotic variance of $n^{1/2}\hat{\theta}_n$, is unknown, and the test statistic $T_n = n^{1/2}\hat{\sigma}_n^{-1}(\hat{\theta}_n - \theta_0)$ follows the distribution H_n when θ_0 is the true value of θ.

    a. Prove that a test of the null hypothesis $H_0 : \theta \leq \theta_0$ against the alternative hypothesis $H_1 : \theta > \theta_0$ that rejects the null hypothesis if $T_n > z_{1-\alpha}$ is a first-order accurate test.

    b. Prove that a test of the null hypothesis $H_0 : \theta \leq \theta_0$ against the alternative hypothesis $H_1 : \theta > \theta_0$ that rejects the null hypothesis if $T_n > z_{1-\alpha} + n^{-1/2}\hat{s}_1(z_{1-\alpha})$ is a second-order accurate test.

33. Let $X_1, \ldots, X_n$ be a sequence of independent and identically distributed random variables that have an EXPONENTIAL(θ) distribution for all $n \in \mathbb{N}$. Consider testing the null hypothesis $H_0 : \theta \leq \theta_0$ against the alternative hypothesis $H_1 : \theta > \theta_0$ for some $\theta \in \mathbb{R}$. This model falls within the smooth function model. When n is large it is often suggested that rejecting the null hypothesis when $T_n = n^{1/2}\hat{\sigma}_n^{-1}(\bar{X}_n - \theta_0) > z_{1-\alpha}$, where $\hat{\sigma}_n^2$ is the unbiased version of the sample variance, is a test that is approximately valid. Find an asymptotic expansion for the accuracy of this approximate test.

34. In the context of the proof of Theorem 10.12, prove that $\Lambda_n = n(\hat{\theta}_n - \theta_0)^2 I(\theta_0) + o_p(n^{-1})$, as $n \to \infty$ where $\theta_{1,n} = \theta_0 + n^{-1/2}\delta$.

35. Under the assumptions outlined in Theorem 10.11, show that *Wald's statistic*, which is given by $Q = n(\hat{\theta}_n - \theta_0)I(\hat{\theta}_n)$ where $I(\hat{\theta}_n)$ denotes the Fisher information number evaluated at the maximum likelihood statistics $\hat{\theta}_n$, has an asymptotic CHISQUARED(1) distribution under the null hypothesis $H_0 : \theta = \theta_0$.

36. Under the assumptions outlined in Theorem 10.11, show that *Rao's efficient score statistic*, which is given by $Q = n^{-1}U_n^2(\theta_0)I^{-1}(\theta_0)$ has an asymptotic CHISQUARED(1) distribution under the null hypothesis $H_0 : \theta = \theta_0$, where

$$U_n(\theta_0) = \sum_{i=1}^{n} \frac{\partial}{\partial \theta} \log[f(X_i; \theta)]\Big|_{\theta=\theta_0}.$$

Does this statistic require us to calculate the maximum likelihood estimator of θ?

37. Under the assumptions outlined in Theorem 10.11, show that Wald's statistic, $Q = n(\hat{\theta}_n - \theta_0)I(\hat{\theta}_n)$, has an asymptotic CHISQUARED$[1, \delta^2 I(\theta_0)]$ distribution under the sequence of alternative hypotheses $\{\theta_{1,n}\}_{n=1}^{\infty}$ where $\theta_{1,n} = \theta_0 + n^{-1/2}\delta$.

38. Under the assumptions outlined in Theorem 10.11, show that Rao's efficient score statistic, which is given by $Q = n^{-1}U_n^2(\theta_0)I^{-1}(\theta_0)$ has an asymptotic CHISQUARED$[1, \delta^2 I(\theta_0)]$ distribution under the sequence of alternative hypothesis $\{\theta_{1,n}\}_{n=1}^{\infty}$ where $\theta_{1,n} = \theta_0 + n^{-1/2}\delta$.

39. Suppose that $X_1, \ldots, X_n$ is a set of independent and identically distributed random variables from a continuous distribution F. Let $\xi \in (0, 1)$ and define $\theta = F^{-1}(\xi)$, the ξth population quantile of F. To compute a confidence region for θ, let $X_{(1)} \leq X_{(2)} \leq \cdots \leq X_{(n)}$ be the order statistics of the sample and let B be a BINOMIAL(n, ξ) random variable. The usual point estimator of θ is $\hat{\theta} = X_{\lfloor np \rfloor + 1}$ where $\lfloor x \rfloor$ is the largest integer strictly less than x. A confidence interval for θ is given by $C(\alpha, \omega) = [X_{(L)}, X_{(U)}]$, where L and U are chosen so that $P(B < L) = \omega_L$ and $P(B \geq U) = 1 - \omega_U$ and $\alpha = \omega_U - \omega_L$. See Section 3.2 of Conover (1980) for examples using this method.

  a. Derive an observed confidence level based on this confidence interval for an arbitrary interval subset $\Psi = (t_L, t_U)$ of the parameter space of θ.

  b. Derive an approximate observed confidence level for an arbitrary interval subset $\Psi = (t_L, t_U)$ of the parameter space of θ that is based on approximating the BINOMIAL distribution with the NORMAL distribution when n is large.

40. Suppose $X_1, \ldots, X_n$ is a set of independent and identically distributed random variables from a POISSON(θ) distribution where $\theta \in \Omega = (0, \infty)$. Garwood (1936) suggests a $100\alpha\%$ confidence interval for θ using the form

$$C(\alpha, \omega) = \left[\frac{\chi^2_{2Y; \omega_L}}{2n}, \frac{\chi^2_{2(Y+1); \omega_U}}{2n} \right],$$

where

$$Y = \sum_{i=1}^{n} X_i,$$

and $\boldsymbol{\omega} = (\omega_L, \omega_U) \in W_\alpha$ where

$$W_\alpha = \{\boldsymbol{\omega} = (\omega_1, \omega_2) : \omega_L \in [0,1], \omega_U \in [0,1], \omega_U - \omega_L = \alpha\}.$$

Therefore $L(\omega_L) = \frac{1}{2}n^{-1}\chi_{2Y;1-\omega_L}^2$ and $U(\omega_U) = \frac{1}{2}n^{-1}\chi_{2(Y+1);1-\omega_U}^2$. Derive an observed confidence level based on this confidence interval for an arbitrary interval subset $\Psi = (t_L, t_U)$ of the parameter space of θ.

41. Suppose $X_1, \ldots, X_n$ is a random sample from an EXPONENTIAL location family of densities of the form $f(x) = \exp[-(x - \theta)]\delta\{x; [\theta, \infty)\}$, where $\theta \in \Omega = \mathbb{R}$.

    a. Let $X_{(1)}$ be the first order-statistic of the sample $X_1, \ldots, X_n$. That is $X_{(1)} = \min\{X_1, \ldots, X_n\}$. Prove that

$$C(\alpha, \boldsymbol{\omega}) = [X_{(1)} + n^{-1}\log(1 - \omega_U), X_{(1)} + n^{-1}\log(1 - \omega_L)],$$

    is a $100\alpha\%$ confidence interval for θ when $\omega_U - \omega_L = \alpha$ where $\omega_L \in [0,1]$ and $\omega_U \in [0,1]$. *Hint: Use the fact that the density of $X_{(1)}$ is $f(x_{(1)}) = n\exp[-n(x_{(1)} - \theta)]\delta\{x_{(1)}; [\theta, \infty)\}$.*

    b. Use the confidence interval given above to derive an observed confidence level for an arbitrary region $\Psi = (t_L, t_U) \subset \mathbb{R}$ where $t_L < t_U$.

42. Suppose $X_1, \ldots, X_n$ is a random sample from a UNIFORM$(0, \theta)$ density where $\theta \in \Omega = (0, \infty)$.

    a. Find a $100\alpha\%$ confidence interval for θ when $\omega_U - \omega_L = \alpha$ where $\omega_L \in [0,1]$ and $\omega_U \in [0,1]$.

    b. Use the confidence interval given above to derive an observed confidence level for an arbitrary region $\Psi = (t_L, t_U) \subset \mathbb{R}$ where $0 < t_L < t_U$.

43. Let $\mathbf{X}_1, \ldots, \mathbf{X}_n$ be a set of independent and identically distributed d-dimensional random vectors from a distribution F with real valued parameter θ that fits within the smooth function model. Let Ψ be an interval subset of the parameter space of θ, which will be assumed to be a subset of the real line. When σ is unknown, a correct observed confidence level for Ψ is given by

$$\alpha_{\text{stud}}(\Psi) = H_n[n^{1/2}\hat{\sigma}_n^{-1}(\hat{\theta} - t_L)] - H_n[n^{1/2}\hat{\sigma}_n^{-1}(\hat{\theta} - t_U)],$$

where H_n is the distribution function of $n^{1/2}\hat{\sigma}_n^{-1}(\hat{\theta}_n - \theta)$. Suppose that H_n is unknown, but that we can estimate H_n using its Edgeworth expansion. That is, we can estimate H_n with $\hat{H}_n(t) = \Phi(t) + n^{-1/2}\hat{v}_1(t)\phi(t)$, where $\hat{v}_1(t) = v_1(t) + O_p(n^{-1/2})$, as $n \to \infty$. The observed confidence level can then be estimated with

$$\tilde{\alpha}_{\text{stud}}(\Psi) = \hat{H}_n[n^{1/2}\hat{\sigma}_n^{-1}(\hat{\theta} - t_L)] - \hat{H}_n[n^{1/2}\hat{\sigma}_n^{-1}(\hat{\theta} - t_U)].$$

Prove that $\tilde{\alpha}_{\text{stud}}$ is second-order accurate.

44. Let $X_1, \ldots, X_n$ be a set of independent and identically distributed random variables from a $N(\theta, \sigma^2)$ distribution conditional on θ, where θ has a $N(\lambda, \tau^2)$ distribution where σ^2, λ and τ^2 are known. Prove that the posterior distribution of θ is $N(\tilde{\theta}_n, \tilde{\sigma}_n^2)$ where

$$\tilde{\theta}_n = \frac{\tau^2 \bar{x}_n + n^{-1}\sigma^2\lambda}{\tau^2 + n^{-1}\sigma^2},$$

and

$$\tilde{\sigma}_n^2 = \frac{\sigma^2\tau^2}{n\tau^2 + \sigma^2}.$$

45. Let X be a single observation from a discrete distribution with probability distribution function

$$f(x|\theta) = \begin{cases} \frac{1}{4}\theta & x \in \{-2, -1, 1, 2\}, \\ 1 - \theta & x = 0, \\ 0 & \text{elsewhere,} \end{cases}$$

where $\theta \in \Omega = \{\frac{1}{5}, \frac{2}{5}, \frac{3}{5}, \frac{4}{5}\}$. Suppose that the prior distribution on θ is a UNIFORM$\{\frac{1}{5}, \frac{2}{5}, \frac{3}{5}, \frac{4}{5}\}$ distribution. Suppose that $X = 2$ is observed. Compute the posterior distribution of θ and the Bayes estimator of θ using the squared error loss function.

46. Let X be a single observation from a discrete distribution with probability distribution function

$$f(x|\theta) = \begin{cases} n^{-1}\theta & x \in \{1, 2, \ldots, n\}, \\ 1 - \theta & x = 0, \\ 0 & \text{elsewhere,} \end{cases}$$

where $\theta \in \Omega = \{(n+1)^{-1}, 2(n+1)^{-1}, \ldots, n(n+1)^{-1}\}$. Suppose that the prior distribution on θ is a UNIFORM$\{(n+1)^{-1}, 2(n+1)^{-1}, \ldots, n(n+1)^{-1}\}$ distribution. Suppose that $X = x$ is observed. Compute the posterior distribution of θ and the Bayes estimator of θ using the squared error loss function.

47. Let $X_1, \ldots, X_n$ be a set of independent and identically distributed random variables from a POISSON(θ) distribution and let θ have a GAMMA(α, β) prior distribution where α and β are known.

a. Prove that the posterior distribution of θ is a GAMMA$(\tilde{\alpha}, \tilde{\beta})$ distribution where

$$\tilde{\alpha} = \alpha + \sum_{i=1}^{n} Y_i,$$

and $\tilde{\beta} = (\beta^{-1} + n)^{-1}$.

b. Compute the Bayes estimator of θ using the squared error loss function. Is this estimator consistent and asymptotically NORMAL in accordance with Theorem 10.15?

48. Let $X_1, \ldots, X_n$ be a set of independent and identically distributed random variables from a POISSON(θ) distribution and let θ have a prior distribution of the form

$$\pi(\theta) = \begin{cases} \theta^{-1/2} & \theta > 0 \\ 0 & \text{elsewhere.} \end{cases}$$

This type of prior is known as an *improper prior* because it does not integrate to one. In particular, this prior is known as the *Jeffrey's prior* for the POISSON distribution. See Section 10.1 of Bolstad (2007) for further information on this prior.

a. Prove that the posterior distribution of θ is a GAMMA$(\tilde{\alpha}, \tilde{\beta})$ distribution where

$$\tilde{\alpha} = \tfrac{1}{2} + \sum_{i=1}^{n} Y_i,$$

and $\tilde{\beta} = n^{-1}$. Note that even though the prior distribution is improper, the posterior distribution is not.

b. Compute the Bayes estimator of θ using the squared error loss function. Is this estimator consistent and asymptotically NORMAL in accordance with Theorem 10.15?

10.7.2 Experiments

1. Write a program in R that will simulate $b = 1000$ samples of size n from a T(ν) distribution. For each sample the compute the sample mean, the sample mean with 5% trimming, the sample mean with 10% trimming, and the sample median. Estimate the mean squared error of estimating the population mean for each of these estimators over the b samples. Use your program to obtain the mean square estimates when $n = 25, 50$, and 100 with $\nu = 3, 4, 5, 10$, and 25.

a. Informally compare the results of these simulations. Does the sample median appear to be more efficient than the sample mean as indicated by the asymptotic relative efficiency when ν equals three and four? Does the trend reverse itself when ν becomes larger? How do the trimmed mean methods compare to the sample mean and the sample median?

b. Now formally compare the four estimators using an analysis of variance with a randomized complete block design where the treatments are taken to be the estimators, the blocks are taken to be the sample sizes, and the observed mean squared errors are taken to be the observations. How do the results of this analysis compare to the results observed above?

2. Write a program in R that simulates 1000 samples of size n from a POISSON(θ) distribution, where n and θ are specified below. For each sample

compute the two estimators of $P(X_n = 0) = \exp(-\theta)$ given by

$$\hat{\theta}_n = n^{-1} \sum_{i=1}^{n} \delta\{X_i; \{0\}\},$$

which is the proportion of values in the sample that are equal to zero, and $\tilde{\theta}_n = \exp(-\bar{X}_n)$. Use the 1000 samples to estimate the bias, standard error, and the mean squared error for each case. Discuss the results of the simulations in terms of the theoretical findings of Exercise 11. Repeat the experiment for $\theta = 1$, 2, and 5 with $n = 5$, 10, 25, 50, and 100.

3. Write a program in R that simulates 1000 samples of size n from a distribution F with mean θ, where both n and F are specified below. For each sample compute two 90% upper confidence limits for the mean: the first one of the form $\bar{X}_n - n^{-1/2}\hat{\sigma}_n z_{0.10}$ and the second of the form $\bar{X}_n - n^{-1/2}\hat{\sigma}_n t_{0.10,n-1}$, and determine whether θ is less than the upper confidence limit for each method. Use the 1000 samples to estimate the coverage probability for each method. How do these estimated coverage probabilities compare for the two methods with relation to the theory presented in this chapter? Formally analyze your results and determine if there is a significant difference between the two methods at each sample size. Use $n = 5$, 10, 25, 50 and 100 for each of the distributions listed below.

a. $N(0,1)$

b. $T(3)$

c. EXPONENTIAL(1)

d. EXPONENTIAL(10)

e. LAPLACE$(0,1)$

f. UNIFORM$(0,1)$

4. Write a program in R that simulates 1000 samples of size n from a $N(\theta, 1)$ distribution. For each sample compute the sample mean given by $\bar{X}_n$ and the Hodges super-efficient estimator $\hat{\theta}_n = \bar{X}_n + (a-1)\bar{X}_n\delta_n$ where $\delta_n = \delta\{|\bar{X}_n|; [0, n^{-1/4})\}$. Using the results of the 1000 simulated samples estimate the standard error of each estimator for each combination of $n = 5$, 10, 25, 50, and 100 and $a = 0.25, 0.50, 1.00$ and 2.00. Repeat the entire experiment once for $\theta = 0$ and once for $\theta = 1$. Compare the estimated standard errors of the two estimators for each combination of parameters given above and comment on the results in terms of the theory presented in Example 10.9.

5. Write a program in R that simulates 1000 samples of size n from a distribution F with mean θ where n, θ and F are specified below. For each sample compute the observed confidence level that θ is in the interval $\Psi = [-1, 1]$ as

$$T_{n-1}[n^{1/2}\hat{\sigma}_n^{-1}(\hat{\theta}_n + 1)] - T_{n-1}[n^{1/2}\hat{\sigma}_n^{-1}(\hat{\theta}_n - 1)],$$

where $\hat{\theta}_n$ is the sample mean and $\hat{\sigma}_n$ is the sample standard deviation. Keep track of the average observed confidence level over the 1000 samples.

Repeat the experiment for $n = 5, 10, 25, 50$ and 100 and comment on the results in terms of the consistency of the method.

    a. F is a $N(\theta, 1)$ distribution with $\theta = 0.0, 0.25, \ldots, 2.0$.

    b. F is a LAPLACE$(\theta, 1)$ distribution with $\theta = 0.0, 0.25, \ldots, 2.0$.

    c. F is a CAUCHY$(\theta, 1)$ distribution $\theta = 0.0, 0.25, \ldots, 2.0$, where θ is taken to be the median (instead of the mean) of the distribution.

6. Write a program in R that simulates 1000 samples of size n from a distribution F with mean θ, where n, F, and θ are specified below. For each sample test the null hypothesis $H_0 : \theta \leq 0$ against the alternative hypothesis $H_1 : \theta > 0$ using two different tests. In the first test the null hypothesis is rejected if $n^{1/2}\hat{\sigma}_n^{-1}\bar{X}_n > z_{0.90}$ and in the second test the null hypothesis is rejected if $n^{1/2}\hat{\sigma}_n^{-1}\bar{X}_n > t_{0.90, n-1}$. Keep track of how many times each test rejects the null hypothesis over the 1000 replications for $\theta = 0.0, 0.10\sigma, 0.20\sigma, \ldots, 2.0\sigma$ where σ is the standard deviation of F. Plot the number of rejections against θ for each test on the same set of axes, and repeat the experiment for $n = 5, 10, 25, 50$ and 100. Discuss the results in terms of the power functions of the two tests.

    a. F is a $N(\theta, 1)$ distribution.

    b. F is a LAPLACE$(\theta, 1)$ distribution.

    c. F is a EXPONENTIAL(θ) distribution.

    d. F is a CAUCHY$(\theta, 1)$ distribution $\theta = 0.0, 0.25, \ldots, 2.0$, where θ is taken to be the median (instead of the mean) of the distribution.

7. The interpretation of frequentist results of Bayes estimators is somewhat difficult because of the sometimes conflicting views of the resulting theoretical properties. This experiment will look at two ways of looking at the asymptotic results of this section.

    a. Write a program in R that simulates a sample of size n from a $N(0, 1)$ distribution and computes the Bayes estimator under the assumption that the mean parameter θ has a $N(0, \frac{1}{2})$ prior distribution. Repeat the experiment 1000 times for $n = 10, 25, 50$ and 100, and make a density histogram of the resulting Bayes estimates for each sample size. Place a comparative plot of the asymptotic NORMAL distribution for $\tilde{\theta}_n$ as specified by Theorem 10.15. How well do the distributions agree, particularly when n is larger?

    b. Write a program in R that first simulates θ from a $N(0, \frac{1}{2})$ prior distribution and then simulates a sample of size n from a $N(\theta, 1)$ distribution, conditional on the simulated value of θ. Compute the Bayes estimator of θ for each sample. Repeat the experiment 1000 times for $n = 10, 25, 50$ and 100, and make a density histogram of the resulting Bayes estimates for each sample size. Place a comparative plot of the asymptotic NORMAL distribution for $\tilde{\theta}_n$ as specified by Theorem 10.15. How well do the distributions agree, particularly when n is larger?

c. Write a program in R that first simulates θ from a $N(\frac{1}{2}, \frac{1}{2})$ prior distribution and then simulates a sample of size n from a $N(\theta, 1)$ distribution, conditional on the simulated value of θ. Compute the Bayes estimator of θ for each sample under the assumption that θ has a $N(0, \frac{1}{2})$ prior distribution. Repeat the experiment 1000 times for $n = 10,\ 25,\ 50$ and 100, and make a density histogram of the resulting Bayes estimates for each sample size. Place a comparative plot of the asymptotic NORMAL distribution for $\tilde{\theta}_n$ as specified by Theorem 10.15. How well do the distributions agree, particularly when n is larger? What effect does the misspecification of the prior have on the results?

CHAPTER 11

Nonparametric Inference

I had assumed you'd be wanting to go to the bank. As you're paying close attention to every word I'll add this: I'm not forcing you to go to the bank, I'd just
assumed you wanted to.

The Trial by Franz Kafka

11.1 Introduction

Nonparametric statistical methods are designed to provide valid statistical estimates, confidence intervals and hypothesis tests under very few assumptions
about the underlying model that generated the data. Typically, as the name
suggests, these methods avoid parametric models, which are models that can
be represented using a single parametric family of functions that depend on
a finite number of parameters. For example, a statistical method that makes
an assumption that the underlying distribution of the data is NORMAL is a
parametric method as the model for the data can be represented with a single
parametric family of densities that depend on two parameters. That is, the
distribution F comes from the family $\mathcal{N}$ where

$$\mathcal{N} = \{f(x) = (2\pi\sigma^2)^{-1/2}\exp[-\tfrac{1}{2}\sigma^{-2}(x-\mu)^2] : -\infty < \mu < \infty, 0 < \sigma^2 < \infty\}.$$

On the other hand, if we develop a statistical method that only makes the assumption that the underlying distribution is symmetric and continuous, then
the method is nonparametric. In this case the underlying distribution cannot be represented by a single parametric family that has a finite number
of parameters. Nonparametric methods are important to statistical theory as
parametric assumptions are not always valid in applications. A general introduction to these methods can be found in Gibbons and Chakraborti (2003),
Hollander and Wolfe (1999), and Sprent and Smeeton (2007).

The development of many nonparametric methods depends on the ability to
find statistics that are functions of the data, and possibly a null hypothesis,
that have the same known distribution over a nonparametric family of distributions. If the distribution is known, then the statistic can be use to develop
a hypothesis test that is valid over the entire nonparametric family.

Definition 11.1. *Let $\{X_n\}_{n=1}^{\infty}$ be a set of random variables having joint* *distribution F where $F \in \mathcal{A}$, a collection of joint distributions in $\mathbb{R}^n$. The function $T(X_1, \ldots, X_n)$ is distribution free over $\mathcal{A}$ is the distribution of T if the same for all $F \in \mathcal{A}$.*

Example 11.1. Let $\{X_n\}_{n=1}^{\infty}$ be a set of independent and identically distributed random variables from a continuous distribution F that has median equal to θ. Consider the statistic

$$T(X_1, \ldots, X_n) = \sum_{k=1}^{n} \delta\{X_k - \theta; (-\infty, 0]\}.$$

It follows that $\delta\{X_k - \theta; (-\infty, 0]\}$ has a BERNOULLI($\frac{1}{2}$) distribution for $k = 1, \ldots, n$, and hence the fact that $X_1, \ldots, X_n$ are independent and identically distributed random variables implies that $T(X_1, \ldots, X_n)$ has a BINOMIAL($n, \frac{1}{2}$) distribution. From Definition 11.1 it follows that $T(X_1, \ldots, X_n)$ is distribution free over the class of continuous distributions with median equal to θ. In applied problems θ is usually unknown, but the statistic $T(X_1, \ldots, X_n)$ can be used to develop a hypothesis test of $H_0 : \theta = \theta_0$ against $H_1 : \theta \neq \theta_0$ when θ is replaced by θ_0. That is,

$$B_n(X_1, \ldots, X_n) = \sum_{k=1}^{n} \delta\{X_k - \theta_0; (-\infty, 0]\}.$$

Under the null hypothesis B_n has a BINOMIAL($n, \frac{1}{2}$) distribution and therefore the null hypothesis $H_0 : \theta = \theta_0$ can be tested at significance level α by rejecting H_0 whenever $B \in R(\alpha)$ where $R(\alpha)$ is any set such that $P[B \in R(\alpha)|\theta = \theta_0] = \alpha$. This type of test is usually called the *sign test*. ∎

Many nonparametric procedures are based on ranking the observed data.

Definition 11.2. *Let $X_1, \ldots, X_n$ be an observed sample from a distribution F and let $X_{(1)}, \ldots, X_{(n)}$ denote the ordered sample. The rank of X_i, denoted by $R_i = R(X_i)$, equals k if $X_i = X_{(k)}$.*

We will assume that the ranks are unique in that there are not two or more of the observed sample values that equal one another. This will be assured with probability one when the distribution F is continuous. Under this assumption, an important property of the ranks is that their joint distribution does not depend on F. This is due to the fact that the ranks always take on the values $1, \ldots, n$ and the assignment of the ranks to the values $X_1, \ldots, X_n$ is a random permutation of the integers in the set $\{1, \ldots, n\}$.

Theorem 11.1. *Let $X_1, \ldots, X_n$ be a set of independent and identically distributed random variables from a continuous distribution F and let $\mathbf{R}' = (R_1, \ldots, R_n)$ be the vector of ranks. Define the set*

$$\mathbf{R}_n = \{\mathbf{r} : \mathbf{r} \text{ is a permutation of the integers } 1, \ldots, n\}. \tag{11.1}$$

Then $\mathbf{R}$ is uniformly distributed over $\mathbf{R}_n$.

For a proof of Theorem 11.1 see Section 2.3 of Randles and Wolfe (1979).

Rank statistics find common use in nonparametric methods. These types of statistics are often used to compare two or more populations as illustrated below.

Example 11.2. Let $X_1, \ldots, X_n$ be a set of independent and identically distributed random variables from a continuous distribution F and let $Y_1, \ldots, Y_m$ be a set of independent and identically distributed random variables from a continuous distribution G where $G(t) = F(t - \theta)$ and θ is an unknown parameter. This type of model for the distributions F and G is known as the *shift model*, and in the special case where the means of F and G both exist, $\theta = E(Y_1) - E(X_1)$. To test the null hypothesis $H_0 : \theta = 0$ against the alternative hypothesis $H_1 : \theta \neq 0$ compute the ranks of the combined sample $\{X_1, \ldots, X_m, Y_1, \ldots, Y_n\}$. Denote this combined sample as $Z_1, \ldots, Z_{n+m}$ and denote the corresponding ranks as $R_1, \ldots, R_{n+m}$. Note that under the null hypothesis the combined sample can be treated as a single sample from the distribution F. Now consider the test statistic

$$M_{m,n} = \sum_{i=1}^{n+m} D_i R_i,$$

where $D_i = 0$ when Z_i is from the sample $X_1, \ldots, X_m$ and $D_i = 1$ when Z_i is from the sample $Y_1, \ldots, Y_n$. Theorem 11.1 implies both that the random vector $\mathbf{D} = (D_1, \ldots, D_{n+m})'$ is distribution free and that $\mathbf{D}$ is independent of the vector of ranks given by $\mathbf{R}$. Hence, it follows that $M_{m,n} = \mathbf{D}'\mathbf{R}$ is distribution free. A test based on $M_{m,n}$ is known as the *Wilcoxon, Mann, and Whitney Rank Sum* test. The exact distribution of the test statistic under the null hypothesis can be derived by considering possible all equally likely configurations of $\mathbf{D}$ and $\mathbf{R}$, and computing the value of $M_{m,n}$ for each. See Chapter 4 of Hollander and Wolfe (1999) for further details. ∎

When considering populations that are symmetric about a shift parameter θ, the sign of the observed value, which corresponds to whether the observed value is greater than, or less than, the shift parameter can also be used to construct distribution free statistics.

Theorem 11.2. *Let $Z_1, \ldots, Z_n$ be a set of independent and identically distributed random variables from a distribution F that is symmetric about zero. Let $R_1, \ldots, R_n$ denote the ranks of the absolute observations $|Z_1|, \ldots, |Z_n|$ with $\mathbf{R}' = (R_1, \ldots, R_n)$ and let $C_i = \delta\{Z_i; (0, \infty)\}$ for $i = 1, \ldots, n$, with $\mathbf{C}' = (C_1, \ldots, C_n)$. Then*

1. *The random variables in $\mathbf{R}$ and $\mathbf{C}$ are mutually independent.*

2. *Then $\mathbf{R}$ is uniformly distributed over the set $\mathbf{R}_n$.*

3. *The components of $\mathbf{C}$ are a set of independent and identically distributed* BERNOULLI$(\frac{1}{2})$ *random variables.*

A proof of Theorem 11.2 can be found in Section 2.4 of Randles and Wolfe (1979).

Example 11.3. Let $(X_1, Y_1), \ldots, (X_n, Y_n)$ be a set of independent and identically distributed paired random variables from a continuous bivariate distribution F. Let G and H be the marginal distributions of X_n and Y_n, respectively. Assume that $H(x) = G(x - \theta)$ for some shift parameter θ and that G is a symmetric distribution about zero. Let $Z_i = X_i - Y_i$ and let $R_1, \ldots, R_n$ denote the ranks of the absolute differences $|Z_1|, \ldots, |Z_n|$. Define $C_i = \delta\{Z_i; (0, \infty)\}$ for $i = 1, \ldots, n$, with $\mathbf{C}' = (C_1, \ldots, C_n)$. The *Wilcoxon signed rank* statistic is then given by $W_n = \mathbf{C}'\mathbf{R}$, which is the sum of the ranks of the absolute differences that correspond to positive differences. When testing the null hypothesis $H_0 : \theta = \theta_0$ the value of θ in W_n is replaced by θ_0. Under this null hypothesis it follows from Theorem 11.2 that W_n is distribution free. The distribution of W_n under the null hypothesis can be found by enumerating the value of W_n over all possible equally likely permutations of the elements of $\mathbf{C}$ and $\mathbf{R}$. For further details see Section 3.1 of Hollander and Wolfe (1999). ∎

The analysis of the asymptotic behavior of test statistics like those studied in Examples 11.1 to 11.3 is the subject of this chapter. The next section will develop a methodology for showing that such statistics are asymptotically NORMAL using the theory of U-statistics.

11.2 Unbiased Estimation and U-Statistics

Let $\{X_n\}_{n=1}^{\infty}$ be a sequence of independent and identically distributed random variables from a distribution F with functional parameter $\theta = T(F)$. Hoeffding (1948) considered a class of estimators of θ that are unbiased and also have a distribution that converges weakly to a NORMAL distribution as the sample size increases. Such statistics are not necessarily directly connected with nonparametric methods. For example, the sample mean and variance are examples of such statistics. However, many of the classical nonparametric test statistics fall within this area. As such, the methodology of Hoeffding (1948) provides a convenient method for finding the asymptotic distribution of such test statistics.

To begin our development we consider the concept of estimability. Estimability in this case refers to the smallest sample size for which an unbiased estimator of a parameter exists.

Definition 11.3. *Let $\{X_n\}_{n=1}^{\infty}$ be a sequence of independent and identically distributed random variables from a distribution $F \in \mathcal{F}$, where $\mathcal{F}$ is a collection of distributions. A parameter $\theta \in \Omega$ is estimable of degree r over $\mathcal{F}$ if r is the smallest sample size for which there exists a function $h^* : \mathbb{R}^r \to \Omega$ such that $E[h^*(X_1, \ldots, X_r)] = \theta$ for every $F \in \mathcal{F}$ and $\theta \in \Omega$.*

Example 11.4. Let $\{X_n\}_{n=1}^{\infty}$ be a sequence of independent and identically distributed random variables from a distribution $F \in \mathcal{F}$ where $\mathcal{F}$ is a collection of distributions that have a finite second moment. Let $\theta = V(X_1)$. We will show that θ is estimable of degree two over $\mathcal{F}$. First, we demonstrate that the degree of θ is not greater than two. To do this we note that if we define $h^*(X_1, X_2) = \frac{1}{2}(X_1 - X_2)^2$, then $E[\frac{1}{2}(X_1 - X_2)^2] = \frac{1}{2}(\theta + \mu^2 - 2\mu^2 + \theta + \mu^2) = \theta$ where $\mu = E(X_1)$. Therefore, the degree of θ is at most two. To show that the degree of θ is not one we must show that there is not a function $h_1^*(X_1)$ such that $E[h_1^*(X_1)] = \theta$ for all $F \in \mathcal{F}$ and $\theta \in \Omega$. Following the suggestion of Randles and Wolfe (1979) we first assume that such a function exists and search for a contradiction. For example, if such an h_1^* exists then

$$E[X_1^2 - h_1^*(X_1)] = \theta + \mu^2 - \theta = \mu^2. \tag{11.2}$$

Now, is it possible that such a function exists? As suggested by Randles and Wolfe (1979) we shall consider what happens when F corresponds to a UNIFORM$(\eta - \frac{1}{2}, \eta + \frac{1}{2})$ distribution where $\eta \in \mathbb{R}$. In this case

$$E[X_1^2 - h_1(X_1)] = \int_{\eta - \frac{1}{2}}^{\eta + \frac{1}{2}} x_1^2 - h_1^*(x_1) dx_1 = \int_{\eta - \frac{1}{2}}^{\eta + \frac{1}{2}} x_1^2 dx_1 - \int_{\eta - \frac{1}{2}}^{\eta + \frac{1}{2}} h_1^*(x_1) dx_1. \tag{11.3}$$

The first integral in Equation (11.3) is given by $\eta^2 + \frac{1}{12}$ and hence the second integral must equal η^2. That is,

$$\int_{\eta - \frac{1}{2}}^{\eta + \frac{1}{2}} h_1^*(x_1) dx_1 = \eta^2.$$

This implies that $h^*(x_1)$ must be a linear function of the form $a + bx_1$ for some constants a and b. However, direct integration implies that

$$\int_{\eta - \frac{1}{2}}^{\eta + \frac{1}{2}} (a + bx_1) dx_1 = a + b\eta$$

which is actually linear in η, which is a contradiction, and hence no such function exists when $\eta \neq 0$, and therefore θ is not estimable of degree one over $\mathcal{F}$. It is important to note that this result depends on the family of distributions considered. See Exercise 1. ∎

Now, suppose that $X_1, \ldots, X_n$ is a set of independent and identically distributed random variables from a distribution $F \in \mathcal{F}$ and let θ be a parameter that is estimable of degree r with function $h^*(X_1, \ldots, X_r)$. Note that $h^*(X_1, \ldots, X_r)$ is an unbiased estimator of θ since $E[h^*(X_1, \ldots, X_r)] = \theta$ for all $\theta \in \Omega$. To compute this estimator the sample size n must be at least as large as r. When $n > r$ this estimator is unlikely to be very efficient since it does not take advantage of the information available in the entire sample. In fact, $h^*(X_{i_1}, \ldots, X_{i_r})$ is also an unbiased estimator of θ for any set of indices $\{i_1, \ldots, i_r\}$ that are selected without replacement from the set $\{1, \ldots, n\}$. The central idea of a U-statistic is to form an efficient unbiased estimator of θ by averaging together all $\binom{n}{r}$ possible unbiased estimators of θ corresponding to the function h^* computed on all $\binom{n}{r}$ possible samples of the form $\{X_{i_1}, \ldots, X_{i_r}\}$ as described above. This idea is much simpler to implement if h^* is a symmetric function of its arguments. That is, if $h^*(X_1, \ldots, X_r) = h^*(X_{i_1}, \ldots, X_{i_r})$ where in this case $\{i_1, \ldots, i_r\}$ is any permutation of the integers in the set $\{1, \ldots, r\}$. In fact, given any function $h^*(X_1, \ldots, X_r)$ that is an unbiased estimator of θ, it is possible to construct a symmetric function $h(X_1, \ldots, X_r)$ that is also an unbiased estimator of θ. To see why this is true we construct $h(X_1, \ldots, X_r)$ as

$$h(X_1, \ldots, X_r) = (r!)^{-1} \sum_{\mathbf{a} \in A_r} h^*(X_{a_1}, \ldots, X_{a_r}),$$

where $\mathbf{a}' = (a_1, \ldots, a_r)$ and A_r is the set that contains all vectors whose elements correspond to the permutations of the integers in the set $\{1, \ldots, r\}$. Note that because $X_1, \ldots, X_n$ are independent and identically distributed it follows that $E[h^*(X_{a_1}, \ldots, X_{a_r})] = \theta$ for all $\mathbf{a} \in A_r$ and hence

$$E[h(X_1, \ldots, X_r)] = (r!)^{-1} \sum_{\mathbf{a} \in A_r} E[h^*(X_{a_1}, \ldots, X_{a_r})] = \theta.$$

Therefore $h(X_1, \ldots, X_r)$ is a symmetric function that can be used in place of $h^*(X_1, \ldots, X_r)$. With the symmetric function $h(X_1, \ldots, X_r)$ we now define a U-statistic as the average of all possible values of the function h computed over all $\binom{n}{r}$ possible selections of r random variables from the set $\{X_1, \ldots, X_n\}$.

Definition 11.4. *Let $X_1, \ldots, X_n$ be a set of independent and identically distributed random variables from a distribution F with functional parameter θ. Suppose the θ is estimable of degree r with a symmetric function h which will be called the kernel function. Then a U-statistic for θ is given by*

$$U_n = U_n(X_1, \ldots, X_n) = \binom{n}{r}^{-1} \sum_{\mathbf{b} \in \mathbf{B}_{n,r}} h(X_{b_1}, \ldots, X_{b_r}), \quad (11.4)$$

where $\mathbf{B}_{n,r}$ is a set that contains all vectors whose elements correspond to unique selections of r integers from the set $\{1, \ldots, n\}$, taken without replacement.

Example 11.5. Let $X_1, \ldots, X_n$ be a set of independent and identically distributed random variables from a distribution F with finite mean θ. Let $h(x) = x$ and note that $E[h(X_i)] = \theta$ for all $i = 1, \ldots, n$. Therefore

$$U_n = n^{-1} \sum_{i=1}^n X_i,$$

is a U-statistic of degree one. ∎

Example 11.6. Let $X_1, \ldots, X_n$ be a set of independent and identically distributed random variables from a continuous distribution F and define $\theta = P(X_n \leq \xi)$ for a specified real constant ξ. Consider the function $h(x) = \delta\{x - \xi; (-\infty, 0]\}$ where $E[h(X_n)] = P(X_n - \xi \leq 0) = P(X_n \leq \xi) = \theta$. Therefore a U-statistic of degree one for the parameter θ is given by

$$B_n = n^{-1} \sum_{i=1}^n \delta\{X_i - \xi; (-\infty, 0]\}.$$

This statistic corresponds to the test statistic used by the sign test for testing hypotheses about a quantile of F. ∎

Example 11.7. Let $X_1, \ldots, X_n$ be a set of independent and identically distributed random variables from a continuous distribution F that is symmetric about a point θ. Let $\mathbf{R}$ be the vector of ranks of $|X_1 - \theta|, \ldots, |X_n - \theta|$ and let $\mathbf{C}$ be an $n \times 1$ vector with i^{th} element $c_i = \delta\{X_i - \theta; (0, \infty)\}$ for $i = 1, \ldots, n$. The signed rank statistic was seen in Example 11.3 to have the form $W_n = \mathbf{C}'\mathbf{R}$. The main purpose of this exercise is to demonstrate that this test statistic can be written as the sum of two U-statistics. To simplify the notation used in this example, let $Z_i = X_i - \theta$ for $i = 1, \ldots, n$ so that $\mathbf{R}$ contains the ranks of $|Z_1|, \ldots, |Z_n|$ and $c_i = \delta\{Z_i; (0, \infty)\}$. Let us examine each term in the statistic

$$W_n = \sum_{i=1}^{n} R_i \delta\{Z_i; (0, \infty)\} = \sum_{i=1}^{n} \tilde{R}_i \delta\{Z_{(i)}; (0, \infty)\}, \qquad (11.5)$$

where $\tilde{R}_i$ is the absolute rank associated with $Z_{(i)}$. There are two possibilities for the ith term in the sum in Equation (11.5). The term will be zero if $Z_i < 0$. If $Z_i > 0$ then the ith term will add $\tilde{R}_i$ to the sum. Suppose for the moment that $\tilde{R}_i = r$ for some $r \in \{1, \ldots, n\}$. This means that there are $r - 1$ values from $|Z_1|, \ldots, |Z_n|$ such that $|Z_j| < |Z_{(i)}|$ along with the one value that equals $|Z_{(i)}|$. Hence, the ith term will add

$$\sum_{j=1}^{n} \delta\{|Z_j|; (0, |Z_{(i)}|]\} = \sum_{j=1}^{n} \delta\{|Z_{(j)}|; (0, |Z_{(i)}|]\},$$

to the sum in Equation (11.5) when $Z_i > 0$. Now let us combine these two conditions. Let $Z_{(1)}, \ldots, Z_{(n)}$ denote the order statistics of $Z_1, \ldots, Z_n$. Let $i < j$ and note that $\delta\{Z_{(i)} + Z_{(j)}; (0, \infty)\} = 1$ if and only if $Z_{(j)} > 0$ and $|Z_{(i)}| < Z_{(j)}$. To see why this is true consider the following cases. If $Z_{(j)} < 0$ then $Z_{(i)} < 0$ since $i < j$ and hence $\delta\{Z_{(i)} + Z_{(j)}; (0, \infty)\} = 0$. Similarly, it is possible that $Z_{(j)} > 0$ but $|Z_{(i)}| > |Z_{(j)}|$. This can only occur when $Z_{(i)} < 0$, or else $Z_{(j)}$ would be larger than $Z_{(j)}$ which cannot occur because $i < j$. Therefore $|Z_{(i)}| > |Z_{(j)}|$ and $Z_{(i)} < 0$ implies that $Z_{(i)} + Z_{(j)} < 0$ and hence $\delta\{Z_{(i)} + Z_{(j)}; (0, \infty)\} = 0$. However, if $Z_{(j)} > 0$ and $|Z_{(i)}| < |Z_{(j)}|$ then it must follow that $\delta\{Z_{(i)} + Z_{(j)}; (0, \infty)\} = 1$. Therefore, it follows that the term in Equation (11.5) can be written as

$$\sum_{j=1}^{n} \delta\{|Z_{(j)}|; (0, Z_{(i)})\} = \sum_{j=1}^{i} \delta\{Z_{(i)} + Z_{(j)}\}; (0, \infty)\},$$

where the upper limit of the sum on the right hand side of the equation reflects the fact that we only add in observations less than or equal to $Z_{(j)}$, which is the signed rank of $Z_{(i)}$. Therefore, it follows that

$$
\begin{aligned}
W &= \sum_{i=1}^{n} \sum_{j=1}^{i} \delta\{Z_{(i)} + Z_{(j)}; (0, \infty)\} \\
&= \sum_{i=1}^{n} \sum_{j=1}^{i} \delta\{Z_i + Z_j; (0, \infty)\} \\
&= \sum_{i=1}^{n} \delta\{2Z_i; (0, \infty)\} + \sum_{i=1}^{n} \sum_{j=i+1}^{n} \delta\{Z_i + Z_j; (0, \infty)\}. \quad (11.6)
\end{aligned}
$$

The first term in Equation (11.6) can be written as $nU_{1,n}$ where $U_{1,n}$ is a U-statistic of the form

$$
U_{1,n} = n^{-1} \sum_{i=1}^{n} \delta\{2Z_i; (0, \infty)\},
$$

and the second term in Equation (11.6) can be written as $\binom{n}{2} U_{2,n}$ where $U_{2,n}$ is a U-statistic of the form

$$
U_{2,n} = \binom{n}{2}^{-1} \sum_{i=1}^{n} \sum_{j=i+1}^{n} \delta\{Z_i + Z_j; (0, \infty)\}.
$$

∎

A key property of U-statistics which makes them an important topic in statistical estimation theory is that they are optimal in that they have the lowest variance of all unbiased estimators of θ.

Theorem 11.3. *Let $X_1, \ldots, X_n$ be a set of independent and identically distributed random variables from a distribution F with parameter θ. Let U_n be a U-statistic for θ and let T_n be any other unbiased estimator of θ, then $V(U_n) \leq V(T_n)$.*

A proof of Theorem 11.3 can be found in Section 5.1.4 of Serfling (1980).

The main purpose of this section is to develop conditions under which a U-statistic is asymptotically normal. In order to obtain such a result, we first need to develop an expression for the variance of a U-statistic. The form of the U-statistic defined in Equation (11.4) suggests that we need to obtain an expression for the variance of the sum

$$
\sum_{\mathbf{b} \in \mathbf{B}_{n,r}} h(X_{b_1}, \ldots, X_{b_r}).
$$

If the terms in this sum where independent of one another then we could exchange the variance and the sum. However, it is clear that if we choose two distinct elements $\mathbf{b}$ and $\mathbf{b}'$ from $\mathbf{B}_{n,r}$, the two terms $h(X_{b_1}, \ldots, X_{b_r})$ and $h(X_{b'_1}, \ldots, X_{b'_r})$ could have as many as $r - 1$ of the random variables from the set $\{X_1, \ldots, X_n\}$ is common, but may have as few as zero of these variables in common. Note that the two terms could not have all r variables in common because we are assuming that $\mathbf{b}$ and $\mathbf{b}'$ are distinct. We also note that if n was not sufficiently large then there may also be a lower bound on the number of random variables that the two terms could have in common. In general we will assume that n is large enough so that the lower limit is always zero. In this case we have that

$$V\left[\sum_{\mathbf{b}\in\mathbf{B}_{n,r}} h(X_{b_1},\ldots,X_{b_r})\right] =$$

$$\sum_{\mathbf{b}\in\mathbf{B}_{n,r}}\sum_{\mathbf{b}'\in\mathbf{B}_{n,r}} C[h(X_{b_1},\ldots,X_{b_r}),h(X_{b'_1},\ldots,X_{b'_r})].$$

To simplify this expression let us consider the case where the sets $\{b_1,\ldots,b_r\}$ and $\{b'_1,\ldots,b'_r\}$ have exactly c elements in common. For example, we can consider the term,

$$C[h(X_1,\ldots,X_c,X_{c+1},\ldots,X_r),h(X_1,\ldots,X_c,X_{r+1},\ldots,X_{2r-c})],$$

where we assume that $n > 2r - c$. Now consider comparing this covariance to another term that also has exactly c variables in common such as

$$C[h(X_1,\ldots,X_c,X_{c+1},\ldots,X_r),h(X_1,\ldots,X_c,X_{r+2},\ldots,X_{2r-c+1})].$$

Note that the two covariances will be equal because the joint distribution of $X_{r+1},\ldots,X_{2r-c}$ is exactly the same as the joint distribution of $X_{r+2},\ldots,$ X_{2r-c+1} because $X_1,\ldots,X_n$ are assumed to be a sequence of independent and identically distributed random variables. This fact, plus the symmetry of the function h will imply that any two terms that have exactly c variables in common will have the same covariance. Therefore, define

$$\zeta_c = C[h(X_1,\ldots,X_c,X_{c+1},\ldots,X_r),h(X_1,\ldots,X_c,X_{r+1},\ldots,X_{2r-c})],$$

for $c = 0,\ldots,r-1$. The number of terms of this form in the sum of covariances equals the number of ways to choose r indices from the set of n, which is $\binom{n}{r}$, multiplied by the number of ways to choose c common indices from the r chosen indices which is $\binom{r}{c}$, multiplied by the number of ways to choose the remaining $r - c$ non-common indices from the set of $n - r$ indices not chosen from the first selection which is $\binom{n-r}{r-c}$. Therefore, it follows that the variance of a U-statistic of the form defined in Equation (11.4) is

$$V(U_n) = \binom{n}{r}^{-2}\sum_{\mathbf{b}\in\mathbf{B}_{n,r}}\sum_{\mathbf{b}'\in\mathbf{B}_{n,r}} C[h(X_{b_1},\ldots,X_{b_r}),h(X_{b'_1},\ldots,X_{b'_r})]$$

$$= \binom{n}{r}^{-2}\sum_{c=0}^{r}\binom{n}{r}\binom{r}{c}\binom{n-r}{r-c}\zeta_c$$

$$= \binom{n}{r}^{-1}\sum_{c=0}^{r}\binom{r}{c}\binom{n-r}{r-c}\zeta_c. \tag{11.7}$$

Note further that

$$\zeta_0 = C[h(X_1,\dots,X_r), h(X_{r+1},\dots,X_{2r})] = 0,$$

since $X_1,\dots,X_r$ are mutually independent of $X_{r+1},\dots,X_{2r}$. Hence, the expression in Equation (11.7) simplifies to

$$V(U_n) = \sum_{c=1}^{r} \binom{r}{c}\binom{n-r}{r-c}\zeta_c. \tag{11.8}$$

Example 11.8. Let $X_1,\dots,X_n$ be a set of independent and identically distributed random variables from a distribution with mean θ and finite variance σ^2. In Example 11.5 it was shown that the sample mean $U_n = \bar{X}_n$ is a U-statistic of degree $r = 1$ for θ with function $h(x) = x$. The variance of this U-statistic can then be computed using Equation (11.8) to find

$$
\begin{aligned}
V(U_n) &= \binom{n}{1}^{-1}\sum_{c=1}^{r}\binom{1}{c}\binom{n-1}{1-c}\zeta_c \\
&= n^{-1}\zeta_1 \\
&= n^{-1}C(X_1,X_1) \\
&= n^{-1}\sigma^2,
\end{aligned}
$$

which is the well-known expression for the variance of the sample mean. ∎

Example 11.9. Let $X_1,\dots,X_n$ be a set of independent and identically distributed random variables from a distribution with mean θ and finite variance σ^2. Consider estimating the parameter θ^2 which can be accomplished with a U-statistic of degree $r = 2$ with symmetric function $h(x_1,x_2) = x_1 x_2$. The variance of this U-statistic can then be computed using Equation (11.8) to find

$$
\begin{aligned}
V(U_n) &= \binom{n}{2}^{-1}\sum_{c=1}^{2}\binom{2}{c}\binom{n-2}{2-c}\zeta_c \\
&= \binom{n}{2}^{-1}\binom{2}{1}\binom{n-2}{1}\zeta_1 + \binom{n}{2}^{-1}\binom{2}{2}\binom{n-2}{0}\zeta_2 \\
&= \frac{2(n-2)}{\frac{1}{2}n(n-1)}\zeta_1 + \frac{1}{\frac{1}{2}n(n-1)}\zeta_2 \\
&= \frac{4(n-2)}{n(n-1)}\zeta_1 + \frac{2}{n(n-1)}\zeta_2.
\end{aligned}
$$

Now

$$
\begin{aligned}
\zeta_1 &= C(X_1 X_2, X_1 X_3) \\
&= E(X_1^2 X_2 X_3) - E(X_1 X_2)E(X_1 X_3) \\
&= \theta^2(\theta^2 + \sigma^2) - \theta^4 \\
&= \theta^2 \sigma^2,
\end{aligned}
$$

and

$$\begin{aligned}
\zeta_2 &= C(X_1 X_2, X_1 X_2) \\
&= E(X_1^2 X_2^2) - E(X_1 X_2)E(X_1 X_2) \\
&= (\theta^2 + \sigma^2)^2 - \theta^4 \\
&= 2\theta^2 \sigma^2 + \sigma^4.
\end{aligned}$$

Therefore, we can conclude that

$$\begin{aligned}
V(U_n) &= \frac{4(n-2)\theta^2 \sigma^2}{n(n-1)} + \frac{4\theta^2 \sigma^2}{n(n-1)} + \frac{2\sigma^4}{n(n-1)} \\
&= \frac{4\theta^2 \sigma^2}{n} + \frac{2\sigma^4}{n(n-1)}.
\end{aligned}$$

∎

The variance of a U-statistic can be quite complicated, but it does turn out that the leading term in Equation (11.8) is dominant from an asymptotic viewpoint.

Theorem 11.4. *Let $X_1, \ldots, X_n$ be a set of independent and identically distributed random variables from a distribution F. Let U_n be an rth-order U-statistic with symmetric kernel function $h(x_1, \ldots, x_r)$. If $E[h^2(X_1, \ldots, X_r)] < \infty$ then $V(U_n) = n^{-1} r^2 \zeta_1 + o(n^{-1})$, as $n \to \infty$.*

Proof. From Equation (11.7) we have that

$$nV(U_n) = \sum_{c=1}^{r} n \binom{n}{r}^{-1} \binom{r}{c} \binom{n-r}{r-c} \zeta_c, \tag{11.9}$$

where $\zeta_1, \ldots, \zeta_r$ are finite because we have assumed that $E[h^2(X_1, \ldots, X_r)] < \infty$. It is the asymptotic behavior of the coefficients of ζ_c that will determine the behavior of $nV(U_n)$. We first note that for $c = 1$ we have that

$$n \binom{n}{r}^{-1} \binom{r}{c} \binom{n-r}{r-c} = n \binom{n}{r}^{-1} \binom{r}{1} \binom{n-r}{r-1} \zeta_c = \frac{r^2 [(n-r)!]^2}{(n-2r+1)!(n-1)!}.$$

Canceling the identical terms in the numerator and the denominator of the right hand side of this equation yields

$$n \binom{n}{r}^{-1} \binom{r}{1} \binom{n-r}{r-1} \zeta_c = \frac{r^2 (n-r) \cdots (n-2r+1)}{(n-1) \cdots (n-r+1)} = r^2 \prod_{i=1}^{r-1} \frac{n-i+1-r}{n-i},$$

where it is instructive to note that the number of terms in the product does not depend on n. Therefore, since

$$\lim_{n \to \infty} r^2 \prod_{i=1}^{r-1} \frac{n-i+1-r}{n-i} = r^2,$$

it follows that the first term in Equation (11.9) converges to $r^2 \zeta_1$ as $n \to \infty$.

For the case where $c \in \{2, \ldots, r\}$, we have that

$$n \binom{n}{r}^{-1} \binom{r}{c} \binom{n-r}{r-c} = \frac{(r!)^2}{c![(r-c)!]^2} \frac{[(n-r)!]^2}{(n-2r+c)!(n-1)!}. \tag{11.10}$$

Again, we cancel identical terms on the numerator and the denominator of the second term on the right hand side of Equation (11.10) to yield

$$\frac{[(n-r)!]^2}{(n-2r+c)!(n-1)!} = \frac{(n-r)\cdots(n-2r+c+1)}{(n-1)\cdots(n-r+1)} =$$

$$\left[\prod_{i=1}^{c-1}(n-r+c-i)^{-1} \right] \left[\prod_{i=1}^{r-c} \frac{n-r-i+1}{n-i} \right],$$

where again we note that the number of terms in each of the products does not depend on n. Therefore, we have that

$$\lim_{n\to\infty} \prod_{i=1}^{c-1}(n-r+c-i)^{-1} = 0,$$

and

$$\lim_{n\to\infty} \prod_{i=1}^{r-c} \frac{n-r-i+1}{n-i} = 1.$$

Therefore, it follows that

$$\lim_{n\to\infty} n \binom{n}{r}^{-1} \binom{r}{c} \binom{n-r}{r-c} \zeta_c = 0,$$

and hence the result follows. $\qquad\square$

Example 11.10. Let $X_1, \ldots, X_n$ be a set of independent and identically distributed random variables from a distribution with mean θ and finite variance σ^2. Continuing with Example 11.9 we consider estimating the parameter θ^2 with a U-statistic of degree $r = 2$ with symmetric kernel function $h(x_1, x_2) = x_1 x_2$. The variance of this U-statistic was computed in Example 11.9 as

$$V(U_n) = \frac{4\theta^2\sigma^2}{n} + \frac{2\sigma^4}{n(n-1)}.$$

One can verify the result of Theorem 11.4 by noting that

$$\lim_{n\to\infty} nV(U_n) = \lim_{n\to\infty} 4\theta^2\sigma^2 + \frac{2\sigma^4}{n-1} = 4\theta^2\sigma^2 = r^2\zeta_1.$$

$\blacksquare$

We are now in a position to develop conditions under which U-statistics are asymptotically normal. We begin with the case where $r = 1$. In this case the U-statistic defined in Equation (11.4) has the form

$$U_n = n^{-1} \sum_{i=1}^{n} h(X_i),$$

which is the sum of independent and identically distributed random variables. Therefore, it follows from Theorem 4.20 (Lindeberg and Lévy) that $n^{1/2}\zeta_1^{-1/2}(U_n - \theta) \xrightarrow{d} Z$ as $n \to \infty$ where Z is a $N(0,1)$ random variable.

For the case when $r > 1$ the problem becomes more complicated as the terms in the sum in a U-statistic are no longer necessarily independent. The approach for establishing asymptotic normality for these types of U-statistics is based on finding a function of the observed data that has the same asymptotic behavior as the U-statistic, but is the sum of independent and identically distributed random variables. Theorem 4.20 can then be applied to this related function, thus establishing the asymptotic normality of the U-statistic. To simplify matters, we will actually first center the U-statistic about the origin. That is, if U_n is a U-statistic of order r, then we will actually be working with the function $U_n - \theta$ which has expectation zero. The method for finding a function of the data that is a sum of independent and identically distributed terms that has the same asymptotic behavior as $U_n - \theta$ is based on finding a projection of $U_n - \theta$ onto the space of functions that are sums of independent and identically distributed random variables.

Recall that a projection of a point in a metric space on a subspace is accomplished by finding a point in the subspace that is closest to the specified point. For example, we can consider the vector space $\mathbb{R}^3$ with vector $\mathbf{x} \in \mathbb{R}^3$. Let $\mathcal{P}$ denote a two-dimensional subspace of $\mathbb{R}^3$ corresponding to a plane. Then the vector $\mathbf{x}$ is projected onto $\mathcal{P}$ by finding a vector $\mathbf{p} \in \mathcal{P}$ that minimizes $\|\mathbf{x} - \mathbf{p}\|$. For our purpose we will consider projecting $U_n - \theta$ onto the space of functions given by

$$\mathcal{V}_n = \left\{ V_n \,\middle|\, V_n = \sum_{i=1}^{n} k(X_i) \right\},$$

where k is a real-valued function. The function k that will result from this projection will usually depend on some unknown parameters of F. However, this does not affect the usefulness of the results since we are not actually interested in computing the function; we only need to establish its asymptotic properties.

Consider a U-statistic of order r given by U_n. We wish to project $U_n - \theta$ onto the space $\mathcal{V}_n$. In order to do this we need a measure of distance between $U_n - \theta$ and functions that are in $\mathcal{V}_n$. For this we will use the expected square distance between the two functions. That is, we use $\|U_n - \theta - V_n\| = E[(U_n - \theta - V_n)^2]$.

Theorem 11.5. *Let U_n be a U-statistic of order r calculated on $X_1, \ldots, X_n$, a set of independent and identically distributed random variables. The projection of $U_n - \theta$ onto $\mathcal{V}_n$ is given by*

$$V_n = rn^{-1} \sum_{i=1}^{n} \{E[h(X_i, X_2, \ldots, X_r)|X_i] - \theta\}. \tag{11.11}$$

Proof. In order to prove this result we must show that $V_n \in \mathcal{V}_n$ and that V_n

minimizes $\|U_n - \theta - V_n\|$. To show that $V_n \in \mathcal{V}_n$, we need only note that the conditional expectation $E[h(X_i, X_2, \ldots, X_r)|X_i]$ is only a function of X_i and hence we can take the function $k(X_i)$ to be defined as

$$\tilde{k}(X_i) = rn^{-1}\{E[h(X_i, X_2, \ldots, X_r)|X_i] - \theta\}.$$

To prove that $V_{1,n}$ minimizes $\|U_n - \theta - V_{1,n}\|$ we let V be an arbitrary member of $\mathcal{V}_n$, and note that

$$
\begin{aligned}
\|U_n - \theta - V\| &= E[(U_n - \theta - V)^2] \\
&= E\{[(U_n - \theta - V_n) + (V_n - V)]^2\} \\
&= E[(U_n - \theta - V_n)^2] + E[(V_n - V)^2] \\
&\quad + 2E[(U_n - \theta - V_n)(V_n - V)].
\end{aligned}
$$

Now, suppose that V has the form

$$V = \sum_{i=1}^{n} k(X_i),$$

where k is a real valued function. Then

$$
\begin{aligned}
E[(U_n - \theta - V_n)(V_n - V)] &= E\left\{(U_n - \theta - V_n)\sum_{i=1}^{n}[\tilde{k}(X_i) - k(X_i)]\right\} \\
&= \sum_{i=1}^{n} E\{(U_n - \theta - V_n)[\tilde{k}(X_i) - k(X_i)]\}.
\end{aligned}
$$

$$(11.12)$$

Evaluating the term in the sum on the right hand side of Equation (11.12) we use Theorem A.17 to find that

$$
\begin{aligned}
E\{(U_n - \theta - V_n)[\tilde{k}(X_i) - k(X_i)]\} = \\
E[E\{(U_n - \theta - V_n)[\tilde{k}(X_i) - k(X_i)]|X_i\}], \quad (11.13)
\end{aligned}
$$

where the outer expectation on the right hand side of Equation (11.13) is taken with respect to X_i. Therefore, it follows that

$$
\begin{aligned}
E[E\{(U_n - \theta - V_n)[\tilde{k}(X_i) - k(X_i)]|X_i\}] = \\
E\{[\tilde{k}(X_i) - k(X_i)]E(U_n - \theta - V_n|X_i)\}. \quad (11.14)
\end{aligned}
$$

To evaluate the conditional expectation on the right hand side of Equation (11.14) we note that

$$
\begin{aligned}
E(U_n|X_i) &= E\left[\binom{n}{r}^{-1}\sum_{b \in \mathbf{B}_{n,r}} h(X_{b_1}, \ldots, X_{b_r})\,\middle|\, X_i\right] \\
&= \binom{n}{r}^{-1}\sum_{b \in \mathbf{B}_{n,r}} E[h(X_{b_1}, \ldots, X_{b_r})|X_i]. \quad (11.15)
\end{aligned}
$$

There are $\binom{n-1}{r}$ terms in the sum on the right hand side of Equation (11.15) where $i \notin \{b_1, \ldots, b_r\}$. For these terms we have that $E[h(X_{b_1}, \ldots, X_{b_r})|X_i] = E[h(X_{b_1}, \ldots, X_{b_r})] = \theta$. The remaining $\binom{n-1}{r-1}$ terms in the sum have the form

$$E[h(X_{b_1}, \ldots, X_{b_r})|X_i] = E[h(X_i, X_1, \ldots, X_{r-1})|X_i] = r^{-1}n\tilde{k}(X_i) + \theta,$$

which follows from the definition of the function $\tilde{k}$ and the fact that $X_1, \ldots, X_n$ are a set of independent and identically distributed random variables. Therefore,

$$E(U_n|X_i) =$$
$$\binom{n}{r}^{-1} \left[\binom{n-1}{r} + \binom{n-1}{r-1} \right] \theta + \binom{n}{r}^{-1} \binom{n-1}{r-1} r^{-1}n\tilde{k}(X_i) =$$
$$\theta + \tilde{k}(X_i).$$

Similarly, we have that

$$E(V_n|X_i) = E\left[\sum_{j=1}^{n} \tilde{k}(X_j) \,\middle|\, X_i \right] = \sum_{j=1}^{n} E[\tilde{k}(X_j)|X_i].$$

Now, when $i = j$ we have that $E[\tilde{k}(X_i)|X_i] = \tilde{k}(X_i)$ and when $i \neq j$, Theorem A.17 implies

$$
\begin{aligned}
E[\tilde{k}(X_j)|X_i] &= E[\tilde{k}(X_j)] \\
&= rn^{-1}E\{E[h(X_j, X_2, \ldots, X_r)|X_j]\} - rn^{-1}\theta \\
&= rn^{-1}E[h(X_j, X_2, \ldots, X_r)] - rn^{-1}\theta \\
&= 0.
\end{aligned}
$$

Therefore, $E(V_n|X_i) = \tilde{k}(X_i)$ and $E(U_n - \theta - V_n|X_i) = 0$, from which we can conclude that $E\{(U_n - \theta - V_n)[\tilde{k}(X_i) - k(X_i)]\} = 0$ for all $i = 1, \ldots, n$. Hence, it follows that $||U_n - \theta - V|| = E[(U_n - V_n)^2] + E[(V_n - V)^2]$. Because both terms are non-negative and the first term does not depend on V, it follows that minimizing $||U_n - \theta - V||$ is equivalent to minimizing the second term $E[(V_n - V)^2]$, which can be made zero by choosing $V = V_n$. Therefore, V_n minimizes $||U_n - \theta - V||$ and the result follows. $\qquad\square$

Now that we have determined that the projection of $U_n - \theta$ is given by V_n, we can now prove that V_n has an asymptotic NORMAL distribution and that $U_n - \theta$ has the same asymptotic properties as V_n. This result, which was proven by Hoeffding (1948), establishes conditions under which a U-statistic is asymptotically normal.

Theorem 11.6 (Hoeffding). *Let $X_1, \ldots, X_n$ be a set of independent and identically distributed random variables from a distribution F. Let θ be a parameter that estimable of degree r with symmetric kernel function $h(x_1, \ldots, x_r)$. If*

$E[h^2(x_1,\ldots,x_r)] < \infty$ and $\zeta_1 > 0$ then

$$n^{1/2}(r^2\zeta_1)^{-1/2}\left[\binom{n}{r}^{-1}\sum_{b\in B_{n,r}} h(X_{b_1},\ldots,X_{b_r}) - \theta\right] \xrightarrow{d} Z,$$

as $n \to \infty$ where Z is a N$(0,1)$ random variable.

Proof. The proof of this result proceeds in two parts. We first establish that the projection of $U_n - \theta$ onto $\mathcal{V}_n$ has an asymptotic NORMAL distribution. We then prove that $\|U_n - \theta - V_n\|$ converges to zero as $n \to \infty$, which will then be used to establish that the two statistics have the same limiting distribution. Let U_n have the form

$$U_n = \frac{1}{\binom{n}{r}}\sum_{b\in B_{n,r}} h(X_{b_1},\ldots,X_{b_r}),$$

and let V_n be the projection of $U_n - \theta$ onto the space $\mathcal{V}_n$. That is

$$V_n = \sum_{i=1}^{n} \tilde{k}(X_i),$$

where $\tilde{k}$ is defined in the proof of Theorem 11.5. Because $X_1,\ldots,X_n$ is a set of independent and identically distributed random variables, it follows that $\tilde{k}(X_1),\ldots,\tilde{k}(X_n)$ is also a set of independent and identically distributed random variables and hence Theorem 4.20 (Lindeberg and Lévy) implies that

$$n^{1/2}\tilde{\sigma}^{-1}\left\{n^{-1}\sum_{i=1}^{n} n\tilde{k}(X_i) - E[n\tilde{k}(X_i)]\right\} \xrightarrow{d} Z, \qquad (11.16)$$

as $n \to \infty$ where Z is a N$(0,1)$ random variable. Now Theorem A.17 implies that

$$\begin{aligned}
E[n\tilde{k}(X_1)] &= E\{rE[h(X_1,\ldots,X_r)|X_1] - r\theta\} \\
&= rE[h(X_1,\ldots,X_r)] - r\theta \\
&= 0. \qquad (11.17)
\end{aligned}$$

Similarly, to find $\tilde{\sigma}^2$ we note that

$$\begin{aligned}
V[n\tilde{k}(X_1)] &= E[n^2\tilde{k}^2(X_1)] \\
&= r^2E\{[E[h(X_1,\ldots,X_r)|X_1] - \theta^2]^2\} \\
&= r^2V\{E[h(X_1,\ldots,X_r)|X_1]\},
\end{aligned}$$

since the result of Equation (11.17) implies that $E\{E[h(X_1,\ldots,X_r)|X_1]\} = \theta$.

Therefore, it follows that

$$
\begin{aligned}
V[n\tilde{k}(X_1)] &= r^2 E\{E^2[h(X_1,\ldots,X_r)|X_1]\} - r^2\theta^2 \\
&= r^2 E\{E[h(X_1,\ldots,X_r)|X_1]E[h(X_1,\ldots,X_r)|X_1]\} - r^2\theta^2 \\
&= r^2 E\{E[h(X_1,\ldots,X_r)|X_1]E[h(X_1,X_{r+1},\ldots,X_{2r-1})|X_1]\} \\
&\quad - r^2\theta^2 \\
&= r^2 E\{E[h(X_1,\ldots,X_r)h(X_1,X_{r+1},\ldots,X_{2r-1})|X_1]\} - r^2\theta^2 \\
&= r^2 E[h(X_1,\ldots,X_r)h(X_1,X_{r+1},\ldots,X_{2r-1})] - r^2\theta^2.
\end{aligned}
$$

Now, note that

$$
\begin{aligned}
\zeta_1 &= C[h(X_1,\ldots,X_r),h(X_1,X_{r+1},\ldots,X_{2r-1})] \\
&= E[h(X_1,\ldots,X_r)h(X_1,X_{r+1},\ldots,X_{2r-1})] \\
&\quad - E[h(X_1,\ldots,X_r)]E[h(X_1,X_{r+1},\ldots,X_{2r-1})] \\
&= E[h(X_1,\ldots,X_r)h(X_1,X_{r+1},\ldots,X_{2r-1})] - \theta^2.
\end{aligned}
$$

Therefore, it follows that

$$
E[h(X_1,\ldots,X_r)h(X_1,X_{r+1},\ldots,X_{2r-1})] = \zeta_1 + \theta^2,
$$

and hence we have shown that $nV[\tilde{k}(X_1)] = r^2\zeta_1$, or equivalently we have shown that $V[\tilde{k}(X_1)] = n^{-2}r^2\zeta_1$. Substituting the expressions for the expectation and variance of $\tilde{k}(X_1)$ into the result of Equation (11.16) yields the result that $r^{-1}(n\zeta_1)^{-1/2}V_n \xrightarrow{d} Z$ as $n \to \infty$. For the next step, we begin by proving that $\|U_n - \theta - V_n\| \to 0$ as $n \to \infty$. We first note that

$$
\|U_n - \theta - V_n\| = E[(U_n - \theta - V_n)^2] =
$$
$$
E[(U_n - \theta)^2] - 2E[V_n(U_n - \theta)] + E[V_n^2]. \quad (11.18)
$$

To evaluate the first term on the right hand side of Equation (11.18) we note that $E(U_n) = \theta$ and hence Theorem 11.4 implies that

$$
E[(U_n - \theta)^2] = V(U_n) = n^{-1}r^2\zeta_1 + o(n^{-1}), \quad (11.19)
$$

as $n \to \infty$. To evaluate the second term on the right hand side of Equation (11.18),

$$
\begin{aligned}
E[V_n(U_n - \theta)] &= E\left\{\left[\sum_{i=1}^{n}\tilde{k}(X_i)\right]\left[\binom{n}{r}^{-1}\sum_{b\in\mathbf{B}_{n,r}}h(X_{b_1},\ldots,X_{b_r}) - \theta\right]\right\} \\
&= \binom{n}{r}^{-1}\sum_{i=1}^{n}\sum_{b\in\mathbf{B}_{n,r}}E\{\tilde{k}(X_i)[h(X_{b_1},\ldots,X_{b_r}) - \theta]\}.
\end{aligned}
$$

Now, if $i \notin \{b_1,\ldots,b_r\}$ then $\tilde{k}(X_i)$ and $h(X_{b_1},\ldots,X_{b_r})$ will be independent and hence

$$
E\{\tilde{k}(X_i)[h(X_{b_1},\ldots,X_{b_r}) - \theta]\} = E[\tilde{k}(X_i)]E[h(X_{b_1},\ldots,X_{b_r}) - \theta] = 0.
$$

For the remaining $\binom{n-1}{r-1}$ terms where $i \in \{b_1,\ldots,b_r\}$, we apply Theorem A.17

to find that

$$
\begin{aligned}
E\{\tilde{k}(X_i)[h(X_{b_1},\ldots,X_{b_r})-\theta]\} &= E[E\{\tilde{k}(X_i)[h(X_{b_1},\ldots,X_{b_r})-\theta]|X_i\}]\\
&= E[\tilde{k}(X_i)E\{[h(X_{b_1},\ldots,X_{b_r})-\theta]|X_i\}]\\
&= nr^{-1}E[\tilde{k}^2(X_i)]\\
&= n^{-1}r\zeta_1.
\end{aligned}
$$

Therefore, it follows that

$$
E[V_n(U_n-\theta)] = \sum_{i=1}^{n} \frac{r\binom{n-1}{r-1}\zeta_1}{n\binom{n}{r}} = r^2 n^{-1}\zeta_1. \tag{11.20}
$$

To evaluate the third term on the right hand side of Equation (11.18), we have that

$$
E(V_n^2) = E\left\{\left[\sum_{i=1}^{n}\tilde{k}(X_i)\right]\left[\sum_{j=1}^{n}\tilde{k}(X_j)\right]\right\} = \sum_{i=1}^{n}\sum_{j=1}^{n} E[\tilde{k}(X_i)\tilde{k}(X_j)].
$$

Now when $i \neq j$, $\tilde{k}(X_i)$ and $\tilde{k}(X_j)$ are independent and $E[\tilde{k}(X_i)\tilde{k}(X_j)] = E[\tilde{k}(X_i)]E[\tilde{k}(X_j)] = 0$. Therefore,

$$
E(V_n^2) = \sum_{i=1}^{n} E[\tilde{k}^2(X_i)] = nE[\tilde{k}^2(X_1)] = n^{-1}r^2\zeta_1. \tag{11.21}
$$

Combining the results of Equations (11.18)–(11.21) yields

$$
\|U_n-\theta-V_n\| = n^{-1}r^2\zeta_1 - 2n^{-1}r^2\zeta_1 + n^{-1}r^2\zeta_1 + o(n^{-1}) = O(n^{-1}),
$$

as $n \to \infty$, so that $\|U_n-\theta-V_n\| \to 0$ as $n \to \infty$. Therefore from Definition 5.1 it follows that $U_n-\theta-V_n \xrightarrow{qm} 0$ as $n \to \infty$. Theorems 5.2 and 4.8 imply then that $U_n-\theta-V_n \xrightarrow{d} 0$ as $n \to \infty$, and therefore $U_n-\theta$ and V_n converge to the same distribution as $n \to \infty$. Hence, Equation (11.16) implies that $r^{-1}(n\zeta_1)^{-1/2}(U_n-\theta) \xrightarrow{d} Z$ as $n \to \infty$, and the result is proven. $\qquad\square$

Example 11.11. Let $X_1,\ldots,X_n$ be a set of independent and identically distributed random variables from a distribution with mean θ and finite variance σ^2. Continuing with Example 11.9 we consider estimating the parameter $E[X_1^2] = \theta^2$ with a U-statistic of degree $r = 2$ that has a symmetric kernel function $h(x_1,x_2) = x_1x_2$ with $\zeta_1 = \theta^2\sigma^2$. The assumption that the variance is finite implies that $\theta < \infty$ and we have that $E[h^2(X_1,X_2)] = E[X_1^2X_2^2] = E[X_1^2]E[X_2^2] = (\theta^2+\sigma^2)^2 < \infty$. Further $\zeta_1 = \theta^2\sigma^2 > 0$ so that Theorem 11.6 implies that $n^{1/2}(2\theta\sigma)^{-1}(U_n-\theta^2) \xrightarrow{d} Z$ as $n \to \infty$ where Z is a N$(0,1)$ random variable. $\qquad\blacksquare$

Example 11.12. Let $X_1,\ldots,X_n$ be a set of independent and identically distributed random variables from a continuous distribution F that is symmetric about a point θ. Let **R** be the vector of ranks of $|X_1-\theta|,\ldots,|X_n-\theta|$, let **C** be an $n \times 1$ vector with i^{th} element $c_i = \delta\{X_i-\theta;(0,\infty)\}$ for $i = 1,\ldots,n$, and

$Z_i = X_i - \theta$ for $i = 1, \ldots, n$. The Wilcoxon signed rank statistic was seen in Example 11.3 to have the form $W = \mathbf{C}'\mathbf{R}$. In Example 11.7 it was shown that $W = nU_{1,n} + \frac{1}{2}n(n-1)U_{2,n}$ where $U_{1,n}$ and $U_{2,n}$ are U-statistics of orders $r = 1$ and $r = 2$, respectively. These U-statistics are given by

$$U_{1,n} = n^{-1} \sum_{i=1}^{n} \delta\{2Z_i; (0, \infty)\},$$

and

$$U_{2,n} = \frac{2}{n(n-1)} \sum_{i=1}^{n} \sum_{j=i+1}^{n} \delta\{Z_i + Z_j; (0, \infty)\}.$$

To find the asymptotic distribution of W, we first note that

$$n^{1/2} \binom{n}{2}^{-1} [W - E(W)] =$$

$$\frac{2n^{1/2}}{n-1}[U_{1,n} - E(U_{1,n})] + n^{1/2}[U_{2,n} - E(U_{2,n})]. \quad (11.22)$$

For the first term on the right hand side of Equation (11.22) we note that $V(U_{1,n}) = n^{-1}V(\delta\{2Z_i; (0, \infty)\})$ where

$$E(\delta\{2Z_i; (0, \infty)\}) = P(\delta\{2Z_i; (0, \infty)\} = 1) = P(Z_i > 0) = \tfrac{1}{2}$$

since Z_i has a symmetric distribution about zero. Since $\delta^2\{2Z_i; (0, \infty)\} = \delta\{2Z_i; (0, \infty)\}$ it follows also that $E(\delta^2\{2Z_i; (0, \infty)\}) = \frac{1}{2}$. Therefore, we have that $V(\delta\{2Z_i; (0, \infty)\}) = \frac{1}{4}$, and hence $V(U_{1,n}) = \frac{1}{4}n^{-1}$. Theorem 3.10 (Weak Law of Large Numbers) then implies that $U_{1,n} \xrightarrow{p} \frac{1}{2}$ as $n \to \infty$. Noting that $2n^{1/2}(n-1)^{-1} \to 0$ as $n \to \infty$ we can then apply Theorem 4.11 (Slutsky) to find that $2n^{1/2}(n-1)^{-1}[U_{1,n} - E(U_{1,n})] \xrightarrow{p} 0$ as $n \to \infty$. Therefore, it follows that the asymtptotic distribution of $n^{1/2}\binom{n}{2}^{-1}[W - E(W)]$ is the same as $n^{1/2}[U_{2,n} - E(U_{2,n})]$. We wish to apply Theorem 11.6 to the second term on the right hand side of Equation (11.22), and therefore we need to verify the assumptions required for Theorem 11.6. Let f be the density of Z_1 where, by assumption, f is symmetric about zero. Noting that the independence between Z_i and Z_j implies that the joint distribution between Z_i and Z_j is $f(z_i)f(z_j)$, it follows that

$$
\begin{aligned}
E[\delta^2\{Z_i + Z_j; (0,\infty)\}] &= E[\delta\{Z_i + Z_j; (0,\infty)\}] \\
&= P(Z_i + Z_j > 0) \\
&= \int_{-\infty}^{\infty} \int_{-z_i}^{\infty} f(z_i)f(z_j)\,dz_j\,dz_i \\
&= \int_{-\infty}^{\infty} f(z_i) \int_{-z_i}^{\infty} f(z_j)\,dz_j\,dz_i \\
&= \int_{-\infty}^{\infty} f(z_i)[1 - F(-z_i)]\,dz_i \\
&= \int_{-\infty}^{\infty} F(z_i)f(z_i)\,dz_i \\
&= \int_{0}^{1} t\,dt = \tfrac{1}{2},
\end{aligned}
$$

where we have used the fact that the symmetry of f implies that $1 - F(-z) = F(z)$. Since $\tfrac{1}{2} < \infty$ we have verified that $E[h^2(x_1,\ldots,x_r)] < \infty$. To verify the second assumption we note that

$$
\begin{aligned}
\zeta_1 &= E[\delta\{Z_i + Z_j; (0,\infty)\}\delta\{Z_i + Z_k; (0,\infty)\}] - \tfrac{1}{4} \\
&= P(\{Z_i + Z_j > 0\} \cap \{Z_i + Z_k > 0\}) - \tfrac{1}{4} \\
&= \int_{-\infty}^{\infty} \int_{-z_i}^{\infty} \int_{-z_i}^{\infty} f(z_i)f(z_j)f(z_k)\,dz_k\,dz_j\,dz_i - \tfrac{1}{4} \\
&= \int_{-\infty}^{\infty} f(z_i) \int_{-z_i}^{\infty} f(z_j)[1 - F(-z_i)]\,dz_j\,dz_i - \tfrac{1}{4} \\
&= \int_{-\infty}^{\infty} f(z_i)F^2(z_i)\,dz_i - \tfrac{1}{4} \\
&= \int_{0}^{1} t^2\,dt - \tfrac{1}{4} = \tfrac{1}{3} - \tfrac{1}{4} = \tfrac{1}{12} > 0.
\end{aligned}
$$

Hence, the second assumption is verified. Theorem 11.6 then implies that $n^{1/2}[U_{2,n} - \tfrac{1}{2}] \xrightarrow{d} Z_2$ where Z_2 has a $N(0, \tfrac{1}{3})$ distribution, and therefore

$$
n^{1/2} \binom{n}{2}^{-1} [W - E(W)] \xrightarrow{d} Z_2,
$$

as $n \to \infty$. Further calculations can be used to refine this result to find that

$$
\frac{W - \tfrac{1}{4}n(n+1)}{[\tfrac{1}{24}n(n+1)(2n+1)]^{1/2}} \xrightarrow{d} Z, \tag{11.23}
$$

as $n \to \infty$ where Z has a $N(0,1)$ distribution. See Exercise 5. This result is suitable for using W to test the null hypothesis $H_0 : \theta = 0$ using approximate rejection regions. Figures 11.1–11.3 plot the exact distribution of W under the null hypothesis for $n = 5$, 7, and 10. It is clear in Figure 11.3 that the

Figure 11.1 *The exact distribution of the signed-rank statistic when $n = 5$.*

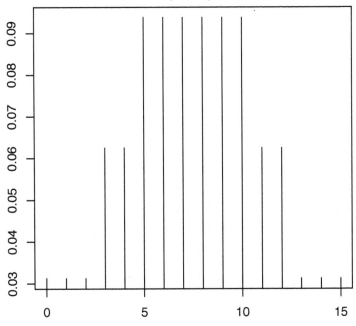

normal approximation should work well for this sample size and larger. Table 11.1 compares some exact quantiles of the distribution of W with some given by the normal approximation. Note that Equation (11.23) implies that the α quantile of the distribution of W can be approximated by

$$\tfrac{1}{4}n(n+1) + z_\alpha[\tfrac{1}{24}n(n+1)(2n+1)]^{1/2}. \tag{11.24}$$

■

The topic of U-statistics can be expanded in many ways. For example, an overview of U-statistics for two or more samples can be found in Section 3.4 of Randles and Wolfe (1979). For other generalizations and asymptotic results for U-statistics see Kowalski and Tu (2008), Lee (1990), and Chapter 5 of Serfling (1980).

11.3 Linear Rank Statistics

Another class of statistics that commonly occur in nonparametric statistical inference are *linear rank statistics*, which are linear functions of the rank vector. As with U-statistics, linear rank statistics are asymtptotically NORMAL under some very general conditions.

Figure 11.2 *The exact distribution of the signed-rank statistic when n = 7.*

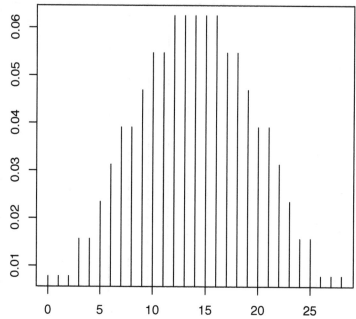

Figure 11.3 *The exact distribution of the signed-rank statistic when n = 10.*

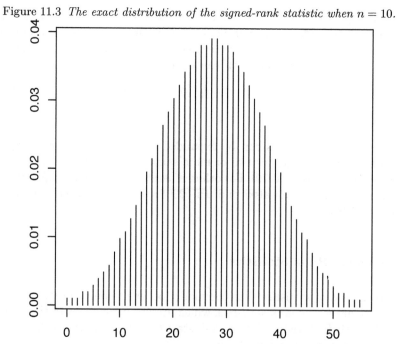

Table 11.1 *A comparison of the exact quantiles of the signed rank statistic against those given by the normal approximation given in Equation (11.24). The approximate quantiles have been rounded to the nearest integer.*

n	Exact Quantiles			Normal Approximation			Relative Error (%)		
---	0.10	0.05	0.01	0.10	0.05	0.01	0.10	0.05	0.01
5	12	14	15	10	11	14	16.7	21.5	6.7
6	17	18	21	14	15	19	17.6	16.7	9.5
7	22	24	27	18	20	24	18.2	16.7	11.1
8	27	30	34	23	26	31	14.8	13.3	8.8
9	34	36	41	29	32	38	14.7	11.1	7.3
10	40	44	49	35	39	45	12.5	11.4	8.16
25	211	224	248	198	211	236	6.2	5.8	4.8
50	771	808	877	745	783	853	3.4	3.1	2.7

Definition 11.5. *Let $X_1, \ldots, X_n$ be a set of independent and identically distributed random variables from a distribution F and let $\mathbf{R}' = (r_1, \ldots, r_n)$ be the vector a ranks associated with $X_1, \ldots, X_n$. Let a and c denote functions that map the set $\{1, \ldots, n\}$ to the real line. Then the statistic*

$$S = \sum_{i=1}^{n} c(i)a(r_i),$$

is called a linear rank statistic. The set of constants $c(1), \ldots, c(n)$ are called the regression constants, and the set of constants $a(1), \ldots, a(n)$ are called the scores of the statistic.

The important ingredients of Definition 11.5 are that the elements of the vector $\mathbf{R}$ correspond to a random permutation of the integers in the set $\{1, \ldots, n\}$ so that $a(r_1), \ldots, a(r_n)$ are random variables, but the regression constants $c(1), \ldots, c(n)$ are not random. The frequent use of ranks in classical nonparametric statistics makes linear rank statistics an important topic.

Example 11.13. Let us consider the Wilcoxon, Mann, and Whitney Rank Sum test statistic from Example 11.2. That is, let $X_1, \ldots, X_m$ be a set of independent and identically distributed random variables from a continuous distribution F and let $Y_1, \ldots, Y_n$ be a set of independent and identically distributed random variables from a continuous distribution G where $G(t) = F(t - \theta)$, where θ is an unknown parameter. Denoting the combined sample as $Z_1, \ldots, Z_{n+m}$, and the corresponding ranks as $R_1, \ldots, R_{n+m}$, the test statistic

$$M_{m,n} = \sum_{i=1}^{n+m} D_i R_i, \tag{11.25}$$

where $D_i = 1$ when Z_i is from the sample $X_1, \ldots, X_m$ and $D_i = 0$ when

Z_i is from the sample $Y_1, \ldots, Y_n$. For simplicity assume that $Z_i = X_i$ for $i = 1, \ldots, m$ and $Z_j = Y_{j-n+1}$ for $j = m+1, \ldots, n+m$. Then the statistic given in Equation (11.25) can be written as a linear rank statistic of the form given in Definition 11.5 with $a(i) = i$ and $c(i) = \delta\{i; \{m+1, \ldots, n+m\}\}$ for all $i = 1, \ldots, n+m$.   ∎

Example 11.14. Once again consider the two-sample setup studied in Example 11.2, except in this case we will use the *median test* statistic proposed by Mood (1950) and Westenberg (1948). For this test we compute the median of the combined sample $X_1, \ldots, X_m, Y_1, \ldots, Y_n$ and then compute the number of values in the sample $Y_1, \ldots, Y_n$ that exceed the median. Note that under the null hypothesis that $\theta = 0$, where the combined sample all comes from the sample distribution, we would expect that half of these would be above the median. If $\theta \neq 0$ then we would expect either greater than, or fewer than, of these values to exceed the median. Therefore, counting the number of values that exceed the combined median provides a reasonable test statistic for the null hypothesis that $\theta = 0$. This test statistic can be written in the form of a linear rank statistic with $c(i) = \delta\{i; \{m+1, \ldots, n+m\}\}$ and $a(i) = \delta\{i; \{\frac{1}{2}(n+m+1), \ldots, n+m\}\}$. The form of the score function is derived from the fact that if the rank of a value from the combined sample exceeds $\frac{1}{2}(n+m+1)$ then the corresponding value exceeds the median. This score function is called the *median score function*.   ∎

Typically, under a null hypothesis the elements of the vector $\mathbf{R}$ correspond to a random permutation of the integers in the set $\{1, \ldots, n\}$ that is uniformly distributed over the set $\mathbf{R}_n$ that is defined in Equation (11.1). In this case the distribution of S can be found by enumerating the values of S over the $r!$ equally likely permutations in $\mathbf{R}_n$. There are also some general results that are helpful in the practical application of tests based on linear rank statistics.

Theorem 11.7. *Let S be a linear rank statistic of the form*

$$S = \sum_{i=1}^{n} c(i) a(r_i).$$

If $\mathbf{R}$ is a vector whose elements correspond to a random permutation of the integers in the set $\{1, \ldots, n\}$ that is uniformly distributed over $\mathbf{R}_n$ then $E(S) = n\bar{a}\bar{c}$ and

$$V(S) = (n-1)^{-1} \left\{ \sum_{i=1}^{n} [a(i) - \bar{a}]^2 \right\} \left\{ \sum_{j=1}^{n} [c(j) - \bar{c}]^2 \right\}$$

where

$$\bar{a} = n^{-1} \sum_{i=1}^{n} a(i),$$

and

$$\bar{c} = n^{-1} \sum_{i=1}^{n} c(i).$$

Theorem 11.7 can be proven using direct calculations. See Exercise 6. More complex arguments are required to obtain further properties on the distribution of linear rank statistics. For example, the symmetry of the distribution of a linear rank statistic can be established under fairly general conditions using arguments based on the composition of permutations. This result was first proven by Hájek (1969).

Theorem 11.8 (Hájek). *Let S be a linear rank statistic of the form*

$$S = \sum_{i=1}^{n} c(i)a(r_i).$$

Let $c_{(1)}, \ldots, c_{(n)}$ and $a_{(1)}, \ldots, a_{(n)}$ denote the ordered values of $c(1), \ldots, c(n)$ and $a(1), \ldots, a(n)$, respectively. Suppose that $\mathbf{R}$ is a vector whose elements correspond to a random permutation of the integers in the set $\{1, \ldots, n\}$ that is uniformly distributed over $\mathbf{R}_n$. If $a_{(i)} + a_{(n+1-i)}$ or $c_{(i)} + c_{(n+1-i)}$ is constant for $i = 1, \ldots, n$, then the distribution of S is symmetric about $n\bar{a}\bar{c}$.

A proof of Theorem 11.8 can be found in Section 8.2 of Randles and Wolfe (1979).

Example 11.15. Let us consider the rank sum test statistic from Example 11.2 which can be written as a linear rank statistic of the form given in Definition 11.5 with $a(i) = i$ and $c(i) = \delta\{i; \{m+1, \ldots, n+m\}\}$ for all $i = 1, \ldots, n+m$. See Example 11.13. Note that $a_{(i)} = a(i)$ and that $a(i) + a(m+n-i+1) = m+n+1$ for all $i \in \{1, \ldots, m+n\}$ so that Theorem 11.8 implies that the distribution of the rank sum test statistic is symmetric when the null hypothesis that $\theta = 0$ is true. Some examples of the distribution are plotted in Figures 11.4–11.6. ∎

The possible symmetry of the distribution of a linear rank statistic indicates that its distribution could be a good candidate for the normal approximation, an idea that is supported by the results provided in Example 11.15. The remainder of this section is devoted to developing conditions under which linear rank statistics have an asymptotic NORMAL distribution. To begin development of the asymptotic results we now consider a sequence of linear rank statistics $\{S_n\}_{n=1}^{\infty}$ where

$$S_n = \sum_{i=1}^{n} c(i, n)a(R_i, n), \qquad (11.26)$$

where we emphasize that both the regression constants and the scores depend on the sample size n. Section 8.3 of Randles and Wolfe (1979) points out that the regression constants $c(1, n), \ldots, c(n, n)$ are usually determined by the type of problem under consideration. For example, in Examples 11.13 and 11.14, the regression constants are used to distinguish between the two samples. Therefore, it is advisable to put as few restrictions on the types of regression constants that can be considered so that as many different types of problems

Figure 11.4 *The distribution of the rank sum test statistic when* $n = m = 3$.

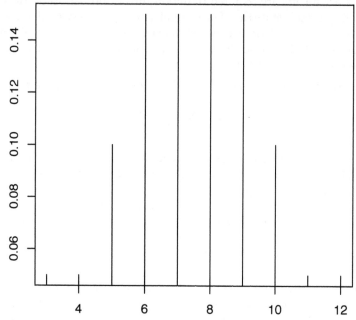

Figure 11.5 *The distribution of the rank sum test statistic when* $n = m = 4$.

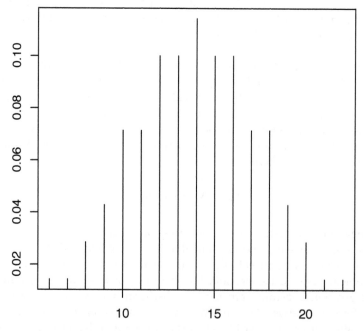

Figure 11.6 *The distribution of the rank sum test statistic when $n = m = 5$.*

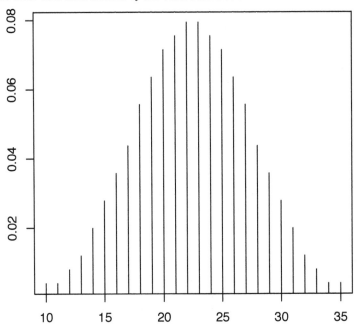

as possible can be addressed by the asymptotic theory. A typical restriction is given by *Noether's condition.*

Definition 11.6 (Noether). *Let $c(1, n), \ldots, c(n, n)$ be a set of regression constants for a linear rank statistic of the form given in Equation (11.26). The regression constants follow Noether's condition if*

$$\lim_{n \to \infty} \frac{\sum_{i=1}^{n} d(i, n)}{\max_{i \in \{1,\ldots,n\}} d(i, n)} = \infty,$$

where

$$d(i, n) = \left[c(i, n) - n^{-1} \sum_{i=1}^{n} c(i, n) \right]^2,$$

for $i = 1, \ldots, n$ and $n \in \mathbb{N}$.

This condition originates from Noether (1949) and essentially keeps one of the constants from dominating the others.

Greater latitude is given in choosing the score function, and hence more restrictive assumptions can be implemented on them. In particular, the usual approach is to consider score functions of the form $a(i, n) = \alpha[i(n + 1)^{-1}]$, where α is a function that does not depend on n and is assumed to have certain properties.

Definition 11.7. *Let α be a function that maps the open unit interval $(0,1)$ to $\mathbb{R}$ such that,*

1. *$\alpha(t) = \alpha_1(t) - \alpha_2(t)$ where α_1 and α_2 are non-decreasing functions that map the open unit interval $(0,1)$ to $\mathbb{R}$.*
2. *The integral*

$$\int_0^1 \left[\alpha(t) - \int_0^1 \alpha(u) du \right]^2 dt$$

   is non-zero and finite.

Then the function $\alpha(t)$ is called a square integrable score function.

Not all of the common score functions can be written strictly in the form $a(i,n) = \alpha[i(n+1)^{-1}]$ where α is a square integrable score function. However, a slight adjustment to this form will not change the asymptotic behavior of a properly standardized linear rank statistic, and therefore it suffices to consider this form.

To establish the asymptotic normality of a linear rank statistic, we will need to develop several properties of the ranks, order statistics, and square integrable score functions. The first result establishes independence between the rank vector and the order statistics for sets of independent and identically distributed random variables.

Theorem 11.9. *Let $X_1, \ldots, X_n$ be a set of independent and identically distributed random variables from a continuous distribution F. Let $R_1, \ldots, R_n$ denote the ranks of $X_1, \ldots, X_n$ and let $X_{(1)}, \ldots, X_{(n)}$ denote the order statistics. Then $R_1, \ldots, R_n$ and $X_{(1)}, \ldots, X_{(n)}$ are mutually independent.*

A proof of Theorem 11.9 can be found in Section 8.3 of Randles and Wolfe (1979).

Not surprisingly, we also need to establish several properties of square integrable score functions. The property we establish below is related to limiting properties of the expectation of the the score function.

Theorem 11.10 (Hájek and Šidák). *Let U be a* UNIFORM$(0,1)$ *random variable and let $\{g_n\}_{n=1}$ be a sequence of functions that map the open unit interval $(0,1)$ to $\mathbb{R}$, where $g_n(U) \xrightarrow{a.c.} g(U)$ as $n \to \infty$, and*

$$\limsup_{n\to\infty} E[g_n^2(U)] \leq E[g^2(U)]. \tag{11.27}$$

Then

$$\lim_{n\to\infty} E\{[g_n(U) - g(U)]^2\} = 0.$$

Proof. We begin by noting that

$$E\{[g_n(U) - g(U)]^2\} = E[g_n^2(U)] - 2E[g_n(U)g(U)] + E[g^2(U)]. \tag{11.28}$$

For the first term on the right hand side of Equation (11.28) we note that since $\{g_n^2(U)\}_{n=1}^{\infty}$ is a sequence of non-negative random variables that converge almost certainly to $g(U)$, it follows from Theorem 5.10 (Fatou) that

$$E[g(U)] \leq \liminf_{n\to\infty} E[g_n(U)].$$

Combining the result with Equation (11.28) yields

$$\limsup_{n\to\infty} E[g_n^2(U)] \leq E[g^2(U)] \leq \liminf_{n\to\infty} E[g_n(U)],$$

so that Definition 1.3 implies that

$$\lim_{n\to\infty} E[g_n^2(U)] = E[g(U)]. \tag{11.29}$$

For the second term on the right hand side of Equation (11.28) we refer to Theorem II.4.2 of Hájek and Šidák (1967), which shows that

$$\lim_{n\to\infty} E[g_n(U)g(U)] = E[g^2(U)]. \tag{11.30}$$

Combining the results of Equations (11.28)–(11.30) yields the result.   □

In most situations the linear rank statistic is used as a test statistic to test a null hypotheses that implies that the rank vector $\mathbf{R}$ is uniformly distributed over the set $\mathbf{R}_n$. In this case each component of $\mathbf{R}$ has a marginal distribution that is uniform over the integers $\{1, \ldots, n\}$. Hence, the expectation $E\{\alpha^2[(n+1)^{-1}R_1]\}$ is equivalent to the expectation $E\{\alpha^2(n+1)^{-1}U_n]\}$ where U_n is a UNIFORM$\{1, \ldots, n\}$ random variable. It can be shown that $U_n \overset{d}{\to} U$ as $n \to \infty$ where U is a UNIFORM$(0,1)$ random variable. The result given in Theorem 11.11 below establishes the fact that the expectations converge as well.

Theorem 11.11. *Let α be a square integrable score function and let U be a* UNIFORM$\{1, \ldots, n\}$ *random variable. Then*

$$\lim_{n\to\infty} E\{\alpha^2[(n+1)^{-1}U_n]\} = \lim_{n\to\infty} n^{-1} \sum_{i=1}^{n} \alpha^2[(n+1)^{-1}i] =$$

$$\int_0^1 \alpha^2(t)dt. \tag{11.31}$$

A complete proof of Theorem 11.11 is quite involved. The interested reader should consult Section 8.3 of Randles and Wolfe (1979) for the complete details. A similar result is required to obtain the asymptotic behavior of the score function evaluated at a BINOMIAL random variable.

Theorem 11.12. *Let α be a square integrable score function and let Y_n be a* BINOMIAL(n, θ) *random variable. Then*

$$\lim_{n\to\infty} E\{\alpha[(n+2)^{-1}(Y_n+1)]\} = \alpha(\theta)$$

for all $\theta \in (0,1) \setminus A$ where the Lebesgue measure of A is zero.

A proof of Theorem 11.12 can be found in Section 8.3 of Randles and Wolfe

(1979). We finally require a result that shows that $\alpha[(n+1)^{-1}R_1]$, where R_1 is the rank of the first observation U_1 from a set of n independent and identically distributed random variables from a UNIFORM$(0,1)$ distribution can be approximated by $\alpha(U_1)$. The choice of the first observation is by convenience, what matters is that we have the rank of an observation from a UNIFORM sample. This turns out to be the key approximation in developing the asymptotic NORMALITY of linear rank statistics.

Theorem 11.13. *Let α be a square integrable score function and let $U_1, \ldots, U_n$ be a set of independent and identically distributed* UNIFORM$(0,1)$ *random variables. Suppose that R_1 is the rank of U_1. Then $|\alpha[(n+1)^{-1}R_1] - \alpha(U_1)| \xrightarrow{qm} 0$ as $n \to \infty$.*

Proof. We begin by noting that

$$E[\{\alpha[(n+1)^{-1}R_1] - \alpha(U_1)\}^2] =$$
$$E\{\alpha^2[(n+1)^{-1}R_1]\} - 2E\{\alpha(U_1)\alpha[(n+1)^{-1}R_1]\} + E[\alpha^2(U_1)]. \quad (11.32)$$

Since U_1 is a UNIFORM$(0,1)$ random variable it follows that

$$E[\alpha^2(U_1)] = \int_0^1 \alpha^2(t)dt. \quad (11.33)$$

The marginal distribution of R_1 is UNIFORM$\{1, \ldots, n\}$ so that Theorem 11.11 implies that

$$\lim_{n \to \infty} E\{\alpha^2[(n+1)^{-1}R_1]\} = \int_0^1 \alpha^2(t)dt. \quad (11.34)$$

To evaluate the second term on the right hand side of Equation (11.32) we note that the rank of U_1 equals the number of values in $\{U_1, \ldots, U_n\}$ that are less than or equal to U_1 (including U_1). Equivalently, the rank of U_1 equals the number of differences $U_1 - U_i$ that are non-negative. Therefore

$$\alpha[(n+1)^{-1}R_1] = \alpha\left\{(n+1)^{-1}\left[\sum_{i=1}^n \delta\{U_1 - U_i; [0, \infty)\}\right]\right\}.$$

Let

$$Y_n = \sum_{i=2}^n \delta\{(U_1 - U_i); [0, \infty)\},$$

and note that Y_n is a BINOMIAL$(n-1, \gamma)$ random variable where $\gamma = P(U_1 - U_i \geq 0)$. Therefore, Theorem A.17 implies that

$$E\{\alpha(U_1)\alpha[(n+1)^{-1}R_1]\} = E\{\alpha(U_1)\alpha[(n+1)^{-1}(Y_n+1)]\} =$$
$$E[E\{\alpha(U_1)\alpha[(n+1)^{-1}(Y_n+1)]\}|U_1]. \quad (11.35)$$

Now, $Y_n|U_1$ is a BINOMIAL$(n-1, \lambda)$ random variable where $\lambda = P(U_1 - U_i \geq 0|U_1) = P(U_i \leq U_1) = U_1$ since U_i is a UNIFORM$(0,1)$ random variable for $i = 2, \ldots, n$. Let $\beta_n(u) = E\{\alpha[(n+1)^{-1}(B+1)]\}$ where B is a BINOMIAL$(n-1, u)$

random variable, then Equation (11.35) implies that

$$E\{\alpha[(n+1)^{-1}R_1]\alpha(U_1)\} = E[\beta_n(U)\alpha(U)],$$

where U is a UNIFORM$(0,1)$ random variable. Theorem 11.12 implies that

$$\lim_{n\to\infty}\beta_n(u) = \lim_{n\to\infty}E\{\alpha[(n+1)^{-1}(B+1)]\} = \alpha(u),$$

so that the next step is then to show that

$$\lim_{n\to\infty}E[\beta_n(U)\alpha(U)] = E[\alpha^2(U)],$$

using Theorem 11.10. To verify the assumption of Equation (11.27) we note that Theorem 2.11 (Jensen) implies that

$$\begin{aligned} E[\beta_n^2(U)] &= E[E^2\{\alpha[(n+1)^{-1}(B+1)]|U\}] \\ &\leq E[E\{\alpha^2[(n+1)^{-1}(B+1)]|U\}] \\ &= E\{\alpha^2[(n+1)^{-1}R_1]\}, \end{aligned}$$

since we have shown that R_1 has the same distribution as $B+1$. Theorem 11.11 implies that

$$\lim_{n\to\infty}E\{\alpha^2[(n+1)^{-1}R_1]\} = E[\alpha^2(U)].$$

Therefore, Theorem 1.6 implies that

$$\limsup_{n\to\infty}E[\beta_n^2(U)] \leq \limsup_{n\to\infty}E\{\alpha^2[(n+1)^{-1}R_1]\} = E[\alpha^2(U)],$$

which verifies the assumption in Equation (11.27). Therefore, Theorem 11.10 implies that

$$\lim_{n\to\infty}E[\beta_n(U)\alpha(U)] = E[\alpha^2(U)] = \int_0^1 \alpha^2(t)dt. \tag{11.36}$$

Conbining the results of Equations (11.32)–(11.36) yields the result. $\qquad\square$

We are now ready to tackle the asymptotic normality of a linear rank statistic when the null hypothesis is true. This result, first proven by Hájek (1961) is proven using a similar approach to Theorem 11.6 (Hoeffding), in that the linear rank statistic is approximated by a simpler statistic whose asymptotic distribution is known. Therefore, the crucial part of the proof is based on showing that the two statistics have the same limiting distribution.

Theorem 11.14 (Hájek). *Let S_n be a linear rank statistic with regression constants $c(1,n),\dots,c(n,n)$ and score function a. Suppose that*

1. *$a(i,n) = \alpha[(n+1)^{-1}i]$ for $i = 1,\dots,n$ where α is a square integrable score function.*

2. *$c(1,n),\dots,c(n,n)$ satisfy Noether's condition given in Definition 11.6.*

3. *$\mathbf{R}$ is a rank vector that is uniformly distributed over the set $\mathbf{R}_n$ for each $n \in \mathbb{N}$.*

Then $\sigma_n^{-1}(S_n - \mu_n) \xrightarrow{d} Z$ *as* $n \to \infty$ *where* Z *is a* $N(0,1)$ *random variable,* $\mu_n = n\bar{c}_n\bar{a}_n$, *and*

$$\sigma_n^2 = (n-1)^{-1} \left\{ \sum_{i=1}^{n} [c(i,n) - \bar{c}_n]^2 \right\} \left\{ \sum_{i=1}^{n} [a(i,n) - \bar{a}_n]^2 \right\}.$$

Proof. Let $U_1, \ldots, U_n$ be a set of independent and identically distributed random variables from a UNIFORM$(0,1)$ distribution and let $U_{(1)}, \ldots, U_{(n)}$ denote the corresponding order statistics, with $\mathbf{U}$ denoting the vector containing the n order statistics.

Conside the linear rank statistic S_n and note that

$$
\begin{aligned}
S_n &= \sum_{i=1}^{n} c(i,n)a(R_i,n) \\
&= \sum_{i=1}^{n} c(i,n)a(R_i,n) - n\bar{c}_n\bar{a}_n + n\bar{c}_n\bar{a}_n \\
&= \sum_{i=1}^{n} c(i,n)a(R_i,n) - \bar{c}_n \sum_{i=1}^{n} a(i,n) + n\bar{c}_n\bar{a}_n \qquad (11.37) \\
&= \sum_{i=1}^{n} [c(i,n) - \bar{c}_n]a(R_i,n) + n\bar{c}_n\bar{a}_n \\
&= \sum_{i=1}^{n} [c(i,n) - \bar{c}_n]\alpha[(n+1)^{-1}R_i] + n\bar{c}_n\bar{a}_n, \qquad (11.38)
\end{aligned}
$$

where we have used the fact that

$$\sum_{i=1}^{n} a(i,n) = \sum_{i=1}^{n} a(R_i,n),$$

to establish Equation (11.37). The key idea to proving the desired result is based on approximating the statistic in Equation (11.38) with one whose asymptotic distribution can easily be found. Let W_n be a UNIFORM$\{1, \ldots, n+1\}$ random variable. Then, it can be shown that $(n+1)^{-1}W_n \xrightarrow{d} U$ as $n \to \infty$ where U is a UNIFORM$(0,1)$ random variable. This suggests approximating $(n+1)^{-1}R_i$ with U, or approximating the asymptotic behavior of S_n using the statistic

$$V_n = \sum_{i=1}^{n} [c(i,n) - \bar{c}_n]\alpha(U_i) + n\bar{c}_n\bar{a}_n.$$

The first step is to find the asymptotic distribution of V_n. Note that

$$
\begin{aligned}
E(V_n) &= E\left[\sum_{i=1}^{n} [c(i,n) - \bar{c}_n]\alpha(U_i) \right] + n\bar{c}_n\bar{a}_n \\
&= \sum_{i=1}^{n} [c(i,n) - \bar{c}_n]E[\alpha(U_i)] + n\bar{c}_n\bar{a}_n.
\end{aligned}
$$

Note that because $U_1, \ldots, U_n$ are identically distributed, it follows that

$$E(V_n) = E[\alpha(U_1)] \sum_{i=1}^{n} [c(i,n) - \bar{c}_n] + n\bar{c}_n\bar{a}_n = n\bar{c}_n\bar{a}_n = \mu_n,$$

since

$$\sum_{i=1}^{n} [c(i,n) - \bar{c}_n] = 0.$$

Similarly, since $U_1, \ldots, U_n$ are mutually independent, it follows that

$$\begin{aligned}
V(V_n) &= V\left[\sum_{i=1}^{n} [c(i,n) - \bar{c}_n]\alpha(U_i)\right] \\
&= \sum_{i=1}^{n} [c(i,n) - \bar{c}_n]^2 V[\alpha(U_i)] \\
&= V[\alpha(U_1)] \sum_{i=1}^{n} [c(i,n) - \bar{c}_n]^2 \\
&= \tilde{\alpha}^2 \sum_{i=1}^{n} [c(i,n) - \bar{c}_n]^2,
\end{aligned}$$

where

$$\tilde{\alpha}^2 = V[\alpha(U_1)] = \int_0^1 [\alpha(t) - \bar{a}]^2 dt.$$

Now,

$$\begin{aligned}
V_n - \mu_n &= \sum_{i=1}^{n} [c(i,n) - \bar{c}_n]\alpha(U_i) \\
&= \sum_{i=1}^{n} [c(i,n) - \bar{c}_n]\alpha(U_i) - \sum_{i=1}^{n} [c(i,n) - \bar{c}_n]\bar{a} \\
&= \sum_{i=1}^{n} \{[c(i,n) - \bar{c}_n]\alpha(U_i) - [c(i,n) - \bar{c}_n]\bar{a}\},
\end{aligned}$$

where

$$\bar{a} = E[\alpha(U_i)] = \int_0^1 \alpha(t)dt.$$

Define $Y_{i,n} = [c(i,n) - \bar{c}_n]\alpha(U_i)$, $\mu_{i,n} = [c(i,n) - \bar{c}_n]\bar{a}$, and $\sigma_{i,n}^2 = [c(i,n) - \bar{c}_n]^2\tilde{\alpha}^2$, for $i = 1, \ldots, n$. Then Theorem 6.1 (Lindeberg, Lévy, and Feller) will imply that

$$Z_n = n^{1/2}\tau_n^{-1/2}(\bar{Y}_n - \bar{\mu}_n) \overset{d}{\to} Z \tag{11.39}$$

as $n \to \infty$ where Z has a $N(0,1)$ distribution, as long as we can show that the associated assumptions hold. In terms of the notation of Theorem 6.1, we have that

$$\bar{\mu}_n = n^{-1}\sum_{i=1}^{n}\mu_{i,n} = n^{-1}\sum_{i=1}^{n}[c(i,n) - \bar{c}_n]\bar{\alpha},$$

and

$$\tau_n^2 = \sum_{i=1}^{n}\sigma_{i,n}^2 = \sum_{i=1}^{n}[c(i,n) - \bar{c}_n]^2\tilde{\alpha}^2. \tag{11.40}$$

For the first condition we must show that

$$\lim_{n\to\infty}\max_{i\in\{1,\dots,n\}}\tau_n^{-2}\sigma_{i,n}^2 = 0. \tag{11.41}$$

Now,

$$\tau_n^{-2}\sigma_{i,n}^2 = \frac{\tilde{\alpha}^2[c(i,n) - \bar{c}_n]^2}{\tilde{\alpha}^2\sum_{i=1}^{n}[c(i,n) - \bar{c}_n]^2} = \frac{[c(i,n) - \bar{c}_n]^2}{\sum_{i=1}^{n}[c(i,n) - \bar{c}_n]^2}.$$

It then follows that the assumption required in Equation (11.41) is implied by Noether's condition given in Definition 11.6. The second condition we need to show is that

$$\lim_{n\to\infty}\tau_n^{-2}\sum_{i=1}^{n}E(|Y_i - \mu_{i,n}|^2\delta\{|Y_i - \mu_{i,n}|; (\varepsilon\tau_n, \infty)\}) = 0,$$

for every $\varepsilon > 0$. Let $\varepsilon > 0$, then we begin showing this condition by noting that

$$E(|Y_i - \mu_{i,n}|^2\delta\{|Y_i - \mu_{i,n}|; (\varepsilon\tau_n, \infty)\}) =$$

$$\int_{L_i(\varepsilon,n)}|[c(i,n) - \bar{c}_n]\alpha(u) - [c(i,n) - \bar{c}_n]\bar{\alpha}|^2 du =$$

$$\int_{L_i(\varepsilon,n)}[c(i,n) - \bar{c}_n]^2[\alpha(u) - \bar{\alpha}]^2 du,$$

where $L_i(\varepsilon, n) = \{u : |c(i,n) - \bar{c}_n||\alpha(u) - \bar{\alpha}| > \varepsilon\tau_n\}$. Now let

$$\Delta_n = \tau_n\left[\max_{1\le j\le n}|c(j,n) - \bar{c}_n|\right]^{-1},$$

and note that

$$\begin{aligned}L_i(\varepsilon, n) &= \{u : |c(i,n) - \bar{c}_n||\alpha(u) - \bar{\alpha}| > \varepsilon\tau_n\} \\ &\subset \left\{u : \max_{1\le j\le n}|c(i,n) - \bar{c}_n||\alpha(u) - \bar{\alpha}| > \varepsilon\tau_n\right\} \\ &= \{u : |\alpha(u) - \bar{\alpha}| > \varepsilon\Delta_n\} \\ &= \bar{L}(\varepsilon, n),\end{aligned}$$

which no longer depends on the index i. Because the integrand is non-negative we have that

$$\int_{L_i(\varepsilon,n)} [c(i,n) - \bar{c}_n]^2 [\alpha(u) - \bar{\alpha}]^2 du \le$$

$$\int_{\bar{L}(\varepsilon,n)} [c(i,n) - \bar{c}_n]^2 [\alpha(u) - \bar{\alpha}]^2 du =$$

$$[c(i,n) - \bar{c}_n]^2 \int_{\bar{L}(\varepsilon,n)} [\alpha(u) - \bar{\alpha}]^2 du,$$

where we note that the integral no longer depends on the index i. Therefore, Equation (11.40) implies that

$$\tau_n^{-2} \sum_{i=1}^n E(|Y_i - \mu_{i,n}|^2 \delta\{|Y_i - \mu_{i,n}|; (\varepsilon\tau_n, \infty)\}) \le$$

$$\tau_n^{-2} \left\{ \int_{\bar{L}(\varepsilon,n)} [\alpha(u) - \bar{\alpha}]^2 du \right\} \left\{ \sum_{i=1}^n [c(i,n) - \bar{c}_n]^2 \right\} =$$

$$\tau_n^{-2} \bar{\alpha}^2 \tilde{\alpha}^{-2} \left\{ \int_{\bar{L}(\varepsilon,n)} [\alpha(u) - \bar{\alpha}]^2 du \right\} \left\{ \sum_{i=1}^n [c(i,n) - \bar{c}_n]^2 \right\} =$$

$$\tilde{\alpha}^{-2} \int_{\bar{L}(\varepsilon,n)} [\alpha(u) - \bar{\alpha}]^2 du.$$

To take the limit we note that

$$\Delta_n = \tau_n \left[\max_{1 \le j \le n} |c(j,n) - \bar{c}_n| \right]^{-1} =$$

$$\tilde{\alpha} \left\{ \sum_{i=1}^n [c(i,n) - \bar{c}_n]^2 \right\}^{1/2} \left[\max_{1 \le j \le n} |c(j,n) - \bar{c}_n| \right]^{-1}.$$

Noether's condition of Definition 11.6 implies that $\Delta_n \to \infty$ as $n \to \infty$ and hence $\varepsilon\Delta_n \to \infty$ as $n \to \infty$ for each $\varepsilon > 0$. Therefore, it follows that

$$\lim_{n\to\infty} \tilde{\alpha}^{-2} \int_{\bar{L}(\varepsilon,n)} [\alpha(u) - \bar{\alpha}]^2 du = 0,$$

and the second condition is proven, and hence the convergence described in Equation (11.39) follows.

We will now consider the mean square difference between S_n and V_n. We begin by noting that

$$S_n - V_n =$$

$$\sum_{i=1}^{n}[c(i,n) - \bar{c}_n]a(R_i, n) + n\bar{c}_n\bar{a}_n - \sum_{i=1}^{n}[c(i,n) - \bar{c}_n]\alpha(U_i) - n\bar{c}_n\bar{a}_n =$$

$$\sum_{i=1}^{n}[c(i,n) - \bar{c}_n][a(R_i, n) - \alpha(U_i)].$$

Therefore, using the fact that the conditional distribution of $\mathbf{R}$ is still uniform over $\mathbf{R}_n$ conditional on $\mathbf{U}$, we have that

$$E[(S_n - V_n)^2|\mathbf{U} = \mathbf{u}] =$$

$$E\left\{\left[\sum_{i=1}^{n}[c(i,n) - \bar{c}_n][a(R_{i,n}) - \alpha(U_i)]\right]^2 \middle| \mathbf{U} = \mathbf{u}\right\}.$$

Denote $c^*(i,n) = c(i,n) - \bar{c}_n$ and $a^*(i,n) = a(i) - \alpha(U_i)$, where we note that conditional on $\mathbf{U} = \mathbf{u}$, $\alpha(U_i)$ will be a constant. Then we have that $E[(S_n - V_n)^2|\mathbf{U} = \mathbf{u}] = E[(S_n^*)^2|\mathbf{U} = \mathbf{u}]$, where S_n^* is a linear rank statistic of the form

$$S_n^* = \sum_{i=1}^{n}c^*(i,n)a^*(R_i, n).$$

Now, note that

$$\bar{c}_n^* = n^{-1}\sum_{i=1}^{n}c^*(i,n) = n^{-1}\sum_{i=1}^{n}[c(i,n) - \bar{c}_n]^2 = 0.$$

Therefore, Theorem 11.7 implies that $E(S_n^*|\mathbf{U} = \mathbf{u}) = 0$, and hence

$$E[(S_n - V_n)^2|\mathbf{U} = \mathbf{u}] = E[(S_n^*)^2|\mathbf{U} = \mathbf{u}] = V(S_n^*|\mathbf{U} = \mathbf{u}).$$

Thus, Theorems 11.7 and 11.9 imply that

$$E[(S_n - V_n)^2|\mathbf{U} = \mathbf{u}] =$$

$$(n-1)^{-1}\left\{\sum_{i=1}^{n}[a^*(i,n) - \bar{a}_n^*]^2\right\}\left\{\sum_{i=1}^{n}[c^*(i,n) - \bar{c}_n^*]^2\right\} =$$

$$(n-1)^{-1}\left\{\sum_{i=1}^{n}[a(i,n) - \alpha(U_i) - \bar{a} + \bar{\alpha}_U]^2\right\}\left\{\sum_{i=1}^{n}[c(i,n) - \bar{c}_n]^2\right\},$$

where the last equality follows from the definitions of $a^*(i,n)$, $c^*(i,n)$, and Equation (11.37). We also note that the order statistic $U_{(i)}$ is associated with rank i, and that

$$\bar{\alpha}_U = n^{-1}\sum_{i=1}^{n}\alpha(U_{(i)}).$$

Recall that for a random variable Z such that $V(Z) < \infty$, it follows that

$V(Z) \leq E(Z^2)$. Applying this formula to the case where Z has a UNI-
FORM$\{x_1, \ldots, x_n\}$ distribution implies that

$$\sum_{i=1}^{n}(x_i - \bar{x}_n)^2 \leq \sum_{i=1}^{n} x_i^2.$$

Therefore, we have that

$$(n-1)^{-1}\left\{\sum_{i=1}^{n}[a(i,n) - \alpha(U_i) - \bar{a} + \bar{a}_U]^2\right\}\left\{\sum_{i=1}^{n}[c(i,n) - \bar{c}_n]^2\right\} \leq$$

$$(n-1)^{-1}\left\{\sum_{i=1}^{n}[a(i,n) - \alpha(U_i)]^2\right\}\left\{\sum_{i=1}^{n}[c(i,n) - \bar{c}_n]^2\right\} =$$

$$n(n-1)^{-1}\left\{n^{-1}\sum_{i=1}^{n}[a(i,n) - \alpha(U_i)]^2\right\}\left\{\sum_{i=1}^{n}[c(i,n) - \bar{c}_n]^2\right\}.$$

Note that

$$E\left\{[a(R_1^*, n) - \alpha(U_1)]^2 | \mathbf{U} = \mathbf{u}]\right\} = \sum_{i=1}^{n}[a(i,n) - \alpha(U_1)]^2 P(R_1^* = i).$$

Theorem 11.9 implies that $P(R_1^* = i | \mathbf{U} = \mathbf{u}) = n^{-1}$ for $i = 1, \ldots, n$ and hence

$$E\left\{[a(R_1^*, n) - \alpha(U_1)]^2 | \mathbf{U} = \mathbf{u}]\right\} = n^{-1}\sum_{i=1}^{n}[a(i,n) - \alpha(U_1)]^2.$$

Therefore,

$$(n-1)^{-1}\left\{\sum_{i=1}^{n}[a(i,n) - \alpha(U_{(i)}) - \bar{a} - \bar{a}_U]^2\right\}\left\{\sum_{i=1}^{n}[c(i,n) - \bar{c}_n]^2\right\} \leq$$

$$n(n-1)^{-1}\left\{\sum_{i=1}^{n}[c(i,n) - \bar{c}_n]^2\right\} E\left\{[a(R_1^*, n) - \alpha(U_1)]^2 | \mathbf{U} = \mathbf{u}]\right\},$$

which in turn implies that

$$E[(S_n - V_n)^2 | \mathbf{U} = \mathbf{u}] \leq$$

$$n(n-1)^{-1}\left\{\sum_{i=1}^{n}[c(i,n) - \bar{c}_n]^2\right\} E\left\{[a(R_1^*, n) - \alpha(U_1)]^2 | \mathbf{U} = \mathbf{u}]\right\}. \quad (11.42)$$

Using Theorem A.17 and taking the expectation of both sides of Equation
(11.42) yields

$$E\{E[(S_n - V_n)^2 | \mathbf{U} = \mathbf{u}]\} = E[(S_n - V_n)^2],$$

and

$$E\{E\left\{[a(R_1^*, n) - \alpha(U_1)]^2 | \mathbf{U} = \mathbf{u}]\right\}\} = E\left\{[a(R_1^*, n) - \alpha(U_1)]^2\right\},$$

so that

$$E[(S_n - V_n)^2] \leq n(n-1)^{-1} \left\{ \sum_{i=1}^{n} [c(i,n) - \bar{c}_n]^2 \right\} E\left\{ [a(R_1^*, n) - \alpha(U_1)]^2 \right\}.$$

Recalling that

$$\tau_n^2 = \tilde{\alpha} \sum_{i=1}^{n} [c(i,n) - \bar{c}_n]^2,$$

we have that

$$\lim_{n \to \infty} E[\tau_n^{-2}(S_n - V_n)^2] \leq$$

$$\lim_{n \to \infty} \tau_n^{-2} n(n-1)^{-1} \left\{ \sum_{i=1}^{n} [c(i,n) - \bar{c}_n]^2 \right\} E\left\{ [a(R_1^*, n) - \alpha(U_1)]^2 \right\} =$$

$$\lim_{n \to \infty} n(n-1)^{-1} \tilde{\alpha}^{-2} E\left\{ [a(R_1^*, n) - \alpha(U_1)]^2 \right\} \leq$$

$$\lim_{n \to \infty} \tilde{\alpha}^{-2} E\left\{ [a(R_1^*, n) - \alpha(U_1)]^2 \right\} = 0,$$

by Theorem 11.13. Hence $\tau_n^{-1}(S_n - V_n) \xrightarrow{q.m.} 0$ as $n \to \infty$. Using the same type of arguments used in proving Theorem 11.6, it then follows that $\sigma_n^{-1}(S_n - \mu_n)$ and $\tau_n^{-1}(V_n - \mu_n)$ have the same limiting distribution, and hence the result is proven.   □

Example 11.16. Consider the rank sum test statistic M which has the form of a linear rank statistic with regression constants $c(i, m, n) = \delta\{i; \{m + 1, \ldots, n + m\}\}$ and score function $a(i) = i$ for $i = 1, \ldots, n + m$. For these regression constants, it follows that

$$\bar{c}_{m,n} = (n+m)^{-1} \sum_{i=1}^{n+m} c(i,m,n) = (n+m)^{-1} \sum_{i=m+1}^{n+m} 1 = n(n+m)^{-1},$$

and

$$\sum_{i=1}^{n+m} [c(i,m,n) - \bar{c}_{m,n}]^2 = \sum_{i=1}^{n+m} [c(i,m,n) - n(n+m)^{-1}]^2$$

$$= \sum_{i=1}^{m} [0 - n(n+m)^{-1}]^2$$

$$+ \sum_{i=m+1}^{n+m} [1 - n(n+m)^{-1}]^2$$

$$= mn^2(n+m)^{-2} + nm^2(n+m)^{-2}$$

$$= mn(n+m)^{-1}.$$

To verify Noether's condition of Definition 11.6, we note that

$$\max_{i\in\{1,\ldots,n+m\}}[c(i,m,n)-\bar{c}_{m,n}]^2 = \max\{\bar{c}_{m,n}^2, (1-\bar{c}_{m,n})^2\}$$

$$= \max\{n^2(n+m)^{-2}, m^2(n+m)^{-2}\}$$

$$= (n+m)^{-2}\max\{n^2, m^2\}$$

$$= (n+m)^{-2}(\max\{n,m\})^2.$$

In order to simplify the condition in Definition 11.6, we must have that

$$N_{m,n} = \left\{\max_{i\in\{1,\ldots,n+m\}}[c(i,m,n)-\bar{c}_{m,n}]^2\right\}^{-1}\sum_{i=1}^{n+m}[c(i,m,n)-\bar{c}_{m,n}]^2 =$$

$$\frac{nm(n+m)^2}{(n+m)(\max\{n,m\})^2} = \frac{nm(n+m)}{(\max\{n,m\})^2} \to \infty$$

as $n+m \to \infty$. Note that if $n > m$ then we have that $N_{m,n} = nm(n+m)n^{-2} = mn^{-1}(n+m)$, and if $m \geq n$ then we have that $N_{m,n} = nm(n+m)m^{-2} = nm^{-1}(n+m)$. Therefore, it follows that

$$N_{m,n} = \frac{(n+m)\min\{m,n\}}{\max\{m,n\}} = \frac{n\min\{m,n\}+m\min\{m,n\}}{\max\{m,n\}}. \qquad (11.43)$$

Note that no matter what the relative sizes of m and n are, one of the terms in the sum on the right hand side of Equation (11.43) will have the form $\min\{m,n\}$ and the other will have the form $\min^2\{m,n\}(\max\{m,n\})^{-1}$. Therefore,

$$N_{m,n} = \min\{m,n\}\{1+\min\{m,n\}[\max\{m,n\}]^{-1}\}.$$

Therefore, if $\min\{m,n\} \to \infty$ as $n+m \to \infty$ then $N_{m,n} \to \infty$ as $n+m \to \infty$ and Noether's condition will hold. The score function can be written as $a(i) = i = (m+n+1)\alpha[(m+n+1)^{-1}i]$ where $\alpha(t) = t$ for $t \in (0,1)$. Therefore,

$$\bar{\alpha} = \int_0^1 t\, dt = \tfrac{1}{2},$$

and

$$\tilde{\alpha}^2 = \int_0^1 (t-\tfrac{1}{2})^2 dt = \tfrac{1}{12}.$$

Definition 11.7 then implies that α is square integrable score function. Now

$$\mu_{m+n} = (m+n)\bar{c}_{m,n}\bar{a}_{m,n} =$$

$$(m+n)n(m+n)^{-1}(m+n)^{-1}\sum_{i=1}^{m+n}i = \tfrac{1}{2}n(m+n+1),$$

and

$$\sigma_{m,n}^2 = (m+n-1)^{-1}\left\{\sum_{i=1}^{m+n}[c(i,n)-\bar{c}_{m,n}]^2\right\}\left\{\sum_{i=1}^{m+n}[a(i,n)-\bar{a}_{m,n}]^2\right\} =$$

$$(m+n-1)^{-1}(m+n)^{-1}mn\sum_{i=1}^{m+n}[i-\tfrac{1}{2}(m+n)(m+n-1)]^2. \quad (11.44)$$

The sum in Equation (11.44) is equal to $(m+n)$ times the variance of a UNIFORM$\{1,2,\ldots,m+n\}$ random variable. Therefore,

$$\sigma_{m,n}^2 = \tfrac{1}{12}(m+n-1)^{-1}(m+n)^{-1}mn(m+n)(m+n+1)(m+n-1) =$$
$$\tfrac{1}{12}mn(m+n+1),$$

and Theorem 11.14 implies that $[M-\tfrac{1}{2}n(m+n+1)][\tfrac{1}{12}mn(m+n+1)]^{-1/2} \xrightarrow{d} Z$ as $n\to\infty$, where Z is a N$(0,1)$ random variable. This verifies the results observed in Figures 11.4–11.6. ∎

11.4 Pitman Asymptotic Relative Efficiency

One of the reasons that many nonparametric statistical methods have remained popular in applications is that few assumptions need to be made about the underlying population, and that this flexibility results in only a small loss of efficiency in many cases. The use of a nonparametric method, which is valid for a large collection of distributions, necessarily entails the possible loss of efficiency. This can manifest itself in larger standard errors for point estimates, wider confidence intervals, or hypothesis tests that have lower power. This is because a parametric method, which is valid for a specific parametric family, is able to take advantage of the structure of the problem to produce a finely tuned statistical method. On the other hand, nonparametric methods have fewer assumptions to rely on and must be valid for a much larger array of distributions. Therefore, these methods, cannot take advantage of this additional structure.

A classic example of this difference can be observed by considering the problem of estimating the location of the mode of a continuous unimodal density. If we are able to reasonably assume that the population is NORMAL, then we can estimate the location of the mode using the sample mean. On the other hand, if the exact parametric form of the density is not known, then the problem can become very complicated. It is worthwhile to note at this point that any potential increase in the efficiency that may be realized by using a parametric method is only valid if the parametric model is at least approximately true. For example, the sample mean will only be a reasonable estimator of the location of the mode of a density for certain parametric models. If one of these models does not hold, then the sample mean may be a particularly unreasonable estimator, and may even have an infinite bias, for example.

To assess the efficiency of statistical hypothesis tests, we must borrow some of the ideas that we encountered in Section 10.4. Statistical tests are usually compared on the basis of their power functions. That is, we would prefer to have a test that rejects the null hypothesis more often when the alternative hypothesis is true. It is important that when two tests are compared on the basis of their power functions that the significance levels of the two tests be the same. This is due to the fact that the power of any test can be arbitrarily increased by increasing the value of the significance level. Therefore, if $\beta_1(\theta)$ and $\beta_2(\theta)$ are the power functions of two tests of the set of hypotheses $H_0 : \theta \in \Omega_0$ and $H_1 : \theta \in \Omega_1$, based on the same sample size we would prefer the test with power function β_1 if $\beta_1(\theta) \geq \beta_2(\theta)$ for all $\theta \in \Omega_1$, where

$$\sup_{\theta \in \Omega_0} \beta_1(\theta) = \sup_{\theta \in \Omega_0} \beta_2(\theta).$$

This view may be too simplistic as it insists that one of the tests be uniformly better than the other. Further, from our discussion in Section 10.4, we know that many tests will do well when the distance between θ and the boundary of Ω_0 is large. Another complication comes from the fact that there are so many parameters which can be varied, including the sample size, the value of θ in the alternative hypothesis, and the distribution. We can remove the sample size from the problem by considering asymptotic relative efficiency using a similar concept encountered for point estimation in Section 10.2. The value of θ in the alternative hypothesis can be eliminated if we consider a sequence of alternative hypotheses that converge to the null hypothesis as $n \to \infty$. This is similar to the idea used in Section 10.4 to compute the asymptotic power of a hypothesis test.

Definition 11.8 (Pitman). *Consider two competing tests of a point null hypothesis $H_0 : \theta = \theta_0$ where θ_0 is a specified parameter value in the parameter space Ω. Let S_n and T_n denote the test statistics for the two tests based on a sample of size n. Let $\beta_{S,n}(\theta)$ and $\beta_{T,n}(\theta)$ be the power functions of the tests based on the test statistics S_n and T_n, respectively, when the sample size equals n.*

1. *Suppose that both tests have size α.*
2. *Let $\{\theta_k\}_{k=1}^{\infty}$ be a sequence of values in Ω such that*

$$\lim_{k \to \infty} \theta_k = \theta_0.$$

3. *Let $\{m(k)\}_{k=1}^{\infty}$ and $\{n(k)\}_{k=1}^{\infty}$ be increasing sequences of positive integers such that both tests have the same limiting significance level and*

$$\lim_{k \to \infty} \beta_{S,m(k)}(\theta_k) = \lim_{k \to \infty} \beta_{T,n(k)}(\theta_k) \in (\alpha, 1).$$

Then, the asymptotic relative efficiency of the test based on the test statistic S_n against the test based on the test statistic T_n is given by

$$\lim_{k \to \infty} m(k)[n(k)]^{-1}.$$

This concept of relative efficiency establishes the relative sample sizes required for the two tests to have the same asymptotic power. This type of efficiency is based on Pitman (1948), and is often called *Pitman relative asymptotic efficiency*.

For simplicity we will assume that both tests reject the null hypothesis $H_0 : \theta = \theta_0$ for large values of the test statistic. The theory presented here can be easily adapted to other types of rejection regions as well. We will also limit our discussion to test statistics that have an asymptotic NORMAL distribution under both the null and alternative hypotheses. The assumptions about the asymptotic distributions of the test statistics used in this section are very similar to those used in the study of asymptotic power in Section 10.4. In particular we will assume that there exist functions $\mu_n(\theta)$, $\eta_n(\theta)$, $\sigma_n(\theta)$ and $\tau_n(\theta)$ such that

$$P\left[\frac{S_{m(k)} - \mu_{m(k)}(\theta_k)}{\sigma_{m(k)}(\theta_k)} \leq t \,\middle|\, \theta = \theta_k\right] \rightsquigarrow \Phi(t),$$

$$P\left[\frac{T_{n(k)} - \eta_{n(k)}(\theta_k)}{\tau_{n(k)}(\theta_k)} \leq t \,\middle|\, \theta = \theta_k\right] \rightsquigarrow \Phi(t),$$

$$P\left[\frac{S_{m(k)} - \mu_{m(k)}(\theta_0)}{\sigma_{m(k)}(\theta_0)} \leq t \,\middle|\, \theta = \theta_0\right] \rightsquigarrow \Phi(t),$$

and

$$P\left[\frac{T_{n(k)} - \eta_{n(k)}(\theta_0)}{\tau_{n(k)}(\theta_0)} \leq t \,\middle|\, \theta = \theta_0\right] \rightsquigarrow \Phi(t),$$

as $k \to \infty$. Let $\alpha \in (0,1)$ be a fixed significance level and let $\{s_{m(k)}(\alpha)\}_{k=1}^{\infty}$ and $\{t_{n(k)}(\alpha)\}_{k=1}^{\infty}$ be sequences of real numbers such that $s_{m(k)}(\alpha) \to z_{1-\alpha}$ and $t_{n(k)}(\alpha) \to z_{1-\alpha}$ as $n \to \infty$ where we are assuming that the null hypothesis $H_0 : \theta = \theta_0$ is rejected when $S_{m(k)} \geq s_{m(k)}(\alpha)$ and $T_{m(k)} \geq t_{m(k)}(\alpha)$, respectively. The tests are assumed to have a limiting significance level α. In particular we assume that

$$\lim_{k\to\infty} P[S_{m(k)} \geq s_{m(k)}(\alpha)|\theta = \theta_0] = \lim_{k\to\infty} P[T_{n(k)} \geq t_{n(k)}(\alpha)|\theta = \theta_0] = \alpha.$$

The power function for the test using the test statistic $S_{m(k)}$ is given by

$$\beta_{S,m(k)}(\theta_1) = P[S_{m(k)} \geq s_{m(k)}(\alpha)|\theta = \theta_1] =$$
$$P\left[\frac{S_{m(k)} - \mu_{m(k)}(\theta_1)}{\sigma_{m(k)}(\theta_1)} \geq \frac{s_{m(k)}(\alpha) - \mu_{m(k)}(\theta_1)}{\sigma_{m(k)}(\theta_1)} \,\middle|\, \theta = \theta_1\right],$$

with a similar form for the power function of the test using the test statistic $T_{n(k)}$. Therefore, the property in Definition 11.8 that requires

$$\lim_{k\to\infty} \beta_{S,m(k)}(\theta_k) = \lim_{k\to\infty} \beta_{T,n(k)}(\theta_k),$$

is equivalent to

$$\lim_{k\to\infty} P\left[\frac{S_{m(k)} - \mu_{m(k)}(\theta_k)}{\sigma_{m(k)}(\theta_k)} \geq \frac{s_{m(k)}(\alpha) - \mu_{m(k)}(\theta_k)}{\sigma_{m(k)}(\theta_k)}\middle|\theta = \theta_k\right] =$$

$$\lim_{k\to\infty} P\left[\frac{T_{n(k)} - \eta_{n(k)}(\theta_k)}{\tau_{n(k)}(\theta_k)} \geq \frac{t_{n(k)}(\alpha) - \eta_{n(k)}(\theta_k)}{\tau_{n(k)}(\theta_k)}\middle|\theta = \theta_k\right],$$

which can in turn be shown to require

$$\lim_{k\to\infty} \frac{s_{m(k)}(\alpha) - \mu_{m(k)}(\theta_k)}{\sigma_{m(k)}(\theta_k)} = \lim_{k\to\infty} \frac{t_{n(k)}(\alpha) - \eta_{n(k)}(\theta_k)}{\tau_{n(k)}(\theta_k)}. \tag{11.45}$$

Similarly, for both tests to have the same limiting significance level we require that

$$\lim_{k\to\infty} P\left[\frac{S_{m(k)} - \mu_{m(k)}(\theta_0)}{\sigma_{m(k)}(\theta_0)} \geq \frac{s_{m(k)}(\alpha) - \mu_{m(k)}(\theta_0)}{\sigma_{m(k)}(\theta_0)}\middle|\theta = \theta_0\right] =$$

$$\lim_{k\to\infty} P\left[\frac{T_{n(k)} - \eta_{n(k)}(\theta_0)}{\tau_{n(k)}(\theta_0)} \geq \frac{t_{n(k)}(\alpha) - \eta_{n(k)}(\theta_0)}{\tau_{n(k)}(\theta_0)}\middle|\theta = \theta_0\right],$$

which in turn requires that

$$\lim_{k\to\infty} \frac{s_{m(k)}(\alpha) - \mu_{m(k)}(\theta_0)}{\sigma_{m(k)}(\theta_0)} = \lim_{k\to\infty} \frac{t_{n(k)}(\alpha) - \eta_{n(k)}(\theta_0)}{\tau_{n(k)}(\theta_0)} = z_{1-\alpha}.$$

Under this type of framework, Noether (1955) shows that the Pitman asymptotic relative efficiency is a function of the derivatives of $\mu_{m(k)}(\theta)$ and $\eta_{n(k)}(\theta)$ relative to $\sigma_{m(k)}(\theta)$ and $\tau_{n(k)}(\theta)$, respectively.

Theorem 11.15 (Noether). *Let S_n and T_n be test statistics based on a sample of size n that reject a null hypothesis $H_0 : \theta = \theta_0$ when $S_n \geq s_n(\alpha)$ and $T_n \geq t_n(\alpha)$, respectively. Let $\{\theta_k\}_{k=1}^{\infty}$ be a sequence of real values greater that θ_0 such that $\theta_k \to \theta_0$ as $k \to \infty$. Let $\{m(k)\}_{k=1}^{\infty}$ and $\{n(k)\}_{k=1}^{\infty}$ be increasing sequences of positive integers. Let $\{\mu_{m(k)}(\theta)\}_{k=1}^{\infty}$, $\{\eta_{n(k)}(\theta)\}_{k=1}^{\infty}$, $\{\sigma_{m(k)}(\theta)\}_{k=1}^{\infty}$, and $\{\tau_{n(k)}(\theta)\}_{k=1}^{\infty}$, be sequences of real numbers that satisfy the following assumptions:*

1. *For all $t \in \mathbb{R}$,*

$$P\left[\frac{S_{m(k)} - \mu_{m(k)}(\theta_k)}{\sigma_{m(k)}(\theta_k)} \leq t\middle|\theta = \theta_k\right] \rightsquigarrow \Phi(t),$$

and

$$P\left[\frac{T_{n(k)} - \eta_{n(k)}(\theta_k)}{\tau_{n(k)}(\theta_k)} \leq t\middle|\theta = \theta_k\right] \rightsquigarrow \Phi(t),$$

as $n \to \infty$.

2. *For all $t \in \mathbb{R}$,*

$$P\left[\frac{S_{m(k)} - \mu_{m(k)}(\theta_0)}{\sigma_{m(k)}(\theta_0)} \leq t\middle|\theta = \theta_0\right] \rightsquigarrow \Phi(t),$$

and

$$P\left[\frac{T_{n(k)} - \eta_{n(k)}(\theta_0)}{\tau_{n(k)}(\theta_0)} \le t \,\middle|\, \theta = \theta_0\right] \rightsquigarrow \Phi(t),$$

as $n \to \infty$.

3.

$$\lim_{k \to \infty} \frac{\sigma_{m(k)}(\theta_k)}{\sigma_{m(k)}(\theta_0)} = 1,$$

and

$$\lim_{k \to \infty} \frac{\tau_{n(k)}(\theta_k)}{\tau_{n(k)}(\theta_0)} = 1.$$

4. *The derivatives of the functions $\mu_{m(k)}(\theta)$ and $\eta_{n(k)}(\theta)$ taken with respect to θ exist, are continuous on an interval $[\theta_0 - \delta, \theta_0 + \delta]$ for some $\delta > 0$, are non-zero when evaluated at θ_0, and*

$$\lim_{k \to \infty} \frac{\mu'_{m(k)}(\theta_k)}{\mu'_{m(k)}(\theta_0)} = \lim_{k \to \infty} \frac{\eta'_{n(k)}(\theta_k)}{\eta'_{n(k)}(\theta_0)} = 1.$$

Define positive constants E_S and E_T as

$$E_S = \lim_{n \to \infty} [n\sigma_n^2(\theta_0)]^{-1/2} \mu'_n(\theta_0)$$

and

$$E_T = \lim_{n \to \infty} [n\tau_n^2(\theta_0)]^{-1/2} \eta'_n(\theta_0).$$

Then the Pitman asymptotic relative efficiency of the test based on the test statistic S_n, relative to the test based on the test statistic T_n is given by $E_S^2 E_T^{-2}$.

Proof. The approach to proving this result is based on showing that the limiting ratio of the sample size sequences $m(k)$ and $n(k)$, when the asymptotic power functions are equal, is the same as the ratio $E_S^2 E_T^{-2}$. We begin by applying Theorem 1.13 (Taylor) to the functions $\mu_{m(k)}(\theta_k)$ and $\eta_{n(k)}(\theta_k)$, where we are taking advantage of Assumption 4, to find that $\mu_{m(k)}(\theta_k) = \mu_{m(k)}(\theta_0) + (\theta_k - \theta_0)\mu'_{m(k)}(\bar{\theta}_k)$ and $\eta_{n(k)}(\theta_k) = \eta_{n(k)}(\theta_0) + (\theta_k - \theta_0)\eta'_{n(k)}(\tilde{\theta}_k)$ where $\bar{\theta}_k \in (\theta_0, \theta_k)$ and $\tilde{\theta}_k \in (\theta_0, \theta_k)$ for all $k \in \mathbb{N}$. Note that even though $\bar{\theta}_k$ and $\tilde{\theta}_k$ are always in the same interval, they will generally not be equal to one another. Now note that

$$\frac{s_{m(k)}(\alpha) - \mu_{m(k)}(\theta_k)}{\sigma_{m(k)}(\theta_k)} =$$

$$\frac{s_{m(k)}(\alpha) - \mu_{m(k)}(\theta_0) + \mu_{m(k)}(\theta_0) - \mu_{m(k)}(\theta_k)}{\sigma_{m(k)}(\theta_0)} \frac{\sigma_{m(k)}(\theta_0)}{\sigma_{m(k)}(\theta_k)}.$$

Assumption 2 and Equation (11.45) imply that

$$\lim_{k \to \infty} \frac{s_{m(k)}(\alpha) - \mu_{m(k)}(\theta_0)}{\sigma_{m(k)}(\theta_0)} = z_{1-\alpha}.$$

Combining this result with Assumption 3 implies that

$$\lim_{k\to\infty} \frac{s_{m(k)}(\alpha) - \mu_{m(k)}(\theta_k)}{\sigma_{m(k)}(\theta_k)} = z_{1-\alpha} + \lim_{k\to\infty} \frac{\mu_{m(k)}(\theta_0) - \mu_{m(k)}(\theta_k)}{\sigma_{m(k)}(\theta_0)}. \quad (11.46)$$

Performing the same calculations with the test based on the test statistic $T_{n(k)}$ yields

$$\lim_{k\to\infty} \frac{t_{n(k)}(\alpha) - \eta_{n(k)}(\theta_k)}{\tau_{n(k)}(\theta_k)} = z_{1-\alpha} + \lim_{k\to\infty} \frac{\eta_{n(k)}(\theta_0) - \eta_{n(k)}(\theta_k)}{\tau_{n(k)}(\theta_0)}. \quad (11.47)$$

Combining these results with the requirement of Equation (11.45) yields

$$\lim_{k\to\infty} \frac{\mu_{m(k)}(\theta_0) - \mu_{m(k)}(\theta_k)}{\sigma_{m(k)}(\theta_0)} \cdot \frac{\tau_{n(k)}(\theta_0)}{\eta_{n(k)}(\theta_0) - \eta_{n(k)}(\theta_k)} = 1.$$

Equations (11.46) and (11.47) then imply that

$$\lim_{k\to\infty} \frac{\mu_{m(k)}(\theta_0) - \mu_{m(k)}(\bar{\theta}_k)}{\sigma_{m(k)}(\theta_0)} \cdot \frac{\tau_{n(k)}(\theta_0)}{\eta_{n(k)}(\theta_0) - \eta_{n(k)}(\tilde{\theta}_k)} =$$

$$\lim_{k\to\infty} \frac{\mu'_{m(k)}(\bar{\theta}_k)\tau_{n(k)}(\theta_0)}{\eta'_{n(k)}(\tilde{\theta}_k)\sigma_{m(k)}(\theta_0)} =$$

$$\lim_{k\to\infty} \frac{m^{1/2}(k)}{n^{1/2}(k)} \cdot \frac{\mu'_{m(k)}(\bar{\theta}_k)}{m^{1/2}(k)\sigma_{m(k)}(\theta_0)} \cdot \frac{n^{1/2}(k)\tau_{n(k)}(\theta_0)}{\eta'_{n(k)}(\tilde{\theta}_k)} =$$

$$\lim_{k\to\infty} \left(\frac{m(k)}{n(k)}\right)^{1/2} E_S E_T^{-1} = 1.$$

Therefore, the Pitman asymptotic relative efficiency, which is given by

$$\lim_{k\to\infty} m(k)[n(k)]^{-1},$$

has the same limit as $E_S^2 E_T^{-2}$. Therefore, the Pitman asymptotic relative efficiency is given by $E_S^2 E_T^{-2}$. $\qquad\qquad\square$

The values E_S and E_T are called the *efficacies* of the tests based on the test statistics S_n and T_n, respectively. Randles and Wolfe (1979) point out some important issues when interpreting the efficacies of test statistics. When one examines the form of the efficacy of a test statistic we see that it measures the rate of change of the function μ_n, in the case of the test based on the test statistic S_n, at the point of the null hypothesis θ_0, relative to σ_n at the same point. Therefore, the efficacy is a measure of how fast the distribution of S_n changes at points near θ_0. In particular, the efficacy given in Theorem 11.15 measures the rate of change of the location of the distribution of S_n near the null hypothesis point θ_0. Test statistics whose distributions change a great deal near θ_0 result in tests that are more sensitive to differences between the θ_0 and the actual value of θ in the alternative hypothesis. A more sensitive test will be more powerful, and such tests will have a larger efficacy. Therefore, if $E_S > E_T$, then the Pitman asymptotic relative efficiency is greater than one, and the test using the test statistic S_n has a higher asymptotic power. Similarly, if $E_S < E_T$ then the relative efficiency is less than one, and the test based on the test statistic T_n has a higher asymptotic power.

In the development of the concept of asymptotic power for individual tests in Section 10.4, a particular sequence of alternative hypotheses of the form $\theta_n = \theta_0 + O(n^{-1/2})$, as $n \to \infty$ was considered. In Theorem 11.15 no explicit form for the sequence $\{\theta_k\}_{k=1}^{\infty}$ is discussed, though there is an implicit form for this sequence given in the assumptions. In particular, it follows that $\theta_k = \theta_0 + O(k^{-1/2})$ as $k \to \infty$, matching the asymptotic form considered in Section 10.4. See Section 5.2 of Randles and Wolfe (1979) for further details on this result.

This section will close with some examples of computing efficacies for the t-test, the signed rank test, and the sign test. In order to make similar comparisons between these tests we will begin by making some general assumptions about the setup of the testing problem that we will consider. Let $X_1, \ldots, X_n$ be a set of independent and identically distributed random variables from a distribution $F(x - \theta)$ that is symmetric about θ. We will assume that F has a density f that is also continuous, except perhaps at a countable number of points. Let θ_0 be a fixed real value and assume that we are interested in testing $H_0 : \theta \leq \theta_0$ against $H_1 : \theta > \theta_0$. In the examples given below we will not concentrate on verifying the assumptions of Theorem 11.15. For details on verifying these assumptions see Section 5.4 of Randles and Wolfe (1979).

Example 11.17. Consider the t-test statistic $T_n = n^{1/2}\sigma^{-1}(\bar{X}_n - \theta_0)$ where the null hypothesis is rejected when $T_n > t_{1-\alpha;n-1}$. Note that $t_{1-\alpha;n-1} \to z_{1-\alpha}$ as $n \to \infty$ in accordance with the assumptions of Theorem 11.15. The form of the test statistic implies that $\mu_T(\theta_0) = \theta_0$ and $\sigma_n(\theta_0) = n^{-1/2}\sigma$ so that the efficacy of the test is given by

$$E_T = \lim_{n \to \infty} \frac{\mu_T'(\theta_0)}{n^{1/2}n^{-1/2}\sigma} = \sigma^{-1}.$$

■

Example 11.18. Consider the signed rank test statistic W_n given by the sum of the ranks of $|X_1 - \theta_0|, \ldots, |X_n - \theta_0|$ that correspond to the cases where $X_i > \theta_0$ for $i = 1, \ldots, n$. Without loss of generality we will consider the case where $\theta_0 = 0$. Following the approach of Randles and Wolfe (1979) we consider the equivalent test statistic $V_n = \binom{n}{2}^{-1} W_n$. Using the results of Example 11.12, it follows that

$$V_n = \binom{n}{2}^{-1} [nU_{1,n} + \tfrac{1}{2}n(n-1)U_{2,n}] = 2(n-1)^{-1}U_{1,n} + U_{2,n},$$

where

$$U_{1,n} = n^{-1} \sum_{i=1}^{n} \delta\{2X_i; (0, \infty)\},$$

and

$$U_{2,n} = 2[n(n-1)]^{-1}\sum_{i=1}^{n}\sum_{j=i+1}^{n}\delta\{X_i + X_j; (0,\infty)\}.$$

The results from Example 11.12 suggest that

$$\mu_n(\theta_1) = E[2(n-1)^{-1}U_{1,n} + U_{2,n}|\theta = \theta_1].$$

When $\theta_1 = \theta_0$ we can use the results directly from Example 11.12, but in this case we need to find the conditional expectation for any $\theta_1 > \theta_0$ so we can evaluate the derivative of $\mu_n(\theta_1)$. Using the shift model it follows that when $\theta = \theta_1$ the distribution is given by $F(x - \theta_1)$ for all $x \in \mathbb{R}$. Therefore,

$$E(U_{1,n}|\theta = \theta_1) = P(X_i > 0|\theta = \theta_1) = \int_0^{\infty} dF(x - \theta_1) =$$

$$\int_{-\theta_1}^{\infty} dF(u) = 1 - F(-\theta_1) = F(\theta_1).$$

Similarly,

$$E(U_{2,n}|\theta = \theta_1) = P(X_i + X_j > 0|\theta = \theta_1) =$$

$$\int_{-\infty}^{\infty}\int_{-x_i}^{\infty} dF(x_j - \theta_1)dF(x_i - \theta_1) = \int_{-\infty}^{\infty}\int_{-x_i-\theta_1}^{\infty} dF(u)dF(x_i - \theta_1) =$$

$$\int_{-\infty}^{\infty}[1 - F(-x_i - \theta_1)]dF(x_i - \theta_1) = \int_{-\infty}^{\infty}[1 - F(-u - 2\theta_1)]dF(u).$$

Therefore,

$$\mu_n(\theta_1) = 2(n-1)^{-1}[1 - F(-\theta_1)] + \int_{-\infty}^{\infty}[1 - F(-u - 2\theta_1)]dF(u). \quad (11.48)$$

Assuming that we can exchange a derivative and the integral in Equation (11.48), it follows that

$$\mu_n'(\theta_1) = 2(n-1)^{-1}f(-\theta_1) + 2\int_{-\infty}^{\infty} f(-u - 2\theta_1)dF(u).$$

Hence, when $\mu_n'(\theta_1)$ is evaluated at $\theta_0 = 0$, we have that

$$\mu_n'(\theta_0) = 2(n-1)^{-1}f(0) + 2\int_{-\infty}^{\infty} f(-u)dF(u) =$$

$$2(n-1)^{-1}f(0) + 2\int_{-\infty}^{\infty} f^2(u)du,$$

where we have used the fact that f is symmetric about zero. To find the variance we note that we can use the result of Example 11.7, which found the variance to have the form $\frac{1}{3}n^{-1}\binom{n}{2}^2$ for the statistic W_n, and hence the variance of V_n is $\sigma_n^2(\theta_0) = \frac{1}{3}n^{-1}$. Therefore, the efficacy of the signed rank

Table 11.2 *The efficacies and Pitman asymptotic relative efficiencies of the t-test (T), the signed rank test (V), and the sign test (B) under sampling from various populations.*

Distribution	E_T^2	E_V^2	E_B^2	$E_V^2 E_T^{-2}$	$E_B^2 E_T^{-2}$	$E_B^2 E_V^{-2}$
$\mathrm{N}(0,1)$	1	$3\pi^{-1}$	$2\pi^{-1}$	$3\pi^{-1}$	$2\pi^{-1}$	$\frac{2}{3}$
$\mathrm{UNIFORM}(-\frac{1}{2},\frac{1}{2})$	12	12	4	1	$\frac{1}{3}$	$\frac{1}{3}$
$\mathrm{LAPLACE}(0,1)$	$\frac{1}{2}$	$\frac{3}{4}$	1	$\frac{3}{2}$	2	$\frac{4}{3}$
$\mathrm{LOGISTIC}(0,1)$	$3\pi^{-2}$	$\frac{1}{3}$	$\frac{1}{4}$	$\frac{1}{9}\pi^2$	$\frac{1}{12}\pi^2$	$\frac{3}{4}$
$\mathrm{TRIANGULAR}(-1,1,0)$	6	$\frac{16}{3}$	4	$\frac{8}{9}$	$\frac{2}{3}$	$\frac{3}{4}$

test is given by

$$\lim_{n\to\infty} \frac{\mu_n'(\theta_0)}{n^{1/2}\sigma_n(\theta_0)} = \lim_{n\to\infty} \frac{\mu_n'(0)}{n^{1/2}\sigma_n(0)} =$$

$$\lim_{n\to\infty} 2(3)^{1/2}(n-1)^{-1}f(0) + 2(3)^{1/2}\int_{-\infty}^{\infty} f^2(u)du =$$

$$2(3)^{1/2}\int_{-\infty}^{\infty} f^2(u)du,$$

and hence

$$E_V^2 = 12\left[\int_{-\infty}^{\infty} f^2(u)du\right]^2. \tag{11.49}$$

The value of the integral in Equation (11.49) has been computed for many distributions. For example see Table B.2 of Wand and Jones (1995). ∎

Example 11.19. Consider the test statistic used by the sign test of Example 11.1, which has the form

$$B = \sum_{i=1}^{n} \delta\{X_i - \theta; (0,\infty)\},$$

where the null hypothesis is rejected when B exceeds a specified quantile of the $\mathrm{BINOMIAL}(\frac{1}{2}, n)$ distribution, which has an asymptotic NORMAL distribution. The efficacy of this test can be shown to be $E_B = 2f(0)$, where we are assuming, as in the previous examples, that $\theta_0 = 0$. See Exercise 9. ∎

One can make several interesting conclusions by analyzing the results of the efficacy calculations from Examples 11.17–11.19. These efficacies, along with the associated asymptotic relative effficiencies, are summarized in Table 11.2 for several distributions. We begin by considering the results for the $\mathrm{N}(0,1)$

distribution. We first note that the efficiencies relative to the t-test observed in Table 11.2 are less than one, indicating that the t-test has a higher efficacy, and is therefore more powerful in this case. The observed asymptotic relative efficiency of the signed rank test is $3\pi^{-1} \simeq 0.955$, which indicates that the signed rank test has about 95% of the efficiency of the t-test when the population is $N(0,1)$. It is not surprising that the t-test is more efficient than the signed rank test, since the t-test is derived under assumption that the population is NORMAL, but what may be surprising is that the signed rank test does so well. In fact, the results indicate that if a sample of size $n = 100$ is required by the signed rank test, then a sample of size $n = 95$ is required by the t-test to obtain the same asymptotic power. Therefore, there is little penalty for using the signed rank test even when the population is normal. The sign test does not fair as well. The observed asymptotic relative efficiency of the sign test is $2\pi^{-1} \simeq 0.637$, which indicates that the sign test has about 64% of the efficiency of the t-test when the population is $N(0,1)$. Therefore, if a sample of size $n = 100$ is required by the sign test, then a sample of size $n = 64$ is required by the t-test to obtain the same asymptotic power. The sign test also does not compare well with the signed rank test.

The UNIFORM$(-\frac{1}{2}, \frac{1}{2})$ distribution is an interesting example because there is little chance of outliers in samples from this distribution, but the shape of the distribution is far from NORMAL. In this case the signed rank test and the t-test perform equally well with an asymptotic relative efficiency of one. The sign test performs poorly in this case with an asymptotic relative efficiency equal to $\frac{1}{3}$, which indicates that the sign test has an asymptotic relative efficiency of about 33%, or that a sample of $n = 33$ is required for the t-test or the signed rank test, then a sample of 100 is required for the sign test to obtain the same asymptotic power.

For the LAPLACE$(0,1)$ distribution the trend begins to turn in favor of the nonparametric tests. The observed asymptotic relative efficiency of the signed rank test is $\frac{3}{2}$, which indicates that the signed rank test has about 150% of the efficiency of the t-test when the population is LAPLACE$(0,1)$. Therefore, if the signed rank test requires a sample of size $n = 100$, then the t-test requires a sample of size $n = 150$ to obtain the same asymptotic power. The sign test does even better in this case with an asymptotic relative efficiency equal to 2 when compared to the t-test. In this case, if the sign test requires a sample of size $n = 100$, then the t-test requires a sample of size $n = 200$ to obtain the same asymptotic power. This is due to the heavy tails of the LAPLACE$(0,1)$ distribution. The sign test, and to a lesser extent, the signed rank test, are robust to the presence of outliers while the t-test is not. An outlying value in one direction may result in failing to reject a null hypothesis even when the remainder of the data supports the alternative hypothesis.

For the LOGISTIC$(0,1)$ distribution we get similar results, but not as drastic. The LOGISTIC$(0,1)$ distribution also has heavier tails than the $N(0,1)$ distribution, but not as heavy as the LAPLACE$(0,1)$ distribution. This is reflected in the asymptotic relative efficiencies. The signed rank test has an asymptotic relative efficiency equal to $\frac{1}{9}\pi^2 \simeq 1.097$ which gives a small advantage to this test over the t-test, while the sign test has an asymptotic relative efficiency equal to $\frac{1}{12}\pi^2 \simeq 0.822$ which gives an advantage to the the the t-test.

For the case when the population follows a TRIANGULAR$(-1,1,0)$ distribution we find that the signed rank test has an asymptotic relative efficiency equal to $\frac{8}{9} \simeq 0.889$, which gives a slight edge to the t-test, and the sign test has an asymptotic relative efficiency equal to $\frac{2}{3} \simeq 0.667$, which implies that the t-test is better than the sign test in this case. This is probably due to the fact that the shape of the TRIANGULAR$(-1,1,0)$ distribution is closer to the general shape of a N$(0,1)$ distribution than many of the other distributions studied here.

Pitman asymptotic relative efficiency is not the only viewpoint that has been developed for comparing statistical hypothesis tests. For example, Hodges and Lehmann (1970) developed the concept of *deficiency*, where expansion theory similar to what was used in this section is carried out to higher order terms. The concept of *Bahadur efficiency*, developed by Bahadur (1960a, 1960b, 1967) considers fixed alternative hypothesis values and power function values and determines the rate at which the significance levels of the two tests converge to zero. Other approaches to asymptotic relative efficiency can be found in Cochran (1952), and Anderson and Goodman (1957).

11.5 Density Estimation

In the most common forms of statistical analysis one observes a sample $X_1,\ldots,X_n$ from a distribution F, and interest lies in performing statistical inference on a characteristic of F, usually denoted by $\boldsymbol{\theta}$. Generally $\boldsymbol{\theta}$ is some function of F and is called a parameter. This type of inference is often performed using a parametric model for F, which increases the power of the inferential methods, assuming the model is true. A more general problem involves estimating F without making any parametric assumptions about the form of F. That is, we wish to compute a nonparametric estimate of F based on the sample $X_1,\ldots,X_n$.

In Section 3.7 we introduced the empirical distribution function as an estimator of F that only relies on the assumption that the sample $X_1,\ldots,X_n$ is a set of independent and identically distributed random variables from F. This estimator was shown to be pointwise consistent, pointwise unbiased, and consistent with respect to Kolmogorov distance. See Theorems 3.16 and 3.18. Overall, the empirical distribution function can be seen as a reliable estimator of a distribution function. However, in many cases, practitioners are not interested in directly estimating F, but would rather estimate the density or probability distribution associated with F, as the general shape of the population is much easier to visualize using the density of probability distribution than with the distribution function.

We first briefly consider the case where F is a discrete distribution where we would be interested in estimating the probability distribution given by $f(x) = P(X = x) = F(x) - F(x-)$, for all $x \in \mathbb{R}$, where X will be assumed to be a random variable following the distribution F. Let $\hat{F}_n$ be the empirical distribution function computed on $X_1, \ldots, X_n$. Since $\hat{F}_n$ is a step function, and therefore corresponds to a discrete distribution, an estimator of F can be derived by computing the probability distribution associated with $\hat{F}_n$. That is, f can be estimated by

$$\hat{f}_n(x) = \hat{F}_n(x) - \hat{F}_n(x-) = n^{-1} \sum_{k=1}^{n} \delta\{X_k; \{x\}\},$$

for all $x \in \mathbb{R}$. Note that $\hat{f}(x)$ is the observed proportion of points in the sample that are equal to the point x. This estimate can be shown to be pointwise unbiased and consistent. See Exercise 14.

In density estimation we assume that F has a continuous density f, and we are interested in estimating the density f based on the sample $X_1, \ldots, X_n$. In this case the empirical distribution function offers little help as $\hat{F}_n$ is a step function corresponding to a discrete distribution and hence has no density associated with it. This section explores the basic development and asymptotic properties of two common estimators of f: the *histogram* and the *kernel density estimator*. We will show how an asymptotic analysis of these estimators is important in understanding the general behavior and application of these estimates.

Let $X_1, \ldots, X_n$ be a set of independent and identically distributed random variables from a distribution F that has continuous density f. For simplicity we will assume that F is differentiable everywhere. We have already pointed out that the empirical distribution function cannot be used directly to derive an estimator of f, due to the fact that $\hat{F}_n$ is a step function. Therefore, we require an estimate of the distribution function that is not a step function, and has a density associated with it. This estimate can then be differentiated to estimate the underlying density.

The *histogram* is a density estimate that is based on using a piecewise linear estimator for F. Let $-\infty < g_1 < g_2 < \cdots < g_d < \infty$ be a fixed grid of points in $\mathbb{R}$. For the moment we will not concern ourselves with how these points are selected, but we will assume that they are selected independent of the sample $X_1, \ldots, X_n$ and that they cover the range of the observed sample in that we will assume that $g_1 < \min\{X_1, \ldots, X_n\}$ and $g_d > \max\{X_1, \ldots, X_n\}$. At each of these points in the grid, we can use the empirical distribution function to

Figure 11.7 *The estimator $\bar{F}_n(t)$ computed on the example set of data indicated by the points on the horizontal axis. The location of the grid points are indicated by the vertical grey lines.*

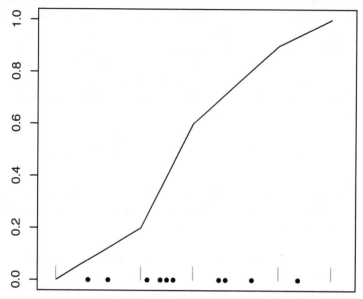

estimate the distribution function F at that point. That is

$$\hat{F}(g_i) = \hat{F}_n(g_i) = n^{-1} \sum_{k=1}^{n} \delta\{X_k; (-\infty, g_i]\},$$

for $i = 1, \ldots, d$. Under our assumptions on the grid of points outlined above we have that $\hat{F}_n(g_1) = 0$ and $\hat{F}_n(g_d) = 1$. Given this assumption, a piecewise linear estimate of F can be obtained by estimating $F(x)$ at a point $x \in [g_i, g_{i+1})$ through linear interpolation between $\hat{F}_n(g_i)$ and $\hat{F}_n(g_{i+1})$. That is, we estimate $F(x)$ with

$$\bar{F}_n(x) = \hat{F}_n(g_i) + (g_{i+1} - g_i)^{-1}(x - g_i)[\hat{F}_n(g_{i+1}) - \hat{F}_n(g_i)],$$

when $x \in [g_i, g_{i+1}]$. It can be shown that $\bar{F}_n$ is a valid distribution function under the assumptions given above. See Exercise 15. See Figure 11.7 for an example of the form of this estimator.

To estimate the density at $x \in (g_i, g_{i+1})$ we take the derivative of $\bar{F}_n(x)$ to obtain the estimator

$$
\begin{aligned}
\bar{f}_n(x) &= \frac{d}{dx}\bar{F}_n(x) \\
&= \frac{d}{dx}\left\{\hat{F}_n(g_i) + (g_{i+1} - g_i)^{-1}(x - g_i)[\hat{F}_n(g_{i+1}) - \hat{F}_n(g_i)]\right\} \\
&= (g_{i+1} - g_i)^{-1}[\hat{F}_n(g_{i+1}) - \hat{F}_n(g_i)] \\
&= n^{-1}(g_{i+1} - g_i)^{-1}\left[\sum_{k=1}^{n}\delta\{X_i; (-\infty, g_{i+1}]\} - \sum_{k=1}^{n}\delta\{X_i; (-\infty, g_i]\}\right] \\
&= (g_{i+1} - g_i)^{-1}n^{-1}\sum_{k=1}^{n}\delta\{X_i,; (g_i, g_{i+1}]\}, \quad\quad (11.50)
\end{aligned}
$$

which is the proportion of observations in the range $(g_i, g_{i+1}]$, divided by the length of the range. The estimator specified in Equation 11.50 is called a *histogram*. This estimator is also often called a *density histogram*, to differentiate it from the *frequency histogram* which is a plot of the frequency of observations within each of the ranges $(g_i, g_{i+1}]$. Note that a *frequency histogram* does not produce a valid density, and is technically not a density estimate. Note that $\bar{f}_n$ will usually not exist at the grid points $g_1, \ldots, g_d$ as $\bar{F}_n$ will usually not be differentiable at these points. In practice this makes little difference and we can either ignore these points, or can set the estimate at these points equal to one of the neighboring estimate values. The form of this estimate is a series of horizontal steps within each range (g_i, g_{i+1}). See Figure 11.8 for an example form of this estimator.

Now that we have specified the form of the histogram we must consider the placement and number of grid points $g_1, \ldots, g_d$. We would like to choose these grid points so that the histogram provides a good estimate of the underlying density, and hence we must develop a measure of discrepancy between the true density f and the estimate $\bar{f}_n$. For univariate parameters we often use the mean squared error, which is the expected square distance between the estimator and the true parameter value, as a measure of discrepancy. Estimators that are able to minimize the mean squared error are considered reasonable estimators of the parameter.

The mean squared error does not directly generalize to the case of estimating a density, unless we consider the pointwise behavior of the density estimate. That is, for the case of the histogram, we would consider the mean squared error of $\bar{f}_n(x)$ as a pointwise estimator of $f(x)$ at a fixed point $x \in \mathbb{R}$ as $\text{MSE}[\bar{f}_n(x), f(x)] = E\{[\bar{f}_n(x) - f(x)]^2\}$. To obtain an overall measure of the performance of this estimator we can then integrate the pointwise mean squared error over the real line. This results in the mean integrated squared

Figure 11.8 *The estimator $\bar{f}_n(t)$ computed on the example set of data indicated by the points on the horizontal axis. The location of the grid points are indicated by the vertical grey lines. These are the same data and grid points used in Figure 11.7.*

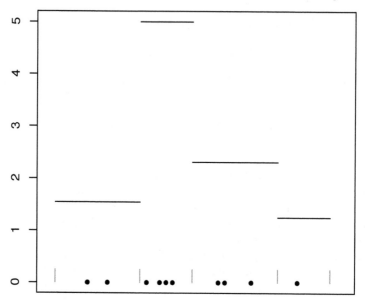

error given by

$$
\begin{aligned}
\text{MISE}(\bar{f}_n, f) &= \int_{-\infty}^{\infty} \text{MSE}[\bar{f}_n(x), f(x)]dx \\
&= \int_{-\infty}^{\infty} E\{[\bar{f}_n(x) - f(x)]^2\}dx \\
&= E\left\{\int_{-\infty}^{\infty} [\bar{f}_n(x) - f(x)]^2 dx\right\},
\end{aligned}
\tag{11.51}
$$

where we have assumed in the final equality that the interchange of the integral and the expectation is permissible. As usual, the mean squared error can be written as the sum of the square bias and the variance of the estimator. The same operation can be performed here to find that the mean integrated squared error can be written as the sum of the integrated square bias and the integrated variance. That is $\text{MISE}(\bar{f}_n, f) = \text{ISB}(\bar{f}_n, f) + \text{IV}(\bar{f}_n)$ where

$$
\text{ISB}(\bar{f}_n, f) = \int_{-\infty}^{\infty} \text{Bias}^2[\bar{f}_n(x), f(x)]dx = \int_{-\infty}^{\infty} \{E[\bar{f}_n(x)] - f(x)\}^2 dx,
$$

and

$$
\text{IV}(\bar{f}_n) = \int_{-\infty}^{\infty} E[\{(\bar{f}_n(x) - E[\bar{f}_n(x)]\}^2]dx.
$$

See Exercise 17. Using the mean integrated squared error as a measure of discrepancy between our density estimate and the true density, we will use an asymptotic analysis to specify how the grid points should be chosen. We will begin by making a few simplifying assumptions. First, we will assume that the grid spacing is even over the range of the distribution. That is, $g_{i+1} - g_1 = h$, for all $i = 1, \ldots, d$ where $h > 0$ is a value called the *bin width*. We will not concern ourselves with the placement of the grid points. We will only focus on choosing h that will minimize an asymptotic expression for the mean integrated squared error of the histogram estimator.

We will assume that f has a certain amount of smoothness. For the moment we can assume that f is continuous, but for later calculations we will have to assume that f' is continuous. In either case it should be clear that f cannot be a step function, which is the form of the histogram. Therefore, if we were to expect the histogram to provide any reasonable estimate asymptotically it is apparent that the bin width h must change with n. In fact, the bins must get smaller as n gets larger in order for the histogram to become a smooth function as n gets large. Therefore, we will assume that

$$\lim_{n \to \infty} h = 0.$$

This is, in fact, a necessary condition for the histogram estimator to be consistent. On the other hand, we must be careful that the bin width does not converge to zero at too fast a rate. If h becomes too small too fast then there will not be enough data within each of the bins to provide a consistent estimate of the true density in that region. Therefore, we will further assume that

$$\lim_{n \to \infty} nh = \infty.$$

For further information on these assumptions, see Scott (1992) and Section 2.1 of Simonoff (1996).

We will begin by considering the integrated bias of the histogram estimator. To find the bias we begin by assuming that $x \in (g_i, g_{i+1}]$ for some $i \in \{1, \ldots, d-1\}$ where $h = g_{i+1} - g_i$ and note that

$$
\begin{aligned}
E[\bar{f}_n(x)] &= E\left[(g_{i+1} - g_i)^{-1} n^{-1} \sum_{k=1}^{n} \delta\{X_i, ; (g_i, g_{i+1}]\} \right] \\
&= (nh)^{-1} \sum_{k=1}^{n} E(\delta\{X_i, ; (g_i, g_{i+1}]\}) \\
&= h^{-1} E(\delta\{X_1, ; (g_i, g_{i+1}]\}) \\
&= h^{-1} \int_{g_i}^{g_{i+1}} dF(t).
\end{aligned}
$$

Theorem 1.15 implies that $f(t) = f(x) + f'(x)(t - x) + \frac{1}{2} f''(c)(t - x)^2$ as $|t - x| \to 0$ where c is some value in the interval $(g_i, g_{i+1}]$ and we have assumed that f has two bounded and continuous derivatives. This implies that

$$\int_{g_i}^{g_{i+1}} f(t)dt = \int_{g_i}^{g_{i+1}} f(x)dt + \int_{g_i}^{g_{i+1}} f'(x)(t-x)dt + \int_{g_i}^{g_{i+1}} \tfrac{1}{2}f''(c)(t-x)^2 dt.$$

$$(11.52)$$

Now

$$\int_{g_i}^{g_{i+1}} f(x)dt = f(x)\int_{g_i}^{g_{i+1}} dt = (g_{i+1} - g_i)f(x) = hf(x),$$

and

$$
\begin{aligned}
\int_{g_i}^{g_{i+1}} f'(x)(t-x)dt &= f'(x)\int_{g_i}^{g_{i+1}} (t-x)dt \\
&= \tfrac{1}{2}f'(x)[g_{i+1}^2 - g_i^2 - 2(g_{i+1} - g_i)x] \\
&= \tfrac{1}{2}f'(x)[(g_{i+1} - g_i)(g_{i+1} + g_i) - 2hx] \\
&= \tfrac{1}{2}hf'(x)(g_{i+1} + g_i - 2x) \\
&= \tfrac{1}{2}hf'(x)[h - 2(x - g_i)].
\end{aligned}
$$

For the final term in Equation (11.52), we will let

$$\xi = \sup_{c \in (g_i, g_{i+1}]} f''(c),$$

where we assume that $\xi < \infty$. Then, assuming that $|t - x| < h$ we have that

$$\left| h^{-3} \int_{g_i}^{g_{i+1}} \tfrac{1}{2}f''(c)(t-x)^2 dt \right| \le \left| \tfrac{1}{2}\xi h^{-3} \int_{g_1}^{g_{i+1}} (t-x)^2 dt \right| \le$$

$$\tfrac{1}{2}h^{-3}\left| \xi \int_{g_i}^{g_{i+1}} h^2 dt \right| = \tfrac{1}{2}h^{-1}\xi(g_{i+1} - g_i) = \tfrac{1}{2}|\xi| < \infty,$$

for all n. Therefore, it follows that

$$\int_{g_i}^{g_{i+1}} \tfrac{1}{2}f''(c)(t-x)^2 dt = O(h^3),$$

as $n \to \infty$. Thus, for $x \in (g_i, g_{i+1}]$ we have that

$$
\begin{aligned}
E[\bar{f}_n(x)] &= h^{-1}\int_{g_i}^{g_{i+1}} f(t)dt \\
&= h^{-1}\{hf(x) + \tfrac{1}{2}hf'(x)[h - 2(x - g_i)] + O(h^3)\} \\
&= f(x) + \tfrac{1}{2}f'(x)[h - 2(x - g_i)] + O(h^2),
\end{aligned}
$$

as $n \to \infty$. Therefore, the pointwise bias is

$$\text{Bias}[\bar{f}_n(x)] = \tfrac{1}{2}f'(x)[h - 2(x - g_i)] + O(h^2),$$

as $h \to 0$, or equivalently as $n \to \infty$. It then follows that the square bias is given by

$$\text{Bias}^2[\bar{f}_n(x)] = \tfrac{1}{4}[f'(x)]^2[h - 2(x - g_i)]^2 + O(h^3).$$

See Exercise 18. The pointwise variance is developed using similar methods.

That is, for $x \in (g_i, g_{i+1}]$, we have that

$$
\begin{aligned}
V[\bar{f}_n(x)] &= (nh)^{-2} \sum_{k=1}^{n} V(\delta\{X_k; (g_i, g_{i+1}]\}) \\
&= n^{-1}h^{-2}V(\delta\{X_1; (g_i, g_{i+1}]\}) \\
&= n^{-1}h^{-2}\left[\int_{g_i}^{g_{i+1}} dF(x)\right]\left[1 - \int_{g_i}^{g_{i+1}} dF(x)\right], \quad (11.53)
\end{aligned}
$$

where we have used the fact that $\delta\{X_k,; (g_i, g_{i+1}]\}$ is a BERNOULLI random variable. To simplify this expression we begin by finding an asymptotic form for the integrals in Equation (11.53). To this end, we apply Theorem 1.15 to the density to find for $x \in (g_i, g_{i+1}]$, we have that

$$
\int_{g_i}^{g_{i+1}} dF(t) = \int_{g_i}^{g_{i+1}} f(x) + f'(x)(t-x) + \tfrac{1}{2}f''(c)(t-x)^2 dt. \quad (11.54)
$$

Using methods similar to those used above, it can be shown that

$$
\int_{g_i}^{g_{i+1}} f'(x)(t-x)dt = O(h^2),
$$

as $n \to \infty$. See Exercise 19. We have previously shown that the last integral in Equation (11.54) is $O(h^3)$ as $n \to \infty$, from which it follows from Theorem 1.19 that

$$
\int_{g_i}^{g_{i+1}} dF(t) = hf(x) + O(h^2),
$$

as $n \to \infty$. Therefore, it follows from Theorems 1.18 and 1.19 that

$$
\begin{aligned}
V[\bar{f}_n(x)] &= n^{-1}h^{-2}[hf(x) + O(h^2)][1 - hf(x) + O(h^2)] \\
&= n^{-1}[f(x) + O(h)][h^{-1} - f(x) + O(h)] \\
&= n^{-1}[h^{-1}f(x) + O(1)] \\
&= (nh)^{-1}f(x) + O(n^{-1}),
\end{aligned}
$$

as $n \to \infty$. Combining the expressions for the pointwise bias and variance yields the pointwise mean squared error, which is given by

$$
\begin{aligned}
\mathrm{MSE}[\bar{f}_n(x)] &= V[\bar{f}_n(x)] + \mathrm{Bias}^2[\bar{f}_n(x)] = \\
&(nh)^{-1}f(x) + \tfrac{1}{4}[f'(x)]^2[h - 2(x - g_i)]^2 + O(n^{-1}) + O(h^3).
\end{aligned}
$$

To obtain the mean integrated squared error, we integrate the pointwise mean squared error separately over each of the grid intervals. That is,

$$
\mathrm{MISE}[\bar{f}_n(x)] = \int_{-\infty}^{\infty} \mathrm{MSE}[\bar{f}_n(x)]dx = \sum_{k=1}^{d} \int_{g_k}^{g_{k+1}} \mathrm{MSE}[\bar{f}_n(x)]dx.
$$

For the grid interval $(g_k, g_{k+1}]$ we have that

$$\int_{g_k}^{g_{k+1}} \mathrm{MSE}[\bar{f}_n(x)]dx = \int_{g_k}^{g_{k+1}} (nh)^{-1}f(x)dx+$$

$$\frac{1}{4}\int_{g_k}^{g_{k+1}} [f'(x)]^2[h - 2(x - g_k)]^2 dx + O(n^{-1}) + O(h^3).$$

Now,

$$\int_{g_k}^{g_{k+1}} (nh)^{-1}f(x)dx = (nh)^{-1}\int_{g_k}^{g_{k+1}} dF(x),$$

and

$$\frac{1}{4}\int_{g_k}^{g_{k+1}} [f'(x)]^2[h - 2(x - g_k)]^2 dx = \frac{1}{4}[f'(\eta_k)]^2 \int_{g_k}^{g_{k+1}} [h - 2(x - g_k)]^2 dx,$$

for some $\eta_k \in (g_k, g_{k+1}]$, using Theorem A.5. Integrating the polynomial within the integral yields

$$\int_{g_k}^{g_{k+1}} [h - 2(x - g_k)]^2 dx = h^3 - 2h^3 + \frac{4}{3}h^3 = \frac{1}{3}h^3,$$

so that

$$\frac{1}{4}\int_{g_k}^{g_{k+1}} [f'(x)]^2[h - 2(x - g_k)]^2 dx = \frac{1}{12}h^3[f'(\eta_k)]^2.$$

Taking the sum over all of the grid intervals gives the total mean integrated squared error,

$$\begin{aligned}
\mathrm{MISE}(\bar{f}, f) &= \sum_{k=1}^{d}(nh)^{-1}\int_{g_k}^{g_{k+1}} dF(x) + \sum_{k=1}^{d}\frac{1}{12}h^3[f'(\eta_k)]^2 \\
&\quad + O(n^{-1}) + O(h^3) \\
&= (nh)^{-1}\int_{-\infty}^{\infty} dF(x) + \frac{1}{12}h^2\sum_{k=1}^{d} h[f'(\eta_k)]^2 + O(n^{-1}) + O(h^3) \\
&= (nh)^{-1} + \frac{1}{12}h^2\sum_{k=1}^{d} h[f'(\eta_k)]^2 + O(n^{-1}) + O(h^3).
\end{aligned}$$

To simplify this expression we utilize the definition of the Riemann integral to obtain

$$\int_{-\infty}^{\infty} [f'(t)]^2 dt = \sum_{k=1}^{d}(h[f'(\eta_k)]^2 + \varepsilon_k),$$

where $\varepsilon_k \le h[f'(g_{k+1}) - f'(g_k)] = O(h)$ as long as f has bounded variation on the interval $(g_k, g_{k+1}]$. Therefore, it follows that

$$\begin{aligned}
\mathrm{MISE}(\bar{f}_n, f) &= (nh)^{-1} + \frac{1}{12}h^2\int_{-\infty}^{\infty} [f'(t)]^2 dt + O(h^3) + O(n^{-1}) \\
&= (nh)^{-1} + \frac{1}{12}h^2 R(f') + O(h^3) + O(n^{-1}),
\end{aligned}$$

Figure 11.9 *This figure demonstrates how a histogram with a smaller bin width is better able to follow the curvature of an underlying density, resulting in a smaller asymptotic bias.*

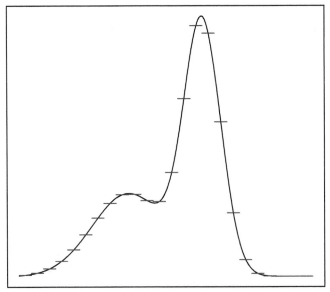

where

$$R(f') = \int_{-\infty}^{\infty} [f'(t)]^2 dt.$$

Note that the mean integrated squared error of the histogram contains the classic tradeoff between bias and variance seen with so many estimators. That is, if the bin width is chosen to be small, the bias will be small since there will be many small bins that are able to capture the curvature of f, as shown by the $\frac{1}{12}h^2 R(f')$ term. But in this case the variance, as shown by the $(nh)^{-1}$ term, will be large, due to the fact that there will be fewer observations per bin. When the bin width is chosen to be large, the bias becomes large as the curvature of f will be not able to be modeled as well by the wide steps in the histogram, while the variance will be small due to the large number of observations per bin. See Figures 11.9 and 11.10.

To find the bin width that minimizes this tradeoff we first truncate the expansion for the mean integrated squared error to obtain the asymptotic mean integrated squared error given by

$$\text{AMISE}(\bar{f}_n, f) = (nh)^{-1} + \tfrac{1}{12}h^2 R(f'). \tag{11.55}$$

Differentiating $\text{AMISE}(\bar{f}_n, f)$ with respect to h, setting the result equal to

Figure 11.10 *This figure demonstrates how a histogram with a large bin width is less able to follow the curvature of an underlying density, resulting in a larger asymptotic bias.*

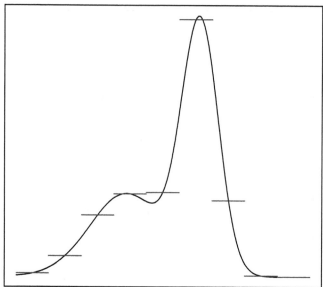

zero, and solving for h gives the asymptotically optimal bin width given by

$$h_{\text{opt}} = n^{-1/3} \left[\frac{6}{R(f')} \right]^{1/3}.$$

See Exercise 20. The resulting asymptotic mean squared error when using the optimal bandwidth is therefore $\text{AMISE}_{\text{opt}}(\bar{f}_n, f) = n^{-2/3}[\frac{9}{16}R(f')]^{1/3}$.

Note the dependence of the optimal bandwidth on the integrated square derivative of the underlying density. This dependence has two major implications. First, it is clear that densities that have smaller derivatives over their range will require a larger bandwidth, and will result in a smaller asymptotic mean integrated square error. This is due to the fact that these densities are more flat and will be easier to estimate with large bin widths in the step function in the histogram. When the derivative is large over the range of the distribution, the optimal bin width is smaller and the resulting asymptotic mean integrated squared error is larger. Such densities are more difficult to estimate using the histogram. The second implication is that the asymptotically optimal bandwidth depends on the form of the underlying density. This means that we must estimate the bandwidth from the observed data, which requires an estimate of the integral of the squared derivative of the density. Wand (1997) suggests using a kernel estimator to achieve this goal, and argues that most of the usual informal rules do not choose the bin width to be small enough. Kernel estimators can also be used to estimate the density itself, and is the second type of density estimator we discuss in this section.

One problem with the histogram is that it is always a step function and therefore does not usually reflect our notion of a continuous and smooth density. As such, there have been numerous techniques developed to provide a smooth density estimate. The estimator we will consider in this section is known as a kernel density estimator, which appears to have been first studied by Fix and Hodges (1951). See Fix and Hodges (1989) for a reprint of this paper. The first asymptotic analysis of this method, which follows along the lines of the developments in this chapter, where studied by Parzen (1962) and Rosenblatt (1956).

To motivate the kernel density estimate, return once again to the problem of estimating the distribution function F. It is clear that if we wish to have a smooth density estimate based on an estimate of F, we must find an estimator for F that itself is smooth. Indeed, we require more than just continuity in this case. As we saw with the histogram, we specified a continuous estimator for F that yielded a step function for the density estimate. Therefore, it seems if we are to improve on this idea we should not only require an estimator for F that is continuous, but it should also be differentiable everywhere. To motivate an approach to finding such an estimate we write the empirical distribution function as

$$\hat{F}_n(x) = n^{-1} \sum_{k=1}^{n} \delta\{X_i; (-\infty, x]\}$$

$$= n^{-1} \sum_{k=1}^{n} \delta\{x - X_i; [0, \infty)\}$$

$$= n^{-1} \sum_{k=1}^{n} K(x - X_i), \qquad (11.56)$$

where $K(t) = \delta\{t; [0, \infty)\}$ for all $t \in \mathbb{R}$. Note that K in this case can be taken to be a distribution function for a degenerate random that concentrates all of its mass at zero, and is therefore a step function with a single step of size one at zero. The key idea behind developing the kernel density estimator is to note that if we replace the function K in Equation (11.56) with any other valid distribution function that is centered around zero, the estimator itself remains a distribution function. See Exercise 21. That is, let K be any non-decreasing right continuous function such that

$$\lim_{t \to \infty} K(t) = 1,$$

$$\lim_{t \to -\infty} K(t) = 0,$$

and

$$\int_{-\infty}^{\infty} t dK(t) = 0.$$

Now define the kernel estimator of the distribution function F to be

$$\tilde{F}_n(x) = n^{-1} \sum_{k=1}^{n} K(x - X_i). \tag{11.57}$$

The problem with the proposed estimator in Equation (11.57) is that the properties of the estimator are a function of the variance of the distribution K. The control this property we introduce a scale parameter h to the function K. That is, define the kernel estimator of the distribution function F as

$$\tilde{F}_{n,h}(x) = n^{-1} \sum_{k=1}^{n} K\left(\frac{x - X_i}{h}\right). \tag{11.58}$$

This scale parameter is usually called a *bandwidth*. Now, if we further assume that K is smooth enough, the kernel estimator given in Equation (11.58) will be differentiable everywhere as we can obtain a continuous estimate of the density f by differentiating the kernel estimator of F to obtain

$$\tilde{f}_{n,h}(x) = (nh)^{-1} \sum_{k=1}^{n} k\left(\frac{x - X_i}{h}\right),$$

the *kernel density estimator* of f with bandwidth h, where $k(t) = K'(t)$. In this case we again have two choices to make about the estimator. We need to decide what *kernel function*, specified by k, should be used, and we also need to decide what bandwidth should be used. As with the histogram, we shall discuss these issues in terms of the mean integrated squared error of the kernel estimate in terms of h and k.

The question of what function should be used for the kernel function k is a rather complicated issue that we will not address in depth. For finite samples the choice obviously makes a difference, but a theoretical quantification of these differences is rather complicated, so researchers have turned to the question of what effect does the choice of the kernel function have asymptotically as $n \to \infty$? It turns out that for large samples the form of k does not affect the *rate* at which the optimal asymptotic mean squared error of the kernel density estimator approaches zero. Therefore, from an asymptotic viewpoint the choice matters little. See Section 3.1.2 of Simonoff (1996). Hence, for the remainder of this section we shall make the following generic assumptions about the form of the kernel function k. We will assume that k is a symmetric continuous density with zero mean finite variance. Given this assumption, we will now show how asymptotic calculations can be used to determine the asymptotically optimal bandwidth.

As with the histogram, we will use the mean integrated squared error given in Equation (11.51) as a measure of the performance of the kernel density estimator. We will assume that f is a smooth density, namely that f'' is continuous and square integrable. As with the histogram, we shall assume that the bandwidth has the properties that

$$\lim_{n\to\infty} h = 0,$$

and

$$\lim_{n\to\infty} nh = \infty.$$

We will finally assume that k is a bounded density that is symmetric and has a finite fourth moment. To simplify notation, we will assume that X is a generic random variable with distribution F. We begin by obtaining an expression for the integrated bias. The expected value of the kernel density estimator at a point $x \in \mathbb{R}$ is given by

$$
\begin{aligned}
E[\tilde{f}_{n,h}(x)] &= E\left[(nh)^{-1}\sum_{k=1}^{n} k\left(\frac{x-X_k}{h}\right)\right] \\
&= (nh)^{-1}\sum_{k=1}^{n} E\left[k\left(\frac{x-X_k}{h}\right)\right] \\
&= h^{-1}E\left[k\left(\frac{x-X}{h}\right)\right] \\
&= h^{-1}\int_{-\infty}^{\infty} k\left(\frac{x-t}{h}\right) dF(t).
\end{aligned}
$$

Now consider the change of variable $v = h^{-1}(t-x)$ so that $t = x + vh$ and $dt = hdv$ to obtain

$$E[\tilde{f}_{n,h}(x)] = \int_{-\infty}^{\infty} k(-v)f(x+vh)dv = \int_{-\infty}^{\infty} k(v)f(x+vh)dv,$$

where the second inequality follows because we have assumed that k is symmetric about the origin. Now, apply Theorem 1.15 (Taylor) to $f(x+vh)$ to find

$$f(x+vh) = f(x) + vhf'(x) + \tfrac{1}{2}(vh)^2 f''(x) + \tfrac{1}{6}(vh)^3 f'''(x) + O(h^4),$$

as $h \to 0$. Therefore, assuming that the integral of the remainder term remains $O(h^4)$ as $h \to 0$, it follows that

$$E[\tilde{f}_{n,h}(x)] = \int_{-\infty}^{\infty} k(v)f(x+vh)dv =$$

$$\int_{-\infty}^{\infty} f(x)k(v)dv + \int_{-\infty}^{\infty} vhf'(x)k(v)dv + \int_{-\infty}^{\infty} \tfrac{1}{2}(vh)^2 f''(x)k(v)dv+$$

$$\int_{-\infty}^{\infty} \tfrac{1}{6}(vh)^3 f'''(x)k(v)dv + O(h^4),$$

as $h \to 0$. Using the fact that k is a symmetric density about the origin, we have that

$$\int_{-\infty}^{\infty} f(x)k(v)dv = f(x)\int_{-\infty}^{\infty} k(v)dv = f(x),$$

$$\int_{-\infty}^{\infty} vhf'(x)k(v)dv = hf'(x)\int_{-\infty}^{\infty} vk(v)dv = 0,$$

$$\int_{-\infty}^{\infty} \tfrac{1}{2}(vh)^2 f''(x)k(v)dv = \tfrac{1}{2}h^2 f''(x)\int_{-\infty}^{\infty} v^2 k(v)dv = \tfrac{1}{2}h^2 f''(x)\sigma_k^2,$$

and

$$\int_{-\infty}^{\infty} \tfrac{1}{6}(vh)^3 f'''(x)k(v)dv = \tfrac{1}{6}h^3 f'''(x)\int_{-\infty}^{\infty} v^3 k(v)dv = 0,$$

where σ_k^2 is the variance of the kernel function k. Therefore,

$$E[\tilde{f}_{n,h}(x)] = f(x) + \tfrac{1}{2}h^2 f''(x)\sigma_k^2 + O(h^4),$$

and the pointwise bias of the kernel density estimator with bandwidth h is

$$\text{Bias}[\tilde{f}_{n,h}(x)] = \tfrac{1}{2}h^2 f''(x)\sigma_k^2 + O(h^4).$$

To compute the mean integrated squared error we require the squared bias. It can be shown that

$$\text{Bias}^2[\tilde{f}_{n,h}(x)] = [\tfrac{1}{2}h^2 f''(x)\sigma_k^2 + O(h^4)]^2 = \tfrac{1}{4}h^4 [f''(x)]^2 \sigma_k^4 + O(h^6).$$

See Exercise 22. To find the pointwise variance of the kernel density estimator we note that

$$V[\tilde{f}_{n,h}(x)] = (nh)^{-2} \sum_{k=1}^{n} V\left[k\left(\frac{x - X_i}{h}\right)\right] = n^{-1}h^{-2} V\left[k\left(\frac{x - X}{h}\right)\right].$$

The variance part of this term can be written as

$$V\left[k\left(\frac{x - X}{h}\right)\right] = \int_{-\infty}^{\infty} k^2\left(\frac{x - t}{h}\right) dF(t) - E^2\left[k\left(\frac{x - X}{h}\right)\right] =$$

$$h\int_{-\infty}^{\infty} k^2(v)f(x + vh)dv - E^2\left[k\left(\frac{x - X}{h}\right)\right].$$

Theorem 1.15 implies that $f(x + vh) = f(x) + O(h)$ as $h \to 0$ so that

$$h\int_{-\infty}^{\infty} k^2(v)f(x + vh)dv = h\int_{-\infty}^{\infty} k^2(v)[f(x) + O(h)]dv$$

$$= hf(x)\int_{-\infty}^{\infty} k^2(v)dv + O(h^2)$$

$$= hf(x)R(k) + O(h^2).$$

From previous calculations, we know that

$$E\left[k\left(\frac{x - X}{h}\right)\right] = hf(x) + O(h^4),$$

so that

$$E^2\left[k\left(\frac{x - X}{h}\right)\right] = h^2 f^2(x) + O(h^4).$$

Thus,

$$
\begin{aligned}
V[\tilde{f}_{n,h}(x)] &= n^{-1}h^{-2}[hf(x)R(k) + O(h^2) - h^2 f(x) + O(h^4)] \\
&= (nh)^{-1}f(x)R(k) + n^{-1}O(1) - n^{-1}f(x) + O(n^{-1}h^2) \\
&= (nh)^{-1}f(x)R(k) + O(n^{-1}).
\end{aligned}
$$

Therefore, the pointwise mean squared error of the kernel estimator with bandwidth h is given by

$$
\mathrm{MSE}[\tilde{f}_{n,h}(x)] = (nh)^{-1}f(x)R(k) + \tfrac{1}{4}h^4[f''(x)]^2\sigma_k^4 + O(h^6) + O(n^{-1}),
$$

as $h \to 0$ and as $n \to \infty$. Integrating we find that the mean integrated squared error is given by

$$
\begin{aligned}
\mathrm{MISE}(\tilde{f}_{n,h}, f) &= (nh)^{-1}R(k)\int_{-\infty}^{\infty} f(x)dx + \tfrac{1}{4}h^4\sigma_k^4\int_{-\infty}^{\infty}[f''(x)]^2 dx \\
&\quad + O(h^6) + O(n^{-1}) \\
&= (nh)^{-1}R(k) + \tfrac{1}{4}h^4\sigma_k^4 R(f'') + O(h^6) + O(n^{-1}),
\end{aligned}
$$

as $h \to 0$ which occurs under our assumptions when $n \to \infty$. As with the case of the histogram, we truncate the error term to get the asymptotic mean integrated squared error, given by

$$
\mathrm{AMISE}(\tilde{f}_{n,h}, f) = (nh)^{-1}R(k) + \tfrac{1}{4}h^4\sigma_k^4 R(f'').
$$

Minimizing the asymptotic mean integrated square error with respect to the bandwidth h yields the asymptotically optimal bandwidth given by

$$
h_{\mathrm{opt}} = n^{-1/5}\left[\frac{R(k)}{\sigma_k^4 R(f'')}\right]^{1/5}.
$$

See Exercise 23. If this bandwidth is known, the asymptotically optimal asymptotic mean integrated squared error using the kernel density estimate with bandwidth h_{opt} is given by

$$
\begin{aligned}
\mathrm{AMISE}_{\mathrm{opt}}(\tilde{f}_{n,h_{\mathrm{opt}}}, f) &= n^{-4/5}R(k)\left[\frac{\sigma_k^4 R(f'')}{R(k)}\right]^{1/5} + \tfrac{1}{4}n^{-4/5}\left[\frac{R(k)}{\sigma_k^4 R(f'')}\right]^{4/5} \\
&= \tfrac{5}{4}[\sigma_k R(k)]^{4/5}[R(f'')]^{1/5}n^{-4/5}. \quad (11.59)
\end{aligned}
$$

Equation (11.59) provides a significant amount of information about the kernel density estimation method. First we note that the asymptotic optimal mean integrated squared error of the kernel density estimate converges to zero at a faster rate than that of the histogram. This indicates, at least for large samples, that the kernel density estimate should provide a better estimate of the underlying density than the histogram from the viewpoint of the mean integrated squared error. We also note that $R(f'')$ plays a prominent role in both the size of the asymptotically optimal bandwidth, and the asymptotic mean integrated squared error. Recall that $R(f'')$ is a measure of the smoothness of the density f. Therefore, we see that if $R(f'')$ is large, then we have a density that is not very smooth. Such a density requires a small bandwidth and at the same time is difficult to estimate as the asymptotically optimal mean integrated squared error becomes larger. On the other hand, if $R(f'')$ is small, a larger bandwidth is required and the density is easier to estimate. It is important to note that the value of $R(f'')$ does not depend on n, and therefore cannot affect the convergence rate of the asymptotic mean squared error.

Aside from the term $R(f'')$ in Equation (11.59), we can also observe that the size of the asymptotic mean integrated squared error is controlled by $\sigma_k R(k)$, which is completely dependent upon the kernel function and is therefore completely within control of the user. That is, we are free to choose the kernel function that will minimize the asymptotic mean integrated squared error. It can be shown that the optimal form of the kernel function is given by $k(t) = \frac{3}{4}(1 - t)^2 \delta\{t; [-1, 1]\}$, which is usually called the *Epanechnikov kernel*, and for which $\sigma_k R(k) = 3/(5^{3/2})$. See Exercise 24. See Bartlett (1963), Epanechnikov (1969) and Hodges and Lehmann (1956). The relative efficiency of other kernel functions can be therefore computed as

$$\text{Relative Efficiency} = \frac{3}{\sigma_k R(k) 5^{3/2}}.$$

For example, it can be shown that the Normal kernel, taken with $\sigma_k = 1$ has a relative efficiency approximately equal to 0.9512. In fact, it can be shown that most of the standard kernel functions used in practice have efficiencies of at least 0.90. See Exercise 25. This, coupled with the fact that the choice of kernel does not have an effect on the rate at which the asymptotic mean integrated square error converges to zero, indicates that the choice of kernel is not as asymptotically important as the choice of bandwidth.

The fact that the optimal bandwidth depends on the unknown density through $R(f'')$ is perhaps not surprising, but it does lend an additional level of difficulty to the prospect of using kernel density estimators in practical situations. There have been many advances in this area; one of the promising early approaches is based on *cross validation* which estimates the risk of the kernel density estimator using a leave-one-out calculation similar to jackknife type estimators. This is the approach proposed by Rudemo (1982) and Bowman (1984). Consistency results for the cross-validation method for estimating the optimal bandwidth can be found in Burman (1985), Hall (1983b) and Stone (1984). Another promising approach is based on attempting to estimate $R(f'')$ directly based on the data. This estimate is usually based on a kernel estimator itself, which also requires a bandwidth that depends on estimating functionals of even higher order derivatives of f. This process is usually carried on for a few iterations with the bandwidth required to estimate the functional of the highest order derivative being computed for a parametric distribution, usually normal. This approach, known as the *plug-in* approach was first suggested by Woodroofe (1970) and Nadaraya (1974). Since then, a great deal of research has been done with these methods, usually making improvements on the rate at which the estimated bandwidth approaches the optimal one. The research is too vast to summarize here, but a good presentation of these methods can be found in Simonoff (1996) and Wand and Jones (1995).

11.6 The Bootstrap

In this section we will consider a very general nonparametric methodology that was first proposed by Efron (1979). This general approach, known as the *bootstrap*, provides a single tool that can be applied to many different problems including the construction of confidence sets and statistical tests, the reduction of estimator bias, and the computation of standard errors. The popularity of the bootstrap arises from its applicability across many different types of statistical problems. The same essential bootstrap technique can be applied to problems in univariate and multivariate inference, regression problems, linear models and time series. The bootstrap differs from the non-parametric techniques like those based on rank statistics in that bootstrap methods are usually approximate. This means that theory developed for the bootstrap is usually asymptotic in nature, though some finite sample results are available. For example, see Fisher and Hall (1991) and Polansky (1999). The second major difference between the bootstrap and nonparametric techniques based on ranks is that in practice the bootstrap is necessarily computationally intensive. While some classical nonparametric techniques also require a large number of computations, the bootstrap algorithm differs in that the computations are usually based on simulations. This section introduces the bootstrap methodology and presents some general asymptotic properties such as the consistency of bootstrap estimates and the correctness and accuracy of bootstrap confidence intervals.

To introduce the bootstrap methodology formally, let $\{X_n\}_{n=1}^{\infty}$ be a sequence of independent and identically distributed random variables following a distribution F. Let $\theta = T(F)$ be a functional parameter with parameter space Ω, and let $\hat{\theta}_n = T(\hat{F}_n)$ be a point estimator for θ where $\hat{F}_n$ is the empirical distribution function defined in Definition 3.5. Inference on the parameter θ typically requires knowledge of the sampling distribution of $\hat{\theta}_n$, or some function of $\hat{\theta}_n$. Let $R_n(\hat{\theta}_n, \theta)$ be the function of interest. For example, when we wish to construct a confidence interval we might take $R_n(\hat{\theta}_n, \theta) = n^{1/2}\sigma^{-1}(\hat{\theta}_n - \theta)$ where σ^2 is the asymptotic variance of $n^{1/2}\hat{\theta}_n$. The distribution function of $R_n(\hat{\theta}_n, \theta)$ is defined to be

$$H_n(t) = P[R_n(\hat{\theta}_n, \theta) \leq t | X_1, \ldots, X_n \sim F]$$

where the notation $X_1, \ldots, X_n \sim F$ is used to represent the situation where $X_1, \ldots, X_n$ is a set of independent and identically distributed random variables from F. In some cases $H_n(t)$ is known; for example when F is known or when $R_n(\hat{\theta}_n, \theta)$ is distribution free. In other cases $H_n(t)$ will be unknown and must either be approximated using asymptotic arguments or must be estimated using the observed data. For example, when F is unknown and θ is the mean, the distribution of $R_n(\hat{\theta}_n, \theta) = n^{1/2}\sigma^{-1}(\hat{\theta}_n - \theta)$ can be approximated by a normal distribution. In the cases where such an approximation is not readily available, or where such an approximation is not accurate enough, one can estimate $H_n(t)$ using the bootstrap.

The bootstrap estimate of the sampling distribution $H_n(t)$ is obtained by estimating the unknown distribution F with an estimate $\hat{F}$ computed using the observed random variables $X_1, \ldots, X_n$. The distribution of $R_n(\hat{\theta}_n, \theta)$ is then found conditional on $X_1, \ldots, X_n$, under the assumption that the distribution is $\hat{F}$ instead of F. That is, the bootstrap estimate is given by

$$\hat{H}_n(t) = P^*[R_n(\hat{\theta}_n^*, \hat{\theta}_n) \le t | X_1^*, \ldots, X_n^* \sim \hat{F}], \qquad (11.60)$$

where $P^*(A) = P^*(A | X_1, \ldots, X_n)$ is the probability measure induced by $\hat{F}$ conditional on $X_1, \ldots, X_n$, $\hat{\theta}_n = T(\hat{F}_n)$, $\hat{\theta}_n^* = T(\hat{F}_n^*)$, and $\hat{F}_n^*$ is the empirical distribution function of $X_1^*, \ldots, X_n^*$. The usual reliance of computing bootstrap estimates using computer simulations is based on the fact that the conditional distribution given in Equation (11.60) is generally difficult to compute in a closed form in most practical problems. An exception is outlined in Section 3 of Efron (1979).

In cases where the bootstrap estimate $\hat{H}_n(t)$ cannot be found analytically one could use a constructive method based on considering all possible equally likely samples from $\hat{F}_n$. Computing $R_n(\hat{\theta}_n, \theta)$ for each of these possible samples and combining the appropriate probabilities would then provide an exact tabulation of the distribution $\hat{H}_n(t)$. This process can become computationally prohibitive even when the sample size is moderate, as the number of possible samples is $\binom{2n-1}{n}$, which equals $92{,}378$ when $n = 10$. See Fisher and Hall (1991). Alternately one can simulate the process that produced the data using $\hat{F}_n$ in place of the unknown distribution F. Using this algorithm, one simulates b sets of n independent and identically distributed random variables from $\hat{F}_n$, conditional on the observed random variables $X_1, \ldots, X_n$. For each set of n random variables the function $R_n(\hat{\theta}_n, \theta)$ is computed. Denote these values as $R_1^*, \ldots, R_b^*$. Note that these observed values are independent and identically distributed random variables from the bootstrap distribution estimate $\hat{H}_n(t)$, conditional on $X_1, \ldots, X_n$. The distribution $\hat{H}_n(t)$ can then be approximated with the empirical distribution function computed on $R_1^*, \ldots, R_b^*$. That is,

$$\hat{H}_n(t) \simeq \tilde{H}_{n,b}(t) = b^{-1} \sum_{i=1}^{b} \delta\{R_i^*; (-\infty, t]\}.$$

Note that the usual asymptotic properties related to the empirical distribution hold, *conditional on* $X_1, \ldots, X_n$. For example, Theorem 3.18 (Glivenko and Cantelli) implies that $\|\hat{H}_n - \tilde{H}_{n,b}\|_\infty \xrightarrow{a.c.} 0$ as $b \to \infty$, where the convergence is relative only to the sampling from $\hat{F}_n$, conditional on $X_1, \ldots, X_n$.

The relative complexity of the bootstrap algorithm, coupled along with the fact that the bootstrap is generally considered to be most useful in the non-parametric framework where the exact form of the population distribution is unknown, means that the theoretical justification for the bootstrap has been typically based on computer-based empirical studies and asymptotic theory. A detailed study of both of these types of properties can be found in Mammen (1992). This section will focus on the consistency of several common boot-strap estimates and the asymptotic properties of several types of bootstrap confidence intervals. We begin by focusing on the consistency of the bootstrap estimate of the distribution $H_n(t)$. There are two ways in which we can view consistency in this case. In the first case we can concern ourselves with the pointwise consistency of $\hat{H}_n(t)$ as an estimator of $H_n(t)$. That is, we can conclude that $\hat{H}_n(t)$ is a pointwise consistent estimator of $H_n(t)$ if $H_n(t) \xrightarrow{p} H_n(t)$ as $n \to \infty$ for all $t \in \mathbb{R}$. Alternatively, we can define $\hat{H}_n$ to be a consistent estimator of H_n if some metric between $\hat{H}_n$ and H_n converges in probability to zero as $n \to \infty$. That is, let d be a metric on $\mathcal{F}$, the space of all distribution functions. The we will conclude that $\hat{H}_n$ is a consistent estimator of H_n if $d(\hat{H}_n, H_n) \xrightarrow{p} 0$ as $n \to \infty$. Most research on the consistency of the bootstrap uses this definition. Both of the concepts above are based on convergence in probability, and in this context these concepts are often referred to as *weak consistency*. In the case where convergence in probability is replaced by almost certain convergence, the concepts above are referred to as *strong consistency*. Because there is more than one metric used on the space $\mathcal{F}$, the consistency of $\hat{H}_n$ as an estimator of H_n is often qualified by the metric that is being used. For example, if $d(\hat{H}_n, H_n) \xrightarrow{p} 0$ as $n \to \infty$ then $\hat{H}_n$ is called a strongly d-consistent estimator of H_n. In this section we will use the supremum metric d_∞ that is based on the inner product defined in Theorem 3.17.

The smooth function model, introduced in Section 7.4, was shown to be a flexible model that contains many of the common types of smooth estimators encountered in practice. The framework of the smooth function model affords us sufficient structure to obtain the strong consistency of the bootstrap estimate of $H_n(t)$.

Theorem 11.16. *Let* $\mathbf{X}_1, \ldots, \mathbf{X}_n$ *be a set of independent and identically distributed d-dimensional random vectors from a distribution F with mean vector* $\boldsymbol{\mu}$ *where* $E(\|\mathbf{X}_n\|^2) < \infty$. *Let* $\theta = g(\boldsymbol{\mu})$ *where* $g : \mathbb{R}^d \to \mathbb{R}$ *is a continuously differentiable function at* $\boldsymbol{\mu}$ *such that*

$$\frac{\partial}{\partial x_i} g(\mathbf{x}) \bigg|_{\mathbf{x} = \boldsymbol{\mu}} \neq 0,$$

for $i = 1, \ldots, d$ *where* $\mathbf{x}' = (x_1, \ldots, x_n)$. *Define* $H_n(t) = P[n^{1/2}(\hat{\theta}_n - \theta) \leq t]$ *where* $\hat{\theta}_n = g(\bar{\mathbf{X}}_n)$. *Then* $d_\infty(\hat{H}_n, H_n) \xrightarrow{a.c.} 0$ *as* $n \to \infty$ *where* $\hat{H}_n(t) = P^*[n^{1/2}(\hat{\theta}_n^* - \hat{\theta}_n) \leq t]$.

For a proof of Theorem 11.16, see Section 3.2.1 of Shao and Tu (1995). The necessity of the condition that $E(||X_n||^2) < \infty$ has been the subject of considerable research; Babu (1984), Athreya (1987), and Knight (1989) have all supplied examples where the violation of this condition results in an inconsistent bootstrap estimate. In the special case where $d = 1$ and $g(x) = x$, the condition has been shown to be necessary and sufficient by Giné and Zinn (1989) and Hall (1990). The smoothness of the function g is also an important aspect of the consistency of the bootstrap. For functions that are not smooth functions of mean vectors there are numerous examples where the bootstrap estimate of the sampling distribution is not consistent. The following example can be found in Efron and Tibshirani (1993).

Example 11.20. Let $X_1, \ldots, X_n$ be a set of independent and identically distributed random variables from a UNIFORM$(0, \theta)$ distribution where $\theta \in \Omega = (0, \infty)$, and let $\hat{\theta}_n = X_{(n)}$, the maximum value in the sample. Suppose that $X_1^*, \ldots, X_n^*$ is a set of independent and identically distributed random variables from the empirical distribution of $X_1, \ldots, X_n$. Then $P^*(\hat{\theta}_n^* = \hat{\theta}_n) = P^*(X_{(n)}^* = X_{(n)})$. Note that $X_{(n)}^*$ will equal $X_{(n)}$ any time that $X_{(n)}$ occurs in the sample at least once. Recalling that, conditional on the observed sample $X_1, \ldots, X_n$, the empirical distribution places a mass of n^{-1} on each of the values in the sample, it follows that $X_{(n)}^* \neq X_{(n)}$ with probability $(1 - n^{-1})^n$. Therefore, the bootstrap estimates the probability $P(\hat{\theta}_n = \theta)$ with $P^*(\hat{\theta}_n^* = \hat{\theta}_n) = 1 - (1 - n^{-1})^n$, and Theorem 1.7 implies that

$$\lim_{n \to \infty} P^*(\hat{\theta}_n^* = \hat{\theta}_n) = \lim_{n \to \infty} 1 - (1 - n^{-1})^n = 1 - \exp(-1).$$

Noting that $X_{(n)}$ is a continuous random variable, it follows that the actual probability is $P(\hat{\theta}_n = \theta) = 0$ for all $n \in \mathbb{N}$. Therefore, the bootstrap estimate of the probability is not consistent. We have plotted the actual distribution of $\hat{\theta}_n$ along with a histogram of the bootstrap estimate of the sampling distribution of $\hat{\theta}_n$ for a set of simulated data from a UNIFORM$(0, 1)$ distribution in Figures 11.11 and 11.12. In Figure 11.12 the observed data are represented by the plotted points along the horizontal axis. Note that since $\hat{\theta}_n^*$ is the maximum of a sample taken from the observed sample $X_1, \ldots, X_n$, $\hat{\theta}_n^*$ will be equal to one of the observed points with probability one with respect to the probability measure P^*, which is a conditional measure on $\hat{\theta}_n$. ∎

The problem with the bootstrap estimate in Example 11.20 is that the parent population is continuous, but the empirical distribution is discrete. The bootstrap usually overcomes this problem in the case where $\hat{\theta}_n$ is a smooth function of the data because the bootstrap estimate of the sampling distribution becomes virtually continuous at a very fast rate as $n \to \infty$. This is due to the large number of atoms in the bootstrap estimate of the sampling distribution. See Appendix I of Hall (1992). However, when $\hat{\theta}_n$ is not a smooth function of the data, such as in Example 11.20, this continuity is never realized and the bootstrap estimate of the sampling distribution of $\hat{\theta}_n$ can fail to be consistent.

Figure 11.11 *The actual density of* $\hat{\theta}_n = X_{(n)}$ *for samples of size* 10 *from a* UNIFORM$(0, 1)$ *distribution.*

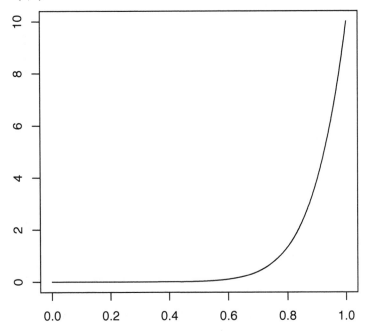

Another class of estimators where the bootstrap estimate of H_n is consistent is for the sampling distributions of certain U-statistics.

Theorem 11.17 (Bickel and Freedman). *Let* $X_1, \ldots, X_n$ *be a set of independent and identically distributed random variables from distribution F with parameter θ. Let U_n be a U-statistic of degree $m = 2$ for θ with kernel function $h(x_1, x_2)$. Suppose that $E[h^2(X_1, X_2)] < \infty$, $E[|h(X_1, X_1)|] < \infty$, and the integral*

$$\int_{-\infty}^{\infty} h(x, y) dF(y),$$

is not a constant with respect to x. The $d_\infty(\hat{H}_n, H_n) \xrightarrow{a.c.} 0$ as $n \to \infty$.

For a proof of Theorem 11.17 see Bickel and Freedman (1981) and Shi (1986). An example from Bickel and Freedman (1981) demonstrates that the condition $E[|h(X_1, X_1)|] < \infty$ cannot be weakened.

Example 11.21. Let $X_1, \ldots, X_n$ be a set of independent and identically distributed random variables from a distribution F. Consider a U-statistic U_n of degree $m = 2$ with kernel function $k(X_1, X_2) = \frac{1}{2}(X_1 - X_2)^2$, which corresponds to the unbiased version of the sample variance. Note that

$$E[h^2(X_1, X_2)] = \int_{-\infty}^{\infty} \int_{-\infty}^{\infty} \frac{1}{4}(x_1 - x_2)^4 dF(x_1) dF(x_2),$$

Figure 11.12 *An example of the bootstrap estimate of* $\hat{\theta}_n = X_{(n)}$ *for a sample of size 10 taken from a* UNIFORM$(0,1)$ *distribution. The simulated sample is represented by the points plotted along the horizontal axis.*

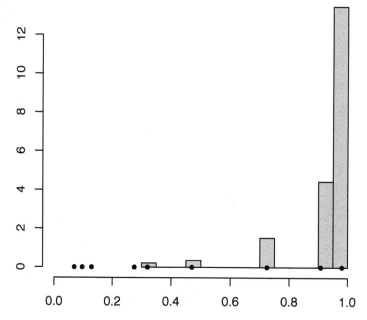

will be finite as long as F has at least four finite moments. Similarly

$$E[|h(X_1, X_1)|] = \int_{-\infty}^{\infty} \tfrac{1}{2}(x_1 - x_1)^2 dF(x_1) = 0 < \infty.$$

Finally, note that

$$\int_{-\infty}^{\infty} h(x_1, x_2) dF(x_2) = \int_{-\infty}^{\infty} \tfrac{1}{2}(x_1 - x_2)^2 dF(x_2) = \tfrac{1}{2}x_1^2 + x_1\mu_1' + \tfrac{1}{2}\mu_2',$$

is not constant with respect to x_1. Therefore, under the condition that F has a finite fourth moment, Theorem 11.17 implies that $d_\infty(\hat{H}_n, H_n) \xrightarrow{a.c.} 0$ as $n \to \infty$ where $H_n(t) = P[n^{1/2}(U_n - \mu_2) \le t]$ and $\hat{H}_n(t)$ is the bootstrap estimate given by $\hat{H}_n(t) = P^*[n^{1/2}(U_n^* - \hat{\mu}_{2,n}) \le t]$. ∎

There are many other consistency results for the bootstrap estimate of the sampling distribution that include results for L-statistics, differentiable statistical functionals, empirical processes, and quantile processes. For an overview of these results see Section 3.2 of Shao and Tu (1995).

Beyond the bootstrap estimate of the sampling distribution, we can also consider the bootstrap estimate of the variance of an estimator, or equivalently the standard error of an estimator. For example, suppose that we take $J_n(t) =$

$P(\hat{\theta}_n \leq t)$ and estimate $J_n(t)$ using the bootstrap to get $\hat{J}_n(t) = P^*(\hat{\theta}_n^* \leq t)$. The bias of $\hat{\theta}_n$ is given by

$$\text{Bias}(\hat{\theta}_n) = E(\hat{\theta}_n) - \theta = \int_{-\infty}^{\infty} t \, dJ_n(t) - \theta, \qquad (11.61)$$

which has bootstrap estimate

$$\widehat{\text{Bias}}(\hat{\theta}_n) = \hat{E}(\hat{\theta}_n) - \hat{\theta}_n = \int_{-\infty}^{\infty} t \, d\hat{J}_n(t) - \hat{\theta}_n. \qquad (11.62)$$

Similarly, the standard error of $\hat{\theta}_n$ equals

$$\sigma_n = \left\{ \int_{-\infty}^{\infty} \left[t - \int_{-\infty}^{\infty} u \, dJ_n(u) \right]^2 dJ_n(t) \right\}^{1/2},$$

which has bootstrap estimate

$$\hat{\sigma}_n = \left\{ \int_{-\infty}^{\infty} \left[t - \int_{-\infty}^{\infty} u \, d\hat{J}_n(u) \right]^2 d\hat{J}_n(t) \right\}^{1/2}.$$

Not surprisingly, the conditions that ensure the consistency of the bootstrap estimate of the variance are similar to what is required to ensure the consistency of the bootstrap estimate of the sampling distribution.

Theorem 11.18. *Let $\mathbf{X}_1, \ldots, \mathbf{X}_n$ be a set of independent and identically distributed d-dimensional random vectors with mean vector $\boldsymbol{\mu}$ and covariance matrix $\boldsymbol{\Sigma}$. Let $\theta = g(\boldsymbol{\mu})$ for a real valued function g that is differentiable in a neighborhood of $\boldsymbol{\mu}$. Define a $d \times 1$ vector $\mathbf{d}(\boldsymbol{\mu})$ to be the vector of partial derivatives of g evaluated at $\boldsymbol{\mu}$. That is, the ith element of $\mathbf{d}(\boldsymbol{\mu})$ is given by*

$$d_i(\boldsymbol{\mu}) = \left. \frac{\partial}{\partial x_i} g(\mathbf{x}) \right|_{\mathbf{x} = \boldsymbol{\mu}},$$

where $\mathbf{x}' = (x_1, \ldots, x_d)$. Suppose that $\mathbf{d}(\boldsymbol{\mu}) \neq \mathbf{0}$. If $E(||\mathbf{X}_1||^2) < \infty$ and

$$\max_{i_1, \ldots, i_n \in \{1, \ldots, n\}} \zeta_n^{-1} |\hat{\theta}_n(\mathbf{X}_{i_1}, \ldots, \mathbf{X}_{i_n}) - \hat{\theta}_n(\mathbf{X}_1, \ldots, \mathbf{X}_n)| \xrightarrow{a.c.} 0, \qquad (11.63)$$

as $n \to \infty$ where

$$\hat{\theta}_n(\mathbf{X}_{i_1}, \ldots, \mathbf{X}_{i_n}) = g\left(n^{-1} \sum_{k=1}^{n} \mathbf{X}_{i_k} \right),$$

and ζ_n is a sequence of positive real numbers such that

$$\liminf_{n \to \infty} \zeta_n > 0,$$

and $\zeta_n = O[\exp(n^\delta)]$ where $\delta \in (0, \frac{1}{2})$, then the bootstrap estimator of $\sigma_n^2 = n^{-1} \mathbf{d}'(\boldsymbol{\mu}) \boldsymbol{\Sigma} \mathbf{d}(\boldsymbol{\mu})$ is consistent. That is $\sigma_n^{-2} \hat{\sigma}_n^2 \xrightarrow{a.c.} 1$ as $n \to \infty$.

A proof of Theorem 11.18 can be found in Section 3.2.2 of Shao and Tu (1995).

The condition given in Equation (11.63) is required because there are cases where the bootstrap estimates of the variance diverges to infinity, a result that is caused by the fact that $|\hat{\theta}_n^* - \hat{\theta}_n|$ may take on some exceptionally large values. Note the role of resampling in this condition. A sample from the empirical distribution $\hat{F}_n$ consists of values from the original sample $\mathbf{X}_1, \ldots, \mathbf{X}_n$. Hence, any particular resample $\hat{\theta}_n$ may be computed on any set of values $\mathbf{X}_{i_1}, \ldots, \mathbf{X}_{i_n}$. The condition given in Theorem 11.18 ensures that none of the values will be too far away from $\hat{\theta}_n$ as $n \to \infty$. An example of a case where the bootstrap estimator in not consistent is given by Ghosh et al. (1984).

As exhibited in Example 11.20, the bootstrap can behave very differently when dealing with non-smooth statistics such as sample quantiles. However, the result given below shows that the bootstrap can still provide consistent variance estimates in such cases.

Theorem 11.19. *Let $X_1, \ldots, X_n$ be a set of independent and identically distributed random variables from a distribution F. Let $\theta = F^{-1}(p)$ and $\hat{\theta}_n = \hat{F}_n^{-1}(p)$ where $p \in (0, 1)$ is a fixed constant. Suppose that $f = F'$ exists and is positive in a neighborhood of θ. If $E(|X_1|^\varepsilon) < \infty$ for some $\varepsilon > 0$ then the bootstrap estimate of the variance $\sigma_n^2 = n^{-1}p(1-p)[f(\theta)]^{-2}$ is consistent. That is, $\hat{\sigma}_n^2 \sigma_n^{-2} \xrightarrow{a.c.} 1$ as $n \to \infty$.*

A proof of Theorem 11.19 can be found in Ghosh et al. (1984). Babu (1986) considers the same problem and is able to prove the result under slightly weaker conditions. It is worthwhile to compare the assumptions of Theorem 11.19 to those of Corollary 4.4, which are used to establish the asymptotic NORMALITY of the sample quantile. These assumptions are required in order to be able to obtain the form of the asymptotic variance of the sample quantile.

The ability of the bootstrap to provide consistent estimates of the sampling distribution and the variance of a statistic is only a small part of the theory that supports the usefulness of the bootstrap in many situations. One of the more surprising results is that under the smooth function model of Section 7.4, the bootstrap automatically performs an Edgeworth type correction. This type of result was first observed in the early work of Babu and Singh (1983, 1984, 1985), Beran (1982), Bickel and Freedman (1980), Hall (1986a, 1986b), and Singh (1981). A fully developed theory appears in the work of Hall (1988a, 1992).

The essential idea is based on the following result. Suppose that $X_1, \ldots, X_n$ is a set of independent and identically distributed d-dimensional random vectors from a distribution F with parameter θ that falls within the smooth function model. Theorem 7.11 implies that the distribution function $G_n(x) = P[n^{1/2}\sigma^{-1}(\hat{\theta}_n - \theta) \leq x]$ has asymptotic expansion

$$G_n(x) = \Phi(x) + \sum_{k=1}^{p} n^{-k/2} r_k(x)\phi(x) + o(n^{-p/2}), \qquad (11.64)$$

as $n \to \infty$, where r_k is a polynomial whose coefficients depend on the moments of F. Theorem 5.1 of Hall (1992) implies that the bootstrap estimate of $G_n(x)$, which is given by $\hat{G}_n(x) = P^*[n^{1/2}\hat{\sigma}_n^{-1}(\hat{\theta}_n^* - \hat{\theta}_n) \le x]$, has asymptotic expansion

$$\hat{G}_n(x) = \Phi(x) + \sum_{k=1}^{p} n^{-k/2}\hat{r}_k(x)\phi(x) + o_p(n^{-p/2}), \qquad (11.65)$$

as $n \to \infty$. The polynomial $\hat{r}_k$ has the same form as r_k, except that the moments of F in the coefficients of the polynomial have been replaced by the corresponding sample moments. One should also note that the error term $o(n^{-p/2})$ in Equation (11.64) has been replaced with the error term $o_p(n^{-p/2})$ in Equation (11.65), which reflects the fact that the error term in the expansion is now a random variable. A proof of this result can be found in Section 5.2.2 of Hall (1992). The same result holds for the Edgeworth expansion of the studentized distribution. That is, if $\hat{H}_n(x) = P^*[n^{1/2}(\hat{\sigma}_n^*)^{-1}(\hat{\theta}_n^* - \hat{\theta}_n) \le x]$ is the bootstrap estimate of the distribution $H_n(x) = P[n^{1/2}\hat{\sigma}_n^{-1}(\hat{\theta}_n - \theta_n) \le x]$ then

$$\hat{H}_n(x) = \Phi(x) + \sum_{k=1}^{p} n^{-k/2}\hat{v}_k(x)\phi(x) + o_p(n^{-p/2}),$$

as $n \to \infty$ where $\hat{v}_k(x)$ is the sample version of $v_k(x)$ for $k = 1, \ldots, p$. Similar results hold for the Cornish–Fisher expansions for the quantile functions of $\hat{G}_n(x)$ and $\hat{H}_n(x)$. Let $\hat{g}_\alpha = \hat{G}_n^{-1}(\alpha)$ and $\hat{h}_\alpha = \hat{H}_n^{-1}(\alpha)$ be the bootstrap estimates of the quantiles of the distributions of $\hat{G}_n$ and $\hat{H}_n$, respectively. Then Theorem 5.2 of Hall (1992) implies that

$$\hat{g}_\alpha = z_\alpha + \sum_{k=1}^{p} n^{-k/2}\hat{q}_k(z_\alpha) + o_p(n^{-p/2}), \qquad (11.66)$$

and

$$\hat{h}_\alpha = z_\alpha + \sum_{k=1}^{p} n^{-k/2}\hat{s}_k(z_\alpha) + o_p(n^{-p/2}),$$

as $n \to \infty$ where $\hat{q}_k$ and $\hat{s}_k$ are the sample versions of q_k and s_k, respectively, for all $k = 1, \ldots, p$. The effect of these results is immediate. Because $\hat{r}_k(x) = r_k(x) + O_p(n^{-1/2})$ and $\hat{v}_k(x) = v_k(x) + O_p(n^{-1/2})$, it follows from Equations (11.64) and (11.65) that

$$\hat{G}_n(x) = \Phi(x) + n^{-1/2}r_k(x)\phi(x) + o_p(n^{-p/2}) = G_n(x) + o_p(n^{-1/2}),$$

and

$$\hat{H}_n(x) = \Phi(x) + n^{-1/2}v_k(x)\phi(x) + o_p(n^{-p/2}) = H_n(x) + o_p(n^{-1/2}),$$

as $n \to \infty$. Therefore, it is clear that the bootstrap does a better job of estimating G_n and H_n than the NORMAL approximation, which would estimate both of these distributions by $\Phi(x)$, resulting in an error term that is $o_p(1)$ as

$n \to \infty$. This effect has far reaching consequences for other bootstrap methods, most notably confidence intervals. Hall (1988a) identifies six common bootstrap confidence intervals and describes their asymptotic behavior. We consider $100\alpha\%$ upper confidence limits using four of these methods.

The *percentile method*, introduced by Efron (1979), estimates the sampling distribution $J_n(x) = P(\hat{\theta}_n \le x)$ with the bootstrap estimate $\hat{J}_n(x) = P^*(\hat{\theta}_n^* \le x)$. The $100\alpha\%$ upper confidence limit is then given by $\hat{\theta}_{\text{back}}^*(\alpha) = \hat{J}_n^{-1}(\alpha)$, where we are using the notation of Hall (1988a) to identify the confidence limit. Note that

$$\hat{J}_n(x) = P^*(\hat{\theta}_n^* \le x) =$$
$$P^*[n^{1/2}\hat{\sigma}_n^{-1}(\hat{\theta}_n^* - \hat{\theta}_n) \le n^{1/2}\hat{\sigma}_n^{-1}(x - \hat{\theta}_n)] = \hat{G}_n[n^{1/2}\hat{\sigma}_n^{-1}(x - \hat{\theta}_n)].$$

Therefore, it follows that $\hat{G}_n\{n^{1/2}\hat{\sigma}_n^{-1}[\hat{J}_n(\alpha) - \hat{\theta}_n]\} = \alpha$, or equivalently that $n^{1/2}\hat{\sigma}_n^{-1}[\hat{J}_n^{-1}(\alpha) - \hat{\theta}_n] = \hat{g}_\alpha$. Hence it follows that $\hat{J}_n(\alpha) = \hat{\theta}_n + n^{-1/2}\hat{\sigma}_n\hat{g}_\alpha$ and therefore $\hat{\theta}_{\text{back}}^*(\alpha) = \hat{\theta}_n + n^{-1/2}\hat{\sigma}_n\hat{g}_\alpha$. The expansion given in Equation (11.66) implies that

$$\hat{\theta}_{\text{back}}^*(\alpha) = \hat{\theta}_n + n^{-1/2}\hat{\sigma}_n[z_\alpha + n^{-1/2}\hat{q}_1(z_\alpha) + n^{-1}\hat{q}_2(z_\alpha) + o_p(n^{-1})],$$

as $n \to \infty$. Noting that $\hat{q}_1(z_\alpha) = q_1(z_\alpha) + O_p(n^{-1/2})$ as $n \to \infty$ implies that

$$\hat{\theta}_{\text{back}}^*(\alpha) = \hat{\theta}_n + n^{-1/2}\hat{\sigma}_n z_\alpha + n^{-1}\hat{\sigma}_n q_1(z_\alpha) + O_p(n^{-3/2}),$$

as $n \to \infty$. In Section 10.3 it was shown that a correct $100\alpha\%$ upper confidence limit for θ has the expansion

$$\hat{\theta}_{\text{stud}}(\alpha) = \hat{\theta}_n - n^{-1/2}\hat{\sigma}_n z_{1-\alpha} - n^{-1}\hat{\sigma}_n s_1(z_{1-\alpha}) + O_p(n^{-1}),$$

as $n \to \infty$. Therefore $|\hat{\theta}_{\text{back}}^*(\alpha) - \hat{\theta}_{\text{stud}}(\alpha)| = O_p(n^{-1})$ as $n \to \infty$ and Defintion 10.7 implies that the backwards, or percentile, method is first-order correct. Calculations similar to those in Section 10.3 can be used to conclude that the method is first-order accurate. Hall (1988a) refers to this limit as the *backwards* limit because the upper confidence limit is based on the upper percentile of the distribution of G_n whereas the form of the correct upper confidence limit is based on the lower tail percentile of H_n. The upper confidence limit will have better performance when $q_1(z_\alpha) = 0$, which occurs when the distribution J_n is symmetric. For the case where θ is the mean of the population, this occurs when the underlying population is symmetric.

A second common upper confidence limit identified by Hall (1988a) is the *hybrid limit* given by $\hat{\theta}_{\text{hyb}}^*(\alpha) = \hat{\theta}_n - n^{-1/2}\hat{\sigma}_n\hat{g}_{1-\alpha}$, which has the expansion

$$\hat{\theta}_{\text{hyb}}^*(\alpha) = \hat{\theta}_n - n^{-1/2}\hat{\sigma}_n[z_{1-\alpha} + n^{-1/2}\hat{q}_1(z_{1-\alpha}) + n^{-1}\hat{q}_2(z_{1-\alpha}) + O_p(n^{-1})],$$

as $n \to \infty$. This confidence limit uses the lower percentile for the upper limit, which is an improvement over the percentile method. However, the percentile is still from the distribution G_n, which assumes that σ is known, rather than the distribution H_n, which takes into account the fact that σ is unknown.

The hybrid method is still first order correct and accurate, though its finite sample behavior has often been shown empirically to be superior to that of the percentile method. See Chapter 4 of Shao and Tu (1995).

The third common interval is the *studentized* bootstrap interval of Efron (1982) which has a $100\alpha\%$ upper confidence limit given by $\hat{\theta}^*_{\text{stud}} = \hat{\theta}_n - n^{-1/2}\hat{\sigma}_n\hat{h}_{1-\alpha}$. In some sense this interval is closest to mimicking the correct interval $\hat{\theta}_{\text{stud}}(\alpha)$, and the advantages of this interval become apparent when we study the asymptotic properties of the confidence limit. We can first note that $\hat{\theta}^*_{\text{stud}}$ has expansion

$$\begin{aligned}\hat{\theta}^*_{\text{stud}}(\alpha) &= \hat{\theta}_n - n^{-1/2}\hat{\sigma}_n[z_{1-\alpha} + n^{-1/2}\hat{s}_1(z_{1-\alpha}) + o_p(n^{-1/2})]\\ &= \hat{\theta}_n - n^{-1/2}\hat{\sigma}_n z_{1-\alpha} - n^{-1}\hat{\sigma}_n s_1(z_{1-\alpha}) + o_p(n^{-1}),\end{aligned}$$

as $n \to \infty$, where it follows that $\hat{\theta}^*_{\text{stud}}(\alpha)$ is second-order correct and accurate. Therefore, from the asymptotic viewpoint this interval is superior to the percentile and hybrid methods. The practical application of this confidence limit can be difficult in some applications. The two main problems are that the confidence limit can be computationally burdensome to compute, and that when n is small the confidence limit can be numerically unstable. See Polansky (2000) and Tibshirani (1988) for further details on stabilizing this method.

In an effort to fix the theoretical and practical deficiencies of the percentile method, Efron (1981, 1987) suggests computing a $100\alpha\%$ upper confidence limit of the form $\hat{\theta}^*_{\text{back}}[\beta(\alpha)]$, where $\beta(\alpha)$ is an adjusted confidence limit that is designed to reduce the bias of the upper confidence limit. The first method is called the *bias corrected* method, and is studied in Exercise 25 of Chapter 10. In this section we will explore the properties of the second method called the *bias corrected and accelerated method*. It is worthwhile to note that Efron (1981, 1987) did not develop these methods based on considerations pertaining to Edgeworth expansion theory. However, the methods can be justified using this theory. The development we present here is based on the arguments of Hall (1988a). Define a function $\hat{\beta}(\alpha) = \Phi[z_\alpha + 2\hat{m} + \hat{a}z_\alpha^2 + O(n^{-1})]$ as $n \to \infty$ where $\hat{m} = \Phi^{-1}[\hat{G}_n(0)]$ is called the *bias correction* parameter and $\hat{a} = -n^{-1/2}z_\alpha^{-2}[2\hat{r}_1(0) - \hat{r}_1(z_\alpha) - \hat{v}_1(z_\alpha)]$ is called the *acceleration constant*. Note that $\hat{\theta}^*_{\text{back}}[\hat{\beta}(\alpha)] = \hat{\theta}_n + n^{-1/2}\hat{\sigma}_n\hat{g}_{\hat{\beta}(\alpha)}$, where $\hat{g}_{\hat{\beta}(\alpha)}$ has a Cornish–Fisher expansion involving $z_{\hat{\beta}(\alpha)}$. Therefore, we begin our analysis of this method by noting that

$$z_{\hat{\beta}(\alpha)} = \Phi^{-1}\{\Phi[z_\alpha + 2\hat{m} + \hat{a}z_\alpha^2 + O(n^{-1})]\} = z_\alpha + 2\hat{m} + \hat{a}z_\alpha^2 + O(n^{-1}),$$

as $n \to \infty$. Therefore, it follows that

$$\hat{g}_{\hat{\beta}(\alpha)} = z_{\hat{\beta}(\alpha)} + n^{-1/2}\hat{q}_1[z_{\hat{\beta}(\alpha)}] + O_p(n^{-1}) =$$
$$z_\alpha + 2\hat{m} + \hat{a}z_\alpha^2 + n^{-1/2}\hat{q}_1[z_{\hat{\beta}(\alpha)}] + O_p(n^{-1}).$$

Theorem 1.13 (Taylor) implies that

$$\hat{q}_1[z_{\hat{\beta}(\alpha)}] = \hat{q}_1[z_\alpha + 2\hat{m} + \hat{a}z_\alpha^2 + O(n^{-1})]$$

$$\hat{q}_1\{z_\alpha + 2\Phi^{-1}[\hat{G}_n(0)] - n^{-1/2}[2\hat{r}_1(0) - \hat{r}_1(z_\alpha) - \hat{v}_1(z_\alpha)]z_\alpha^2 + O(n^{-1})\} =$$

$$\hat{q}_1\{z_\alpha + 2\Phi^{-1}[\hat{G}_n(0)]\} + O_p(n^{-1/2}),$$

as $n \to \infty$. Note that

$$\hat{G}_n(0) = \Phi(0) + n^{-1/2}\hat{r}_1(0)\phi(0) + O(n^{-1}) = \Phi(0) + n^{-1/2}r_1(0)\phi(0) + O_p(n^{-1}),$$

as $n \to \infty$. Using the expansion from Example 1.29, we have that

$$\Phi^{-1}[\Phi(0) + n^{-1/2}r_1(0)\phi(0) + O_p(n^{-1})] =$$

$$\Phi^{-1}[\Phi(0)] + n^{-1/2}r_1(0)\phi(0)[\phi(0)]^{-1} + O(n^{-1}) =$$

$$n^{-1/2}r_1(0) + O(n^{-1}) = O(n^{-1/2}),$$

as $n \to \infty$. Therefore,

$$\hat{q}_1\{z_\alpha + 2\Phi^{-1}[\hat{G}_n(0)]\} + O_p(n^{-1/2}) = \hat{q}_1(z_\alpha) + O_p(n^{-1/2}) =$$

$$q_1(z_\alpha) + O_p(n^{-1/2}).$$

Therefore, it follows that

$$\begin{aligned}
\hat{g}_{\hat{\beta}(\alpha)} &= z_\alpha + 2\Phi^{-1}[\hat{G}_n(0)] - n^{-1/2}[2\hat{r}_1(0) - \hat{r}_1(z_\alpha) - \hat{v}_1(z_\alpha)] \\
&\quad + n^{-1/2}q_1(z_\alpha) + O_p(n^{-1}) \\
&= z_\alpha + 2n^{-1/2}r_1(0) - 2n^{-1/2}r_1(0) + n^{-1/2}r_1(z_\alpha) \\
&\quad + n^{-1/2}v_1(z_\alpha) + n^{-1/2}q_1(z_\alpha) + O_p(n^{-1}) \\
&= z_\alpha + n^{-1/2}r_1(z_\alpha) + n^{-1/2}v_1(z_\alpha) + n^{-1/2}q_1(z_\alpha) + O_p(n^{-1}),
\end{aligned}$$

$$(11.67)$$

as $n \to \infty$. Recall that $q_1(z_\alpha) = -r_1(z_\alpha)$ and $s_1(z_\alpha) = -v_1(z_\alpha)$ so that the expansion in Equation (11.67) becomes $\hat{g}_{\hat{\beta}}(\alpha) = z_\alpha - n^{-1/2}s_1(z_\alpha) + O_p(n^{-1})$ as $n \to \infty$, and hence the upper confidence limit of the accelerated and bias corrected percentile method is

$$\begin{aligned}
\hat{\theta}^*_{\text{bca}}(\alpha) &= \hat{\theta}_n + n^{-1/2}\hat{\sigma}_n[z_\alpha - n^{-1/2}s_1(z_\alpha) + O_p(n^{-1})] \\
&= \hat{\theta}_n + n^{-1/2}\hat{\sigma}_n z_\alpha - n^{-1}\hat{\sigma}_n s_1(z_\alpha) + O_p(n^{-3/2}) \\
&= \hat{\theta}_n - n^{-1/2}\hat{\sigma}_n z_{1-\alpha} - n^{-1}\hat{\sigma}_n s_1(z_\alpha) + O_p(n^{-3/2}),
\end{aligned}$$

as $n \to \infty$, which matches $\hat{\theta}_{\text{stud}}(\alpha)$ to order $O(n^{-3/2})$, and hence it follows that the bias corrected and accelerated method is second-order accurate and correct.

As with the confidence intervals studied in Section 10.3, the asymptotic accuracy of two-sided bootstrap confidence intervals are less effected by correctness. It is also worthwhile to take note that bootstrap confidence intervals may behave quite differently outside the smooth function model. See, for example, Hall and Martin (1991). There are also many other methods for constructing bootstrap confidence intervals that are at least second-order correct. For examples, see Hall and Martin (1988), Polansky and Schucany (1997), Beran (1987), and Loh (1987).

11.7 Exercises and Experiments

11.7.1 Exercises

1. Let $\{X_n\}_{n=1}^{\infty}$ be a sequence of independent and identically distributed random variables from a distribution $F \in \mathcal{F}$ where $\mathcal{F}$ is the collection of a distributions that have a finite second moment and have a mean equal to zero. Let $\theta = V(X_1)$ and prove that θ is estimable of degree one over $\mathcal{F}$.

2. Suppose that $X_1, \ldots, X_n$ is a set of independent and identically distributed random variables from a distribution F with finite variance θ. It was shown in Example 11.4 that θ is estimable of degree two.

   a. Prove that $h(X_i, X_j) = \frac{1}{2}(X_i - X_j)^2$ is a symmetric function such that $E[h(X_i, X_j)] = \theta$ for all $i \neq j$.

   b. Using the function $h(X_i, X_j)$, form a U-statistic for θ and prove that it is equivalent to the unbiased version of the sample standard deviation.

   c. Using Theorem 11.5, find the projection of this U-statistic.

   d. Find conditions under which Theorem 11.6 (Hoeffding) applies to this statistic, and specify its weak convergence properties.

3. Let $X_1, \ldots, X_n$ be a set of independent and identically distributed random variables from a distribution F with mean θ. In Example 11.9 the U-statistic

$$U_n = 2[n(n-1)]^{-1} \sum_{i=1}^{n} \sum_{j=1}^{i} X_i X_j,$$

was considered as an unbiased estimator of θ^2.

   a. Using Theorem 11.5, find the projection of this U-statistic.

   b. Find conditions under which Theorem 11.6 (Hoeffding) applies to this statistic, and specify its weak convergence properties.

4. Let $X_1, \ldots, X_n$ be a set of independent and identically distributed random variables from a distribution F with mean θ. Suppose that we are interested in estimating $\theta(1 - \theta)$ using a U-statistic.

   a. Find a symmetric kernel function for this parameter, and develop the corresponding U-statistic.

   b. Using Theorem 11.5, find the projection of this U-statistic.

c. Find conditions under which Theorem 11.6 (Hoeffding) applies to this statistic, and specify its weak convergence properties.

5. Let $X_1, \ldots, X_n$ be a set of independent and identically distributed random variables from a continuous distribution F that is symmetric about a point θ. Let $\mathbf{R}$ be the vector of ranks of $|X_1 - \theta|, \ldots, |X_n - \theta|$ and let $\mathbf{C}$ be an $n \times 1$ vector with i^{th} element $c_i = \delta\{X_i - \theta; (0, \infty)\}$ for $i = 1, \ldots, n$. The signed rank statistic was seen in Example 11.3 to have the form $W = \mathbf{C}'\mathbf{R}$. In Example 11.12 is was further shown that

$$n^{1/2} \binom{n}{2}^{-1} [W - E(W)] \xrightarrow{d} Z_2,$$

as $n \to \infty$, where Z_2 has a $N(0, \frac{1}{3})$ distribution.

a. Using direct calculations, prove that under the null hypothesis that $\theta = 0$ it follows that $E(W) = \frac{1}{4}n(n+1)$.

b. Using direct calculations, prove that under the null hypothesis that $\theta = 0$ it follows that $V(W) = \frac{1}{24}n(n+1)(2n+1)$.

c. Prove that

$$\frac{W - \frac{1}{4}n(n+1)}{[\frac{1}{24}n(n+1)(2n+1)]^{1/2}} \xrightarrow{d} Z,$$

as $n \to \infty$ where Z has a $N(0, 1)$ distribution.

6. Let S be a linear rank statistic of the form

$$S = \sum_{i=1}^{n} c(i)a(r_i).$$

If $\mathbf{R}$ is a vector whose elements correspond to a random permutation of the integers in the set $\{1, \ldots, n\}$ then prove that $E(S) = n\bar{a}\bar{c}$ and

$$V(S) = (n-1)^{-1} \left\{ \sum_{i=1}^{n} [a(i) - \bar{a}]^2 \right\} \left\{ \sum_{j=1}^{n} [c(j) - \bar{c}]^2 \right\}$$

where

$$\bar{a} = n^{-1} \sum_{i=1}^{n} a(i),$$

and

$$\bar{c} = n^{-1} \sum_{i=1}^{n} c(i).$$

7. Consider the rank sum test statistic from Example 11.2, which is a linear rank statistic with $a(i) = i$ and $c(i) = \delta\{i; \{m+1, \ldots, n+m\}\}$ for all $i = 1, \ldots, n+m$. Under the null hypothesis that the shift parameter θ is zero, find the mean and variance of this test statistic.

8. Consider the median test statistic described in Example 11.14, which is a linear rank statistic with $a(i) = \delta\{i; \{\frac{1}{2}(m+n+1),\ldots,m+n\}\}$ and $c(i) = \delta\{i; \{m+1,\ldots,n+m\}\}$ for all $i = 1,\ldots,n+m$.

   a. Under the null hypothesis that the shift parameter θ is zero, find the mean and variance of the median test statistic.

   b. Determine if there are conditions under which the distribution of the median test statistic under the null hypothesis is symmetric.

   c. Prove that the regression constants satisfy Noether's condition.

   d. Define $a(t)$ such that $a(i) = a[(m+n+1)^{-1}i]$ for all $i = 1,\ldots,m+n$ and show that a is a square integrable function.

   e. Prove that the linear rank statistic

$$D = \sum_{i=1}^{n} a(i,n)c(i,n),$$

   converges weakly to a $N(0,1)$ distribution when it has been properly standardized.

9. Let $X_1,\ldots,X_n$ be a set of independent and identically distributed random variables from a distribution F with continuous and bounded density f, that is assumed to be symmetric about a point θ. Consider testing the null hypothesis $H_0 : \theta = \theta_0$ against the alternative hypothesis $H_1 : \theta > \theta_0$, by rejecting the null hypothesis when the test statistic

$$B = \sum_{i=1}^{n} \delta\{X_i - \theta_0; (0,\infty)\},$$

   is too large. Let α denote the desired significance level of this test. Without loss of generality assume that $\theta_0 = 0$.

   a. Show that the critical value for this test converges to $z_{1-\alpha}$ as $n \to \infty$.

   b. In the context of Theorem 11.15 show that we can take $\mu_n(\theta) = F(\theta)$ and $\sigma_n^2 = \frac{1}{4}n$.

   c. Using the result derived above, prove that the efficacy of this test is given by $2f(0)$.

10. Prove that

$$\int_{-\infty}^{\infty} f^2(x)dx$$

   equals $\frac{1}{2}\pi^{-1/2}, 1, \frac{1}{4}, \frac{1}{6}$, and $\frac{2}{3}$ for the $N(0,1)$, UNIFORM$(-\frac{1}{2},\frac{1}{2})$, LAPLACE$(0,1)$, LOGISTIC$(0,1)$, and TRIANGULAR$(-1,1,0)$ densities, respectively.

11. Prove that the square efficacy of the t-test equals $1, 12, \frac{1}{2}, 3\pi^{-2}$, and 6 for the $N(0,1)$, UNIFORM$(-\frac{1}{2},\frac{1}{2})$, LAPLACE$(0,1)$, LOGISTIC$(0,1)$, and TRIANGULAR$(-1,1,0)$ densities, respectively.

12. Prove that the square efficacy of the sign test equals $2\pi^{-1}$, 4, 1, $\frac{1}{4}$, and 4 for the $N(0,1)$, UNIFORM$(-\frac{1}{2}, \frac{1}{2})$, LAPLACE$(0,1)$, LOGISTIC$(0,1)$, and TRIANGULAR$(-1,1,0)$ densities, respectively.

13. Consider the density $f(x) = \frac{3}{20}5^{-1/2}(5-x^2)\delta\{x; (-5^{1/2}, 5^{1/2})\}$. Prove that $E_V^2 E_T^{-2} \simeq 0.864$, which is a lower bound for this asymptotic relative efficiency established by Hodges and Lehmann (1956). Comment on the importance of this lower bound is statistical applications.

14. Let $X_1, \ldots, X_n$ be a set of independent and identically distributed random variables from a discrete distribution with distribution function F and probability distribution function f. Assume that F is a step function with steps at points contained in the countable set D. Consider estimating the probability distribution function as

$$\hat{f}_n(x) = \hat{F}_n(x) - \hat{F}_n(x-) = n^{-1}\sum_{k=1}^{n}\delta\{X_k; \{x\}\},$$

for all $x \in \mathbb{R}$. Prove that $\hat{f}_n(x)$ is an unbiased and consistent estimator of $f(x)$ for each point $x \in \mathbb{R}$.

15. Let $X_1, \ldots, X_n$ be a set of independent and identically distributed random variables from a discrete distribution with distribution function F and probability distribution function f. Let $-\infty < g_1 < g_2 < \cdots < g_d < \infty$ be a fixed grid of points in $\mathbb{R}$. Assume that these points are selected independent of the sample $X_1, \ldots, X_n$ and that $g_1 < \min\{X_1, \ldots, X_n\}$ and $g_d > \max\{X_1, \ldots, X_n\}$. Consider the estimate of F given by

$$\bar{F}_n(x) = \hat{F}_n(g_i) + (x - g_i)[\hat{F}_n(g_{i+1}) - \hat{F}_n(g_i)],$$

when $x \in [g_i, g_{i+1}]$. Prove that this estimate is a valid distribution function conditional on $X_1, \ldots, X_n$.

16. Let $X_1, \ldots, X_n$ be a set of independent and identically distributed random variables from a distribution F with continuous density f. Prove that the histogram estimate with fixed grid points $-\infty < g_1 < \cdots < g_d < \infty$ such that $g_1 < \min\{X_1, \ldots, X_n\}$ and $g_d > \max\{X_1, \ldots, X_n\}$ given by

$$\bar{f}_n(x) = (g_{i+1} - g_i)^{-1}n^{-1}\sum_{k=1}^{n}\delta\{X_i, ; (g_i, g_{i+1}]\},$$

is a valid density function, conditional on $X_1, \ldots, X_n$.

17. Prove that the mean integrated squared error can be written as the sum of the integrated square bias and the integrated variance. That is, prove that $\text{MISE}(\bar{f}_n, f) = \text{ISB}(\bar{f}_n, f) + \text{IV}(\bar{f}_n)$, where

$$\text{ISB}(\bar{f}_n, f) = \int_{-\infty}^{\infty} \text{Bias}^2(\bar{f}_n(x), f(x))dx = \int_{-\infty}^{\infty}[E[\bar{f}_n(x)] - f(x)]^2 dx,$$

and

$$\text{IV}(\bar{f}_n) = \int_{-\infty}^{\infty} E\{[(\bar{f}_n(x) - E(\bar{f}_n(x))]^2\}dx.$$

18. Using the fact that the pointwise bias of the histogram is given by,

$$\text{Bias}[\bar{f}_n(x)] = \tfrac{1}{2}f'(x)[h - 2(x - g_i)] + O(h^2),$$

as $h \to 0$, prove that the square bias is given by

$$\text{Bias}^2[\bar{f}_n(x)] = \tfrac{1}{4}[f'(x)]^2[h - 2((x - g_i)]^2 + O(h^3).$$

19. Let f be a density with at least two continuous and bounded derivatives and let $g_i < g_{i+1}$ be grid points such that $h = g_{i+1} - g_i$.

$$\int_{g_i}^{g_{i+1}} f'(x)(t - x)dt = O(h^2),$$

as $n \to \infty$, where

$$\lim_{n \to \infty} h = 0.$$

20. Given that the asymptotic mean integrated squared error for the histogram with bin width h is given by

$$\text{AMISE}(\bar{f}_n, f) = (nh)^{-1} + \tfrac{1}{12}h^2 R(f'),$$

show that the value of h that minimizes this function is given by

$$h_{\text{opt}} = n^{-1/3}\left(\frac{6}{R(f')}\right)^{1/3}.$$

21. Let K be any non-decreasing right-continuous function such that

$$\lim_{t \to \infty} K(t) = 1,$$

$$\lim_{t \to -\infty} K(t) = 0,$$

and

$$\int_{-\infty}^{\infty} t\,dK(t) = 0.$$

Define the kernel estimator of the distribution function F to be

$$\tilde{F}_n(x) = n^{-1}\sum_{k=1}^{n} K(X_i - x).$$

Prove that $\tilde{F}_n$ is a valid distribution function.

22. Use the fact that the pointwise bias of the kernel density estimator with bandwidth h is given by

$$\text{Bias}[\tilde{f}_{n,h}(x)] = \tfrac{1}{2}h^2 f''(x)\sigma_k^2 + O(h^4),$$

as $h \to 0$ to prove that the square bias is given by

$$\text{Bias}^2[\tilde{f}_{n,h}(x)] = \tfrac{1}{4}h^4[f''(x)'^2\sigma_k^4 + O(h^6).$$

23. Using the fact that the asymptotic mean integrated squared error of the kernel estimator with bandwidth h is given by,

$$\text{AMISE}(\tilde{f}_{n,h}, f) = (nh)^{-1}R(k) + \tfrac{1}{4}h^4\sigma_k^4 R(f''),$$

show that the asymptotically optimal bandwidth is given by

$$h_{\text{opt}} = n^{-1/5} \left[\frac{R(k)}{\sigma_k^4 R(f'')} \right]^{1/5}.$$

24. Consider the Epanechnikov kernel given by $k(t) = \frac{3}{4}(1-t)^2 \delta\{t; [-1,1]\}$. Prove that $\sigma_k R(k) = 3/(5\sqrt{5})$.

25. Compute the efficiency of each of the kernel functions given below relative to the Epanechnikov kernel.

   a. The *Biweight* kernel function, given by $\frac{15}{16}(1-t^2)^2 \delta\{t; [-1,1]\}$.
   b. The *Triweight* kernel function given by $\frac{35}{32}(1-t^2)^3 \delta\{t; [-1,1]\}$.
   c. The *Normal* kernel function given by $\phi(t)$.
   d. The *Uniform* kernel function given by $\frac{1}{2}\delta\{t; [-1,1]\}$.

26. Let $\hat{f}_n(t)$ denote a kernel density estimator with kernel function k computed on a sample $X_1, \ldots, X_n$. Prove that,

$$E\left[\int_{-\infty}^{\infty} \hat{f}_n(t) f(t) dt \right] = h^{-1} E\left[\int_{-\infty}^{\infty} k\left(\frac{t-X}{h}\right) f(t) dt \right],$$

where the expectation on the right hand side of the equation is taken with respect to X.

27. Let $X_1, \ldots, X_n$ be a set of independent and identically distributed random variables from a distribution F with functional parameter θ. Let $\hat{\theta}_n$ be an estimator of θ. Consider estimating the sampling distribution of $\hat{\theta}_n$ given by $J_n(t) = P(\hat{\theta}_n \leq t)$ using the bootstrap resampling algorithm described in Section 11.6. In this case let $\hat{\theta}_1^*, \ldots, \hat{\theta}_b^*$ be the values of $\hat{\theta}_n$ computed on b resamples from the original sample $X_1, \ldots, X_n$.

   a. Show that the bootstrap estimate of the bias given by Equation (11.61) can be approximated by $\widehat{\text{Bias}}(\hat{\theta}_n) = \bar{\theta}_n^* - \hat{\theta}_n$, where

$$\bar{\theta}_n^* = b^{-1} \sum_{i=1}^{n} \hat{\theta}_i^*.$$

   b. Show that the standard error estimate of Equation (11.62) can be approximated by

$$\tilde{\sigma}_n = \left\{ b^{-1} \sum_{i=1}^{n} (\hat{\theta}_i^* - \bar{\theta}^*)^2 \right\}^{1/2}.$$

28. Let $X_1, \ldots, X_n$ be a set of independent and identically distributed random variables from a distribution F with mean θ with $\hat{\theta}_n = \bar{X}_n$. Consider estimating the sampling distribution of $\hat{\theta}_n$ given by $J_n(t) = P(\hat{\theta}_n \leq t)$ using the bootstrap resampling algorithm described in Section 11.6. Find closed expressions for the bootstrap estimates of the bias and standard error of $\hat{\theta}_n$. This is an example where the bootstrap estimates have closed forms and do not require approximate simulation methods to compute the estimates.

29. Let $X_1, \ldots, X_n$ be a set of independent and identically distributed random variables from a distribution F with mean θ. Let $R_n(\hat{\theta}, \theta) = n^{1/2}(\hat{\theta}_n - \theta)$ where $\hat{\theta}_n = \bar{X}_n$, the sample mean. Let $H_n(t) = P[R_n(\hat{\theta}_n, \theta) \leq t]$ with bootstrap estimate $\hat{H}_n(t) = P^*[R_n(\hat{\theta}_n^*, \hat{\theta}_n) \leq t]$. Using Theorem 11.16, under what conditions can we conclude that $d_\infty(\hat{H}_n, H_n) \xrightarrow{a.c.} 0$ as $n \to \infty$?

30. Let $X_1, \ldots, X_n$ be a set of independent and identically distributed random variables from a distribution F with mean μ. Define $\theta = g(\mu) = \mu^2$, and let $R_n(\hat{\theta}, \theta) = n^{1/2}(\hat{\theta}_n - \theta)$ where $\hat{\theta}_n = g(\bar{X}_n) = \bar{X}_n^2$. Let $H_n(t) = P[R_n(\hat{\theta}_n, \theta) \leq t]$ with bootstrap estimate $\hat{H}_n(t) = P^*[R_n(\hat{\theta}_n^*, \hat{\theta}_n) \leq t]$. Using the theory related by Theorem 11.16, under what conditions can we conclude that $d_\infty(\hat{H}_n, H_n) \xrightarrow{a.c.} 0$ as $n \to \infty$?

31. Let $\mathbf{X}_1, \ldots, \mathbf{X}_n$ be a set of two-dimensional independent and identically distributed random vectors from a distribution F with mean vector $\boldsymbol{\mu}$. Let $g(\mathbf{x}) = x_2 - x_1^2$ where $\mathbf{x}' = (x_1, x_2)$. Define $\theta = g(\boldsymbol{\mu})$ with $\hat{\theta}_n = g(\bar{\mathbf{X}}_n)$. Let $R_n(\hat{\theta}, \theta) = n^{1/2}(\hat{\theta}_n - \theta)$ and $H_n(t) = P[R_n(\hat{\theta}_n, \theta) \leq t]$ with bootstrap estimate $\hat{H}_n(t) = P^*[R_n(\hat{\theta}_n^*, \hat{\theta}_n) \leq t]$. Using Theorem 11.16, under what conditions can we conclude that $d_\infty(\hat{H}_n, H_n) \xrightarrow{a.c.} 0$ as $n \to \infty$? Explain how this result can be used to determine the conditions under which the bootstrap estimate of the sampling distribution of the sample variance is strongly consistent.

32. Use Theorem 11.16 to determine the conditions under which the bootstrap estimate of the sampling distribution of the sample correlation is strongly consistent.

33. Let $X_1, \ldots, X_n$ be a set of independent and identically distributed random variables from a distribution F with mean θ. Consider a U-statistic of degree two with kernel function $h(X_1, X_2) = X_1 X_2$ which corresponds to an unbiased estimator of θ^2. Describe under what conditions, if any, Theorem 11.17 could be applied to this U-statistic to obtain a consistent bootstrap estimator of the sampling distribution.

34. Let $X_1, \ldots, X_n$ be a set of independent and identically distributed random variables from a distribution F with mean θ with estimator $\hat{\theta}_n = \bar{X}_n$.

   a. Use Theorem 11.18 to find the simplest conditions under which the bootstrap estimate of the standard error of $\hat{\theta}_n$ is consistent.

   b. The standard error of $\hat{\theta}_n$ can also be estimated without the bootstrap by $n^{-1/2}\hat{\sigma}_n$, where $\hat{\sigma}_n$ is the sample standard deviation. Under what conditions is this estimator consistent? Compare these conditions to the simplest conditions required for the bootstrap.

35. For each method for computing bootstrap confidence intervals detailed in Section 11.6, describe an algorithm for which the method could be implemented practically using simulation for an observed set of data. Discuss the computational cost of each method.

36. Prove that the bootstrap hybrid method is first-order accurate and correct.

37. Using the notation of Chapter 7, show that the acceleration constant used in the bootstrap accelerated and bias corrected bootstrap confidence interval as the form

$$\hat{a} = \tfrac{1}{6} n^{-1} \sigma^{-3} \sum_{i=1}^{d} \sum_{j=1}^{d} \sum_{k=1}^{d} \hat{a}_i \hat{a}_j \hat{a}_k \hat{\mu}_{ijk},$$

under the smooth function model where $\hat{a}_i$ and $\hat{\mu}_{ijk}$ have the same form as a_i and μ_{ijk} except that the moments of F have been replaced with the corresponding sample moments.

38. In the context of the development of the bias corrected and accelerated bootstrap confidence interval, prove that

$$\hat{m} + (\hat{m} + z_\alpha)[1 - \hat{a}(\hat{m} + z_\alpha)]^{-1} = z_\alpha + 2\hat{m} + z_\alpha^2 \hat{a} + O_p(n^{-1}),$$

as $n \to \infty$. This form of the expression is of the same form given by Efron (1987).

11.7.2 Experiments

1. Write a program in R that simulates 1000 samples of size n from a distribution F with mean θ, where n, F and θ are specified below. For each sample compute the U-statistic for the parameter θ^2 given by

$$U_n = 2[n(n-1)]^{-1} \sum_{i=1}^{n} \sum_{j=1}^{i} X_i X_j.$$

For each case make a histogram of the 1000 values of the statistic and evaluate the results in terms of the theory developed in Exercise 3. Repeat the experiment for $n = 5, 10, 25, 50$, and 100 for each of the distributions listed below.

a. F is a $\mathrm{N}(0,1)$ distribution.
b. F is a UNIFORM$(0,1)$ distribution.
c. F is a LAPLACE$(0,1)$ distribution.
d. F is a CAUCHY$(0,1)$ distribution.
e. F is a EXPONENTIAL(1) distribution.

2. Write a program in R that simulates 1000 samples of size n from a distribution F with mean θ, where n, F and θ are specified below. For each sample compute the U-statistic for the parameter θ^2 given by

$$U_n = 2[n(n-1)]^{-1} \sum_{i=1}^{n} \sum_{j=1}^{i} X_i X_j.$$

In Exercise 3, the projection of U_n was also found. Compute the projection on each sample as well, noting that the projection may depend on population parameters that you will need to compute. For each case make a histogram of the 1000 values of each statistic and comment on how well each one does with estimating θ^2. Also construct a scatterplot of the pairs of each estimate computed on the sample in order to study how the two statistics relate to one another. If the projection is the better estimator of θ, what would prevent us from using it in practice? Repeat the experiment for $n = 5, 10, 25, 50$, and 100 for each of the distributions listed below.

a. F is a $N(0, 1)$ distribution.

b. F is a UNIFORM$(0, 1)$ distribution.

c. F is a LAPLACE$(0, 1)$ distribution.

d. F is a CAUCHY$(0, 1)$ distribution.

e. F is a EXPONENTIAL(1) distribution.

3. Write a program in R that simulates 1000 samples of size n from a distribution F with location parameter θ, where n, F and θ are specified below. For each sample use the sign test, the signed rank test, and the t-test to test the null hypothesis $H_0 : \theta \leq 0$ against the alternative hypothesis $H_1 : \theta > 0$ at the $\alpha = 0.10$ significance level. Over the 1000 samples keep track of how often the null hypothesis is rejected. Repeat the experiment for $n = 10, 25$ and 50 with $\theta = 0.0, \frac{1}{20}\sigma, \frac{2}{20}\sigma, \ldots, \sigma$, where σ is the standard deviation of F. For each sample size and distribution plot the proportion of rejections for each test against the true value of θ. Discuss how the power of these tests relate to one another in terms of the results of Table 11.2.

a. F is a $N(\theta, 1)$ distribution.

b. F is a UNIFORM$(\theta - \frac{1}{2}, \theta + \frac{1}{2})$ distribution.

c. F is a LAPLACE$(\theta, 1)$ distribution.

d. F is a LOGISTIC$(\theta, 1)$ distribution.

e. F is a TRIANGULAR$(-1 + \theta, 1 + \theta, \theta)$ distribution.

4. Write a program in R that simulates five samples of size n from a distribution F, where n and F are specified below. For each sample compute a histogram estimate of the density $f = F'$ using the bin-width bandwidth estimator of Wand (1997), which is supplied by R. See Section B.2. For each sample size and distribution plot the original density, along with the five histogram density estimates. Discuss the estimates and how well they are able to capture the characteristics of the true density. What type of characteristics appear to be difficult to estimate? Does there appear to be areas where the kernel density estimator has more bias? Are there areas where the kernel density estimator appears to have a higher variance? Repeat the experiment for $n = 50, 100, 250, 500$, and 1000.

a. F is a $N(\theta, 1)$ distribution.

b. F is a UNIFORM$(0, 1)$ distribution.

c. F is a CAUCHY$(0, 1)$ distribution.

    d. F is a TRIANGULAR$(-1, 1, 0)$ distribution.

    e. F corresponds to the mixture of a N$(0, 1)$ distribution with a N$(2, 1)$ distribution. That is, F has corresponding density $\frac{1}{2}\phi(x) + \frac{1}{2}\phi(x - 2)$.

5. Write a program in R that simulates five samples of size n from a distribution F, where n and F are specified below. For each sample compute a kernel density estimate of the density $f = F'$ using the plug-in bandwidth estimator supplied by R. See Section B.6. Plot the original density, along with the five kernel density estimates on the same set of axes, making the density estimate a different line type than the true density for clarity. Discuss the estimates and how well they are able to capture the characteristics of the true density. What type of characteristics appear to be difficult to estimate? Does there appear to be areas where the kernel density estimator has more bias? Are there areas where the kernel density estimator appears to have a higher variance? Repeat the experiment for $n = 50, 100, 250,$ 500, and 1000.

    a. F is a N$(\theta, 1)$ distribution.

    b. F is a UNIFORM$(0, 1)$ distribution.

    c. F is a CAUCHY$(0, 1)$ distribution.

    d. F is a TRIANGULAR$(-1, 1, 0)$ distribution.

    e. F corresponds to the mixture of a N$(0, 1)$ distribution with a N$(2, 1)$ distribution. That is, F has corresponding density $\frac{1}{2}\phi(x) + \frac{1}{2}\phi(x - 2)$.

6. Write a function in R that will simulate 100 samples of size n from the distributions specified below. For each sample use the nonparametric bootstrap algorithm based on b resamples to estimate the distribution function $H_n(t) = P[n^{1/2}(\hat{\theta}_n - \theta) \le t]$ for parameters and estimates specified below. For each bootstrap estimate of $H_n(t)$ compute $d_\infty(\hat{H}_n, H_n)$. The function should return the sample mean of the values of $d_\infty(\hat{H}_n, H_n)$ taken over the k simulated samples. Compare these observed means for the cases specified below, and relate the results to the consistency result given in Theorem 11.16. Also comment on the role that the population distribution and b have on the results. Treat the results as a designed experiment and use an appropriate linear model to find whether the population distribution, the parameter, b and n have a significant effect on the mean value of $d_\infty(\hat{H}_n, H_n)$. For further details on using R to compute bootstrap estimates, see Section B.4.16.

    a. N$(0, 1)$ distribution, θ is the population mean, $\hat{\theta}_n$ is the usual sample mean, $b = 10$ and 100, and $n = 5, 10, 25,$ and 50.

    b. T(2) distribution, θ is the population mean, $\hat{\theta}_n$ is the usual sample mean, $b = 10$ and 100, and $n = 5, 10, 25,$ and 50.

    c. POISSON(2) distribution, θ is the population mean, $\hat{\theta}_n$ is the usual sample mean, $b = 10$ and 100, and $n = 5, 10, 25,$ and 50.

d. N$(0, 1)$ distribution, θ is the population variance, $\hat{\theta}_n$ is the usual sample variance, $b = 10$ and 100, and $n = 5, 10, 25$, and 50.

e. T(2) distribution, θ is the population variance, $\hat{\theta}_n$ is the usual sample variance, $b = 10$ and 100, and $n = 5, 10, 25$, and 50.

f. POISSON(2) distribution, θ is the population variance, $\hat{\theta}_n$ is the usual sample variance, $b = 10$ and 100, and $n = 5, 10, 25$, and 50.

7. Write a function in R that will simulate 100 samples of size n from the distributions specified below. For each sample use the nonparametric bootstrap algorithm based on $b = 100$ resamples to estimate the standard error of the sample median. For each distribution and sample size, make a histogram of these bootstrap estimates and compare the results with the asymptotic standard error for the sample median given in Theorem 11.19. Comment on how well the bootstrap estimates the standard error in each case. For further details on using R to compute bootstrap estimates, see Section B.4.16. Repeat the experiment for $n = 5, 10, 25$, and 50.

a. F is a N$(0, 1)$ distribution.

b. F is an EXPONENTIAL(1) distribution.

c. F is a CAUCHY$(0, 1)$ distribution.

d. F is a TRIANGULAR$(-1, 1, 0)$ distribution.

e. F is a T(2) distribution.

f. F is a UNIFORM$(0, 1)$ distribution.

APPENDIX A

Useful Theorems and Notation

A.1 Sets and Set Operators

Suppose Ω is the *universal set*, that is, the set that contains all of the elements of interest. Membership of an element to a set is indicated by the $\in$ relation. Hence, $a \in A$ indicates that the element a is contained in the set A. The relation $\notin$ indicates that an element is not contained in the indicated set. A set A is a *subset* of Ω if all of the elements in A are also in Ω. This relationship will be represented with the notation $A \subset \Omega$. Hence $A \subset \Omega$ if and only if $a \in A$ implies $a \in \Omega$. If A and B are subsets of Ω then $A \subset B$ if all the elements in A are also in B, that is $A \subset B$ if and only if $a \in A$ implies $a \in B$. The *union* of two sets A and B is a set that contains all elements that are either in A or B or both sets. This set will be denoted by $A \cup B$. Therefore $A \cup B = \{\omega \in \Omega : \omega \in A \text{ or } \omega \in B\}$. The *intersection* of two sets A and B is a set that contains all elements that are common to both A and B. This set will be denoted by $A \cap B$. Therefore $A \cap B = \{\omega \in \Omega : \omega \in A \text{ and } \omega \in B\}$. The *complement* of a set A is denoted by A^c and is defined as $A^c = \{\omega \in \Omega : \omega \notin A\}$, which is the set that contains all the elements in Ω that are not in A. If $A \subset B$ then the elements of A can be subtracted from A using the operation $B \backslash A = \{\omega \in B : \omega \notin A\} = A^c \cap B$.

Unions and intersections distribute in much the same way as sums and products do.

Theorem A.1. *Let $\{A_k\}_{k=1}^{n}$ be a sequence of sets and let B be another set. Then*

$$B \cup \left(\bigcap_{k=1}^{n} A_k \right) = \bigcap_{k=1}^{n} (A_k \cup B),$$

and

$$B \cap \left(\bigcup_{k=1}^{n} A_k \right) = \bigcup_{k=1}^{n} (A_k \cap B).$$

Taking complements over intersections or unions changes the operations to unions and intersections, respectively. These results are usually called De Morgan's Laws.

Theorem A.2 (De Morgan). *Let $\{A_k\}_{k=1}^n$ be a sequence of sets. Then*

$$\left(\bigcap_{k=1}^n A_k\right)^c = \bigcup_{k=1}^n A_k^c,$$

and

$$\left(\bigcup_{k=1}^n A_k\right)^c = \bigcap_{k=1}^n A_k^c.$$

The number systems have their usual notation. That is, $\mathbb{N}$ will denote that natural numbers, $\mathbb{Z}$ will denote the integers, and $\mathbb{R}$ will denote the real numbers.

A.2 Point-Set Topology

Some results in this book rely on some fundamental concepts from metric spaces and point-set topology. A detailed review of this subject can be found in Binmore (1981). Consider a space Ω and a metric δ. Such a pairing is known as a *metric space*.

Definition A.1. *A metric space consists of a set Ω and a function $\rho : \Omega \times \Omega \to \mathbb{R}$, where ρ satisfies*

1. *$\rho(x, y) \geq 0$ for all $x \in \Omega$ and $y \in \Omega$.*

2. *$\rho(x, y) = 0$ if and only if $x = y$.*

3. *$\rho(x, y) = \rho(y, x)$ for all $x \in \Omega$ and $y \in \Omega$.*

4. *$\rho(x, z) \leq \rho(x, y) + \rho(y, z)$ for all $x \in \Omega$, $y \in \Omega$, and $z \in \Omega$.*

Let $\omega \in \Omega$ and $A \subset \Omega$, then we define the distance from the point ω to the set A as

$$\delta(\omega, A) = \inf_{a \in A} d(\omega, a).$$

It follows that for any non-empty set A, there exists at least one point in Ω such that $d(\omega, A) = 0$, noting that the fact that $d(\omega, A) = 0$ does not necessarily imply that $\omega \in A$. This allows us to define the *boundary* of a set.

Definition A.2. *Suppose that $A \subset \Omega$. A boundary point of A is a point $\omega \in \Omega$ such that $d(\omega, A) = 0$ and $d(\omega, A^c) = 0$. The set of all boundary points of a set A is denoted by ∂A.*

Note that $\partial A = \partial A^c$ and that $\partial \emptyset = \emptyset$. The concept of boundary points now makes it possible to define *open* and *closed* sets.

Definition A.3. *A set $A \subset \Omega$ is open if $\partial A \subset A^c$. A set $A \subset \Omega$ is closed if $\partial A \subset A$.*

It follows that A is closed if and only if $\delta(\omega, A) = 0$ implies that $\omega \in A$. It also follows that $\emptyset$ and Ω are open, the union of the collection of open sets is open, and that the intersection of any finite collection of open sets is open. On the other hand, $\emptyset$ and Ω are also closed, the intersection of any collection of closed sets is closed, and the finite union of any set of closed sets is closed.

Definition A.4. *Let A be a subset of Ω in a metric space. The interior of A is defined as $A° = A \setminus \partial A$. The closure of A is defined as $A^- = A \cup \partial A$.*

It follows that for any subsets A and B of Ω that $\omega \in A^-$ if and only if $\delta(\omega, A) = 0$, if $A \subset B$ then $A^- \subset B^-$, and that A^- is the smallest closed set containing A. For properties about interior sets, one needs only to note that $[(A^c)^-]^c = A°$.

A.3 Results from Calculus

The results listed below are results from basic calculus referred to in this book. These results, along with their proofs, can be found in Apostol (1967).

Theorem A.3. *Let f be an integrable function on the interval $[a, x]$ for each $x \in [0, b]$. Let $c \in [a, b]$ and define*

$$F(x) = \int_c^x f(t)dt$$

for $x \in [a, b]$. Then $F'(x)$ exists at each point $x \in (a, b)$ where f is continuous and $F'(x) = f(x)$.

Theorem A.4. *Suppose that f and g are integrable functions with at least one derivative on the interval $[a, b]$. Then*

$$\int_a^b f(x)g'(x)dx = f(b)g(b) - f(a)g(a) - \int_a^b f'(x)g(x)dx.$$

Theorem A.5. *Suppose that f and w are continuous functions on the interval $[a, b]$. If w does not change sign on $[a, b]$ then*

$$\int_a^b w(x)f(x)dx = f(\xi) \int_a^b w(x)dx,$$

for some $\xi \in [a, b]$.

Theorem A.6. *Suppose that f is any integrable function and $R \subset \mathbb{R}$. Then*

$$\left| \int_R f(t) \right| \le \int_R |f(t)|dt.$$

Theorem A.7. *If both f and g are integrable on the real interval $[a, b]$ and $f(x) \le g(x)$ for every $x \in [a, b]$ then*

$$\int_a^b f(x)dx \le \int_a^b g(x)dx.$$

It is helpful to note that the results from integral calculus transfer to expectations as well. For example, if X is a random variable with distribution F, and f and g are Borel functions such that $f(X(\omega)) \leq g(X(\omega))$ for all $\omega \in \Omega$, then

$$E[g(X)] = \int g(X(\omega)) dF(\omega) \leq \int f(X(\omega)) dF(\omega) = E[f(X)]. \qquad (A.1)$$

In many cases when we are dealing with random variables, we may use results that have slightly weaker conditions than that from classical calculus. For example, the result of Equation (A.1) remains true under the assumption that $P[f(X(\omega)) \leq g(X(\omega))] = 1$.

A.4 Results from Complex Analysis

While complex analysis does not usually play a large role in the theory of probability and statistics, many of the arguments in this book are based on characteristic functions, which require a basic knowledge of complex analysis. Complete reviews of complex analysis can be found in Ahlfors (1979) and Conway (1975). In this section x will denote a complex number of the form $x_1 + ix_2 \in \mathbb{C}$ where $i = (-1)^{1/2}$.

Definition A.5. *The absolute value or modulus of $x \in \mathbb{C}$ is $|x| = [x_1^2 + x_2^2]^{1/2}$.*

Definition A.6 (Euler). *The complex value $\exp(iy)$, where $y \in \mathbb{R}$, can be written as $\cos(y) + i \sin(y)$.*

Theorem A.8. *Suppose that $x \in \mathbb{C}$ such that $|x| \leq \frac{1}{2}$. Then $|\log(1-x)+x| \leq |x^2|$.*

Theorem A.9. *Suppose that $x \in \mathbb{C}$ and $y \in \mathbb{C}$, then $|\exp(x) - 1 - y| \leq (|x - y| + \frac{1}{2}|y|^2) \exp(\gamma)$ where $\gamma \geq |x|$ and $\gamma \geq |y|$.*

The following result is useful for obtaining bounds involving characteristic functions.

Theorem A.10. *For $x \in \mathbb{C}$ and $y \in \mathbb{C}$ we have that $|x^n - y^n| \leq n|x-y|z^{n-1}$ if $|x| \leq z$ and $|y| \leq z$, where $z \in \mathbb{R}$.*

Theorem A.11. *Let $n \in \mathbb{N}$ and $y \in \mathbb{R}$. Then*

$$\left| \exp(iy) - \sum_{k=0}^{n} \frac{(iy)^k}{k!} \right| \leq \frac{2|y|^n}{n!},$$

and

$$\left| \exp(iy) - \sum_{k=0}^{n} \frac{(iy)^k}{k!} \right| \leq \frac{|y|^{(n+1)}}{(n+1)!}.$$

A proof of Theorem A.11 can be found in the Appendix of Gut (2005) or in Section 2.3 of Kolassa (2006).

The integration of complex functions is simplified in this book because we are always integrating with respect to the real line, and not the complex plane.

Theorem A.12. *Let f be an integrable function that maps the real line to the complex plane. Then*

$$\left| \int_{-\infty}^{\infty} f(x)dx \right| \leq \left| \int_{-\infty}^{\infty} |f(x)|dx. \right.$$

A.5 Probability and Expectation

Theorem A.13. *Let $\{B_n\}_{n=1}^{k}$ be a partition of a sample space Ω and let A be any other event in Ω. Then*

$$P(A) = \sum_{n=1}^{k} P(A|B_n)P(B_n).$$

Theorem A.14. *Let X be a random variable such that $P(X = 0) = 1$. Then $E(X) = 0$.*

Theorem A.15. *Let X be a random variable such that $P(X \geq 0) = 1$. If $E(X) = 0$ then $P(X = 0) = 1$.*

Theorem A.16. *Let X and Y be random variables such that $X \leq Y$ with probability one. Then $E(X) \leq E(Y)$.*

Theorem A.17. *Let X and Y be any two random variables. Then $E(X) = E[E(X|Y)]$.*

A.6 Inequalities

Theorem A.18. *Let x and y be real numbers, then $|x + y| \leq |x| + |y|$.*

Theorem A.19. *Let x and y be positive real numbers. Then $|x + y| \leq 2\max\{x, y\}$.*

The following inequalities can be considered an extension of the Theorem A.18 to powers of sums. The results can be proven using the properties of convex functions.

Theorem A.20. *Suppose that x, y and r are positive real numbers. Then*

1. $(x + y)^r \leq 2^r (x^r + y^r)$ *when $r > 0$.*
2. $(x + y)^r \leq x^r + y^r$ *when $0 < r \leq 1$.*
3. $(x + y)^r \leq 2^{r-1}(x^r + y^r)$ *when $r \geq 1$.*

A proof of Theorem A.20 can be found in Section A.5 of Gut (2005).

Theorem A.21. *For any $x \in \mathbb{R}$, $\exp(x) \geq x + 1$ and for $x > 0$ $\exp(-x) - 1 + x \leq \frac{1}{2}x^2$.*

A proof of Theorem A.21 can be found in Section A.1 of Gut (2005).

A.7 Miscellaneous Mathematical Results

Theorem A.22. *Let x and y be real numbers and $n \in \mathbb{N}$. Then*

$$(x+y)^n = \sum_{i=0}^{n} \binom{n}{i} x^i y^{n-i} = \sum_{i=0}^{n} \frac{n!}{(n-i)!i!} x^i y^{n-i}.$$

Theorem A.23. *Suppose that x is a real value such that $|x| < 1$. Then*

$$\sum_{k=0}^{\infty} x^k = (1+x)^{-1}.$$

Theorem A.24. *If a and b are positive real numbers, then $x^b \exp(-ax) \to 0$ as $x \to \infty$.*

A.8 Discrete Distributions

A.8.1 The Bernoulli Distribution

A random variable X has a BERNOULLI(p) distribution if the probability distribution function of X is given by

$$f(x) = \begin{cases} p^x (1-p)^{1-x} & \text{for } x \in \{0,1\} \\ 0 & \text{otherwise,} \end{cases}$$

where $p \in (0,1)$. The expectation and variance of X are p and $p(1-p)$, respectively. The moment generating function of X is $m(t) = 1 - p + p\exp(t)$ and the characteristic function of X is $\psi(t) = 1 - p + p\exp(it)$.

A.8.2 The Binomial Distribution

A random variable X has a BINOMIAL(n,p) distribution if the probability distribution function of X is given by

$$f(x) = \begin{cases} \binom{n}{x} p^x (1-p)^{n-x} & \text{for } x \in \{0,1,\dots,n\} \\ 0 & \text{otherwise,} \end{cases}$$

where $n \in \mathbb{N}$ and $p \in (0,1)$. The expectation and variance of X are np and $np(1-p)$, respectively. The moment generating function of X is $m(t) = [1-p+p\exp(t)]^n$ and the characteristic function of X is $\psi(t) = [1-p+p\exp(it)]^n$. A BERNOULLI$(p)$ random variable is a special case of a BINOMIAL(n,p) random variable with $n = 1$.

A.8.3 The Geometric Distribution

A random variable X has a GEOMETRIC(θ) distribution if the probability distribution function of X is given by

$$f(x) = \begin{cases} \theta(1-\theta)^{x-1} & x \in \mathbb{N} \\ 0 & \text{elsewhere.} \end{cases}$$

The expectation and variance of X are θ^{-1} and $\theta^{-2}(1-\theta)$, respectively. The moment generating function of X is $m(t) = [1 - (1-\theta)\exp(t)]\theta \exp(t)$ and the characteristic function of X is $\psi(t) = [1 - (1-\theta)\exp(it)]\theta \exp(it)$.

A.8.4 The Multinomial Distribution

A d-dimensional random vector $\mathbf{X}$ has a MULTINOMIAL($n, d, \mathbf{p}$) distribution if the joint probability distribution function of $\mathbf{X}$ is given by

$$f(\mathbf{x}) = \begin{cases} n! \prod_{k=1}^{n}(n_k!)^{-1}p_k^{n_k} & \text{for } \sum_{k=1}^{n} n_k = n \text{ and} \\ & n_k > 0 \text{ for } k \in \{1,\ldots,d\}, \\ 0 & \text{otherwise.} \end{cases}$$

The mean vector of $\mathbf{X}$ is $n\mathbf{p}$ and the covariance matrix of $\mathbf{X}$ has $(i,j)^{\text{th}}$ element $\Sigma_{ij} = np_i(\delta_{ij} - p_j)$, for $i = 1,\ldots,d$ and $j = 1,\ldots,d$.

A.8.5 The Poisson Distribution

A random variable X has a POISSON(λ) distribution if the probability distribution function of X is given by

$$f(x) = \begin{cases} \frac{\lambda^x \exp(-\lambda)}{x!} & \text{for } x \in \{0, 1, \ldots\} \\ 0 & \text{otherwise,} \end{cases}$$

where λ, which is called the *rate*, is a positive real number. The expectation and variance of X are λ. The moment generating function of X is $m(t) = \exp\{\lambda[\exp(t) - 1]\}$, the characteristic function of X is $\psi(t) = \exp\{\lambda[\exp(it) - 1]\}$, and the cumulant generating function of X is

$$c(t) = \lambda[\exp(t) - 1] = \lambda \sum_{k=1}^{\infty} \frac{t^k}{k!}.$$

A.8.6 The (Discrete) Uniform Distribution

A random variable X has a UNIFORM$\{1, 2, \ldots, u\}$ distribution if the probability distribution function of X is given by

$$f(x) = \begin{cases} u^{-1} & \text{for } x \in \{1, 2, \ldots, u\}, \\ 0 & \text{otherwise,} \end{cases}$$

where u is a positive integer. The expectation and variance of X are $\frac{1}{2}(u+1)$ and $\frac{1}{12}(u+1)(u-1)$. The moment generating function of X is

$$m(t) = \frac{\exp(t)[1 - \exp(ut)]}{u[1 - \exp(t)]},$$

and the characteristic function of X is

$$\psi(t) = \frac{\exp(it)[1 - \exp(uit)]}{u[1 - \exp(it)]}.$$

A.9 Continuous Distributions

A.9.1 The Beta Distribution

A random variable X has an BETA(α, β) distribution if the density function of X is given by

$$f(x) = \begin{cases} \frac{1}{B(\alpha,\beta)} x^{\alpha-1}(1-x)^{\beta-1} & \text{for } x \in (0, 1) \\ 0 & \text{elsewhere,} \end{cases}$$

where α and β are positive real numbers and $B(\alpha, \beta)$ is the beta function given by

$$B(\alpha, \beta) = \frac{\Gamma(\alpha)\Gamma(\beta)}{\Gamma(\alpha+\beta)} = \int_0^1 x^{\alpha-1}(1-x)^{\beta-1}dx.$$

The expectation and variance of X are $\alpha/(\alpha+\beta)$ and

$$\frac{\alpha\beta}{(\alpha+\beta)^2(\alpha+\beta+1)},$$

respectively.

A.9.2 The Cauchy Distribution

A random variable X has a CAUCHY(α, β) distribution if the density function of X is given by

$$f(x) = \left\{ \pi\beta \left[1 + \left(\frac{x - \alpha}{\beta} \right)^2 \right] \right\}^{-1},$$

for all $x \in \mathbb{R}$. The moments and cumulants of X do not exist. The moment generating function of X does not exist. The characteristic function of X is $\exp(it\alpha - |t|\beta)$.

A.9.3 The Chi-Squared Distribution

A random variable X has an CHISQUARED(ν) distribution if the density function of X is given by

$$f(x) = \frac{x^{(\nu-2)/2} \exp(-\frac{1}{2}x)}{2^{\nu/2} \Gamma(\frac{1}{2}\nu)},$$

where $x > 0$ and ν is a positive integer. The mean of X is ν and the variance of X is 2ν. The moment generating function of X is $m(t) = (1 - 2t)^{-\nu/2}$, the characteristic function of X is $\psi(t) = (1 - 2it)^{-\nu/2}$, and the cumulant generating function is $c(t) = -\frac{1}{2}\nu \log(1 - 2t)$.

A.9.4 The Exponential Distribution

A random variable X has an EXPONENTIAL(β) distribution if the density function of X is given by

$$f(x) = \begin{cases} \beta^{-1} \exp(x/\beta) & \text{for } x > 0 \\ 0 & \text{otherwise,} \end{cases}$$

where $\beta > 0$. The expectation and variance of X are β and β^2, respectively. The moment generating function of X is $m(t) = (1 - t\beta)^{-1}$, the characteristic function of X is $\psi(t) = (1 - it\beta)^{-1}$, and the cumulant generating function is $c(t) = -\log(1 - t\beta)$.

A.9.5 The Gamma Distribution

A random variable X has a GAMMA(α, β) distribution if the density function of X is given by

$$f(x) = \begin{cases} \frac{1}{\Gamma(\alpha)\beta^\alpha} x^{\alpha-1} \exp(-x/\beta) & \text{for } x > 0 \\ 0 & \text{otherwise} \end{cases}$$

where $\alpha > 0$ and $\beta > 0$. The expectation and variance of X are $\alpha\beta$ and $\alpha\beta^2$, respectively. The moment generating function of X is $m(t) = (1 - t\beta)^{-\alpha}$ and the characteristic function of X is $\psi(t) = (1 - it\beta)^{-\alpha}$.

A.9.6 The LaPlace Distribution

A random variable X has a $\mathrm{LAPLACE}(\alpha, \beta)$ distribution if the density function of X is given by

$$f(x) = \begin{cases} \frac{1}{2} \exp\left(\frac{x-\alpha}{\beta}\right), & x < \alpha \\ 1 - \frac{1}{2} \exp\left[-\left(\frac{x-\alpha}{\beta}\right)\right], & x \geq \alpha. \end{cases}$$

The expectation and the variance of X are α and $2\beta^2$, respectively. The moment generating function of X is $m(t) = (1 - t^2\beta^2)^{-1} \exp(t\alpha)$ and the characteristic function of X is $\psi(t) = (1 + t^2\beta^2)^{-1} \exp(it\alpha)$.

A.9.7 The Logistic Distribution

A random variable X has an $\mathrm{LOGISTIC}(\mu, \sigma)$ distribution if the density function of X is given by

$$f(x) = \sigma^{-1}\{1 + \exp[-\sigma^{-1}(x-\mu)]\}^{-2} \exp[-\sigma^{-1}(x-\mu)],$$

for all $x \in \mathbb{R}$. The expectation and the variance of X are μ and $\frac{1}{3}\pi^2\sigma^2$, respectively.

A.9.8 The Lognormal Distribution

A random variable X has an $\mathrm{LOGNORMAL}(\mu, \sigma)$ distribution if the density function of X is given by

$$f(x) = \begin{cases} \frac{1}{x\sigma(2\pi)^{1/2}} \exp\left\{-\frac{[\log(x)-\mu]^2}{2\sigma^2}\right\} & \text{for } x > 0 \\ 0 & \text{elsewhere}, \end{cases}$$

where μ is a real number and σ is a positive real number. The expectation and variance of X are $\exp(\mu + \frac{1}{2}\sigma^2)$ and $\exp(2\mu + \sigma^2)[\exp(\sigma^2) - 1]$, respectively.

A.9.9 The Multivariate Normal Distribution

A d-dimensional random vector $\mathbf{X}$ has a $\mathbf{N}(d, \boldsymbol{\mu}, \boldsymbol{\Sigma})$ distribution if the density function of $\mathbf{X}$ is given by

$$f(\mathbf{x}) = (2\pi)^{-d/2} |\boldsymbol{\Sigma}|^{-1/2} \exp[-\tfrac{1}{2}(\mathbf{x} - \boldsymbol{\mu})'\boldsymbol{\Sigma}^{-1}(\mathbf{x} - \boldsymbol{\mu})]$$

where $\boldsymbol{\mu}$ is a d-dimensional real vector and $\boldsymbol{\Sigma}$ is a $d \times d$ covariance matrix. The expectation and covariance of $\mathbf{X}$ are $\boldsymbol{\mu}$ and $\boldsymbol{\Sigma}$, respectively. The moment generating function of $\mathbf{X}$ is $\phi(\mathbf{t}) = \exp(\boldsymbol{\mu}'\mathbf{t} + \frac{1}{2}\mathbf{t}'\boldsymbol{\Sigma}\mathbf{t})$. The characteristic function of $\mathbf{X}$ is $\psi(\mathbf{t}) = \exp(i\boldsymbol{\mu}'\mathbf{t} + \frac{1}{2}\mathbf{t}'\boldsymbol{\Sigma}\mathbf{t})$.

A.9.10 The Non-Central Chi-Squared Distribution

A random variable X has an CHISQUARED(ν, δ) distribution if the density function of X us given by

$$f(x) = \begin{cases} 2^{-v/2} \exp[-\frac{1}{2}(x + \delta)] \sum_{k=0}^{\infty} \frac{x^{\nu/2+k-1}\delta^k}{\Gamma(\frac{\nu}{2}+k)2^{2k}k!} & \text{for } x > 0, \\ 0 & \text{elsewhere.} \end{cases}$$

The expectation and variance of X are $\nu + \delta$ and $2(\nu + 2\delta)$, respectively. When $\delta = 0$ the distribution is equivalent to a CHISQUARED(ν) distribution.

A.9.11 The Normal Distribution

A random variable X has a N(μ, σ) distribution if the density function of X us given by

$$f(x) = (2\pi\sigma^2)^{-1/2} \exp[-(x - \mu)^2/2\sigma] \text{ for } x \in \mathbb{R},$$

where $\mu \in \mathbb{R}$ and $\sigma > 0$. The expectation and variance of X are μ and σ^2, respectively. The moment generating function of X is $m(t) = \exp(t\mu + \frac{1}{2}\sigma^2 t^2)$ and the characteristic function of X is $\psi(t) = \exp(it\mu + \frac{1}{2}\sigma^2 t^2)$. A *standard normal* random variable is a normal random variable with $\mu = 0$ and $\sigma^2 = 1$.

A.9.12 Student's t Distribution

A random variable X has a T(ν) distribution if the density function of X is given by

$$f(x) = \frac{\Gamma[\frac{1}{2}(\nu + 1)]}{\Gamma(\frac{1}{2}\nu)}(\pi\nu)^{-1/2}(1 + x^2\nu^{-1})^{-(\nu+1)/2},$$

where $x \in \mathbb{R}$ and ν is a positive integer known as the *degrees of freedom*. Assuming the $\nu > 1$, the mean of X is 0 and if $\nu > 2$ then the variance of X is $\nu(\nu - 2)^{-1}$. The moment generating function of X does not exist.

A.9.13 The Triangular Distribution

A random variable X has a TRIANGULAR(α, β, γ) distribution if the density function of X is given by

$$f(x) = \begin{cases} 2[(\beta - \alpha)(\gamma - \alpha)]^{-1}(x - \alpha) & x \in (\alpha, \gamma) \\ 2[(\beta - \alpha)(\beta - \gamma)]^{-1}(\beta - x) & x \in [\gamma, \beta) \\ 0 & \text{elsewhere,} \end{cases}$$

where $\alpha \in \mathbb{R}$, $\beta \in \mathbb{R}$, $\gamma \in \mathbb{R}$ such that $\alpha \leq \gamma \leq \beta$. The expectation and the variance of X are $\frac{1}{3}(\alpha + \beta + \gamma)$ and $\frac{1}{18}(\alpha^2 + \beta^2 + \gamma^2 - \alpha\beta - \alpha\gamma - \beta\gamma)$, respectively. The moment generating function of X is

$$m(t) = \frac{2[(\beta - \gamma)\exp(t\alpha) - (\beta - \alpha)\exp(t\gamma) + (\gamma - \alpha)\exp(t\beta)]}{(\beta - \alpha)(\gamma - \alpha)(\beta - \gamma)t^2},$$

and the characteristic function of X is

$$\psi(t) = \frac{-2[(\beta - \gamma)\exp(it\alpha) - (\beta - \alpha)\exp(it\gamma) + (\gamma - \alpha)\exp(it\beta)]}{(\beta - \alpha)(\gamma - \alpha)(\beta - \gamma)t^2}.$$

A.9.14 The (Continuous) Uniform Distribution

A random variable X has a UNIFORM(α, β) distribution if the density function of X is given by

$$f(x) = \begin{cases} (\beta - \alpha)^{-1} & \text{for } \alpha < x < \beta \\ 0 & \text{otherwise,} \end{cases}$$

where $\alpha \in \mathbb{R}$ and $\beta \in \mathbb{R}$ such that $\alpha < \beta$. The expectation and variance of X are $\frac{1}{2}(\alpha + \beta)$ and $\frac{1}{12}(\beta - \alpha)^2$. The moment generating function of X is $m(t) = [t(\beta - \alpha)]^{-1}[\exp(t\beta) - \exp(t\alpha)]$ and the characteristic function of X is $\psi(t) = [it(\beta - \alpha)]^{-1}[\exp(it\beta) - \exp(it\alpha)]$.

A.9.15 The Wald Distribution

A random variable X has a WALD(μ, λ) distribution if the density function of X is given by

$$f(x) = \begin{cases} \lambda^{1/2}(2\pi)^{-1/2}x^{-3/2}\exp[-\frac{1}{2}\lambda\mu^{-2}x^{-1}(x - \mu)^2] & x > 0 \\ 0 & \text{elsewhere,} \end{cases}$$

where $\mu > 0$ and $\lambda > 0$. The expectation and variance of X are μ and $\mu^3\lambda^{-1}$. The moment generating function of X is $m(t) = \exp(\lambda\mu^{-1})[1 - (1 - 2\lambda^{-1}\mu^2t)^{1/2}]$ and the characteristic function of X is $\psi(t) = \exp(\lambda\mu^{-1})[1 - (1 - 2\lambda^{-1}\mu^2it)^{1/2}]$.

Using R for Experimentation

B.1 An Introduction to R

The statistical software package R is a statistical computing environment that is similar in implementation to the S package developed at Bell Laboratories by John Chambers and his colleagues. The R package is a GNU project and is available under a free software license. Open source code for R, as well as compiled implementations for Unix, Linux, OS X, and Windows are available from www.r-project.org. A major strength of the R statistical computing environment is the ability to simulate data from numerous distributions. This allows R to be easily used for simulations and other types of statistical experiments.

This appendix provides some useful tips for using R as a tool for visualizing asymptotic results and is intended to be an aid to those wishing to solve the experimental exercises in the book. This appendix does assume a basic working knowledge of R, though this section will cover many of the basic ideas used in R.

B.2 Basic Plotting Techniques

The results of simulations are most effectively understood using visual representations such as plots and histograms. The basic mechanism for plotting a set of data pairs in R is the plot function:

```
plot(x,y,type,xlim,ylim,main,xlab,ylab,lty,pch,col)
```

Technically, the plot function has the header plot(x,y,...) where the optional arguments type, main, xlab, ylab, and lty are passed to the par function. The result of the plot function is to send a plot if the pairs given by the arguments x and y to the current graphics device, usually a separate graphics window, depending on the specific way your version of R has been set up. If no optional arguments are used then the plot is a simple scatterplot and the labels for the horizontal and vertical axes are taken to be the names of the objects passed to the function for the arguments x and y.

The optional argument **type** specifies what type of plot should be constructed. Some of the possible values of **type**, along with the resulting type of plot produced are

"p", which produces a scatterplot of individual points;

"l", which connects the specified points with lines, but does not plot the points themselves;

"b", which plots both the lines and the points as described in the two options given above; and

"n", which sets up the axes for the plot, but does not actually plot any values.

The specification **type="n"** can be used to set up a pair of axes upon which other objects will be plotted later. If the **type** argument is not used, then the plot will use the value stored by the **par** command, which in most cases corresponds to the option **type="p"**. The current settings can be viewed by executing the **par** command without any arguments.

The arguments **xlim** and **ylim** specify the range of the horizontal and vertical axes, respectively. The range is expressed by an array of length two whose first element corresponds to the minimum value for the axis and whose second component corresponds to the maximum value for the axis. For example, the specification **xlim=c(0,1)** specifies that the axis should have a range from zero to one. If these arguments are not specified then R uses a specific algorithm to compute what these ranges should be based on the ranges of the specified data. In most cases R does a good job of selecting ranges that make the visually appealing. However, when many sets of data are plotted on a single set of axes, as will be discussed later in this section, the ranges of the axes for the initial plot may need to be specified so that the axes are sufficient to contain all of the objects to be plotted. If ranges are specified and points lie outside the specified ranges, then the plot is still produced with the specified ranges, and R will usually return a warning for each point that it encounters that is outside the specified ranges.

The arguments **main, xlab,** and **ylab** specify the labels used for the main title, the label for the horizontal axis, and the vertical axis, respectively.

The argument **lty** specifies the type of line used when the argument **type="l"** or **type="b"** is used. The line types can either be specified as an integer or as a character string. The possible line types include

Integer	Character String	Line Type Produced
0	"blank"	No line is drawn
1	"solid"	Solid line
2	"dashed"	Dashed line
3	"dotted"	Dotted line
4	"dotdash"	Dots alternating with dashes
5	"longdash"	Long dashes
6	"twodash"	Two dashes followed by blank space

Alternatively, a character string of up to 8 characters may be specified, giving the length of line segments which are alternatively drawn and skipped. Consult the help pages for the **par** command for further details on this specification.

The argument **pch** specifies the symbol or character to be used for plotting points when the argument **type="p"** or **type="b"** is used. This can be either specified by a single character or by an integer. The integers 1–18 specify the set of plotting symbols originally used in the S software package. In addition, there is a special set of R plotting symbols which can be obtained by specified integers between 19 and 25. These specifications produce the following symbols:

Argument	Symbol
pch=19	solid circle
pch=20	small circle
pch=21	circle
pch=22	square
pch=23	diamond
pch=24	triangle point-up
pch=25	triangle point down

Other options are also available. See the R help page for the **points** command for further details.

The **col** argument specifies the color used for plotting. Colors can be specified in several different ways. The simplest way is to specify a character string that contains the name of the color. Some examples of common character strings that R understands are **"red"**, **"blue"**, **"green"**, **"orange"**, and **"black"**. A complete list of the possible colors can be obtained by executing the function **colors** with no arguments. Another option is to use one of the many color specification functions provided by R. For example, a gray color can be specified using the **gray(level)** function where the argument **level** is set to a number between 0 and 1 that specifies how dark the gray shading should be. Alternatively, colors can be specified directly in terms of their red-green-blue

(RGB) components with a character string of the form "#RRGGBB", where each of the pairs RR, GG, BB are of two digit hexadecimal numbers giving a value between 00 and FF. For example, specifying the argument col="#FF0000" is equivalent to using the argument col="red".

Many of the results shown in this book consist of plots of more than one set of data on a single plot. There are many ways that this can be achieved using R, but probably the simplest approach is to take advantage of the two functions points and lines. The points function adds points to the current plot at points specified by two arguments x and y as in the plot function. The optional arguments pch and col can also be used with this function to change the plotting symbol or the color of the plotting symbol. The lines function adds lines to the current plot that connect the points specified by the two arguments x and y. The optional arguments lty and col can also be used with this function.

It is important to note that when plotting multiple sets of data on the same set of axes, the range of the two axes, which are set in the original plot function, should be sufficient to handle all of the points specified in the multiple plots. For example, suppose that we wish to plot linear, quadratic, and cubic functions for a range of x values between 0 and 2, all on the same set of axes, using different line types for each function. If we execute the commands

```
x <- seq(0,2,0.001)
y1 <- x
y2 <- x^2
y3 <- x^3
plot(x,y1,type="l",lty=1)
lines(x,y2,lty=2)
lines(x,y3,lty=3)
```

the resulting plot will cut off the quadratic and cubic functions because the original plot command set up the vertical axis based on the range of y1. To fix this problem we need only find the minimum and maximum values before we execute the original plot command, and then specify this range in the plot command. That is

```
x <- seq(0,2,0.001)
y1 <- x
y2 <- x^2
y3 <- x^3
yl <- c(min(y1,y2,y3),max(y1,y2,y3))
plot(x,y1,type="l",lty=1,ylim=yl)
lines(x,y2,lty=2)
lines(x,y3,lty=3)
```

The final plotting function that will generally be helpful in performing the experiments suggested in this book is the hist function, which plots a histogram of a set of data. The usage of the hist function is given by

```
hist(x, breaks = "Sturges", freq = NULL, right = T, col = NULL,
     border = NULL, main, xlim, ylim, xlab, ylab)
```

The arguments of the function are

x is a vector of values for which the histogram will be plotted.

breaks specifies how many cells to be used when plotting the histogram. This argument can also specify the location of the cell endpoints. breaks can be one of the following:

- A vector giving the location of the endpoints of the histogram cells.
- A single number specifying the number of cells to be used when plotting the histogram.
- A character string naming an algorithm to compute the number of cells to be used when plotting the histogram.
- A function to compute the number of cells to be used when plotting the histogram.

In all but the first case, R uses the specified number as a suggestion only. The only way to force R to use the number you wish is to specify the endpoints of the cells. The default values of breaks specifies the algorithm "Sturges". Other possible algorithm names are "FD" and "Scott". Consult the R help page for the hist function for further details on these algorithms.

freq is a logical argument that specifies whether a frequency or density histogram is plotted. If freq=T is specified, then a frequency histogram is plotted. If freq=F is specified then a density histogram is plotted. In this case the histogram has a total area of one. When comparing a histogram to a known density on the same plot, using a density histogram usually gives better results.

right is a logical argument that specifies whether the endpoints of the cells are included on the left or right hand side of the cell interval. If right=T is specified then the cells include the right endpoint but not the left. If right=F is specified then the cells include the left endpoint but not the right.

col specifies the color of the bars plotted on the histogram. The default value of NULL yields unfilled bars.

border specifies the color of the border around the bars. The default is to use the color used for plotting the axes.

main, xlab, ylab can be used to specify the main title, the label for the horizontal axis, and the label for the vertical axis, respectively.

xlim and ylim can be used to specify the ranges of the horizontal and vertical axes. These options are useful when overlaying a histogram with a plot of a density for comparison.

Wand (1997) argues that many schemes for selecting the number of bins implemented in standard statistical software usually uses too few bins. A more reasonable method for selecting the bin width of a histogram is provided in the `KernSmooth` library and has the form `dpih(x)`, where `x` is the observed vector of data and we have omitted many technical arguments which have reasonable default values. In practice this function estimates the optimal width of the bins based on the observed data, and not the number of bins. Therefore, the endpoints of the histogram classes must be calculated to implement this function. For example, if we wish to simulate a sample of size 100 from a $N(0,1)$ distribution and create a histogram of the data based on the methodology of Wand (1997), then we can use the following code:

```
x <- rnorm(100)
h <- dpih(x)
bins <- seq(min(x)-h,max(x)+2*h,by=h)
hist(x,breaks=bins)
```

B.3 Complex Numbers

The standard R package has the ability to handle complex numbers in a native format. The basic function used for creating complex valued vectors is the `complex` function. The real and imaginary parts of a complex vector can be recovered using the functions `Re` and `Im`. The modulus of a complex number can be obtained using the `mod` function. The usage of these functions is summarized below. Further information about these functions, including some optional arguments not summarized below, can be found at www.r-project.org.

`complex(length.out=0, real, imaginary)` creates a vector whose length equals `length.out` that contains complex numbers whose real parts are stored in `real` and imaginary parts `imaginary`.

`Re(x)`: returns a vector of the same size as `x`, whose elements correspond to the real parts of the complex vector `x`.

`Im(x)`: returns a vector of the same size as `x`, whose elements correspond to the imaginary parts of the complex vector `x`.

`mod(x)`: returns a vector of the same size as `x`, whose elements correspond to modulus of the complex vector `x`.

B.4 Standard Distributions and Random Number Generation

The R package includes functions that can compute the density (or probability distribution function), distribution function, and quantile function for many standard distributions including nearly all of the distributions used in this book. There are also functions that will easily generate random samples from these distributions as well. This section provides information about the R functions for each of the distributions used in this book. Further information about these functions can be found at `www.r-project.org`.

B.4.1 The Bernoulli and Binomial Distributions

`dbinom(x, size, prob, log=F)`
Calculates the probability $P(X = x)$ where X has a BINOMIAL distribution based on `size` independent BERNOULLI experiments with success probability `prob`. The optional argument `log` indicates whether the logarithm of the probability should be returned. The BERNOULLI distribution is implemented by specifying `size=1`.

`pbinom(q, size, prob, lower.tail=T, log.p=F)`
Calculates the cumulative probability $P(X \leq q)$ where X has a BINOMIAL distribution based on `size` independent BERNOULLI experiments with success probability `prob`. The optional argument `lower.tail` indicates whether $P(X \leq q)$ should be returned (the default) or if $P(X > q)$ should be returned. The optional argument `log.p` indicates whether the natural logarithm of the probability should be returned. The BERNOULLI distribution is implemented by specifying `size=1`.

`qbinom(p, size, prob.lower.tail=T, loq.p=F)`
Returns the pth quantile of a BINOMIAL distribution based on `size` independent BERNOULLI experiments with success probability `prob`. The optional argument `lower.tail` indicates whether $p = P(X \leq x)$ (the default) or if $p = P(X > x)$. The optional argument `log.p` indicates whether p is the natural logarithm of the probability. The BERNOULLI distribution is implemented by specifying `size=1`.

`rbinom(n, size, prob)`
Generates a random sample of size n from a BINOMIAL distribution based on `size` independent BERNOULLI experiments with success probability `prob`. The BERNOULLI distribution is implemented by specifying `size=1`.

B.4.2 The Beta Distribution

`dbeta(x, shape1, shape2, log = F)`
Calculates the density function $f(x)$ of a random variable X that has a BETA(α, β) distribution with the α parameter equal to `shape1` and the β parameter equal to `shape2`. The optional argument `log` indicates whether the logarithm of the density should be returned.

`pbeta(q, shape1, shape2, lower.tail = T, log.p = F)`
Calculates the cumulative probability $P(X \leq q)$ where X has a BETA(α, β) distribution with the α parameter equal to `shape1` and the β parameter

equal to `shape2`. The optional argument `lower.tail` indicates whether $P(X \leq q)$ should be returned (the default) or if $P(X > q)$ should be returned. The optional argument `log.p` indicates whether the natural logarithm of the probability should be returned.

`qbeta(p, shape1, shape2, lower.tail=T, log.p = F)`
Returns the pth quantile of a BETA(α, β) distribution with the α parameter equal to `shape1` and the β parameter equal to `shape2`. The optional argument `lower.tail` indicates whether $p = P(X \leq x)$ (the default) or if $p = P(X > x)$. The optional argument `log.p` indicates whether p is the natural logarithm of the probability.

`rbeta(n, shape1, shape2)`
Generates a random sample of size n from a BETA(α, β) distribution with the α parameter equal to `shape1` and the β parameter equal to `shape2`.

B.4.3 The Cauchy Distribution

`dcauchy(x, location=0, scale=1, log=F)`
Calculates the density function $f(x)$ of a random variable X that has a CAUCHY(α, β) distribution with the α parameter equal to `location` and the β parameter equal to `scale`. The optional argument `log` indicates whether the logarithm of the density should be returned.

`pcauchy(q, location=0, scale=1, lower.tail=T, log.p=F)`
Calculates the cumulative probability $P(X \leq q)$ where X has a CAUCHY(α, β) distribution with the α parameter equal to `location` and the β parameter equal to `scale`. The optional argument `lower.tail` indicates whether $P(X \leq q)$ or $P(X > q)$ should be returned. The optional argument `log.p` indicates whether the natural logarithm of the probability should be returned.

`qcauchy(p, location=0, scale=1, lower.tail=T, log.p=F)`
Returns the pth quantile of a CAUCHY(α, β) distribution with the α parameter equal to `location` and the β parameter equal to `scale`. The optional argument `lower.tail` indicates whether p equals $P(X \leq q)$ or $P(X > q)$. The optional argument `log.p` indicates whether p is the natural logarithm of the probability.

`rcauchy(n, location = 0, scale = 1)`
Generates a random sample of size n from a CAUCHY(α, β) distribution with the α parameter equal to `location` and the β parameter equal to `scale`.

B.4.4 The Chi-Squared Distribution

`dchisq(x, df, log = F)`
Calculates the density function $f(x)$ of a random variable X that has a

CHISQUARED(η) distribution with the η parameter equal to df. The optional argument log indicates whether the logarithm of the density should be returned.

pchisq(q, df, lower.tail = T, log.p = F)

Calculates the cumulative probability $P(X \leq q)$ where X is a random variable with a CHISQUARED(η) distribution with the η parameter equal to df. The optional argument lower.tail indicates whether $P(X \leq q)$ or $P(X > q)$ should be returned. The optional argument log.p indicates whether the natural logarithm of the probability should be returned.

qchisq(p, df, lower.tail = T, log.p = F)

Returns the pth quantile of a CHISQUARED(η) distribution with the η parameter equal to df. The optional argument lower.tail indicates whether $p = P(X \leq q)$ (the default) or if $p = P(X > q)$. The optional argument log.p indicates whether p is the natural logarithm of the probability.

rchisq(n, df)

Generates a sample of size n from a CHISQUARED(η) distribution with the η parameter equal to df.

B.4.5 The Exponential Distribution

dexp(x, rate = 1, log = F)

Calculates the density function $f(x)$ of a random variable X that has an EXPONENTIAL(θ) distribution with θ^{-1} equal to rate. The optional argument log indicates whether the logarithm of the density should be returned.

pexp(q, rate = 1, lower.tail = T, log.p = F)

Calculates the cumulative probability $P(X \leq q)$ where X is a random variable with an EXPONENTIAL(θ) distribution with θ^{-1} equal to rate. The optional argument lower.tail indicates whether $P(X \leq q)$ or $P(X > q)$ should be returned. The optional argument log.p indicates whether the natural logarithm of the probability should be returned.

qexp(p, rate = 1, lower.tail = T, log.p = F)

Returns the pth quantile of a EXPONENTIAL(θ) distribution with θ^{-1} equal to rate. The optional argument lower.tail indicates whether $p = P(X \leq q)$ (the default) or if $p = P(X > q)$. The optional argument log.p indicates whether p is the natural logarithm of the probability.

rexp(n, rate = 1)

Generates a sample of size n from an EXPONENTIAL(θ) distribution with θ^{-1} equal to rate.

B.4.6 The Gamma Distribution

dgamma(x, shape, rate = 1, scale = 1/rate, log = F)

Calculates the density function $f(x)$ of a random variable X that has a

GAMMA(α, β) distribution with α equal to shape and β^{-1} equal to rate. Alternatively, one can specify the scale parameter instead of the rate parameter where scale is equal to β. The optional argument log indicates whether the logarithm of the density should be returned.

pgamma(q, shape, rate=1, scale=1/rate, lower.tail=T, log.p=F)
    Calculates the cumulative probability $P(X \leq q)$ where X is a random variable with a GAMMA(α, β) distribution with α equal to shape and β^{-1} equal to rate. Alternatively, one can specify the scale parameter instead of the rate parameter where scale is equal to β. The optional argument lower.tail indicates whether $P(X \leq q)$ or $P(X > q)$ should be returned. The optional argument log.p indicates whether the logarithm of the probability should be returned.

qgamma(p, shape, rate=1, scale=1/rate, lower.tail=T, log.p = F)
    Returns the pth quantile of a GAMMA(α, β) distribution with α equal to shape and β^{-1} equal to rate. Alternatively, one can specify the scale parameter instead of the rate parameter where scale is equal to β. The optional argument lower.tail indicates whether $p = P(X \leq q)$ (the default) or if $p = P(X > q)$. The optional argument log.p indicates whether p is the natural logarithm of the probability.

rgamma(n, shape, rate = 1, scale = 1/rate)
    Generates a sample of size n from a GAMMA(α, β) distribution with α equal to shape and β^{-1} equal to rate. Alternatively, one can specify the scale parameter instead of the rate parameter where scale is equal to β.

B.4.7 The Geometric Distribution

dgeom(x, prob, log = F)
    Calculates the probability $P(X = x)$ where X has a GEOMETRIC(θ) distribution with θ specified by prob. The optional argument log indicates whether the logarithm of the probability should be returned.

pgeom(q, prob, lower.tail = T, log.p = F)
    Calculates the cumulative probability $P(X \leq q)$ where X has a GEOMETRIC(θ) distribution with θ specified by prob. The optional argument lower.tail indicates whether $P(X \leq q)$ or $P(X > q)$ should be returned. The optional argument log.p indicates whether the logarithm of the probability should be returned.

qgeom(p, prob, lower.tail = T, log.p = F)
    Returns the pth quantile of a GEOMETRIC(θ) distribution with θ specified by prob. The optional argument lower.tail indicates whether $p = P(X \leq q)$ (the default) or if $p = P(X > q)$. The optional argument log.p indicates whether p is the natural logarithm of the probability.

```
rgeom(n, prob)
```
Generates a sample of size n from a GEOMETRIC(θ) distribution with θ specified by prob.

B.4.8 The LaPlace Distribution

There does not appear to be a standard R library at this time that supports the LAPLACE(a, b) distribution. The relatively simple form of the distribution makes it fairly easy to work with, however. For example, a function for the density function can be programmed as

```
dlaplace <- function(x, a=0, b=1)
    return(exp(-1*abs(x-a)/b)/(2*b))
```

A function for the distribution function can be programmed as

```
plaplace <- function(q, a=0, b=1)
{
    if(x<a) return(0.5*exp(x-a)/b)
    else return(1-0.5*exp((a-x)/b)/(2*b))
}
```

and a function for the quantile function can be programmed as

```
qlaplace <- function(p, a=0, b=1)
    return(a-b*sign(p-0.5)*log(1-2*abs(p-0.5)))
```

A function to generate a sample of size n from a LAPLACE(a, b) distribution can be programmed as

```
rlaplace <- function(n, a=0, b=1)
{
    u <- runif(n,-0.5,0.5)
    return(a-b*sign(u)*log(1-2*abs(u)))
}
```

Please note that these are fairly primitive functions in that they do no error checking and may not be the most numerically efficient methods. They should be sufficient for performing the experiments in this book as long as one is careful with their use.

B.4.9 The Lognormal Distribution

```
dlnorm(x, meanlog = 0, sdlog = 1, log = F)
```
Calculates the density function $f(x)$ of a random variable X that has a LOGNORMAL(μ, σ^2) density with μ equal to logmean and σ equal to sdlog. The optional argument log indicates whether the logarithm of the density should be returned.

plnorm(q, meanlog = 0, sdlog = 1, lower.tail = T, log.p = F)
Calculates the cumulative probability $P(X \leq q)$ where X has a LOGNOR-MAL(μ, σ^2) density with μ equal to logmean and σ equal to sdlog. The optional argument lower.tail indicates whether $P(X \leq q)$ (the default) or if $P(X > q)$ should be returned. The optional argument log.p indicates whether the natural logarithm of the probability should be returned.

qlnorm(p, meanlog = 0, sdlog = 1, lower.tail = T, log.p = F)
Returns the pth quantile of a LOGNORMAL(μ, σ^2) density with μ equal to logmean and σ equal to sdlog. The optional argument lower.tail indicates whether $p = P(X \leq q)$ (the default) or if $p = P(X > q)$. The optional argument log.p indicates whether p is the natural logarithm of the probability.

rlnorm(n, meanlog = 0, sdlog = 1)
Generates a sample of size n from a LOGNORMAL(μ, σ^2) density with μ equal to logmean and σ equal to sdlog.

B.4.10 The Multinomial Distribution

rmultinom(n, size, prob)
Generates n $k \times 1$ independent random vectors. Each random random vector follows a MULTINOMIAL distribution where size outcomes are classified into k categories that have associated probabilities specified by the $k \times 1$ vector prob. The value of k is determined by the function from the length of prob.

dmultinom(x, size=NULL, prob, log=F)
Returns the probability $P(\mathbf{X} = \mathbf{x})$ where $\mathbf{x}$ is specified by the $k \times 1$ vector x. The random vector $\mathbf{X}$ has a MULTINOMIAL where size outcomes are classified into k categories that have associated probabilities specified by the $k \times 1$ vector prob. By default size is set equal to sum(x) and need not be specified. The optional argument log indicates whether the logarithm of the probability should be returned.

B.4.11 The Normal Distribution

dnorm(x, mean = 0, sd = 1, log = F)
Calculates the density function $f(x)$ of a random variable X that has a N(μ, σ^2) distribution where μ is specified by mu and σ is specified by sigma. The optional argument log indicates whether the logarithm of the density should be returned.

pnorm(q, mean = 0, sd = 1, lower.tail = T, log.p = F)
Calculates the cumulative probability $P(X \leq q)$ where X has a N(μ, σ^2) distribution where μ is specified by mu and σ is specified by sigma. The

optional argument `lower.tail` indicates whether $P(X \leq q)$ (the default) or if $P(X > q)$ should be returned. The optional argument `log.p` indicates whether the natural logarithm of the probability should be returned.

`qnorm(p, mean = 0, sd = 1, lower.tail = T, log.p = F)`
Returns the pth quantile of a $N(\mu, \sigma^2)$ distribution where μ is specified by `mu` and σ is specified by `sigma`. The optional argument `lower.tail` indicates whether $p = P(X \leq q)$ (the default) or if $p = P(X > q)$. The optional argument `log.p` indicates whether p is the natural logarithm of the probability.

`rnorm(n, mean = 0, sd = 1)`
Generates a sample of size n from a $N(\mu, \sigma^2)$ distribution where μ is specified by `mu` and σ is specified by `sigma`.

B.4.12 The Multivariate Normal Distribution

Samples can be simulated from a $\mathbf{N}(\boldsymbol{\mu}, \boldsymbol{\Sigma})$ distribution using the `mvrnorm(n = 1, mu, Sigma)` function which can be found in the `MASS` library. The argument `n` specifies the size of the sample to be generated, `mu` is the mean vector and `Sigma` is the covariance matrix. The function returns a $n \times d$ matrix object where d is the dimension of the vector `mu`.

B.4.13 The Poisson Distribution

`dpois(x, lambda, log = F)`
Calculates the probability $P(X = x)$ where X has a POISSON distribution with rate specified by `lambda`. The optional argument `log` indicates whether the logarithm of the probability should be returned.

`ppois(q, lambda, lower.tail = T, log.p = F)`
Calculates the cumulative probability $P(X \leq q)$ where X has a POISSON distribution with rate specified by `lambda`. The optional argument `lower.tail` indicates whether $P(X \leq q)$ should be returned (the default) or if $P(X > q)$ should be returned. The optional argument `log.p` indicates whether the natural logarithm of the probability should be returned.

`qpois(p, lambda, lower.tail = T, log.p = F)`
Returns the pth quantile of a POISSON distribution with rate specified by `lambda`. The optional argument `lower.tail` indicates whether $p = P(X \leq q)$ (the default) or if $p = P(X > q)$. The optional argument `log.p` indicates whether p is the natural logarithm of the probability.

`rpois(n, lambda)`
Generates a random sample of size n from a POISSON distribution with rate specified by `lambda`.

B.4.14 Student's t Distribution

`dt(x, df, log = F)`
Calculates the density function $f(x)$ of a random variable X that has a $T(\nu)$ distribution with ν equal to df. The optional argument log indicates whether the logarithm of the density should be returned.

`pt(q, df, lower.tail = T, log.p = F)`
Calculates the cumulative probability $P(X \le q)$ where X has a $T(\nu)$ distribution with ν equal to df. The optional argument lower.tail indicates whether $P(X \le q)$ should be returned (the default) or if $P(X > q)$ should be returned. The optional argument log.p indicates whether the natural logarithm of the probability should be returned.

`qt(p, df, lower.tail = T, log.p = F)`
Returns the pth quantile of a $T(\nu)$ distribution with ν equal to df. The optional argument lower.tail indicates whether $p = P(X \le q)$ (the default) or if $p = P(X > q)$. The optional argument log.p indicates whether p is the natural logarithm of the probability.

`rt(n, df)`
Generates a random sample of size n from a $T(\nu)$ distribution with ν equal to df.

B.4.15 The Continuous Uniform Distribution

`dunif(x, min=0, max=1, log = F)`
Calculates the density function $f(x)$ of a random variable X that has a UNIFORM(α, β) distribution with α equal to min and β equal to max. The optional argument log indicates whether the logarithm of the density should be returned.

`punif(q, min=0, max=1, lower.tail = T, log.p = F)`
Calculates the cumulative probability $P(X \le q)$ where X has a UNIFORM(α, β) distribution with α equal to min and β equal to max. The optional argument lower.tail indicates whether $P(X \le q)$ should be returned (the default) or if $P(X > q)$ should be returned. The optional argument log.p indicates whether the natural logarithm of the probability should be returned.

`qunif(p, min=0, max=1, lower.tail = T, log.p = F)`
Returns the pth quantile of a UNIFORM(α, β) distribution with α equal to min and β equal to max. The optional argument lower.tail indicates whether $p = P(X \le q)$ (the default) or if $p = P(X > q)$. The optional argument log.p indicates whether p is the natural logarithm of the probability.

`runif(n, min=0, max=1)`

Generates a random sample of size n from a UNIFORM(α, β) distribution with α equal to min and β equal to max.

B.4.16 The Discrete Uniform Distribution

There is not full support for the UNIFORM$(x_1, x_2, \ldots, x_n)$ distribution in R, though sampling from this distribution can be accomplished using the sample function which has the following form:

```
sample(x, size, replace = F, prob = NULL)
```

where x is a vector that contains the units to be sampled, size is a non-negative integer that equals the sample size, replace is a logical object that specifies whether the sampling should take place with or without replacement, and prob is a vector of probabilities for each of the elements in the event that non-uniform sampling is desired. Therefore, if we wish to generate a sample of size 10 from a UNIFORM$(1, 2, 3, 4, 5, 6)$ distribution then we can use the command sample(1:6,10,replace=T). Specification of the vector prob allows us to simulate sampling from any discrete distribution. For example, if we wish to generate a sample of size 10 from a discrete distribution with probabilty distribution function

$$f(x) = \begin{cases} \frac{1}{4} & x = -1, 1 \\ \frac{1}{2} & x = 0 \end{cases}$$

then we can use the command

```
sample(c(-1,0,1),10,replace=T,prob=c(0.25,0.50,0.25))
```

Finally, if we wish to simulate resamples from a sample x in order to perform bootstrap calculations, then we can use the command

```
sample(x,length(x),replace=T)
```

In fact, a rudimentary bootstrap function can be specified as

```
bootstrap <- function(x,b,fun,...)
{
    n <- length(x)
    xs <- matrix(sample(x,n*b,replace=T),b,n)
    return(apply(xs,1,fun,...))
}
```

The function returns the b values of the statistic specified by fun computed on the b resamples. For example, the following command simulates a sample of size 25 from a $N(0, 1)$ distribution, generates 1000 bootstrap resamples, calculates the 10% trimmed mean of each, and makes a histogram of the resulting values.

```
hist(bootstrap(rnorm(25),1000,mean,trim=0.10))
```

A more sophisticated implementation of the bootstrap can be found in the boot library.

B.4.17 The Wald Distribution

Support for the Wald distribution can be found in the library SuppDists, which is available from most official R internet sites.

dinvGauss(x, nu, lambda, log=F)
  Calculates the density of X at x where X has a WALD(μ, λ) distribution where μ is given by nu and λ is specified by lambda. The optional argument log indicates whether the natural logarithm of the density should be returned.

pinvGauss(q, nu, lambda, lower.tail=T, log.p=F)
  Calculates the cumulative probability $P(X \leq q)$ where X has a WALD(μ, λ) distribution where μ is given by nu and λ is specified by lambda. The optional argument lower.tail indicates whether $P(X \leq q)$ should be returned (the default) or if $P(X > q)$ should be returned. The optional argument log.p indicates whether the natural logarithm of the probability should be returned.

qinvGauss(p, nu, lambda, lower.tail=T, log.p=F)
  Returns the pth quantile of a textscWald(μ, λ) distribution where μ is given by nu and λ is specified by lambda. The optional argument lower.tail indicates whether $p = P(X \leq q)$ (the default) or if $p = P(X > q)$. The optional argument log.p indicates whether p is the natural logarithm of the probability.

rinvGauss(n, nu, lambda)
  Generates a random sample of size n from a textscWald(μ, λ) distribution where μ is given by nu and λ is specified by lambda.

B.5 Writing Simulation Code

Most of the experiments suggested in this book should run in a reasonable amount of time on a standard Mac or PC. If some of the simulations seem to be taking too long, then some of the parameters can be changed to make them run faster, many times without too much loss of information about the concept under consideration. Instructors are encouraged to try out some of the simulations on their local network before assigning some of the larger experiments in order to establish whether the simulation parameters are reasonable for their students complete in a reasonable amount of time.

There are also some basic R concepts that will keep the simulations efficient as possible. In general, it is usually advised that even though R does support looping, most notably using the `for` loop, loops in general should be avoided if possible. This can be accomplished in R using several helpful concepts that are built into R.

The first way to avoid loops is to note that R can do many calculations on an entire vector or matrix at once. For example, suppose that A and B are two matrix objects with n rows and m columns. One could multiply the corresponding elements of the matrices using the double loop:

```
C <- matrix(0,n,m)
for(i in 1:n} for(j in 1:m) C[i,j] <- A[i,j] * B[i,j]
```

However, it is much more efficient to use the simple command C <- A*B, which multiplies the corresponding elements. This also works with addition, subtraction and division. One should note that the matrix product of A and B, assuming they are conformable is given by the command A%*%B. Many standard R functions will also work on the elements of a vector. For example, cos(A) will return a matrix object whose elements correspond to the cosine of the elements of A. Similarly, if one wishes to compute a vector of $N(0,1)$ quantiles for $p = 0.01, \ldots, 0.99$, one can avoid using a loop by simply executing the command qnorm(seq(0.01,0.99,0.01)).

One should also note that R offers vector versions of many functions that have the option of returning vectors. This is particularly useful when simulating a set of data. For example, one could simulate a sample of size 10 from a $N(0,1)$ distribution using the commands:

```
z <- matrix(0,10,1)
for(i in 1:10) z[i] <- dnorm(1)
```

It is much more efficient to just use the command z <- dnorm(10). In fact, many of the experiments in this book specify simulating 1000 samples of size 10, for example. If all the samples are from the same distribution, and are intended to be independent of one another, then all of the samples can be simulated at once using the command z <- matrix(dnorm(10*1000),1000,10). Each sample will then correspond to a row of the object z.

The final suggested method for avoiding loops in R is to use the `apply` command whenever possible. The `apply` command allows one to compute many R functions row-wise or column-wise on an array in R without using a loop. The `apply` command has the following form:

```
apply(X, MARGIN, FUN, ...)
```

where X is the object that a function will be applied to, MARGIN indicates whether the function will be applied to rows (MARGIN=1), columns (MARGIN=2), or both (MARGIN=c(1,2)), and FUN is the function that will be applied. Optional arguments for the function that is being applied can be passed after these three arguments have been specified. For example, suppose that we wish to compute a 10% trimmed mean on the rows of a matrix object X. This can be accomplished using the command `apply(X,1,mean,trim=0.10)`.

B.6 Kernel Density Estimation

Kernel density estimation is supported in R through several add-on libraries. Two of these libraries are particularly useful for kernel density estimation. The sm library is a companion to the book of Bowman and Azzalini (1997), and provides a variety of functions for several types of smoothing methods. The second library is the `KernSmooth` library that is a companion to the book of Wand and Jones (1995). The `KernSmooth` library is specific to kernel smoothing methods, and we will briefly demonstrate how to obtain a kernel density estimate using this library. There are two basic functions that we require from this library. The first function estimates the optimal bandwidth based on an iterated plug-in approach. The form of this function is `dpik(x, level=2, kernel="normal")` where x is the observed data, `level` corresponds to how many iterations are used, and `kernel` is the type of kernel function being used. To keep matters simple, we have purposely left out many more technical arguments which can be left with their default value. To compute the kernel estimator the function `bkde(x, kernel = "normal", bandwidth)` is used, where x is the observed data, `kernel` is the type of kernel function being used, and `bandwidth` is the bandwidth to be used. The output from this function is a list object that contains vectors x and y that correspond to a grid on the range of the data, and the corresponding value of the kernel density estimator at each of the grid points. The output from this function can be passed directly to the `plot` function. For example, to simulate a set of 100 observations from a N(0, 1) distribution, estimate the optimal bandwidth, and plot the corresponding estimate we can use the commands:

```
x <- rnorm(100)
plot(bkde(x,bandwidth=dpik(x)),type="l")
```

B.7 Simulating Samples from Normal Mixtures

Some of the experiments in this book require the reader to simulate samples from normal mixtures. There are some libraries which offer some techniques, but for the experiments in this book a simple approach will suffice. Suppose that we wish to simulate a sample of size n from a normal mixture density of the form

$$f(x) = \sum_{i=1}^{p} \omega_i \sigma_i^{-1} \phi[\sigma_i^{-1}(x - \mu_i)], \qquad (B.1)$$

where $\omega_1, \ldots, \omega_p$ are the weights of the mixture which are assumed to add to one, $\mu_1, \ldots, \mu_p$ are the means of the normal densities, and $\sigma_1^2, \ldots, \sigma_p^2$ are the variances of the normal densities. The essential behind simulating a sample from such a density lies behind the fact that if X has the density given in Equation (B.1), then X has the same distribution as $Z = \mathbf{Y'W}$ where $\mathbf{Y}$ has a MULTIONOMIAL$(1, p, \boldsymbol{\omega})$ distribution and $\mathbf{W}$ has a $\mathbf{N}(\boldsymbol{\mu}, \boldsymbol{\Sigma})$ distribution where $\boldsymbol{\omega'} = (\omega_1, \ldots, \omega_p)$, $\boldsymbol{\mu'} = (\mu_1, \ldots, \mu_p)$, and $\boldsymbol{\Sigma} = \mathrm{Diag}\{\sigma_1^2, \ldots, \sigma_p^2\}$. Therefore, suppose we wish to simulate an observation from the density

$$f(x) = \tfrac{1}{4}\phi(x - 1) + \tfrac{1}{2}\phi(x) + \tfrac{1}{4}\phi(x + 1),$$

then we could use the commands

```
omega <- c(0.25,0.50,0.25)
mu <- c(-1,0,1)
Sigma <- diag(1,2,2)
W <- mvrnorm(1, mu, Sigma)
Y <- rmultinom(1,3,omega)
Z <- t(W)%*%Y
```

B.8 Some Examples

B.8.1 Simulating Flips of a Fair Coin

In this simulation, we consider flipping a fair coin n times. For each flip we wish to keep track of the proportion of flips that are heads. The experiment is repeated b times and the resulting proportions are plotted together on a single plot that demonstrates how the proportion converges to $\tfrac{1}{2}$ as n gets large. For the example we have used $n = 100$ and $b = 5$, but these parameters are easily changed.

```
n <- 100
b <- 5
p <- matrix(0,b,n)

plot(c(0,100),c(0.5,0.5),type="l",lty=2,ylim=c(0,1),
    xlab="Flip Number",ylab="Proportion Heads")

for(i in 1:b) lines(seq(1,n,1),cumsum(rbinom(n=100,size=1,
    prob=0.5))/seq(1,n,1))
```

The resulting output should be a plot similar to the one in Figure B.1.

B.8.2 Investigating the Central Limit Theorem

In this simulation we simulate 100 samples of size 10 from a EXPONENTIAL(1) distribution and compute the mean of each sample. A histogram of the resulting sample means is then plotted, along with a NORMAL density for comparison. The mean and variance of the normal density that is plotted are

Figure B.1 *Example output for simulating flips of a fair coin.*

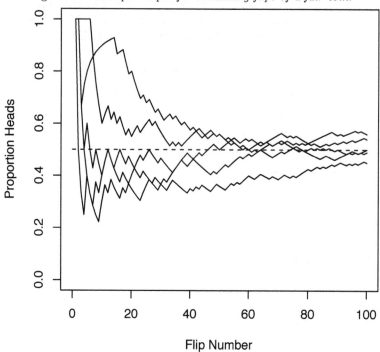

computed to match the mean and variance of the observed sample means. Note that this simulation sets up the samples in a 100×10 matrix and uses the apply function to compute the sample mean of each row. This simulation also sets up the ranges of the horizontal and vertical axes so that the overlay of the density function can observed. In setting up the range of the vertical axis we use the hist function with the argument plot=F. Using this argument causes the histogram not to be plotted, but does return a list the contains the calculated heights of the density bars. Therefore, we are able to calculate the maximum value of the density curve, contained in the y object, along with the maximum density from the histogram contained in the object returned by the command hist(obs.means,plot=F)$density).

```
x <- matrix(rexp(1000,1),100,10)
obs.means <- apply(x,1,mean)
norm.mean <- mean(obs.means)
norm.sd <- sd(obs.means)
xl <- c(norm.mean-4*norm.sd,norm.mean+4*norm.sd)
x.grid <- seq(xl[1],xl[2],length.out=1000)
y <- dnorm(x.grid,norm.mean,norm.sd)
yl <- c(0,max(y,hist(obs.means,plot=F)$density))
```

Figure B.2 *Example output from the simulation that investigates the Central Limit Theorem.*

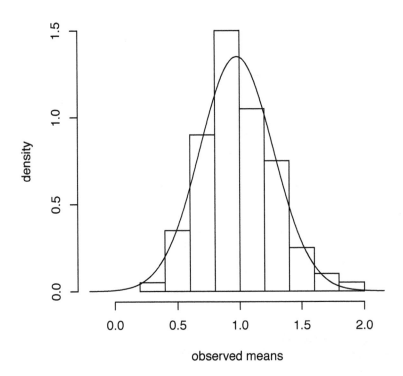

The Central Limit Theorem

```
hist(obs.means,freq=F,xlab="observed means",ylab="density",
     main="The Central Limit Theorem", xlim=xl,ylim=yl)
lines(x.grid,y)
```

The resulting output should be a plot similar to the one in Figure B.2.

B.8.3 Plotting the Normal Characteristic Function

The characteristic function of a $N(\mu, \sigma^2)$ random variable is difficult to visualize because it is a complex valued function of t given by $\psi(t) = \exp(-it\mu - \frac{1}{2}t^2\sigma^2)$. One solution to this problem is to use a three-dimensional scatterplot to visualize the function. In the following example code, we use the function scatterplot3d from the package scatterplot3d. Using the native complex data type in R makes this type of plot relatively easy to produce.

```
library(scatterplot3d)
```

Figure B.3 *Example output for plotting the normal characteristic function.*

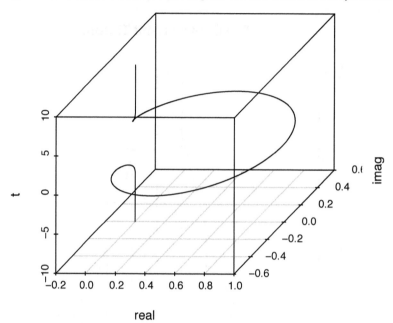

```
t <- seq(-10,10,0.001)
cf <- exp(complex(1,0,1)*t-0.5*t*t)

scatterplot3d(Re(cf),Im(cf),t,type="l",xlab="real",ylab="imag")
```

The resulting output should be a plot similar to the one in Figure B.3.

B.8.4 Plotting Edgeworth Corrections

In the R program below we demonstrate how to plot Edgeworth corrected density for the standardized distribution of the sample mean when the population is a translated GAMMA distribution. In this case we have used a sample size equal to three. See Example 7.1 for further details.

```
n <- 3
x <- seq(-3,4,0.01)
y1 <- dnorm(x)
y2 <- dgamma(x+sqrt(n),n,sqrt(n))
y3 <- y1 + y1/(3*sqrt(n))*(x^3-3*x)
y4 <- y3 + y1/n*((1/18)*(x^6-15*x^4+45*x^2-15)+
```

Figure B.4 *Example output for plotting Edgeworth corrections.*

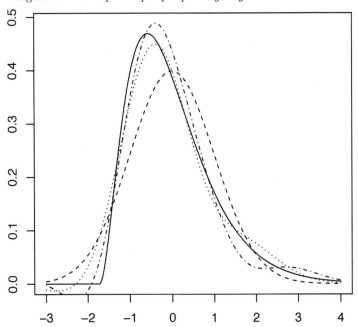

```
      (3/8)*(x^4-6*x^2+3))
plot(x,y1,type="l",xlab="",ylab="",ylim=c(0, max(y1,y2,y3,y4)),
     lty=2)
lines(x,y2)
lines(x,y3,lty=3)
lines(x,y4,lty=4)
```

The resulting output should be a plot similar to the one in Figure B.4.

B.8.5 Simulating the Law of the Iterated Logarithm

Theorem 3.15 (Hartman and Wintner) provides a result on the extreme fluctuations of the sample mean. The complexity of this result makes it difficult to visualize. The code below was used to produce Figure 3.6.

```
ss <- seq(5,500,1)
x <- rnorm(max(ss))
sm <- matrix(0,length(ss),1)
lf <- matrix(0,length(ss),1)
uf <- matrix(0,length(ss),1)
```

```
ul <- matrix(0,length(ss),1)
ll <- matrix(0,length(ss),1)
for(i in seq(1,length(ss),1))
{
    sm[i] <- sqrt(ss[i])*mean(x[1:ss[i]])
    uf[i] <- max(sm[1:i])
    lf[i] <- min(sm[1:i])
    ul[i] <- sqrt(2*log(log(ss[i])))
    ll[i] <- -1*ul[i]
}
yl <- c(min(sm,uf,lf,ul,ll),max(sm,uf,lf,ul,ll))
plot(ss,sm,type="l",ylim=yl)
lines(ss,uf,lty=2)
lines(ss,lf,lty=2)
lines(ss,ll,lty=3)
lines(ss,ul,lty=3)
```

References

Ahlfors, L. (1979). *Complex Analysis*. New York: McGraw-Hill.

Akhiezer, N. I. (1965). *The Classical Moment Problem*. New York: Hafner Publishing Company.

Anderson, T. W. and Goodman, L. A. (1957). Statistical inference about Markov chains. *The Annals of Mathematical Statistics*, **28**, 89–109.

Apostol, T. M. (1967). *Calculus: Volume I*. 2nd Ed. New York: John Wiley and Sons.

Apostol, T. M. (1974). *Mathematical Analysis*. 2nd Ed. Menlo Park, CA: Addison-Wesley.

Arnold, B. C., Balakrishnan, N., and Nagaraja, H. N. (1993). *A First Course in Order Statistics*. New York: John Wiley and Sons.

Athreya, K. B. (1987). Bootstrap of the mean in the infinite variance case. *The Annals of Statistics*, **14**, 724–731.

Averbukh, V. I. and Smolyanov, O. G. (1968). The various definitions of the derivative in linear topological spaces. *Russian Mathematical Surveys*, **23**, 67–113.

Babu, G. J. (1984). Bootstrapping statistics with linear combinations of chi-square as weak limit. *Sankhyā, Series A*, **46**, 85–93.

Babu, G. J. (1986). A note on bootstrapping the variance of the sample quantile. *The Annals of the Institute of Statistical Mathematics*, **38**, 439–443.

Babu, G. J. and Singh, K. (1983). Inference on means using the bootstrap. *The Annals of Statistics*, **11**, 999–1003.

Babu, G. J. and Singh, K. (1984). One-term Edgeworth correction by Efron's bootstrap. *Sankhyā Series A*, **46**, 219–232.

Babu, G. J. and Singh, K. (1985). Edgeworth expansions for sampling without replacement from finite populations. *Journal of Multivariate Analysis*, **17**, 261–278.

Bahadur, R. R. (1958). Examples of inconsistency of maximum likelihood estimates. *Sahnkya*, **20**, 207–210.

Bahadur, R. R. (1960a). Asymptotic efficiency of tests and estimators. *Sankhyā*, **20**, 229–252.

Bahadur, R. R. (1960b). Stochastic comparison of tests. *The Annals of Mathematical Statistics*, **31**, 276–295.

Bahadur, R. R. (1964). On Fisher's bound for asymptotic variances. *The Annals of Mathematical Statistics*, **35**, 1545–1552.

Bahadur, R. R. (1964). Rates of convergence of estimates and test statistics. *The Annals of Mathematical Statistics*, **38**, 303–324.

Barnard, G. A. (1970). Discussion on paper by Dr. Kalbfleisch and Dr. Sprott. *Journal of the Royal Statistical Society, Series B*, **32**, 194–195.

Barndorff-Nielsen, O. E. (1978). *Information and Exponential Families in Statistical Theory*. New York: John Wiley and Sons.

Barndorff-Nielsen, O. E. and Cox, D. R. (1989). *Asymptotic Techniques for Use in Statistics*. London: Chapman and Hall.

Bartlett, M. S. (1963). Statistical estimation of density functions. *Sankhyā, Series A*, **25**, 245–254.

Basu, D. (1955). An inconsistency of the method of maximum likelihood. *The Annals of Mathematical Statistics*, **26**, 144–145.

Beran, R. (1987). Prepivoting to reduce level error of confidence sets. *Biometrika*, **74**, 457–468.

Beran, R. (1982). Estimated sampling distributions: The bootstrap and competitors. *The Annals of Statistics*, **10**, 212–225.

Beran, R. and Ducharme, G. R. (1991). *Asymptotic Theory for Bootstrap Methods in Statistics*. Montréal: Centre De Recherches Mathematiques.

Berry, A. C. (1941). The accuracy of the Gaussian approximation to the sum of independent variates. *Transactions of the American Mathematical Society*, **49**, 122–136.

Bhattacharya, R. N. and Ghosh, J. K. (1978). On the validity of the formal Edgeworth expansion. *The Annals of Statistics*, **6**, 434–451.

Bhattacharya, R. N. and Rao, C. R. (1976). *Normal Approximation and Asymptotic Expansions*. New York: John Wiley and Sons.

Bickel, P. J. and Freedman, D. A. (1980). *On Edgeworth Expansions and the Bootstrap*. Unpublished manuscript.

Bickel, P. J. and Freedman, D. A. (1981). Some asymptotic theory for the bootstrap. *The Annals of Statistics*, **9**, 1196–1217.

Billingsley, P. (1986). *Probability and Measure*. New York: John Wiley and Sons.

Billingsley, P. (1999). *Convergence of Probability Measures*. New York: John Wiley and Sons.

Binmore, K. G. (1981). *Topological Ideas*. Cambridge: Cambridge University Press.

Bolstad, W. M. (2007). *Introduction to Bayesian Statistics*. New York: John Wiley and Sons.

Bowman, A. W. (1984). An alternative method of cross-validation for the smoothing of density estimates. *Biometrika*, **71**, 353–360.

Bowman, A.W. and Azzalini, A. (1997). *Applied Smoothing Techniques for Data Analysis: the Kernel Approach with S-Plus Illustrations*. Oxford: Oxford University Press.

Bratley, P., Fox, B. L., and Schrage, L. E. (1987). *A Guide to Simulation*. New York: Springer-Verlag.

Buck, R. C. (1965). *Advanced Calculus*. New York: McGraw-Hill.

Burman, P. (1985). A data dependent approach to density estimation. *Zeitschrift für Wahrscheinlichkeitstheorie und Verwandte Gebiete*, **69**, 609–628.

Butler, R. W. (2007). *Saddlepoint Approximations with Applications*. Cambridge: Cambridge University Press.

Cantelli, F. P. (1933). Sulla determinazione empirica delle leggi di probabilitia. *Giorn. Inst. Ital. Attuari*, **4**, 421–424.

Casella, G. and Berger, R. L. (2002). *Statistical Inference*. Pacific Grove, CA: Duxbury.

Chen, P.-N. (2002). *Asymptotic refinement of the Berry-Esseen constant*. Unpublished manuscript.

Christensen, R. (1996). *Plane Answers to Complex Questions*. New York: Springer.

Chow, Y. S. and Teicher, H. (2003). *Probability Theory: Independence, Interchangeability, Martingales*. New York: Springer.

Chung, K. L. (1974). *A Course in Probability*. Boston, MA: Academic Press.

Cochran, W. G. (1952). The χ^2 test of goodness of fit. *The Annals of Mathematical Statistics*, **23**, 315–345.

Conway, J. B. (1975). *Functions of One Complex Variable*. New York: Springer-Verlag.

Copson, E. T. (1965). *Asymptotic Expansions*. Cambridge: Cambridge University Press.

Cornish, E. A. and Fisher, R. A. (1937). Moments and cumulants in the specification of distributions. *International Statistical Review*, **5**, 307–322.

Cramér, H. (1928). On the composition of elementary errors. *Skandinavisk Aktuarietidskrift*, **11**, 13–74, 141–180.

Cramér, H. (1946). *Mathematical Methods of Statistics*. Princeton, NJ: Princeton University Press.

Cramér, H. (1970). *Random Variables and Probability Distributions*. 3rd Ed. Cambridge: Cambridge University Press.

Daniels, H. E. (1954). Saddlepoint approximations in statistics. *The Annals of Mathematical Statistics*, **25**, 631–650.

De Bruijn, N. G. (1958). *Asymptotic Methods in Analysis*. New York: Dover.

Dieudonné, J. (1960). *Foundations of Modern Analysis*. New York: John Wiley and Sons.

Edgeworth, F. Y. (1896). The asymmetrical probability curve. *Philosphical Magazine*, Fifth Series, **41**, 90–99.

Edgeworth, F. Y. (1905). The law of error. *Proceedings of the Cambridge Philosophical Society*, **20**, 36–65.

Edgeworth, F. Y. (1907). On the representation of a statistical frequency by a series. *Journal of the Royal Statistical Society, Series A*, **70**, 102–106.

Efron, B. (1979). Bootstrap methods: Another look at the jackknife. *The Annals of Statistics*, **7**, 1–26.

Efron, B. (1981). Nonparametric standard errors and confidence intervals. *Canadian Journal of Statistics*, **9**, 139–172.

Efron, B. (1982). *The Jackknife, the Bootstrap, and Other Resampling Plans*. Philadelphia, PA: Society for Industrial and Applied Mathematics.

Efron, B. (1987). Better bootstrap confidence intervals. *Journal of the American Statistical Association*, **82**, 171–200.

Efron, B. and Gong, G. (1983). A leisurely look at the bootstrap, the jackknife, and cross validation. *The American Statistician*, **37**, 36–48.

Efron, B., Holloran, E., and Holmes, S. (1996). Bootstrap confidence levels for phylogenetic trees. *Proceedings of the National Academy of Sciences of the United States of America*, **93**, 13492–13434.

Efron, B. and Tibshirani, R. J. (1993). *An Introduction to the Bootstrap*. London: Chapman and Hall.

Efron, B. and Tibshirani, R. J. (1998). The problem of regions. *The Annals of Statistics*, **26**, 1687–1718.

Epanechnikov, V. A. (1969). Non-parametric estimation of a multivariate probability density. *Theory of Probability and its Applications*, **14**, 153–158.

Erdélyi, A. (1956). *Asymptotic Expansions*. New York: Dover Publications.

Esseen, C.-G. (1942). On the Liapounoff limit of error in the theory of probability. *Arkiv för Matematik, Astronomi och Fysik*, **28A**, 1–19.

Esseen, C.-G. (1945). Fourier analysis of distribution functions. A mathematical study of the Laplace-Gaussian law. *Acta Mathematica*, **77**, 1–125.

Esseen, C.-G. (1956). A moment inequality with an application to the central limit theorem. *Skandinavisk Aktuarietidskrift*, **39**, 160–170.

Feller, W. (1935). Über den zentralen grenzwertsatz der wahrscheinlichkeitsrechnung. *Mathematische Zeitschrift*, **40**, 521–559.

Feller, W. (1971). *An Introduction to Probability Theory and its Application, Volume 2*. 2nd Ed. New York: John Wiley and Sons.

Felsenstein, J. (1985). Confidence limits on phylogenies: An approach using the bootstrap. *Evolution*, 783–791.

Fernholz, L. T. (1983). *von Mises Calculus for Statistical Functionals*. New York: Springer-Verlag.

Finner, H. and Strassburger, K. (2002). The partitioning principle. *The Annals of Statistics*, **30**, 1194–1213.

Fisher, R. A. (1915). Frequency distribution of the values of the correlation coefficient in samples from an indefinitely large population. *Biometrika*, **10**, 507–521.

Fisher, N. I. and Hall, P. (1991). Bootstrap algorithms for small sample sizes. *Journal of Statistical Planning and Inference*, **27**, 157–169.

Fisher, R. A. and Cornish, E. A. (1960). The percentile points of distributions having known cumulants. *Technometrics*, **2**, 209–226.

Fix, E. and Hodges, J. L. (1951). Discriminatory analysis - nonparametric discrimination: consistency properties. *Report No. 4, Project No. 21-29-004*. Randolph Field, TX: USAF School of Aviation Medicine.

Fix, E. and Hodges, J. L. (1989). Discriminatory analysis - nonparametric discrimination: consistency properties. *International Statistical Review*, **57**, 238–247.

Fréchet, M. (1925). La notion de differentielle dans l'analyse generale. *Annales Scientifiques de l'École Normale Supérieure*, **42**, 293–323.

Fristedt, B. and Gray, L. (1997). *A Modern Approach to Probability Theory*. Boston, MA: Birkhäuser.

Garwood, F. (1936). Fiducial limits for the Poisson distribution. *Biometrika*, **28**, 437–442.

Ghosh, M. (1994). On some Bayesian solutions of the Neyman-Scott problem. *Statistical Decision Theory and Related Topics*, Volume V. J. Berger and S. S. Gupta, eds. New York: Springer-Verlag. 267–276.

Ghosh, M., Parr, W. C., Singh, K., and Babu, G. J. (1984). A note on bootstrapping the sample median. *The Annals of Statistics*, **12**, 1130–1135.

Gibbons, J. D. and Chakraborti, S. (2003). *Nonparametric Statistical Inference*. Boca Raton, FL: CRC Press.

Giné, E. and Zinn, J. (1989). Necessary conditions for the bootstrap of the mean. *The Annals of Statistics*, **17**, 684–691.

Glivenko, V. (1933). Sulla determinazione empirica delle leggi di probabilitia. *Giornate dell'Istituto Italiano degli Attuari*, **4**, 92–99.

Gnedenko, B. V. (1962). *The Theory of Probability*. New York: Chelsea Publishing Company.

Gnedenko, B. V. and Kolmogorov, A. N. (1968). *Limit Distributions for Some Sums of Independent Random Variables*. Reading, MA: Addison-Wesley.

Graybill, F. A. (1976). *Theory and Application of the Linear Model*. Pacific Grove, CA: Wadsworth and Brooks.

Gut, A. (2005). *Probability: A Graduate Course*. New York: Springer.

Hájek, J. (1969). *Nonparametric Statistics*. San Francisco, CA: Holden-Day.

Hájek, J. (1972). Local asymptotic minimax and admissibility in estimation. *Proceedings of the Sixth Berkeley Symposium on Mathematical Statistics and Probability*, Volume I, 175–194.

Hájek, J. and Šidák, Z. (1967). *Theory of Rank Tests*. New York: Academic Press.

Hall, P. (1983a). Inverting an Edgeworth expansion. *The Annals of Statistics*, **11**, 569–576.

Hall, P. (1983b). Large sample optimality of least squares cross-validation in density estimation. *The Annals of Statistics*, **11**, 1156–1174.

Hall, P. (1986a). On the bootstrap and confidence intervals. *The Annals of Statistics*, **14**, 1431–1452.

Hall, P. (1986b). On the number of bootstrap simulations required to construct a confidence interval. *The Annals of Statistics*, **14**, 1453–1462.

Hall, P. (1988a). Theoretical comparison of bootstrap confidence intervals. *The Annals of Statistics*, **16**, 927–953.

Hall, P. (1988b). *Introduction to the Theory of Coverage Processes*. New York: John Wiley and Sons.

Hall, P. (1990). Asymptotic properties of the bootstrap for heavy-tailed distributions. *The Annals of Probability*, **18**, 1342–1360.

Hall, P. (1992). *The Bootstrap and Edgeworth Expansion*. New York: Springer.

Hall, P. and Martin, M. A. (1988). On bootstrap resampling and iteration. *Biometrika*, **75**, 661–671.

Hall, P. and Martin, M. A. (1991). One the error incurred using the bootstrap variance estimator when constructing confidence intervals for quantiles. *Journal of Multivariate Analysis*, **38**, 70–81.

Halmos, P. R. (1958). *Finite-Dimensional Vector Spaces*. 2nd Ed. Princeton, NJ: Van Nostrand.

Halmos, P. R. (1974). *Measure Theory*. New York: Springer-Verlag.

Hardy, H. G. (1949). *Divergent Series*. Providence, RI: AMS Chelsea Publishing.

Heyde, C. C. (1963). On a property of the lognormal distribution. *Journal of the Royal Statistical Society, Series B*, **25**, 392–393.

Hochberg, Y. and Tamhane, A. C. (1987). *Multiple Comparison Procedures*. New York: John Wiley and Sons.

Hodges, J. L. and Lehmann, E. L. (1956). Thee efficiency of some nonparametric competitors of the *t*-test. *The Annals of Mathematical Statistics*, **27**, 324–335.

Hodges, J. L. and Lehmann, E. L. (1970). Deficiency. *The Annals of Mathematical Statistics*, **41**, 783–801.

Hoeffding, W. (1948). A class of statistics with asymptotically normal distribution. *The Annals of Mathematical Statistics*, **19**, 293–325.

Hollander, M. and Wolfe, D. A. (1999). *Nonparametric Statistical Methods*. 2nd Ed. New York: John Wiley and Sons.

Hsu, P. L. and Robbins, H. (1947). Complete convergence and the law of large numbers. *Proceedings of the National Academy of Sciences of the United States of America*, **33**, 25–31.

Huber, P. J. (1966). Strict efficiency excludes superefficiency. *The Annals of Mathematical Statistics*, **37**, 1425.

Jensen, J. L. (1988). Uniform saddlepoint approximations. *Advances in Applied Probability*, **20**, 622–634.

Johnson, N. L., Kotz, S., and Balakrishnan, N. (1994). *Distributions in Statistics: Continuous Univariate Distributions*. Volume I. 2nd Ed. New York: John Wiley and Sons.

Keller, H. H. (1974). *Differential Calculus on Locally Convex Spaces*. Lecture Notes in Mathematics Number 417. Berlin: Springer-Verlag.

Kendall, M. and Stuart, A. (1977). *The Advanced Theory of Statistics, Volume 1: Distribution Theory*. 4th Ed. New York: Macmillan Publishing Company.

Khuri, A. I. (2003). *Advanced Calculus with Applications in Statistics*. New York: John Wiley and Sons.

Knight, K. (1989). One the bootstrap of the sample mean in the infinite variance case. *The Annals of Statistics*, **17**, 1168–1175.

Koenker, R. W. and Bassett, G. W. (1984). Four (Pathological) examples in asymptotic statistics. *The American Statistician*, **38**, 209–212.

Kolassa, J. E. (2006). *Series Approximation Methods in Statistics*. 3rd Ed. New York: Springer.

Kolmogorov, A. N. (1956). *Foundations of the Theory of Probability*. New York: Chelsea Publishing Company.

Kowalski, J. and Tu, X. M. (2008). *Modern Applied U-Statistics*. New York: John Wiley and Sons.

Landau, E. (1974). *Handbuch der Lehre von der Verteilung der Primzahlen*. Providence, RI: AMS Chelsea Publishing.

Le Cam, L. (1953). On some asymptotic properties of maximum likelihood estimates and related Bayes' estimates. *University of California Publications in Statistics*, **1**, 277–330.

Le Cam, L. (1979). *Maximum Likelihood Estimation: An Introduction*. Lecture Notes in Statistics Number 18. University of Maryland, College Park, MD.

Lee, A. J. (1990). *U-Statistics: Theory and Practice*. New York: Marcel Dekker.

Lehmann, E. L. (1983). *Theory of Point Estimation*. New York: John Wiley and Sons.

Lehmann, E. L. (1986). *Testing Statistical Hypotheses*. Pacific Grove, CA: Wadsworth and Brooks/Cole.

Lehmann, E. L. (1999). *Elements of Large-Sample Theory*. New York: Springer.

Lehmann, E. L. and Casella, G. (1998). *Theory of Point Estimation*. New York: Springer.

Lehmann, E. L. and Shaffer, J. (1988). Inverted distributions. *The American Statistician*, **42**, 191–194.

Lévy, P. (1925). *Calcul des Probabilités*. Paris: Gauthier-Villars.

Lindeberg, J. W. (1922). Eine neue herleitung des exponentialgezetzes in der wahrscheinlichkeitsrechnung. *Mathematische Zeitschrift*, **15**, 211–225.

Loéve, M. (1977). *Probability Theory*. 4th Ed. New York: Springer-Verlag.

Loh, W.-Y. (1987). Calibrating confidence coefficients. *Journal of the American Statistical Association*, **82**, 155–162.

Lukacs, E. (1956). On certain periodic characteristic functions. *Compositio Mathematica*, **13**, 76–80.

Mammen, E. (1992). *When Does Bootstrap Work? Asymptotic Results and Simulations*. New York: Springer.

Miller, R. G. (1981). *Simultaneous Statistical Inference*. 2nd Ed. New York: Springer-Verlag.

Mood, A. M. (1950). *Introduction to the Theory of Statistics*. 3rd Ed. New York: McGraw-Hill.

Nadaraya, E. A. (1974). On the integral mean squared error of some non-parametric estimates for the density function. *Theory of Probability and Its Applications*, **19**, 133–141.

Nashed, M. Z. (1971). Differentiability and related properties of nonlinear operators: some aspects of the role of differentials in nonlinear functional analysis. *Nonlinear Functional Analysis and Applications*. L. B. Rall, ed. New York: Academic Press. 103–309.

Neyman, J. and Scott. E. (1948). Consistent estimates based on partially consistent observations. *Econometrica*, **16**, 1–32.

Noether, G. E. (1949). On a theorem by Wald and Wolfowitz. *The Annals of Mathematical Statistics*, **20**, 455–458.

Noether, G. E. (1955). On a theorem of Pitman. *The Annals of Mathematical Statistics*, **26**, 64–68.

Parzen, E. (1962). On the estimation of a probability density function and the mode. *The Annals of Mathematical Statistics*, **33**, 1065–1076.

Petrov, V. V. (1995). *Limit Theorems of Probability Theory: Sequences of Independent Random Variables*. New York: Oxford University Press.

Petrov, V. V. (2000). Classical-type limit theorems for sums of independent random variables. *Limit Theorems of Probability Theory, Encyclopedia of Mathematical Sciences*. Number 6, 1–24. New York: Springer.

Phanzagl, J. (1970). On the asymptotic efficiency of median unbiased estimates. *The Annals of Mathematical Statistics*, **41**, 1500–1509.

Pitman, E. J. G. (1948). *Notes on Non-Parametric Statistical Inference*. Unpublished notes from Columbia University.

Polansky, A. M. (1995). *Kernel Smoothing to Improve Bootstrap Confidence Intervals*. Ph.D. Dissertation. Dallas, TX: Southern Methodist University.

Polansky, A. M. (1999). Upper bounds on the true coverage of bootstrap percentile type confidence intervals. *The American Statistician*, **53**, 362–369.

Polansky, A. M. (2000). Stabilizing bootstrap-t confidence intervals for small samples. *Canadian Journal of Statistics*, **28**, 501–516.

Polansky, A. M. (2003a). Supplier selection based on bootstrap confidence regions of process capability indices. *International Journal of Reliability, Quality and Safety Engineering*, **10**, 1–14.

Polansky, A. M. (2003b). Selecting the best treatment in designed experiments. *Statistics in Medicine*, **22**, 3461–3471.

Polansky, A. M. (2007). *Observed Confidence Levels: Theory and Application*. Boca Raton, FL: Chapman Hall/CRC Press.

Polansky, A. M. and Schucany, W. R. (1997). Kernel smoothing to improve bootstrap confidence intervals. *Journal of the Royal Statistical Society, Series B*, **59**, 821–838.

Pollard, D. (2002). *A User's Guide to Measure Theoretic Probability*. Cambridge: Cambridge University Press.

Putter, H. and Van Zwet, W. R. (1996). Resampling: consistency of substitution estimators. *The Annals of Statistics*, **24**, 2297–2318.

Randles, R. H. and Wolfe, D. A. (1979). *Introduction to the Theory of Nonparametric Statistics*. New York: John Wiley and Sons.

Rao, C. R. (1963). Criteria for estimation in large samples. *Sankhyā*, **25**, 189–206.

Reeds, J. A. (1976). *On the Definition of von Mises Functionals*. Ph.D. Dissertation. Cambridge, MA: Harvard University.

Rosenblatt, M. (1956). Remarks on some nonparametric estimates of a density function. *The Annals of Mathematical Statistics*, **27**, 832–837.

Royden, H. L. (1988). *Real Analysis*. 3rd Ed. New York: Macmillan.

Rudemo, M. (1982). Empirical choice of histograms and kernel density estimators. *Scandinavian Journal of Statistics*, **9**, 65–78.

Scott, D. W. (1992). *Multivariate Density Estimation: Theory, Practice and Visualization*. New York: John Wiley and Sons.

Sen, P. K. and Singer, J. M. (1993). *Large Sample Methods in Statistics*. London: Chapman and Hall.

Serfling, R. J. (1980). *Approximation Theorems of Mathematical Statistics*. New York: John Wiley and Sons.

Severini, T. A. (2005). *Elements of Distribution Theory*. Cambridge: Cambridge University Press.

Shao, J. and Tu, D. (1995). *The Jackknife and Bootstrap*. New York: Springer.

Shi, X. (1986). A note on bootstrapping U-statistics. *Chinese Journal of Applied Probability and Statistics*, **2**, 144–148.

Shiganov, I. S. (1986). Refinement of the Upper Bound of the constant in the central limit theorem. *Journal of Soviet Mathematics*, **35**, 2545–2550.

Shohat, J. A. and Tamarkin, J. D. (1943). *The Problem of Moments*. 4th Ed. Providence, RI: American Mathematical Society.

Silverman, B. W. (1986). *Density Estimation*. London: Chapman and Hall.

Simmons, G. (1971). Identifying probability limits. *The Annals of Mathematical Statistics*, **42**, 1429–1433.

Simonoff, J. S. (1996). *Smoothing Methods in Statistics*. New York: Springer.

Singh, K. (1981). One the asymptotic accuracy of Efron's bootstrap. *The Annals of Statistics*, **9**, 1187–1195.

Slomson, A. B. (1991). *An Introduction to Combinatorics*. Bocca Raton, FL: CRC Press.

Sprecher, D. A. (1970). *Elements of Real Analysis*. New York: Dover.

Sprent, P. and Smeeton, N. C. (2007). *Applied Nonparametric Statistical Methods*. 4th Ed. Boca Raton, FL: Chapman and Hall/CRC Press.

Stefansson, G., Kim, W.-C., and Hsu, J. C. (1988). On confidence sets in multiple comparisons. *Statistical Decision Theory and Related Topics IV*. S.S. Gupta and J. O. Berger, Eds. New York: Academic Press. 89–104.

Stone, C. J. (1984). An asymptotically optimal window selection rule for kernel density estimates. *The Annals of Statistics*, **12**, 1285–1297.

Tchebycheff, P. (1890). Sur duex theéorèmes relatifs aux probabilités. *Acta Mathematica*, **14**, 305–315.

Tibshirani, R. (1988). Variance stabilization and the bootstrap. *Biometrika*, **75**, 433–444.

van Beek, P. (1972). An application of Fourier methods to the problem of sharpening of the Berry-Esseen inequality. *Zeitschrift für Wahrscheinlichkeitstheorie und Verwandte Gebiete*, **23**, 183–196.

van der Vaart, A. W. (1998). *Asymptotic Statistics*. Cambridge: Cambridge University Press.

von Mises, R. (1947). On the asymptotic distribution of differentiable statistical functionals. *The Annals of Mathematical Statistics*, **18**, 309–348.

Wand, M. P. (1997). Data-based choice of histogram bin width. *The American Statistician*, **51**, 59–64.

Wand, M. P. and Jones, M. C. (1995). *Kernel Smoothing*. London: Chapman and Hall.

Westenberg, J. (1948). Significance test for median and interquartile range in samples from continuous populations of any form. *Proceedings Koningklijke Nederlandse Akademie van Wetenschappen*, **51**, 252–261.

Westfall, P. H. and Young, S. S. (1993). *Resampling-Based Multiple Testing*. New York: John Wiley and Sons.

Winterbottom, A. (1979). A note on the derivation of Fisher's transformation of the correlation coefficient. *The American Statistician*, **33**, 142–143.

Withers, C. S. (1983). Asymptotic expansions for the distribution and quantiles of a regular function of the empirical distribution with applications to nonparametric confidence intervals. *The Annals of Statistics*, **11**, 577–587.

Withers, C. S. (1984). Asymptotic expansions for distributions and quantiles with power series cumulants. *Journal of The Royal Statistical Society, Series B*, **46**, 389–396.

Wolfowitz, J. (1965). Asymptotic efficiency of the maximum likelihood estimator. *Theory of Probability and its Applications*, **10**, 247–260.

Woodroofe, M. (1970). On choosing a delta-sequence. *The Annals of Mathematical Statistics*, **41**, 1665–1671.

Yamamuro, S. (1974). *Differential Calculus in Topological Linear Spaces*. Lecture Notes in Mathematics, Number 374. Berlin: Springer-Verlag.

Young, N. (1988). *An Introduction to Hilbert Space*. Cambridge: Cambridge University Press.

Zolotarev, V. M. (1986). *Sums of Independent Random Variables*. New York: John Wiley and Sons.

Author Index

Subject Index

almost certain convergence, 107
    alternate definition, 108
    continuous mapping theorem, 122–124, 154, 155
    random vectors, 118, 119, 154
    relation to complete convergence, 114–116, 154
    relation to convergence in r^{th} mean, 250
    relation to convergence in probability, 110, 111
    sum of random variables, 128, 129
alternative hypothesis, 425
approximate local linearization, 348
asymptotic chi-squared distribution, 276
asymptotic expansion, 28
    integration by parts, 33
    inversion, 39
asymptotic normality, 195
    functional parameter, 375, 376
    multivariate, 201
    non-identical distributions, 255, 261
    relation to asymptotic standard error, 198
    sample moments, 214
    sample quantiles, 215, 218, 219, 226
    transformation, 266, 268, 270, 274
    triangular arrays, 263, 264, 279
asymptotic order notation, 30
    properties, 35, 37, 38, 49
asymptotic relative efficiency
    mean and median, 392
    point estimation, 392
    sample proportion, 393

asymptotically equivalent, 31

Backmann-Landau notation, 31
backwards confidence interval, 464
Bahadur efficiency, 524
Bayes estimator, 447, 448
    absolute error loss, 448
    asymptotic optimality, 456
    binomial distribution, 449, 450, 459
    consistency, 456
    exponential family, 458
    normal mean, 458, 469
    Poisson mean, 469, 470
    squared error loss, 448
Bayes risk, 448
Bayes' theorem, 448
Bayesian methods, 447
    posterior distribution, 447
    prior distribution, 447
Berry and Esseen theorem, 208, 212
    universal constant, 212
bias, 384
    asymptotic expansion, 387
bias-corrected confidence interval, 464
bin width, 529
binomial expansion, 570
Bonferroni inequality, 59
bootstrap, 541
    acceleration constant, 551, 560
    backwards confidence limit, 550
    bias corrected and accelerated confidence limit, 551
    bias corrected confidence limit, 551

9 780367 383138